Human Biology

Jones & Bartlett Learning Titles in Biological Science

Human Biology

EIGHTH EDITION

Daniel D. Chiras
The Evergreen Institute
Gerald, Missouri

JONES & BARTLETT
LEARNING

World Headquarters
Jones & Bartlett Learning
5 Wall Street
Burlington, MA 01803
978-443-5000
info@jblearning.com
www.jblearning.com

Jones & Bartlett Learning books and products are available through most bookstores and online booksellers. To contact Jones & Bartlett Learning directly, call 800-832-0034, fax 978-443-8000, or visit our website, www.jblearning.com.

Substantial discounts on bulk quantities of Jones & Bartlett Learning publications are available to corporations, professional associations, and other qualified organizations. For details and specific discount information, contact the special sales department at Jones & Bartlett Learning via the above contact information or send an email to specialsales@jblearning.com.

Production Credits

Chief Executive Officer: Ty Field
President: James Homer
Chief Product Officer: Eduardo Moura
Executive Publisher: William Brottmiller
Publisher: Cathy L. Esperti
Senior Acquisitions Editor: Erin O'Connor
Editorial Assistant: Rachel Isaacs
Production Manager: Louis C. Bruno, Jr.
Marketing Manager: Lindsay White

Manufacturing and Inventory Control Supervisor: Amy Bacus
Composition: Circle Graphics, Inc.
Cover Design: Michael O'Donnell
Photo Research and Permissions Coordinator: Lauren Miller
Cover Image: © Eduard Kyslynsky/ShutterStock, Inc.
Dedication Page Image: © Photos.com
Printing and Binding: Courier Companies
Cover Printing: Courier Companies

To order this product, use ISBN: 978-1-284-03181-2

Library of Congress Cataloging-in-Publication Data
Chiras, Daniel D.
  Human biology / Daniel D. Chiras.—Eighth edition.
    pages cm
  ISBN 978-1-284-02770-9
  1. Human biology—Textbooks. I. Title.
  QP34.5.C4853 2014
  612—dc23   2013008541

6048

Printed in the United States of America
17 16 15 14 13   10 9 8 7 6 5 4 3 2 1

The Jones & Bartlett Learning
Commitment to the Environment

As a book publisher, Jones & Bartlett Learning is committed to reducing its impact on the environment. Many Jones & Bartlett Learning titles are printed using recycled, post-consumer paper. We purchase our paper from manufacturers committed to sustainable, environmentally sensitive processes. We employ the Internet and office computer network technology in our effort toward sustainable solutions. New communication technology provides opportunities to reduce our use of paper and other resources through online delivery of instructors' materials and educational information—and, of course, we recycle in the office.

Dedicated to my two sons, Skyler and Forrest,
who continue to amaze me with their love, talent, and wisdom.

Brief Contents

Contents

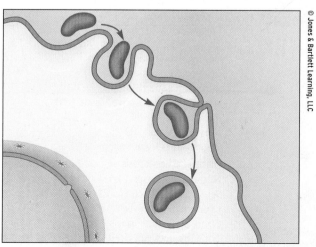

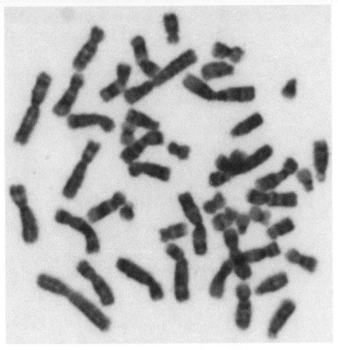

© Photos.com

PART 2   Structure and Function of the Human Body

© Polo Productions/Brand X Pictures/Alamy Images

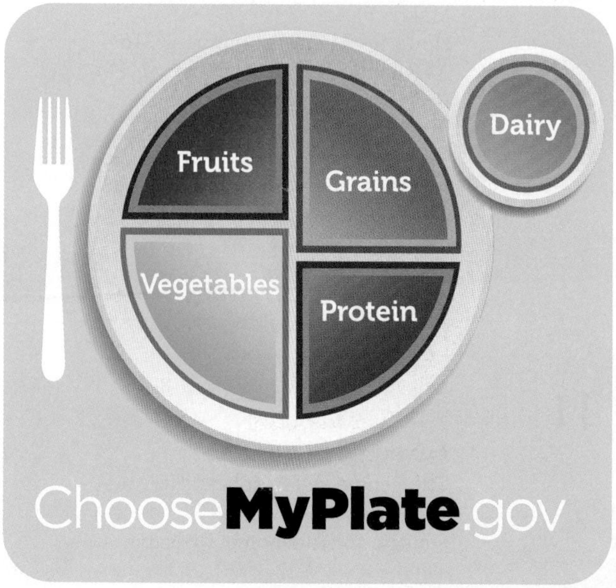

Courtesy of USDA

Fruits
Grains
Dairy
Vegetables
Protein

Choose**MyPlate**.gov

© Rob Marmion/ShutterStock, Inc.

Courtesy of CDC

18 How Genes Work and How Genes Are Controlled 391

19 Genetic Engineering and Biotechnology 411

20 Cancer 429

PART 5   Human Reproduction and Development

21 Human Reproduction 442

© Jones & Bartlett Learning, LLC

© LiquidLibrary

Preface

Human Biology, Eighth Edition, is written specifically with the undergraduate, non-science major in mind. The text contains an abundance of timely and important information about the structure, function, health, and disease of the human body, exploring the world from the cellular level, then the level of tissues and organs, and finally pulling back to view humans as organisms within a complex evolutionary and ecological environment.

The central organizing theme of the book—homeostasis—illustrates the internal balancing act that has evolved over the history of humankind, allowing our bodies to regulate conditions integral to our survival. For our bodies to achieve homeostasis and maintain health, they must regulate their temperature, levels of chemicals in the blood, as well as a host of cellular functions. Homeostasis and health also depend on our bodies' external environment—the quality of the air we breathe, the water we drink, and the foods we eat.

When I began writing this book, many editions ago, I had several goals in mind. The first is to educate you about your body and the ways to take good care of it. The more you know about how your body works, the more able you are to make healthy decisions now and in the future. The book is also meant to teach you about scientific literacy—how scientists gather information, analyze it, and come to conclusions about it—which will make you a better consumer of media. When you read about a scientific discovery in the news, you will be able to make better decisions about how reliable the information might be and how it might affect your life, if you understand how the information came to be known. This knowledge will also help you understand difficult and controversial political issues, like stem cell research (why it's controversial and why many people feel it is important). Finally, this book encourages you to think critically about the human body and health, looking at the body as a system—a collection of organs and molecules working together, influenced by stimuli inside and out.

Organization

Human Biology is divided into six logical parts. Part 1 outlines the fundamental biological and chemical principles you need to know to understand the human body, as well as introducing the cell, the unit of all life. You will also read about the scientific method and critical thinking, both of which are applied throughout the book.

Part 2 examines the fascinating structure and function of the human body—how organ systems work, and what happens when they don't work properly. Homeostasis is emphasized in these chapters as a unifying biological principle, applicable to the function and dysfunction of all life.

In Part 3 you'll learn about the human immune system and how it responds to disease, as well as some diseases that affect it directly. A chapter on infectious diseases (a collaboration with Dr. Jeffrey Pommerville of Glendale Community College) and

an *A Closer Look* mini-chapter on HIV/AIDS will illustrate how the immune system defends the body from disease and what can happen when the immune system is compromised. We'll also discuss how infectious diseases spread, the emergence of new and re-emerging diseases, and the threat of bioterrorism.

Part 4 dives down to the level of your cells and beyond, investigating cells and the materials of heredity they contain and how cells and genes function and reproduce themselves. We'll also take a look at what happens when cell division gets out of control (cancer) and what humans can and *have* done by manipulating genetic material (genetic engineering).

Part 5 examines human reproduction, development, and aging. We'll highlight everyday practices that will help you achieve healthy development and aging. An *A Closer Look* mini-chapter also covers common sexually transmitted infections—how they spread, the effects they can have, and how you can prevent them.

Finally, in Part 6 we look at humans from a big-picture point of view. We'll discuss evolution—how we got here and why we got here the way we did—as well as the basic ideas behind ecology and healthy, natural environments. The final chapter surveys some of the problems of our society, those that we have created in the natural world that affect us every day. We'll discuss possible solutions for redirecting human society onto a healthier and more sustainable long-term course.

An Integrated Approach to Teaching Chemistry

Human Biology uses a bold method of teaching chemistry: rather than discussing all of the chemistry you need to know in one chapter early in the book, I introduce the basics of chemistry in Chapter 2, with a brief overview of major biological molecules. Then, throughout the book, more specific information on important molecules (carbohydrates, lipids, proteins, amino acids, and nucleic acids) is given as you need it. For example, you'll learn about proteins and lipids when we discuss the plasma membrane in Chapter 3. Later on in that chapter, you'll learn about carbohydrates when we cover energy and metabolism, and we'll go over even more detail on protein, lipids, and carbohydrates in Chapter 7, Nutrition and Digestion. Nucleic acids are presented in detail in the discussion of molecular genetics in Part 4.

This approach has two chief benefits: (1) It prevents you from becoming overwhelmed by a deluge of seemingly unconnected facts that many students find difficult to grasp and impossible to remember if not given in proper context, and (2) it shows you that the chemistry you are learning is relevant to your understanding of human biology.

New to the *Eighth Edition*

The primary emphasis of this edition, as in previous editions, is to help you better understand your body and to highlight the steps you can take to improve your health and fitness. Essential to this goal is the understanding of the body's systems and the many ways in which they are affected by environmental conditions. This edition has undergone extensive revision and updating. I have reorganized the book, moving the chapters on the circulatory system and blood before the nutrition and digestion chapter. I've added examples, analogies, and figures to illustrate new and existing concepts. There are new Point/ Counterpoints on current topics of great interest including vaccination against common diseases, vaccination against human papilloma virus, the safety of genetically modified organisms, and the safety of milk from cows treated with growth hormone. The in-chapter key concepts are expanded and are now at the end of sections, making this a more useful learning tool. I pose questions in the text that require you to stop and think to assimilate information you are learning, and I pose research questions in the end-of-chapter Concept Review for independent study. Coverage of all diseases has been expanded to include new information on causes, treatment, and prevention. Chapter 21's mini companion chapter, *A Closer Look: Sexually Transmitted Diseases*, has been revised and updated with the latest statistics and information.

Some new topics I have added are primary cilia, white and brown fat, energy production by fat through beta-oxidation, gastro-esophageal reflux disease (GERD), use of pig organs in repairing damaged human ones, photorefractive keratectomy (PRK) and laser epithelial keratomileusis (LASEK) operations, natural killer cells, and the importance of a nutrient-rich diet containing phytochemical antioxidants and anticancer agents. Chapter 7 includes a new section on controlling hunger and the latest Department of Agriculture diet guidelines called MyPlate. Chapter 14 has a new section on immunocompetence.

I continue to stress the importance of critical thinking throughout the book, helping you learn to evaluate judiciously information on your own. With this critical thinking approach in mind, the *Eighth Edition* tackles a variety of subjects found in the news today, such as *trans* fats, overconsumption and obesity, organic foods and farming, vitamins, vaccination, cancer prevention, and even global climate change. The more you learn about these topics, the wiser you'll be as a citizen, consumer of media, and manager of your own health and well-being.

Special Features

I hope you find this to be a user-friendly book. In this edition, as in previous ones, I have tried hard to ensure that the material is approachable and easy to understand. I've simplified complex subjects to some degree and have used examples and analogies to make some concepts easier to understand. For the most part, *Human Biology* concentrates on basic information, the key facts and concepts that you need to know. I've clearly defined key terms and added pronunciations of some of the more difficult ones. Additionally, the following features are here to make your learning experience fun, interesting, and memorable.

Practical Pointers to Increase Health and Fitness

The *Health Tips* included in each chapter offer you sound advice on some of the many ways you can improve your life,

often requiring very little change in your daily routine. The tips offer nutrition information and suggestions for increasing your physical fitness. A brief description of the reasoning and scientific evidence behind each tip is also presented. All Health Tips were derived from the latest research published in reputable scientific journals.

Health Tip 10-5

Get in the habit of exercise now and exercise as often as you can to maintain muscle mass.

Why?

Starting in midlife, adults lose as much as one-third to one-fourth of a pound of muscle mass per year if they don't do something to prevent it. Adopting a healthy lifestyle now, which includes plenty of exercise, makes it easier to stay active as you get older.

Learning the Process of Science

When studying science, there is no way to get around the body of facts that make up our knowledge. But just as important as these facts is the way we came to know them. To highlight some of the key discoveries that have been made in the field of human biology, I've included numerous essays called *Scientific Discoveries that Changed the World*. They discuss breakthroughs such as Robert Hooke's first description of cells in 1665; William Harvey's seventeenth-century studies of the circulation of blood; Louis Pasteur, Robert Koch, and the germ theory of disease; and Watson, Crick, Wilkins, and Franklin's work on the structure and function of DNA. In this way, we are able to highlight the work of some of history's most important scientists *and* to illustrate how scientific discoveries can drastically change our view of the world. These essays will also illuminate the scientific method and demonstrate the fact that scientific advances usually require the efforts of many people, sometimes working in what seem like very different areas. Cooperation and the exchange of information are key to scientific progress.

Scientific Discoveries that Changed the World

3-1 The Discovery of Cells
Featuring the Work of Hooke, Leeuwenhoek, Brown, Schleiden, Schwann, and Virchow

One of the fundamental principles of biology is known as the *cell theory*. The cell theory comprises three parts: (1) the cell is the basic unit of structure of all organisms, (2) all organisms consist of one or more cells, and (3) all cells arise from pre-existing cells. Although this may seem rather elementary, it was not so obvious to early scientists who labored with relatively crude instruments and without the benefit of many facts we now take for granted.

One of those scientific pioneers who opened our eyes to the world of cells was Robert Hooke, a seventeenth-century British mathematician, inventor, and scientist. Equipped with a relatively crude microscope, Hooke observed just about everything he could lay his hands on—which he described in his book *Micrographia*, published in 1665.

One especially useful description was that made on a thin slice of cork (Figure 1). Peering through his microscope, Hooke beheld a network of tiny, boxlike compartments that reminded him of a honeycomb. He called these compartments *cellulae*, meaning "little rooms." Today, we know them as cells.

Hooke did not really see cells but, rather, cell walls, the structures that surround the cell membranes of plant cells. The cytoplasm and cellular organelles had been destroyed when he prepared the tissues for slicing.

Hooke's work was complemented a few years later by Antony van Leeuwenhoek, a Dutch shopkeeper who spent much of his free time designing simple microscopes. Like Hooke, Leeuwenhoek examined just about everything he could find and wrote extensively on his observations. Wayne Becker, a cell biologist at the University of Wisconsin writes, "His detailed reports attest to both the high quality of his lenses and his keen powers of observation." Becker continues, "They also reveal an active imagination

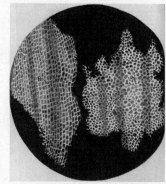

FIGURE 1 **Thin Slice of Cork** (Courtesy of Library of Congress, Prints & Photographs Division [reproduction number LC-USZ62-95187].)

How to Live a Healthier Life

Health Notes are essays emphasizing vital information for your current and future health and wellness. They present practical advice on proper diet, exercise, and stress management—always important, and especially while you're in school!—as well as hair loss, heart disease, breast augmentation, cancer prevention, the dangers of recreational drug use, and aging. This is a book to hold onto and refer back to throughout your life!

11–1 Recreational Drug Use: Is It a Good Idea?

Jason Wiley was a popular young man in his college dormitory. He was friendly, generous, and fun to be with. Like others of his age, Jason sometimes used recreational drugs like Special K or ecstasy. But one day, Jason didn't return from his night of partying. When his friends found him, he was dead in the back seat of another friend's car. He had died from an overdose of Special K and methamphetamine.

Jason is just one of millions of youths the world over who uses recreational drugs. The most commonly used are (1) methamphetamine, also known as meth, crystal meth, speed, ice, and crystal; (2) phencyclidine, also called PCP, angel dust, supergrass, killer weed, embalming fluid, and rocket fuel; (3) ecstasy; (4) ketamine, also referred to as Special K or simply k; (5) cocaine; (6) LSD; (7) gamma hydroxybutyrate or GHB, liquid x, Georgia home boy, goop, gamma-oh, and grievous body harm; and (8) Ritalin, also known as Vitamin R.

Many of these drugs, such as methamphetamine, ketamine, ecstasy, GHB, cocaine, and Ritalin, are stimulants. They give the user a sense of euphoria and increased well being. These drugs increase heart rate, blood pressure

FIGURE 1 A rave. (© dwphotos/ShutterStock, Inc.)

GHB, also commonly used at raves with ketamine and ecstasy, can cause seizures, severe respiratory depression, and coma. It is also called the date rape drug, because men have used it to incapacitate women, leaving them vulnerable to sexual assault.

Getting Both Sides of the Story

Many current biological discoveries come hand in hand with weighty ethical challenges, resulting in considerable debate, both in academic circles and in the news media. Genetic engineering and human embryonic stem cell research, for example, have sparked some very lively ongoing discussions. As such, *Human Biology* presents a number of modern controversies from differing perspectives in *Point/Counterpoint* boxes. Each

consists of two brief essays written by distinguished writers and thinkers, each advocating an opposing view on important issues such as genetic engineering, health care spending, cancer screening, physician-assisted suicide, vaccination, and the current decline in biodiversity.

These essays will not only help inform you about many current issues but also offer you a great opportunity to use your critical thinking skills as you weigh each argument and come to your own conclusions. Your professor may decide to use the Point/Counterpoints to spark a class discussion, so come ready with your own views and the evidence to back them up. Additional ideas, questions, and links to web pages supporting different views on the issues at hand are available at the Jones & Bartlett Learning website for *Human Biology* (go.jblearning.com/HumanBio8ecw).

Learning How to Think Critically

Chapter 1 of *Human Biology* presents a number of guidelines for improving your critical thinking skills, making you a more discerning thinker in all of your classes and a more informed citizen. While critical thinking is encouraged throughout the text, I emphasize it at the start and finish of each chapter with a *Thinking Critically* exercise. The chapter begins with the description of a problem or the conclusion of a scientific study to which I ask you to apply your critical thinking. At the end of the chapter, I present my own analysis of the problem or study.

Each of these exercises underscores one or two of the critical thinking guidelines outlined in Chapter 1. Critical thinking questions also appear after every Point/Counterpoint and in the Concept Review questions at the end of each chapter.

THINKING CRITICALLY

Researchers at a major medical college in your area have discovered that cells live longer when cultured (grown) in petri dishes containing a certain vitamin. They suggest that this vitamin might help people live longer. A local newspaper runs with the story, and one of your friends is thinking of taking the vitamin supplement in hope of living longer. What advice would you give him before he embarks on this course of action?

Understanding Nature's Balancing Act

Human health depends on homeostasis. Homeostasis, in turn, depends on many other factors, which are described in the *Health and Homeostasis* sections at the conclusion of many chapters. These passages will help you gain a broader understanding of how your diet, stress levels, and level of physical activity affect your health and your body's attempts to maintain balance.

Boiling It Down to Essentials

Nearly all chapter sections are followed by summary statements that stress the key concepts covered in that section.

13-7 Health and Homeostasis

Hormones orchestrate an incredible number of body functions and help to create a dynamic balance that's vital for human health. Hormones influence homeostasis primarily by controlling the rate of various metabolic reactions and by regulating ionic balance. When this balance is altered, our health suffers (Table 13-3).

The endocrine system, like other systems, is sensitive to outside factors. Stress, for example, can lead to an imbalance

These *Key Concepts* are a great study tool to review major concepts as you read and prepare for exams.

> **KEY CONCEPTS**
>
> Much of our basic understanding of heredity comes from the work of one man, Gregor Mendel, who studied genetics of garden peas.

Summing It All Up

An extensive *Study Guide* appears at the end of every chapter, including a chapter summary, an analysis of the *Thinking Critically* exercise for that chapter, a list of key terms and concepts, a numbered review of all key topics covered, a set of review questions, and a self-quiz so you can test yourself and see how much of the chapter's contents you've retained.

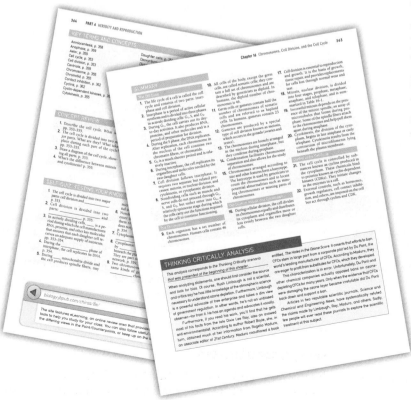

Making Each Concept Visible

Human Biology contains a remarkable collection of illustrations and photographs to bring each concept to life before your eyes. In this edition, we've added numerous new figures and refined some existing ones to enhance the visual and educational experience you'll have with the book.

Enhancement Using the Web

The eighth edition of *Human Biology* is designed to interact with an extensive website, go.jblearning.com/HumanBio8ecw, designed exclusively for this book by Jones & Bartlett Learning. The site offers you unprecedented opportunities to integrate this text with the online world, connecting you with independent websites covering the latest biology news and information on Point/Counterpoint debates, cancer and infectious diseases, Health Note topics, and Health Tips. Through these linkages, *Human Biology* is able to provide you with up-to-the minute information throughout the life of the edition. When there is a link to an outside website, a brief discussion places the site in context so you can get the most out of the information contained in each link.

The *Human Biology* website also contains Tools for Learning, an online review guide, designed specifically to help you study for exams. You'll find outlines of each chapter, review questions with which to test your knowledge, hundreds of animated key-term flashcards, figure labeling exercises, and links to a number of interactive tutorials.

Ancillaries

Jones & Bartlett Learning offers an impressive variety of traditional print and interactive multimedia supplements to assist instructors and to aid students in mastering human biology. Additional information and review copies of any of the following items are available through your Jones & Bartlett Learning account specialist or by going to go.jblearning.com/Biology.

For Instructors

The **Instructor's Media CD**, compatible with Windows® and Macintosh® platforms, provides instructors with the following critical ancillaries:

PowerPoint® Lecture Outline Slides. This presentation package provides lecture notes and images from the text for each chapter of *Human Biology, Eighth Edition*. A PowerPoint viewer is provided on the CD. Instructors with the Microsoft® PowerPoint software can customize the outlines, images, and order of presentation.

PowerPoint Image Bank. This presentation provides the illustrations, photographs, and tables (to which Jones & Bartlett Learning holds the copyright or has permission to reproduce digitally) inserted into PowerPoint slides. With the Microsoft PowerPoint program, instructors can quickly and easily copy individual images into their existing lecture slides.

The following online instructor's resources are available for qualified instructors to download from go.jblearning.com/HumanBio8e.

Instructor's Manual. This guide is provided as a text file containing chapter outlines, learning objectives, key terms, concept questions, and teaching tips, including ways to encourage discussion in class using the Point/Counterpoint features.

Test Bank. This reserve of over 2200 exam questions is available as text files.

For Students

Case Studies for Understanding the Human Body, Second Edition, by Stanton Braude, Deena Goran, and Alexander P. Miceli of Washington University. This supplementary case studies workbook is available both in print and as a customizable electronic product. It provides exercises that lend themselves to a cooperative learning setting where students work together to review and solve open-ended questions pertaining to the human body in health and disease. The case studies link directly back to content covered in *Human Biology, Eighth Edition*, making these two books a truly integrative teaching and learning tool.

Human Biology Lab Manual by Charles Welsh of LaRoche College. This laboratory manual contains 18 exercises that may be taught in any order, offering instructors flexibility to mold the text to the specific needs of their students.

Guide to Infectious Diseases by Body System, Second Edition, by Jeffrey C. Pommerville of Glendale Community College. This excellent tool guides students through the microbial diseases that affect the human body. Each of the 16 units offers a brief introduction to the human anatomical systems and the bacterial, viral, fungal, or parasitic agents that infect each system, as well as the diseases they cause and therapies that may be used to treat them. Anatomical illustrations are captioned with the diseases' signs and symptoms.

Human Anatomy Flash Cards: Skeletal and Muscular Systems. This set of flash cards is a valuable and convenient tool designed to test and reinforce students' knowledge of the skeletal and muscular systems described in the text.

Digital Learning Solutions

With this eighth edition we are excited to offer **Navigate Human Biology**, a completely customizable, comprehensive, and interactive courseware solution for your undergraduate introductory human biology course. **Navigate Human Biology** transforms how students learn and instructors teach by bringing together authoritative and interactive content aligned to course objectives with student practice activities and assessments and learning analytics reporting tools. **Navigate Human Biology** empowers faculty and students with easy-to-use web-based curriculum solutions that optimize student success, identify retention risks, and improve completion rates.

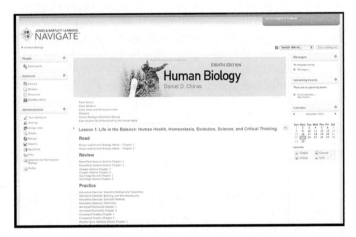

Incorporated into **Navigate Human Biology** is **Navigate eFolio**, an interactive eBook with enhanced activities and engaging learning tools. **Navigate eFolio** is an exciting new choice for students and instructors looking for a more interactive learning experience. **Navigate eFolio** reinforces important concepts through exercises, enhanced diagrams, activities, videos, and animations—bringing key concepts to life! Students can practice what they read and see immediate feedback on their performance. **Navigate eFolio: Human Biology** is also available for standalone purchase by visiting go.jblearning.com/NavigateEfolioHumanBiology8e.

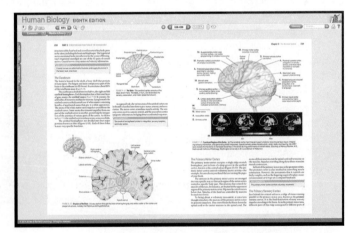

Daniel D. Chiras
Gerald, Missouri

Acknowledgments

A project of this magnitude is the fruit of a great many people. I wish to thank the thousands of scientists and teachers who have contributed to our understanding of human biology. A special thanks to the extraordinary teachers who have made tremendous contributions to my education, especially the late Weldon Spross, Ed Evans, the late Dr. H. T. Gier, the late Dr. Gilbert Greenwald, the late Dr. Howard Matzke, and Dr. Douglas Poorman. Their teaching techniques made me a better teacher and textbook author, and for that I'm extremely grateful.

I am also deeply indebted to many people for their assistance during the writing of this book. A great debt of gratitude goes to the folks at Jones & Bartlett Learning. Many thanks to those who assisted with the production of the book, including my publisher Cathy Esperti, talented editor Erin O'Connor, editorial assistant Rachel Isaacs, production editor Lou Bruno, photo researcher Lauren Miller, and copyeditor Shellie Newell. I greatly appreciate the efforts of the Jones & Bartlett Learning account specialists who have helped make this book a success. It has been a pleasure and an honor to have worked with such a fine and talented group of people.

Thanks also to the many authors who contributed the Point/Counterpoints in this book. Your work will make this a more exciting journey for students as they begin to appreciate different perspectives of crucial issues.

Throughout this time, my two delightful sons, Skyler and Forrest, have offered considerable support and a counterbalance to the stresses and strains of a project of this magnitude. You're the light of my life, guys. A world of thanks to my sweetheart, Linda, for her continuous support, inspiration, guidance, and unwavering love.

Finally, a special thanks to all the reviewers on this and previous editions who offered many useful comments throughout this project. Their insight and attention to detail have been greatly appreciated. Below is a list of those who have reviewed the manuscripts.

D. Darryl Adams, Mankato State University
Donald K. Alford, Metropolitan State College
David R. Anderson, Pennsylvania State University–Fayette Campus
Felix Baerlocher, Mount Allison University
Jack Bennett, Northern Illinois University
Holly Boettger-Tong, Wesleyan College
Charles E. Booth, Eastern Connecticut State University
J. D. Brammer, North Dakota State University
Judith Byrnes-Enoch, Empire State College
Vic Chow, City College of San Francisco
Ann Christensen, Pima Community College
Tony Contento, Iowa State University
Peter Colverson, Mohawk Valley Community College
Francoise Cossette, University of Ottawa
John D. Cowlishaw, Oakland University
Penni Croot, SUNY Potsdam
Richard Crosby, Treasure Valley Community College

John Cummings, Waynesburg College
Jeffrey Dean, Vanderbilt University
Debby Dempsey, Northern Kentucky University
Melanie DeVores, Sam Houston State University
Stephen Freedman, Loyola University of Chicago
Sheldon R. Gordon, Oakland University
Martin Hahn, William Paterson College
John P. Harley, Eastern Kentucky University
Robert R. Hollenbeck, Metropolitan State College
Carl Johnson, Vanderbilt University
Wendel J. Johnson, University of Wisconsin, Marinette
Florence Juillerat, Indiana University–Purdue University
Martin Kapper, Central Connecticut State University
Ruth Logan, Santa Monica College
Charles Mays, DePauw University

David Mork, St. Cloud State University
Donald J. Nash, Colorado State University
Emily C. Oaks, State University of New York–Oswego
Lewis Peters, Northern Michigan University
Richard E. Richards, University of Colorado at Colorado Springs
Lynette Rushton, South Puget Sound Community College
Miriam Schocken, Empire State College
Richard Shippee, Vincennes University
Beverly Silver, James Madison University
Mary Vetter, University of Regina
David Weisbrot, William Paterson College
Richard Weisenberg, Temple University
Terrance O. Weitzel, Jr., Jacksonville University
Roberta Williams, University of Nevada–Las Vegas
Tommy Wynn, North Carolina State University

About the Author

Dr. Chiras received his Ph.D. in reproductive physiology from the University of Kansas Medical School in 1976. In September 1976, Dr. Chiras joined the Biology Department at the University of Colorado in Denver in a teaching and research position. Since then, he has taught numerous undergraduate and graduate courses, including general biology, cell biology, histology, endocrinology, and reproductive biology. Dr. Chiras also has a strong interest in environmental issues and has taught a variety of courses on the subject. He is currently a visiting professor at Colorado College in Colorado Springs, Colorado. He is also the founder and director of The Evergreen Institute Center for Renewable Energy and Green Building, where he teaches classes on residential solar electricity, wind energy, home energy efficiency, natural building, and more (www.evergreeninstitute.org).

Dr. Chiras is the author of numerous technical publications on ovarian physiology, critical thinking, sustainability, environmental education, green building, and renewable energy. He has also written numerous articles for newspapers and magazines on environmental issues. Dr. Chiras wrote the environmental pollution section for World Book's annual publication *Science Year in Review* from 1993 through 2012 and has written numerous articles on human biology and environmental issues for *World Book Encyclopedia, Encyclopedia Americana,* and *Grolier's Multimedia Encyclopedia.*

Dr. Chiras has published five college and high school textbooks, including *Environmental Science* (Jones & Bartlett Learning) and *Natural Resource Conservation: Management for a Sustainable Future* (with John P. Reganold and Oliver S. Owen; Prentice Hall).

Dr. Chiras's books include *Lessons from Nature: Learning to Live Sustainably on the Earth* (Island Press), *Study Skills for Science Students* (West) and *Essential Study Skills* (Brooks Cole). His books also include *The Natural House* (Chelsea Green), *The Solar House* (Chelsea Green), *The New Ecological Home* (Chelsea Green), *Superbia! 31 Ways to Create Sustainable Neighborhoods* (New Society), *EcoKids: Raising Children Who Care for the Earth* (New Society), *The Homeowner's Guide to Renewable Energy* (New Society), and *Games for the Classroom*, which he wrote with Ed Evans, Jr., his 9th grade physics and chemistry science teacher. His newest books include *Power from the Wind, Power from the Sun; Green Home Improvement; Wind Energy Basics;* and *Solar Electricity Basics.*

Dr. Chiras enjoys biking, organic gardening, hiking, canoeing, and music. He lives on a 60-acre farm in Gerald, Missouri, in a passive solar home supplied by solar and wind power. He raises grass-fed beef with his wife Linda.

Study Skills

College is a demanding time. For many students, term papers, tests, reading assignments, and classes require a new level of commitment to their education. At times, the workload can become overwhelming.

Fortunately, there are many ways to lighten the load and make time spent in college more profitable. This section offers some helpful tips on ways to enhance your study skills. It teaches you how to improve your memory, how to become a better note taker, and how to get the most out of what you read. It also helps you prepare better for tests and become a better test taker.

Mastering these study skills will require some work, mostly to break old, inefficient habits. In the long run, though, the additional time you spend now learning to become a better learner will pay huge dividends. Over the long haul, improved study skills will save you lots of time and help you improve your knowledge of facts and concepts. That will, no doubt, lead to better grades and very likely a more fruitful life.

General Study Skills

- Study in a quiet, well-lighted space. Avoid noisy, distracting environments.
- Turn off televisions and radios.
- Turn off your cell phone or return calls and text messages after your scheduled study time.
- Work at a desk or table. Don't lie on a couch or bed.
- Establish a specific time each day to study, and stick to your schedule.
- Study when you are most alert. Many students find that they retain more if they study in the evening a few hours before bedtime.
- Take frequent breaks—one every hour or so. Exercise or move around during your study breaks to help you stay alert.
- Reward yourself after a study session with a mental pat on the back or a snack.
- Study each subject every day to avoid cramming for tests. Some courses may require more hours than others, so adjust your schedule accordingly.
- Look up new terms or words whose meaning is unclear to you in the glossaries in your textbooks or in a dictionary.

Improving Your Memory

You can improve your memory by following the PMC method. The PMC method involves three simple learning steps: (1) paying attention, (2) making information memorable, (3) correlating new information with facts you already know.

Step 1 Paying attention means taking an active role in your education—taking your mind out of neutral. Eliminate distractions when you study. Review what you already know and formulate questions about what you are going to learn *before* a lecture or *before* you read a chapter in the text. Reviewing and questioning help prime the mind.

Step 2 Making information memorable means finding ways to help you retain information in your memory. Repetition, mnemonics, and rhymes are three helpful tools.

- Repetition can help you remember things. The more you hear or read something, the more likely you are to remember it, especially if you're paying attention. Jot down important ideas and facts while you read or study to help involve all of the senses.
- Mnemonics are useful learning tools to help remember lists of things. I use the mnemonic CARRRP to remember the biological principles of sustainability: conservation, adaptability, recycling, renewable resources, restoration, and population control.
- Rhymes and sayings can also be helpful when trying to remember lists of facts.
- If you're having trouble remembering key terms, look up their roots in the dictionary. This often helps you remember their meaning.
- You can also draw pictures and diagrams of processes to help remember them.

Step 3 Correlating new information with the facts and concepts you already know helps tie facts together, making sense out of the bits and pieces you are learning.

- Instead of filling your mind with disjointed facts and figures, try to see how they relate with what you already know. When studying new concepts, spend some time tying information together to get a view of the big picture.
- After studying your notes or reading your textbook, go back and review the main points. Ask yourself how this new information affects your view of life or critical issues and how you may be able to use it.

Becoming a Better Note Taker

- Spend 5 to 10 minutes before each lecture reviewing the material you learned in the previous lecture. This is extremely important!
- Know the topic of each lecture before you enter the class and spend a few minutes reflecting on facts you already know about the subject about to be discussed.
- If possible, read the text before each lecture. If not, at least look over the main headings in the chapter, read the topic sentence of each paragraph, and study the figures. If your chapter has a summary, read it too.
- Develop a shorthand system of your own to facilitate note taking. Symbols such as = (equals), > (greater than), < (less than), w/ (with), and w/o (without) can

save lots of time so you don't miss the main points or key facts.

- Develop special abbreviations to cut down on writing time. "M" might stand for "muscle," "T" might be used for "trachea," and "NI" could be used to signify "nerve impulse."
- Omit vowels and abbreviate words to decrease writing time (for example: omt vwls & abbrvte wrds to dcrs wrtng tme). This will take some practice.
- Don't take down every word your professor says, but be sure your notes contain the main points, supporting information, and important terms.
- Watch for signals from your professor indicating important material that might show up on the next test (for example, "This is an extremely important point . . .").
- If possible, sit near the front of the class to avoid distractions.
- Politely ask your professor to repeat key points you didn't have a chance to jot down in your notes.
- Review your notes soon after the lecture is over while they're still fresh in your mind. Be sure to leave room in your notes written during class so you can add material you missed. If you have time, recopy your notes after each lecture.
- Compare your notes with those of your classmates to be sure you understood everything and did not miss any important information.
- Attend all lectures.
- Use a tape recorder if you have trouble keeping up.
- If your professor talks too quickly, politely ask him or her to slow down.
- If you are unclear about a point, ask during class. Chances are other students are confused as well. If you are too shy, go up after the lecture and ask, or visit your professor during his or her office hours.

How to Get the Most Out of What You Read

- Before you read a chapter or other assigned readings, preview the material by reading the main headings or outline to see how the material is organized.
- Pause over each heading and ask a question about it.
- Next, read the first sentence of each paragraph. When you have finished, turn back to the beginning of the chapter and read it thoroughly.
- Take notes in the margin or on a separate sheet of paper. Underline or highlight key points. You can save a lot of time by simply putting a check mark at the beginning of key sentences.
- Don't skip terms that are confusing to you. Look them up in the glossary or in a dictionary. Make sure you understand each term before you move on.
- Use the study aids in your textbook, including summaries and end-of-chapter questions. Don't just look over the questions and say, "Yeah, I know that." Write out the answer to each question as if you were turning

it in for a grade, and save your answers for later study. Look up answers to questions that confuse you. This book has questions that test your understanding of facts and concepts. Critical thinking questions are also included to sharpen your skills.

Preparing for Tests

- Don't fall behind on your reading assignments.
- Review lecture notes as often as possible.
- If you have the time, you may want to outline your notes and assigned readings.
- Space your study to avoid cramming. One week before your exam, go over all of your notes. Study for two nights, then take a day off from that subject. Study again for a couple of days. Take another day off from that subject. Then make one final push before the exam, being sure to study not only the facts and concepts but also how the facts are related. Unlike cramming, which puts a lot of information into your brain for a short time, spacing will help you retain information for the test and for the rest of your life.
- Be certain you can define all terms and give examples of how they are used.
- You may find it useful to write flash cards to review terms and concepts.
- After you have studied your notes and learned the material, look at the big picture—the importance of the knowledge and how the various parts fit together.
- You may want to form a study group to discuss what you are learning and to test one another.
- Attend review sessions offered by your instructor or by your teaching assistant, but study before the session and go to the session with questions.
- See your professor or class teaching assistant with questions as they arise.
- Take advantage of free or low-cost tutoring offered by your school or, if necessary, hire a private tutor to help you through difficult material. Get help quickly, though. Don't wait until you are hopelessly lost. Remember that learning is a two-way street. A tutor won't help unless you are putting in the time.
- If you are stuck on a concept, it may be that you have missed an important point in earlier material. Look back over your notes or ask your tutor or professor what facts might be missing and causing you to be confused.
- If you have time, write and take your own tests. Include all types of questions.
- Study tests from previous years, if they are available legally.
- Determine how much of a test will come from lecture notes and how much will come from the textbook.

Taking Tests

- Eat well and get plenty of exercise and sleep before tests.

- Remain calm during the test by deep breathing.
- Arrive at the exam on time or early.
- If you have questions about the wording of a question, ask your professor.
- Skip questions you can't answer right away, and come back to them at the end of the session if you have time.
- Read each question carefully and be sure you understand its full meaning before answering it.
- For essay questions and definitions, organize your thoughts first on the back of the test before you start writing.

Now take a few moments to go back over the list. Check off those things you already do. Then, mark the new ideas you want to incorporate into your study habits. Make a separate list, if necessary, and post it by your desk or on the wall and keep track of your progress.

(© Deklofenak/Shutterstock, Inc.)

chapter 1

Life in the Balance: Human Health, Homeostasis, Evolution, Science, and Critical Thinking

Three to four million years ago, humanlike organisms roamed the grasslands of Africa (Figure 1-1). Scientists dubbed them *Australopithecus afarensis* (aus-TRAL-owe-PITH-a-CUSS A-far-EN-suss). Standing only three feet tall and walking upright, these creatures, one of our early ancestors, subsisted in large part on a diet of roots, seeds, nuts, and fruits. Studies suggest that they supplemented their primarily vegetarian diet with carrion, animals that had been killed by predators or that had died from other causes. Our early ancestors also may have captured and killed animals for meat.

THINKING CRITICALLY

Your local newspaper reports the results of an experiment a local high school student performed to test the effects of a special healthy diet on the cholesterol content of chicken eggs. He obtained 20 chickens from two different breeders. Chickens from breeder A were fed his special diet; the other birds from breeder B were fed a diet of standard chicken feed, which he purchased at the local livestock feed store. The boy found that the eggs from the chickens fed his healthy diet had lower levels of cholesterol than the eggs from the other group. The story created quite a stir in the local media—so much so that the boy's father is trying to acquire funding to market his son's new feed. Do you see any potential problems with this study?

FIGURE 1-1 *Australopithecus afarensis* Current scientific evidence suggests that *Australopithecus afarensis* was the first humanlike ape. Its skeletal remains indicate that it walked upright.

Weak and slow compared to other large animals, our early ancestors could have easily ended up as an evolutionary dead end. Fortunately, though, they possessed several anatomical features that tipped the scales in their favor. Undoubtedly, one of the most important characteristics was their brain.

Thanks in large part to our brains, human beings have not only survived but flourished. Today, humans inhabit a world of marvelous technologies, which make our lives easier, more convenient, and more fun. Rather than roaming in small bands, as our early ancestors did, the majority of the world's people today live in cities and towns that offer amenities our ancestors never would have dreamed possible. Rather than collecting nuts and berries from the plants around us, most people in the modern world purchase their food from grocery stores supplied by highly mechanized farms. Many farmers are now using satellites and remote sensing devices, as well as computerized machinery to produce food for the seven billion human inhabitants. Today, thanks to advances in genetics, scientists have begun to alter the genetic material

of the cells of plants and animals to increase food production. They have even developed tissues and organs that could be transplanted into human beings, replacing worn out or damaged body parts. Scientists have even begun to manipulate our hereditary material in an attempt to cure diseases long thought to be untreatable.

For better or for worse, humans have become a major player in evolution, a process of genetic change that results in structural, functional, and behavioral changes in groups of organisms known as populations. These changes, in turn, result in organisms better equipped to cope with their environment—that is, better able to survive and reproduce.

From most perspectives, the human experiment has been an overwhelming success. However, a growing body of evidence shows that human success has its price. As society advances, we are also causing considerable damage to the life-support system of the planet upon which we—and all other species—depend. Such changes, scientist warn us, could have dramatic effects on life on Earth, even our own survival.

Like other texts on human biology, the bulk of this book will take you on a very important journey through the human body with many practical applications. On this journey, you will learn a great deal about yourself—how you got here, how you inherited certain characteristics from your parents, and how your body functions. You will learn how broken bones mend and study the basics of nutrition. You will discover how your immune and nervous systems operate. You will learn about many common diseases and—perhaps more important—how to prevent them. In Health Tips like the one on this page, you will learn about the many things you can do to live a long and healthy life.

As you will soon see, the information you learn from this book will prove useful to you in many ways. It will also help you understand important political debates over issues such as genetic engineering, stem cell research, vaccination, and even energy and pollution. It will help you understand injuries and diseases from which you and loved ones may suffer.

Human biology is also a fascinating subject. As you proceed through this book, be sure to take time to marvel at the human body—the intricate details of the cell, the structure and function of organs, and the intriguing manner in which the various parts work together.

Health Tip 1-1

To learn more quickly, perform better in school, live a more emotionally stable life, and reduce your chances of getting sick, be sure to get plenty of sleep.

Why?

Inadequate sleep not only impairs memory, it makes it harder to learn. Inadequate sleep can also result in emotional instability and lowers immune defenses, increasing the likelihood that you will contract the common cold or the flu. Lack of sleep is one reason why college students often suffer from one cold after another.

1-1 Health and Homeostasis

Most of us want to live a long, healthy life. Human health depends on numerous biological mechanisms that evolved over many millions of years. These internal biological processes help to maintain a fairly constant internal condition within our bodies, a state often referred to as *homeostasis* (home-e-oh-STAY-siss).

What Is Homeostasis?

The term **homeostasis** comes from two Greek words, homeo, which means "the same," and stasis, which means "standing." Literally translated, homeostasis means "staying the same." Some people refer to homeostasis as a state of internal constancy. In reality, however, homeostasis is not a static state; it is a dynamic (ever-changing) state. To understand what I mean, consider a familiar example, body temperature.

Humans are warm-blooded creatures. What that means is that we generate body heat internally and thus maintain a fairly constant body temperature—about 98.68° F. If you were to measure your body temperature through the day, however, you would find that it varies. Body temperature, for instance, falls slightly at night when you are asleep and rises during daylight hours. It increases even more—sometimes a lot more—when you participate in strenuous physical activity.

Like many other internal conditions, then, body temperature fluctuates within a range. This is what is meant when we say that body temperature is in homeostasis: It changes a bit from time to time, but remains more or less constant.

Homeostasis is achieved through a variety of automatic mechanisms that compensate for internal bodily changes and external changes—changes in our environment (Figure 1-2). Homeostatic mechanisms require sensors, structures that detect internal and external change—for example, temperature sensors in the skin that keep track of air temperature. These sensors elicit a response that offsets the change, helping to maintain a fairly constant state. On very cold days, for example, sensors

in our skin detect the cold, chilly air. If it is cold enough, they stimulate shivering, a rhythmic contraction of muscles that generates body heat, compensating for low temperatures. This is just one of many homeostatic mechanisms in our bodies.

Homeostatic mechanisms also maintain fairly constant levels of nutrients as well as other chemicals like salts and hormones in our blood. Maintaining constant levels of dozens of chemicals inside our bodies is vital for maintaining human health, survival, and reproduction.

Homeostatic mechanisms also exist in ecosystems. An ecosystem is a biological system consisting of organisms and their environment. Homeostatic mechanisms help achieve balance in ecosystems.

A highly simplified example illustrates the point. In the grasslands of Kansas, rodent populations generally remain

FIGURE 1-2 Keeping Warm The human body is remarkably able to tolerate a wide variety of conditions thanks to internal mechanisms that maintain relatively constant internal conditions. (© Losevsky Pavel/ ShutterStock, Inc.)

FIGURE 1-3 Predator Control Snakes play an important role in controlling rodent populations, thus helping to maintain ecosystem homeostasis. (© MikeE/ShutterStock, Inc.)

fairly constant from one year to the next. This phenomenon results, in part, from predators, animals that hunt and kill other organisms. Predators such as snakes, coyotes, foxes, and hawks feed on rodents and, thus, help to control rodent populations (Figure 1-3).

Although predators are a crucial element in maintaining environmental homeostasis in these grasslands and virtually all other natural systems, a host of other factors also contribute to it, such as weather and food supplies. It is the net effect of these factors that determines population sizes.

In this book, the term *homeostasis* is used to refer to the balance that occurs at all levels of biological organization—from cells to organisms to ecosystems. The abundance of homeostatic mechanisms in nature suggests their importance to life on Earth. These mechanisms are just one of the many positive outcomes of evolution.

Maintaining "balance" is essential to the health and welfare of all organisms, humans included at many levels. Without it, cells would fall into disarray, organisms would perish, and ecosystems would collapse.

> **KEY CONCEPTS**
>
> Homeostasis is a state of relative constancy that helps ensure human health; it is achieved automatically by numerous physiological processes in the body that respond to internal and external changes.

Healthy Environments

As you might suspect, the health of our environment and the health of the organisms that live in the environment, including us human beings, are closely linked. Changes in the chemical composition of the air we breathe caused by pollution, for instance, can have significant negative impacts on human health. Only recently, scientists discovered that certain emissions from trucks, buses, cars, and coal-fired power plants (called polycyclic aromatic hydrocarbons or PAHs) can result in lower birth weight and impaired mental development in young children who were exposed while still in their mother's womb.

The health of an organism is also affected by less tangible, but very real, changes in our social and psychological environment. Highly stressful environments, for example, can lead to more frequent colds and other serious ailments. **Health and Homeostasis** sections in this text outline some of the connections between human health and the health of our environment.

Although humans are the central focus of this text, it is important to note that many of the species that share this world with us are affected by the condition of the environment. Scientists, for instance, are finding that many drugs that people take like those chemicals found in birth control pills are excreted in their urine and end up in the effluent of sewage treatment plants. From there they enter rivers, lakes, and streams. These chemicals are having profound effects on the growth, reproduction, and survival of aquatic species, especially fish. Scientists are concerned that one class of chemicals, antibiotics, in our waterways such as lakes and streams could result in antibiotic resistant bacteria. Humans who ingest these bacteria in drinking water could become deathly ill. Doctors worry that they won't have antibiotics to treat the resistant strains. An even greater problem may be the extensive use of antibiotics—and lots of them—in the production of poultry and livestock for human consumption.

> **KEY CONCEPTS**
>
> Human health depends on healthy chemical, physical, and psychological environment.

Dimensions of Health

For many years, human health was defined as the absence of disease (Figure 1-4a). As long as a person had no obvious symptoms of a disease, that person was considered healthy. Although such a person may have had clogged arteries from a lifetime of fatty hamburgers and snack foods, it wasn't until symptoms of heart disease—for example, chest pain—became apparent that the patient was considered unhealthy.

> **Health Tip 1-2**
>
> To lose or maintain weight, eat larger portions of vegetables and whole grains and cut way back on meat and fats.
>
> *Why?*
>
> Vegetables and whole grains contain a lot less fat—and, therefore, fewer calories—than meats. Whole grains, surprisingly enough, also contain proteins that your body needs. Vegetables provide many nutrients not found in meats that the body requires for long-term health. Vegetables also contain fiber that is important for health. Remember, only a small portion of meat is needed to satisfy the body's need for protein.

Today, health experts rely on a more comprehensive definition of health. It takes into account *both* physical and emotional well-being.

(a) The old concept

Poor health ←————————————————————————→ Good health

Obvious disease or illness	No obvious disease or illness

(b) The new concept

Poorest health ←—— Poor health —— Good health ——→ Best health

Obvious disease or illness	More risk factors	A few risk factors	No obvious disease or illness
Many risk factors			No risk factors
Poor fitness			Good fitness
Poor mental health			Good mental health

FIGURE 1–4 **Old and New Concepts of Health** (left, © Patrick Sheandell O'Carroll/PhotoAlto/PictureQuest; right, © Jim Boorman/Pixland/age fotostock.)

Physical health refers to the state of the body—how well it is working. Physical health can be measured by checking temperature, blood pressure, blood sugar levels, and a number of other variables. Abnormalities in these measurements may be a signal that one's physical health is in jeopardy, even though there are no obvious symptoms of illness. Medical scientists use the term *risk factor* to refer to abnormal conditions such as high blood pressure or high blood cholesterol levels that put a person at risk for disease. The presence of one or more risk factors is a sign of less-than-perfect health. Obviously, the more risk factors there are, the worse one's physical health is (Figure 1-4b).

As shown in Figure 1-4b, the absence of risk factors indicates in the best health. A few risk factors indicate that your health is less than optimum—just good. More risk factors indicate your health is poor. Even in poor health, you may not exhibit any symptoms—at least not yet. A friend of mine who exhibited no signs of heart disease at age 50 put her children to bed, went downstairs to the couch, then suffered a fatal heart attack. Under the new and more realistic concept of health, then, even though you may feel okay and not exhibit obvious signs of disease, such as a failing heart, the presence of risk factors indicates that your health is compromised.

Scientists also use the term risk factor to refer to activities that make an individual more likely to develop diseases. Smoking, lack of exercise, and a fatty diet, for example, are risk factors for heart and artery disease. Many people today lack exercise and eat poorly, putting them at risk for a wide assortment of diseases, including heart attack and late-onset diabetes.

Health Tip 1-3

Regular exercise is essential to good health, so, try to get at least 30 minutes of aerobic exercise (running, riding a bike, or swimming) at least 3 times a week. More is better.

Why?

Exercise helps us maintain a healthy weight. Being overweight increases the risk of many diseases, from heart attack and stroke to late-onset diabetes to breast cancer in women. It is important to remember that exercise burns calories directly, but also builds muscle mass. Muscle has a higher metabolic rate than fat, and it continues to burn calories after we're done exercising, providing prolonged benefits! The cool thing about staying in shape is you may be able to eat more and not gain weight!

Physical health is also measured by one's level of physical fitness. If you can't walk up a set of stairs without gasping for air, you're not considered very physically fit. You're more likely to have other problems later in life—for example, heart disease.

Emotional well-being also factors into an assessment of a person's health. Especially relevant is your ability to cope with stress. Inability to cope may lead to physical problems, such as high blood pressure and heart disease.

Mental and physical fitness are measures of our abilities to meet the demands of life. Fit people are able to cope with daily psychological stresses and are able to move about without becoming short of breath. They're also better employees, less likely to take days off because of illness. For that and other reasons, some companies are now paying employees to adopt more healthy lifestyles. IBM, for instance, offers

TABLE 1-1	Healthy Habits

Sleep seven to eight hours per day[*]

Eat a healthy breakfast every day

Eat a healthy diet with lots of fruits and vegetables

Avoid snacking on junk food (sweets or fatty foods) between meals

Maintain ideal weight

Do not smoke

Avoid alcohol or use it moderately

Exercise regularly

Manage stress in your life

[*] Not all people need this much sleep. If you're one of them, don't try to force yourself to sleep more than you need.

financial incentives to employees who exercise, lose weight, and stop smoking. Some employees receive an extra $600 a year in incentives for pursuing a path to a healthier life. Why would a company do this?

IBM estimates that for every dollar it spends on promoting wellness, it saves $3 in health care costs. Employees require much less sick time, too, which enhances productivity. Currently, more than half of IBM's workforce has signed up for the program. Maintaining good health is a lifelong job that is best begun early in life, but it is never too late to steer onto a healthy path. Table 1-1 lists numerous healthy habits.

By incorporating these habits into your lifestyle, you can increase your chances of living a long, healthy life. You may want to start slowly, incorporating one idea after another.

> **KEY CONCEPTS**
>
> Human health is a state of physical and mental well-being characterized by absence of disease and risk factors that could lead to problems in the future.

Health and Homeostasis

As just pointed out, your mental and physical health depends on homeostatic mechanisms. When these mechanisms are out of whack or break down completely, illness results. Persistent stress, for example, can disrupt several of the body's homeostatic mechanisms, leading to disease. It also weakens the immune system, making us more prone to viruses and bacteria. According to a recent study, people under stress are twice as likely to suffer from colds and the flu as those who are not. If it is prolonged, stress can increase the risk of diseases of the heart and arteries. Fortunately, stress can be dramatically reduced by exercise, relaxation training, massage, acupuncture, and a number of other measures discussed in Health Note 1-1.

> **KEY CONCEPTS**
>
> Human health is dependent on properly functioning homeostatic systems; damage to these systems lead to many common diseases.

1-2 Evolution and the Characteristics of Life

Homeostasis is a central theme of this book because it is so essential to maintaining health and to the continuation of life. It is also important because so many human activities upset homeostatic mechanisms—to the detriment of humans and all living beings. Another key concept of biology and a subtheme of this book is evolution. A few words on the subject are essential to your understanding of human biology.

All life forms alive today exist because of evolution. In fact, every cell and every organ in the human body is a product of millions of years of evolution. As just noted, even the intricate homeostatic mechanisms evolved over long periods. How did life evolve?

Figure 1-5 shows the five major groups or kingdoms of organisms that exist today. This diagram also illustrates evolutionary relationships—how these kingdoms are related. The simplest, bacteria-like organisms, belonging to a group called the monerans, were the first to evolve. They gave rise to a more complex set of organisms, known as the protistans. The protistans consist of single-celled organisms such as amoebas and paramecia. During the course of evolution, the protistans gave rise to three additional major groups, or kingdoms: plants, fungi, and animals. We humans belong to the animal kingdom.

As evolutionary biologists have discovered, this common lineage is responsible for the striking similarities among the Earth's organisms. One common feature of all living organisms is that they rely on the same type of genetic material. Other similarities also exist. A comparison of certain anatomical features, such as the bones in a person's arm and in the wings of birds, reveals a remarkable resemblance that speaks of a common ancestry. Let's take a closer look at the common characteristics of living things.

Common Characteristics of Organisms

An analysis of organisms reveals eight common features, referred to as the characteristics of life. A brief look at these characteristics not only shows our evolutionary connection with other organisms but also helps us understand human beings better.

The first characteristic of life is that all organisms, including humans, are made of cells. Cells are tiny structures that are the fundamental building blocks of living organisms. Cells, in turn, consist of molecules, nonliving particles composed of smaller units called atoms. Glucose molecules, for instance, contain carbon, hydrogen, and oxygen atoms. Molecules, in turn, combine to form the parts of the cell.

The second characteristic of life is that all organisms grow and maintain their complex organization by taking in molecules and energy from their surroundings. As you will see in subsequent chapters, humans must expend considerable amounts of energy to maintain their complex internal organization. In fact, 70% to 80% of the energy adults need is used just to maintain their bodies—to transport molecules in and out of cells, to make proteins in cells, and to perform other basic body functions. The rest is used for activity—walking, running, talking, and so on.

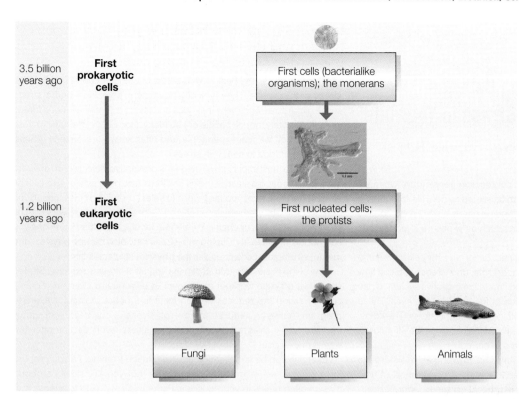

3.5 billion years ago

First prokaryotic cells

First cells (bacterialike organisms); the monerans

1.2 billion years ago

First eukaryotic cells

First nucleated cells; the protists

Fungi

Plants

Animals

FIGURE 1-5 Evolution of the Pro-karyotic and Eukaryotic Cells This diagram illustrates the evolutionary history of life. Organisms fall into one of five major groupings, or kingdoms. The first life-forms were the monerans, which are single-celled prokaryotic organisms. They gave rise to the protistans, single-celled eukaryotic organisms. Eukaryotic protistans, in turn, gave rise to plants, fungi, and animals. (Cells, courtesy of CDC; amoeba, courtesy of Sutherland Maciver, School of Biomedical Sciences, University of Edinburgh; mushroom, © Photos .com; flower, © Photodisc; fish, courtesy of Eric Engbretson/U.S. Fish & Wildlife Service.)

The third characteristic of life is that all living things exhibit a feature called metabolism. Metabolism refers to the chemical reactions in the cells and tissues of organisms. These reactions consist of two types—those in which food substances are built up into living tissue (anabolic reactions) and those in which food is broken down into simpler substances, often releasing energy (catabolic reactions). In human body cells, millions of reactions occur each second to maintain life.

The fourth characteristic of life is homeostasis. All organisms depend heavily on homeostatic mechanisms. They ensure health, reproduction, and the survival of individual organisms and entire species (groups of similar organisms).

The fifth characteristic of living things is that they sense and respond to changes in their environment. The ability to perceive stimuli and respond to them is important in the day-to-day survival of all organisms and the survival of all species over time.

The sixth characteristic of life is reproduction and growth. All organisms are capable of reproduction and growth. Humans produce offspring by combining sex cells, sperm and eggs from males and females, respectively.

The seventh characteristic of life is evolution. All life is capable of evolving, that is, changing with changes in their environment. Over time, these changes may lead to the emergence of entirely new species. The mechanism by which species change is explored elsewhere in the text.

The eighth characteristic of life is that all organisms are part of the Earth's ecosystems. Humans, like every other plant, animal, and microorganism, are dependent on the Earth's ecosystems. They depend on them for food, fiber, water, oxygen, and a host of free services such as natural flood control. The Earth's ecosystems are not only the life-support system of the planet, they are essential to the human economy. All wealth ultimately springs from the Earth and its ecosystems.

KEY CONCEPTS

Human beings share many similarities with other living beings, like metabolism, reproduction, and homeostasis.

What Makes Humans Unique?

Human beings are one of evolution's many products. Although we are like many other species, we are a unique form of life. Several features distinguish us from other species.

One of the key differences between humans and other animals is our ability to acquire and use complex languages. Another distinguishing feature is our culture. Culture has been defined in many ways. Humorist Will Cuppy remarked that culture is anything we do that monkeys don't. On a more serious note, culture can be defined as the ideas, values, customs, skills, and arts of a people. While other species may have some rudiments of culture—for instance, some communication skills—humans are unique in the biological world because of the complexity of our cultural achievements (Figure 1-6a).

Humans also differ from other animals in our ability to plan for the future. Although a few other animals seem to share this ability, most "planning"—like a bird's nest-building activities—is probably the result of instinct programmed by the animal's genetic material. In contrast, building skyscrapers requires an extraordinary amount of forethought.

Another unique characteristic of humans is our unrivaled ability to shape the environment. Although other species can alter their environment, we possess an extraordinary capacity in this regard. We erect huge dams in steep-sided gorges to protect downstream areas against floods and to create a year-round supply of water. We till the soil to grow food. We extract valuable minerals from the depths of the Earth. We build islands to provide land to make new cities.

health**note**

1-1   Maintaining Balance: Reducing Stress in Your Life

Stress is a normal occurrence in everyday life. But just what is it?

Stress is a psychological and physical reaction people have when we are exposed to certain stimuli. Stress can be elicited by non–life-threatening situations such as a blind date or a final exam. It also can be caused by potentially life-threatening situations such as exposure to dangerous machinery in a factory.

Stressful stimuli can be real or imagined. Either way, they elicit the same response in the body: an increase in heart rate, an increase in blood flow to the muscles and a decrease in blood flow to the digestive system, a rise in blood sugar (glucose), and a dilation of the pupils of the eyes. These physical changes in the body help us to respond to the stress. They may, for instance, help us flee from a stressful stimulus. Once the stimulus is gone, though, the body typically returns to normal.

How stress affects us, however, depends on how long we're exposed to it. If the stressful stimulus is short-lived, our bodies recover nicely. Some argue that a little stress may actually improve a person's performance.

Long-term exposure to stressful stimuli, however, can have serious consequences. As a result, a prolonged period of stress may lead to disease. One reason this happens is that the body's immune system is often depressed by stress. The immune system protects us from bacteria and viruses that cause colds and flu and other diseases. Prolonged stress also results in changes in the blood vessels. These changes may accelerate the accumulation of cholesterol, which clogs the arteries and may eventually result in strokes and heart attacks.

Stress doesn't affect all people in the same way. Some people recognize the stress they're feeling and channel its energies into productive work. They are better able to cope with stress. Psychologists believe that people who handle stress the best have a sense of being in control, despite the stress of their work. They typically have clear objectives and a strong sense of purpose. They view their jobs and life as a challenge, not a threat.

Unfortunately, not all people are so lucky. Many of us are not in a position of control; we feel expendable and often view ourselves as victims. What can be done to deal with stress?

One of the most important strategies is preventive: selecting an environment and creating a lifestyle that is as stress-free as possible. As a college student, for instance, you may want to select a realistic class load. If you must work to pay your way through college or if you're taking very difficult courses like physics or chemistry, sign up for a lighter class load—one you can handle more easily. Coping with stress may also require physical and mental strategies. Let's consider the physical strategies first.

One of the easiest ways to lessen the impact of stress is exercise. Studies show that a single workout at the gym, a bike ride, a swim, or a cross-country skiing trip reduces tension for 2 to 5 hours. A regular exercise program, however, reduces the overall stress in one's life. Individuals who are easily stressed usually find that stress levels decline after two weeks of exercise.

Exercise can be supplemented by relaxation training. As you prepare for a difficult test or get ready for a date that you are nervous about, tension often builds in your muscles. Periodically stopping to release that tension helps to reduce physical stress. Stretching or taking a walk can help. Some people find it useful to tighten their muscles forcefully and then let them relax. Massage therapy and acupuncture can also be used to reduce stress. Stress-reduction programs on CDs, videos, and DVDs can teach relaxation methods, as can trained therapists.

Psychological factors may also play a role in stress. In many instances, people have an exaggerated response to a stressful stimulus. Dealing with the thoughts that exaggerate your stress can help you better control your stress. Start paying attention to the thoughts that provoke anxiety in your life. Are they exaggerated? If so, why? For example, are you nervous before exams? Why? Would better preparation reduce your anxiety? You might

(a)

(b)

FIGURE 1–6   Human Culture: Progress and Setbacks   (a) Architecture is one of our greatest achievements. But in reshaping our environment, we sometimes create enormous backlashes. (b) Earthen walls, or levees, along the Mississippi River, for example, have been used to control flooding in the past 100 years but have led to more frequent flooding in downstream areas. (a, © David Acosta Allely/ ShutterStock, Inc.; b, courtesy of Andrea Booher/FEMA Photo News.)

remind yourself that you've been through stressful situations before and always triumphed. This could help you better cope with stress.

Finding the source of your anxiety and taking positive action to alleviate it are helpful ways of reducing stress. However, stress reduction is not always easy. Test anxiety, for example, may be deeply rooted in feelings of inadequacy. Many people struggle with low self-esteem. A trained psychologist can help you find the causes and assist you in learning to feel better about yourself. Psychological help is as important as medical help these days, given the complexity and pace of our society. There is no shame in seeking counseling.

Biofeedback is another form of stress relief. A trained healthcare worker places sensors on you and connects them to a machine that monitors heart rate, breathing, muscle tension, or some other physiological indicator of stress (Figure 1). During a biofeedback session, your trainer first will help you relax, and then discuss a stressful situation. When one of the indicators shows that you are suffering from stress, a signal is given off. Your goal is to consciously reduce the frequency of the signal. For example, if your heart started beating faster when you thought about taking an exam, the machine will make a clicking sound. By breathing deeply and relaxing, you consciously slow down your heart rate; at that point the clicking sound slows down and then disappears.

Learning to recognize the symptoms of stress and to counter them is the goal of biofeedback. Eventually, you should be able to do it without the aid of a machine.

You can also reduce tension by managing your time and your workload efficiently. Numerous books on this subject can help you learn to budget your time more effectively. See the Study Skills section at the beginning of this book for ideas on ways to be a more efficient learner.

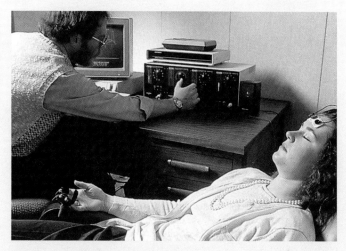

FIGURE 1 **Biofeedback** Student in a biofeedback session. (© Will and Deni McIntyre/Science Source, Inc.)

If these techniques don't work, you may want to see a doctor. He or she can prescribe medications that relieve anxiety and muscle tension, help you sleep, or combat depression. Herbal remedies such as valerian root are also available.

Relieving stress in our lives helps us reduce the risk of disease and enables us to relax and enjoy life. It also makes us more pleasant to be around. All in all, it is best to start learning early in life how to reduce or cope with stress. Lessons learned now will be useful for many years to come.

 Visit Human Biology's Internet site for links to websites offering more information on this topic.

Despite the benefits of our remarkable technologies, attempts to control nature sometimes backfire, creating larger problems. For instance, studies show that efforts to control upstream flooding on the Mississippi River by building levees along sections of the river have led to more frequent floods and more damage downstream (Figure 1-6b). Levees prevent water from spilling over the banks of rivers. While this protects upstream communities, the result is a larger slug of water delivered to downstream communities and hence increased flooding in such areas—unless they are protected by even higher levees.

KEY CONCEPTS

Although human beings are similar in many ways to all other life forms, we do have some unique characteristics, including our ability to acquire and use complex language and to think and plan ahead.

1-3 Understanding Science

The systematic study of the universe and its many parts falls into the realm of science. The term science comes from the Latin word *scientia*, which means "to know" or "to discern." Today, science is defined as a body of knowledge derived from observation and experimentation. It also involves ways of learning facts. In other words, it requires methods of acquiring information.

Many people view science as an uninteresting endeavor best left in the hands of a select brainy few (Figure 1-7). In reality, science is an exciting endeavor that often involves enormous creative energy. Because it teaches us about the workings of the world around us, science can be a source of great fascination. As the paleontologist Robert Bakker, a consultant to the company that made the dinosaurs for the movie *Jurassic Park*, once noted, science is "fun for the mind."

Science also has a practical side. It provides information that can improve our lives. It helps us understand important phenomena such as the weather and the spread of disease. Knowledge of science makes us better voters, better able to understand many complex issues. And an understanding of

FIGURE 1-7 **Not Just for Scientists** Science is essential to human progress and a benefit to all of us. It is a process that we all engage in. (© JHDT Stock Images LLC/ShutterStock, Inc.)

science, notably human biology, can help us make informed decisions about our lifestyle—what we eat, how much we eat, how much we sleep, and the amount of exercise we get.

The Scientific Method

Scientists employ a technique called the **scientific method** to obtain information. As shown in Figure 1-8, the scientific method generally begins with observations and measurements. In some cases, these may be part of carefully conducted experiments. Others may occur more haphazardly. A scientist on vacation in the tropics, for instance, may notice a phenomenon that sparks her curiosity that, in turn, leads to experimentation.

You may not realize that most of us use the scientific method in our day-to-day lives. To see how the scientific method works, suppose you sat down at a computer, turned it on but nothing happened. You might also have noticed that the lights in the room hadn't come on. These two observations would lead to a **hypothesis** (high-POTH-eh-siss), a

tentative explanation of the phenomenon. From your observations, you might hypothesize that the electricity in your house was off.

You could test your hypothesis by performing an experiment. An experiment is a procedure designed to test some idea. In this case, all you would need to do would be to try the light switch in the kitchen. If the lights in the kitchen worked, you would reject your original hypothesis and form a new one. Perhaps, you hypothesize, the circuit breaker to your study had been tripped. To test this idea, you would "perform" another experiment, locating the circuit breaker to see if it was turned off. If the circuit breaker was off, you would conclude that your second hypothesis were correct. To substantiate your conclusion, you would throw the switch and see if your computer worked.

This process involving observation, hypothesis, and experimentation forms the foundation of the scientific method. Although scientific experimentation may be much more complicated than discovering the reason for a computer failure, the process itself is the same.

Experimentation in biology usually requires two groups: experimental and control. The **experimental group** is the one that is tested or manipulated in some way. The **control group** is not tested or manipulated. Both groups are treated identically except in one way. The difference in treatment is known as the **experimental variable**. For example, they are fed the same diet, given water freely, housed under identical lighting conditions in cases of the same size. One group, however, is exposed to a potentially toxic pollutant. Valid conclusions come from comparing the two groups. Consider an example to illustrate this point. In order to test the effect of a new drug on laboratory mice, a good scientist would start with a group of mice of the same age, sex, weight, genetic composition, and so on (Figure 1-9). These animals would be divided into two groups, the experimental and control groups. Both groups would be treated the same throughout the experiment, receiving the same diet and being housed in the same type of cage at the same temperature. The only

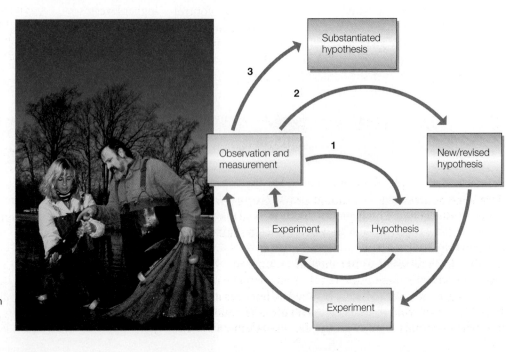

FIGURE 1-8 **Scientific Method** Scientific study begins with observation and measurement. These activities lead to hypotheses that can be tested by experiments. New and revised hypotheses are derived from experimentation. (Courtesy of Scott Bauer/USDA.)

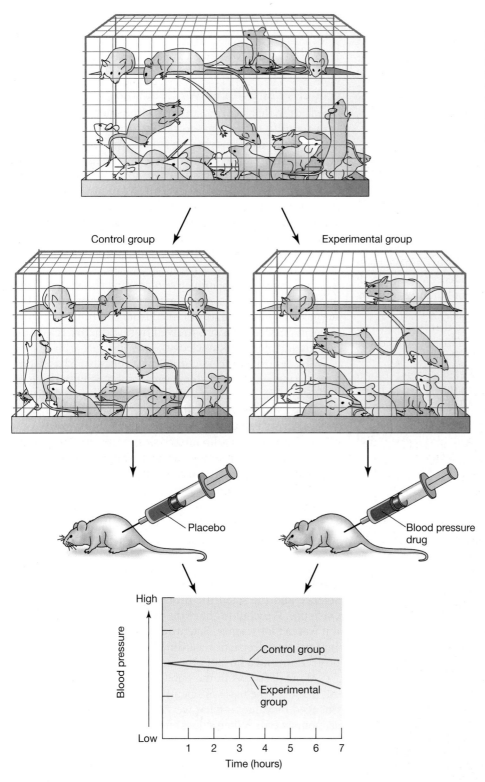

FIGURE 1–9 **Experimental Design** A good experiment uses a large number of identical subjects. They are placed into two groups: control and experimental. Both groups are treated the same, except for one variable. In this case, the experimental group is given an experimental blood pressure-lowering drug. The control group is given an injection of saline, the solution in which the drug is dissolved.

will be—because of possible variation in response. In laboratory experiments, at least 10 test animals are required for both control and experimental groups for reliable statistical analysis; groups larger than 10 are even better. For human health studies, much larger groups are generally used because of the wider genetic variability among people. Because it is often unethical to experiment on humans—for example, to intentionally expose them to pollutants to determine their effect—researchers often extrapolate results from animal studies to humans. That is, they assume that if a chemical proves to be toxic to mice it will be toxic to humans. To further their study, they also use epidemiological studies, that is, studies on populations exposed unintentionally to toxic pollutants, for example, at work. If the researchers notice a higher than anticipated incidence of a disease, they can attribute it to the exposure. Although such studies don't necessarily prove causation, they can help us draw conclusions, especially if other researchers find similar results or if studies with laboratory animals support the findings.

In nutrition, researchers use epidemiological studies to assess the effect of nutrients or foods on human health. In such studies, researchers often follow a large group of subjects over a long period. The diets of those who develop a certain disease—say heart disease—are compared to those who do not. Differences in diet can be associated with the development of various diseases.

difference between the two should be the drug given to the experimental group. Consequently, any observed differences between the groups could be attributed to the treatment (the experimental variable).

Besides having an experimental group and a control group, good experimentation requires an adequate number of subjects to ensure that any observed differences are real. Individual variation is natural. As a rule, the smaller the number of animals in each group, the less reliable the data

KEY CONCEPTS

Scientific knowledge is acquired and refined through the scientific method, an orderly procedure that often begins with observations or measurements of natural phenomena that lead to hypotheses that are tested in carefully designed experiments.

What Is a Theory?

Scientific method leads to the accumulation of scientific facts (really, well-tested hypotheses). Over time, as knowledge accumulates, scientists are able to gain a broader understanding of the way the world works. These broad generalizations are known as theories.

Theories are supported by numerous facts established by careful observation, measurement, and experimentation. Unlike hypotheses, theories cannot be tested by single experiments because they encompass many bits of information. Atomic theory, for instance, explains the structure of the atom and fits numerous observations made in different ways over many decades.

A theory commands respect in science because it has stood the test of time. This does not mean that theories always stand forever. As history bears out, numerous theories have been modified or discarded as new scientific evidence accumulated. Even widely held theories that persisted for hundreds of years have been overturned. In 140 AD, for example, the Greek astronomer Ptolemy (TAL-eh-me) proposed a theory that placed Earth at the center of our solar system. This was called the geocentric theory. For nearly 1,500 years, the geocentric view held sway. Many astronomers vigorously defended this position while ignoring observations that did not fit the theory. In 1580, the astronomer Nicolaus Copernicus proposed a new theory—the heliocentric view—which placed the sun at the center of the solar system. His work stimulated considerable controversy and eventually prevailed because it fit the observations.

Because theories may require occasional modifications or outright rejection, scientists must be willing to analyze new evidence that throws into question their most cherished beliefs. For the most part, though, theories are talked about as if they were fact. Some people even object to calling a theory a "theory" for fear that it sounds tentative. That's not always the case, however. Many theories are extremely sound—based on a great deal of research.

Be careful how you use the word "theory." In physics, the term *theory* tends to be used a bit loosely. As noted elsewhere in the text, for instance, physicists speculate that the fundamental unit of all matter, even light, is a weird string of energy, known as a superstring or string, for short. These strings are also believed to form subatomic particles like electrons and protons. Interestingly, there is no scientific evidence to support this view, only complex mathematical equations that suggest that these strings might exist. As physicist Joseph D. Lykken notes in an article on superstrings in *Science Year* (a publication of World Book), "Physicists often think and work in mathematical terms before their ideas can be translated into concepts that can be tested in the laboratory." Based on this work, they construct theories, such as the superstring theory. They're more appropriately referred to as hypotheses.

The word *theory* is commonly used in everyday conversation. A friend, for instance, might say, "My theory about why Jane missed the party is that she didn't want to see her ex-boyfriend." Jane's feelings aside, this is hardly worthy of the status of a theory. What your friend really meant was his "hypothesis," for his explanation was truly a tentative explanation that could be tested by experimentation—in this case, a phone call to Jane. It was not a theory in the strictest sense.

> **KEY CONCEPTS**
>
> Theories are generalizations about the physical, chemical, and biological world we live in based on many experimental observations.

Inductive and Deductive Reasoning

One feature that sets humans apart from most other animals is our ability to reason—to think about things and draw conclusions based on what we see and hear or are told. Two types of reasoning are commonly used. The first is known as inductive reasoning. Inductive reasoning occurs when one makes a generalization based on observations. For example, suppose that, based on your observations of fellow students, you concluded that students tend to be a lot healthier (for example, suffer fewer colds) if they get a good night's sleep every night. This is inductive reasoning. You are relying on observations to reach a general conclusion.

Scientists practice inductive reasoning all the time. They infer general laws based on facts and observations. When supported by lots of data obtained from carefully controlled experiments, conclusions derived from inductive reasoning help us advance our knowledge.

The opposite process—drawing a conclusion based on a general rule—is called deductive reasoning. Scientists engage in this kind of thinking as well, as do nonscientists. Consider the sleep example once again. The general conclusion that a good night's sleep means better health might be used to explain why a student who misses sleep is always ill. "I think Marshall is sick all the time because he doesn't sleep much" is an example of deductive reasoning.

> **KEY CONCEPTS**
>
> Inductive reasoning occurs when one arrives at generalizations based on numerous observations; deductive reasoning occurs when one draws a conclusion based on a general rule.

Science and Human Values

Science and the scientific process are essential to modern existence. We wouldn't have the MP3 player or the cell phone if it weren't for science. Scientific knowledge can also influence political decisions regarding health care, environmental protection, and a host of other issues.

Many decisions in the public-policy arena, however, are not made on the basis of scientific facts. Rather, they are influenced by values—what we view as right or wrong—and economic needs. When values are framed in the absence of scientific knowledge, however, they can lead to less effective, even self-defeating, policies. How does science influence values?

Science can influence human values in many ways. The study of ecology, for example, helps us understand the importance of the ecosystems to human well-being as the source of all our resources, even the oxygen we breathe. Ecology also helps uncover relationships that are not obvious to most people, such as the role of bacteria in recycling nutrients. Understanding the importance of the natural world to human well-being, even economic well-being, shapes our values (what we think is important) and gives us good reason to protect the Earth's ecosystems.

> **KEY CONCEPTS**
>
> Science provides an incredible amount of useful information that informs many aspects of our lives and can even shape our values.

Vivisection: A Medieval Legacy *by Elliot M. Katz*

Vivisection, an outdated and extremely cruel form of biomedical research, is the purposeful burning, drugging, blinding, infecting, irradiating, poisoning, shocking, addicting, shooting, freezing, and traumatizing of healthy animals. In psychological studies, baby monkeys are separated from their mothers and driven insane; in smoking research, dogs have tobacco smoke forced into their lungs; in addiction studies, chimpanzees, monkeys, and dogs are addicted to cocaine, heroin, and amphetamines, then forced into convulsions and painful withdrawal symptoms; in vision research, kittens and monkeys are blinded; in spinal cord studies, kittens and cats have their spinal cords severed; in military research, cats, dogs, monkeys, goats, pigs, mice, and rats die slow, agonizing deaths after being exposed to deadly radiation, chemical, and biological agents.

Started at a time when the scientific community did not believe animals felt pain, vivisection has left a legacy of animal suffering of unimaginable proportions. Descartes, the father of vivisection, asserted that the cries of a laboratory animal had no more meaning than the metallic squeak of an overwound clock spring. Though the research community considers vivisection a "necessary evil," a growing number of scientists and health professionals see vivisection as simply evil.

As a veterinarian, I was taught that vivisection was essential to human health. My eyes were finally opened to the full horrors and futility of vivisection when years later, faculty members and campus veterinarians at the University of California informed me that animals were dying by the thousands from severe neglect and abuse; that vivisectors and campus officials were denying and concealing the abuses; and that experiments were conducted whose only benefit was to the school's finances and researchers' careers. I discovered that animal "care" committees, typically cited as assurance that animals are used responsibly, are in fact "rubber stamp" committees composed mostly of vivisectors who routinely approve each other's projects. Over the years, I have witnessed an ongoing pattern of university officials denying documented charges of misconduct, attempting instead to discredit critics of vivisection, ultimately defending even the most ludicrous and cruel experiments as necessary and humane.

I discovered that assertions by the biomedical community that vivisection is an essential and indispensable part of protecting the public's health are simply untrue. Vivisection can and should be ended. It is scientifically outdated and morally wrong. There is a plethora of modern biomedical technology that can be used to improve society's health without harming animals. The advent of sophisticated scanning technologies, including computerized tomography (CT), positron emission tomography (PET), and magnetic resonance imaging (MRI), has given scientists the ability to examine people and animals noninvasively. This technology has isolated abnormalities in the brains of patients with Alzheimer's disease, epilepsy, and autism, revolutionizing diagnosis and treatment of these diseases. Tissue and cell cultures are being increasingly used to screen cancer and AIDS drugs. Progress with AIDS has come from areas entirely unrelated to animal experimentation. Human skin cell cultures are used to test new products and drugs for toxicity and irritancy.

Elliot M. Katz, DVM (here with companion, Manco) is a graduate of the Cornell University School of Veterinary Medicine. He is president and founder of In Defense of Animals, a national animal rights organization. (© Elliot M. Katz)

Why, then, is vivisection so entrenched and defended with an almost religious fervor? Dr. Murry Cohen summed it up when he stated, "Change is difficult for most people, but it is particularly painful for scientific and medical bureaucracies, which fight to maintain the status quo, especially if required change might imply admission of previously held incorrect ideas or flawed axioms." Vivisection continues today because of vested interests, habit, economics, and legal considerations, not for the real advancement of science and public health.

When presented with the facts, members of the public almost unanimously express their desire to see an end to the horrors of vivisection. Thousands of professionals have reevaluated the sense, efficacy, and worth of vivisection and have formed or joined organizations working to end this outdated and cruel form of research. The ending of vivisection will lead to improved public health and restore to medicine and science much needed excellence and compassion for all beings, human and nonhuman alike.

Sharpening Your Critical Thinking Skills

1. Summarize the main points of each author.
2. Do these authors use data or ethical, anecdotal (stories or experiences) arguments to make their cases?
3. Do you have a view on this issue? What factors weigh most heavily in making up your mind?

Visit Human Biology's Internet site for links to websites offering more information on this topic.

fits into the category of science by the way the experiment was conducted. Did the experiment have a control group? Were the control and experimental groups treated identically except for the experimental variable? Did the experimenters use an adequate number of subjects?

Even if all of these conditions are met, beware. In science, one experiment is rarely sufficient to permit one to draw firm conclusions. Careful scrutiny may reveal significant design flaws: perhaps mice being tested in a drug study were resistant to the drug under examination or were hypersensitive to it. As a rule of thumb, wait for scientific verification of the results. A second researcher may repeat the experiment with similar results. In some cases, a new researcher may find different results.

Nowhere is caution more necessary than when you encounter announcements of scientific breakthroughs on the television news and in magazine and newspaper articles. Ever eager to showcase new scientific studies, the media sometimes does a grave disservice to the advancement of scientific knowledge; further study may show that findings were invalid. Unfortunately, the media often fails to publish contradictory results from follow-up studies. Scientific journals often favor studies that have positive results, too. Researchers who write up a study that shows no effect often have a difficult time publishing their results. Ultimately, the public—even the scientific community—is left with a false impression of reality.

The fourth rule of critical thinking is: question the conclusions derived from facts. Surprisingly, even if an experiment is run correctly, there's no guarantee that the conclusions drawn from the results are correct. How can that be? The answer may lie in bias, ignorance, and error. Bias refers to personal beliefs that taint the interpretation of results. Ignorance is a lack of full knowledge. This, in turn, may lead a scientist to misinterpret his or her results. Finally, error does occur, in spite of our best efforts.

Two questions should be asked when one analyzes the conclusions of an experiment: (1) Do the facts support the conclusions? and (2) Are there alternative interpretations? Consider an example.

One of the earliest studies on lung cancer showed that people who consumed large quantities of table sugar (sucrose) had a higher incidence of lung cancer than those who ate table sugar in moderate amounts. The researchers concluded that lung cancer was caused by sugar consumption. Many people had trouble believing this conclusion, which forced a re-examination of the study. It, in turn, showed that the group with a higher incidence of lung cancer included a higher percentage of cigarette smokers. Subsequent studies showed that smoking, not sugar consumption, is the culprit. Smokers, it seems do consume more sugar, but it is smoking that causes lung cancer, not sugar intake.

The health effects of coffee consumption may have fallen victim to a similar false correlation. A study of data from the British National Health Service, for instance, uncovered a link between coffee consumption and heart disease. The initial research showed that heavy coffee drinkers had a higher incidence of heart disease. Further study of the data, however, suggested that the link was probably between smoking and

heart disease. Heavy coffee drinkers, it turns out, tend also to engage in a number of unhealthy activities, especially smoking and alcohol consumption. A study published in 2006 that eliminated the influence of smoking showed that there was no link between coffee consumption and heart disease.

Besides showing the importance of scrutinizing the conclusions of a scientific study, this example illustrates a key principle of medical research: correlation does not necessarily mean causation. In other words, two factors that appear to be related (correlated) may, in fact, not be linked at all.

The fifth rule of critical thinking is: look for assumptions and biases. This rule is related to the previous rule, questioning conclusions, but is so important that it warrants closer examination. Biases and hidden assumptions are to thinking what cyanide is to food—a poison. Unfortunately, biases and hidden assumptions run rampant in today's society.

In many contemporary debates over a wide range of issues, proponents often present information that supports their point of view. This selective inclusion of supportive data and exclusion of contradictory information is often an expression of a hidden agenda. What happens is that people make up their mind about an issue and then seek out information that supports their point of view. Sadly, some people even lie about the facts to support their point of view.

The sixth rule of critical thinking is: question the source of the facts—that is, who is telling them. Closely related to the previous rule, this one calls on us to learn about the people who performed various research studies or analyses. It asks us to familiarize ourselves with the people taking various positions. Bias can influence scientific researchers. Be especially wary of people with political motives. An association with a partisan group may be a red flag, a warning that bias may have influenced their conclusions. Their penchant for "putting a spin" on things often results in doctored truths.

Sometimes a study of the biographies of the people delivering the information is as instructive as an examination of their conclusions. Beware of "experts" who have a hidden agenda. Experts from industry who swear under oath about the safety of their product may be biased or even deceitful. Environmental experts may also slant the data to support their view.

Also beware of people who may not know as much as you think they should. Although we think of physicians as experts on human health, most of them received little or no training in nutrition in medical school. Many medical students still graduate without a full understanding of the role of nutrition in preventing disease and promoting good health. For questions about diet, you may be better off consulting a registered dietitian.

Scientists can be biased, too. If a scientist is testing the effects of an experimental drug on humans, but financial support for the research comes from a pharmaceutical company that stands to make billions of dollars from the new product, his results may be unintentionally—or sometimes intentionally—biased. Fortunately, the scientific community has built-in mechanisms to weed out bad science. When a researcher submits a paper for publication, it is reviewed by his or her peers. If the peers find fault in the study, it may be rejected. If the study and its conclusions appear valid, it is published in scientific journals. After it is published, however, it is further scrutinized and sometimes criticized by other scientists. Additional work may substantiate or refute the findings, helping us slowly but surely advance our understanding of the world around us.

The seventh rule of critical thinking is: understand your own biases, hidden assumptions, and areas of ignorance. So far, this discussion on critical thinking has concentrated on ways you can uncover mistakes in reasoning that other people make. But what about your own biases, hidden assumptions, and areas of ignorance? Do they affect your ability to think critically? Do you continually seek out information that only supports your beliefs? Are you seeking information from biased sources?

Uncomfortable as it may be, it's essential to uncover—and correct—your own biases. Only then can you become a critical thinker. Remember, too, that despite the vast amount of information human society has accumulated over the years, we do not have all of the answers. We're constantly adding to and refining our knowledge. Nowhere is this more evident

than in the field of nutrition where new discoveries are continually throwing into question previous ideas and dietary recommendations. Bear in mind that occasional reversals don't mean that researchers aren't making progress in understanding health and nutrition. And, above all, be patient. Maybe in another 1,000 years, we will have all of the answers! In the meantime, we'll have to content ourselves with an incomplete and evolving knowledge base.

As you read this text, you will be presented with examples to help you sharpen your critical thinking skills. Each chapter starts with a critical thinking exercise. Even if your teacher doesn't assign these exercises, it is a good idea to take time to read and study them. The Point/Counterpoints in the book will also help you hone your critical thinking skills.

> **KEY CONCEPTS**
>
> Critical thinking is an orderly process of thought that enables one to discern fact from fallacy; it requires careful gathering of facts and analysis of those facts, including the ways in which they were derived and the source of those facts.

SUMMARY

Health and Homeostasis

1. Humans, like all other organisms, have evolved mechanisms that ensure relative internal constancy (homeostasis). These homeostatic mechanisms are vital to our health, survival, and reproduction.
2. Homeostatic mechanisms exist at all levels of biological organization, from cells to organisms to ecosystems.
3. The health of all species and ecosystems is dependent on the functioning of homeostatic mechanisms. When these mechanisms break down, illnesses often result.
4. Human health has traditionally been defined as the absence of disease, but a broader definition of health is now emerging. Under this definition, good health implies a state of physical and mental well-being.
5. Physical well-being is characterized by an absence of disease or symptoms of disease, a lack of risk factors that lead to disease, and good physical fitness.
6. Mental health is also characterized by a lack of mental illness and a capacity to deal effectively with the normal stresses and strains of life.
7. Human health and the health of the many species that share this planet with us depend on a properly functioning, healthy ecosystem. Thus, alterations of the environment can have severe repercussions for all species, including humans.

Evolution and the Characteristics of Life

8. Evolution results in structural, functional, and behavioral changes in populations. These changes, in turn, result in organisms better equipped to cope with their environment—that is, better able to survive and reproduce.

9. All life forms alive today exist because of evolution. In fact, every cell and every organ and function, including homeostasis, in the human body is a product of millions of years of evolution.
10. Living organisms belong to five major groups or kingdoms. Humans belong to the animal kingdom.
11. Evolution is responsible for the great diversity of life forms. However, because the Earth's organisms evolved from early cells that arose over 3.5 billion years ago, all organisms, including humans, share many common characteristics. Items 12–18 list the common characteristics of all life forms.
12. All organisms, including humans, are made up of cells.
13. All other organisms grow and maintain their complex organization by taking in molecules and energy from their surroundings.
14. All living things house many chemical reactions. These reactions are collectively referred to as metabolism.
15. All organisms possess homeostatic mechanisms.
16. All organisms exhibit the capacity to perceive and respond to stimuli.
17. All organisms are capable of reproduction and growth.
18. All organisms are the product of evolutionary development and are subject to evolutionary change.
19. All organisms are part of the Earth's ecosystems.
20. Although humans are similar to many other organisms, we also possess many unique abilities and features: culture, our ability to plan for the future, and an enormous ability to reshape the Earth through ingenuity and technology.

Understanding Science

21. Science is both a systematic method of discovery and a body of information about the world around us.
22. Scientists gather information and test ideas through the scientific method. The scientific method begins with observations and measurements, often made during experiments. Observations and measurements may lead to hypotheses, tentative explanations of natural phenomena that can be tested in experiments. The results of experiments help scientists support or refute their hypotheses.
23. The body of scientific knowledge also contains theories, broad generalizations about the way the world works, based on numerous studies. Theories can change over time as new information becomes available, though, this is rare.

Critical Thinking

24. Critical thinking is a useful tool in science and life and is defined as careful analysis that helps us distinguish knowledge from beliefs or judgments.
25. Critical thinking provides a way to analyze issues and virtually information.
26. Table 1-2 summarizes the critical thinking rules.

THINKING CRITICALLY ANALYSIS

This analysis corresponds to the Thinking Critically scenario that was presented at the beginning of this chapter.

This experiment has several major flaws. First, the boy used chickens from two different farms, so the chickens could have been genetically dissimilar. Differences in genetics could have been responsible for the differences in cholesterol content in the eggs. Second, although differences were found in the cholesterol content of the eggs of the two groups, we don't know if they were statistically significant. Good statistical analysis is necessary to determine whether measured differences are substantial enough to be attributable to the treatment. Another problem is the small sample size. Before I donated any money to this new venture, I'd want to see it performed on a larger group of genetically similar chickens. The fourth problem is that no mention was made of the differences between the two feeds. A careful analysis is essential to solidify one's confidence.

Clearly, this simple experiment has some flaws, but there's an even larger problem that we haven't addressed yet—notably the underlying supposition that cholesterol in eggs raises blood cholesterol levels. As it turns out, the liver produces lots of cholesterol. It's responsible for most of the cholesterol floating around in our bloodstreams. Furthermore, recent studies have shown that eggs don't contain as much cholesterol as was once thought and that eating foods rich in saturated fat (like hamburgers and bacon) is linked to elevated cholesterol levels in your bloodstream. (You'll learn more about this elsewhere in the text.) Because of these findings, the American Heart Association has raised its recommendation for egg consumption from three a week to one a day for healthy people. To control cholesterol, you're better off cutting back on foods that have a high content of saturated fat like hamburgers and bacon.

KEY TERMS AND CONCEPTS

Carrion, p. 1
Control group, p. 10
Critical thinking, p. 13
Deductive reasoning, p. 12
Ecosystem, p. 3
Evolution, p. 2
Experiment, p. 10

Experimental group, p. 10
Experimental variable, p. 10
Homeostasis, p. 3
Hypothesis, p. 10
Inductive reasoning, p. 12
Kingdom, p. 6
Metabolism, p. 7

Predators, p. 4
Pseudoscience, p. 13
Risk factor, p. 5
Science, p. 9
Scientific method, p. 10
Sensors, p. 3
Theory, p. 12

CONCEPT REVIEW

1. How would you define life? What are the characteristics of all life forms? p. 6
2. In what ways are a rock and a living organism (for example, a bird) similar, and in what ways are they different? p. 6
3. Describe the concept of homeostasis. How does it apply to humans? How does it apply to ecosystems? Give examples. p. 3
4. Using the definition of health and the list of healthy habits in Table 1-1, assess your own health. In what areas do you excel? What areas need improvement? pp. 4–6
5. What is a risk factor, and what are the risk factors that detract from the health of you and your friends? p. 5

6. In what ways are humans different from other animals? In what ways are they similar? p. 7
7. Describe the scientific method, and give some examples of how you have used it recently in your own life. p. 10
8. How do a hypothesis and a theory differ? How is the word theory misused in everyday language? pp. 10–12
9. List the critical thinking skills presented in this chapter. Describe each one and what it means to you. Which skills seem to be most important to you? pp. 13–19

10. A graduate student injects 10 mice with a pollutant commonly found in the environment and finds that all of his animals die within a few days. Eager to publish his results, the student comes to you, his adviser. What would you suggest the student do before publishing his results? pp. 10–11
11. Given your knowledge of scientific method and critical thinking, make a list of reasons why scientists might disagree on a particular issue or research finding. pp. 13–19

biology.jbpub.com/chiras/8e/

The site features eLearning, an online review area that provides quizzes, chapter outlines, and other tools to help you study for your class. You can also follow useful links for in-depth information, research the differing views in the Point/Counterpoints, or keep up on the latest health news.

SELF-QUIZ: TESTING YOUR KNOWLEDGE

1. _____ results in structural, functional, and behavioral changes in populations of organisms that make them better able to survive and reproduce. p. 2

2. _____ is referred to as a state of relative internal constancy. p. 3

3. Abnormal conditions in the human body such as high blood pressure that may lead to ill health or serious conditions are called _____ factors. p. 5

4. Human health is dependent on physical health and _____ well-being. p. 4

5. Organisms are organized into five major groups called _____. p. 6

6. Chemical reactions in the body are known as _____. p. 7

7. Scientists use a process called the _____ method to learn new information and test ideas. p. 10

8. A(n) _____ is a testable assertion. p. 10

9. In many scientific experiments, two groups of subjects are used. One group, the _____ group is treated identically to the experimental group except that it is not exposed to the experimental variable. p. 10

10. _____ _____ is a process that allows us to analyze problems, assertions, conclusions, and research, distinguishing beliefs from knowledge. p. 13

(© iStockphoto.com/agsandrew.)

2

The Chemistry of Life

Chemistry is a field very few people know much about, yet it has many useful applications. In recent decades, for example, chemists have developed thousands of new drugs. These chemicals have helped humankind combat a wide range of often deadly or debilitating diseases. One of the most notable examples is antibiotics, chemicals that kill harmful bacteria. Not only have antibiotics reduced suffering, they have also greatly reduced mortality, especially among infants.

THINKING CRITICALLY

Imagine that you are a researcher at a major medical school. You are considered a leading authority on a disease you have been studying for the past 25 years. Because of this, a major pharmaceutical company hires you to help them determine whether a new drug they're developing will cure this disease. Being busy with your own research, you ask a couple of your new graduate students to devise an experiment to test the potential effectiveness of the drug. The students quickly come up with an idea: they suggest that you treat 100 patients with the drug at two different dosages, and then follow the patients for a year to see how many recover. Do you think this is a good experiment? Why or why not?

Advances in chemistry have also led to countless improvements in household products that make our lives safer. House paints are a case in point. At one time, most paints contained lead-based pigments. Because lead is toxic to the nervous system, chips of paint flaking off walls and ceilings pose a health hazard to young children who routinely put foreign objects in their mouths as they explore their new world. Modern paints lacking lead-based pigments are a definite improvement, although many contain potentially harmful chemicals known as VOCs (volatile organic chemicals) and antimicrobial agents that could cause health problems (Figure 2-1). To address this problem, chemists have created even safer paints that lack these potentially harmful chemicals. They are known as low- or no-VOC paints.

The truth of the matter is that almost everything we touch or look at in our daily lives has been brought to us or improved by chemists. But chemistry is important in other ways, too. A little knowledge of chemistry, for example, can help us understand environmental problems such as global warming, ozone depletion, urban air pollution, and acid rain. On a more personal level, a knowledge of chemistry can also help us live long, healthy lives by enabling us to make informed decisions about our diets.

But why study chemistry at the beginning of a biology course? The answer is simple. All organisms are made up of chemicals. These chemicals undergo many reactions that are the basis of life. In order to understand organisms, you must understand some basic principles of chemistry.

This chapter introduces you to several key principles of chemistry, which form the foundation for the study of human biology. We begin with a discussion of atoms and molecules, and then look at water, acids, bases, and buffers. The major biological molecules like proteins and lipids are briefly

FIGURE 2-1 **Better Living Through Chemistry** A new line of paint, stains, and finishes, like those being applied here from AFM in San Diego, California, help reduce exposure of workers and residents to toxic chemicals once contained in many paint products. (Courtesy of AFM. All rights reserved.)

mentioned here but will be discussed more fully in subsequent chapters as they become important to your understanding of life and the human body.

2-1 Atoms and Subatomic Particles

The world in which we live is composed of a wide variety of physical matter such as wood, plastics, and metal. **Matter** is defined as anything that has **mass** and occupies space. Technically, mass is a measure of the amount of matter in an object. Thus, a water balloon has a greater mass than an air-filled balloon.

Atoms

Matter is composed of tiny particles called **atoms**. The word atom comes from the Greek *atomos*, which means "incapable of division." The ancient Greek philosophers who proposed the existence of atoms believed that all matter consisted of small particles that lacked any internal structure. Over time, though, scientists have found that atoms consist of even smaller particles, called **subatomic particles**. The subatomic particles are electrons, protons, and neutrons.

Although scientists once considered the subatomic particles to be the elementary particles of which all atoms are made, physicists now believe that they are made of even smaller components. Each proton and neutron is composed

of three smaller particles, known as quarks. They, in turn, are made up of smaller components called superstrings, or strings for short (Figure 2-2). **Electrons** are made of superstrings as well. What are superstrings?

"Superstrings are incredibly tiny and unimaginably weird loops of energy," says physicist Joseph D. Lykken. Superstrings form everything in the cosmos from planets to atoms in our body to rays of sunlight. "If you could see these superstrings, they would appear to be vibrating, merging, breaking apart, disappearing, and reappearing right before your eyes." Lykken adds, however, "Physicists have yet to find experimental evidence that superstrings exist, but mathematical calculations suggest that superstrings may be real."

In this chapter, we focus our attention on the subatomic particles: electrons, protons, and neutrons (Figure 2-3). **Protons** (PRO-tauns) are located in the center of each atom, in a region known as the **nucleus** (Figure 2-3). Each proton has a positive charge. **Neutrons** (NEW-trauns) are also found in the nuclei of atoms. Unlike protons, they are uncharged particles. Protons and neutrons are very similar in mass. (Neutrons are

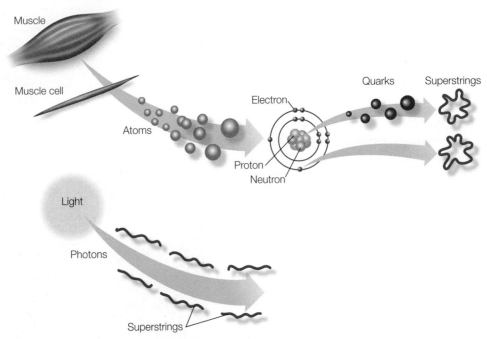

FIGURE 2-2 **Superstrings** Physicists believe the subatomic particles are actually composed of smaller units known as superstrings, or strings, as shown here.

slightly heavier.) Because protons are positively charged, the nucleus of an atom is positively charged.

Surrounding the positively charged nucleus is a region called the *electron cloud*. It contains tiny, negatively charged particles, called **electrons** (ee-LECK-traun). They move rapidly through a large volume of space. They weigh about 1/1800 as much as protons.

As shown in Figure 2-3, most of the volume atoms occupy consists of electron cloud. As the world-renowned astronomer Carl Sagan once remarked, atoms are mostly "electron fluff."

Although most of the space an atom occupies is made up of the electron cloud, most of the mass of an atom can be attributed to the protons and neutrons packed into the tiny nucleus. These particles are so heavy, in fact, that a single cubic centimeter (about ⅓ teaspoon) would weigh 100 million tons.

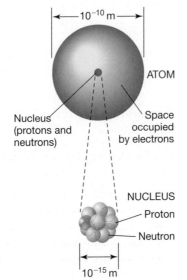

FIGURE 2–3 **The Atom** The atom consists of two regions. The central nucleus contains protons and neutrons and makes up 99.9% of an atom's mass. Surrounding the nucleus is the electron cloud, where the electrons orbit around the nucleus. (Figure not drawn to scale.)

Atoms are electrically neutral because they contain the same number of protons and electrons. The positive charges, thus, cancel out the negative ones.

> **KEY CONCEPTS**
>
> All matter is made of atoms, which are made of three subatomic particles: protons, neutrons, and electrons.

The Elements

Early scientists accounted for the diversity of matter by assuming that there must exist a number of pure substances and that these substances could be combined into an infinite variety of different forms. They called these pure substances **elements**. Today, modern chemists define an element as a pure substance (for example, gold or lead) that contains only one type of atom.

At this writing, 92 naturally occurring elements are known. More than a dozen additional elements can be made in the laboratory. Of the 92 naturally occurring elements, only about 20 are found in organisms. Four elements in this group—carbon, oxygen, hydrogen, and nitrogen (remember: COHN)—comprise 98% of the atoms of all living things.

Elements are listed on a chart called the periodic table of elements (Table 2-1). The periodic table is a summary of vital statistics on each element. As illustrated in Table 2-1, most elements are represented on the table by a one- or two-letter symbol. The element hydrogen, for example, is indicated by H. Oxygen is designated by O, and carbon is represented by C. Aluminum is designated by Al.

Above each symbol on the periodic table is its atomic number. The **atomic number** of an element is the number of protons in the nuclei of an element's atoms. Carbon's atomic number is six. It contains six protons. Oxygen contains eight protons. Its atomic number is eight. Because all atoms are electrically neutral, the atomic number also equals the number of electrons in the atoms of each element.

TABLE 2-1	A Simplified Periodic Table						
I	II	III	IV	V	VI	VII	VIII
1 H Hydrogen 1.0 ← Atomic number / Atomic symbol / Mass number							
3 Li Lithium 7.0	4 Be Beryllium 9.0	5 B Boron 11.0	6 C Carbon 12.0	7 N Nitrogen 14.0	8 O Oxygen 16.0	9 F Fluorine 19.0	10 Ne Neon 20.2
11 Na Sodium 23.0	12 Mg Magnesium 24.3	13 Al Aluminum 27.0	14 Si Silicon 28.1	15 P Phosphorus 31.0	16 S Sulfur 32.1	17 Cl Chlorine 35.5	18 A Argon 40.0
19 K Potassium 39.1	20 Ca Calcium 40.1						

Note that each element is represented by a one- or two-letter symbol. Elements are listed according to their atomic number (the number of protons). Their atomic mass is also shown.

As you can see, below the chemical symbol for each element is another number, called the *mass number*. The mass number is the mass of an individual atom, mostly determined by the number of protons and neutrons in the nucleus of a chemical element. That's because electrons have virtually no mass, and therefore contribute very little to the mass of an atom. As shown in Table 2-1, carbon, which contains six protons and six neutrons, has an atomic mass of 12. Oxygen atoms contain eight protons and eight neutrons. The mass number of oxygen is 16.

KEY CONCEPTS

Pure substances are called elements and each element is made up of one type of atom.

Isotopes

Although elements are made of identical atoms, there are very slight differences in some of the atoms of most elements. These different forms of atoms are called **isotopes** (EYE-so-topes). Isotopes of an element differ only in the number of neutrons they contain. Hydrogen, for example, has three slightly different atomic forms. The most common hydrogen atom contains one proton and one electron, but no neutrons. The next most common form has one proton, one electron, and one neutron. The least common hydrogen atom has one proton, one electron, and two neutrons. The three forms, or isotopes, of hydrogen are chemically identical—that is, they all react the same. Changes in the number of neutrons, therefore, have very little effect on atoms.

While additional neutrons don't change the chemical nature of atoms of a given element, their presence makes some atoms of an element unstable. To achieve a more stable state, many isotopes release small bursts of energy, or tiny energetic particles from their nuclei. These emissions are known as **radiation**. Radiation carries energy and mass away from an atom's nucleus, which helps atoms achieve more stable states. These are called **radioactive isotopes** or **radionuclides** (ray-dee-oh-NEW-klides).

Radiation has provided numerous benefits to humankind. For example, scientists have discovered a wide variety of uses for radiation in medicine, including X-rays to study bones. Radioactive chemicals are also used in medical tests such as thyroid function. One of the most important advancements is the use of radioactive isotopes as markers in experiments (Scientific Discoveries 2-1). Another extremely important application is the use of naturally occurring radioactive isotopes to date rocks and fossils. Information from this technique has helped scientists pinpoint significant evolutionary events and has given us important insights into the history of life on Earth.

Despite the many benefits of radiation, its use has not occurred without some risks. For example, radiation exposure damages molecules in body cells and causes changes in the genetic material called **mutations**. Mutations can result in birth defects and can lead to cancer. Despite these and other problems, radiation is widely used. It is now even being used to sterilize foods, thus eliminating the need for refrigeration. The **Point/Counterpoint** in this chapter examines the pros and cons of food irradiation.

KEY CONCEPTS

Although we say that all elements are made up of identical atoms, there are very slight differences in some atoms of an element, notably in the number of neutrons in the nuclei. These alternative forms are called isotopes.

Scientific Discoveries that Changed the World

2-1 The Discovery of Radioactive Chemical Markers
Featuring the Work of Schoenheimer

Biochemist Rudolf Schoenheimer (1898–1941) and his colleagues offered science a chemical tool that would yield important new information and result in significant scientific and medical advances: radioactively labeled molecules. A radioactively labeled molecule is a molecule containing one or more radioactive atoms (isotopes).

A radioactive atom incorporated in a biological molecule, such as an **amino acid** used to make proteins, emits tiny bursts of radioactivity. Radioactive markers permit scientists to follow marked molecules as they progress through the body's maze of chemical reactions. That is, they allow scientists to see where they go and how they are changed. Before the emergence of this technique, substances such as fatty acids or amino acids administered to an animal were lost almost immediately, because they were indistinguishable from molecules in the body.

Radioactive markers also permit scientists to determine where chemicals are stored and how they are removed from the body. In fact, many of the physiological processes you will read about in this book were investigated by using radioactively labeled molecules.

Before Schoenheimer's discovery, the inability to track molecules resulted in a huge gap in our knowledge about metabolism and body functions. Futile attempts to "tag" molecules with nonradioactive molecular markers generally failed, because the altered molecules differed so much from their natural substances that they were treated differently by the body.

Through experimentation, Schoenheimer showed that radioactively labeled molecules were so similar to the unlabeled molecule that they reacted identically. This made them ideal markers and opened the door to many additional discoveries.

From Schoenheimer's brilliant experiments, a new picture of life has emerged. He and his colleagues showed that molecules of the body such as DNA, which previously had been considered stable, are actually in a continuous state of flux. Even apparently dormant fat deposits in the body were ever-changing.

Since their introduction in the 1930s, radioactive markers have also enjoyed great popularity in medicine, where they're often used to determine the functional status of various organs. Radioactive iodide ions, for instance, can be administered to a patient to assess the function of the thyroid gland, a hormone-producing gland in the neck.

Radioactively labeled molecules are helping scientists unravel the mysteries of body functions. Labeled glucose molecules, for instance, can be used to determine the relative activity of different parts of the brain, permitting scientists to map brain functions accurately.

These are but a few examples of the tremendous contribution of radioactively labeled molecules to medicine and science.

2-2   The Making of a Molecule

The diversity of matter in the world around us is a result of the presence of a wide range of elements, each with its own unique properties. This diversity also arises from the fact that atoms of various elements can combine to form compounds and molecules. The term **compound** refers to a substance made up of two or more atoms. Each compound has a unique chemical formula that indicates the ratio of the atoms found in it. For instance, CO_2 is the chemical formula of the greenhouse gas carbon dioxide. It contains one carbon atom and two oxygen atoms.

The term **molecule** refers to the smallest particle of a compound that still retains the properties of that compound. Methane gas given off by rotting vegetation, for instance, is a chemical compound consisting of methane molecules. The chemical formula of this flammable gas that we cook with is CH_4. CH_4 contains one carbon atom for every four hydrogen atoms. Hydrogen gas is a chemical compound consisting of two hydrogen atoms (H_2).

> **KEY CONCEPTS**
> Atoms bond to one another to form molecules.

Chemical Bonds

Atoms join together by chemical bonds to form organic and inorganic compounds. There are two types of bonds that form between atoms: ionic and covalent. These are known as intramolecular bonds. As you shall soon see, electrons of atoms are responsible for creating both of these bonds.

> **KEY CONCEPTS**
> Atoms bond to form stable configurations.

Ionic Bonds

Ionic bonds are formed between two atoms when one atom loses an electron and the other atom gains an electron. This reaction creates two charged particles, known as **ions**

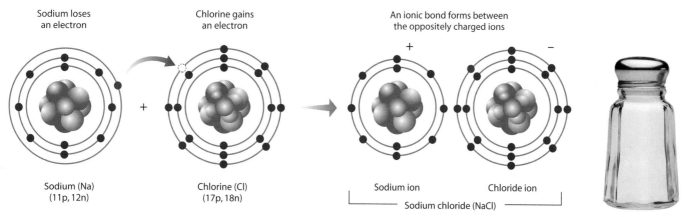

Chlorine gains an electron

An ionic bond forms between the oppositely charged ions

+   −

Sodium (Na)
(11p, 12n)

Chlorine (Cl)
(17p, 18n)

Sodium ion   Chloride ion

Sodium chloride (NaCl)

FIGURE 2-4 Ions and Ionic Bonds Sodium (Na) and chlorine (Cl) atoms differ in their electron affinity. (Photo © Photodisc.)

(EYE-ons). In this reaction, the atom that loses its electron becomes positively charged, and the one that gains an electron is negatively charged. The positively charged ion is attracted to the negatively charged electron by a weak electrostatic force. This force holds them together and is referred to as an ionic bond. Figure 2-4 shows how an ionic bond forms between a chlorine atom and a sodium atom in table salt (sodium chloride).

> **KEY CONCEPTS**
> Ionic bonds are electrostatic attractions between two oppositely charged ions.

Covalent Bonds

In contrast to ionic bonds, covalent bonds occur when two atoms "share" electrons. That is to say, when two atoms are bound to each other by a **covalent bond**, they are held together by the act of sharing electrons. How's that? As shown in Figure 2-5, hydrogen gas, H_2, consists of two hydrogen atoms. In this molecule, the electrons of each hydrogen atom orbit around both atoms, holding them together. This arrangement known as a hydrogen bond usually produces a much stronger union than an ionic bond.

Methane is another compound whose molecules are held together by covalent bonds. As noted above, methane (CH_4) is a gas that consists of four hydrogen atoms and one carbon atom. The four hydrogen atoms are covalently bonded to the carbon atom. To understand how this molecule forms, we turn first to the central atom, carbon.

Electrons in atoms orbit around the nucleus at different distances from the nucleus. Some electrons are found in orbits close to the nucleus. Others orbit in regions more distant. It's the outermost electrons that participate in chemical bonds. Consider an example: Carbon atoms have six electrons, two in a region or shell close to the nucleus and four in an outer shell. It is the four outer-shell electrons that participate in covalent bonds. That is, they can be shared with other atoms to form covalent bonds. In the methane molecule, for instance, each of the four hydrogen atoms contains one electron. To form methane, the four outer shell electrons of carbon are shared by the four hydrogen atoms' electrons. This arrangement fills the outer shell of carbon. (It is full when it contains eight electrons.)

Many other atoms react similarly, filling the outer shell during covalent bonding. Thus, chemists have coined the term *octet rule* (meaning a group of eight) to explain this behavior. You can use it to predict how atoms will react.

Covalent bonding can be represented by using dots and X's for the outer-shell electrons of an atom. Figure 2-6 uses this technique to show the electron sharing in a molecule of methane. In this example, carbon is represented by the letter C surrounded by four outer-shell electrons, indicated by dots. Each hydrogen atom is represented by H, and the electron of each hydrogen atom is indicated by an X. As illustrated, each hydrogen atom shares its electron with the carbon atom, giving the carbon atom four additional electrons and thus creating a full outer shell for carbon. The shell contains eight electrons.

Another important molecule is water, H_2O. As the formula indicates, a water molecule consists of a single atom of oxygen bonded to two atoms of hydrogen. Oxygen atoms contain eight electrons, two in the inner shell and six in the outer shell. Oxygen reacts with two hydrogen atoms to form water. By combining with two hydrogen atoms, oxygen fills its outer shell.

The bonds between the carbon atom and each of the hydrogen atoms in methane and the oxygen atom and each of the hydrogen atoms in water are formed as a result of the sharing of one pair of electrons—one from each atom. These bonds are referred to as *single covalent bonds*. In chemical nomenclature, they are often indicated by a single line joining two atoms, as in H—H.

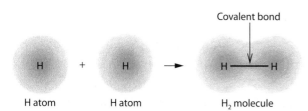

Covalent bond

H   +   H   →   H——H

H atom   H atom H_2 molecule

FIGURE 2-5 Covalent Bond Nonpolar covalent bonds are characterized by a sharing of electrons.

Point/Counterpoint   Controversy Over Food Irradiation

Food Irradiation: Too Many Questions *by Donald B. Louria, MD*

The debate about the safety of irradiating foods raises six issues:

1. The safety issue. Irradiated food does not become radioactive, but the radiation does cause chemical changes in food molecules. The major concerns are genetic damage in those consuming the foods and nutrient loss after irradiation.

Donald B. Louria, MD teaches at the UMDNJ–New Jersey Medical School, where he is Chairman Emeritus of the Department of Preventive Medicine and Community Health. (© Donald B. Louria)

A highly controversial, but flawed, study on small numbers of malnourished children in India suggested that consumption of freshly irradiated wheat could produce chromosomal abnormalities. However, a study of a much larger group of healthy adults in China showed no such abnormalities. The proponents of food irradiation have criticized the Indian study and pointed to the larger study they feel showed no chromosome damage. But, that does not answer the concerns. If food irradiation becomes an accepted technique, millions of malnourished people will eat irradiated foods. What is needed is a careful study of possible genetic damage involving adequate numbers of children and adults. I am prepared to believe that consumption of irradiated foods is unlikely to produce significant harm, but such a belief must be supported by proper data.

2. The nutrition issue. Irradiated food loses some of its nutritional value; the extent of loss of vitamin content depends on the type of food and the dose of radiation—the higher the radiation dose, the greater the loss. Irradiation proponents suggest the effects of irradiation on vitamin content are not different from that of conventional food processing (such as heating). But, some evidence suggests that when irradiated foods are processed (frozen, thawed, heated), there is an accelerated loss of vitamins. Furthermore, some foods that would be irradiated (fruits, vegetables) are eaten raw so the arguments about heating would not apply. Proponents also maintain that the diet in the United States contains redundant vitamins, so some loss through irradiation would not be of concern. Of course, this would not be true for millions of people in other countries, as well as those below the poverty line in the United States and those over age 60 who often have low blood vitamin concentrations.

Arguing that our diet contains adequate vitamins and that destroying the vitamin content of foods by irradiation is of no importance is not likely to be viewed well by many Americans.

3. The necessity for this technology. Proponents say food irradiation will reduce diarrheal illnesses. That is true, but the potential benefits have been exaggerated. There are 200 million diarrheal episodes a year in the United States; of these, only about one-third are food borne and, in most cases, the cause is unknown so food irradiation is not the answer to diarrheal disease in the United States. But, it would reduce substantially diarrheal illnesses due to several bacteria (*Salmonella, Listeria,* and *E. coli*). Proper cooking procedures are just as effective as food irradiation and that, combined with more intensive surveillance of food production and distribution, reduces the need to use irradiation.

4. Helping to solve world hunger. Will irradiating foods prolong their shelf lives and thus, feed the world? Shelf lives will definitely be increased; that is an important benefit for the less-developed world, but the proponents have provided no adequate data on the extent of the benefit. If the trade-off is longer shelf life, but less nutritious foods, that would be unacceptable to many people.

5. The issue of competing technologies. During the next decades, scientists will develop foods that have longer shelf lives and are more nutritious. Advances in biotechnology are likely to give us useful alternatives to food irradiation. Other technologic advances will greatly improve our ability to detect microbial contamination of foods before they are allowed to reach the marketplace.

6. The environmental pollution issue. Food irradiation is a technology that will result in numerous irradiation plants in the United States. The few irradiation plants operating in the United States have, in some cases, contaminated their workers and the environment. Imagine the potential for contamination if there were hundreds of such plants using radioactive materials.

Food irradiation does have potential benefits, but it also raises substantial concerns. Whether we should adopt the technology is obviously a matter for continuing debate. Certainly, the technology should not be adopted until the issue of potential genetic damage has been resolved and there is agreement that the public will always be informed not only that a given food has been irradiated, but also the extent, if any, of nutritional damage.

Food Irradiation: Safe and Sound by George G. Giddings

Irradiated foods are safe and wholesome. The ionizing radiation process applied to foods offers certain proven public health and economic benefits without significant public health risks when carried out according to well-established principles and procedures. Decades of worldwide research and testing by competent, knowledgeable, objective, and responsible scientists have led to this conclusion. This conclusion is also supported by some 30 years of experience in the radiation sterilization of medical devices and other health care products to prevent infections, plus a growing list of industrial and consumer products, including foods and their raw materials, ingredients, and packaging materials. The technology is so pollution-free that the EPA exempts it from environmental impact statements.

There has been organized political opposition to food irradiation by a network of special-interest activists serving various political agendas, notably the antinuclear/anti-irradiation one. This network includes a handful of scientists and medical professionals from other fields who act as "expert witnesses" against food irradiation to serve their hidden agendas. For example, they still point to an old Indian study on irradiated wheat and a few children as suggestive of risk even though it was dismissed in the mid-1970s. Opponents point to minor losses of certain vitamins where in fact irradiation is gentler toward vitamins and other nutrients than comparable food processes and even cooking. Their campaign is doomed to failure in the face of the unshakable facts, including a growing appreciation for public health and other proven benefits of food irradiation, and its growing worldwide regulatory approval, industrial usage, and public acceptance. The American Medical Association and World Health Organization are among the many professional bodies that have endorsed food irradiation.

Food irradiation is undoubtedly the most versatile physical process yet applied to food substances in terms of the range and variety of objectives it can accomplish. Radiation can:

George G. Giddings, Ph.D., has been involved in food irradiation research and development since 1963 and has written and spoken extensively on the subject. (© George G. Giddings)

- inhibit the sprouting of foods, such as potatoes and onions, and delay spoilage.
- rid fruits, vegetables, and grains of insect pests and prevent parasites from infecting consumers of fish and meats.
- rid foods of microbial pathogens such as the salmonellae.
- delay microbial spoilage of a wide variety of animal and plant products by reducing microbe levels.
- sterilize packaging materials, eliminating microorganisms that would otherwise contaminate products.
- sterilize a wide variety of prepared or cooked foods such as meat, poultry, and fishery products, which have already been used to feed astronauts in space and immune-compromised patients.

All of these beneficial effects and more can be, and are, readily accomplished by the application of ionizing (gamma, electron, and X-ray) radiation according to well-established principles and procedures. Nevertheless, irradiation must compete with a number of other new technologies. As a result, it is not likely to be applied to all, or even a high percentage, of the national and world food supply. It will therefore be used in cases in which it is clearly the best all-around choice.

Sharpening Your Critical Thinking Skills

1. What is food irradiation? Why is it used?

2. List and summarize the key points of each author.

3. Using your critical thinking skills, analyze each author's position. Are you inclined to agree with either one on all issues? Why or why not?

4. In your opinion, what is the best course for the development and implementation of this technology?

5. What facts would you need to know to form a personal conclusion about this issue?

Visit Human Biology's Internet site for links to websites offering more information on this topic.

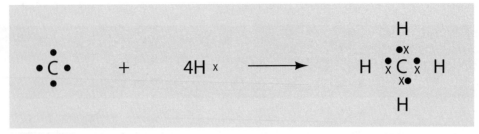

FIGURE 2-6 Covalent Bonding Simplified You can keep track of the electrons being shared in covalent bonds by representing the outer-shell electrons as dots or Xs. See the text for an explanation.

Some atoms share two pairs of electrons with other atoms, forming double covalent bonds. Oxygen, for instance, can share two of its outer-shell electrons with a carbon atom, forming a *double covalent bond* (C=O), as shown in Figure 2-7. Other atoms can share three pairs of electrons, forming *triple covalent bonds* (Figure 2-7).

Health Tip 2-1

Human health depends on an adequate intake of water. Be sure to drink plenty of liquid every day.

How much?

Despite what you may have been told, you don't need to drink eight glasses (8 fluid ounces each) of water a day to be healthy. Recent studies also show that most people only need about four 8-ounce (one cup) glasses of water or some other liquid refreshment each day to stay adequately hydrated. And there's generally no reason to force yourself to drink. According to new research, fluid intake driven by thirst will, for most people, ensure that they are well-hydrated. Remember, too, that some water comes in the food you eat. Sauces, fruits, and vegetables, for instance provide a fair amount of water.

> **KEY CONCEPTS**
>
> Covalent bonds are formed between two atoms that share electrons; this sharing binds the atoms together to form molecules.

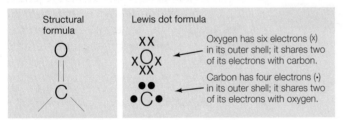

(a) Double covalent bond

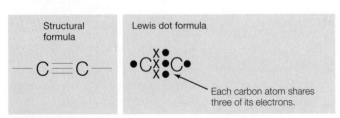

(b) Triple covalent bond

FIGURE 2-7 Double and Triple Covalent Bonds (a) When atoms share two pairs of electrons, a double covalent bond is formed. (b) On rare occasions, atoms share three pairs of electrons, creating triple covalent bonds.

Polar Covalent Bonds

One special type of covalent bond is the polar covalent. **Polar covalent bonds** occur any time there is an unequal sharing of electrons by two atoms. In other words, the electrons spend more time orbiting around one atom than the other. Why does this occur?

Not all atoms have the same affinity (attraction) for electrons. Some attract electrons more than others. For example, oxygen and hydrogen join to form a single covalent bond, but oxygen has a slightly higher affinity for electrons than hydrogen. Consequently, the electron of the hydrogen atom when bonded to oxygen tends to spend more time around the oxygen atom than around the hydrogen atom. Because of this, the oxygen atom has a slightly negative charge. The hydrogen atom is visited less frequently by the electron and has a slightly positive charge (Figure 2-8a). The result is a polar covalent bond. A polar covalent bond is a covalent bond whose atoms bear a slight charge—either positive or negative. (Note that oxygen has a slightly negative charge because of the asymmetry of the water molecule.)

The presence of polar covalent bonds in molecules makes entire molecules polar. That is, it gives them positively and negatively charged regions. Consider one of the most important of all polar molecules, water. As noted earlier, each hydrogen atom forms a covalent bond with the oxygen atom of water. Each covalent bond is polar. As Figure 2-8a illustrates, water has three slightly charged atoms. The slightly charged atoms are often attracted to oppositely charged atoms on neighboring water molecules (Figure 2-8b).

Polar molecules are also formed when hydrogen atoms covalently bond to atoms such as nitrogen and fluorine, which, like oxygen, have a slightly higher affinity for electrons than hydrogen.

> **KEY CONCEPTS**
>
> Polar covalent bonds occur any time there is an unequal sharing of electrons by two atoms; in such cases the shared electron spends more time around one atom, giving it a slightly positive charge.

Hydrogen Bonds

As just noted, chemists recognize two types of chemical bonds: ionic and covalent. These are known as *intramolecular bonds* because they occur between the atoms of compounds.

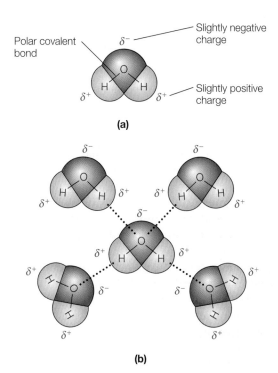

FIGURE 2-8 **Hydrogen Bonding** (a) Slightly unequal sharing of electrons in the water molecule creates a polar molecule. Electrons tend to spend more time around the oxygen nucleus, making it slightly negative. The hydrogen atoms are, therefore, slightly positive, as indicated by the symbols δ^+. (b) Because of the polarity, hydrogen atoms of one molecule are attracted to oxygen atoms of another. The attraction is called a hydrogen bond. (© Photos.com.)

These bonds hold the atoms of ionic and covalent compounds together. Neighboring molecules can also be attracted to one another. This attractive force is known as an *intermolecular bond.* Consider water.

In a glass of water, positively charged hydrogen atoms of water molecules attract negatively charged oxygen atoms of other water molecules. The electrostatic attractions between the positively charged hydrogen atoms of one water molecule and the negatively charged oxygen atoms of another are intermolecular bonds known as **hydrogen bonds**.

Hydrogen bonds are responsible for many of the unique properties of water and are essential to life on Earth (discussed shortly). Hydrogen bonds are also found in other biologically important molecules like the hereditary material DNA and proteins discussed elsewhere in the text.

> **KEY CONCEPTS**
>
> Hydrogen bonds often occur between oppositely slightly charged atoms of different molecules.

Organic and Inorganic Compounds

Chemists classify compounds as either organic or inorganic. **Organic compounds** contain molecules that are made primarily of carbon atoms. Proteins, for instance, are organic compounds, as are sugars.

Many organic molecules are quite large. That's because carbon atoms can be joined to one another by multiple covalent bonds much like the beads of a necklace. These long chains of carbon atoms form the "backbone" of many organic compounds. Attached to the backbone are a number of other atoms, most often hydrogen, oxygen, and

nitrogen. In organic molecules, all atoms are covalently bonded.

Inorganic compounds are defined primarily by exclusion—they are those compounds that are not organic. That is, they are not made primarily of carbon. As a rule, inorganic compounds are generally small molecules. And, they usually consist of atoms joined by ionic bonds, as opposed to organic compounds, which are always covalently bonded. Sodium chloride and magnesium chloride are examples of inorganic compounds. As you will see in the study of biology, not all rules apply 100% of the time. This is true here: not all inorganic compounds contain ionic bonds. Water, for example, is an inorganic compound whose atoms are joined by covalent bonds. Carbon dioxide is another exception. Even though carbon forms the core of this molecule, it is considered an inorganic compound. Table 2-2 summarizes the main features of each type of compound.

> **KEY CONCEPTS**
>
> Chemical compounds fit into two categories: organic and inorganic.

TABLE 2-2	Comparison of Organic and Inorganic Compounds	
Organic		**Inorganic**
Consists primarily of molecules containing carbon and hydrogen		Usually consists of positively and negatively charged ions
Atoms linked by covalent bonds		Usually consists of atoms joined by ionic bonds
Often consists of large molecules that contain many atoms		Always contains small numbers of atoms

2-3   Water, Acids, Bases, and Buffers

Now that you've learned a few basics of chemistry, let's look at water, a biologically important molecule.

Water

Scientific historian and naturalist Loren Eiseley once wrote that, "if there is magic on the planet, it has to be water." This remarkable substance produces billowy clouds, icicles that hang from the limbs of trees, and elegant waterfalls. Water is also of great practical importance to life on Earth—to you and me and all other living things.

Water is a major component of all cells and organisms. In fact, nearly two-thirds of the human body is water. Thus, if you weigh 100 pounds, nearly 70 pounds of your body weight is water. Water is an important biological solvent, a fluid that dissolves other chemical substances. Human blood, for example, contains large amounts of water that dissolve and transport nutrients, hormones, and wastes throughout our bodies.

Water also participates in many chemical reactions in the body—for example, in the breakdown of protein (described later). Water is also essential to cellular energy production. That's why dehydration can significantly lower your energy levels. A drink of water on a hot summer afternoon can give you a boost of energy.

Water serves as a lubricant. Saliva, which is largely water, lubricates food in the mouth and esophagus, the tube that transports food from our mouths to our stomachs. Saliva also helps to keep the teeth cleaned. Watery fluid in the joints enables bones to slide over one another, thus facilitating body motion.

Finally, water helps us regulate body temperature. Perspiration, for example, rids our bodies of heat, for when it evaporates, water draws off heat (Figure 2-9). People sometimes suffer heatstroke in hot, humid climates because the excess water

FIGURE 2-9 Perspiration Cools the Body
Perspiration is an automatic homeostatic response that helps rid the body of heat. (© LiquidLibrary.)

in the atmosphere reduces evaporation. This causes heat in the body to build up, which can cause a person to collapse.

> **KEY CONCEPTS**
> Water is an inorganic molecule that is essential to life for many reasons.

The Dissociation of Water

Water molecules are fairly stable; nonetheless, some molecules break apart, forming two charged units known as ions, as shown in the following reaction:

$$H_2O \quad \leftrightarrow \quad H^+ \quad + \quad OH^-$$

water   hydrogen ion   hydroxide ion

Because the hydrogen and hydroxide ions formed in this reaction can react with one another to reform water molecules, this reaction is said to be reversible. The double arrow in the reaction formula indicates this fact.

Although water can dissociate, the ratio of water molecules to the ions, H^+ and OH^-, in the human body is about 500 million to 1. Nevertheless, even the slightest change in the hydrogen ion concentration can alter cells and organisms, shutting down biochemical pathways and sometimes killing organisms. The human body contains several homeostatic mechanisms to ensure a constant level of these ions.

> **KEY CONCEPTS**
> Some water molecules split into positively charged hydrogen and negatively charged hydroxide ions.

Acids

In pure water, the concentrations of H^+ and OH^- ions are equal. A solution containing an equal number of these ions is said to be chemically neutral.

Chemical neutrality can be upset by adding or removing H^+ and OH^-. For example, when hydrochloric acid (HCl) is added to water, it dissociates into hydrogen ions (H^+) and chloride ions (Cl^-). Adding hydrochloric acid increases the hydrogen ion concentration of the solution.

Scientists use the term **acids** to refer to substances like HCl that add hydrogen ions to a solution. Solutions with proportionately more H^+ and OH^- are said to be acidic.

Bases are substances that remove H^+ from solution. Sodium hydroxide (NaOH) is an example of a base. When added to aqueous (water-based) solutions, it dissociates into Na^+ and OH^-. The hydroxide ions react with hydrogen ions (formed by the breakdown or dissociation of water), forming new water molecules. As sodium hydroxide is added to a solution of pure water, the hydrogen ion concentration declines, and the hydroxide ion concentration increases. A solution with a greater concentration of OH^- than H^+ ions is said to be basic.

Acidity is measured on the **pH scale** (Figure 2-10). As illustrated, the pH scale ranges from 0 to 14. Neutral substances have a pH of 7. Basic substances have a pH greater than 7, and acidic substances have a pH less than 7.

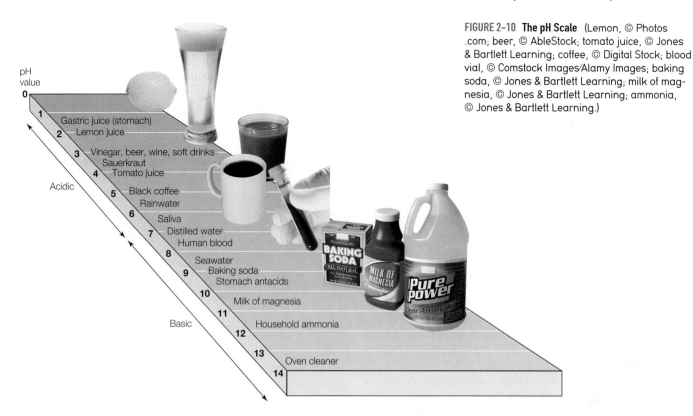

FIGURE 2–10 **The pH Scale** (Lemon, © Photos .com; beer, © AbleStock; tomato juice, © Jones & Bartlett Learning; coffee, © Digital Stock; blood vial, © Comstock Images/Alamy Images; baking soda, © Jones & Bartlett Learning; milk of magnesia, © Jones & Bartlett Learning; ammonia, © Jones & Bartlett Learning.)

pH value

0
1 — Gastric juice (stomach)
2 — Lemon juice
3 — Vinegar, beer, wine, soft drinks
    Sauerkraut
4 — Tomato juice
5 — Black coffee
    Rainwater
6 — Saliva
7 — Distilled water
    Human blood
8 — Seawater
9 — Baking soda
    Stomach antacids
10 — Milk of magnesia
11 — Household ammonia
12
13 — Oven cleaner
14

Acidic

Basic

On the pH scale, a change in one pH unit represents a tenfold change in acidity. Thus, a solution with a pH of 3 is 10 times more acidic than one with a pH of 4 and 100 times more acidic than one with a pH of 5.

Most biochemical reactions occur at pH values between 6 and 8. Human blood, for example, has a pH of 7.4. So important is this to normal body function that a slight shift in the pH of the blood for even a short period can be fatal.

> **KEY CONCEPTS**
> Acids are substances that add hydrogen ions to solution, decreasing the pH; bases remove hydrogen ions and increase the pH.

Maintaining pH: Buffers

Biological systems operate within a narrow pH range, as noted above. This narrow range is maintained by naturally occurring chemicals called buffers. **Buffers** (BUFF-firs) occur in organisms and even ecosystems, particularly soil and water, where they help to protect against potentially dangerous shifts in pH. How do they work?

Buffers can be likened to hydrogen-ion sponges. They help maintain a constant pH by removing hydrogen ions from solution when levels increase. Buffers give back the hydrogen ions when levels fall.

One of the most common buffers is carbonic acid. Found in the blood of animals, carbonic acid is formed from water and carbon dioxide, a waste product of cellular energy production. In blood, carbonic acid dissociates into bicarbonate and hydrogen ions:

$$H_2CO_3 \leftrightarrow H^+ \leftrightarrow HCO_3^-$$

carbonic acid (weak acid)   hydrogen ion   bicarbonate (weak base)

In our blood, and in lakes and rivers where carbonic acid is also present, this chemical reaction shifts back and forth in response to changing levels of hydrogen ions. Thus, when hydrogen ions are added to water, they combine with bicarbonate ions, driving the reaction to the left. But when the hydrogen-ion concentration falls, the reaction is driven to the right. In either case, the pH of the water remains constant.

> **KEY CONCEPTS**
> Buffers are chemical substances in the environment and in our bodies that help maintain pH within a narrow range essential for life.

2-4 Overview of Other Biologically Important Molecules

This introduction to chemistry is intended to give you a foundation that will help you understand human biology. After all, the cells, tissues, and organs of the human body are composed of numerous chemicals. Many processes vital to life are nothing more than chemical reactions involving organic and inorganic compounds. Lets take a look at the major organic molecules.

In your study of biology, you will encounter four major groups of organic molecules: (1) carbohydrates; (2) lipids; (3) amino acids, peptides, and proteins; and (4) nucleic acids.

Rather than describe each one in detail here, I will present a few important facts. These molecules are discussed in more detail in subsequent chapters, as you encounter them in your study of human biology.

Carbohydrates are organic molecules that range in size from very small molecules such as the blood sugar glucose to very large ones such as starch. Large biological molecules like starch are actually composed of many small molecules.

Carbohydrates serve many functions in our bodies. Small carbohydrates like glucose, for instance, are the main source of cellular energy. You and I obtain the carbohydrates we need to produce energy by eating foods rich in starch, such as crackers, pastas, rice, and cereals. Many of us also ingest lots of carbohydrate in beverages like soft drinks (Health Tip 2-2). Unfortunately, our bodies convert excess sugar we ingest into fat, which leads to weight gain.

Lipids (LIP-ids) are a rather diverse group of biologically important molecules that includes fats and steroids such as cholesterol. Many lipids we consume are a source of energy; others are important structural components of cells. For example, the membranes that enclose the cells of our bodies are mostly made of lipids. Lipids play other roles as well. Some are chemical precursors that react to form important hormones like testosterone. Deposits of lipids beneath the skin of animals form an insulating layer that keeps us and others animals like whales warm in cold environments.

Lipids are found in many foods, especially meat, nuts, fried foods, and many snack foods like chips, donuts, and many chocolate products. Although lipids are essential to a healthy diet, most people consume too much lipid in fatty foods like French fries, hamburgers, and a wide assortment of snack foods. Excess intake of lipids leads to excessive weight gain and a whole host of debilitating, even fatal, health problems like heart disease, diabetes, and premature breakdown of joints.

Most readers have heard the terms amino acids, proteins, and peptides. **Proteins** are long molecules consisting of many smaller molecules, the amino acids. Molecules consisting of many smaller molecules are known as *polymers*. The subunits are known as *monomers*. Found in many foods, especially meats and milk products, proteins are important structural elements of cells, as you will see elsewhere in the text. Hair and fingernails are made from protein, too. Many proteins in the human body function as enzymes, special molecules that speed up chemical reactions in the body. **Peptides** are very small proteins—chains of amino acids. Contrary to popular belief proteins are not a source of energy except when a person is starving. (Then the body will break down protein to make energy.)

Health Tip 2-2

Watch what you drink! Select water or unsweetened tea or coffee over sweetened or calorie-rich drinks such as sodas, fruit juices, and beer.

Why?

Many people either forget—or don't realize—that beverages are often packed with calories (mostly sugars). These calories add up quickly. Excess sugar is converted to fat. What is more, most sweetened or calorie-rich drinks offer few, if any, essential nutrients other than sugar and water. Nutritionists call them "empty calories." Liquid calories also don't alleviate hunger as effectively as solid foods.

Health Tip 2-3

To lose or maintain weight, substitute low-fat dairy products like low-fat yogurt and low-fat milk and lean meat like chicken breast and fish for fattier whole milk, meats, and cheeses.

Why?

Low-fat dairy foods and low-fat meats provide the protein you need without all of the calories of their higher-fat cousins.

The fourth group of biologically important molecules is the **nucleic** (new-CLAY-ick) **acids**. This structurally complex group includes DNA, the genetic material found in cells. These molecules are long chains of smaller molecules called **nucleotides** (NEW-klee-oh-tides). As many students know, the genetic information in DNA is determined by the sequence of nucleotides. Another nucleic acid is RNA. It plays a key role in protein synthesis, as you shall see elsewhere in the text.

> **KEY CONCEPTS**
>
> Numerous organic compounds play an important role in human biology; the most important are carbohydrates, lipids, nucleic acids, proteins, and peptides.

2-5   Health and Homeostasis

Elsewhere in the text you learned that human health is dependent on a healthy environment. Our environment includes the social and psychological worlds we inhabit and well as the physical environment we inhabit. Exposure to an unhealthy environment can result in disease. Stress from a poor social or psychological environment can be as dangerous as pollutants in a contaminate air and water. One disease that can result from an unhealthy environment is cancer.

Cancer is a disease in which body cells begin to divide uncontrollably, producing growths called tumors. Unless destroyed, many tumors eventually kill the person. Like many other diseases that afflict humans, cancer sometimes results from imbalances, for example, excesses of cancer-causing agents in our environment, as the following story reveals.

For 2,000 years, the people in Lin Xian, China, 250 miles south of Beijing, have been dying in record numbers from cancer of the esophagus, the muscular tube that transports food to the stomach. So prevalent is the disease that one of every four persons once succumbed to this ruthless killer, whose incidence is higher there than anywhere else in the world.

In 1959, scientists began a systematic study of the people living in the valley around Lin Xian to discover the origins of the disease and put an end to it. Scientists first found that esophageal cancer in Lin Xian was the result of a group of chemicals called nitrosamines (NYE-trose-ah-MEANS).

Nitrosamines, they discovered, were being produced in the stomachs of the residents from two other chemical compounds: nitrites and amines. But where did they come from?

Research showed that residents had abnormally high levels of nitrites, which came from abnormally high levels of nitrate in the vegetables the residents consumed (Figure 2-11). The amines were present in moldy bread, a delicacy in the region. But why the excess of nitrates in vegetables? And why the higher-than-normal levels of nitrites in people?

Chemical analyses of the soil in which the region's crops were grown showed that the topsoil around Lin Xian was deficient in molybdenum (meh-LIB-deh-num), an element required by plants in minute quantities (Figure 2-11). This deficiency caused the plants to concentrate nitrates. It also reduced vitamin C production by plants. Nitrates from plants were apparently being converted to nitrites in the stomachs of the residents. Low levels of vitamin C in the plants the people ate promoted the conversion of nitrates to nitrites. When the residents were given vitamin C tablets, their nitrite levels dropped.

Ultimately, then, a soil deficiency, caused either by intensive agriculture, which depletes the soil, or by a natural deficiency, put the local residents who ate moldy bread rich in amines at high risk for esophageal cancer.

To reduce the risk of esophageal cancer, the villagers now coat wheat and corn seeds with molybdenum. As a result, nitrite levels in vegetables have dropped 40%, and vitamin C levels have increased 25%.

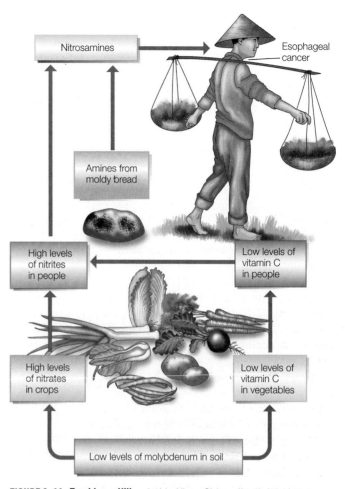

FIGURE 2-11 **Tracking a Killer** In Lin Xian, China, the link between esophageal cancer and the soil was uncovered by medical researchers.

KEY CONCEPTS

Chemical imbalances in our environment can affect our food supply and our health in profound ways.

SUMMARY

1. An understanding of chemistry is essential to understanding biology because all organisms are composed of chemicals and many life processes are nothing more than chemical reactions.

Atoms and Subatomic Particles

2. The physical world we live in is composed of matter. Matter is anything that has mass and occupies space. All matter is composed of tiny particles called atoms.
3. Atoms contain subatomic particles: electrons, protons, and neutrons.
4. Negatively charged particles, electrons, are the smallest of these three subatomic particles; they are found in the electron cloud around the dense, central nucleus.
5. Protons are positively charged particles that are located in the nucleus with neutrons, which are uncharged.
6. Atoms are electrically neutral because the number of electrons always equals the number of protons.

7. A pure substance such as gold or lead that contains only one type of atom is known as an element. The elements are listed on the periodic table by atomic number, which is the number of protons in the nucleus.
8. Most elements are made up of mixtures of two or more isotopes, atoms with slightly different atomic masses. This is the result of additional neutrons. Additional neutrons in the nuclei of atoms often make isotopes unstable. To achieve a more stable state, many isotopes release radiation—small bursts of energy, or tiny energetic particles—from their nuclei.
9. Radiation has proved to be a useful tool in medicine and science, but it can also damage cells of organisms, leading to birth defects and cancer.

The Making of a Molecule

10. Atoms react with one another to form organic and inorganic compounds.

Their reactivity is a result of the interaction of electrons.
11. Two types of bonds are found between atoms, ionic and covalent.
12. During some atomic reactions, electrons are transferred from one atom to another. Transferring an outer-shell electron from one atom to another creates two oppositely charged ions. An electrostatic attraction forms between them and is referred to as an ionic bond.
13. Covalent bonds form between atoms that share one or more pairs of electrons. This sharing of electrons holds the atoms together.
14. Unequal sharing of electrons results in the formation of a polar covalent bond, which results in the formation of polar molecules. Polar molecules are frequently attracted to one another by hydrogen bonds, weak electrostatic attractions between the oppositely charged atoms of neighboring molecules.

15. Organic molecules are compounds made up primarily of carbon and hydrogen. The atoms of these generally large molecules are held together by covalent bonds.
16. Inorganic molecules are much smaller molecules whose atoms are often linked by ionic bonds.

Water, Acids, Bases, and Buffers

17. Water is an important inorganic molecule and a major component of all body cells. It serves as a solvent, which transports many substances in the blood, body tissues, and cells. It also participates in many chemical reactions, serves as a lubricant, and helps regulate body temperature.

18. Although fairly stable, water molecules can dissociate, forming hydrogen and hydroxide ions. In pure water, the concentration of these ions is equal and water has a pH of 7.
19. Adding acidic substances increases the concentration of hydrogen ions, causing the pH to fall.
20. Substances that remove hydrogen ions from solutions cause the pH to climb and are called bases. A solution with a pH less than 7 is acidic; a solution with a pH greater than 7 is basic.
21. Biological systems contain buffers, chemical compounds that offset changes in the concentration of hydrogen and hydroxide ions. Buffers are therefore an important component of chemi-cal homeostasis in organisms and the environment.

Overview of Other Biologically Important Molecules

22. Carbohydrates, lipids, amino acids, proteins, and nucleic acids are biologically important molecules. They vary considerably in structure and function. Details of their structure and function are covered in later chapters.

Health and Homeostasis

23. Tiny imbalances of chemicals in the environment can cause upsets in agricultural crops, which, combined with other factors, can lead to cancer.

THINKING CRITICALLY ANALYSIS

This analysis corresponds to the Thinking Critically scenario that was presented at the beginning of this chapter.

If you said no, you'd be correct. If you said that it was a bad idea because the experiment lacked a control group, you'd also be correct. Interestingly, some diseases show spontaneous remission. That is, they get better on their own without treatment. If you had performed the study as the student suggested and 60% of the patients recovered, could you conclude that the drug was

effective? No. You'd need a control group. That is, you would need a group who received no treatment. If 60% of the patients in the control group got better, then you would conclude that the drug was not effective. If only 10% got better and 60% of the experimental group recovered, then you might be able to conclude that the drug was effective. But don't be too quick to rush to judgment. In science, certainty comes from additional studies—just to be sure the observed effects are real.

KEY TERMS AND CONCEPTS

Acid, p. 32
Amino acid, p. 34
Atom, p. 23
Atomic number, p. 24
Base, p. 32
Buffer, p. 33
Carbohydrate, p. 34
Compound, p. 26
Covalent bond, p. 27
Electron, pp. 23–24
Element, p. 24
Hydrogen bond, p. 31

Inorganic compound, p. 31
Ion, p. 26
Ionic bond, p. 26
Isotope, p. 25
Lipid, p. 34
Mass, p. 23
Matter, p. 23
Molecule, p. 26
Mutation, p. 25
Neutron, p. 23
Nucleic acid, p. 34
Nucleotide, p. 34

Nucleus, p. 23
Organic compound, p. 31
Peptide, p. 34
pH scale, p. 32
Polar covalent bond, p. 30
Protein, p. 34
Proton, p. 23
Radiation, p. 25
Radioactive isotopes, p. 25
Radionuclide, p. 25
Subatomic particle, p. 23

CONCEPT REVIEW

1. Why is the study of chemistry so important to the study of human biology? pp. 22–23.
2. Describe the structure of an atom, using a diagram to illustrate your answer. pp. 23–24.
3. Why is the matter on Earth so varied in its appearance? p. 24.

4. Define each of the following: electrons, protons, and neutrons. Which of these subatomic particles is involved in chemical reactions? pp. 23–24, 26.
5. Define the following terms: atomic number, mass number, and isotope. pp. 24–25.
6. Why do many isotopes emit radiation? p. 25.

7. How are ionic and covalent bonds different? Describe what holds the atoms together in both cases. In what ways are the bonds similar? pp. 26–27.
8. Describe how polar covalent bonds form and why they result in the formation of hydrogen bonds. p. 30.

9. Temperature is a measure of the speed of molecules: the higher the temperature, the higher the speed. With this information and your knowledge of water and hydrogen bonds, why does water have a higher boiling point than a nonpolar liquid like alcohol? p. 31.

10. Describe the biological significance of water. In other words, what functions does water play in the human body? p. 32.

11. How does the body regulate H⁺ levels in the blood? Describe the process. Why is it so important that the acidity of the blood remain constant? (You may have to do some research to answer this question.) pp. 32–33.

12. Which of the four groups of organic molecules is responsible for heredity? p. 34.

13. What is the main source of energy in the body? Where does it come from? p. 34.

14. Describe how the environment we live in affects human health. What is meant by the environment? Give specific examples, if you can. In what ways is the environment you live in healthy and in what ways is it unhealthy? pp. 34–35.

SELF-QUIZ: TESTING YOUR KNOWLEDGE

1. Each atom has a nucleus made up of _____ and neutrons. p. 23.

2. In atoms the number of protons is usually equal to the number of _____. p. 24.

3. The atomic number of an atom is equal to the number of _____ in the nucleus. p. 24.

4. An alternative form of an atom is called an _____; it differs in the number of neutrons in the nucleus. p. 25.

5. A bond formed between two atoms that share electrons is called a(n) _____ bond. p. 27.

6. _____ bonds form between oppositely charged atoms of polar covalent bonds of different molecules. pp. 26–27.

7. Proteins and carbohydrates consist primarily of carbon, hydrogen, and oxygen; these two molecules are types of _____ compounds. p. 31.

8. On the pH scale, a pH of ___ is neutral; a solution with a pH less than this is _____ and a solution with a pH greater than this is _____. p. 32.

9. Chemicals like bicarbonate that resist pH change are called _____. p. 33.

10. When water molecules dissociate, they form ___ and _____ ions. p. 32.

biology.jbpub.com/chiras/8e/

The site features eLearning, an online review area that provides quizzes, chapter outlines, and other tools to help you study for your class. You can also follow useful links for in-depth information, research the differing views in the Point/Counterpoints, or keep up on the latest health news.

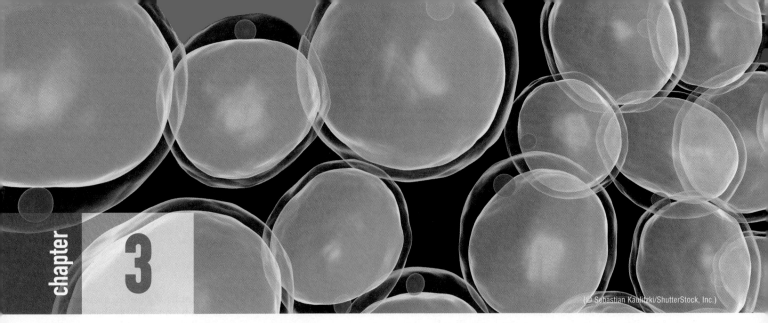

The Life of the Cell

THINKING CRITICALLY

Researchers at a major medical college in your area have discovered that cells live longer when cultured (grown) in petri dishes containing a certain vitamin. They suggest that this vitamin might help people live longer. A local newspaper runs with the story, and one of your friends is thinking of taking the vitamin supplement in hope of living longer. What advice would you give him before he embarks on this course of action?

For many years, medical researchers have been performing promising experiments with embryonic stem cells. Stem cells are undifferentiated cells taken from very early human embryos only four to five days after fertilization. These cells have the potential to develop in a variety of directions, for example, into muscle, nerve cells, or any of the other 220 cell types found in the human body. Scientists hope that stem cells can be used to replace diseased cells, allowing the reversal of a number of chronic, debilitating diseases such as juvenile diabetes, Parkinson's disease, and others. They could also be used to treat cancer and genetic diseases. Proponents of this research believe that it may someday make a significant contribution to medical science, reducing pain and suffering in tens of thousands of people worldwide—people who otherwise have little hope of a cure. Already, studies are showing that stem cell transplantation successfully cures some patients who have the debilitating nervous system disease known as Parkinson's disease, which currently afflicts millions of people worldwide, including actor Michael J. Fox of *Back to the Future* movie fame and the famous boxer Muhammad Ali.

Although the research results are promising, embryonic stem cell research poses a number of technical challenges. It also poses several significant moral and political dilemmas, although new discoveries may alleviate much of the concern. One such discovery was that embryonic stem cells could be removed from early embryos without destroying them. Another key discovery is that through genetic engineering certain adult cells can be converted back to stem cells. This stem cell research controversy is discussed in the Point/Counterpoint in this chapter.

This chapter focuses on the cell and some key biological molecules. The information presented here forms a foundation on which a good understanding of human biology can be built. This material will also help you understand the details of stem cell research.

3-1 Microscopes: Illuminating the Structure of Cells

Cells are exceedingly small and, with few exceptions, cannot be seen with the naked eye. Our knowledge of them has depended on the **microscope**, a device consisting of a lens or combination of lenses used to enlarge tiny objects. (For a discussion of the discovery of cells, see Scientific Discoveries 3-1.)

Microscopes fit into two categories: (1) *light microscopes*, which use ordinary visible light to illuminate the specimen under study; and (2) *electron microscopes*, which use a beam of electrons to create a visual image of a specimen. Both types of microscope enlarge tiny images and allow us to see the minute details of a cell. Light microscopes magnify objects from 100 to 400 times their original size (Figure 3-1). Electron microscopes enlarge objects 100,000 times their original size. One type of electron microscope, known as a scanning electron microscope, allows scientists to obtain three-dimensional views of cells (Figure 3-2). Photographs from both light and electron microscopes are used throughout this book.

> **KEY CONCEPTS**
>
> Microscopes allow us to study the structure of cells; two types are commonly used: light microscopes and electron microscopes.

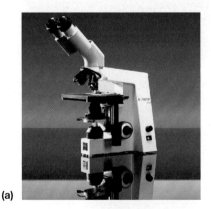

(a)

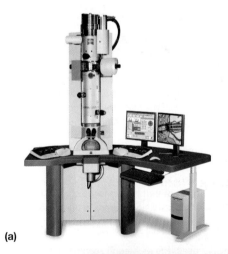

(a)

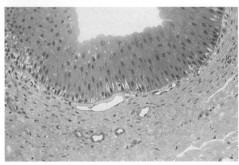

(b)

FIGURE 3-1 The Light Microscope (a) In the ordinary light microscope, the image is illuminated by light that is cast from a mirror or (in this case) a light bulb located below the object. Two lenses (one in the eyepiece and one just above the object) magnify the object. (Courtesy of Carl Zeiss MicroImaging, Inc., Thornwood, NY.) (b) Photo of a section of the lining of the urinary bladder taken through a light microscope. (© Biodisc/Visuals Unlimited.)

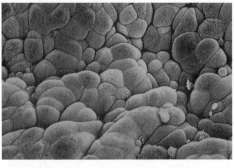

(b)

FIGURE 3-2 The Electron Microscope (a) Scanning electron microscope. (Courtesy of Carl Zeiss, NTS) (b) Three-dimensional image (known as a scanning electron micrograph) of the lining of the urinary bladder. (© Biodisc/Visuals Unlimited.)

Scientific Discoveries that Changed the World

3-1 The Discovery of Cells
Featuring the Work of Hooke, Leeuwenhoek, Brown, Schleiden, Schwann, and Virchow

One of the fundamental principles of biology is known as the *cell theory*. The cell theory comprises three parts: (1) the cell is the basic unit of structure of all organisms, (2) all organisms consist of one or more cells, and (3) all cells arise from pre-existing cells. Although this may seem rather elementary, it was not so obvious to early scientists who labored with relatively crude instruments and without the benefit of many facts we now take for granted.

One of those scientific pioneers who opened our eyes to the world of cells was Robert Hooke, a seventeenth-century British mathematician, inventor, and scientist. Equipped with a relatively crude microscope, Hooke observed just about everything he could lay his hands on—which he described in his book *Micrographia*, published in 1665.

One especially useful description was that made on a thin slice of cork (Figure 1). Peering through his microscope, Hooke beheld a network of tiny, boxlike compartments that reminded him of a honeycomb. He called these compartments *cellulae*, meaning "little rooms." Today, we know them as cells.

Hooke did not really see cells but, rather, cell walls, the structures that surround the cell membranes of plant cells. The cytoplasm and cellular organelles had been destroyed when he prepared the tissues for slicing.

Hooke's work was complemented a few years later by Antony van Leeuwenhoek, a Dutch shopkeeper who spent much of his free time designing simple microscopes. Like Hooke, Leeuwenhoek examined just about everything he could find and wrote extensively on his observations. Wayne Becker, a cell biologist at the University of Wisconsin writes, "His detailed reports attest to both the high quality of his lenses and his keen powers of observation." Becker continues: "They also reveal an active imagination, since at one point he reported seeing a 'homunculus' (little man) in the nucleus of a human sperm cell."

Leeuwenhoek discovered bacteria and protozoans (single-celled eukaryotic organisms). Although his microscopes were superior to any others around at that time, they were still crude by modern standards. This fact and the tendency of scientists at that time to focus only on the observation and description of life forms stalled progress in our understanding of cells for some time. In fact, more than a century passed before cell biology moved significantly forward.

Aided by improved lenses, the eighteenth-century English botanist Robert Brown noted that every plant cell he studied contained a centrally placed structure, now called a nucleus. A German colleague, Matthias Schleiden, concluded that all plant tissues consisted of cells. One year

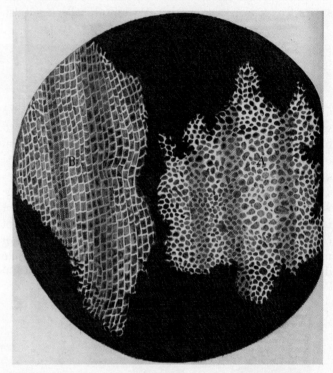

FIGURE 1 **Thin Slice of Cork** (Courtesy of Library of Congress, Prints & Photographs Division [reproduction number LC-USZ62-95187].)

later, Theodor Schwann, a German scientist, arrived at a similar conclusion regarding animal cells. This discovery laid to rest an earlier hypothesis that plants and animals were structurally different.

Based on earlier research and his own work, Schwann proposed the first two parts of the cell theory—the cell is the basic unit of structure for all organisms and all organisms consist of one or more cells. Less than 20 years after the discovery of cell division, the German physiologist Rudolf Virchow added the third tenet of the cell theory: all cells arise from pre-existing cells.

Like other discoveries, the cell theory is the work of many people over many years. Although only a few people are credited with this important discovery, the credit really belongs to an entire line of scientists who, through careful observation and experimentation, have changed our view of the world.

3-2   An Overview of Cell Structure

The human body contains 220 different types of cells. These cells perform very different tasks. Because the function of a cell is frequently reflected in its structure, very few cells look alike. Although cells vary in structure and function, we'll focus our attention on similarities as we examine a fictional entity, the typical animal cell, shown in Figure 3-3. Don't look for it under the microscope; it only exists on the pages of biology textbooks. You may want to refer to Table 3-1 as you study the following material. It's also a useful resource when you review this material for tests.

The Nuclear and Cytoplasmic Compartments
Human cells are the product of millions of years of evolution. As illustrated in Figure 3-3, human cells consist of two basic compartments, the nuclear and the cytoplasmic. The nuclear compartment, or **nucleus**, is the control center of the cell. That is, it contains the genetic information that regulates the structure and function of the cell. As shown in Figure 3-3, the nucleus is surrounded by a double membrane, the **nuclear envelope**.

The cytoplasmic compartment lies between the nucleus and the **plasma membrane**, the outermost structure of the cell. It contains a substance known as cytoplasm. **Cytoplasm** is a semifluid material that consists of numerous molecules, including water, protein, ions, nutrients, vitamins, dissolved gases, and waste products. It forms a nutrient pool from which the cell draws chemicals it needs for metabolism. It is also a dumping ground for wastes, which are later removed from cells.

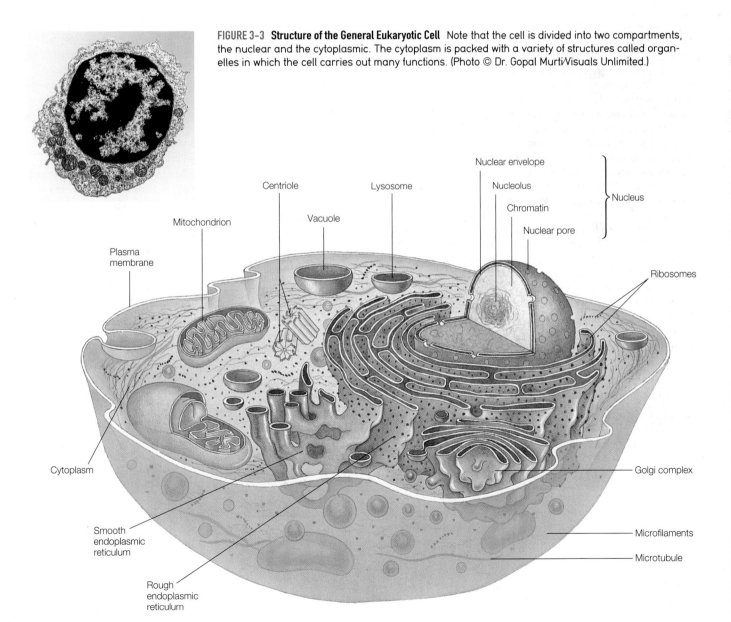

FIGURE 3-3 Structure of the General Eukaryotic Cell Note that the cell is divided into two compartments, the nuclear and the cytoplasmic. The cytoplasm is packed with a variety of structures called organelles in which the cell carries out many functions. (Photo © Dr. Gopal Murti/Visuals Unlimited.)

Point/Counterpoint   Fetal Cell Transplantation

Fetal Tissue Transplants: Auschwitz Revisited *by Tom Longua*

Persons who advocate medical research using tissue from aborted babies claim it has genuine medical value. But that premise is just plain false. On April 22, 1999, both the *New York Times* and *Washington Post* ran stories about a four-year trial by the National Institute of Neurological Disorder, in which a group of Parkinson's patients had fetal cells implanted into their brains, while others had a "sham" operation in which no cells were implanted.

Tom Longua has been an educator for 39 years and is currently instructor of anatomy and physiology at the Denver Academy of Court Reporting. He has been active in the pro-life movement since 1974 and is a member of the board of directors of Colorado Right to Life. (© Thomas J. Longua)

In patients over 60 (those most susceptible to Parkinson's), no difference was observed between the two groups, either on the neurologists' examinations or on the patients' own perception of their condition. Even among younger patients—for whom the treatment allegedly "worked"—there was no difference between the two groups in initiating walking, walking balance, the number of falls, tremors, "freezing" (sudden loss of all movement), and dyskinesia.

But even if the claims were true, should a civilized society allow research on human beings who may be killed, albeit legally, even though they are innocent of any crime? After World War II, Nazi doctors who experimented on prisoners in concentration camps were tried at Nuremberg for "crimes against humanity." The prosecutor opened his case with this statement:

> The defendants in this case are charged with. . . . Atrocities committed in the name of medical science. . . . The wrongs which we seek to condemn and punish have been so calculated, so malignant, and so devastating, that civilization cannot tolerate their being ignored, because it cannot survive their being repeated.

Yet with research involving tissue from elective abortions, we are committing the same "crimes against humanity." Even the claimed justification is the same.

- Nazi doctor August Hirt: "Wouldn't it be ridiculous to send the bodies to the crematory oven without giving them an opportunity to contribute to the progress of society?"
- Lawrence Lawn, a doctor involved in research with tissue from elective abortions: "We are simply using something which is destined for the incinerator to benefit mankind."

Proponents of such research claim that it is acceptable if it is "separated" from abortions that are freely chosen anyway. Again, this argument

echoes the Nazi doctors' defense at Nuremberg: "We caused no deaths. They [the victims] were all consigned to death by legal authorities; with . . . professional correctness we tried to salvage some good from their plight."

But "separation" is a myth. As the Auschwitz research depended on the "acceptability" of killing prisoners, fetal tissue research depends on the "acceptability" of killing helpless, innocent human beings in the first place. (Certainly there is nothing wrong with using tissue from ectopic pregnancies or miscarriages; but this debate concerns babies who are purposely killed simply because they are unwanted.) Moreover, new abortion techniques have been contrived specifically for the purpose of securing usable (and salable) tissue. So much for "separation."

But besides the perverseness of the practice itself, such research will have other detrimental effects:

- It will certainly increase the number of abortions. Some medical journals claim that fetal tissue could be used to treat a vast array of diseases. In a 1988 guest editorial in the *Wall Street Journal*, Dr. Emanuel Thorne estimated that the 1.6 million annual abortions in the United States would not be enough to keep up with the demand for such tissue. While there have been well-publicized cases of women becoming pregnant to create organ or tissue sources for their parents and other children, it is unlikely that many women would get pregnant to create tissue that might be "useful" for others (although stranger things have happened). But with money to be made, the abortion industry will certainly pressure women who are undecided about having abortions, using the argument that the tissue will serve some "benevolent" purpose. Indeed, retired Colorado abortionist James Parks bragged, "[My patients] say, 'Thank God, some good is going to come out of this.'" (*The New York Times*, 11/19/89)
- In a 1991 article on this topic, this writer said, "[This research] will lead to trafficking in human 'spare parts.'" Well, eight years later, Congress launched an investigation into exactly that grisly practice, finding that it was occurring, and that it was extremely lucrative. As the Associated Press reported on October 10, 1999, Congressman Tom Tancredo (R-CO), "displayed a brochure . . . that lists prices for fetal parts, including '$50 for eyes, $150 for lungs and hearts and $999 for an eight-week brain.'"
- Advocacy of fetal tissue research will hinder other—and ethical—research to find cures for some diseases. There are many potential treatments being studied, such as autografts and use of patients' own stem cells, but some researchers have abandoned those possibilities because of the much-touted (albeit dishonest) fetal tissue approach.

Advocates of fetal tissue research like to depict themselves as "humanitarians." But then, so did the doctors on trial at Nuremberg.

Human Fetal Tissue Should Be Used to Treat Human Disease *by Curt R. Freed*

Despite its legalization, abortion remains a controversial issue and will continue to stir debate in the future. In the United States, the debate centers on whether a woman has the right to control her reproduction. The future developments of this political debate are uncertain, but recent elections suggest that pro-choice candidates have been victorious when elections are based on the abortion issue.

Currently, over 1 million legal abortions are performed in the United States each year; most abortions are performed in the first trimester. For nearly all women, having an abortion is an anguishing choice filled with regret and ambivalence. Nonetheless, the difficult personal decision to terminate a pregnancy is made. After the abortion, fetal tissue is usually discarded.

As an alternative to throwing this tissue away, research has shown that fetal tissue may be useful for treating patients with disabling diseases. For over 50 years, research in animals has demonstrated that fetal tissue has a unique capacity to replace certain cellular deficiencies and so may be useful for treating some chronic diseases of humans. These diseases include Parkinson's disease, diabetes, and some immune system disorders.

Cadaver fetal tissue offers the promise of helping large numbers of Americans with crippling diseases. Parkinson's disease, for example, affects hundreds of thousands of Americans. By reducing the ability to move, the disease can end careers and turn people into invalids. The disease is caused by the death of a small number of critically important nerve cells that produce a chemical called dopamine. Experiments in animals and early experiments in humans indicate that fetal dopamine cells transplanted into the brains of these patients may restore a patient's capacity to move and may even eliminate the disease. Patients whose minds work perfectly well and whose bodies are otherwise normal may become healthy and productive citizens once again.

Using cadaver tissue to treat humans has been debated for nearly 40 years. As kidney, cornea, and other organ transplants were developed in the 1950s, many objected to recovering organs from cadavers. In the intervening decades, opinion has changed so that the practice of recovering these organs from cadavers has gone from a provocative and controversial practice to an accepted policy endorsed by most states. In fact, in most states, a check-off box on the back of driver's licenses is used to give permission for organ donation in the event of the death of the driver. Because abortions are induced, some argue that fetal tissue should be regarded differently from other cadaver tissue. Given the facts that abortion is legal and that fetal tissue is ordinarily discarded, there should be no moral dilemma in using fetal tissue for therapeutic purposes. As with the use of all human tissue for transplant, specific informed consent by the woman donating the tissue must be obtained.

Curt R. Freed, M.D., is a professor of medicine and pharmacology at the University of Colorado School of Medicine in Denver. He has written some 70 articles on medical topics as well as numerous abstracts, chapters, and reviews. (© Curt R. Freed)

Some have proposed that using fetal tissue for therapeutic purposes will increase the number of abortions. This is preposterous. It strains the imagination to think that a woman would get pregnant and have an abortion simply on the chance that the aborted fetal tissue might be used to treat a patient unknown to her. An unwanted pregnancy is an intimate and deeply personal crisis; it is inconceivable that the pregnancy would be seen primarily as a philanthropic opportunity.

Politics and medicine have frequently mixed in the past and will continue to do so in the future. As a physician, I think it is important to try to improve the health of patients with serious diseases. Legally acquired fetal cadaver tissue that would otherwise be discarded should be used to treat humans with disabling diseases.

Sharpening Your Critical Thinking Skills

1. Summarize the positions of each author.

2. Using your critical thinking skills, analyze the view of each author. Is each stand well substantiated? Do the author's biases play a role in each argument?

3. Do you see this debate as scientific or ethical? Explain your answer.

4. Each position is based on at least one key argument. Can you pinpoint them?

5. Which viewpoint do you agree with? Why? What factors (biases) affect your decision?

Visit Human Biology's Internet site for links to websites offering more information on this topic.

TABLE 3-1	Overview of Cell Organelles

Organelle	Structure	Function
Nucleus	Round or oval body; surrounded by nuclear envelope.	Contains the genetic information necessary for control of cell structure and function; DNA contains hereditary information.
Nucleolus	Round or oval body in the nucleus consisting of DNA and RNA.	Produces ribosomal RNA.
Endoplasmic reticulum	Network of membranous tubules in the cytoplasm of the cell. Smooth endoplasmic reticulum contains no ribosomes. Rough endoplasmic reticulum is studded with ribosomes.	Smooth endoplasmic reticulum (SER) is involved in the production of phospholipids and has many different functions in different cells; round endoplasmic reticulum (RER) is the site of the synthesis of lysosomal enzymes and proteins for extracellular use.
Ribosomes	Small particles found in the cytoplasm; made of RNA and protein.	Aid in the production of proteins on the RER and polysomes.
Polysome	Molecule of mRNA bound to ribosomes.	Site of protein synthesis.
Golgi complex	Series of flattened sacs usually located near the nucleus.	Sorts, chemically modifies, and packages proteins produced on the RER.
Secretory vesicles	Membrane-bound vesicles containing proteins produced by the RER and repackaged by the Golgi complex; contain protein hormones or enzymes.	Store protein hormones or enzymes in the cytoplasm awaiting a signal for release.
Food vacuole	Membrane-bound vesicle containing material engulfed by the cell.	Stores ingested material and combines with lysosome.
Lysosome	Round, membrane-bound structure containing digestive enzymes.	Combines with food vacuoles and digests materials engulfed by cells.
Mitochondria	Round, oval, or elongated structures with a double membrane. The inner membrane is thrown into folds.	Complete the breakdown of glucose, producing NADH and ATP.
Cytoskeleton	Network of microtubules and microfilaments in the cell.	Gives the cell internal support, helps transport molecules and some organelles inside the cell, and binds to enzymes of metabolic pathways.
Cilia	Small projections of the cell membrane containing microtubules; found on a limited number of cells.	Propel materials along the surface of certain cells.
Flagella	Large projections of the cell membrane containing microtubules; found in humans only on sperm cells.	Provide motive force for sperm cells.
Centrioles	Small cylindrical bodies composed of microtubules arranged in nine sets of triplets; found in animal cells, not plants.	Help organize spindle apparatus necessary for cell division.

As shown in Figure 3-3, the cell contains numerous structures called **organelles**. Organelles ("little organs") carry out ... functions (Table 3-1). Many organelles are bounded ... therefore subcompartments within the ...

... is essential to the smooth func... ...cause it permits cells to segregate ... This, in turn, increases the effi... ...complex organisms perform many ...ly. In a way, a cell is like a modern ...hysically isolated departments, each

performing specific functions, but all working together toward one purpose.

The cytoplasm of the cell also contains organelles that are not membrane-bound. These organelles perform many important functions, too (see Table 3-1).

Giving shape to the cell is a network of protein tubules and filaments in the cytoplasm called the **cytoskeleton** (Figure 3-4). Besides giving support, the cytoskeleton helps organize the cell's activities, which greatly increases the efficiency with which it carries out its many functions. One way the cytoskeleton increases efficiency is by binding with enzymes.

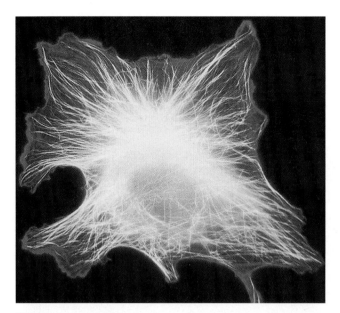

FIGURE 3-4 **The Cytoskeleton** Photomicrograph of the cytoskeleton of a human fibroblast (connective tissue cell). The microtubules are yellow, and the microfilaments are red. (© M. Schliwa/Visuals Unlimited.)

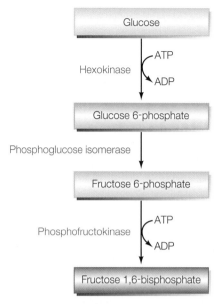

FIGURE 3-5 **Metabolic Pathway** Reactions in the cell occur as parts of larger pathways where the product of one reaction becomes the reactant of another, as in this biochemical pathway. This illustrates the early steps in glycolysis, the breakdown of glucose in the cytoplasm of eukaryotic cells. Note that each reaction has its own enzyme (labeled in red). Also note that the names of these, and all other enzymes, end in -ase.

Enzymes are proteins that increase the rate of chemical reactions in cells. As you will see later in the chapter, many chemical reactions in cells of the body occur in series, with one reaction giving rise to a substance that's used in the next.

A series of linked reactions is called a *metabolic pathway* and is the cellular equivalent of an assembly line in a factory. The enzymes of the metabolic pathway are the cell's equivalent of assembly line workers.

Molecules enter one end of a metabolic pathway and are modified along their course by enzymes (Figure 3-5). Eventually, a finished product comes off the line. In many cases, the enzymes of metabolic pathways are arranged along the cytoskeleton in proper order; the product of one reaction is conveniently situated for the next reaction.

> **KEY CONCEPTS**
>
> Cells are divided into compartments that help segregate functions, leading to more efficient performance; human cells consist of two major compartments, the cytoplasmic and nuclear.

3-3 The Structure and Function of the Plasma Membrane

The plasma membrane is the outermost part of a cell. It holds the cell together but also controls what goes in and out of the cell. In so doing, the plasma membrane determines the contents of a cell's internal chemical environment—a function essential to cellular homeostasis and survival.

The Structure of the Plasma Membrane

The plasma membrane and all of the other internal membranes of the cell consist of three biological molecules: lipids, proteins, and a small amount of carbohydrate (Figure 3-6). **Lipids** or fats are a chemically diverse group of biochemicals that form waxes, grease, and oils. All are insoluble in water.

Lipids serve a variety of functions in the body. Some provide energy. Others serve as insulation and energy storage. Still others play a structural role. Three different lipids are discussed in this book. The first, which is found in the membranes of cells, is the **phospholipid** (FOSS-foh-lip-id). Shown in Figure 3-7, phospholipids form lollypop-shaped mole-cules with a polar (charged) head and a nonpolar (uncharged) tail. Most of the lipid molecules in the plasma membrane are phospholipids.

Experimental studies suggest that the phospholipid molecules of the plasma membrane are arranged in a double layer—or bilayer. Figure 3-6 shows the most widely accepted model (theory) of the plasma membrane's structure, the **fluid mosaic model**—so named because the membrane consists of many different components like a mosaic and the membrane is an oily liquid that allows many materials to pass freely through it. As shown, the polar heads of the outer layer of phospholipid molecules face outward. They are in contact with a watery solution that bathes all cells. It is known as **extracellular fluid**. You will also note that the polar heads of the inner layer of molecules are directed inward toward the watery cytoplasm.

Interspersed in the lipid bilayer of human cells are large protein molecules known as *integral proteins*. **Proteins** are

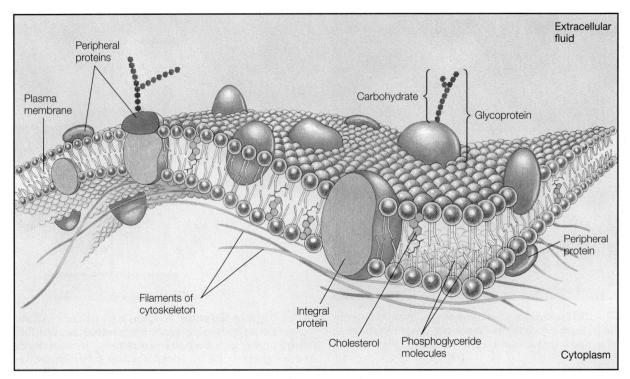

FIGURE 3-6 **Fluid Mosaic Model of the Plasma Membrane of Animal Cells** The fluid mosaic model is the most widely accepted theory of the plasma membrane. Phospholipids are the chief lipid component. They are arranged in a bilayer. Integral proteins float like icebergs in a sea of lipids.

large molecular weight compounds made of many smaller molecules, known as **amino acids**. Amino acids contain a central carbon atom, attached to which are an amino group (NH_2) and a carboxylic acid group (COOH)—from which they derive their name (Figure 3-8). As shown in the figure, the central carbon atom also bonds to a hydrogen and a variable group (indicated by the letter R). About 20 amino acids are found in the proteins of humans. Each one has a different R group.

Proteins serve many different functions. One important class of proteins, the enzymes, accelerates chemical reactions in the body. Without enzymes, few reactions would occur. Other proteins play a structural role. Keratin (CARE-ah-tin), for example, is a protein in human hair and nails. Collagen (coll-AH-gin), the most abundant protein in the human body, forms fibers that are found in ligaments, tendons, and bones.

Proteins are synthesized in the body one amino acid at a time. Although you would think that it takes a long time to produce a protein containing a couple hundred amino acids, it actually occurs very rapidly. A large protein containing hundreds of amino acids is assembled in under a minute.

Figure 3-9 shows how amino acids bind. The bond between two amino acids is called a **peptide bond**. As a protein is synthesized, the chain of amino acids begins to twist and bend (Figure 3-10). Amino acid chains may form random coils or very precise coils that resemble phone cords. This structure is known as an *alpha helix*. Amino acid chains also may form folded or pleated sheets. Take a look at Figure 3-10 for details.

Many amino acid chains bend and twist to form elaborate three-dimensional structures that are often globular in nature. The integral proteins of the plasma membrane are globular structures that float freely like giant icebergs in their sea of lipid. As shown in Figure 3-6, some of them completely penetrate the plasma membrane; others penetrate only partway. Several integral proteins may join to form pores that permit the movement of molecules into and out of the cell. Others may attach to the underlying cytoskeleton, anchoring the plasma membrane in place.

Another group of proteins, the **peripheral proteins**, is found in the plasma membrane, often attached to integral proteins. On the cytoplasmic side of the membrane, they may also bind to the cytoskeleton.

Plasma membranes contain small but significant amounts of carbohydrate. Most carbohydrates consist of short chains of simple sugars known as *oligosaccharides*. (Oligo is Greek for having few.) Shown in Figure 3-6, these segments attach to integral proteins that protrude into the extracellular fluid of cells. A protein combined with a carbohydrate is a **glycoprotein**. The glycoproteins of the plasma membrane are vital to cell function.

Another type of lipid is also found in the plasma membrane. This lipid is cholesterol. **Cholesterol** is a steroid. Its chemical formula is shown in Figure 3-11. Cholesterol and other steroids consist of four ring structures that are joined together to form a single molecule. Cholesterol molecules are found in the plasma membranes of cells wedged in between the phospholipid molecules (Figure 3-7). Their importance is not well-understood. They appear to make the membrane more fluid and are also involved in cell-to-cell communication and other processes.

KEY CONCEPTS

The plasma membrane of cells consists of lipid, protein, and carbohydrates; the membrane helps control the movement of materials in and out of a cell, which is vital for normal operation of cells and homeostasis.

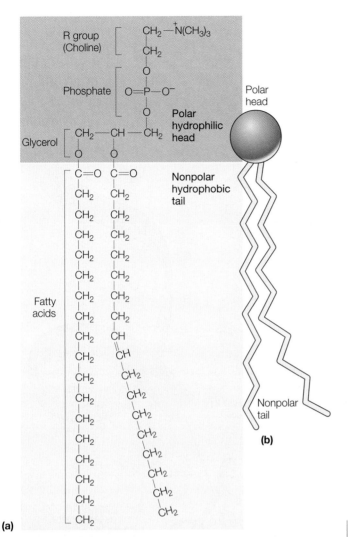

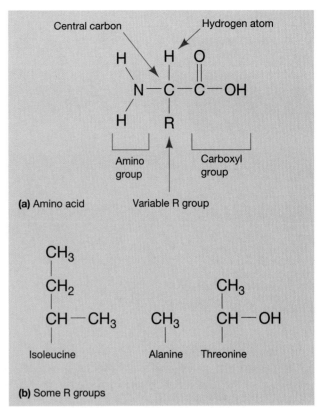

FIGURE 3-8 **Structure of Amino Acids** (a) The amino acid is a small organic molecule with four groups, one of which is variable and is designated by the letter R. (b) Some representative R groups.

FIGURE 3-7 **Phospholipids** (a) Each phospholipid consists of a glycerol backbone, two fatty acids, a phosphate, and a variable group generally designated by the letter R. In this case, the R group is choline, which is polar. (b) Because of the R group, the molecule has a polar head. The nonpolar tail region is formed by the two fatty acid chains.

Maintaining Homeostasis

The plasma membrane regulates the flow of molecules and ions into and out of the cell. This helps maintain the precise chemical concentrations inside and outside the cell—both essential for homeostasis and optimal cell function.

Because the plasma membrane regulates what molecules can enter and leave a cell, it is said to be selectively permeable. The plasma membrane also plays a vital role in cellular communication. Communication is defined very broadly here to include the transmission or receipt of any kind of message. Some hormones in the tissue fluid surrounding cells, for example, attach to specific integral proteins (membrane receptors) in the plasma membranes of cells. Attachment to the membrane receptor, in turn, sends a message to the cell's interior that triggers important changes in the cell's structure and function. Consequently, hormones are said to "communicate" with the cell through its membrane.

The plasma membrane is also part of an elaborate cellular identification system. The cells in each of us have a unique protein "cellular fingerprint." Because the protein

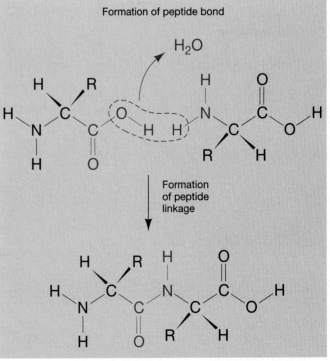

FIGURE 3-9 **Formation of the Peptide Bond** The carboxyl and amino groups react in such a way that a covalent bond is formed between the carbon of the carboxyl group of one amino acid and the nitrogen of the amino group of another. This bond is called a peptide bond.

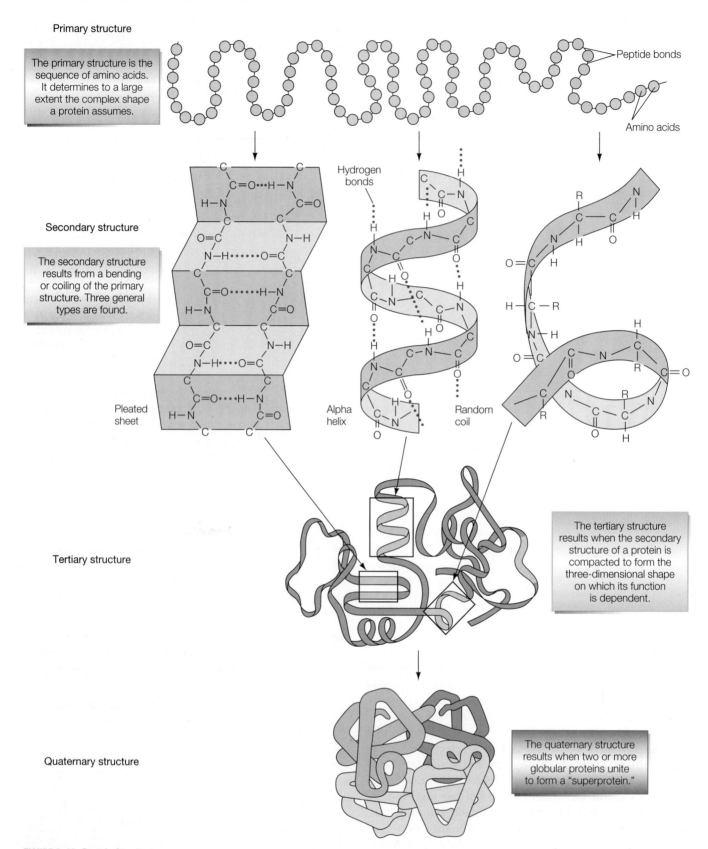

Primary structure

The primary structure is the sequence of amino acids. It determines to a large extent the complex shape a protein assumes.

Peptide bonds

Amino acids

Secondary structure

The secondary structure results from a bending or coiling of the primary structure. Three general types are found.

Hydrogen bonds

Pleated sheet

Alpha helix

Random coil

Tertiary structure

The tertiary structure results when the secondary structure of a protein is compacted to form the three-dimensional shape on which its function is dependent.

Quaternary structure

The quaternary structure results when two or more globular proteins unite to form a "superprotein."

FIGURE 3–10 Protein Structure

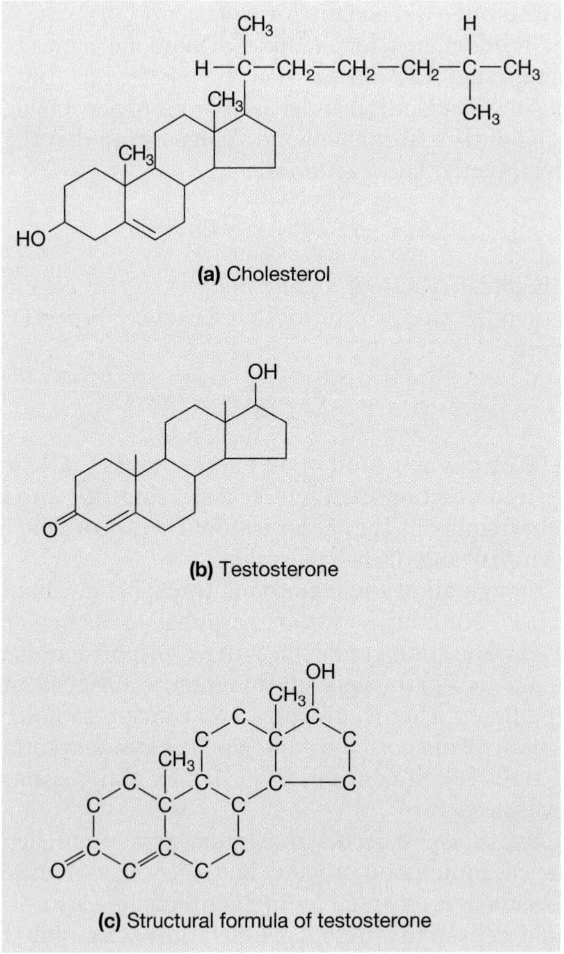

(a) Cholesterol

(b) Testosterone

(c) Structural formula of testosterone

FIGURE 3-11 **Steroids** Steroids, such as (a) cholesterol and (b) testosterone, consist of four rings joined together. Parts (a) and (b) are a shorthand way of drawing (c) the structural formula.

composition of the plasma membrane of the cells of each individual is unique, a person's immune system can recognize its own cells and determine when foreign cells—such as bacteria—invade. It mounts an attack directed only at the foreign cells.

The immune system can also identify cancer cells. Cancer cells arise from normal healthy body cells that undergo genetic mutations. Cancer cells often proliferate wildly within an organism and eventually kill the host for reasons discussed elsewhere in the text.

Although cancer cells pose a threat to healthy humans, something unusual happens when they become cancerous: their cellular fingerprint changes. The immune system views them as foreign invaders and mounts an attack, often successfully eliminating them. Without this marvelous ability to recognize foreign invaders or altered cells, we would very likely perish in rapid order from infections or tumors.

> **KEY CONCEPTS**
>
> The plasma membrane is vital to cellular homeostasis and therefore the health and welfare of all living organisms.

Membrane Transport

As noted earlier, cells are bathed in a liquid called extracellular fluid, also known as **interstitial fluid** (in-ter-STICH-al).

This liquid, which fills the spaces between cells of the body, consists primarily of water, as does the cytoplasm of cells. It also contains a variety of chemical substances that are dissolved or suspended in the water. Some of these substances pass through the membrane with ease; others must be transported across the membrane with the aid of special molecules. All told, five basic mechanisms exist for moving molecules and ions across plasma membranes (Table 3-2).

> **KEY CONCEPTS**
>
> Molecules move through membranes either passively, flowing down concentration gradients, or actively, being pumped in or out of cells.

Diffusion

Some molecules move passively across the plasma membrane, flowing from regions of high concentration to regions of low concentration. Movement such as this is known as **diffusion**.

Although diffusion may be a new concept to some, it is really quite familiar. You experience diffusion when a friend puts on perfume or cologne in the bathroom. Soon the smell permeates the air of neighboring rooms. The chemicals that cause odor moved from the bathroom into neighboring rooms by diffusion.

Diffusion across the plasma membrane is influenced by the chemical and physical properties of the membrane itself. Because the plasma membrane is primarily made of lipid, lipid-soluble substances (such as alcohols) can diffuse through the plasma membrane with relative ease. If you

TABLE 3-2	Summary of Plasma Membrane Transport
Process	**Description**
Simple diffusion	Flow of ions and molecules from high concentrations to low. Water-soluble ions and molecules probably pass through pores; water-insoluble molecules pass directly through the lipid layer.
Facilitated diffusion	Flow of ions and molecules from high concentrations to low concentrations with the aid of protein carrier molecules in the membrane.
Active transport	Transport of molecules from regions of low concentration to regions of high concentration with the aid of transport proteins in the cell membrane and ATP.
Endocytosis	Active incorporation of liquid and solid materials outside the cell by the plasma membrane. Materials are engulfed by the cell and become surrounded in a membrane.
Exocytosis	Release of materials packaged in secretory vesicles.
Osmosis	Diffusion of water molecules from regions of high water (low solute) concentration to regions of low water (high solute) concentrations.

FIGURE 3–12 Simple Diffusion Lipid-soluble substances pass through the membrane directly via simple diffusion.

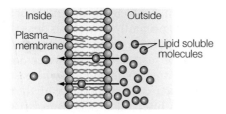

have studied chemistry, you know that chemical compounds readily dissolve in chemical compounds of a similar nature. Chemists are fond of saying, "Likes dissolve likes." Steroid hormones are lipid-soluble and therefore also pass directly through the plasma membrane by diffusion (Figure 3-12).

The plasma membrane also permits small molecules to pass with relative ease. Gases such as oxygen and carbon dioxide, for example, pass through the plasma membrane via diffusion. Water, another small molecule, passes through the membrane with relative ease, too.

Although diffusion refers to any movement of molecules or ions down a concentration gradient—that is, from high to low concentration—two types of diffusion exist: simple and facilitated. Movement of lipid-soluble chemicals, water, and gases through the membrane without assistance, as described above, is simple diffusion.

> **KEY CONCEPTS**
> The flow of molecules and ions across cell membranes from a region of high concentration to low concentration is called diffusion.

Carrier Proteins and Diffusion

Although water passes through the plasma membrane by diffusion, water molecules also pass through pores in the plasma membrane. This speeds up the process. Larger water-soluble molecules such as glucose and amino acids also pass through the membrane through pores, or openings, formed by integral proteins. This process is passive, that is, it requires no energy from ATP (defined below), and is known as **facilitated diffusion**, as shown in Figure 3-13a. (The protein molecules of the pore aid or facilitate diffusion.)

The proteins that form the pores are known as **carrier proteins**. They transport small water-soluble molecules like amino acids and ions across the membrane from areas of high concentration to areas of low concentration. How they operate remains a bit of a mystery.

> **KEY CONCEPTS**
> Water-soluble molecules pass through the cell membrane with the aid of proteins inside the membrane, known as carrier proteins.

Active Transport

Another transport mechanism in plasma membranes is known as active transport. **Active transport** is the movement of molecules across membranes with the aid of protein carrier molecules in the plasma membrane and with energy supplied by a special molecule called ATP (Figure 3-13b). ATP or **adenosine triphosphate** (ah-DEN-oh-seen try-FOSS-fate) consists of three smaller molecules, shown in Figure 3-14: a nitrogen-containing organic base, adenine; a five-carbon sugar, ribose; and three phosphate groups.

ATP is often likened to a form of energy currency. When the cell needs energy, for example, to move molecules across its membrane, ATP splits off a phosphate, forming ADP (adenosine diphosphate) and one molecule of phosphate. The breaking of the phosphate bond yields energy that the cell can use directly. The reaction is:

$$ATP \rightarrow ADP + P_i + Energy$$

Chemical reactions in the body that give off energy often turn the energy over to ADP. That way, new ATP can be formed.

$$ADP + P_i + Energy \rightarrow ATP$$

ATP serves as a kind of an energy shuttle, picking up energy from reactions that release it and shuttling it to reactions that require it. The energy is stored in the covalent bond between ADP and the last phosphate.

Although all of the membrane transport mechanisms we've been studying so far are responsible for movement down concentration gradients, active transport moves molecules and ions in the opposite direction: against concentration gradients. That is, during active transport, molecules and ions are transported from regions of low concentration to regions of high concentration. It's for this reason that energy is needed.

Active transport occurs in cells that concentrate chemical substances to function properly and, like the various forms of diffusion, it is essential for maintaining homeostasis. For example, cells in the thyroid gland in the neck require large quantities of the iodide ion (I^-) to manufacture the hormone thyroxine (thigh-ROX-in). But because levels of I^- in the bloodstream are fairly low, thyroid cells must actively transport iodide ions into the cytoplasm where the iodide concentration is many times higher. Cells also use active transport to move materials out of their cytoplasm against concentration gradients.

> **KEY CONCEPTS**
> Some molecules are pumped across cell membranes from regions of low concentration to high concentration via active transport, a process that requires energy.

Endocytosis

Cells also incorporate materials by **endocytosis** (en-doe-sigh-TOE-siss; literally, "into the cell"), as shown in Figures 3-15a and b. Endocytosis requires ATP and consists of two related activities: phagocytosis (fag-oh-sigh-toe-siss) and pinocytosis. **Phagocytosis** ("cell eating") occurs when cells engulf larger particles, such as bacteria and viruses. In humans, phagocytosis is limited to relatively few types of cells—those involved in protecting the body against foreign invaders. **Pinocytosis** ("cell drinking") occurs when cells engulf extracellular fluids and dissolved materials. Most, if not all, cells are capable of pinocytosis.

> **KEY CONCEPTS**
> Large molecules like proteins and entire cells like bacteria can be engulfed by cells; this process is known as endocytosis.

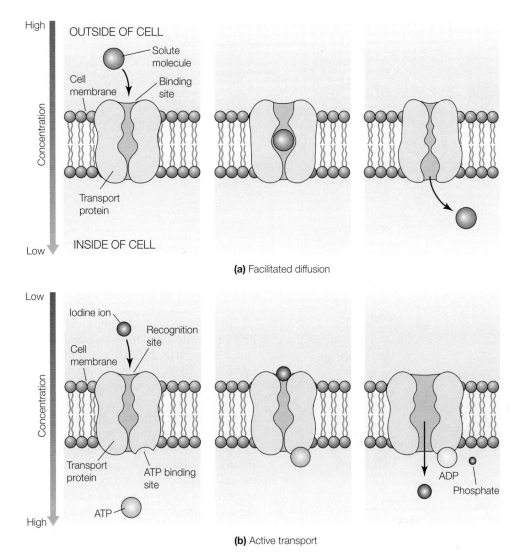

FIGURE 3-13 **Facilitated Diffusion and Active Transport** (a) Water-soluble molecules can also diffuse through membranes with the assistance of proteins in facilitated diffusion. (b) Other proteins use energy from ATP to move against concentration gradients in a process called active transport.

Exocytosis

The reverse of endocytosis is **exocytosis** ("out of the cell"). This process allows cells to release large molecules held in their cytoplasm (Figure 3-15b). For example, in many endocrine glands, protein hormones are packaged into tiny membrane-bound sacs, called *vesicles*, inside the cells. These vesicles

migrate to the plasma membranes of the cells and fuse with them. At the point of fusion, the membranes break down, and the protein hormone is released into the extracellular fluid. This process is exocytosis.

> **KEY CONCEPTS**
>
> Cells regurgitate undigested materials in a process known as exocytosis.

Osmosis

Like any other small molecule, water moves from one side of a plasma membrane to the other by diffusion. The diffusion of water across a selectively permeable membrane, however, is given a special name, **osmosis** (oss-MOE-siss; this word comes from a Greek word meaning "to push"). To understand osmosis, consider a simplified example in Figure 3-16.

In this example, we begin with a large bag made of a special plastic that's selectively permeable. We fill the bag with water and table sugar (sucrose) and then submerge it in a large beaker of distilled water (Figure 3-16). Distilled

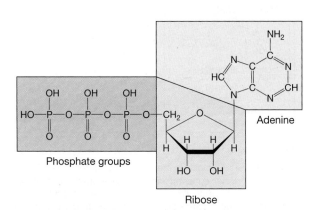

FIGURE 3-14 **Molecular Structure of the Nucleotide ATP** ATP is often called the energy currency of the cell.

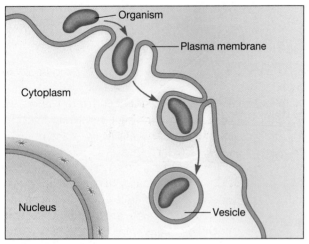

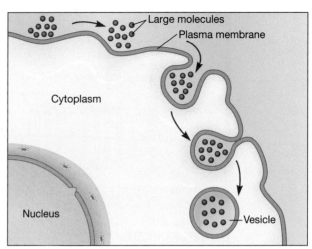

Pinocytosis

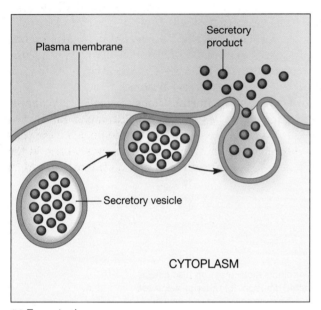

(b) Exocytosis

FIGURE 3-15 **Endocytosis and Exocytosis** (a) Cells can engulf large particles, cell fragments, and even entire cells via endocytosis. Two types exist: phagocytosis and pinocytosis. (b) Exocytosis, the reverse process, rids the cell of large particles.

In humans and other multicellular organisms, osmosis helps regulate the concentration of fluid surrounding the cell, keeping it at the same concentration as that of the cell's cytoplasm. In such cases, the extracellular fluid is said to be isotonic (having the same strength).

If the fluid surrounding a cell is not isotonic, serious problems can be expected. For example, if red blood cells are immersed in a solution that is more concentrated than their cytoplasm, water will move out of the cells, and the cells will shrink. A solution with a solute concentration higher than the cell's cytoplasm is said to be hypertonic (having a greater strength). A solution with a solute concentration lower than the cell's cytoplasm is said to be hypotonic (having a lesser strength). If red blood cells are placed in such a solution, water will rush in, causing the cells to swell and burst.

Combined with blood pressure and other forces, osmotic pressure plays an important role in the filtering of blood in the kidney—a function essential to homeostasis.

KEY CONCEPTS

Osmosis is the diffusion of water across a cell membrane from regions of high water concentration to regions of low water concentration.

water contains no impurities. The plastic is permeable only to water. Soon after immersion in the distilled water, water molecules will begin to diffuse inward—that is, down the concentration gradient—and the bag will begin to swell, as indicated by the arrows in Figure 3-16b. In this system, water moves from a region of high water concentration (distilled water) to a region of low water concentration (inside the bag). The concentration difference across the membrane "drives" the water across a selectively permeable membrane.

Whenever two fluids with different concentrations of solute (dissolved substance) are separated by a selectively permeable membrane, water will flow from one to the other, moving down the concentration gradient. The driving force is called **osmotic pressure**. The greater the difference in water concentration, the greater the osmotic pressure, and the more quickly the water moves.

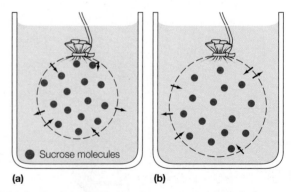

(a)   **(b)**

FIGURE 3-16 **Osmosis** Osmosis is the diffusion of water molecules from a region of higher water concentration (or low solute concentration) to one of lower water concentration (or high solute concentration) across a selectively permeable membrane. (a) To demonstrate the process, immerse a bag of sugar water in a solution of pure water. (b) Water diffuses into the bag (toward the lower concentration), causing it to swell.

3-4   Cellular Compartmentalization: Organelles

Now that we've looked at the plasma membrane, let's turn our attention to the organelles—structures that carry out many other key functions of the cell. We begin with the nucleus.

The Nucleus

The nucleus is one of the cell's most important organelles (Figure 3-17). It is often likened to a cellular command center. The nucleus is contained by a double membrane, the nuclear envelope, which separates the nuclear material from the cytoplasm. Minute channels in the envelope, the **nuclear pores**, allow materials to pass into and out of the nucleus (Figure 3-17b).

The bulk of the nucleus contains long, threadlike fibers of DNA and protein, known as **chromatin** (CHROME-eh-tin). It appears as fine granules in transmission electron micrographs (Figure 3-17a). The DNA in the nucleus is the genetic material. It houses all of the information a cell needs for proper development and functioning—hence, the description of the nucleus as a central command center.

A strand of chromatin with its protein is known as a **chromosome**. Just before cell division, the chromosomes coil to form short, compact bodies, highly visible through the light microscope (Figures 3-18a and b). This compaction greatly facilitates cell division, making it much easier to divide the nuclear contents. A detailed discussion of DNA can be found elsewhere in the text.

Proteins, water, and other small molecules and ions are also found in the nucleus, forming a semifluid material known as the **nucleoplasm** (new-KLEE-oh-plazem)—the nuclear equivalent of cytoplasm.

Also found in the nuclei of cells are structures known as *nucleoli* ("little nuclei"; singular, **nucleolus**). Nucleoli appear as small, clear, oval structures in light micrographs and as dense bodies in transmission electron micrographs. They are found only between cell divisions (Figure 3-3). Nucleoli are regions of the DNA that are actively engaged in the production of a biological molecule known as RNA. RNA, or ribonucleic acid, is involved in protein synthesis, as you shall see elsewhere in the text. Nucleoli are involved in the production of one of the three types of RNA, **ribosomal RNA** (also known as rRNA; RYE-bow-zomal). Ribosomal RNA gets its name from the fact that it combines with certain proteins to form cellular organelles known as ribosomes.

Ribosomes are nonmembranous organelles that appear as small, dark granules in electron micrographs. Ribosomes consist of two subunits. Produced in the nucleus, the subunits enter the cytoplasm through the nuclear pores. In the cytoplasm, the subunits unite to play an important part in

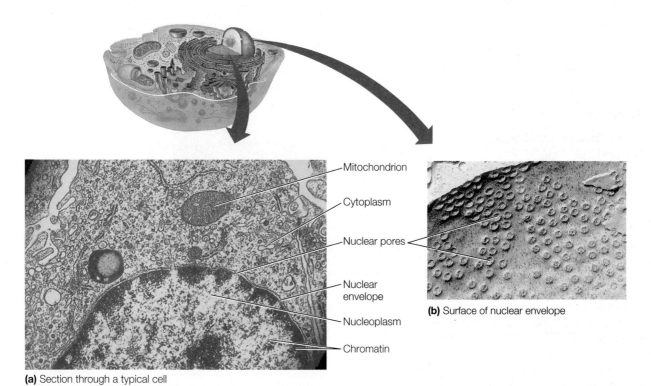

Mitochondrion
Cytoplasm
Nuclear pores
Nuclear envelope
Nucleoplasm
Chromatin

(a) Section through a typical cell

(b) Surface of nuclear envelope

FIGURE 3-17 The Nucleus (a) The nucleus houses the genetic information that controls the structure and function of the cell. The nuclear envelope, made of lipids and proteins like the plasma membrane, actually consists of two membranes, separated by a space. Pores in the membrane allow the movement of molecules into the nucleus, providing raw materials for the synthesis of DNA and RNA. They also allow RNA molecules to travel into the cytoplasm, where they participate in the production of protein. Chromatin may be densely packed or loosely arranged in the nucleus. (© K. G. Murty/Visuals Unlimited.) (b) Colorized scanning electron micrograph of the nuclear membrane, showing numerous pores. (© Don W. Fawcett/Visuals Unlimited.)

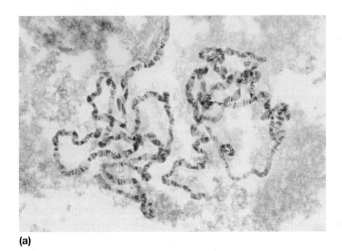

(a)

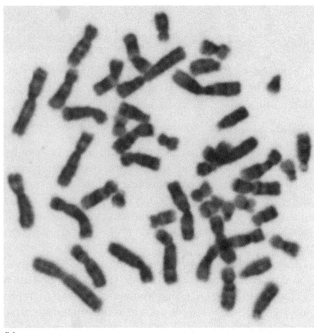

(b)

FIGURE 3-18 Chromosomes (a) The threadlike chromosomes, made of chromatin fibers consisting of protein and DNA, must condense (b) before the nucleus can divide. (© Photos.com.) (b) Condensed chromosomes. Can you think of any advantages to this strategy? (© Photos.com.)

protein synthesis, described elsewhere in the text along with a detailed discussion of RNA.

> **KEY CONCEPTS**
>
> The nucleus houses the DNA that contains the genetic information that, through an elaborate process, controls the structure and function of the cell.

Mitochondria

Like a factory, a cell requires energy and a special structure to generate it. This structure is a cellular organelle known

as the **mitochondrion** (MY-toe-CON-dree-on) shown in Figure 3-19. Although mitochondria vary considerably in form and number from cell to cell, they "share" several common characteristics. All mitochondria contain outer and inner membranes. The inner membrane is thrown into folds or *cristae* (CHRIS-tee; Figure 3-19b). This membrane therefore divides the mitochondrion into two distinct compartments, the inner compartment and the outer compartment, each with its own function. The cristae also increase the surface area on which chemical reactions take place.

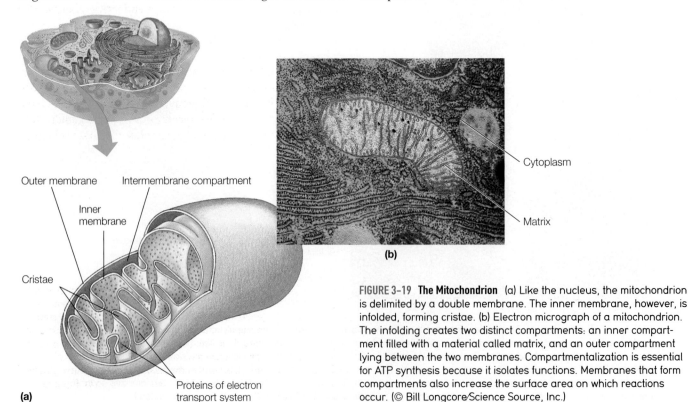

Outer membrane   Intermembrane compartment

Inner membrane

Cristae

Cytoplasm

Matrix

(b)

Proteins of electron transport system

(a)

FIGURE 3-19 The Mitochondrion (a) Like the nucleus, the mitochondrion is delimited by a double membrane. The inner membrane, however, is infolded, forming cristae. (b) Electron micrograph of a mitochondrion. The infolding creates two distinct compartments: an inner compartment filled with a material called matrix, and an outer compartment lying between the two membranes. Compartmentalization is essential for ATP synthesis because it isolates functions. Membranes that form compartments also increase the surface area on which reactions occur. (© Bill Longcore/Science Source, Inc.)

Although the mitochondrion is often likened to a power plant, it is important to note that this organelle doesn't make energy. That's impossible. You can't make energy. All energy comes from pre-existing forms of energy. Even the energy released by the combustion of coal comes from another source, the sun. It was captured by ancient plants and converted to coal. Burning coal releases ancient solar energy.

Just like a coal-fired power plant, the mitochondrion simply liberates energy held within the chemical bonds of organic molecules, usually sugars and fats. Mitochondria use the energy they extract from these molecules to make ATP. Cells, in turn, use the energy of ATP to synthesize chemical substances, transport molecules across membranes, divide, contract, and move about.

Also like a coal-fired power plant, mitochondria are not 100% efficient. That is, mitochondria only capture approximately one-third of the energy stored in their fuel source, mostly the simple sugar glucose. This energy is captured by the cell to produce ATP. The rest of the energy is given off as heat. This "waste" product creates body heat. It, in turn, is necessary for maintaining normal enzymatic function in warm-blooded animals like you and me. Without it, many processes would cease, and we would soon die.

> **KEY CONCEPTS**
> Mitochondria are membrane-bound organelles that house the reactions in which most of cellular energy production occurs.

Protein Production

Many cells synthesize a variety of molecules vital for growth, reproduction, and day-to-day maintenance. A number of these molecules are used within the cell; others, such as hormones, are released from the cell into the bloodstream where they travel to distant cells, eliciting specific responses. This section examines three organelles involved in cellular synthesis: the endoplasmic reticulum, ribosomes, and the Golgi complex.

> **KEY CONCEPTS**
> Proteins used inside and outside cells are produced one amino acid at a time inside the cytoplasm of cells thanks to three organelles.

Endoplasmic Reticulum

Coursing throughout the cytoplasm of many cells is a branched network of membranous channels, the **endoplasmic reticulum** (END-oh-PLAZ-mick rah-TICK-u-lum; meaning intracellular network; Figure 3-20a). These membranous channels are derived chiefly from the nuclear envelope and may be coated with ribosomes (Figure 3-20c). Ribosome-studded endoplasmic reticulum is referred to as the **rough endoplasmic reticulum (RER)**. The RER produces a variety of proteins, such as plasma membrane proteins. In the pancreas, it produces digestive enzymes. In certain endocrine glands, RER produces protein hormones.

Proteins produced by the RER are made on the outside surface of the organelle but are quickly transferred into its interior, possibly to protect them from damage. The interior of the RER not only protects the proteins, but also contains 30 to 40 enzymes that chemically convert some proteins into glycoproteins. Proteins and glycoproteins are then transferred to yet another organelle, the Golgi complex, for final packaging.

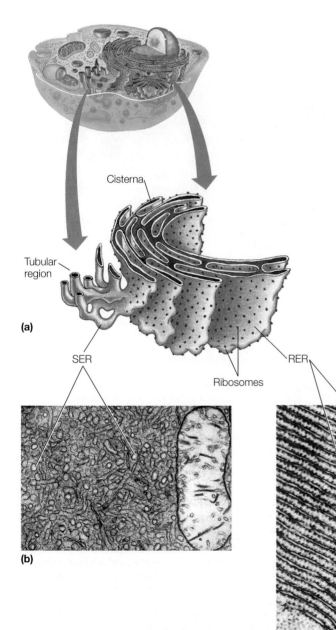

Cisterna

Tubular region

(a)

SER

Ribosomes

RER

(b)

(c)

FIGURE 3–20 **The Endoplasmic Reticulum** (a) Created by a network of membranes in the cytoplasm, the endoplasmic reticulum is often studded with small particles, the ribosomes, forming rough endoplasmic reticulum. This is the site of the synthesis of protein for lysosomes and extracellular use (digestive enzymes and hormones). (b) Electron micrographs of SER and (c) RER. (b and c, © Don W. Fawcett/Visuals Unlimited.)

Cells also contain endoplasmic reticulum lacking ribosomes (Figure 3-20b). Known as the **smooth endoplasmic reticulum (SER)**, this organelle specializes in the production of phospholipids. They are needed to replenish the plasma membrane. The SER performs a variety of functions in different cells. In the human liver, for instance, it detoxifies certain drugs, such as barbiturates, one type of sedative. In the stomach of humans and other mammals, the SER of certain cells produces hydrochloric acid, which coagulates protein in the stomach. This is necessary for subsequent protein digestion by enzymes in the small intestine. In the adrenal glands, the SER produces steroid hormones.

KEY CONCEPTS

The cell contains two types of endoplasmic reticulum, smooth and rough. Protein is produced on the rough endoplasmic reticulum; the smooth endoplasmic reticulum helps rebuild plasma membrane and produces a number of other functions in specialized cells.

Ribosomes

As noted earlier, **ribosomes** are tiny particles made of protein and ribosomal RNA (rRNA). They are synthesized in the nuclei of cells and enter the cytoplasm through the nuclear pores. In the cytoplasm, ribosomes attach to other strands of RNA molecules, known as **messenger RNA (mRNA)**. Messenger RNA molecules are also produced in the nuclei of cells. As you shall see elsewhere in the text, mRNA contains information required to synthesize protein.

In the cytoplasm, several ribosomes attach to a single strand of mRNA, forming a structure known as a **polyribosome** or **polysome**. Together, mRNA and its attached ribosomes produce protein. If it is a protein for intracellular use, like those of the cytoskeleton or cytoplasmic enzymes, the polysome remains free within the cytoplasm. If the protein is intended for extracellular use—that is, it is a hormone that will be released from the cell and circulated in the blood—the polysome attaches to the endoplasmic reticulum, forming RER. The protein they produce is transferred into the interior of the RER for packaging, as described earlier.

KEY CONCEPTS

Ribosomes are cellular organelles that play a key role in protein assembly on the rough endoplasmic reticulum.

The Golgi Complex

Proteins and glycoproteins manufactured on the surface of the RER move from it to the Golgi complex. The **Golgi complex** (GOL-gee) consists of a series of membranous sacs (Figures 3-21a and b), whose function is discussed shortly. How to proteins move from the RER to the Golgi?

Proteins and glycoproteins produced on the surface of the RER are first transferred to the inside of this structure. They are then transported internally to the ends of membranous channels (Figure 3-21a). Here, the proteins accumulate. As they do, the ends bulge and eventually pinch off, forming tiny membrane-bound sacs known as **transport vesicles** (VESS-sick-kuls). The transport vesicles protect the proteins and glycoproteins from cytoplasmic enzymes and also carry them to the Golgi complex.

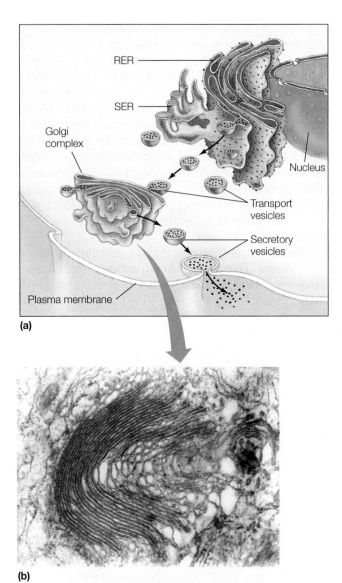

(a)

(b)

FIGURE 3–21 **Protein Synthesis and Secretion** (a) Protein packed in lysosomes and secretory granules (for later export) is synthesized on the RER and then transferred in tiny transport vesicles to the Golgi complex. Protein is sometimes chemically modified in the cisternae of both the RER and the Golgi complex. The Golgi complex separates protein by destination and repackages it into secretory granules, which remain in the cytoplasm until secreted by exocytosis. (b) Transmission electron micrograph of the Golgi complex. (Courtesy of Prof. Constantin Craciun from Electron Microscopy Center, Babes-Bolyai University, Cluj-Napoca, Romania.)

The Golgi complex performs three functions. First, it sorts the molecules it receives. That is, it separates proteins by destination, ensuring that products end up in the right place. Second, like the RER, the Golgi complex chemically modifies many proteins, often adding carbohydrates or other small molecules. Finally, the Golgi complex repackages certain proteins into lysosomes or secretory vesicles (or secretory granules; Figure 3-21a).

Secretory vesicles are membranous organelles in the cytoplasm. They serve as temporary storage sites for hormones and digestive enzymes that will eventually be excreted by the cells. When the signal for release comes, secretory vesicles migrate to the plasma membrane and fuse with it. The fused membranes soon dissolve, releasing the contents of the secretory vesicle into the extracellular space. This is an example of exocytosis.

Lysosomes

Lysosomes (LIE-so-ZOMES; literally, digestive bodies) are membrane-bound organelles that contain numerous digestive enzymes produced by the RER and packaged by the Golgi complex (Figure 3-22). These organelles break down materials phagocytized by cells.

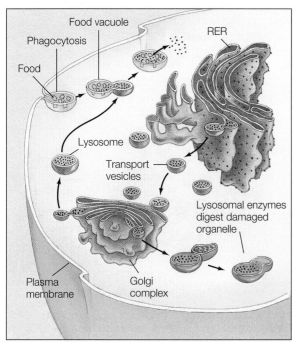

(a)

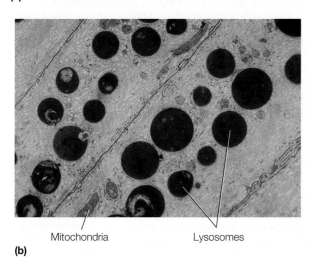

Mitochondria   Lysosomes

(b)

FIGURE 3-22 **Lysosomes**   (a) The digestive enzymes of the lysosome are produced on the RER and transported to the Golgi complex for repackaging. Lysosomes produced by the Golgi complex fuse with food vacuoles; this allows their enzymes to mix with the contents of the food vacuole. The enzymes digest the contents, which diffuse through the membrane into the cytoplasm, where they are used. The membrane surrounding the lysosome helps protect the cell from digestive enzymes. (b) Electron micrograph of lysosomes within a macrophage. The dark-staining bodies are the lysosomes, filled with digestive enzymes that are used by these cells to break down ingested material. (© Don W. Fawcett/Visuals Unlimited.)

As illustrated in the top of Figure 3-22, material engulfed by the cell by endocytosis is enclosed in a membrane derived from the plasma membrane. The resultant structure is called a **food vacuole**. Lysosomes attach to food vacuoles and release their enzymes into them. The enzymes then break down the molecules inside the food vacuole into smaller molecules. These, in turn, diffuse into the cytoplasm, where they are used by the cell. The undigested material left behind is expelled from the cell by exocytosis.

Besides providing the cell with nutrients, lysosomes destroy aged or defective cellular organelles, such as mitochondria. They simply bind to them, release their destructive enzymes, which break down the organelle, releasing chemicals into the cytoplasm for reuse. Lysosomes may also digest protein and other molecules that have become defective. This process helps cells maintain their structure and function.

Lysosomes also rid the body of injured or aged cells. As a cell reaches the end of its life, lysosomes within it break open and release their enzymes, destroying the cell. Lysosomal enzymes are also released during certain disease states. During a heart attack, for example, heart muscle cells die because of a lack of oxygen. Muscle cells are then destroyed from within by their own lysosomes. As the cells disintegrate, the lysosomes are released into the bloodstream. Their presence can be used to confirm a physician's diagnosis of a heart attack. Several other diseases can also be detected by blood enzyme levels.

Lysosomes also play an important role in embryonic development. In humans, for example, the fingers of an early embryo are webbed (Figure 3-23). The cells of the webbing, however, are genetically programmed to self-destruct at a certain point in development. This process is caused by the release of lysosomal enzymes. Destruction of the cells removes the webbing.

Most cells in the human body only contain a few lysosomes; these are used primarily to recycle aged organelles or perhaps to kill the cell when its useful lifetime is over. However, certain cell types contain hundreds of lysosomes. An example is a blood cell known as a *neutrophil* (NEW-trowe-FIL). Neutrophils are white blood cells that act as a

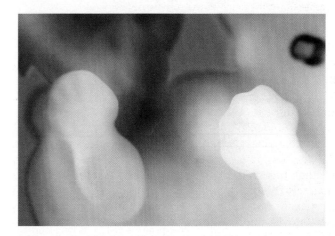

FIGURE 3-23 **The Human Embryonic Hand**   The webs between the fingers present during the fetal stage are normally destroyed by lysosomes during development. (© Cabisco/Visuals Unlimited.)

police force. They rid the body of harmful bacteria and viruses that enter through cuts and scratches or other routes. As they circulate in the blood and body tissues, neutrophils phagocytize and digest bacteria and viruses they encounter. This elimination of potentially infectious microorganisms helps maintain homeostasis and our health.

> **KEY CONCEPTS**
>
> Lysosomes are membrane-bound organelles that contain digestive enzymes used to remove aged organelles, destroy cells, or, in specialized cells, digest material engulfed by them.

Flagella

Most cells in the human body are part of a tissue or organ and thus remain in the same place throughout their life. One exception is the mammalian sperm cell. This highly mobile cell moves with the aid of a special organelle known as a **flagellum** (fla-GEL-um; plural, flagella; Latin for "whip").

As shown in Figure 3-24, a flagellum is a long, whiplike extension of the plasma membrane of the sperm cell. It contains numerous small tubules, the **microtubules**, which are shown in Figure 3-25a. Microtubules produce the motive force. As illustrated, the nine pairs of microtubules of the flagellum are arranged around a central pair. Biologists typically refer to this as the "9 + 2" arrangement. The flagella in sperm, however, have additional fibers outside the central 9 + 2 array, which provide extra strength.

At the base of each flagellum is an anchoring structure, the **basal body** (Figure 3-25b). The basal body gives rise to the flagellum during development and contains nine sets of microtubules (with three microtubules each), but no central pair.

> **KEY CONCEPTS**
>
> Flagella are cellular organelles that are found in sperm cells; they permit sperm to move.

Cilia

Several fixed cells in the body, such as those lining the trachea (TRAY-key-ah) or wind pipe, contain an organelle called the **cilium** (SILL-ee-um). Cilia (plural, SILL-ee-ah) are small

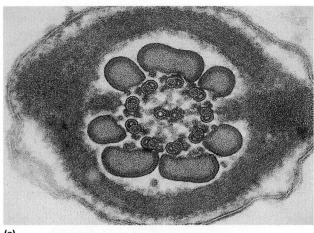

(a)

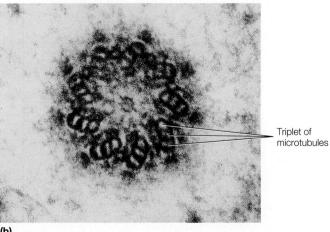

Triplet of microtubules

(b)

FIGURE 3–25 **Flagella** (a) A sperm flagellum, showing the 9 + 2 arrangement and additional fibers thought to provide support and strength to the vigorously beating tail. (b) A basal body is found at the base of each cilium and flagellum. It consists of nine sets of triplets that give rise to the microtubule doublets of the flagellum and cilium. (a and b, © David M. Phillips/Visuals Unlimited.)

extensions of the cell membrane (Figure 3-26). Although cilia contain microtubules like flagella, cilia have only a central core of microtubules in a 9 + 2 arrangement. Hundreds, even thousands, of cilia can be found on a single cell (Figure 3-26).

Cilia perform extremely important functions in the human body. In the trachea, for example, they beat toward the mouth. This moves mucus containing trapped dust particles away from the lungs. In the mouth, the mucus is either swallowed or expectorated (spit out). Cilia are therefore part of an automatic cleansing mechanism, vital for protecting the lungs and maintaining our health.

At the base of each cilium is a basal body identical to those found in flagella. The basal body gives rise to the cilium during development and anchors the organelle in place.

Cilia and flagella are structurally similar, but flagella are much longer than cilia and far less numerous. Normal human sperm cells, for example, have one flagellum each, whereas each of the ciliated cells lining the respiratory system and parts of the female reproductive system (notably the Fallopian tubes) contains thousands of cilia. Flagella and cilia also beat differently. Cilia perform a stiff rowing motion with a

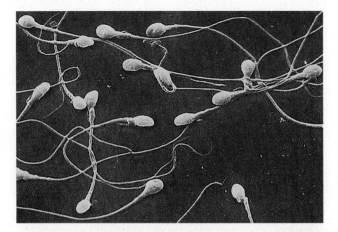

FIGURE 3–24 **The Sperm Cell** The sperm cell is a marvel of architecture, uniquely "designed" to streamline the cell for its long journey to fertilize the ovum. The nuclear material is compacted into the sperm head. The flagellum helps propel the sperm through the female reproductive tract. (© David M. Phillips/Visuals Unlimited.)

FIGURE 3-26 **Cilia** Cilia on cells lining the trachea of the human respiratory tract. (© David M. Phillips/Visuals Unlimited.)

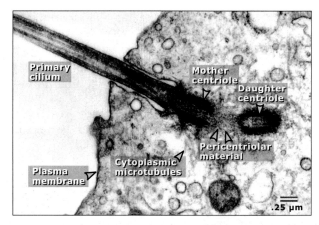

FIGURE 3-28 **Primary Cilium** (© Conly L. Rieder, Wadsworth Center.)

flexible return stroke, whereas flagella beat more like whips, undulating in a continuous motion (Figure 3-27). Because cilia and flagella often beat continuously, they require a more or less constant supply of energy, provided by ATP.

> **KEY CONCEPTS**
>
> Certain cells are covered with cilia, tiny organelles that move substances along the surface of the cells.

Primary Cilia

Cell biologists have also found that most cells of the body contain a single, prominent nonmotile cilium, a **primary cilium** (Figure 3-28). These cilia, discovered over 100 years ago, resemble ordinary cilia but lack the central pair. They were long thought to be nonfunctional. Evidence now suggests that they may play an important role in cellular communication. Primary cilia respond to an assortment of chemical and physical stimuli.

They contain receptors in their membranes to which chemicals attach, signaling a cascade of changes in the genes of a cell. These, in turn, lead to physiological changes inside the cell.

As one science writer noted, "a single cilium is a cell's eyes and nose, GPS receiver, and even weather vane." Research shows that the loss of the primary cilium makes it impossible for a cell to interact with the complex environment it inhabits.

Studies show that primary cilia are necessary for proper development, required throughout life for normal cell function, and are vital throughout the entire body. Numerous studies even show that malfunctioning primary cilia may be related to numerous diseases and birth defects, among them high blood pressure, heart disease, hardened arteries, cleft palate, kidney failure, and cancer.

> **KEY CONCEPTS**
>
> Almost all cells have a single nonmotile cilium, the primary cilium, that plays an important role in communication.

Ameboid Movement

Another type of movement is **ameboid motion** (ah-ME-boyd). During amoeboid movement, the cell sends out many small cytoplasmic projections, or **pseudopodia** (SUE-dough-POW-dee-ah; "false feet"). Pseudopodia attach to solid surfaces "in front" of the cell (Figure 3-29). The cytoplasm then flows into the pseudopodia, moving the cell forward.

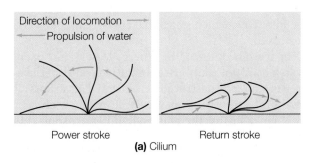

Power stroke   Return stroke
(a) Cilium

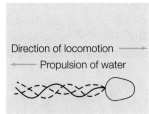

Continuous propulsion
(b) Flagellum

FIGURE 3-27 **Beating of Cilia and Flagella** (a) Cilia beat with a stiff power stroke something like a rowing motion. During the return stroke, the cilium is flexible. (b) The flagellum beats like a whip, over and over, propelling the sperm cell forward.

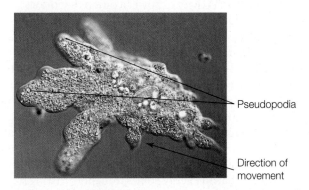

FIGURE 3-29 **Ameboid Movement** Some cells in the human body move by amoeboid movement. Studies of the amoeba (shown here) show that cells that move by amoeboid movement send out minute extensions called pseudopodia (false feet), which attach to solid surfaces. Cytoplasm then flows into the pseudopodia, moving the cell forward. (© M. Abbey/Visuals Unlimited.)

Although most cells in the human body are stationary, there are a few very important cells that can move. One of them is the neutrophil, mentioned earlier. The neutrophil is one type of white blood cell. It is transported in the blood throughout the bloodstream. These cells, however, also escape through the walls of thin-walled blood vessels called capillaries in body tissues. Inside the tissues, the neutrophils move about by ameboid motion, gobbling up bacteria and damaged cells by phagocytosis, thus helping rid the body of potentially harmful organisms and removing cells that are no longer functioning.

> **KEY CONCEPTS**
> Some cells of the human body migrate throughout the body by ameboid motion, a kind of cellular crawling.

3-5 Energy and Metabolism

Humans need energy to function. That energy is consumed by the cells. It allows them to carry out the many functions they perform. Cells obtain energy primarily from two sources: the breakdown of glucose, a type of carbohydrate, and the breakdown of triglycerides, a type of fat. Under certain circumstances, notably during starvation, they can even acquire energy from amino acids released by the breakdown of important body proteins like those in muscles. In this section, we'll look briefly at how cells in humans acquire energy from the simple sugar glucose. Before we examine this process, however, a few words about energy are in order. We'll then look at glucose, the body's main source of energy, and other carbohydrates.

Energy is a mystery to most of us. To understand energy, we must first recognize that it comes in many forms. For example, humans in many countries rely today on fossil fuels such as coal and natural gas. In other countries, wood is still heavily used. Sunlight, wind, and hydropower are also forms of energy. Even a cube of sugar contains energy! Touch a match to it, and it will burn, giving off heat and light energy, two additional forms of energy. Consume that cube of sugar and the cells of your body will extract its energy to run many internal processes.

There's more to energy than its diverse forms. Even the casual observer can tell you that energy can be converted from one form to another. Natural gas, for example, when burned, is converted to heat and light. Heat energy can be used to produce electricity. Even wind and solar energy can be converted to electricity to power a pump that draws water from the ground.

Not only can energy be converted to other forms, but it has to in order for us to derive benefit from it. It is not raw forms of energy that we need. It is the byproducts of energy that are unleashed when we "process" them in various energy-liberating technologies or energy-converting processes in our cells that meet our needs.

Another fact you need to know about energy is that it can be neither created nor destroyed. Physicists call this the **First Law of Thermodynamics** or, simply, the **First Law**. The First Law says that all energy comes from pre-existing forms. Even though you may think cells are "creating" energy when they process sugars we eat, all they are doing is unleashing energy contained in the sugar molecules—specifically, the energy locked in the chemical bonds of the sugar molecules. It, in turn, came from sunlight. The Sun's energy came from the fusion of hydrogen atoms in the Sun's interior.

More important to us, however, is the **Second Law of Thermodynamics**. The **Second Law**, says, quite simply, that any time one form of energy is converted to another—for example, when sugar is converted into cellular energy—it is degraded. Translated, that means energy conversions transform high-quality energy resources to low-quality energy. Natural gas, for instance, consists primarily of methane (CH_4). Natural gas contains a huge amount of energy in a small volume; it's locked up in the covalent bonds that attach the carbon atom to the four hydrogen atoms of the methane molecules. When these bonds are broken, the stored chemical energy is released. Light and heat energy are the products. Both light and heat, however, are less concentrated or lower quality forms of energy. Hence, we say that natural gas, a concentrated form of energy, is "degraded." In electric power plants, only about 50% of the energy contained in natural gas is converted to electrical energy. The rest is "lost" as heat and is dissipated into the environment. The same thing occurs in the body. It captures approximately one-third of the energy contained in the food molecules we consume; the rest is lost as heat.

What all this means is that no energy conversion is 100% efficient. But let's get something straight. Some of you may be wondering if this discussion of "energy losses" is a violation of the First Law, which states that energy cannot be created or destroyed.

The truth be known, the "energy losses" during energy conversion are not really losses in the true sense of the word. Energy is not really destroyed; it is released in various forms, some useful and others, such as heat, not so useful. In the body, energy lost in the conversion of glucose to cellular energy is lost as heat. It warms our bodies and eventually escapes into space, and is no longer available to us.

Now that you understand a bit about energy, let's look at it the way a physicist does. To a physicist, energy is defined as "the ability to do work." More accurately, energy is that elusive something that allows us to do work; we and our machines, that is.

Any time you lift an object, for example, or slide an object across the floor, you are performing work. The energy you consume allows your cells and you to do work. The same holds for our machines. Any time a machine lifts something or moves it from one place to another, it performs work.

Energy, quite simply, is valuable because it allows us to perform work. It powers our bodies. It powers our homes. It powers our society. We cannot exist without energy.

Now that you understand energy, let's take a look at those molecules that we consume to generate energy.

KEY CONCEPTS

Cells contain enzymes that break down food stuffs yielding energy that is required for a wide assortment of processes vital to life.

Energy Production in Cells

Humans are powered by two organic chemicals in the foods we eat, carbohydrates and lipids. Carbohydrates are a general class of biological molecules. **Carbohydrates** are organic compounds that range in size from simple sugars such as glucose to very large molecules such as starch. The molecular structures of two simple carbohydrates—ribose, a five-carbon sugar, and glucose, a six-carbon sugar—are shown in Figure 3-30. These are also known as **monosaccharides**. It gets its name, like so many things, from the Greek language. *Monos* means single and *sacchar* means sugar.

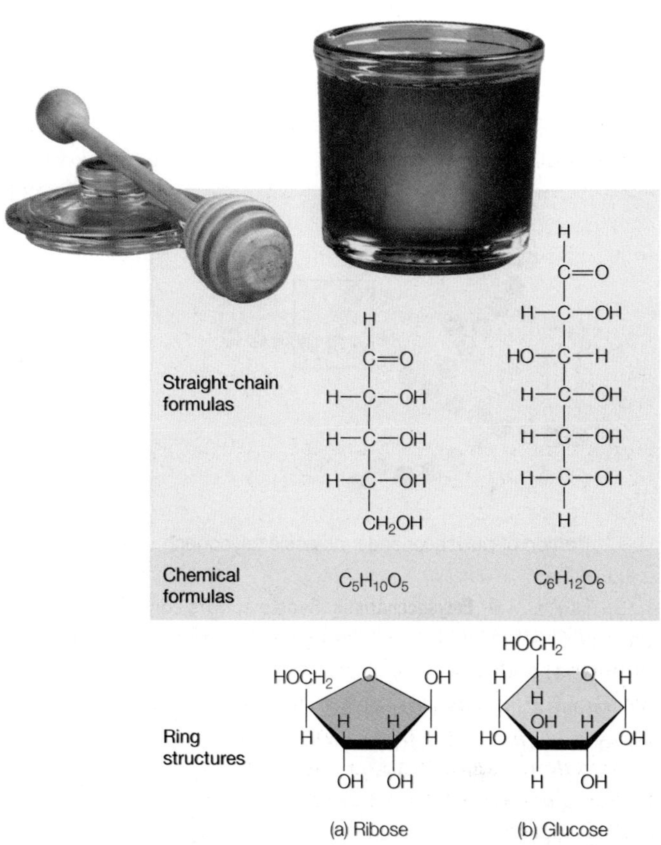

Health Tip 3-1

To lose or maintain weight, drink water and other non-caloric beverages with meals.

Why?

Soda and fruit drinks provide valuable water we need to stay healthy, but also supply an enormous number of calories in the form of sugar (sucrose or fructose). The extra calories can add up quickly; if you consume more calories than you need, they will be converted to body fat.

FIGURE 3-30 Two Monosaccharides Honey is rich in glucose. (Photo © Digital Stock.)

Simple sugars such as glucose and fructose (another six-carbon sugar that is found in fruits) can unite by covalent bonds. When glucose and fructose combine, they form sucrose, commonly known as *table sugar* (Figure 3-31). Sucrose is a **disaccharide**, because it consists of two monosaccharides.

As noted earlier, monosaccharides can unite to form long, branching chains (Figure 3-32). These are known as **polysaccharides** and include such important molecules as starch, glycogen, and cellulose. These molecules will be discussed further elsewhere in the text.

Glucose molecules, like other organic molecules, contain energy—a fact that you can easily demonstrate by holding a spoonful of sugar over a candle. The sugar will soon catch fire, releasing energy in the forms of light and heat. This energy is stored in the covalent bonds that bind the atoms of the sucrose molecule. Cells get their energy from glucose. They tap into this energy by breaking glucose molecules apart in an orderly series of chemical reactions. Energy is released in steps, a little at a time, unlike combustion, which releases it very quickly. This series of chemical reactions that liberates energy from glucose begins in the cytoplasm and is completed in the mitochondria. The complete breakdown of glucose is referred to as **cellular respiration**.

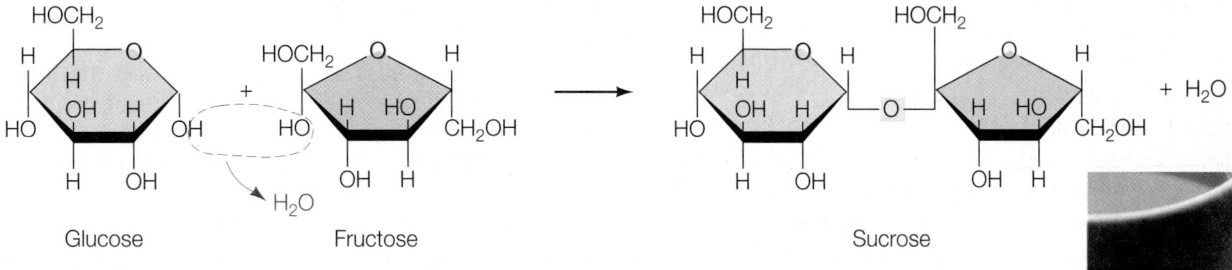

Formation of sucrose

FIGURE 3-31 Disaccharides Simple sugars such as glucose and fructose can combine to form the disaccharide known as sucrose. Table sugar is an example. (Photo © Photos.com.)

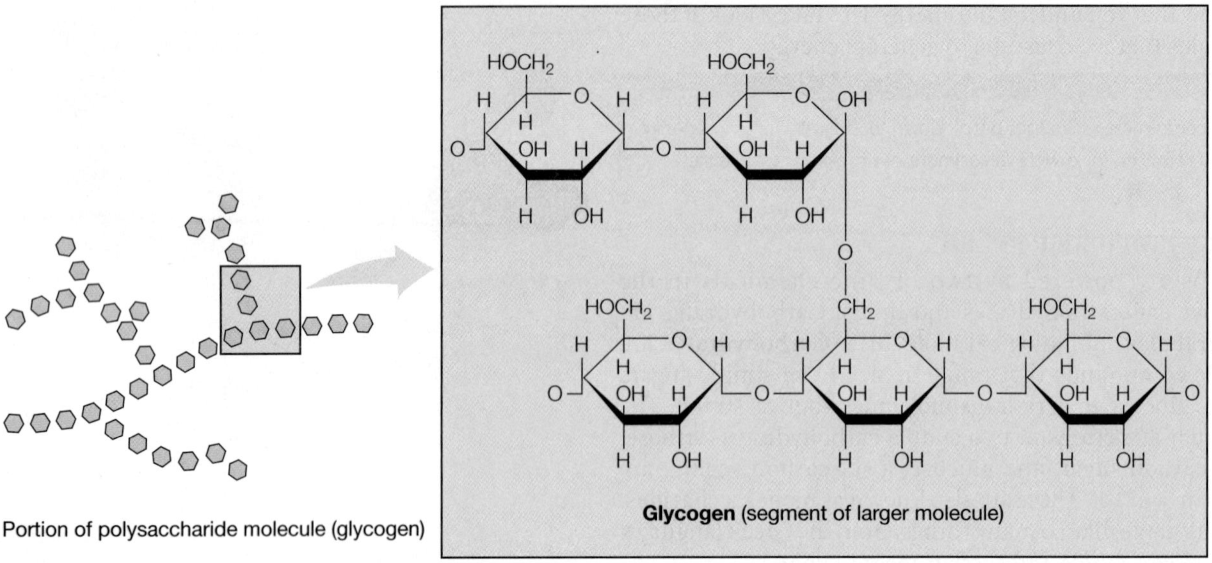

Portion of polysaccharide molecule (glycogen)

Glycogen (segment of larger molecule)

FIGURE 3-32 **Polysaccharides** Simple sugars can also combine to form polysaccharides such as glycogen, which is a storage form of glucose in animals and is found in muscle and liver.

During cellular respiration, glucose, which contains six carbon atoms, is broken down into six molecules of carbon dioxide and six molecules of water. The overall reaction is:

$$\text{glucose} + \text{oxygen} \rightarrow \frac{\text{carbon}}{\text{dioxide}} + \text{water} + \text{energy}$$

$$C_6H_{12}O_6 + 6\,O_2 \rightarrow 6\,CO_2 + 6\,H_2O + 36\,ATP$$

As shown, the complete breakdown of glucose requires oxygen. Without oxygen, glucose can be only partially broken down, and only a fraction of its energy can be captured. The products of the complete breakdown of glucose are carbon dioxide, water, and ATP.

The breakdown of glucose in cellular respiration gives off energy and is therefore known as an *exergonic reaction*. During cellular respiration, cells use energy in glucose to synthesize ATP from inorganic phosphate (P_i) and ADP. ATP is virtually the only kind of energy cells can use.

ATP synthesis is an *endergonic reaction* (energy-requiring). In order for this reaction to occur, it must receive energy. It gets this energy from cellular respiration.

In living things, exergonic and endergonic reactions are often linked so that energy from one reaction is transferred to the other. The complete breakdown of glucose generates enough energy to produce 36 molecules of ATP from ADP and inorganic phosphate.

Cells of the human body are also powered by lipids we eat, that is, various fats. The fats that provide energy are known as *triglycerides*. They are found in oils in nuts and seeds and animal fats.

To obtain energy, fatty acids are first removed from the glycerol backbone of the triglyceride molecules. The fatty acids are broken down two carbons at a time (Figure 3-33). This process, known as beta oxidation, produces lots of ATP, which powers our bodies at rest, but especially when we walk, run, row, or perform other aerobic (low-intensity exercise).

KEY CONCEPTS

Carbohydrates and lipids, notably triglycerides, are the major sources of cellular energy.

Cellular Respiration

In cells, the complete breakdown of glucose in cellular respiration requires numerous chemical reactions that occur during four distinct stages: (1) glycolysis, (2) the transition reaction, (3) the citric acid cycle, and (4) the electron transport system (Figure 3-34). Table 3-3 summarizes the stages of cellular respiration and lists their products. Although these reactions may seem overwhelming at first glance, the logic is pretty simple: In a series of chemical reactions, the cell breaks the six-carbon glucose molecule apart, yielding six carbon dioxide molecules and lots of energy primarily in the form of ATP. The following paragraphs provide a brief summary of each part of this process.

KEY CONCEPTS

The complete breakdown of glucose in cells is known as cellular respiration and occurs in four steps in the cytoplasm and mitochondria of cells.

Glycolysis

The first phase of cellular respiration in cells is glycolysis. **Glycolysis** (gly-COLL-eh-siss) is a metabolic pathway that occurs in the cytoplasm of cells. As Figure 3-35 shows, during glycolysis, glucose is split in half, forming two three-carbon molecules known as pyruvic acid or pyruvate. The energy released during this process nets the cell a modest two molecules of ATP. It also yields two molecules of NADH. **NADH** stands for nicotine adenine dinucleotide. It exists in two forms, **NAD⁺** and NADH.

As illustrated in Figure 3-35, NADH is formed from NAD⁺. NAD⁺ is a chemical substance that picks up two electrons given off by chemical reactions that occur during cellular respiration.

As you shall soon see, NADH gives up those energy-rich electrons to make more ATP.

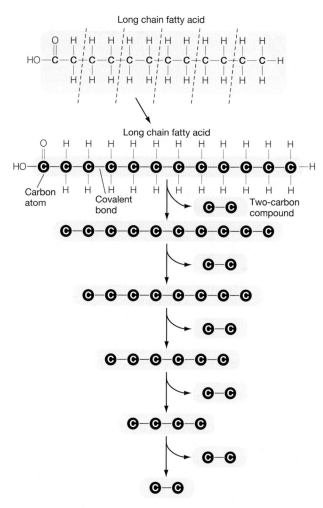

FIGURE 3-33 Beta Oxidation A two-carbon unit is split off at each step.

Glycolysis is the breakdown of the six-carbon compound glucose into two three-carbon compounds known as pyruvate. It occurs in the cytoplasm.

The Transition Reaction

In the second stage of cellular respiration, the **transition reaction**, pyruvate generated in glycolysis diffuses from the cytoplasm into the inner compartment of the mitochondrion. Here it reacts with a large molecule called coenzyme A (CoA), as shown in Figure 3-36. During this reaction, one carbon atom is removed from each pyruvate molecule. This reaction yields a two-carbon compound (acetyl group) and carbon dioxide. The two-carbon compound binds to coenzyme A (Figure 3-36). The product of this reaction is acetyl CoA (ah-SEAT-ill). Two electrons lost during the chemical reaction are picked up by two molecules of NAD^+. This causes NAD^+ to be converted to NADH.

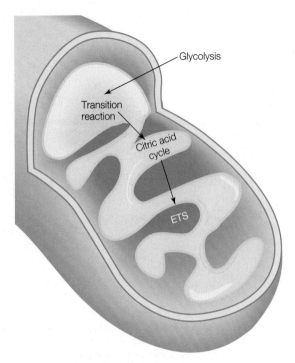

FIGURE 3-34 **Cellular Respiration** This process involves four steps: glycolysis, the transition reaction, the citric acid cycle, and the electron transport system. Glycolysis occurs in the cytoplasm; the remaining steps take place in the mitochondrion. Most of the ATP is produced in the electron transport system (ETS) from the energy stripped from electrons given off by the previous three stages.

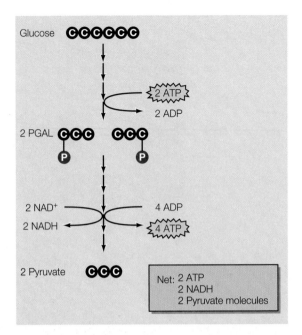

FIGURE 3-35 **Overview of Glycolysis** During this process, glucose is broken down into two molecules of PGAL, which are converted to pyruvate. The cell nets two ATP and two NADH molecules.

> **KEY CONCEPTS**
>
> In the transition reaction, one carbon is cleaved off each pyruvate molecule.

The Citric Acid Cycle

Two molecules of acetyl CoA produced during the transition reaction then enter the third stage of cellular respiration. This stage is a cyclic metabolic pathway located inside the inner compartment of the mitochondrion. This elaborate pathway is called the **Krebs cycle**, after the scientist who worked out many of its details, or the **citric acid cycle**, after the very first product formed in the reaction sequence.

During the cycle, each two-carbon compound produced in the transition reaction binds to a four-carbon com-

pound found inside the mitochondrion known as *oxaloacetate* (ox-AL-oh-ass-eh-tate). The result is a six-carbon product, citric acid. Citric acid proceeds through a series of chemical reactions, each catalyzed (defined shortly) by its own enzyme. Along the way, the molecule is modified many times. During the complex molecular rearrangements, two molecules of carbon dioxide are removed from the six-carbon chain, so that oxaloacetate is regenerated. This ensures the continuation of the cycle.

As shown in Figure 3-36, one of the chemical reactions of the citric acid cycle yields an ATP molecule. Because two molecules of pyruvate enter the citric acid cycle for each glucose molecule broken down by glycolysis, the citric acid cycle nets the cell two ATP molecules. It is important to note that several other reactions in the citric acid cycle release electrons. They are "funneled" into the fourth, and final, part of cellular respiration, the electron transport system. The electrons are transported to the electron transport system by NADH and a closely related compound, **FADH** (flavin adenine dinucleotide). Both of these molecules are electron carriers, because they shuttle the electrons from the citric acid cycle to the electron

TABLE 3-3	Overview of Cellular Energy Production	
Reaction	**Location**	**Description and Products**
Glycolysis	Cytoplasm	Breaks glucose into two three-carbon compounds, pyruvate; nets two ATP and two NADH molecules
Transition reaction	Mitochondrion	Removes one carbon dioxide from each pyruvate, producing two acetyl CoA molecules; produces two NADH molecules
Citric acid cycle	Inner compartment of mitochondrion	Completes the breakdown of acetyl cycle CoA; produces two ATPs per glucose; produces numerous NADH and FADH2 molecules
Electron transport	Inner membrane of mitochondrion	Accepts electrons from NADH and FADH2, generated by previous system reactions; produces 34 ATPs

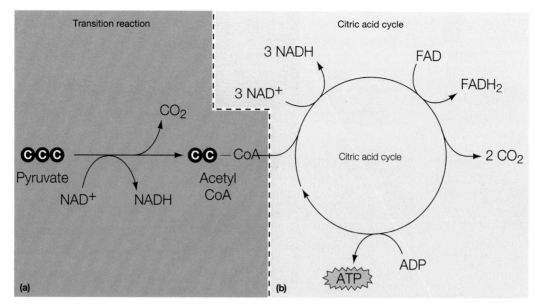

FIGURE 3–36 Overview of the Transition Reaction and the Citric Acid Cycle (a) The transition reaction cleaves off one carbon dioxide and binds the remaining two-carbon compound temporarily to a large molecule called coenzyme A. The result is a molecule called acetyl-CoA. It then enters the citric acid cycle. (b) Occurring in the matrix of the mitochondrion, the citric acid cycle liberates two carbon dioxides and produces one ATP per pyruvate molecule. Its main products, however, are NADH and $FADH_2$, bearing high-energy electrons that are transferred to the electron transport system.

transport system. The energy of the electrons they deliver to the electron transport system is used to produce ATP.

The Electron Transport System

The last stage of cellular respiration is the **electron transport system**. It produces the bulk of the cell's energy in the form of ATP from electrons liberated during chemical reactions occurring in glycolysis and the citric acid cycle. As just noted, these electrons are delivered to the electron transport system by NADH and FADH. They give up their electrons to protein carrier molecules embedded on the inner surface of the inner membrane of the mitochondria. The carrier proteins constitute the electron transport system (ETS).

Electrons entering the ETS are passed from one protein to another and are eventually given over to oxygen molecules inside the matrix. During their journey along this chain, the electrons lose energy, much like a hot potato passed along by a line of people. The energy lost along the way is used to make ATP.

All told, the electron transport system produces 32 ATP molecules per glucose molecule. Because two ATPs are generated by glycolysis and two are produced in the citric acid cycle, the total output for cellular respiration is 36 ATP molecules per glucose molecule. The "de-energized" electrons from the electron transport system eventually combine with hydrogen ions and oxygen to produce water.

Enzymes

Enzymes play a key role in cellular respiration, and many other cellular processes, so let's take some time to study them. *Enzymes* are proteins that serve as catalysts. A *catalyst* is a substance that facilitates chemical reactions.

Enzymes are protein molecules in the cells of the body that speed up chemical reactions, allowing them to occur very rapidly—at rates necessary to maintain life. Without them, the cellular metabolism would grind to a halt. So important are enzymes that a missing enzyme in a metabolic pathway will shut the pathway down—in much the same way that a missing worker on an assembly line interrupts production.

Each enzyme is a large, globular protein with a small region known as the **active site** (Figure 3-37a). The active site is an indentation, or pocket, where the chemical reaction occurs. Its shape corresponds to that of the **substrate**, the molecule undergoing reaction. Because each enzyme has its own uniquely shaped active site capable of binding to one or at most a few substrates, enzymes are said to be *specific*. This feature allows the cells to regulate chemical reactions very precisely.

Although enzymes play an active role in the metabolism of cells, they are unchanged by the reaction. As a result, they can be used over and over again.

Enzymes can also be controlled by the cells themselves, which provides additional control over cellular metabolism. The most common regulators of enzyme activity are the end products of chemical reactions. When produced in excess, the end products often bind to specific control regions on enzymes. The control regions are called **allosteric sites** (al-oh-STAIR-ick). Allosteric sites are the molecular equivalent of switches. When an end product binds to one, it turns the enzyme off, shutting down the entire metabolic pathway (Figure 3-37b). In some metabolic pathways, end products may bind directly to the active site of an enzyme. This

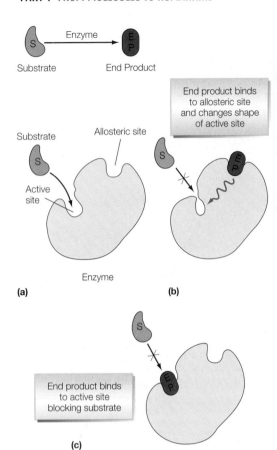

FIGURE 3-37 Enzyme Structure and Function Many enzymes are like factory workers on an assembly line. (a) Two-dimensional drawing of an enzyme. The active site conforms to the shape of the reacting molecule(s). (b) The site is a molecular switch that regulates enzyme activity. In some metabolic pathways, the end products bind to the site. This may turn the active site on or off, depending on the enzyme. (c) In others, the end products of a reaction bind to the active site itself, blocking it entirely. (Photo © Photodisc.)

binding also blocks substrates from entering the site, thereby inhibiting the enzyme and shutting down the metabolic pathway (Figure 3-37c).

Cellular respiration can be controlled by the buildup of chemicals produced in the cycle, such as citrate and ATP. Each one shuts down a specific enzyme in the cycle.

It is important to point out that enzymes are controlled by many other mechanisms, beyond the scope of this text. These control mechanisms give the cell the capacity to regulate the complex chemical reactions that constitute metabolism, one of the characteristics of all living things.

> **KEY CONCEPTS**
>
> Enzymes are protein molecules found in the cells of our bodies; they speed up many chemical reactions essential to life.

3-6 Fermentation

What do tired muscles, cheese, and wine have in common? The answer is that they're all products of a process called fermentation. **Fermentation** is a chemical reaction in which pyruvic acid produced from glycolysis is converted to one of several other end products (Figure 3-38). Although fermentation by certain microorganisms is used to produce bread and alcoholic beverages from grains and other materials, it also occurs in human cells when oxygen levels run low—for example, during prolonged and strenuous exercise. The end product is lactic acid, not alcohol. Fermentation in human cells also produces a small amount of energy. Let's look briefly at this process in body cells.

During strenuous exercise, oxygen consumption in muscle cells may exceed replenishment. When this occurs, oxygen levels inside the cells fall. (The cells are said to be anaerobic.) The lack of oxygen shuts down the electron transport system. As noted above, oxygen accepts electrons in the electron transport system. When it is absent, the electron transport system can't function. When the electron transport system falters, so does the citric acid cycle. Fortunately, glycolysis still continues. This ensures a small amount of energy production. In the absence of oxygen, however, the end product of glycolysis, pyruvate, is converted to lactic acid as shown in Figure 3-38a.

Rather than shutting energy production down completely, muscle cells thus continue to generate energy via glycolysis and fermentation. Unfortunately, fermentation is exceedingly inefficient, netting the cell only 2 ATPs for each glucose molecule, compared with 36 during the complete breakdown.

In humans, heavy exercise, such as weight lifting and running, depletes muscle oxygen, making cells become anaerobic. This, in turn, results in the buildup of lactic acid in muscle cells.

Lactic acid causes the muscle fatigue and soreness you feel after you've exercised heavily. Lactic acid, however, diffuses out of the muscle within hours and is carried by the blood to the liver, where it is converted first to pyruvate and then back to glucose in a series of chemical reactions that is essentially the reverse of glycolysis. Some of the glucose is used to syn-

Principles of Structure and Function

4 chapter

Jenny Hawn has cystic fibrosis, a deadly disease that affects the respiratory system and pancreas. One of the main symptoms of the disease is the production of excess mucus by the cells lining the respiratory tract. Mucus traps bacteria and viruses, resulting in frequent lung infections.

THINKING CRITICALLY

Healthcare costs are skyrocketing. Some critics argue that part of the reason is that our public healthcare dollars are being wasted on procedures such as organ transplants for needy people, which cost taxpayers $200,000 or more each. Such expenditures, they say, are draining dollars that could be invested in preventive medicine, such as prenatal care.

Prenatal care consists of visits to doctors for checkups and advice on nutrition and other matters that are crucial to the development of a healthy fetus. If problems arise, they can be dealt with immediately, not when it's too late and too costly. One organ transplant costs as much as prenatal care for 1,300 to 1,400 pregnant women, assistance that itself could prevent ailments that might even necessitate costly transplant procedures later in life.

Should Americans spend more on preventive medicine (including prenatal care) and less on dramatic life-saving measures, such as organ transplants? Why or why not?

healthnote

4-1 The Truth about Herbal Remedies

A male patient is treated by his physician for a painful infection of the prostate gland, one of the glands that produce a large portion of the semen. The doctor prescribes an antibiotic and advises the patient to take an herbal remedy, saw palmetto, to help prevent prostatic enlargement, a common problem in older men. A young woman suffering from restlessness and minor problems with sleep is advised by a friend to try some valerian root, another type of herbal remedy.

Herbal remedies are derived from roots, bark, seeds, fruit, flowers, leaves, and branches of plants. In China, some herbal remedies even contain animal parts. An estimated 3,000 different medicinal herbs are currently used throughout the world.

Herbal remedies are sold in the United States and in other countries by herbalists and may be mixed on site. More commonly, people get their herbs in pill, capsule, or liquid form from health food stores, grocery stores, large discount stores, or through mail order or online suppliers.

Although herbs have become very popular in the United States in recent years, the Asians have been using them for thousands of years. In China, for instance, the history of herbal remedies dates back 5,000 years. Unbeknownst to many doctors, certain herbs were used routinely in the United States and other developed nations before the advent of modern medicines. In the United States, for instance, Echinacea (ek-eh-NAH-shah) was widely used to treat colds prior to the introduction of sulfa drugs (the first generation of synthetic antibiotics). In Germany, herbal remedies are considered prescription medicines and are covered by health insurance.

Herbal treatments have a long history of use. In addition, the plants they are made from have been the source of a surprisingly large proportion of the prescription and over-the-counter (OTC) drugs in use today. That is, many so-called "modern medicines" contain active ingredients originally extracted from plants. Aspirin, for instance, is derived from the inner bark of willow trees. But are herbal treatments effective?

There is no blanket answer to this question. While research shows that many claims about the effectiveness of numerous common herbal preparations appear to be unfounded, not all are. Echinacea, for instance, does appear to enhance immune system function, and it can be used as a preventative and for treatment of colds and flus. Bilberry, dried blueberries, helps maintain eyesight and improves night vision. It was even used in World War II by members of the Royal Air Force who flew night bombing missions. Valerian root is an effective, nonaddicting sleep aid and mild

---

Jenny's disease is genetic. Until recently, there was little hope of curing this disease, which claims most of its victims before they reach the age of 20. Today, however, there may be some hope. Geneticists are experimenting with novel ways to repair the genetic defects responsible for excessive mucus production.

To correct this genetic defect, scientists are experimenting with certain cold viruses that infect the cells lining the respiratory tract. Normally, these viruses inject their DNA into the lining cells, causing cold symptoms. Researchers, however, are cutting out the genes of the virus that cause colds and splicing in new genes—ones that could lead to normal mucus production, hopefully creating a lasting cure for this deadly disease. So far, results of studies on small numbers of patients have been promising.

This research is just one of many forms of "body work" scientists are now performing. Other researchers are working at a level slightly higher—that is, at the level of tissues and organs (defined shortly). Some scientists, for instance, are developing genetically engineered pigs whose cells contain human genes. Their hope is that the pig's organs containing human genes could someday be transplanted into people to replace failed livers or kidneys. They even speculate that an organ could be specially made with a person's own genes to prevent tissue rejection. In addition, this organ would be available in a fraction of the time it

now takes to find a genetically compatible organ from a suitable donor.

Other researchers are attempting to grow tissues in the laboratory that can be used to repair diseased or damaged organs. For example, researchers successfully implanted a biodegradable material into the walls of urinary bladders of young patients to repair badly damaged organs. This material was populated with cells from the patients' damaged bladders. Five years after the transplant, the bladders were operating as well as traditionally repaired bladders.

Researchers are also surgically removing tendons, heart valves, and other tissues from pigs and using them to replace damaged organs in humans. The tissues are stripped of the pig cells, leaving only an underlying layer of connective tissue. It would then be placed where needed and populated by cells of the human recipient, thus avoiding tissue rejection. The tissues would be able to grow if, for example, the transplant was made in a child.

Health Note 4-1 describes other remarkable medical procedures used to repair or rebuild tissues and organs. This chapter presents some basic information about body structure and function, information you will need to understand the human body and many of the exciting research projects you will hear or read about in the news. This chapter also revisits homeostasis, elaborating on what you've already learned and setting the stage for your study of human organ systems, the subject of much of this text.

relaxant. Saw palmetto appears to be effective in reducing enlargement of the prostate (although a recent study has cast some doubt on this conclusion). Ginger is effective in the treatment of nausea and morning sickness, although it should not be used for an extended time during pregnancy. Garlic lowers serum cholesterol and may have antibacterial properties. Feverfew seems to help prevent migraine headaches. Hawthorn fights hypertension. The list goes on.

The main advantage of herbs, say some proponents, is that they have no side effects and contain no chemicals; they are also purportedly safe and effective. Unfortunately, this claim is far from true. Herbal remedies contain plant chemicals, some of which may have very serious side effects. A number of herbs, for instance, damage the liver, including comfrey—an herb that reportedly promotes bone healing. When taken for long periods, its active chemical ingredient is toxic to the liver.

David Kroll, a pharmacologist from the University of Colorado Health Sciences Center, notes further problems with herbal medicine. First and foremost, he argues that U.S. labeling requirements are inadequate. Herbs are sold as dietary supplements, and as such, no claims about their use are available. Individuals often must rely on advice from sales clerks, friends, or articles they read. Second, very little is known about the long-term effects of the wide variety of herbal remedies or their potential interactions with conventional medicines, although our knowledge is expanding.

Others note that without federal regulations that require strict standards for drug content in various herbal remedies, individuals can't know the exact dose they are receiving. Studies have shown that some herbal remedies have none of the active ingredient or less than the required amounts. In Europe, however, stricter laws require much better controls on product content. Canada has also established standards for the purity and content of herbal remedies.

When given a choice between an over-the-counter remedy and an herbal one, Dr. Kroll recommends the over-the-counter treatment because these products are standardized. Even then, some of these medicines may prove ineffective. Common cough medications, for instance, were recently shown to be ineffective. See **Health Tip 4-1**.

Kroll also notes that treatment with herbal remedies, as with OTC preparations, may delay the diagnosis of more serious conditions. The rule: See a doctor first to be certain that you have what you think you have and buy herbal products that are standardized from reputable companies.

The important lesson in all of this is buyer beware. Study an herb thoroughly before you take it, or better yet, see a qualified physician who practices both traditional and herbal medicine.

4-1 From Cells to Organ Systems

The human body is made up of trillions of cells—perhaps as many as 50 trillion, which is 50,000 million cells. These cells are frequently bound together by extracellular fibers and other extracellular materials, forming **tissues** (from the Latin "to weave"). Extracellular materials may be liquid (as in blood), semisolid (as in cartilage), or solid (as in bone). Tissues, in turn, combine to form organs, discrete structures in the body that carry out specific functions.

> **KEY CONCEPTS**
>
> In the human body, cells of like kind make up tissues; tissues are combined to form organs; organs combine to form organ systems.

Primary Tissues

Four major tissue types are found in humans: (1) epithelial, (2) connective, (3) muscle, and (4) nervous. Table 4-1 lists them. Each tissue consists of two or three subtypes. These tissues exist in all organs, but in varying amounts. The lining of the stomach, for example, consists of a single layer of epithelial (ep-eh-THEEL-ee-ill) cells that protects underlying layers (Figure 4-1).

Just beneath the epithelial lining is a layer of connective tissue. It holds tissues together. Beneath that is a thick sheet of smooth muscle cells, which forms the bulk of the stomach wall. When they contract, these muscles help to mix the contents of the stomach. They also help propel food into the small intestine. Smooth muscle cells are also found in blood vessels supplying the tissues of the stomach. Nerves enter with the blood vessels and control the flow of blood.

> **KEY CONCEPTS**
>
> Four primary tissue types are found in the human body; they are called primary tissues.

TABLE 4-1	The Primary Tissues and Their Subtypes
Epithelial tissue	Connective tissue
Membranous	Connective tissue proper
Glandular	Loose connective tissue
Muscle tissue	Dense connective tissue
Cardiac	Specialized connective tissue
Skeletal	Blood
Smooth	Bone
Nervous tissue	Cartilage
Conductive	
Supportive	

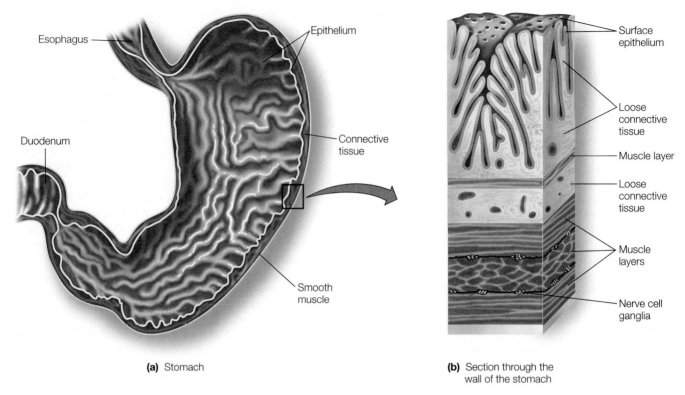

(a) Stomach

(b) Section through the wall of the stomach

FIGURE 4–1 **Human Stomach** (a) The stomach, like all organs, contains all four primary tissues. (b) These are shown here in cross section.

Epithelium

Epithelial tissue exists in two basic forms: membranous and glandular. The membranous epithelia consist of sheets of cells tightly packed together, forming the external coverings or internal linings of organs. Membranous epithelia come in a variety of forms, each specialized to protect underlying tissues. To simplify matters, the membranous epithelia are divided into two broad categories: simple epithelia, consisting of single layers of cells, and stratified epithelia, consisting of many layers of cells (Figure 4-2). Epithelial layers are often

underlain by a basement membrane, a layer of glycoprotein that fuses with the underlying connective tissue, holding the epithelium in place.

The glandular epithelia consist of clumps of cells that form many of the glands of the body. Epithelial glands arise during embryonic development from tiny "ingrowths" of membranous epithelia, as illustrated in Figure 4-3. Some glands remain connected to the epithelium by hollow ducts and are called **exocrine glands** (EX-oh-crin; glands of external secretion); products of the exocrine glands empty by way of the

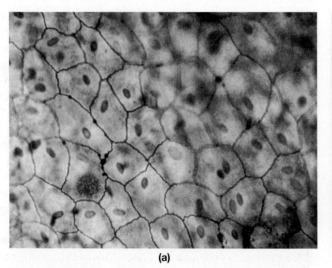

(a)

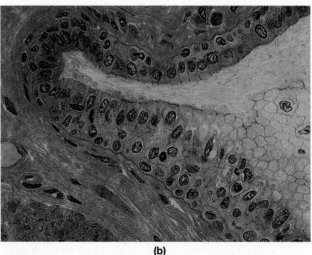

(b)

FIGURE 4–2 **Membranous Epithelia** (a) Single-celled (simple) epithelia and (b) stratified epithelia exist in different parts of the body. (a, © Biophoto Associates/Science Source, Inc.; b, © Donna Beer Stolz, Ph.D., Center for Biologic Imaging, University of Pittsburgh Medical School.)

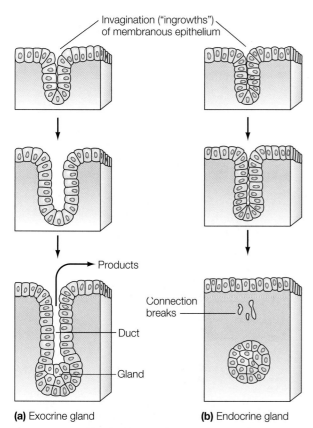

Invagination ("ingrowths") of membranous epithelium

→ Products

Connection breaks

Duct

Gland

(a) Exocrine gland   **(b)** Endocrine gland

FIGURE 4–3 Formation of Endocrine and Exocrine Glands
(a) Exocrine glands arise from invaginations of membranous epithelia that retain their connection. (b) Endocrine glands lose this connection and thus secrete their products into the bloodstream.

The Marriage of Structure and Function

One of the basic rules of architecture is that form (the structure of a building) often follows function—in other words, architectural design reflects underlying function. This rule also applies to living things. As you study human biology, you will find many examples of this rule.

The membranous epithelia provide a good example of the form-follows-function rule. Consider the outer layer of the human skin, the **epidermis** (ep-eh-DERM-iss). Shown in Figure 4-4a, the epidermis consists of numerous cell layers, known as a **stratified squamous epithelium** (SQUAW-mus). The cells of this epithelium are tightly joined by special connections and flatten toward the surface and die. Together, the thickness of the epidermis, the adhesion of one cell to another, and the dry protective layer of dead cells reduce water loss and protect us from dehydration. They also present a formidable barrier to microorganisms. All of these structural features help maintain homeostasis.

> **KEY CONCEPTS**
>
> In the human body, the structure of cells, tissues, and organs often reflects their function.

Connective Tissue

As the name implies, **connective tissue** is the body's glue. It binds cells and other tissues together and is present in all organs in varying amounts.

The body contains several types of connective tissue, each with specific functions (Figure 4-5). Despite the differences, all connective tissues consist of two basic components: cells and varying amounts of extracellular material. Two types of connective tissue will be discussed here: connective tissue proper and specialized connective tissues—bone, cartilage, and blood.

> **KEY CONCEPTS**
>
> Connective tissues of the body bind cells and various structures together, helping to form tissues organs.

Connective Tissue Proper

Connective tissue proper consists of two types: dense connective tissue and loose connective tissue. The chief difference between them lies in the ratio of cells to extracellular fibers, as shown in Figure 4-5a and b and Figures 4-6a and b.

As the name implies, **dense connective tissue** (DCT) consists primarily of densely packed fibers. The fibers are produced by cells interspersed between the fibers. DCT is found in ligaments and tendons (Figure 4-6b). Ligaments join bones to bones at joints and provide support for joints. Tendons join muscle to bone and aid in body movement.

DCT is also found at other sites, such as the layer of the skin underlying the epidermis, known as the **dermis**. Although this is DCT, the fibers are less regularly arranged than those in ligaments and tendons.

As shown in Figure 4-6b, **loose connective tissue** (LCT) contains cells in a loose network of protein fibers—specifically collagen and elastic fibers. Loose connective tissue forms around blood vessels in the body and in skeletal muscles, where it binds the muscle cells together. It

ducts (Figure 4-3a). In humans, sweat glands in the skin are exocrine glands. They produce a clear, watery fluid that is released onto the surface of the skin by small ducts. This fluid evaporates from the skin and cools the body. The salivary glands and pancreas are also exocrine glands.

Some glandular epithelial cells break off completely from their embryonic source, as shown in Figure 4-3b, to form endocrine glands (EN-doh-crin; glands of internal secretion). The **endocrine glands** produce hormones that are released into the bloodstream, where they travel to other parts of the body.

> **KEY CONCEPTS**
>
> Epithelium is a primary tissue; it forms linings of organs and during embryological development some types of epithelial tissue forms glands of internal and external secretion.

Health Tip 4-1

Got a cough? You may want to skip the over-the-counter cough medicines. *Why?*

There is no clinical evidence to show that over-the-counter cough medicines actually work. Does anything work?

Yes. Both Benadryl (an antihistamine) and Sudafed (a decongestant) appear to provide relief.

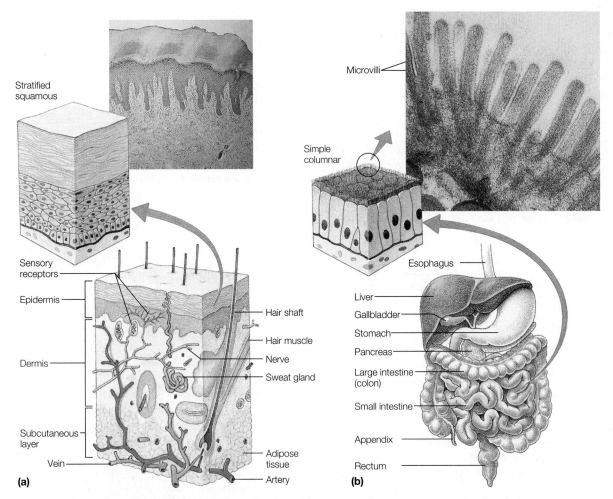

FIGURE 4-4 **Comparison of Two Epithelia with Different Functions** (a) A cross section of the skin showing the stratified squamous epithelium of the epidermis (above), which protects underlying skin from sunlight and desiccation. (Photo © Fred Hossler/ Visuals Unlimited.) (b) The simple columnar epithelium of the lining of the small intestine, which is specialized for absorption. The plasma membranes of the cells lining the intestine are thrown into folds (microvilli) that greatly increase the surface area for absorption. (Photo © David M. Phillips/Visuals Unlimited.)

also lies beneath epithelial linings of the intestines and trachea, anchoring them to underlying structures.

The extracellular fibers found in dense and loose connective tissue are produced by a connective tissue cell known as the **fibroblast** (FIE-bro-blast). Fibroblasts also repair tissue damage. When skin is cut or scraped, for example, fibroblasts in the dermis migrate into the injured area. There they begin producing collagen fibers that begin to fill the wound. Skin cells then move in, creating new skin. If the wound is small, the area may be overgrown entirely by the epidermis—so much so that it leaves no scar. In larger wounds, however, the epidermal cells may be unable to cover the entire wound, leaving some of the underlying collagen exposed, which produces a visible scar.

Besides binding tissues together, loose connective tissue houses several cell types that protect us against bacterial and viral infections. One of the most important of these protective cells is the **macrophage** (MACK-row-FAYGE; "big eater"). Macrophages engulf (phagocytize) microorganisms that penetrate the skin and underlying loose connective tissues after an injury. This prevents bacteria from spreading to other parts of the body. Inside the macrophage, microorganisms are destroyed by lysosomes, an enzyme-containing organelle, discussed elsewhere in the text. Macrophages also play a role in immune protection, a topic also discussed elsewhere in the text.

> **KEY CONCEPTS**
>
> Connective tissue proper consists of dense connective tissue like that found in tendons and loose connective tissue like that found beneath the skin.

Adipose Tissue

Some LCT contains fat cells. The **fat cell** is one of the most distinctive of all body cells (Figure 4-5d). Fat cells occur singly or in groups of varying size. Large numbers of fat cells in a given region form a modified type of LCT known as **adipose tissue** (AD-eh-poze), or, less glamorously, fat.

Scientists recognize two types of adipose tissue in the body. In the first, **white fat**, each cell contains a single large and highly conspicuous fat droplet (Figure 4-7a).

The other type of fat, **brown fat**, contains numerous small fat globules (Figure 4-7b). The cells of brown fat also contain a large number of iron-containing mitochondria, which give it its brownish color. In both types of fat cell, fat globules occupy virtually the entire cell and therefore "press" the cytoplasm and the nucleus to the periphery. Besides containing many smaller fat globules and iron-containing mitochondria, brown fat cells are surrounded by more small blood vessels called

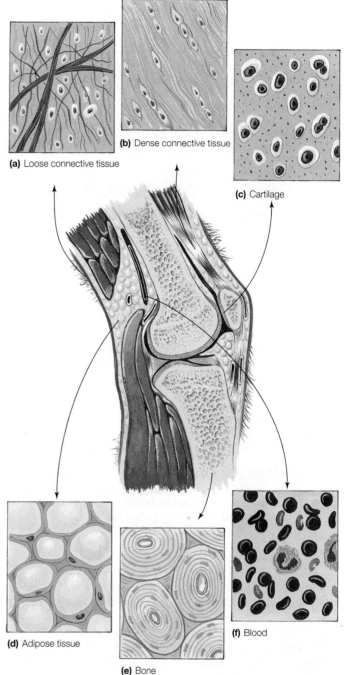

(a) Loose connective tissue

(b) Dense connective tissue

(c) Cartilage

(d) Adipose tissue

(e) Bone

(f) Blood

FIGURE 4-5 **Connective Tissue** Connective tissue consists of many diverse subtypes. (Photo © Photodisc.)

capillaries than are found in white fat. They are there to provide oxygen and nutrients to this more metabolically active fat.

White fat is widely distributed throughout the body. It serves as an important storage depot for triglycerides, a type of lipid, which are used as an energy source by the body. Fatty deposits also provide insulation for humans, many other mammals, and some birds, especially those that live in aquatic environments where heat loss can be substantial. Some fat deposits help cushion organs like the kidney, helping to protect them from physical damage. White fat cells also produce a hormone, known as *leptin*. As you will learn elsewhere in the text, leptin helps regulate hunger—in fact, it is a naturally occurring hunger suppressant that normally keeps us from overeating.

Brown fat is found in newborns. In fact, it is quite abundant in newborns and for good reason—it helps them combat cold temperatures. It is found along the back, along the upper part of the spine, and in the shoulders. How?

Brown fat is extremely metabolically active. It produces a substantial amount of heat. This is helpful to newborns because they are not yet capable of shivering when cold or taking other measures to stay warm like donning clothing or covering themselves with blankets. How does the brown fat cell produce so much heat?

Enzymes in the cell's mitochondria break glucose down, capturing energy locked in the covalent bonds. They store the energy in ATP molecules for later use. Because this process is not 100% efficient, waste heat is generated. It's what produces our body temperature. In brown fat cells, however, the mitochondria make proportionately less ATP and proportionately more heat—a lot more heat.

For many years, scientists thought that brown fat was found only in newborn infants. Recent studies have debunked this belief, showing that brown fat is also found in adults, although it is restricted to the neck and large blood vessels of the chest.

For humans, fatty deposits often become unsightly. To rid the body of these deposits, nutritionists advise exercise and reduced food intake. Some individuals opt for a surgical measure called *liposuction* (LIE-poh-suck-shun; Figure 4-8). In this procedure, a small incision is made in the skin through which surgeons insert a device they use to suck out unwanted fat deposits under the skin in various locations such as the buttocks and abdomen. The fat cells extracted from one region can even be transferred to other regions, such as the breast and lips, to resculpt the human body. Liposuction is a relatively safe technique but, like most surgery, not free from risk.

KEY CONCEPTS

The human body contains two types of adipose tissue or fat: white and brown; they differ significantly in their function.

Health Tip 4-2

To lose or maintain weight, eat a salad with low-fat dressing before lunch and dinner. Then reduce portion sizes of your main course.
Why?
Salads fill you up, yet contain few calories. This strategy will not only help you lose weight, it will provide the valuable vitamins, minerals, fiber, and other nutrients your body needs to maintain your health.

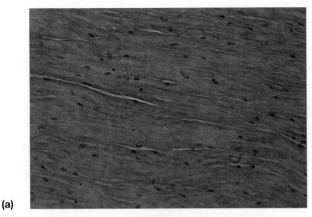

(a)

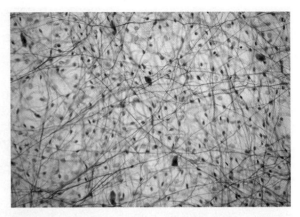

(b)

FIGURE 4-6   Connective Tissue as Seen through the Light Microscope   (a) Dense connective tissue. (b) Loose connective tissue. (a and b © Donna Beer Stolz, Ph.D., Center for Biologic Imaging, University of Pittsburgh Medical School.)

Specialized Connective Tissue

The body contains three types of specialized connective tissue: cartilage, bone, and blood.

Cartilage. **Cartilage** consists of cells embedded in an abundant and rather impenetrable extracellular material, the *matrix* (Figure 4-9). Surrounding virtually all types of cartilage is a layer of dense, irregularly packed connective tissue. This layer contains capillaries that supply nutrients to cartilage cells through diffusion. No blood vessels penetrate the cartilage itself. Because cartilage cells are nourished by diffusion from capillaries, damaged cartilage heals very slowly. Therefore, joint injuries that involve the cartilage often take years to repair or may not heal at all.

Three types of cartilage are found in humans: hyaline, elastic, and fibrocartilage. The most prevalent type is

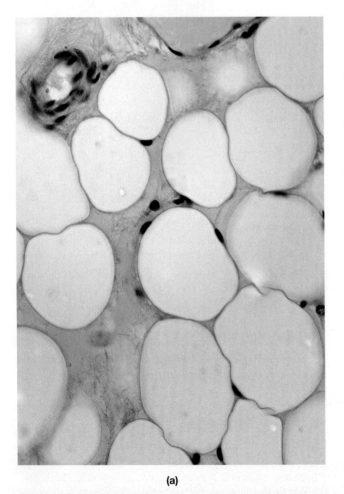

(a)

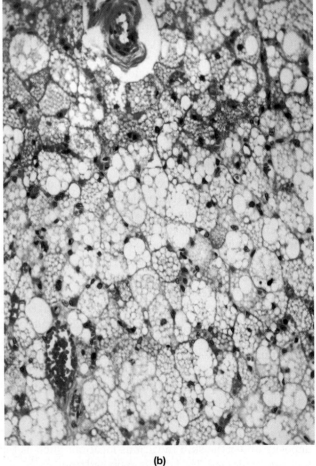

(b)

FIGURE 4-7   (a) Light micrograph of white fat. (© Donna Beer Stolz, Ph.D., Center for Biologic Imaging, University of Pittsburgh Medical School.) (b) Light micrograph of brown fat. (© Biophoto Associates/Science Source.)

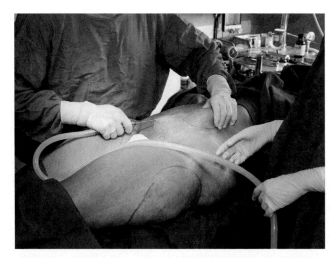

FIGURE 4-8 **Liposuction** During liposuction surgery, the physician aspirates fat from deposits lying beneath the skin, helping to reduce unsightly accumulations. (© Girish Menor/ShutterStock, Inc.)

hyaline cartilage (HIGH-ah-lynn; Figure 4-9a). Hyaline cartilage contains numerous collagen fibers, which appear white to the naked eye. Found on the ends of many bones in joints, hyaline cartilage in such locations greatly reduces friction, so bones can move over one another with ease. Hyaline cartilage also makes up the bulk of the nose and is found in the larynx (voice box) and the trachea or wind pipe. The ends of the ribs join to the sternum (breastbone) via hyaline cartilage. In embryonic development, the first skeleton is made of hyaline cartilage. It is later converted to bone.

Elastic cartilage contains many wavy elastic fibers, which give it flexibility (Figure 4-9b). Elastic cartilage is therefore found in regions where support and flexibility are required—for example, in the ears.

Fibrocartilage is a rare form of cartilage. It contains strong collagen fibers and is found in areas that must withstand tension and pressure (Figure 4-9c). Fibrocartilage is found in the shock absorbers, that is, the **intervertebral disks**

Matrix

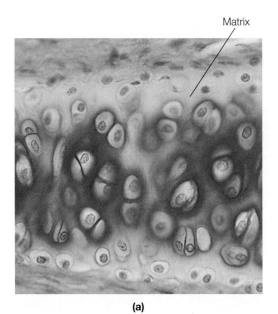

(a)

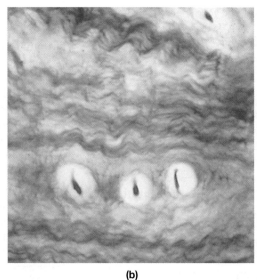

(b)

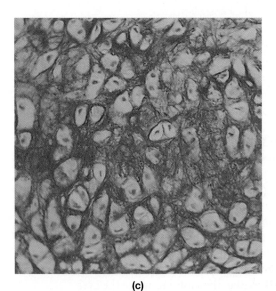

(c)

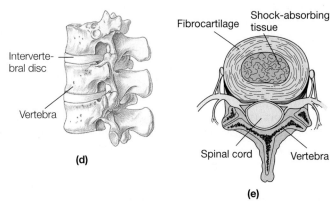

(d)

(e)

FIGURE 4-9 **Cartilage as Seen through the Light Microscope** (a) Hyaline. (© Fred Hossler/ Visuals Unlimited.) (b) Elastic. (© John D. Cunningham/Visuals Unlimited.) (c) Fibro- cartilage. (© Cabisco/Visuals Unlimited.) (d) Intervertebral disk showing location and (e) arrangement of fibrocartilage in a protective ring around the soft, spongy part of the disk that absorbs shock.

(a)   (b)

FIGURE 4-10 **Protecting Your Back** Bend your knees as the man in
(a) is doing, not your back, when you're lifting. (b) With your back
straight, stand up and lift with your legs. (a and b, © Jones & Bartlett
Learning, LLC.)

(in-ter-VER-tah-braul), which lie between the bones of the
spine, the vertebrae (Figure 4-9d). An intervertebral disk
consists of a soft, cushiony central region that absorbs shock.
Fibrocartilage forms a ring around the central portion of the
disk, holding it in place (Figure 4-9e).

The fibrocartilage ring may weaken and tear, especially
in overweight individuals. This causes the central part of
the disk to bulge outward, pressing against nearby nerves.
This condition is referred to as a *slipped* or *herniated disk*
and usually occurs in the neck or lower back, resulting in a
significant amount of pain in the neck, back, or one or both
legs, depending on the location of the damaged disk.

This problem can be corrected by surgery, although physical therapy is often the first plan of attack. Lasers are also being
used to remove herniated disks. In some cases, the entire disk
is removed and the vertebrae are fused. You can reduce your
chances of "slipping" a disk in the first place by watching your
weight, keeping in shape, sitting upright (not slouching) in a
chair, and lifting heavy objects carefully (Figure 4-10).

Bone. Bone is a form of specialized connective tissue that
provides internal support as well as protection to internal
organs such as the brain. Bone also plays an important role in
maintaining blood calcium levels and is, therefore, a homeostatic organ. Calcium is required for many body functions,
including muscle contraction and normal nerve functioning.

Like all connective tissues, bone consists of cells embedded in an abundant extracellular matrix (Figure 4-11b). Bone
matrix consists primarily of collagen fibers (which give
bone its strength and resiliency), interspersed with numerous needlelike salt crystals containing calcium, phosphate,
and hydroxide ions. They give bone its hardness.

Two types of bone tissue are found in the body: compact
bone and spongy bone (Figure 4-11a). **Compact bone** is, as its
name implies, dense and hard. As illustrated in Figure 4-11b,
the cells in compact bone are located in concentric rings of
calcified matrix. These cells surround a central canal through
which the blood vessels and nerves pass. Each bone cell, or
osteocyte, has numerous processes that course through tiny
canals in the bony matrix, known as *canaliculi* (CAN-al-ICK-
u-LIE; literally "little canals"). The canaliculi provide a route
for nutrients and wastes to flow to and from the bone cells.

Inside most bones of the body is a tissue known as spongy
bone. **Spongy bone** consists of an irregular network of calcified collagen plates (spicules). As shown in Figure 4-11c, on
the surface of the plates are numerous osteoblasts (OSS-tee-oh-
BLASTS). **Osteoblasts** are a type of bone cell that produces the
collagen, which later becomes calcified. Once these cells are surrounded by calcified matrix, they are referred to as **osteocytes**.

Spongy bone also contains large cells known as **osteoclasts** (Figure 4-11c and d). These cells digest bony matrix,
releasing calcium, when blood calcium levels fall. They also
help to resculpt bones internally to help meet changing forces
that occur as a person's activity levels change.

Blood. Blood is another specialized form of connective tissue—
although some biologists prefer to consider it a separate tissue
type called *vascular tissue*. Blood consists of formed elements
and a large amount of extracellular material. The formed elements of blood consist of the red and white blood cells and
the platelets. The formed elements are responsible for 45% of
the blood volume. The extracellular material is a fluid known
as **plasma**. It makes up the remaining 55%.

Red blood cells transport oxygen and small amounts
of carbon dioxide to and from the lungs and body tissues.
White blood cells rid the body of foreign organisms such
as viruses and bacteria that can cause infections. **Platelets**
are fragments of large cells (megakaryocytes) located in the
red bone marrow, the principal site of blood cell formation.
Platelets play a key role in blood clotting.

> **KEY CONCEPTS**
>
> Specialized connective tissue performs specific functions vital to
> homeostasis.

Muscle Tissue

Muscle tissue is found in virtually every organ in the body.
Muscle gets its name from the Latin word for "mouse" (*mus*).
Early observers likened the contracting muscle of the biceps
to a mouse moving under a carpet.

Muscle is an excitable tissue. When stimulated, it contracts, producing mechanical force. Muscle cells working in
large numbers can create enormous forces. Muscles of the jaw,
for instance, create a pressure of 200 pounds per square inch,
forceful enough to snap off a finger. (Don't try this at home.)
Muscle also moves body parts, propels food along the digestive
tract, and expels the fetus from the uterus during birth. Heart
muscle contracts and pumps blood through the 50,000 miles
of blood vessels in the human body. Acting in smaller numbers,
muscle cells are responsible for intricate movements, such as
those required to play the violin or move the eyes.

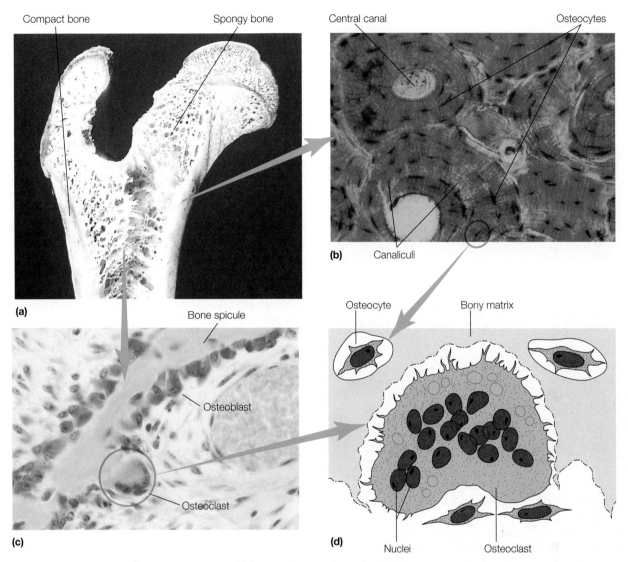

FIGURE 4-11 **Bone** (a) Compact and spongy bone, shown in a section of the humerus. (© R. Calentine/Visuals Unlimited.) (b) Light micrograph of compact bone. (© John D. Cunningham/Visuals Unlimited.) (c) Photomicrograph of spongy bone, showing osteoblasts and osteoclasts. (© R. Kessel/Visuals Unlimited.) (d) Osteoclast digesting surface of bony spicule.

Three types of muscle tissue are found in humans: skeletal, cardiac, and smooth. The cells in each type of muscle tissue contain two contractile protein filaments, actin and myosin. These very same fibers were first encountered in your study of the microfilamentous network lying beneath the plasma membrane of cells. This layer of fibers plays a key role in cell division. When stimulated, actin and myosin filaments of muscle cells slide over one another, shortening and causing contraction.

Skeletal Muscle. The majority of the body's muscle is called **skeletal muscle**—so named because it is frequently attached to the skeleton. When skeletal muscle contracts, it causes body parts (arms and legs, for instance) to move. Most skeletal muscle in the body is under voluntary, or conscious, control. Signals from the brain cause the muscle to contract. A notable exception is the skeletal muscle of the upper esophagus, which contracts automatically during swallowing. (The initial act of swallowing is voluntary—pushing the food to the back of the oral cavity—but from that point on, the action is involuntary.)

Skeletal muscle cells are long cylinders formed during embryonic development by the fusion of many embryonic muscle cells. Because of this, skeletal muscle cells are usually referred to as *muscle fibers*. Each muscle fiber contains many nuclei (Figure 4-12a). Because of the dense array of contractile fibers in the cytoplasm of the muscle fiber, the nuclei become pressed against the plasma membrane. As a result, muscle fibers cannot divide, and damaged muscle cells cannot be replaced.

Skeletal muscle fibers appear banded, or striated, when viewed under the light microscope (Figure 4-12a). The striations result from the unique arrangement of actin and myosin filaments inside muscle cells.

Cardiac Muscle. Like skeletal muscle, **cardiac muscle** is banded (Figure 4-12b). Unlike skeletal muscle, cardiac muscle is involuntary; that is, it contracts without conscious control. Found only in the walls of the heart, each cardiac muscle cell contains a single nucleus. Cardiac muscle cells also branch and interconnect freely, and individual cells are tightly connected

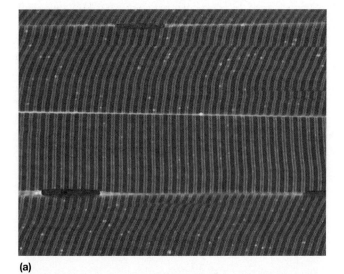

(a)

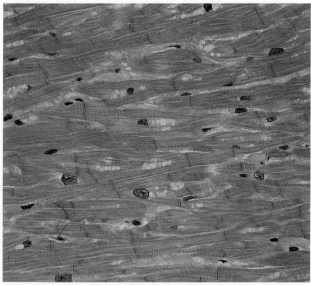

(b)

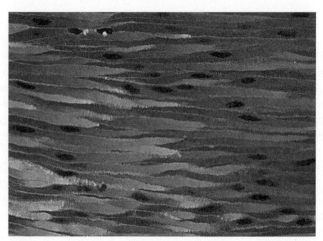

(c)

FIGURE 4-12 **Three Types of Muscle as Seen through the Light Microscope** (a) Skeletal (252x). (© Dr. Gladden Willis/Visuals Unlimited, Inc.) (b) Cardiac (400x). (© Donna Beer Stolz, Ph.D., Center for Biologic Imaging, University of Pittsburgh Medical School.) (c) Smooth (398x). (© M. I. Walker/Science Source, Inc.)

to one another. This adaptation helps maintain the structural integrity of the heart, an organ subject to incredible strain as it pumps blood through the body day and night. The points of connection also provide pathways for electrical impulses to travel from cell to cell, allowing the muscle of the heart chambers to contract uniformly when stimulated.

Smooth Muscle. **Smooth muscle**, so named because it lacks visible striations, is involuntary. Actin and myosin filaments are present in smooth muscle cells but are not arranged in an organized fashion like those found in other types (Figure 4-12c). Smooth muscle cells typically occur in large groups, specifically bands or sheets—for example, in the walls of hollow organs such as the intestine, stomach, and uterus (see Figure 4-1).

Smooth muscle cells in the wall of the stomach churn the food, mixing the stomach contents, and force tiny spurts of liquefied food into the small intestine. Smooth muscle contractions also propel the food along the intestinal tract. Smooth muscle cells may also occur in small groups, for example, in small rings that surround tiny blood vessels. When these cells contract, they reduce the supply of blood to tissues. Smooth muscle cells may also occur in the connective tissue of certain organs such as the prostate gland of boys and men. They help propel fluid from this gland during ejaculation.

> **KEY CONCEPTS**
>
> Muscle consists of highly specialized cells that contract when stimulated. Three types of muscle tissue are found in the body: cardiac, skeletal, and smooth.

Nervous Tissue

Last but not least of the primary tissues is **nervous tissue**. It consists of two types of cells: conducting cells or neurons and nonconductive cells or neuroglia (literally, nerve glue). Together, the neurons and the neuroglial cells form the brain, spinal cord, and nerves, described elsewhere in the text.

The conducting cells or **neurons** are sometimes modified to respond to specific stimuli, such as pain or temperature. Stimulation results in nerve impulses, which the neuron transmits from one region of the body to another.

The nonconductive cells of the nervous system are called **neuroglia** and were long thought to be merely a kind of nervous system connective tissue. They are by far the most prominent type of cell in the brain and spinal cord, outnumbering nerve cells by about nine to one. Although nonconducting cells provide physical support for nerve cells, we now know that they also perform other functions. One type of neuroglial cell, for instance, engulfs bacteria, helping reduce brain infections. Another type of neuroglial cell transports nutrients from blood vessels to neurons. This cell also produces a hormone that stimulates regrowth of nerve cells. Researchers hope that one day this hormone can be used to treat diseases of the nervous system resulting from the degeneration of neurons, such as Parkinson's disease. The hormone-producing cells also guard against toxins by creating a barrier to many potentially harmful substances. Although neuroglial cells have long been considered nonconductive, new research suggests that they communicate with each other as well as with nerve cells.

Neurons. At least three distinct types of neurons are found in the body. Despite obvious structural differences, they share several common features. We will study these similarities by looking at one of the most common (and complex) nerve cells, the **multipolar neuron**, shown in Figures 4-13a and b.

The multipolar neuron contains a prominent cell body to which are attached several short, highly branched processes, known as *dendrites* (DEN-drites). (It is the presence of these processes that gives this neuron its name, multipolar.) The

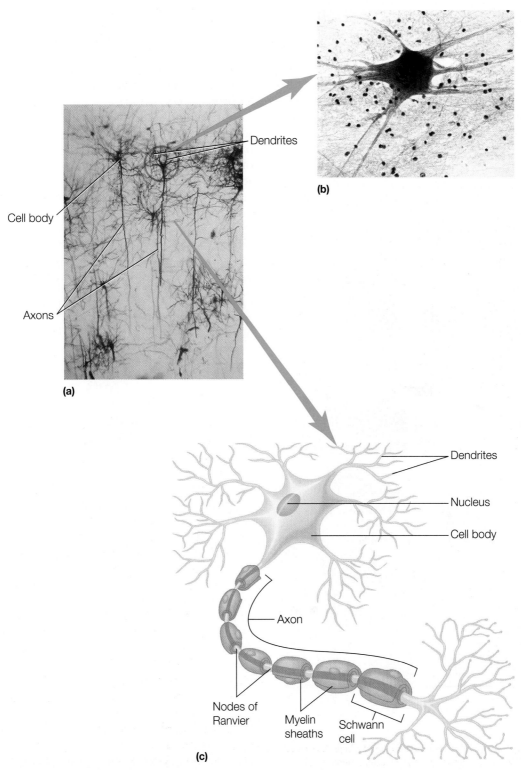

FIGURE 4-13 Multipolar Neurons (a) Attached to the cell body of the multipolar neuron are many highly branched dendrites (b), which deliver impulses to the cell body. (a, © John D. Cunningham/Visuals Unlimited; b, © Manfred Kage/Science Source.) (c) Multipolar neurons have one long, unbranching fiber called the axon, which transmits impulses away from the cell body.

dendrites receive impulses from receptors (discussed next) or other neurons and transmit them to the cell body. Also attached to the cell body of the multipolar neuron is a large, fairly thick process, known as the *axon* (AXE-on). It transports bioelectric impulses away from the nerve cell body.

Nervous tissue plays an important role in body function. It monitors internal and external conditions through various sensors called *receptors*. Nerve cells use this information to bring about internal changes that ensure our survival and well-being. Input from various sensors is fed into the brain and spinal cord via nerves. Incoming information is then compiled and appropriate responses are mounted. Some may be automatic responses, ones we're not consciously aware of. Others involve conscious responses. Both conscious and unconscious responses help maintain homeostasis.

> **KEY CONCEPTS**
>
> Nervous tissues consists of highly specialized cells called neurons that are capable of conducting bioelectric impulses from part of the body to another vital to control many body functions.

Organs and Organ Systems

Cells in your body contain organelles ("little organs") that carry out many of its functions in isolation from the biochemically active cytoplasm. As pointed out elsewhere in the text, compartmentalization such as this is an important evolutionary adaptation. Compartmentalization also occurs in organisms in their internal organs.

Organs are discrete structures that have evolved to perform specific functions, such as digestion, enzyme production, and hormone production. Most organs, however, do not function alone. Instead, they are part of groups of co-operative organs, called **organ systems**. The brain, spinal cord, and nerves are all organs that belong to an organ system known as the nervous system.

As you will see, components of an organ system are sometimes physically connected—as in the digestive system (Figure 4-14a). In other cases, they are dispersed throughout the body—as in the endocrine system (Figure 4-14b). In addition, some organs belong to more than one system. The pancreas, for example, produces digestive enzymes used to break down food in the small intestine. The pancreas is

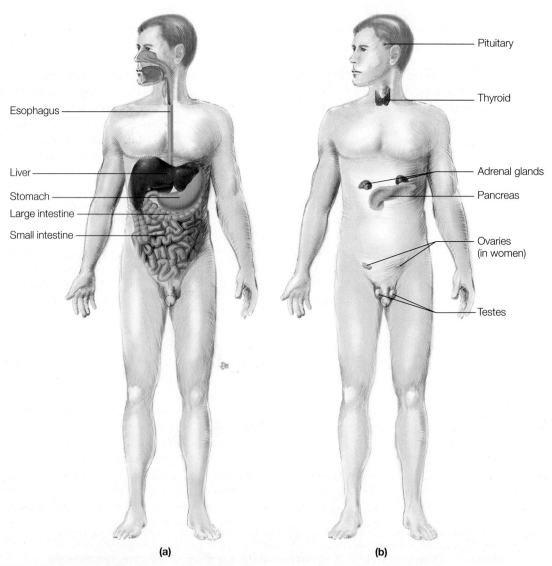

Esophagus

Liver

Stomach

Large intestine

Small intestine

Pituitary

Thyroid

Adrenal glands

Pancreas

Ovaries (in women)

Testes

(a)   **(b)**

FIGURE 4-14   **Comparison of the Endocrine and Digestive Systems**   (a) The digestive system. (b) The endocrine system.

therefore part of the digestive system. However, some cells of the pancreas produce hormones that are released into the bloodstream where they circulate to body tissues to help control blood glucose levels. Thus, the pancreas also belongs to the endocrine system.

Many of the chapters in this text describe the organ systems of the human body and the functions they perform, paying special attention to their role in homeostasis. Figure 4-15 summarizes the functions of each organ system and lists the

role each plays. You may want to take a moment to read the descriptions in the boxes describing the organ systems before proceeding.

> **KEY CONCEPTS**
>
> Organs are structures that evolved to perform very specific functions like digestion and respiration; organs are part of organ systems consisting of numerous organs that function together to carry out their functions.

4-2 Principles of Homeostasis

Organ systems perform many functions, virtually all of which are involved in **homeostasis**. Defined as a state of relative internal constancy, homeostasis occurs on a variety of levels—in cells, tissues, organs, organ systems, organisms, and even the environment. Homeostatic systems at all levels of biological organization have several common features.

> **KEY CONCEPTS**
>
> The organs of the body perform a multitude of functions, all essential for maintaining homeostasis.

Negative Feedback

Most of the processes you will be studying in this course are controlled by a process biologists refer to as feedback mechanisms. The most common type is called **negative feedback**. To understand how negative feedback systems work, consider a familiar example, the heating system of a house. It contains many of the features present in biological feedback mechanisms (Figure 4-16a).

In the winter, the heating system of a house maintains a constant internal temperature, even though the outside temperature fluctuates considerably. Heat lost through ceilings, walls, and windows is replaced by heat generated from the combustion of natural gas or oil in the furnace (Figure 4-16a).

The furnace is controlled by a thermostat. It monitors room temperature. When indoor temperature falls below the setting, the thermostat sends a signal to the furnace, turning it on. Heat from the furnace is then distributed through the house, raising the room temperature. When the room temperature reaches the desired setting, the thermostat shuts the furnace off. Like all negative feedback mechanisms, the product of the system (heat) "feeds back" on the process, shutting it down.

A graph of a hypothetical house temperature is shown in Figure 4-16b and illustrates another important principle of homeostasis: Homeostatic systems do not maintain absolute constancy. Rather, they maintain conditions (such as body temperature) within a given range.

All homeostatic mechanisms in humans operate in a similar fashion, maintaining conditions within a narrow range around an operating point. The operating point is akin to the setting on a thermostat. As you will see in later chapters, our bodies maintain fairly constant levels of a great many chemical components, including hormones, nutrients, wastes, and ions. We also maintain physical conditions such as body temperature, blood pressure, blood flow, and others. Most of this is done subconsciously and automatically.

> **KEY CONCEPTS**
>
> Homeostatic mechanisms in the body are controlled primarily by negative feedback loops.

Sensors and Effectors

Biological homeostatic mechanisms contain **sensors** that detect change and **effectors** that correct conditions. In your home, the thermostat is the sensor and the furnace is the effector.

The human body contains many sensors. For example, the skin contains nerve cell endings that detect outside temperature. When temperatures change, they send signals to the brain, alerting it to outside conditions. The brain then sends signals to the body to adjust for temperature, for example, to generate more heat internally if it is too cold. The options for generating more heat are many.

One of the main sources of heat in the human body is the breakdown (catabolism) of glucose and other molecules. For example, cells of the body break down glucose to make ATP. During this process, heat is given off. Each cell, then, is a tiny furnace whose heat radiates outward and is distributed throughout the body by the blood. Unlike the furnace in your home, the cellular "furnaces" cannot be turned up very quickly. They respond much more slowly than a furnace and are part of a delayed response to low temperature. As winter progresses, metabolism increases and the body produces more heat.

In order to respond to sudden changes in outside temperature, the body must rely on more rapid mechanisms. If you walked outdoors on a cold winter night dressed only in a light sweater and blue jeans, for example, receptors in your skin would sense the cold and send signals to the brain. The brain, in turn, would send signals to blood vessels in the skin, causing them to constrict, reducing the flow of blood in the skin. The restriction of blood flow through the skin reduces heat loss. If it is cold enough outside, the brain may also send signals to the muscles, causing them to undergo rhythmic contractions, known as *shivering*. Shivering burns additional glucose, releasing additional heat. Many voluntary actions may also be "ordered" by the brain to reduce heat loss or generate more heat. For example, you might turn around

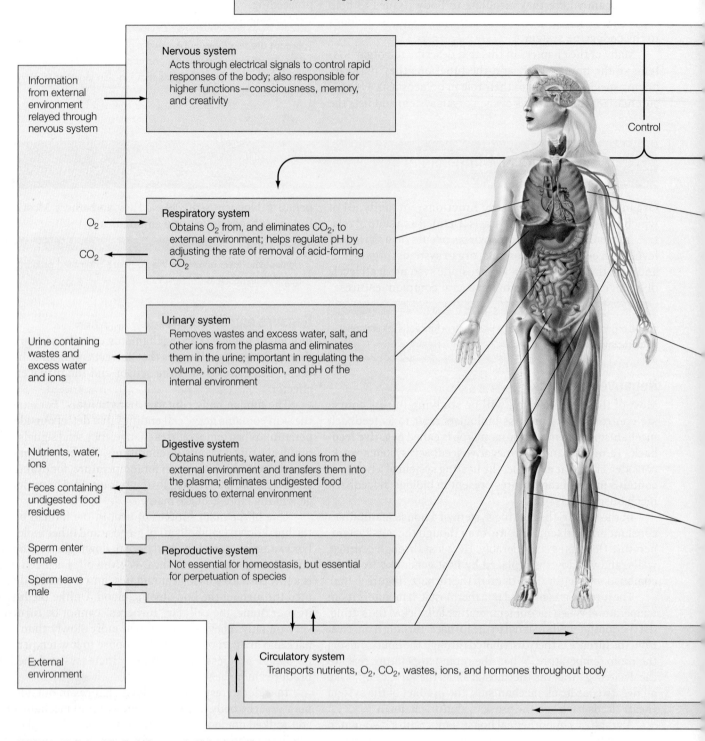

BODY SYSTEMS
Made up of cells organized by specialization to maintain homeostasis

Nervous system
Acts through electrical signals to control rapid responses of the body; also responsible for higher functions—consciousness, memory, and creativity

Information from external environment relayed through nervous system

Control

O_2

CO_2

Respiratory system
Obtains O_2 from, and eliminates CO_2, to external environment; helps regulate pH by adjusting the rate of removal of acid-forming CO_2

Urinary system
Removes wastes and excess water, salt, and other ions from the plasma and eliminates them in the urine; important in regulating the volume, ionic composition, and pH of the internal environment

Urine containing wastes and excess water and ions

Digestive system
Obtains nutrients, water, and ions from the external environment and transfers them into the plasma; eliminates undigested food residues to external environment

Nutrients, water, ions

Feces containing undigested food residues

Sperm enter female

Sperm leave male

Reproductive system
Not essential for homeostasis, but essential for perpetuation of species

External environment

Circulatory system
Transports nutrients, O_2, CO_2, wastes, ions, and hormones throughout body

FIGURE 4–15 **Role of the Body Systems in Maintaining Homeostasis**

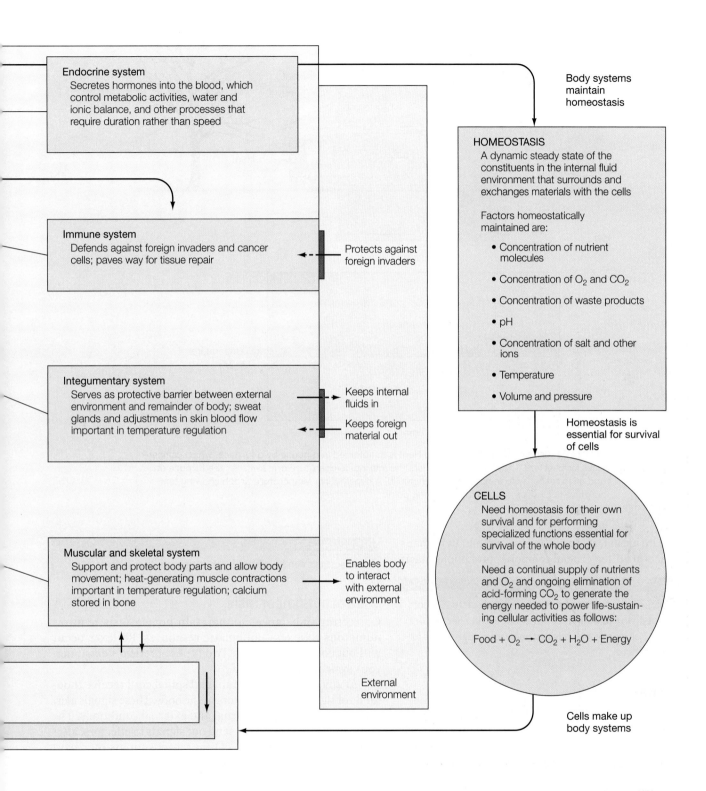

Endocrine system
Secretes hormones into the blood, which control metabolic activities, water and ionic balance, and other processes that require duration rather than speed

Immune system
Defends against foreign invaders and cancer cells; paves way for tissue repair

Protects against foreign invaders

Integumentary system
Serves as protective barrier between external environment and remainder of body; sweat glands and adjustments in skin blood flow important in temperature regulation

Keeps internal fluids in

Keeps foreign material out

Muscular and skeletal system
Support and protect body parts and allow body movement; heat-generating muscle contractions important in temperature regulation; calcium stored in bone

Enables body to interact with external environment

External environment

Body systems maintain homeostasis

HOMEOSTASIS
A dynamic steady state of the constituents in the internal fluid environment that surrounds and exchanges materials with the cells

Factors homeostatically maintained are:

- Concentration of nutrient molecules
- Concentration of O_2 and CO_2
- Concentration of waste products
- pH
- Concentration of salt and other ions
- Temperature
- Volume and pressure

Homeostasis is essential for survival of cells

CELLS
Need homeostasis for their own survival and for performing specialized functions essential for survival of the whole body

Need a continual supply of nutrients and O_2 and ongoing elimination of acid-forming CO_2 to generate the energy needed to power life-sustaining cellular activities as follows:

$$Food + O_2 \rightarrow CO_2 + H_2O + Energy$$

Cells make up body systems

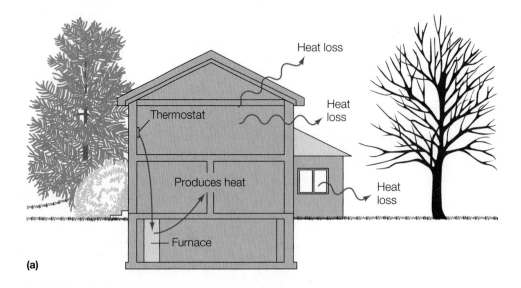

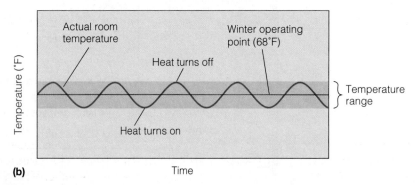

FIGURE 4-16 **Homeostasis in Our Homes** (a) Heat is maintained in a house by a furnace, which compensates for heat loss. The thermostat monitors the internal temperature and switches the furnace on and off in response to temperature changes. (b) A hypothetical temperature graph showing temperature fluctuation around the operating point.

and go back inside at least to grab a coat and hat. In humans, conscious acts are often crucial components of homeostasis.

> **KEY CONCEPTS**
>
> All homeostatic systems require sensors and effectors. Sensors are specialized structures that detect change and effectors are the glands and organs of the body that help make adjustments for those changes.

Upsetting Homeostasis

Homeostasis can be thrown out of balance by many factors. On a hot day, for example, water escapes our bodies very rapidly through perspiration. Although perspiration helps cool us down, if water loss is severe, it can be quite damaging. Severe water loss leads to dehydration, which may upset homeostasis—so much so that death may occur. Going without water for a prolonged period (approximately 3 days) can kill a human being.

Just as changes in the output of a substance—for example, perspiration—drastically alter homeostasis, so do changes in input. Excess salt intake, for example, can result in high blood pressure in some people that can lead to other problems like stroke, kidney disease, and hardening of the arteries. Whatever the cause, imbalances in homeostasis can have dramatic impacts on human health.

> **KEY CONCEPTS**
>
> Homeostasis can be altered by changing inputs or outputs.

Control of Homeostasis

Correcting imbalances to maintain homeostasis requires numerous reflexes—automatic responses. Reflexes occur without conscious control. Two types of reflexes exist: nervous system and hormonal.

At any one time, the brain and spinal cord receive thousands of signals from receptors in the body. These signals alert the nervous system to internal and external conditions. The brain and spinal cord then send out signals to effectors, after integrating (making sense of) the various inputs they have received. The main effectors are the muscles and glands of the body. These are nervous system reflexes.

Hormonal reflexes also help us respond to changes. Hormones are chemical substances produced by the endocrine glands. Released into the blood, hormones travel to distant sites, where they effect change. Hormones don't just pour out of endocrine glands in a constant stream, however. They are released only when needed and usually as part of a chemical reflex essential for maintaining homeostasis.

Nervous and endocrine feedback mechanisms generally occur over considerable distances. Nervous system responses

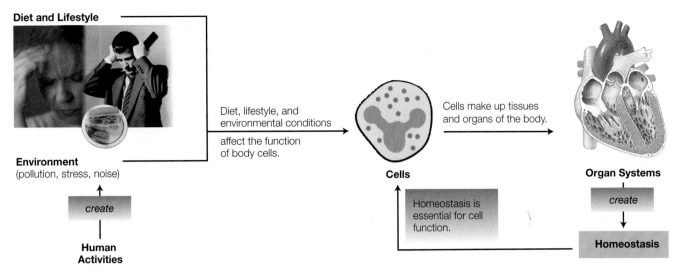

FIGURE 4-17 Human Health Is Dependent on Maintaining Homeostasis Homeostasis, however, is affected by the condition of our environment. Stress, pollution, noise, and other environmental factors upset the function of cells and body systems, thus upsetting homeostasis and human health. As a result, human health is dependent on a healthy environment. (Left photo © Nick Rowe/Photodisc/Getty Images; right photo © Jonnie Miles/Photodisc/Getty Images; sandwich plate © C Squared Studios/Getty Images.)

tend to occur more quickly, however, as the responses are elicited by nerve impulses traveling along nerves at a high rate of speed, as opposed to hormones that must circulate in the bloodstream to reach their effectors.

The body also possesses chemical feedback mechanisms that operate over very short distances. They usually involve chemicals called **paracrines** (PEAR-ah-crins). Produced by individual cells, paracrines diffuse to neighboring cells, where they elicit an effect. One example is epidermal growth factor (EGF), a paracrine produced by skin cells. It is released when skin cells are damaged or lost. EGF stimulates cell division in remaining cells, helping to fix the wound and thus provides a degree of local control.

One of the best-known paracrines is a group of chemical substances known as *prostaglandins* (PROSS-tah-GLAND-ins). Prostaglandins comprise a rather large group of molecules with diverse functions. Some stimulate blood clotting; others stimulate smooth muscle contraction.

> **KEY CONCEPTS**
> The nervous and endocrine systems control virtually all homeostatic processes in the body.

The Link between Health and Homeostasis

This book emphasizes a theme that is important to all of us—notably, that human health is dependent on homeostasis within the cells, tissues, and organs of the body. Homeostasis, in turn, requires a healthy social, psychological, and physical environ-

ment and freedom from disease-causing organisms. The health of the physical environment, our air and water, for instance, is also dependent in part on properly functioning homeostatic systems in nature. These systems evolved to help maintain conditions conducive to life. Unfortunately, they can be damaged by human activities like burning coal in coal-fired power plants, which release a number of harmful chemical pollutants.

The relationship between homeostasis and health is summarized in Figure 4-17. Take a moment to study this diagram. Notice that the organ systems in your body (right side of diagram) help to maintain homeostasis. That's vital for proper cell function. Cells, of course, make up the tissues and organs of the body. Cells also contain homeostatic mechanisms that are vital to their own function as well as to the overall economy of the organism.

On the left side of the diagram, you see that environmental factors, such as pollution and disease organisms, affect the function of cells and body systems in ways that can upset the body's internal balance, thus altering human health. Stress and other factors can also affect homeostasis, detracting from our health. Figure 4-15 also summarizes some key points about homeostasis that you may want to review when studying this material. For a discussion of herbal remedies that help restore health, see Health Note 4-1.

> **KEY CONCEPTS**
> Human health is dependent on the proper functioning of homeostatic mechanisms of the body.

4-3 Biological Rhythms

The previous discussion may give the impression that homeostasis establishes an unwavering stability. In truth, some key bodily processes undergo rhythmic change.

Understanding Biological Rhythms

Body temperature, for instance, varies during a 24-hour period by as much as 0.5°C. Blood pressure may change by as much as

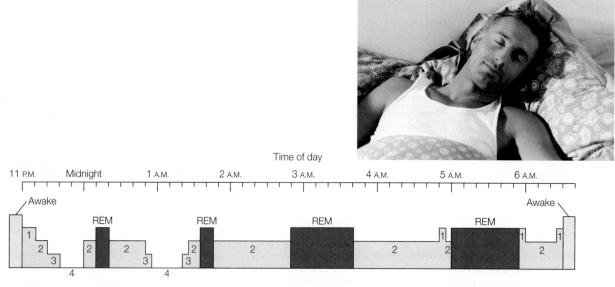

FIGURE 4-18 **Stages of Sleep** Numbers indicate the stages of sleep: the higher the number, the deeper the sleep. Note that around midnight sleep is deepest. As morning approaches, a person's sleep is lighter, and REM sleep, or dream sleep, occurs in longer increments. (Photo © Photos.com.)

20%, and the number of white blood cells, which fight infection, can vary by 50% during the day. Alertness also varies considerably. About 1:00 PM each day, for instance, most people go through a slump. For most of us, activity and alertness peak early in the evening, making this an excellent time to study. Daily cycles, such as these, are called **circadian rhythms** (sir-KADE-ee-an; "about a day"). These also are referred to as **biological rhythms** or **biorhythms**. Circadian rhythms are natural cycles in body function linked to the 24-hour day-night cycle.

Many hormones follow daily rhythmic cycles. The male sex hormone testosterone, for example, follows a 24-hour cycle. The highest levels occur in the night, particularly during dream sleep, also known as REM sleep. Dream sleep occurs primarily in the early morning hours—the later the hour, the longer the periods of dream sleep (Figure 4-18).

Not all cycles occur over 24 hours, however. Some can be much longer. The menstrual cycle, for instance, is a recurring series of events in the reproductive functions of women that lasts, on average, 28 days. During the menstrual cycle, levels of the female sex hormone estrogen undergo dramatic but predictable shifts.

The important point is that although many chemical substances are held within a fairly narrow range by homeostatic mechanisms, others fluctuate widely in normal cycles. These cycles are part of the body's dynamic balance.

KEY CONCEPTS

Many processes in the body fluctuate slightly during the day or over longer periods; these natural fluctuations are called biological rhythms.

Control of Biological Rhythms

Research suggests that the brain controls many internal rhythms. One region in particular, the suprachiasmatic nucleus

(SUE-pra-ki-as-MAT-tick), is thought to play a major role in coordinating several key rhythms.

The suprachiasmatic nucleus (SCN) is a clump of nerve cells located at the base of the brain in a region called the *hypothalamus*. The SCN is believed to regulate other control centers. As a result, the suprachiasmatic nucleus is often referred to as the *master clock* or, more commonly, the *biological clock*. Like the automatic timer of a sprinkler system, the SCN ticks off the minutes, faithfully imposing its control on the body, turning body functions on and off.

Ultimate control of the SCN is thought to reside in a gland in the brain known as the *pineal gland* (PIE-knee-al). It secretes a hormone thought to keep the suprachiasmatic nucleus in sync with the 24-hour day-night cycle. The release of this hormone is controlled by light. Scientists have recently found that the biological clock responds differently to different colors of light. Their work shows that the biological clock (presumably via the pineal gland) is particularly responsive to blue light. Physicians are even using blue light to treat patients who suffer from depression during the winter, when days are shorter and often gloomy. This condition is known as *seasonal-affective disorder* (SAD). In one study, a 45-minute exposure to blue light each morning reduced SAD in 60% of all patients.

Exposure to blue light perks up drowsy people, too, so some businesses in Europe are installing lights whose color can be adjusted. In the morning, when workers first arrive, the lighting is automatically set to provide more blue light. Some experts are suggesting that schools install blue lighting to increase student alertness and performance. Interestingly, however, the problem of drowsy students may be better addressed by making changes in activity at home. Some studies suggest that exposure of teens to blue light from computer screens and television late at night may be throwing off their sleep patterns. By artificially stimulating students at night, this light makes it harder for them to get to sleep . . . so they arrive at school in a state of near slumber. To prevent this, some researchers suggest that students wear yellow-orange glasses that filter out blue light at night and thus help them get to bed on time.

The study of biological rhythms has yielded some important information and insights. One practical application is a better understanding of jet lag—that drowsy, uncomfortable feeling people get from the disruption of sleeping patterns caused by long-distance airplane travel. Jet lag occurs when the body's biological clock becomes out of sync with the day-night cycle of a traveler's new surroundings. A businesswoman who travels from Los Angeles to New York, for instance, may be wide awake at 10:00 PM New York time because her body is still on Los Angeles time—3 hours earlier. If she normally goes to bed at 10:00 PM, she may not feel tired until 1:00 AM in New York. When the alarm goes off at 6:00 AM, our weary traveler crawls out of bed exhausted, because as far as her body is concerned, it is 3:00 AM Los Angeles time. Making matters worse, she may not have fallen asleep until 1:00 AM.

To avoid the discomfort associated with travel, sleep researchers suggest that you abide by your internal clock, following "home time" when in a new time zone, or, barring that, try to reset your biological clock before you get there. You can do this by going to bed an hour or two earlier when you are traveling from west to east. Do just the opposite for east-to-west travel. Insomnia is another disruption of sleeping patterns. More on this topic is in Health Note 4-2.

Research on body rhythms has also shown that people respond to drugs differently at different times of the day and night. Certain drugs that combat high blood pressure, for instance, are more effective if taken at night than in the morning. By administering drugs at the body's most receptive times, physicians may be able to reduce the doses, thereby lowering toxic side effects and fighting diseases more effectively.

> **KEY CONCEPTS**
>
> The brain controls many natural body rhythms.

4-4 Health and Homeostasis

Human health is dependent on maintaining homeostasis, but homeostasis can be upset by many factors. Even stress created by an unhealthy environment can upset our internal balance and our health. The work environment can be a particularly stressful place for many reasons. Difficult bosses, noise, exposure to dangerous equipment, heavy workloads, grouchy coworkers, and shiftwork all contribute to stress that upset homeostasis and damage our health.

Variable work schedules are particularly troublesome. Dr. Richard Restak, a neurologist and author, notes that the "usual rhythms of wakefulness and sleep . . . seem to exert a stabilizing effect on our physical and psychological health." The greatest disrupter of our natural circadian rhythms, he says, is the variable work schedule.

Today, one out of every four working men and one out of every six working women are on a variable work schedule, alternating frequently between day and night shifts. Why? In many industries, factories are kept running 24 hours a day to make optimal use of equipment and buildings. Workers must be on the job day and night. As a result, more restaurants and stores are finding it profitable to be open 24 hours a day, and more healthcare workers must be on duty at night to care for accident victims.

Workers on alternating shifts suffer from a higher incidence of ulcers, insomnia, irritability, depression, and tension than workers on regular shifts. These are all signs that homeostasis is upset. Making matters worse, tired, irritable workers whose judgment is impaired by fatigue pose a threat not only to themselves but also to society. Why?

Humans typically sleep at night and are awake during the day. Make a person work at night when he or she normally sleeps, say sleep experts, and you can expect more accidents and lower productivity. The two most significant accidents at nuclear power plants (Chernobyl in the former Soviet Union and Three Mile Island in the United States) were, many experts believe, caused in part by sleepy workers who missed important signals or made serious errors in judgment because they were working late at night, a time unsuitable for clear thinking. Many experts believe that plane crashes, auto accidents, and acts of medical malpractice can also be traced to judgment errors made by individuals who were working against their natural body rhythms.

Thanks to studies of biological rhythms, researchers are finding ways to reset the body's clock. These efforts could lessen the problems shift workers face and improve safety and productivity.

> **KEY CONCEPTS**
>
> The body's natural rhythms can be disrupted by shift work and by changing time zones.

Health Tip 4-3

To improve your chances of a good night's sleep, the National Sleep Foundation recommends regular exercise.

Why?

Exercise helps decrease stress and strengthens muscles that can reduce aches and pains that disturb sleep, especially as we get older. It also improves balance in old age. If you exercise late in the day, be sure to complete your exercise a couple of hours before bedtime.

health**note**

4-2 Wide Awake at 3 AM: Causes and Cures of Insomnia

It's Thursday evening. The final exam in your human biology course is tomorrow. Forty percent of your grade depends on your score. You've studied hard, but it's well past midnight and you can't get to sleep. It's happened before and you know your ability to think during the test, which is scheduled to start at 8:00 AM, will be impaired. That puts more pressure on you, making it more difficult to sleep.

This frustrating problem, which affects almost everyone at some time in life, is just one of several sleep disorders that physicians call *insomnia* (in-SOM-knee-ah). Another common complaint of insomniacs is waking up in the middle of the night and not being able to fall asleep again for several hours, if at all. In other instances, individuals find that they wake up too early; still others complain of sleeping through the night but waking up feeling groggy and unrefreshed.

Insomnia is best defined as a condition of inadequate or poor quality sleep. That is, you don't get enough sleep or you sleep poorly. It's a problem that affects many people. In a recent study in Canada, for example, two of every five people surveyed said they had trouble sleeping at least once a week. Insomnia affects both men and women but is more common in women and the elderly.

Insomnia may be transient (short term), intermittent (on and off), or chronic (persistent). Short-term and intermittent insomnia are generally of less concern than chronic insomnia, which can last for years. Chronic insomniacs complain of tiredness, a lack of energy, difficulty concentrating, and irritability. Insomnia can make people more susceptible to common illnesses, such as colds.

What causes insomnia? Perhaps the most common cause of insomnia is a change in one's daily routine. Starting a new semester, sleeping in a strange environment, moving into a new home, and anxiety over job interviews or tests are some possible causes. But these are generally short-lived stimuli.

Chronic insomnia is often related to more serious problems. Certain diseases that cause pain and nausea, for example, may affect sleep. Chronic anxiety, for example, caused by a project at work that may take 5 months to complete, with many deadlines along the way, can keep one from falling asleep.

Another very common cause of chronic insomnia is depression. Depression usually manifests itself in the wide-awake-at-3:00-AM syndrome. A person who is depressed may fall asleep without trouble but may wake up every morning at 2:00 or 3:00 AM and may stay awake for a few hours or through the entire night. Depression comes in a variety of shapes and sizes, too—and can last for many years. Insomnia can also be caused by alcohol and caffeine consumption.

What can be done to treat chronic insomnia? If you have a medical problem, treatment of that problem can be helpful. If you are feeling anxious all of the time, the stress-relieving techniques described elsewhere in the text can be helpful, including psychological help. Depression can be treated by psychotherapy, which helps eliminate sources of the problem. Antidepressant drugs are often prescribed for insomniacs, although it may take a few weeks for symptoms of depression to begin to subside. Unfortunately, some antidepressants such as Prozac and Paxil can cause sleeplessness in about one-third of the patients.

Besides these important steps, experts recommend the following dos and don'ts for dealing with insomnia.

- Go to bed and get up at the same time each day.
- Maintain a comfortable temperature in your bedroom—not too hot or too cold.
- Sleep in a quiet, darkened room.
- Exercise regularly during the week to relieve stress. Exercise no later than the early afternoon.
- Engage in quiet tasks such as reading or listening to music in the hour or two before bedtime.
- Try a light bedtime snack with protein just before going to bed.
- Avoid caffeine, which is found in coffees, teas, and caffeinated sodas, especially late in the day.
- Don't use alcohol to stimulate sleep.
- If you can't fall asleep within 30 minutes, get out of bed, engage in some relaxing activity, and then go to bed when you feel tired.
- If you need a nap, take one early in the day, before 3:00 PM.

Finally, insomnia can be treated by over-the-counter sleeping pills such as Tylenol PM or Advil PM. They're supposedly safe and nonaddictive. Prescription medicines can also be taken. One popular prescription is Ambien. Ambien is used to treat short-term insomnia. It is usually not taken for more than 2 weeks because it can worsen insomnia. Several newer medications have also been introduced in recent years.

Another effective treatment is melatonin. Melatonin is a naturally produced hormone that comes from a small gland in the brain, the pineal gland. Melatonin comes in tablets and can be purchased in health food stores or health food sections of grocery stores or other outlets.

SUMMARY

From Cells to Organ Systems

1. The basic structural unit of all organisms is the cell. In humans, cells and extracellular material form tissues, and tissues, in turn, combine to form organs. Organs combine to form organ systems. The human body is made up of ten organ systems.

2. Four basic tissues are found in the bodies of most multicellular animals such as humans: epithelial, connective, muscle, and nervous. Organs contain all four primary tissues in varying proportions.

3. Epithelial tissues consist of two types: membranous epithelia, which form the coverings or linings of organs, and glandular epithelia, which form exocrine (glands of external secretion) and endocrine glands (glands of internal secretion).

4. Connective tissues bind other tissues together, provide protection, and support body structures. All connective tissues consist of cells and extracellular fibers.

5. Two types of connective tissue are found in the body: connective tissue proper (dense and loose connective tissue) and specialized connective tissue (cartilage, bone, and blood).

6. Cartilage consists of specialized cells embedded in a matrix of extracellular fibers and other extracellular material.

7. Bone is a dynamic tissue that provides internal support, protects organs such as the brain, and helps regulate blood calcium levels.

8. Bone consists of bone cells (osteocytes) and a calcified cartilage matrix. Two types of bone tissue exist: spongy and compact.

9. Blood consists of numerous blood cells and platelets and an extracellular fluid, called plasma.

10. Muscle is an excitable tissue that contracts when stimulated. Three types of muscle tissue are found in the human body: skeletal, cardiac, and smooth muscle.

11. Skeletal muscle is under voluntary control, for the most part, and forms the muscles that attach to bones. Cardiac muscle is located in the heart and is involuntary. Smooth muscle is involuntary and forms sheets of varying thickness in the walls of organs and blood vessels.

12. Nervous tissue consists of two types of cells: conducting and nonconducting (supportive). The conducting cells, called neurons, transmit impulses from one region of the body to another. The nonconducting cells are a type of nervous system connective tissue but also play many other important roles.

Principles of Homeostasis

13. The nervous and endocrine systems coordinate the functions of organ systems, helping the body maintain homeostasis.

14. Homeostatic systems do not maintain absolute constancy and can be upset by alterations in input or output. Imbalances can seriously affect an individual's health.

15. Homeostatic mechanisms are reflexes, occurring without conscious control. They operate by negative feedback through the nervous and endocrine systems.

Biological Rhythms

16. Many physiological processes undergo definite rhythmic changes. These natural rhythms may take place over a 24-hour period or over much longer or shorter periods.

17. The brain controls many biological cycles. A clump of nerve cells (the suprachiasmatic nucleus) is thought to play a major role in coordinating these cycles.

Health and Homeostasis

18. The physical, social, and psychological environments we live in greatly affect homeostasis and our health.

19. One of the greatest disrupter of our homeostasis is the variable work schedule, which is surprisingly common among industrialized nations.

20. Workers on alternating shifts suffer from a higher incidence of ulcers, insomnia, irritability, depression, and tension—all indications of homeostasis that's out of whack—than workers on regular shifts. Making matters worse, tired, irritable workers whose judgment is impaired by fatigue pose a threat not only to themselves, but also to society.

21. Researchers are finding ways to reset the biological clock, which could lessen the problems shift workers face.

THINKING CRITICALLY ANALYSIS

This Analysis corresponds to the Thinking Critically scenario that was presented at the beginning of this chapter.

Many of us feel tremendous sympathy for the needy. Are you one of them? Why or why not?

Those who do feel sympathy often want to help in any way they can, even if it means spending hundreds of thousands of dollars on an organ transplant.

That said, most of us recognize that there are limits to the amount of money that is available for public health. We must prioritize our spending. Do you agree or disagree?

After you have given your opinion on this matter, put yourself in the position of a needy parent whose child needs a kidney transplant to survive. How does this perspective affect your position on this issue? What lessons can be learned from this (your) self-examination?

KEY TERMS AND CONCEPTS

Adipose tissue, p. 76
Biological rhythms, p. 90
Biorhythms, p. 90
Blood, p. 80
Bone, p. 80
Brown fat, p. 76
Cardiac muscle, p. 81
Cartilage, p. 78

Circadian rhythms, p. 90
Compact bone, p. 80
Connective tissue, p. 75
Connective tissue proper, p. 75
Dense connective tissue, p. 75
Dermis, p. 75
Effector, p. 85
Elastic cartilage, p. 79

Endocrine gland, p. 75
Epidermis, p. 75
Exocrine gland, p. 74
Fat cells, p. 76
Fibroblast, p. 76
Fibrocartilage, p. 79
Homeostasis, p. 85
Hyaline cartilage, p. 79

CONCEPT REVIEW

1. Define the following terms: tissues, extracellular material, organs, and organ systems. pp. 73–85.
2. List the four basic tissue types and their subtypes. p. 73.
3. Discuss some biological examples showing how structure reflects function. p. 75.
4. Describe the two broad types of connective tissue and how they function. pp. 75–80.
5. Why do cartilage injuries repair so slowly or not at all? Bone is repaired much more easily than cartilage. Why? pp. 78–80.

6. In what way is bone part of a homeostatic system? What bone cells play a part in homeostasis? p. 80.
7. Describe skeletal muscle. What is a muscle fiber? Why do muscle fibers appear banded? What are the contractile proteins found in muscle? Are they found in all types of muscle, or just skeletal muscle? p. 81.
8. Define nervous tissue. What types of cells are found in nervous tissue? pp. 82–84.
9. What is a neuron? Describe its parts? What does each part do? pp. 82–83.
10. What is an organ system? List some examples. pp. 84–85.

11. Define homeostasis, and describe the major principles of homeostasis presented in this chapter. Use an example to illustrate your points. pp. 85–87.
12. Why do all homeostatic systems require sensors and effectors? p. 85.
13. Make a list of homeostatically controlled body functions, p. 85.
14. What are biorhythms? Are biological rhythms an exception to the principle of homeostasis? pp. 89–90.
15. Describe the biological clock. Where is it located? How does shift work upset the biological clock? How can these problems be lessened? pp. 90–91.
16. Why might time of day affect the effectiveness of a medicine you take? p. 91.

SELF-QUIZ: TESTING YOUR KNOWLEDGE

1. The internal and external linings of organs are a type of tissue known as _____. p. 74.
2. A simple epithelium consists of ___ layer(s) of cells. p. 74.
3. A _____ epithelium consists of multiple layers of cells. p. 74.
4. Endocrine glands are made from _____ epithelium during embryonic development. p. 75.
5. Ligaments and tendons are made up of _____ connective tissue. p. 75.
6. The cell responsible for making fibers in connective tissue is known as the _____. p. 75.
7. Accumulations of fat cells in the body are known as _____ tissue. p. 76.
8. _____ is poorly supplied with blood and therefore does not heal readily after injury. p. 78.

9. The most common type of cartilage is known as _____ cartilage. p. 79.
10. Bone cells or _____ are formed when osteoblasts become surrounded by bony matrix. p. 80.
11. The _____ is a type of bone cell that digests bone, releasing calcium into the bloodstream. p. 80.
12. _____ muscle is the type of muscle that is under voluntary control and is responsible for moving body parts like the arms and legs. p. 81.
13. Many organs like the stomach and uterus contain thick sheets of involuntary muscle known as _____ muscle. p. 82.
14. The nerve cell or _____ is a conducting cell found in the nervous system. p. 82.

15. The nonconducting cells in the nervous system are known collectively as _____. p. 82.
16. A collection of organs that function together is called a(n) _____. p. 84.
17. Homeostatic mechanisms in the body are typically controlled by _____ feedback. p. 85.
18. All homeostatic mechanisms rely on _____ to detect changes and _____ to respond to them. p. 85.
19. A biological rhythm lasting 24 hours is known as a _____ rhythm. p. 90.
20. Control of daily body rhythms is believed to be housed primarily in the _____ gland. p. 90.

biology.jbpub.com/chiras/8e/

The site features eLearning, an online review area that provides quizzes, chapter outlines, and other tools to help you study for your class. You can also follow useful links for in-depth information, research the differing views in the Point/Counterpoints, or keep up on the latest health news.

5 | chapter

The Circulatory System

David McMahon, a 48-year-old New York attorney, collapsed in his office one morning and was rushed to a nearby hospital. There a team of cardiologists discovered a blood clot lodged in a narrowed section of his right coronary **artery**, which supplies blood to the heart. The physicians began immediate action to prevent further damage to the oxygen-starved heart muscle. Through a small incision in McMahon's groin, they inserted a tiny plastic catheter into the artery that carries blood to the leg. They then threaded the catheter up through arteries to his heart (Figure 5-1). Once they reached the heart, his physicians guided the catheter

THINKING CRITICALLY

High blood pressure, or hypertension, is a disease that afflicts many people the world over. African Americans, however, are twice as likely to contract the disease as Caucasians. Some scientists believe that this anomaly may be the result of genetic differences in the population and that these differences arose when many blacks were captured in Africa and transported to plantations where they were forced into slave labor. More specifically, they believe that many African Americans are hypertensive today because of a rare genetic trait that helped a small subset of their ancestors survive the inhumane conditions of slavery. That trait is an inherited tendency to conserve salt within the body.

According to the hypothesis, the ancestors of African Americans who could conserve water were more likely to survive the grueling trip to the United States and the arduous conditions on the plantations. Conserving water meant they were less likely to die of dehydration. This trait, which proved to be an asset for slaves, may now be a deadly attribute among their modern descendants. As a scientist, how could you test this hypothesis?

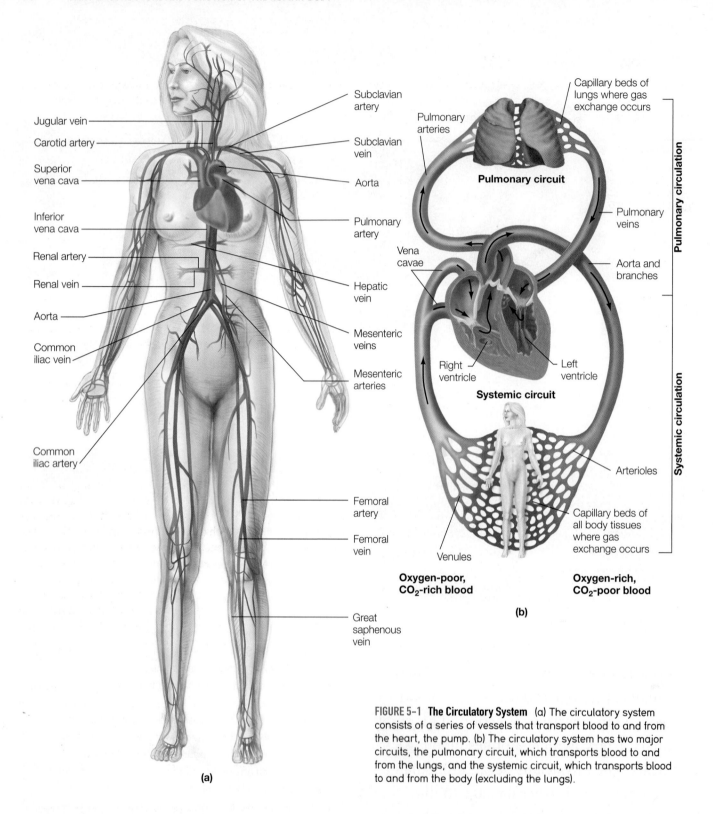

Jugular vein

Carotid artery

Superior
vena cava

Inferior
vena cava

Renal artery

Renal vein

Aorta

Common
iliac vein

Common
iliac artery

Subclavian
artery

Subclavian
vein

Aorta

Pulmonary
artery

Hepatic
vein

Mesenteric
veins

Mesenteric
arteries

Femoral
artery

Femoral
vein

Great
saphenous
vein

(a)

Capillary beds of
lungs where gas
exchange occurs

Pulmonary
arteries

Pulmonary circuit

Pulmonary
veins

Aorta and
branches

Vena
cavae

Right
ventricle

Left
ventricle

Systemic circuit

Arterioles

Capillary beds of
all body tissues
where gas
exchange occurs

Venules

**Oxygen-poor,
CO$_2$-rich blood**

**Oxygen-rich,
CO$_2$-poor blood**

(b)

Pulmonary circulation

Systemic circulation

FIGURE 5–1 The Circulatory System (a) The circulatory system consists of a series of vessels that transport blood to and from the heart, the pump. (b) The circulatory system has two major circuits, the pulmonary circuit, which transports blood to and from the lungs, and the systemic circuit, which transports blood to and from the body (excluding the lungs).

into the coronary artery, where they injected an enzyme, called *streptokinase* (strep-toe-KI-nace), through the catheter. Streptokinase dissolved the blood clot, restoring blood flow to McMahon's heart muscle.

Although David McMahon survived, many others who suffer similar attacks are not so lucky. They either arrive at the hospital too late or, if they do make it to the hospital in a timely fashion, do not receive blood-clot-dissolving agents or other treatments that prevent extensive damage to their heart muscle or, more commonly, do not receive them quickly enough.

Heart attacks strike thousands of Americans each year. They are one of a handful of diseases of the circulatory system caused by the stressful conditions of modern life, smoking, lack of exercise, poor eating habits, and other factors that upset homeostasis. This chapter examines the circulatory system—in sickness and in health. It also discusses the lymphatic system, which functions in both circulation and immune protection.

5-1 The Circulatory System's Function: An Overview

The human circulatory system consists of two parts, a muscular pump, the heart, and a network of vessels (Figure 5-1). These vessels transport blood throughout the body. Blood carries oxygen, which is taken in via the respiratory system, and nutrients, which are absorbed by the digestive system, to the cells, tissues, and organs of the body. The circulation of blood throughout the body also helps to distribute and release heat produced by cells. Besides transporting nutrients and heat, the circulatory system picks up waste from cells in the body's tissues and transports them to the organs of excretion such as the kidneys.

> **KEY CONCEPTS**
>
> The circulatory system transports blood throughout the body, distributing oxygen and nutrients and picking up cellular waste.

5-2 The Heart

The **heart** is a muscular pump in the thoracic (chest) cavity. Often referred to as the workhorse of the cardiovascular system, the heart propels blood through the 50,000 miles of blood vessels in the body. Each day, the heart beats approximately 100,000 times. If you had a dollar for every heartbeat, you would be a millionaire in 10 days. The heart doesn't beat at a steady rate, however. The heart can adjust its rate to meet the changing needs of the body, for example, during exercise.

The heart, shown in Figure 5-2, is a fist-sized organ whose walls are composed of three layers, the pericardium, the myocardium, and the endocardium. The **pericardium** (pear-ah-CARD-ee-um) forms a thin, closed sac that surrounds the heart and the bases of large vessels that enter and leave the heart. The pericardial sac is filled with a clear, slippery watery fluid that reduces the friction produced by the heart's repeated contraction. The middle layer, the **myocardium** (my-oh-CARD-ee-um), is the thickest part of the wall and is composed chiefly of cardiac muscle cells (a type of muscle cell found only in the heart). The inner layer, the **endocardium** (end-oh-CARD-ee-um), is the endothelial layer, which lines the heart chambers.

> **KEY CONCEPTS**
>
> The heart is a muscular organ that propels blood through thousands of miles of blood vessels.

The Pulmonary and Systemic Circuits

To understand how the heart functions, it is important to first know a bit about circulation. As shown in Figure 5-1b, the circulatory system consists of two distinct circuits, the **pulmonary circuit**, which carries blood to and from the lungs, and the **systemic circuit**, which transports blood to and from the rest of the body.

As you shall soon see, the pulmonary circuit has two main functions: (1) to deliver blood to the lungs so that it can be enriched with oxygen, and (2) to help the body get rid of carbon dioxide, a waste product of energy production in cells. Carbon dioxide is released from the lungs.

As shown in Figure 5-1b, the heart consists of four hollow chambers—two on the right side of the heart and two on the left. Blood is pumped through the pulmonary circuit (to and from the lungs) by the right side of the heart—the right atrium and right

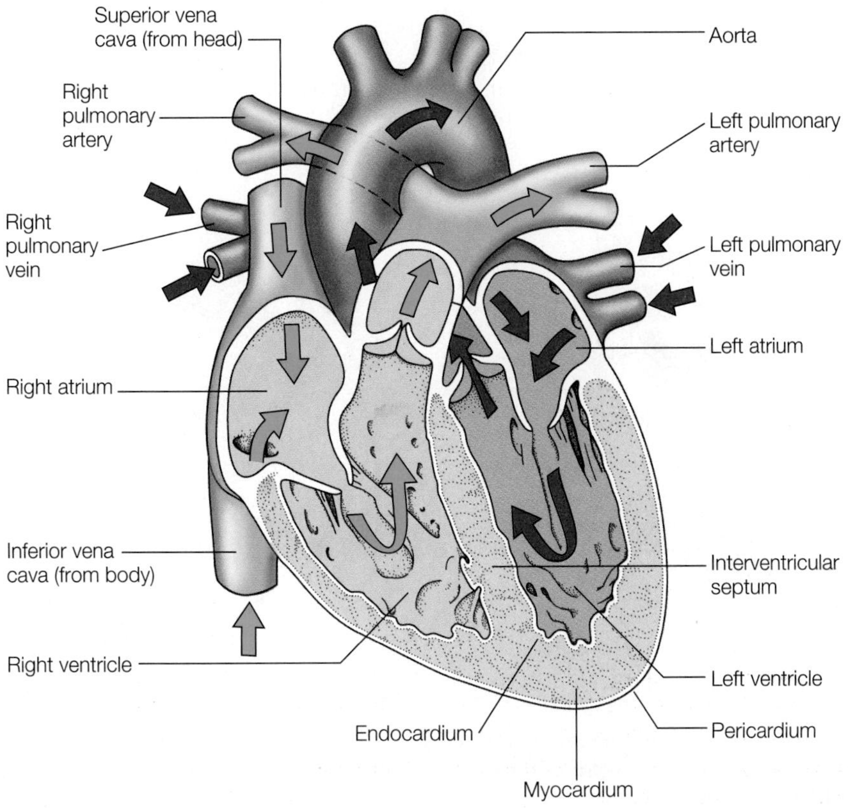

Superior vena cava (from head)
Right pulmonary artery
Right pulmonary vein
Right atrium
Inferior vena cava (from body)
Right ventricle
Endocardium
Myocardium
Aorta
Left pulmonary artery
Left pulmonary vein
Left atrium
Interventricular septum
Left ventricle
Pericardium

FIGURE 5-2 **Blood Flow through the Heart** Deoxygenated (carbon-dioxide-enriched) blood (blue arrows) flows into the right atrium from the systemic circulation and is pumped into the right ventricle. The blood is then pumped from the right ventricle into the pulmonary artery, which delivers it to the lungs. In the lungs, the blood releases its carbon dioxide and absorbs oxygen. Reoxygenated blood (red arrows) is returned to the left atrium, then flows into the left ventricle, which pumps it to the rest of the body through the systemic circuit.

ventricle. Blood is pumped through the systemic circuit by the left side of the heart—the left atrium and left ventricle. Take a moment to study the figure to confirm this fact.

Figure 5-2 illustrates the course that blood takes through the heart. Drawn in blue, blood low in oxygen and rich in carbon dioxide enters the right side of the heart from the **superior** and **inferior venae cavae** (VEEN-ah CAVE-ee, the plural of vena cava). They are part of the systemic circulation. These **veins** empty directly into the **right atrium** (A-tree-um), the uppermost chamber of the heart. The blood is pumped from here into the **right ventricle** (VEN-trick-el), the lower chamber on the right side. When the right ventricle is full, the muscles in its wall contract, forcing blood into the pulmonary arteries, which lead to the lungs.

In the lungs, this blood is oxygenated, and then returned to the heart via the pulmonary veins. The pulmonary veins, in turn, empty directly into the **left atrium**, the upper chamber on the left side of the heart. It's the first part of the systemic circuit.

Next, the oxygen-rich blood is pumped to the **left ventricle**. When it's full, the left ventricle's thick, muscular walls contract and propel the blood into the aorta (a-OR-tah). The **aorta** is the largest artery in the body. It carries the oxygenated blood away from the heart, delivering it to the cells and tissues of the entire body, including the head.

This simplified description of blood flow into, through, and out of the heart is designed to help you understand circulation. Figure 5-3 illustrates how it really works. As shown, both atria fill simultaneously. They also contract simultaneously, propelling blood into the right and left ventricles. The right and left ventricles therefore fill simultaneously, and when both ventricles are full, they contract in unison, pumping the blood into the systemic and pulmonary circuits. The coordinated contraction of heart muscle is brought about by an internal timing device, or pacemaker (described later).

> **KEY CONCEPTS**
> Blood in the human body flows through two distinct circuits: the pulmonary and the systemic.

Heart Valves

The human heart contains four valves that control the direction blood can travel in the heart, and thus ensure a steady flow from the **atria** to the **ventricles** and from the ventricles to the large vessels leading away from the heart (Figure 5-4a).

The valves between the atria and ventricles are known as **atrioventricular valves** (AH-tree-oh-ven-TRICK-u-ler). Each valve consists of two or three flaps of tissue. They are anchored to the inner walls of the ventricles by slender tendinous cords, the **chordae tendineae** (CORD-ee ten-DON-ee-ee), resembling the strings of a parachute (Figures 5-4a and b). The right atrioventricular valve, between the right atrium and right ventricle, is called the *tricuspid valve* (try-CUSS-pid), because it contains three flaps. The left atrioventricular valve is the *bicuspid valve* (bye-CUSS-pid); it contains two flaps. To remember the valves, imagine you are wearing a jersey with the number 32 on the front. This reminds you that the tricuspid valve is on the right side and the bicuspid valve is on the left.

Between the right and left ventricles and the arteries into which they pump blood are the *semilunar valves* (SEM-eye-LUNE-ir) (Figures 5-4a and 5-4c). The semilunar valves (literally, "half moon") consist of three semicircular flaps of tissue (Figures 5-4c and 5-4d).

All of the valves just discussed are one-way valves. They open when blood pressure builds on one side and close when it increases on the other. When the ventricles contract, for example, blood forces the semilunar valves open. Blood flows out of the ventricles into the large arteries. The backflow of blood causes the valve to close, preventing blood from draining back into the ventricles. The atrioventricular valves function in similar fashion.

> **KEY CONCEPTS**
> Four valves inside the human heart control the flow of blood.

Heart Sounds

When physicians listen to their patients' hearts, they are actually listening to the sounds of the heart valves closing. The

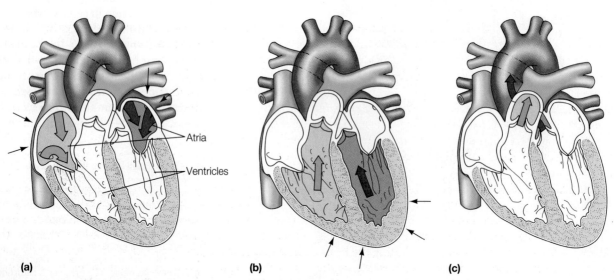

(a)   **(b)**   **(c)**

FIGURE 5-3 Blood Flow through the Heart (a) Blood enters both atria simultaneously from the systemic and pulmonary circuits. When full, the atria pump their blood into the ventricles. (b) When the ventricles are full, they contract simultaneously, (c) delivering the blood to the pulmonary and systemic circuits.

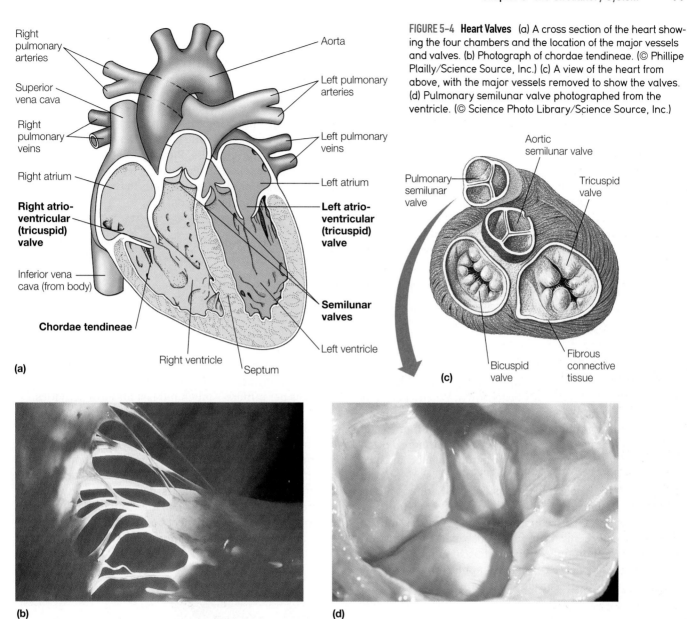

Right pulmonary arteries

Superior vena cava

Right pulmonary veins

Right atrium

Right atrio-ventricular (tricuspid) valve

Inferior vena cava (from body)

Chordae tendineae

Right ventricle

Septum

(a)

Aorta

Left pulmonary arteries

Left pulmonary veins

Left atrium

Left atrio-ventricular (tricuspid) valve

Semilunar valves

Left ventricle

FIGURE 5-4 **Heart Valves** (a) A cross section of the heart showing the four chambers and the location of the major vessels and valves. (b) Photograph of chordae tendineae. (© Phillipe Plailly/Science Source, Inc.) (c) A view of the heart from above, with the major vessels removed to show the valves. (d) Pulmonary semilunar valve photographed from the ventricle. (© Science Photo Library/Science Source, Inc.)

Aortic semilunar valve

Pulmonary semilunar valve

Tricuspid valve

Bicuspid valve

Fibrous connective tissue

(c)

(b)

(d)

noises they hear are called the *heart sounds* and are often described as "LUB-dupp." The first heart sound (LUB) results from the closure of the atrioventricular valves. It is longer and louder than the second heart sound (dupp), produced when the semilunar valves shut.

Interestingly, the right and left atrioventricular valves do not close at precisely the same time. Nor do the semilunar valves. Thus, by careful placement of the stethoscope, a physician can listen to each valve individually to determine whether it is functioning properly.

In some individuals, diseases alter the function of the valves. This, in turn, may dramatically decrease the efficiency of the heart and the circulation of blood. Rheumatic (RUE-mat-tick) fever, for example, is caused by a bacterial infection that affects many parts of the body, including the heart. Although it is rare in the more affluent countries, rheumatic fever is a significant problem in the less developed countries. Rheumatic fever begins as a sore throat caused by certain types of streptococcus (STREP-toe-COCK-iss) bacteria. The

sore throat—known as *strep throat*—is usually followed by general illness. During this infection, the immune system forms antibodies to combat the bacteria. Antibodies are proteins that circulate in the bloodstream seeking out and destroying invading bacteria. Unfortunately, these antibodies bind to proteins on the cells of the tissue lining of heart valves. This damages the valves, preventing them from closing completely. This, in turn, causes blood to leak back into the atria and ventricles after contraction and results in a distinct "sloshing" sound, called a *heart murmur*. This condition reduces the efficiency of the heart and causes the organ to work harder to make up for the inefficient pumping. Increased activity, in turn, causes the muscular walls of the heart to enlarge. In severe cases, it can wear out the heart, resulting in a condition called *heart failure*, The heart does not stop beating, it just can't pump enough blood to keep up with the body's need for oxygen. Patients suffer from a number of symptoms including lung congestion—the buildup of fluid in the lungs that results in shortness of breath—and fluid and water retention throughout much of the rest of the body that leads to swollen ankles, legs, and abdomen. Reduced blood flow to vital organs and muscles makes patients feel tired and weak. Reduced blood flow to the brain leads to dizziness and confusion.

Fortunately, damaged valves can be replaced by artificial implants.

Tumors and scar tissue on the valves have an opposite effect—that is, they reduce blood flow through the heart valves. This condition is known as *valvular stenosis* (sten-OH-siss; from the Greek word *steno*, meaning narrow). Valvular stenosis prevents the ventricles from filling completely. As in valvular incompetence or heart murmur, the heart must beat faster to ensure an adequate supply of blood to the body's tissues. This acceleration also puts additional stress on the organ.

> **KEY CONCEPTS**
> The closing of the four heart valves produce distinct sounds that can be used to diagnose heart disease.

The Heart's Pacemaker

The human heart functions at different rates under different conditions. At rest, it generally beats slowly. When a person is working hard, it beats faster. This variation in heart rate helps the body adjust for differences in the oxygen requirements of muscle cells. The more active a muscle cell is, the more oxygen it needs. Heart rate is controlled by a number of mechanisms.

One of the most important mechanisms is an internal pacemaker, the **sinoatrial (SA) node** (SIGN-oh-A-tree-ill noad; Figure 5-5). Located in the wall of the right atrium, the SA node is composed of a clump of specialized cardiac muscle cells. These cells contract spontaneously and rhythmically. Each contraction produces a tiny electrical impulse, akin to those produced by nerve cells. This impulse spreads rapidly from the SA node to the cardiac muscle and then from muscle cell to muscle cell in both atria. Because cardiac muscle cells are tightly joined, and because the impulse travels quickly, the two atria contract almost simultaneously.

The electrical impulse transmitted throughout the atria next passes to the ventricles. Its passage, however, is briefly slowed by a barrier of unexcitable tissue that separates the atria from the ventricles. The impulse is delayed approximately 1/10 second, giving the atria time to contract and the ventricles time to fill. After this brief delay, the impulse is channeled through a second mass of specialized muscle cells,

the **atrioventricular (AV) node**, shown in Figure 5-5. From the AV node, the impulse travels along a tract of specialized cardiac muscle cells, known as the **atrioventricular bundle**.

As Figure 5-5 shows, the atrioventricular bundle divides into two branches (called the bundle branches) that travel on either side of the wall separating the ventricles. They carry the electrical impulse into the ventricles. Along their course, the bundle branches give off smaller branches that terminate on specialized muscle cells, **Purkinje fibers** (per-KIN-gee), named after the scientist who discovered them. These fibers terminate on the cardiac muscle cells in the walls of the ventricles, stimulating them to contract.

> **KEY CONCEPTS**
> Heart rate is controlled by an internal pacemaker that is, in turn, controlled by the nervous system.

Controlling Heart Rate

Left to their own devices, cardiac muscle cells would contract independently, creating a disorderly, ineffective pumping action. The SA node, however, imposes a single rhythm on all of the heart muscle cells. If the SA node were isolated from outside influences, the human heart would beat at about 100 times per minute—much too fast for most activities.

To bring the heart rate in line with body demand, the SA node must be slowed down. The SA node is curbed by nerve impulses transmitted by nerves arriving from a control center in the brain. At rest or during nonstrenuous activity, these impulses slow the heart to about 70 beats per minute, all that's needed to supply the tissues of the body with oxygen and rid them of wastes. During exercise or stress, the heart rate must increase to meet body demands. In order for this to occur, the decelerating impulses from the brain must

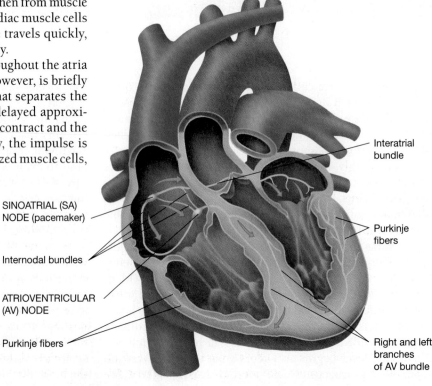

FIGURE 5-5 Conduction of Impulses in the Heart The sinoatrial node is the heart's pacemaker. Located in the right atrium, it sends timed impulses into the atrial heart muscles, coordinating muscle contraction. The impulse travels from cell to cell in the atria, then passes to the atrioventricular node and into the ventricles via the atrioventricular bundle and its two branches, which terminate on the Purkinje fibers.

SINOATRIAL (SA) NODE (pacemaker)

Internodal bundles

ATRIOVENTRICULAR (AV) NODE

Purkinje fibers

Interatrial bundle

Purkinje fibers

Right and left branches of AV bundle

FIGURE 5-6 **The Electrocardiogram** (a) This patient taking a treadmill test to check his heart's performance is wired to a meter that detects electrical activity produced by the heart. (© Ken Sherman/Medichrome/Visuals Unlimited.) (b) An electrocardiogram.

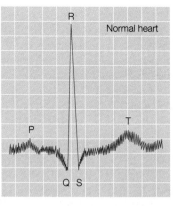

P = atrial depolarization, which triggers atrial contraction.

QRS = depolarization of AV node and conduction of electrical impulse through ventricles. Ventricular contraction begins at R.

T = repolarization of ventricles.

P to R interval = time required for impulses to travel from SA node to ventricles.

(a)   **(b)**

be reduced. (We'll explore this important mechanism shortly.)

Other nerves also influence heart rate. These nerves carry impulses that can speed up the heart. They allow the heart to achieve rates of 180 beats or more, for example, during intense exercise when the body cells' demand for oxygen is great.

Several hormones also play a role in controlling heart rate. One of these is **epinephrine** (EP-eh-NEFF-rin), also known as **adrenalin**. This hormone is produced by the adrenal glands located on top of the kidneys. They secrete this hormone during stress or exercise. Epinephrine boosts heart rate, increasing the flow of blood through the body.

> **KEY CONCEPTS**
>
> The pacemaker, the SA node, operates on its own, but is slowed during periods of inactivity or accelerated during periods of activity by impulses from the brain.

Measuring the Heart's Electrical Activity

When the electrical impulses from the SA node that cause heart muscle cells to contract arrive, they cause electrical changes in the cardiac muscle cells. The changes can be detected by electrodes, small metal plates connected to wires and a voltage meter (Figure 5-6a). The electrodes are placed on a person's chest. The resulting reading on a voltmeter is called an **electrocardiogram** (ECG or, more commonly, EKG; Figure 5-6b).

For a normal person, the electrical tracing has three distinct waves (Figure 5-6b). The first wave, the P wave, represents the electrical changes occurring in the atria of the heart. The second wave, the QRS wave, is a record of the electrical activity taking place during ventricular contraction, and the third wave, the T wave, is a recording of electrical activity occurring as the ventricles relax. Certain diseases of the heart may disrupt one or more waves of the EKG, making the EKG a valuable diagnostic tool for heart doctors (cardiologists).

> **KEY CONCEPTS**
>
> Electrical activity in the heart can be measured on the surface of the chest; this helps doctors determine the health of the heart.

5-3 The Blood Vessels

The heart pumps blood throughout the body in a series of connected blood vessels belonging to either the pulmonary or systemic circuits described earlier. (For a discussion of some of the discoveries that led to our understanding of circulation, see Scientific Discoveries 5-1.)

Arteries transport blood away from the heart and branch many times as they course through the body, forming smaller and smaller vessels that serve all of the organs of the body. The smallest of all arteries is the **arteriole** (are-TEAR-ee-ol). As shown in Figures 5-7a and b, arterioles empty into **capillaries** (CAP-ill-air-ees), tiny, thin-walled vessels that allow nutrients to enter tissues and wastes to escape with ease.

Blood flows out of the capillaries into the smallest of all veins, the **venules**. Venules, in turn, converge to form small

veins. They unite with other small veins, in much the same way that small streams unite to form a river. Blood in veins flows toward the heart. Veins and arteries generally run side by side throughout the body.

Figure 5-8 shows a cross section of an artery and a vein. As illustrated, these two vessels are fairly similar, although veins tend to be smaller and have thinner walls. Both consist of three layers: (1) an external layer of connective tissue, which binds the vessel to surrounding tissues; (2) a middle layer, which is primarily made of smooth muscle; and (3) an internal layer, which is composed of a layer of flattened cells, the endothelium. It also contains a thin, nearly indiscernible layer of connective tissue that binds the endothelium to the middle layer. Figure 5-9 shows these layers and the scientific names assigned to them. With this

Scientific Discoveries that Changed the World

5-1 The Circulation of Blood in Animals
Featuring the Work of Harvey and Hales

The seventeenth-century British physician William Harvey is generally credited with the discovery of the circulation of blood in animals (Figure 1). Harvey is known as a scientist with a short temper who wore a dagger in the fashion of the day, which he reportedly brandished at the slightest provocation. He was probably not the kind of professor you might "argue" with over grades.

Temperament aside, Harvey is generally honored for his work on the role of the heart and the flow of blood in animals and is often praised as a pioneer of scientific method. His application of quantitative procedures to biology, some say, ushered in the modern age of this science.

In Harvey's medical school days, anatomists thought the intestines produced a substance called *chyle*. Chyle, they thought, was a fluid derived from the food people ate. It was passed from the intestines to the liver. The liver, in turn, converted the chyle to venous blood, and then distributed the blood through arteries and veins.

As a medical student, Harvey was told that blood oozing through arteries and veins supplied organs and tissues with nourishment. He was also told that the blood merely ebbed back to the heart and lungs, where impurities were removed. In other words, there was no form of circulation, just an ebb and flow similar to the tides. These ideas had been proposed by the Greek physician Galen fourteen centuries earlier and persisted nearly without challenge until Harvey's time.

As a teacher in the Royal College of Physicians in 1616, Harvey began to describe the circulation of blood, based on his experiments on and observations of animals. He described the muscular character of the heart and the origin of the heartbeat. He also demonstrated that the pulse felt in arteries resulted from blood being pumped through arteries by the heart. Furthermore, he described the pulmonary and systemic circuits and proposed that blood flowed to the tissues and organs of the body via the arteries and returned via the veins.

A brief examination of one of his experiments illustrates that even though Harvey is a key figure in the history of biological science, some of his work was based on poor assumptions and inaccurate observations. As an example, consider the work he used to rebut Galen's hypothesis that the blood was produced by the food people ate.

Harvey first approximated the amount of blood the heart ejected with each heartbeat (stroke volume), then determined the pulse rate. He called on earlier observations of a heart from a human cadaver to determine stroke volume. At that time, he had noted that the left ventricle contained more than 2 ounces of blood, and then, for reasons not entirely clear to historians of science, he hypothesized that the ventricle ejected "a fourth,

a fifth, a sixth, or only an eighth" of its contents. (Today, studies indicate that the heart ejects nearly all of its contents.) Based on this assumption, Harvey estimated that the stroke volume was about 3.9 grams of blood per beat. Modern estimates put it at 89 grams per beat.

Harvey also made a grave error in determining pulse. His value of 33 beats per minute is about half of the actual rate in humans. No one knows how he could have been so wrong. Armed with two erroneous measurements, Harvey derived a figure for the amount of blood that circulated through the body that was 1/36 of the lowest value accepted today.

Regardless, Harvey "proved" his point—that each half-hour the blood pumped by the heart far exceeds the total weight of blood in the body. From this he concluded that blood must be circulated. It is not, as Galen proposed, produced by the food we eat. The amount of food one eats could not produce blood in such volume.

Harvey debunked another falsehood perpetrated through the centuries—the Galenic myth that blood flowed into the extremities in both arteries and veins. Harvey first wrapped a bandage around an extremity. This obstructed the flow of blood through the veins but not the arteries. He noted that the veins swelled because, as he conjectured, blood was being pumped into them via underlying arteries and there was nowhere for the blood to go. Tightening the bandage further cut the blood flow in the arteries as well and thus prevented the veins from swelling. From these observations, Harvey correctly surmised that the arteries deliver blood to the extremities and the veins return it to the heart.

Harvey's work laid the foundation for modern cardiovascular physiology, but left many questions unanswered. Many of these were addressed by the highly industrious English biologist Stephan Hales, who was born a century later. In a long series of scientifically rigorous experiments on horses, dogs, and frogs, Hales explored many aspects of the cardiovascular system. After settling many of the unanswered questions left by Harvey, Hales went on to study plant physiology and is perhaps best known for his work on the circulation of sap in plants.

FIGURE 5-1 William Harvey This flamboyant scientist greatly advanced our knowledge of circulation in animals. (© National Library of Medicine.)

overview in mind, let's examine each type of blood vessel in a little more detail.

> **KEY CONCEPTS**
> Blood flows out of the heart in arteries that branch many times, eventually supplying thin-walled capillaries in tissues of the body and then returns to the heart via veins.

Arteries and Arterioles

As mentioned above, the largest of all arteries is the aorta, a massive vessel that carries oxygenated blood from the left ventricle of the heart to the rest of the body. The aorta loops over the back of the heart, before descending through the chest and abdomen. It gives off large branches along its way from the heart. These branches carry blood to the head, the

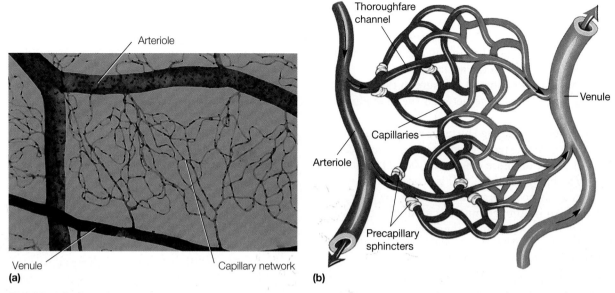

FIGURE 5-7 Capillary Network A network of capillaries between the arteriole and the venule delivers blood to the cells of body tissues (not shown). (photo © D. W. Fawcett-Kuwabara/Visuals Unlimited.)

arms and legs, and organs of the body (Figure 5-1). The very first branches of the aorta are the coronary arteries. They deliver blood to the heart.

The aorta and many of its chief branches contain numerous wavy elastic fibers interspersed among the smooth muscle cells of the wall. For this reason, they're called *elastic arteries*. As blood pulses out of the heart, these arteries expand to accommodate it. Like a stretched rubber band, the elastic fibers cause the arterial walls to recoil. This helps push the blood along the arterial tree, maintaining an even flow of blood through the capillaries.

The elastic arteries branch to form smaller vessels, the *muscular arteries*. Muscular arteries contain fewer elastic fibers but still expand and contract with the flow of blood.

You can feel this expansion and contraction in the arteries lying near the skin's surface in your wrist and neck. It's the pulse that health care workers use to measure heart rate.

The smooth muscle of the muscular arteries responds to a number of stimuli, including nerve impulses, hormones, carbon dioxide, and lactic acid. These stimuli cause the blood vessels to open or close to varying degrees. Opening of blood vessels allows the body to adjust blood flow through its tissues like muscle to meet increased demands for nutrients and oxygen. Arterioles in muscles, for instance, open up or dilate (DIE-late) when a person exercises. This increases blood flow to the muscle, providing oxygen and glucose needed for greater physical exertion. Closing restricts blood flood. Arterioles in the skin close down on cold days to reduce heat loss through the skin.

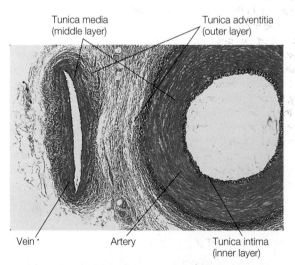

FIGURE 5-8 Artery and Vein A cross section through a vein shows that the muscular layer, the tunica media, is much thinner than it is in an artery. Veins typically lie alongside arteries and, in histological sections such as these, usually have irregular lumens (cavities). (© Cabisco/Visuals Unlimited.)

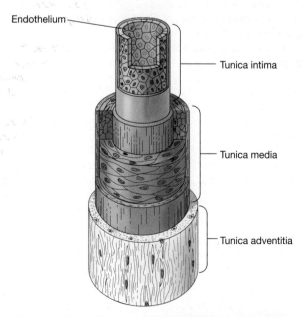

FIGURE 5-9 General Structure of the Blood Vessel The artery shown here consists of three major layers, the tunica intima, tunica media, and tunica adventitia.

Blood Pressure. The force that blood applies to the walls of a blood vessel is known as the **blood pressure**. Blood pressure changes in relation to a person's activity and stress levels. Blood pressure also varies throughout the cardiovascular system, being the highest in the aorta and the lowest in the capillaries and veins. Low blood pressure in the capillaries increases the rate of exchange between the blood and the tissues.

Blood pressure is measured by using an inflatable device with the tongue-twisting name of *sphygmomanometer* (SFIG-mo-man-OM-a-ter), or, more commonly, *blood pressure cuff* (Figure 5-10a). The blood pressure cuff is wrapped around the upper arm. A stethoscope is positioned over the artery just below the cuff. Air is pumped into the cuff until the pressure stops the flow of blood through the artery (Figure 5-10b). The pressure in the cuff is then gradually reduced as air is released. When the blood pressure in the artery exceeds the external pressure of the cuff, the blood starts flowing through the vessel once again. This point represents the systolic pressure (sis-STOL-ick), the peak pressure at the moment the ventricles contract. **Systolic pressure** is the higher of the two numbers in a blood pressure reading (120/70, for example). The pressure at the moment the heart relaxes to let the ventricles fill again is the **diastolic pressure** (DIE-ah-STOL-ick) and is the lower of the two readings. It is determined by continuing to release air from the cuff until no arterial pulsa-

tion is audible. At this point, blood is flowing continuously through the artery. Newer models of blood pressure cuffs no longer require the use of a stethoscope (Figure 5-10a).

A typical reading for a young, healthy adult is about 120/70, although readings vary considerably from one person to the next. As a person ages, blood pressure tends to rise. Thus, a healthy 65-year-old might have a blood pressure reading of 140/90.

> **KEY CONCEPTS**
>
> Arteries branch many times forming smaller and smaller vessels known as arterioles; they transport blood into capillaries that surround the cells of the body.

Capillaries

As described above, the heart and arteries transport blood to the capillaries. Capillaries form branching networks, known as **capillary beds**, among the cells of tissues and organs of the body. The extensive branching of capillaries brings them in close proximity to body cells. It is in these extensive capillary beds that wastes and nutrients are exchanged between the cells of the body and the blood. The walls of the capillaries consist of flattened endothelial cells. These cells permit dissolved substances to pass through them with ease.

If you could remove all of the capillaries from one human and line them up end to end, they would extend over

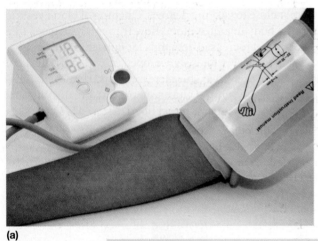

(a)

FIGURE 5–10   Blood Pressure Reading (a) A sphygmomanometer (blood pressure cuff) is used to determine blood pressure. (© Paul Maguire/ShutterStock, Inc.) (b) As shown, the blood pressure (indicated by the red line) rises and falls with each contraction of the heart. When the pressure in the cuff exceeds the arterial peak pressure, blood flow stops (1). No sound is heard. Cuff pressure is gradually released. When pressure in the cuff falls below the arterial pressure, blood starts flowing through the artery once again. This is the systolic pressure (2). The first sound will be heard. Cuff pressure continues to drop. When cuff pressure is equal to the lowest pressure in the artery, the artery is fully open and no sound is heard (3). This is the diastolic pressure.

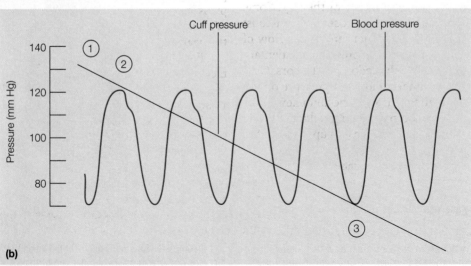

(b)

80,000 kilometers (50,000 miles)—enough to circle the globe at the equator two times.

As blood flows into a capillary bed, nutrients, gases, water, and hormones carried in the blood immediately begin to diffuse out of the tiny thin-walled vessels. Meanwhile, water-dissolved wastes in the tissues, such as carbon dioxide, begin to diffuse inward.

Blood flow through capillaries is controlled by the expansion and contraction of the arterioles that "feed" capillary beds. When the arterioles open, more blood flows into the capillaries. When they constrict, blood flow is restricted. The expansion and contraction of the arterioles feeding capillary beds also helps to regulate body temperature. On a cold winter day, for example, the arterioles close down, restricting blood flow through the capillaries to conserve body heat (that's one reason why your skin feels colder on a cold day). Just the reverse happens on a warm day. The flow of blood through the skin increases, releasing body heat and often creating a pink flush. Capillaries therefore help to maintain homeostasis.

> **KEY CONCEPTS**
>
> Capillaries forms branched networks called capillary beds; these thin-walled vessels allow nutrients to flow out of the blood and waste products to flow in.

Veins and Venules

Blood leaves the capillary beds stripped of its nutrients and loaded with cellular wastes. As it drains from the capillaries, the blood enters the smallest of all veins, the venules. As noted earlier, venules converge to form small veins, which join with others to form larger and larger vessels. Eventually, all blood returning to the heart in the systemic circuit enters the superior or inferior vena cava, the two main veins that drain into the right atrium of the heart (Figure 5-2). These vessels drain the upper and lower parts of the body, respectively.

As noted earlier, blood pressure in the veins is low, and veins have relatively thin walls with fewer smooth muscle cells than arteries (Figure 5-8). Because the veins' walls are so thin, obstructions can cause blood to pool in them. When blood pools in the obstructed veins for extended periods, it forms bulges called **varicose veins** (VEAR-uh-cose; Figure 5-11). The veins may be bluish color, red, or skin colored. They appear on the thighs, calves, and inside of the legs. Some people inherit a tendency to develop varicose veins, but most cases can be attributed to factors that reduce the flow of blood back to the heart such as obesity, pregnancy, sedentary lifestyles, or, much less commonly, abdominal tumors. Obesity, pregnancy, and abdominal tumors put pressure on the inferior vena cava as it courses through the abdomen, slowing blood flow. A sedentary lifestyle simply reduces muscle activity that helps pump blood from the legs up to the heart, described shortly.

Varicose veins can result in considerable discomfort. The restriction of blood flow, for example, may result in throbbing, aching pains, and considerable discomfort. They also result in the buildup of fluid, **edema** (uh-DEEM-ah; swelling), in the ankles and legs. Varicose veins that form in the veins in the wall of the anal canal, the internal hemorrhoidal veins (hem-eh-ROID-il), result in a condition known as *hemor-*

FIGURE 5-11 Varicose Veins Any restriction of venous blood flow to the heart causes veins to balloon out, creating bulges commonly known as varicose veins. (© Audie/ShutterStock, Inc.)

rhoids (hem-eh-ROIDS). Because these veins are supplied by numerous pain fibers, this condition can be painful.

Varicose veins occur in women and to a slightly lesser extent in men. They affect about half of all people 50 years of age and older. Varicose veins can be treated by a number of techniques, including lifestyle changes (increased activity and weight loss), tight-fitting stockings, and sclerotherapy (SKLER-o-ther-a-pee). Sclerotherapy is the most common treatment. During this procedure, a doctor injects a chemical into the vein. This chemical causes the walls of the vein to swell. The venous walls soon come together, sealing shut. This terminates the flow of blood through the vein, and the vein is turned into scar tissue. Small varicose veins can also be blasted with lasers, a treatment that destroys them, or surgically removed. Deep varicose veins can be sealed internally using lasers.

How Veins Work. Blood pressure caused by the beating heart propels blood in the arteries into the capillaries. But blood pressure is extremely low in veins, as shown in Figure 5-12. With such low pressure, how do the veins return blood to the heart?

For blood in veins above the heart, gravity is the chief means of propulsion. For veins below the heart, return flow depends on blood pressure and the movement of body parts, notably muscles, which "squeezes" blood upward. As you walk to class, for example, the contraction of muscles in your legs helps to pump the blood in the veins back to the heart.

Blood flow in veins also relies on valves in the veins. Valves are flaps of tissue that span the veins and prevent the backflow of blood. As illustrated in Figure 5-13, the flaps of the veins resemble those found in the heart. Just as in the valves of the heart, blood pressure, however slight, pushes the flaps open (Figure 5-13). This allows the blood to move forward. As the blood fills the segment of the vein in front of the valve, it pushes back on the valve flaps and forces them shut.

health**note**

5-1 Modern Medicine Turns to Leeches

Donna Amark is a college professor at a Midwestern university. A few years ago, Donna lost her thumb in an unfortunate accident involving a neighbor's dog. Donna stopped by to visit the new neighbor. When no one answered the door, she went to the back yard. There she was confronted by the neighbor's dog, which was chained

to a large maple tree. The dog seemed friendly, but as Donna approached he lunged toward her. She side-stepped the dog and grabbed the chain. Unfortunately, the chain had wrapped around her thumb. When the dog lunged at her a second time, the sudden jerk of the chain severed her thumb, ripping it from her hand, tendons and all.

Paramedics who rushed Donna to the hospital were unable to find the severed thumb, so she elected to have a new procedure. Doctors removed her big toe, tendons and all, and inserted it into the hand where her thumb had once been.

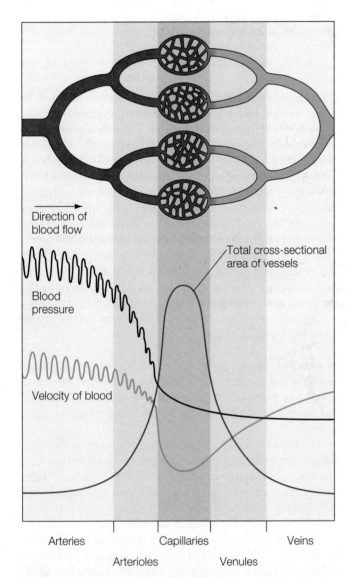

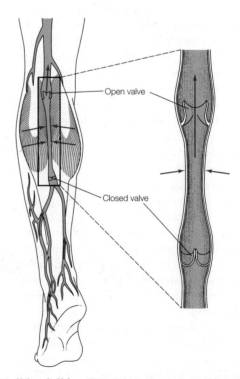

To locate a valve, hold your arm out in front of you and make a fist. The veins should stick out or at least be apparent beneath the skin of your forearm. To locate a valve in the superficial veins on your forearm, press gently on a vein, then run your finger along the vein toward your wrist. You will note that the vein collapses behind your finger until it crosses a valve. (See Health Note 5-1 for a discussion of one novel way to restore blood flow interrupted by injury.)

> **KEY CONCEPTS**
>
> Venules drain capillary beds, then join to form larger vessels, the veins, that return blood to the heart.

FIGURE 5-12 Blood Pressure in the Circulatory System Blood pressure declines in the circulatory system as the vessels branch. Arterial pressure pulses because of the heartbeat, but pulsation is lost by the time the blood reaches the capillary networks, creating an even flow through body tissues. Blood pressure continues to decline in the venous side of the circulatory system.

FIGURE 5-13 Valves in Veins The slight hydrostatic pressure in the veins and the contraction of skeletal muscles propel the blood along the veins back toward the heart. The one-way valves stop the blood from flowing backward.

Her surgery went well. However, despite the surgeon's tedious work of reconnecting blood vessels, Donna's translocated thumb showed no signs of restored blood flow. Soon after, it became apparent that she was going to lose her new thumb. The surgeons in charge made what seemed at first to be a wild suggestion: use leeches to restore the blood flow.

Leeches, it turns out, have become one of modern medicine's most effective tools. They're routinely used in Great Britain and the United States to re-establish blood flow in reattached ears, fingers, and toes. A leech has a wicked looking mouth with 100 teeth. When it is attached to the skin, it bites in. But you don't feel any pain because the leech releases a tiny amount of natural anesthetic that deadens the sensory nerves in the area. The leech also releases a natural anticoagulant into

the wound so blood flows uninterrupted into its mouth. When placed on a reattached finger or toe, the germ-free leech attaches and begins sucking. The patient feels nothing, but in short order oxygen-rich blood begins to flow through the finger or toe, bringing the body part back to life better. The leech is left in place for some time to ensure adequate flow. After it is removed, blood continues to flow through the reattached body part. If further treatment is needed, the leech is placed on the re-attached part again.

Leeches are raised in captivity in sterile environments and shipped to hospitals when needed. In Donna's case, the leeches worked miracles. She's alive and well with functional hands, thanks to the lowly leech, modern medicine's latest and greatest new tool.

5-4   The Lymphatic System

The lymphatic system consists of a branching network of vessels throughout the body that helps transport fluids from tissues back to the circulatory system; it also contains lymph nodes that play an important role in protecting the body against microbial invasion.

The **lymphatic system** is an extensive network of vessels and glands (Figure 5-14). It is functionally related to two systems: the circulatory system and the immune system. This section examines the role it plays in circulation.

You may recall that the cells of the body are bathed in a liquid called *interstitial fluid* (in-ter-STISH-il). Interstitial fluid provides a medium through which nutrients, gases, and wastes can diffuse between the capillaries and the cells.

FIGURE 5-14 **The Lymphatic System** (a) The lymphatic system consists of vessels that transport lymph, excess tissue fluid, back to the circulatory system. (b) Lymph is picked up by lymphatic capillaries that drain into larger vessels. Like the veins, the lymphatic vessels contain valves that prohibit backflow. Lymph nodes are interspersed along the vessels and serve to filter the lymph.

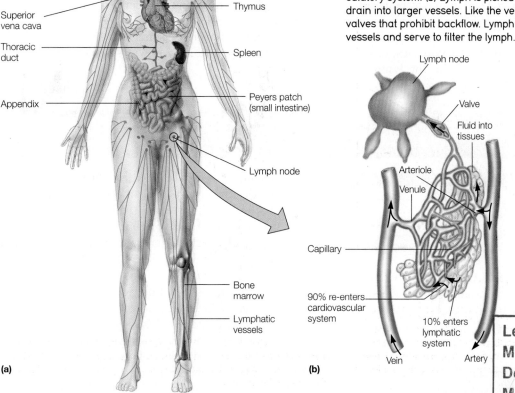

(a)

Right lymphatic duct
Tonsil
Lymph node
Subclavian vein
Superior vena cava
Thymus
Thoracic duct
Spleen
Appendix
Peyers patch (small intestine)
Lymph node
Bone marrow
Lymphatic vessels

(b)

Lymph node
Valve
Fluid into tissues
Arteriole
Venule
Capillary
90% re-enters cardiovascular system
10% enters lymphatic system
Vein
Artery

Tissue fluid is constantly replenished by the capillaries. The flow of water out of the capillaries, however, normally exceeds the return flow by about 3 liters per day. The "excess" water is picked up by small **lymph capillaries** in tissues. Like the capillaries of the circulatory system, these vessels have thin, highly permeable walls through which water and other substances pass with ease.

Lymph drains from the capillaries into larger ducts. These vessels, in turn, merge with others, creating larger and larger ducts that empty into the large veins at the base of the neck.

Lymph moves through the vessels of the lymphatic system in much the same way that blood is transported in veins. In the upper parts of the body, it flows by gravity. In areas below the heart, it is propelled largely by muscle contraction. Breathing, for example, pumps lymph out of the chest. Walking pumps the lymph out of the extremities. Lymphatic flow is also assisted by valves similar to those in the veins.

The lymphatic system also consists of several lymphatic organs: the lymph nodes, the spleen, the thymus, and the tonsils. The lymphatic organs function primarily in immune protection. Lymph nodes, however, play a role worth considering here.

Varying in size and shape, **lymph nodes** are found in association with lymphatic vessels in small clusters in the armpits, groin, neck, and other locations (Figure 5-14). A lymph node consists of a network of fibers and irregular channels that slow down the flow of lymph. As lymph passes through the lymph nodes, the fibers filter out bacteria, viruses, and cellular debris. Lining the channels are numerous cells (macrophages) that engulf (phagocytize) microorganisms and other materials, helping cleanse the body and maintain healthy homeostasis.

Normally, lymph is removed from tissues at a rate equal to its production. In some instances, however, lymph production exceeds the capacity of the system. A burn, for example, may cause extensive damage to blood capillaries, causing them to leak excessively and overwhelm the lymphatic

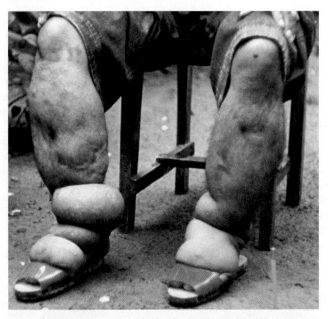

FIGURE 5-15 Elephantiasis A parasitic worm that invades the body stimulates the production of scar tissue, which blocks the flow of lymph through the nodes, causing tissue fluid to build up. This condition is known as elephantiasis. (Courtesy of CDC.)

capillaries. This "flood" results in a buildup of fluid in tissues, that is, *edema*.

Lymphatic vessels may also become blocked. One of the most common causes of blockage is an infection by tiny parasitic worms transmitted to humans by mosquitoes in the tropics. The worm larvae (an immature form) enter lymph vessels and take up residence in the lymph nodes. An inflammatory reaction causes the buildup of scar tissue in the nodes that blocks the flow of lymph. After several years, the lymphatic drainage of certain parts of the body may be almost completely obstructed. A leg may swell so much, in fact, that it weighs as much as the rest of the body (Figure 5-15). This condition is known as *elephantiasis* (el-uh-fun-TIE-ah-sis).

5-5 Cardiovascular Diseases: Causes and Cures

KEY CONCEPTS

The heart and blood vessels are subject to a variety of potentially life-threatening diseases.

Myocardial Infarction

Each year, more than one million Americans suffer from a heart attack. Heart attacks are caused by blood clots that block one or more of the coronary arteries, usually those narrowed by atherosclerotic plaque (discussed next). Coronary arteries, you may recall, are the very first branches of the aorta and supply blood directly to the heart. Blood clots in the coronary arteries restrict the flow of blood to the heart muscle, cutting off the supply of oxygen and nutrients. This

deprivation can damage and even kill the heart muscle cells. The condition is known as a **myocardial infarction**.

As noted in Health Note 5-1, the formation of atherosclerotic plaque results from a combination of factors: stress, overeating, poor diet, lack of exercise, smoking, heredity, and several others. The narrowing of a coronary artery by plaque does not usually block the flow of blood enough to cause a heart attack, however, unless it is quite severe—around 80% or 90% blockage. Nonetheless, less severe narrowing does make the vessel more susceptible to blood clots. Clots often form inside the vessel at the site of narrowing. Clots may also form in other parts of the body. They break loose and flow through the cardiovascular system where they can lodge in the narrowed coronary arteries, obstructing blood flow to heart muscle.

Heart muscle begins to heal soon after a heart attack and takes about eight weeks to return to normal. As in other injuries, however, scar tissue forms in the damaged area. Unfortunately, scar tissue does not contract like cardiac muscle. This reduces the heart's ability to pump blood. How much is lost depends on the severity of the heart attack—the amount of scarring and its location. The outcome of heart attacks varies. If the size of the damaged area is small and if the change in electrical activity of the heart is minor and transient, a heart attack is usually not fatal. If the damage is great or electrical activity is severely disrupted, a heart attack can be fatal. Getting help quickly lessens the chances of severe damage to the heart. Giving a person an aspirin during a heart attack also lessens the damage (it reduces blood clotting).

Heart attacks can occur quite suddenly, without warning, or may be preceded by several weeks of **angina** (an-GINE-ah), pain felt when the supply of oxygen to the myocardium is reduced. Angina is a dull, heavy, constricting pain that appears when an individual is active but disappears when he or she ceases the activity. Angina appears in the center of the chest and can spread to a person's throat, upper jaw, back, and arms (usually just the left one).

Health Tip 5-1

To cut down on what you eat, use smaller plates and glasses.

Why?

Studies show that Americans currently consume, on average, 27% more calories than they did in 1970. One reason for this is larger portion sizes. The larger portion size is partly the result of our using larger glasses, bowls, and plates. Studies show that when using a larger plate, people serve up and consume larger portions. You'll eat about 25% more with a larger plate.

Angina also may be caused by stress and exposure to carbon monoxide, a pollutant that reduces the oxygen-carrying capacity of the blood. Angina begins to show up in men at age 30 and nearly always is caused by coronary artery disease. In women, angina tends to occur at a much later age.

KEY CONCEPTS
Blockages of the arteries supplying heart muscle can lead to a heart attack known as a myocardial infarction.

Preventing Heart Attacks

Heart attacks are a result of many factors, among them poor diet, overeating, and a lack of exercise, all of which can and often do lead to excessive weight gain and obesity. Stress also plays a role in this often fatal disease. Reducing one's risk of heart disease requires proper diet, exercise, and stress management, as explained in Health Note 5-1. The sooner one starts, the better. Many doctors also recommend taking half an aspirin a day to prevent heart attacks. New research, however, shows that this strategy only works in individuals who have already had a heart attack. It doesn't reduce chances of a heart attack in individuals who have never had one.

Prevention is the most effective way to control heart attacks. It could save Americans billions of dollars each year in medical bills, lost work time, and decreased productivity and could save tens of thousands of lives. But given human nature, the fast pace of modern life, the availability of high-calorie foods, and our inattentiveness to our need for exercise and proper diet, heart disease will be around for a long time. To reduce the death rate, physicians therefore also look for ways to treat patients after they have had a heart attack.

KEY CONCEPTS
As always, prevention is the best way to address disease; heart attacks can be prevented by a lifetime of healthy eating and exercise.

Treating Heart Attacks

One promising development is the use of blood-clot-dissolving agents such as streptokinase (mentioned in the introduction of this chapter), urokinase, and tissue plasminogen activator (TPA). When administered within a few hours of the onset of a heart attack, these agents destroy blood clots that block the coronary arteries and can therefore greatly reduce the damage to heart muscle and accelerate a patient's recovery.

Health Tip 5-2

To reduce your chances of a heart attack, avoid *trans* fats in your diet. *Trans* fats occur in many foods made with hydrogenated vegetable oil, especially baked goods and snack foods. Check out labels on foods and avoid products that contain *trans* fat.

Why?

Trans fats increase the risk of coronary heart disease (the buildup of plaque in the arteries of the heart that can lead to heart attacks). Studies show that even when *trans* fats constitute only 1% to 3% of a person's total daily caloric intake, they significantly increase the risk of heart disease.

In mild cases, physicians can open clogged blood vessels by inserting a small catheter with a tiny balloon attached to its tip into the affected artery. After chemical clot dissolvers are administered to a patient, the balloon is inflated, forcing the artery open and loosening the plaque from the wall. This procedure is called *balloon angioplasty* (AN-gee-oh-PLAST-ee). Scientists are also experimenting with lasers that burn away plaque in artery walls. Unfortunately, as in other techniques, within a few months cholesterol builds up again in the walls of arteries.

Yet another treatment used more and more often these days is a tiny device known as a *stent* (Figure 5-16). This tubular wire is placed in the narrowed artery and then is expanded with a tiny balloon. After the balloon is extracted, the stent remains expanded to hold the narrowed artery open. Unfortunately, doctors are finding that, over time, blood clots can form on the stent. Accordingly, they are working on special coatings for stents to reduce or eliminate this problem.

In severe cases where the coronary arteries are completely blocked by atherosclerotic plaque, heart surgeons often perform a procedure known as *coronary bypass surgery* to restore blood flow to the heart muscle. In this procedure, surgeons transplant small segments of veins from the arm or leg into the heart (Figure 5-17). Venous bypasses shunt blood around the clogged coronary arteries, restoring blood flow to the heart muscle. Unfortunately, venous grafts often fill fairly quickly with plaque. The recurrence of plaque in grafts has

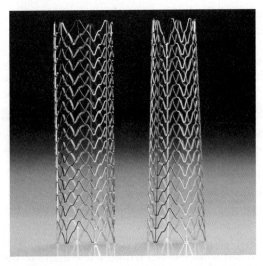

FIGURE 5-16 Stent These small metal devices are inserted into clogged arteries and expanded to restore blood flow. (Photo of RX Acculink® Carotid Stent System provided by Abbot. Used with permission.)

led researchers to turn to marginally important arteries, such as the internal mammary artery, for this procedure. Arteries will prove more resistant to plaque buildup than veins.

> **KEY CONCEPTS**
>
> Many medications and surgical procedures are available to treat heart attacks, including coronary bypass surgery, balloon angioplasty, and stents.

FIGURE 5-17 Coronary Bypass Surgery (a) Atherosclerotic plaque in coronary arteries can block the flow of blood to heart muscle. (© William Ober/Visuals Unlimited.) (b) Venous grafts bypass coronary arteries blocked by atherosclerotic plaque.

(a)

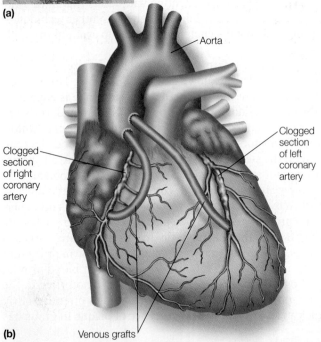

Aorta

Clogged section of left coronary artery

Clogged section of right coronary artery

(b)   Venous grafts

Cardiac Arrest

Heart attacks may lead to sudden cardiac death (SCD), death caused by loss of heart function, known as **sudden cardiac arrest**. SCD is the largest cause of natural death in the United States, according to the Centers for Disease Control and Prevention (CDC). It results in about 350,000 deaths in the United States every year and is responsible for half of all deaths from heart disease.

Sudden cardiac arrest should not be mistaken for a heart attack—that is, a myocardial infarction. Sudden cardiac arrest occurs as a result of a malfunction in the electrical system that controls the regular beating of the heart. Heart beat quickens and becomes extremely irregular.

During an episode, the SA node, discussed above, loses control of the heart muscle. With the SA node no longer in charge, the cardiac muscle cells beat independently, a condition known as **fibrillation**. The lack of coordination converts the heart into an ineffective, quivering mass that pumps little, if any, blood. The heart can stop beating altogether, a condition known as **cardiac arrest**.

The greatest concern during fibrillation is that blood flow to the brain will be so significantly reduced that a person will lose consciousness and die. Emergency treatment is required immediately to prevent death.

Cardiopulmonary resuscitation (CPR) must be administered immediately. CPR involves repetitive compression of the chest cavity over the sternum (breastbone) above the heart and artificial respiration—forcing air into a person's lungs. CPR and mouth-to-mouth resuscitation ensures adequate oxygen and blood flow to the brain until the heart starts beating normally. Normal rhythm may occur during CPR but is more often restored by applying a strong electric shock to the chest by a physician or emergency personnel. This procedure is known as *defibrillation*. The electrical current passes through the wall of the chest and is often sufficient to restore normal electrical activity and heartbeat.

About half of all cases of cardiac arrest occur without prior symptoms. In the remainder, patients complain of dizziness or tachycardia—accelerated heart beat.

> **KEY CONCEPTS**
>
> Malfunction of the electrical system of the heart results in fibrillation, an uncoordinated beating of heart muscle cells that, if not corrected, can lead to death.

High Blood Pressure, Hypertension

Hypertension (also known as high blood pressure) is a prolonged elevation in blood pressure. It is one of the most prevalent health problems of our time, affecting adults as well as children and teens. Nearly a third of all adults have high blood pressure.

Like other cardiovascular diseases, it has many causes, including obesity, high salt intake, heredity, smoking, and kidney disease. Nearly always a symptom-free disease early on, hypertension is characterized by a gradual increase in blood pressure over time. A person may feel fine and display no physical problems whatsoever for years. Symptoms such as headaches, rapid heartbeat, and a general feeling of ill health usually occur only when blood pressure is dangerously high. If left untreated, high blood pressure can lead to serious problems, including heart attacks.

Hypertension is more common in men than in women and is twice as prevalent in African Americans as it is in Caucasians. The increased pressure in the circulatory system forces the heart to work harder. Elevated blood pressure may also damage the lining of arteries, creating a site at which atherosclerotic plaque forms. Atherosclerosis increases the risk of heart attack: a hypertensive person is six times more likely to have a heart attack than an individual with normal blood pressure. Hypertension also increases the chances of plaque buildup in the arteries supplying the brain, which can result in strokes. In fact, hypertension is the leading cause of strokes in America.

High blood pressure can be treated by dramatic improvements in diet—lowering fatty meat consumption and increasing consumption of fruits and vegetables—as well as exercise, cessation of smoking, and numerous types of medications.

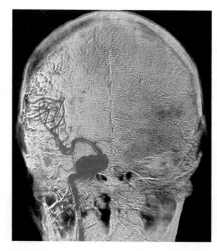

FIGURE 5-18 Aneurysm This X-ray shows a ballooning of one of the arteries in the brain. If untreated, an aneurysm can break, causing a stroke. (© CNRI/Science Source, Inc.)

Health Tip 5-3

Be sure to get plenty of sleep each night, as it greatly lowers your chances of developing hypertension (high blood pressure) that can lead to heart attacks and strokes later in life.

Why?

Many factors contribute to high blood pressure. Obesity is one of the main causes, but so is lack of sleep. Studies of individuals in the 32- to 59-year-old age group show that inadequate sleep doubles the chance of developing high blood pressure.

> **KEY CONCEPTS**
>
> High blood pressure is one of the most common cardiovascular diseases. It develops gradually over time and often goes undetected until too late.

Atherosclerosis

Another common problem of modern times is **atherosclerosis**, the buildup of cholesterol plaque in the walls of arteries. Arteries clogged with cholesterol force the heart to work harder, straining it. Perhaps the most significant problem arises from blood clots that lodge or form in narrowed coronary arteries, reducing the flow of blood to the heart muscle.

> **KEY CONCEPTS**
>
> Atherosclerosis is the build up of cholesterol-containing deposits in the arteries of the body that can lead to serious, often fatal blockages.

Aneurysm

Arteries not only can clog, they can rupture. Certain infectious diseases (such as syphilis), atherosclerosis, and hypertension all can weaken the wall of arteries. This weakening causes arteries to balloon outward, a condition known as an **aneurysm** (Figure 5-18). Like a worn spot on a tire, an aneurysm can rupture when pressure builds inside or when the wall becomes too thin.

When an aneurysm breaks, blood pours out of the circulatory system. Because they happen so quickly, most aneurysm ruptures lead to death. As in most diseases, the first line of defense against aneurysms is prevention. By reducing or eliminating the two main causes—atherosclerosis and high blood pressure—individuals can greatly lower their risk.

Physicians recommend a number of steps to reduce your chances of developing atherosclerosis and hypertension. If you smoke, stop. If you are overweight, exercise and lose weight. If you're one of those people for whom salt raises your blood pressure, cut back. If you suffer from stress at work and at home, find ways to reduce your stress levels. If you drink alcohol, consume it in moderation or quit altogether. If you are fond of foods that are rich in fats and cholesterol, cut down on them and consume foods containing water-soluble fiber such as apples, bananas, citrus fruits, carrots, barley, and oats. These reduce cholesterol uptake by the intestine.

Fish oils also protect the cardiovascular system. Fish oil contains omega-3 fatty acids that have many beneficial functions. For example, it increases high density lipoprotein (HDL), the so-called "good cholesterol." HDL binds to cholesterol in the blood and transports it to the liver for removal. This may decrease plaque buildup in arteries. Omega-3 fatty acids also have been shown to reduce blood clots. Grass-fed beef is naturally higher in omega-3-fatty acids, so if you eat beef, consider purchasing cuts from grass-fed cattle.

Health Tip 5-4

If you are not pregnant or thinking about getting pregnant, be sure to include fish in your diet, on a fairly regular basis.

Why?

Fish are an excellent source of omega-3 fatty acids, especially cold water species such as salmon, lake trout, tuna, mackerel, and herring. You can also obtain omega-3 fatty acids from grass-fed beef. Omega-3 fatty acids help protect against heart disease and stroke.

Pregnant women or women thinking about getting pregnant should avoid fish, however, because it may contain small amounts of mercury that can cause neurological damage to a developing fetus. Some fish are especially likely to have higher levels of mercury such as shark, swordfish, and king mackerel.

Eating fish once a week will provide minimal benefit. Three days a week provides maximum benefit. The highest amounts of omega-3 fatty acids are found in cold water species including salmon, mackerel, lake trout, and sardines. Fish oil capsules apparently provide the same benefit. As explained in the nearby Health Tip, pregnant women should avoid certain types of fish that often contain excess mercury, which arise primarily from the emissions of coal-fired power plants.

Early detection and treatment also help combat cardiovascular disease. As noted earlier, hypertension can be detected by regular blood pressure readings. Atherosclerosis can be detected by blood tests that measure cholesterol. Once any of these diseases is detected, physicians have many options.

> **KEY CONCEPTS**
>
> Weakening of the walls of arteries can lead to ruptures in the main arteries of the body, known as aneurysms. They often occur in the aorta and blood vessels of the brain and can be fatal.

5-6   Health and Homeostasis

The heart and blood vessels carry out many very important functions. They distribute nutrients throughout the body. They also distribute heat throughout the body and remove wastes from cells. These functions, in turn, help to keep the cells of your body—and you—alive and well.

Each of these functions serves a higher purpose: to maintain the relatively constant internal conditions necessary for normal cellular function. That is, they play a key role in homeostasis. Furthermore, by keeping cells, tissues, and organs of the body alive and functioning well, the cardiovascular system ensures that homeostatic mechanisms provided by them work properly. By maintaining the functions of the kidney, for instance, the cardiovascular system ensures that it continues to perform its functions that help ensure homeostasis—for example, its removal of wastes. By maintaining the cells of the liver, the cardiovascular system ensures that it carries out its many important homeostatic roles—for example, making blood-clotting proteins and ridding the body of toxic chemicals.

Cardiovascular diseases upset homeostasis, however. These diseases are caused by many factors, such as stress, overeating, improper diet, and lack of exercise—the so-called "lifestyle factors." By altering the functions of the cardiovascular system, these factors threaten the broader welfare of a person. Keeping your cardiovascular system healthy is not just about preventing heart attacks, it's about pursuing a healthy life—living long and well.

> **KEY CONCEPTS**
>
> The circulatory system is vital to cellular function and therefore to homeostasis.

SUMMARY

The Circulatory System's Function: An Overview

1. The circulatory system is one of the body's chief homeostatic systems.

The Heart

2. The circulatory system consists of a pump, the heart, and two circuits, the pulmonary circuit, which transports blood to and from the lungs, and the systemic circuit, which delivers blood to the body and returns it to the heart.

3. The human heart consists of four chambers: two atria and two ventricles. The right atrium and right ventricle service the pulmonary circuit. The left atrium and left ventricle pump blood into the aorta and are part of the systemic circuit.

4. The right atrium receives blood from the superior and inferior venae cavae. This blood, returning from body tissues, is low in oxygen and rich in carbon dioxide.

5. From the right atrium, blood is pumped into the right ventricle, then to the lungs, where it is resupplied with oxygen and stripped of most of its carbon dioxide. Blood returns to the heart via the pulmonary veins, which empty into the left atrium.

6. Blood is pumped from the left atrium to the left ventricle, then out the aorta to the body, where it supplies cells of tissues and organs with oxygen and picks up cellular wastes.

7. Heart valves help control the direction of blood flow in the heart.

8. Cardiac muscle cells' contraction is coordinated by the heart's pacemaker, the sinoatrial (SA) node.

9. Left on its own, the SA node would produce 100 contractions per minute. Nerve impulses, however, reduce the heart rate to about 70 beats per minute when a person is inactive. During exercise or stress, the heart rate increases to meet body demands.

10. Changes in electrical activity in the heart muscle can be detected by surface electrodes. The tracing produced on a voltage meter attached to the electrodes is called an electrocardiogram (ECG). Certain diseases of the heart may disrupt one or more of the waves, making the ECG a valuable diagnostic tool.

The Blood Vessels

11. The heart pumps blood into the arteries, which distribute the blood to capillary beds. Blood is returned to the heart in the veins.

12. The aorta, which carries blood away from the heart, and many of its chief branches are elastic arteries. As blood pulses out of the heart, the elastic arteries expand to accommodate the blood, then contract, helping pump the blood and ensuring a steady flow through the capillaries. As they course through the body, the elastic arteries branch to form muscular arteries, which also expand and contract with the flow of blood.

13. The smooth muscle in the walls of muscular arteries responds to a variety of stimuli. These stimuli cause the blood vessels to open or close to varying degrees, controlling the flow of blood through body tissues depending on their needs at any moment.

14. Blood pressure and flow rate are highest in the aorta and drop considerably as the arteries branch. By the time the blood reaches the capillaries, its flow and pressure are greatly reduced. This dramatic

decline enhances the rate of exchange between the blood and the tissues.

15. Capillaries are thin-walled vessels that form branching networks, or capillary beds, among the cells of body tissues. Cellular wastes and nutrients are exchanged between the cells of body tissues and the blood in capillary networks.

16. Blood flow through a capillary network is regulated by constriction and relaxation of the arterioles that supply them.

17. In the veins above the heart, blood drains by gravity. Veins below the heart, however, rely on the movement of body parts to squeeze the blood upward and on valves, flaps of tissue that span the veins and prevent the backflow of blood.

The Lymphatic System

18. The lymphatic system is a network of vessels that drains excess interstitial fluid from body tissues and transports it to the bloodstream.

19. The lymphatic system also consists of several lymphatic organs, such as the lymph nodes, the spleen, the thymus, and the tonsils, which function primarily in immune protection.

20. Distributed along the system of lymphatic vessels are small nodular organs called lymph nodes, which filter the lymph, removing bacteria and viruses.

21. Under normal circumstances, lymph is removed from tissues at a rate equal to its production, keeping tissues from swelling. In some cases, however, lymph production exceeds the capacity of the lymphatic capillaries, resulting in swelling (edema).

Cardiovascular Diseases: Causes and Cures

22. Heart attacks are also known as myocardial infarctions. They are caused by blood clots that either form in narrowed coronary arteries or arise elsewhere and breaks loose, only to become lodged in coronary arteries. The obstruction decreases the flow of blood and oxygen to heart muscle, sometimes killing the cells.

23. Heart attacks can occur quite suddenly without warning, or they may be preceded by several weeks of angina, pain felt when blood flow to heart muscle is reduced.

24. The risk of heart attack can be reduced by proper diet and exercise. Numerous treatments are available for patients who have had a heart attack.

25. Another serious heart problem is cardiac arrest. This occurs when the sinoatrial node loses its control over heart muscle. Muscle contraction becomes uncoordinated. Muscle cells beat independently. Heart beat often speeds up and the heart becomes an ineffective, quivering mass. This condition, known as fibrillation, can result in a loss of blood flow to the brain and death.

26. Hypertension or high blood pressure is a prolonged elevation in blood pressure that typically develops over a long period without noticeable signs until it has reached a critical stage. Elevated blood pressure increases one's risk of heart attack and stroke.

27. Atherosclerosis results from the buildup of cholesterol plaque in arteries, which can become clogged by clots, leading to heart attacks, strokes, and rupture of the walls of arteries.

Health and Homeostasis

28. The cardiovascular system ensures homeostasis by distributing nutrients and ridding the body of waste. This helps to maintain cells, tissues, and organs of the body that carry out other vital homeostatic functions.

THINKING CRITICALLY ANALYSIS

This analysis corresponds to the Thinking Critically scenario that was presented at the beginning of this chapter.

This hypothesis suggests that a selective survival process was in operation—that is, that salt-retaining (dehydration-resistant) slaves were more likely to survive the rigors of slavery than those who lacked this trait. To test this idea, one might begin by looking at hypertensive African Americans to see whether they actually have the capacity to retain excess salt and, if so, whether this capacity leads to hypertension.

Another study might involve tests of blacks who reside in the regions of Africa from which slaves were taken. Finding a low rate of hypertension in this group would suggest that a selective survival mechanism was in effect. One such study showed that, in rural parts of Nigeria, blacks did not suffer from hypertension despite the fact that they consumed large amounts of salt. The researchers concluded that if these people are descendants of the ancestors of slaves, as are African-American blacks, then the hypothesis supporting selective survival seems valid.

KEY TERMS AND CONCEPTS

Adrenalin, p. 101
Aneurysm, p. 111
Angina, p. 109
Aorta, p. 98
Arteriole, p. 101
Artery, p. 95
Atherosclerosis, p. 111
Atrioventricular bundle, p. 100
Atrioventricular (AV) node, p. 100
Atrioventricular valve, p. 98
Atrium, p. 98
Blood pressure, p. 104
Capillary, p. 101
Capillary bed, p. 104
Cardiac arrest, p. 110
Chordae tendineae, p. 98

Diastolic pressure, p. 104
Edema, p. 105
Electrocardiogram, p. 101
Endocardium, p. 99
Epinephrine, p. 101
Fibrillation, p. 110
Heart, p. 97
Hypertension, p. 110
Inferior vena cava, p. 98
Left atrium, p. 98
Left ventricle, p. 98
Lymph capillaries, p. 108
Lymph node, p. 108
Lymphatic system, p. 107
Myocardial infarction, p. 108

Myocardium, p. 97
Pericardium, p. 97
Pulmonary circuit, p. 97
Purkinje fibers, p. 100
Right atrium, p. 98
Right ventricle, p. 98
Sinoatrial (SA) node, p. 100
Sudden cardiac arrest, p. 110
Superior vena cava, p. 98
Systemic circuit, p. 97
Systolic pressure, p. 104
Varicose veins, p. 105
Vein, p. 98
Ventricle, p. 98
Venule, p. 101

CONCEPT REVIEW

1. List and describe several ways in which the circulatory system functions in homeostasis. p. 96.
2. Describe how the pacemaker, the sinoatrial node, coordinates muscular contractions of the heart. pp. 100–101.
3. Define the pulmonary and systemic circuits. pp. 97–98.
4. Describe how the valves control the direction of blood flow in the heart. What are valvular incompetence and valvular stenosis? What causes these conditions, and how do they affect the heart? pp. 98–100.
5. Explain how blood moves through the heart? What path does it take? p. 97.

6. How does atherosclerosis of the coronary arteries affect the heart? p. 111.
7. Describe the general structure of the arteries and veins. How are they different? How are they similar? pp. 102, 103, 105.
8. How do the elastic fibers in the major arteries help ensure a continuous blood flow through the capillaries? p. 103.
9. Capillaries illustrate the remarkable correlation between structure and function. Do you agree with this statement? Why or why not? p. 105.
10. Describe how the movement of blood through capillary beds is controlled. Why is this homeostatic mechanism so important? Give some examples. pp. 104–105.

11. Explain how the blood is returned to the heart via the veins. pp. 105–106.
12. Explain the role of the lymphatic system in circulation. pp. 107–108.
13. What is a heart attack? What are the most common symptoms? What are the causes of heart attacks? How can they be prevented? How can heart attacks be treated? pp. 108–109.
14. What is fibrillation? How is it treated? p. 110.
15. What are the common causes of virtually all cardiovascular diseases, including atherosclerosis, heart attacks, and hypertension? p. 108.

SELF-QUIZ: TESTING YOUR KNOWLEDGE

1. The _____ circuit delivers blood from the heart to the rest of the body, excluding the lungs. pp. 97–98.
2. The _____ circuit delivers blood to the lungs. p. 97.
3. The muscle of the heart is known as _____. p. 97.
4. The largest artery in the body is the _____. p. 98.
5. The _____ node is the pacemaker of the heart. p. 100.
6. The ventricles of the heart contract just after the atria because the nerve impulse is delayed by the _____ node. p. 100.

7. Electrical signals are transmitted into the heart muscle of the ventricles via the bundle branches; they terminate on conductive cells among the cardiac muscle cells known as _____ cells. p. 100.
8. A(n) _____ is a recording of the electrical activity in the heart when it contracts. p. 101.
9. Blood pressure is lowest in the _____. p. 104.
10. Nutrient and waste exchange in tissues occurs in _____. p. 104.
11. _____ pressure is the blood pressure recorded when the ventricles contract. p. 104.

12. A _____ infarction results from a blood clot in the coronary artery, which blocks blood flow to heart muscle. p. 108.
13. _____ is a condition that develops slowly and results in elevated blood pressure. This condition does not become detectable until a critical state is achieved. p. 110.
14. Uncontrolled beating of the heart muscle known as _____ may occur during a heart attack. p. 110.
15. Label the drawing. p. 97.

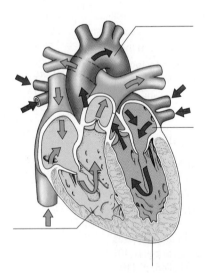

biology.jbpub.com/chiras/8e/

The site features eLearning, an online review area that provides quizzes, chapter outlines, and other tools to help you study for your class. You can also follow useful links for in-depth information, research the differing views in the Point/Counterpoints, or keep up on the latest health news.

(©Warren Goldswain/ShutterStock, Inc.)

The Blood

In 1966, Dr. Leland C. Clark of the University of Cincinnati's College of Medicine immersed a live laboratory mouse in a clear fluorocarbon solution saturated with oxygen. To the amazement of the audience, the mouse did just fine, "breathing" the oxygen-rich liquid as if it were air. Clark then extracted the mouse from the solution. After a moment, the rodent began to move about, apparently unharmed by the ordeal.

THINKING CRITICALLY

Imagine that you are a researcher at a major medical school. You are an expert on a particular blood disease and are asked by a pharmaceutical company to test an experimental drug. You and your students devise an experiment to see if the drug was effective. You find that in 100 patients, 60% who received the drug got better. Only 10% of the control group (of equal size) reported improvement. From these data, you concluded that the drug was effective. But are you correct? What are your thoughts? Can you think of anything that might have thrown off your results?

The point of this demonstration was not to show that an animal could "breathe" this fluid into its lungs and survive—a process called *liquid breathing*. Rather, it was to show that the solution held enormous amounts of oxygen, so much so that it could possibly be used as a substitute for blood. Clark and his colleagues hoped that their "artificial blood" would someday be a boon to medical science, helping paramedics keep accident victims who have lost substantial amounts of blood alive while they are being transported to hospitals. In rural America, where a trip to a hospital takes considerable time, artificial blood could save thousands of lives a year. Today, that dream has yet to be realized.

The artificial blood Dr. Clark tested is a member of group of chemicals known as *perfluorochemicals (PFC)*. To make artificial blood, perfluorochemicals are first mixed with water. This forms an emulsion—tiny droplets of PFC in water. The droplets are about 1/40 the size of blood cells. This emulsion is then mixed with salts, vitamins, nutrients, and antibiotics. The final mix contains about 80 different chemicals and performs much like natural blood. Clinical trials are now underway in the United States and Europe. In this chapter, you will explore the structure and function of an extremely important body tissue, blood.

6-1 Blood: An Overview

Human blood is a far cry from Clark's artificial substitute and the more modern versions. The blood in our circulatory system, for example, is a watery fluid, not a fluorocarbon, and consists of two basic components: plasma and formed elements. **Plasma** is the liquid portion of the blood and is about 90% water. It contains many dissolved substances and three types of formed elements: (1) white blood cells, (2) red blood cells, and (3) platelets. Blood accounts for about 8% of our total body weight. A man weighing 70 kilograms (150 pounds) has about 5.6 kilograms (12 pounds) of blood. That's 5 to 6 liters (1.3 to 1.5 gallons). On average, women have about 1 liter less.

Figure 6-1 shows that the plasma makes up about 55% of the blood volume and formed elements are about 45%. The bulk of the latter consists of red blood cells. White blood cells account for less than 1% of the total volume of the blood.

The blood volume occupied by blood cells is known as the **hematocrit**. The hematocrit varies in individuals living at different altitudes. In people living in Denver, Colorado, a mile above sea level, for example, hematocrits are typically about 5% higher than in people living at sea level. The increase in the blood cell count is the result of a homeostatic mechanism that compensates for the slightly lower level of oxygen in the atmosphere at higher altitudes. The body compensates for lower oxygen levels by increasing its production of oxygen-carrying red blood cells. This allows people living at higher elevations to function normally—that is, it allows the blood to carry a sufficient amount of oxygen. The higher concentration of red blood cells also means that their blood transports more oxygen when an individual is at lower altitudes.

FIGURE 6-1 **Blood Composition** Blood removed from a person can be centrifuged to separate plasma from the cellular component. Red blood cells constitute about 45% of the blood volume, except at higher altitudes where they make up about 50% of the volume to compensate for the lower oxygen levels.

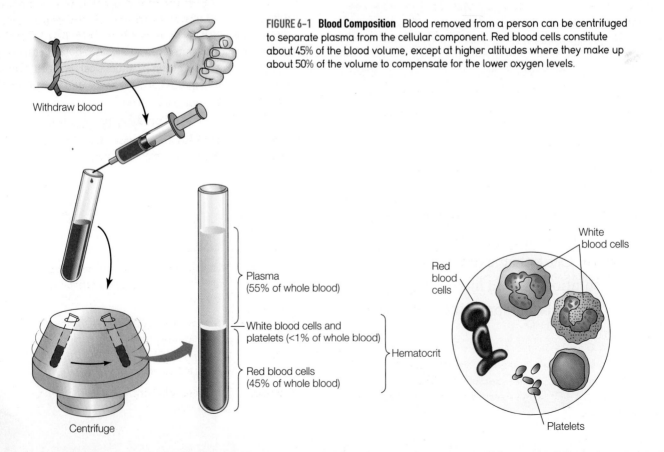

Withdraw blood

Centrifuge

Plasma
(55% of whole blood)

White blood cells and
platelets (<1% of whole blood)

Red blood cells
(45% of whole blood)

Hematocrit

White
blood cells

Red
blood
cells

Platelets

In this day of high-stakes athletic competition, where athletes look for any competitive edge they can find, many athletes train at high altitudes. Athletes at lower elevations can take advantage of the body's ability to produce more RBCs, by sleeping in specially designed chambers that expose them to air with slightly lower oxygen content. The rationale behind high-altitude training and low-oxygen sleep chambers is that an athlete whose blood carries more oxygen will perform better—and win more competitions. In fact, it is partly for this reason that the U.S. Summer Olympic team is currently headquartered in Colorado Springs, Colorado, which is also about a mile above sea level.

> **KEY CONCEPTS**
>
> Blood contains a liquid portion, plasma, and two types of formed elements, blood cells and platelets that perform many crucial tasks from transporting nutrients and wastes, and fighting infection.

6-2 Blood Plasma

Plasma is a light yellow (straw-colored) fluid that plays many important functions in maintaining homeostasis.

Blood as a Transport Medium

Cells of the body require many different substances to survive and carry out their functions. They also produce many wastes that must be removed from the body to ensure good health. Blood plasma helps deliver vital nutrients and other chemicals like glucose and hormones to the cells of the body but also helps to remove a variety of waste products produced by our cells, thus helping maintain homeostasis. A chemical analysis of blood plasma reveals the following substances dissolved or suspended within this watery fluid: (1) gases such as nitrogen, carbon dioxide, and oxygen; (2) ions such as sodium, chloride, and calcium; (3) nutrients such as glucose and amino acids; (4) hormones such as estrogen in women and testosterone in men; (5) proteins such as antibodies; (6) chemical wastes such as carbon dioxide; and (7) lipids such as cholesterol. Most of these chemicals are water soluble, so they are dissolved by water that makes up the bulk of blood plasma. Lipids are water insoluble so are either suspended in tiny globules in plasma or bound to certain plasma proteins, which allows them to be transported through the bloodstream.

> **KEY CONCEPTS**
>
> Blood transports heat throughout the body as well as numerous substances to and from body cells helping ensure proper function.

Regulating pH and Osmotic Pressure

Proteins are the most abundant of all dissolved substances in the plasma. Blood proteins play many roles, depending on the particular protein. All blood proteins contribute to osmotic pressure. As you may recall, osmotic pressure helps regulate the flow of materials in and out of capillaries. It therefore is essential to the proper distribution of wastes and nutrients in the body, which is vital to homeostasis.

Blood proteins also help to regulate the pH of the blood. They do so by binding to hydrogen ions, thus preventing the H^+ concentration in the blood from rising, a process that is also essential to homeostasis.

> **KEY CONCEPTS**
>
> Proteins in the blood help to regulate and maintain its pH (acid content) of the blood, essential to cellular metabolism, and osmotic pressure, which in turn helps regulate the flow of materials in and out of capillaries.

Transport Molecules

Human blood contains three types of protein: (1) albumins; (2) globulins; and (3) fibrinogen (Table 6-1). Albumins (al-BEW-mins) and two types of globulins (GLOB-u-lins) are carrier proteins. **Carrier proteins** are large, water-soluble molecules. They bind to certain hormones and ions and small lipid molecules, as just noted. The binding of a water-insoluble lipid molecule to a carrier protein facilitates its transport in the blood, which consists largely of water. Carrier proteins also protect smaller molecules from destruction by the liver. These proteins therefore are essential to maintaining chemical balance in the blood and tissue fluids.

Not all globulins are carrier proteins. For example, one group of globulins consists of antibodies. **Antibodies** are blood proteins that "neutralize" viruses and bacteria or target them for destruction by phagocytic cells in body tissues known as *macrophages*.

Still another important blood protein is **fibrinogen** (fie-BRIN-oh-gin). At sites of injury, fibrinogen is converted into a filament known as **fibrin** (FIE-brin). Fibrin molecules form a meshwork that traps blood cells forming a blood clot in the walls of injured blood vessels. Clots form quickly and are a homeostatic mechanism to prevent blood loss.

> **KEY CONCEPTS**
>
> Some molecules cannot be readily transported through the bloodstream without the aid of transport proteins.

TABLE 6-1	Summary of Plasma Proteins
Protein	**Function**
Albumins	Maintain osmotic pressure and transport smaller molecules, such as hormones and ions
Globulins	Alpha and beta globulins transport hormones and fat-soluble vitamins; gamma globulins (antibodies) bind to foreign substances
Fibrinogen	Converted into fibrin network that form blood clots

6-3   Red Blood Cells

With this overview of blood plasma in mind, let's turn to the formed elements, blood cells, beginning with the red blood cell.

The **red blood cell** (RBC) or **erythrocyte** is the most abundant cell in human blood. In fact, 20 drops of blood (equal to 1 milliliter) contain approximately 5 billion RBCs. If RBCs were people, a single milliliter of blood would contain slightly more than 70% of the world population!

Structure of RBCs

Human RBCs are highly flexible biconcave disks (concave on both sides), as shown in Figure 6-2b. These cells transport oxygen and, to a lesser degree, carbon dioxide in the blood (Figure 6-2; Table 6-2). The unique biconcave shape of the human RBC is a structural adaptation that increases the cells surface area. The greater the surface area, the more readily gases like oxygen move in and out of RBCs. Like so many other structures encountered in biology, the human RBC is a remarkable example of how its form mirrors function.

Swept along in the bloodstream, RBCs travel many times through the circulatory system each day. As they pass through the capillaries, the RBCs bend and twist. This permits them to pass through the many miles of tiny capillaries whose internal diameters are often smaller than those of the RBCs.

The flexibility of RBCs is an adaptation that serves us well. It allows blood cells to deliver their vital load, oxygen, to the cells of the body. However, not all people are so fortunate. According to various estimates, approximately 1 of every 500 to 1,000 African Americans suffers from a disease called **sickle-cell disease**. This disease results in a marked decrease in the flexibility of RBCs and is caused by a genetic mutation—a slight alteration in the hereditary material (DNA). This mutation, in turn, results in the production of hemoglobin with one incorrect amino acid. This seemingly insignificant defect dramatically alters the three-dimensional structure of the hemoglobin molecule, causing RBCs to transform from biconcave disks into sickle-shaped cells when they encounter low levels of oxygen in capillaries (Figure 6-3). Sickle-shaped cells are considerably less flexible than normal RBCs and are therefore unable to bend and twist. As a result, sickle cells often collect at branching points in capillary beds like logs in a logjam. This blocks blood flow, disrupting the supply of nutrients and oxygen to tissues and organs. The lack of oxygen causes considerable pain and can kill cells. Blockages in the lungs, heart, and brain can also be life-threatening. Blockages in the heart and brain, for instance, often lead to heart attacks and brain damage. Many people who have sickle-cell disease die in their late twenties and thirties; some die even earlier.

> **KEY CONCEPTS**
> Red blood cells are highly flexible cells that transport oxygen throughout the body.

Making New RBCs

On average, a human RBC lives about 120 days. At the end of an RBC's life span, the liver and spleen remove the aged cells from circulation. Not to waste valuable nutrients, much of the iron contained in the hemoglobin of the RBCs (discussed shortly) is captured and then used to produce new RBCs in the red bone marrow. (As you shall soon see, iron in hemoglobin binds to oxygen and is therefore essential to hemoglobin's function.) The recycling of iron is not 100% efficient,

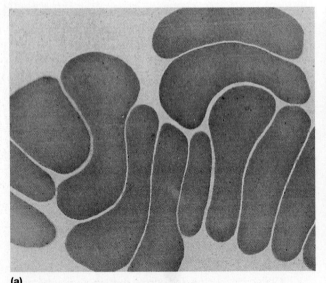

(a)

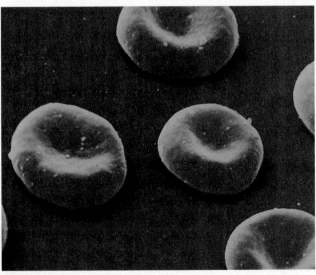

(b)

FIGURE 6-2 **Red Blood Cells** (a) Transmission electron micrograph of human RBCs showing their flexibility. (© David M. Phillips/Visuals Unlimited.)(b) Scanning electron micrograph of human RBCs. (© Stanley Fleger/Visuals Unlimited.)

TABLE 6-2 Summary of Blood Cells

Name	Light Micrograph	Description	Concentration (Number of Cells/mm³)	Life Span	Function
Red blood cells (RBCs)		Biconcave disk; no nucleus	4 to 6 million	120 days	Transports oxygen and carbon dioxide
White blood cells (WBCs) Neutrophil		Approximately twice the size of RBCs; multi-lobed nucleus; clear-staining cytoplasm	3,000 to 7,000	6 hours to a few days	Phagocytizes bacteria
Eosinophil		Approximately same size as neu-trophil; large pink-staining gran-ules; bilobed nucleus	100 to 400	8 to 12 days	Phagocytizes antigen-antibody complex; attacks parasites
Basophil		Slightly smaller than neutrophil; contains large, purple cytoplasmic granules; bilobed nucleus	20 to 50	Few hours to a few days	Releases histamine during inflammation
Monocyte		Larger than neutrophil; cytoplasm grayish-blue; no cytoplasmic gran-ules; U- or kidney-shaped nucleus	100 to 700	Lasts many months	Phagocytizes bacte-ria, dead cells, and cellular debris
Lymphocyte		Slightly smaller than neutrophil; large, relatively round nucleus that fills the cell	1,500 to 3,000	Can persist many years	Involved in immune protection, either attacking cells directly or producing antibodies
Platelets		Fragments of megakaryocytes; appear as small dark-staining granules	250,000	5 to 10 days	Play several key roles in blood clotting

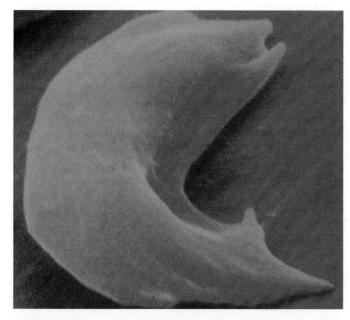

FIGURE 6-3 **Sickle-Cell Anemia** Scanning electron micrograph of a sickle cell. Compare this to the RBCs in Figure 6-2. (© Stanley Fleger/Visuals Unlimited.)

however, so small amounts of iron must be ingested each day in the diet. Loss of blood from an injury or, in women, during menstruation, increases the body's demand for iron. Without adequate iron intake, oxygen transport may become impaired. To make up for lost iron, consider eating all-bran cereal, spinach, tofu, and clams—these are all good sources of dietary iron. Vitamins containing iron can also help you maintain normal iron levels.

Human RBCs are highly specialized "cells" that lose their nuclei and organelles during their formation (Figure 6-2a). Because of this, RBCs cannot divide to replace themselves as they age. In humans, new RBCs are produced inside bones in a tissue known as **bone marrow**. In the blood-producing bone marrow, known as **red marrow**, RBCs are produced by **stem cells**, undifferentiated cells that trace their origins back to embryonic development. These cells give rise to an astounding 2 million RBCs per second!

In infants and children, almost all of the bone marrow is involved in the production of RBCs. As growth slows, however, the red marrow of many bones becomes inactive and gradually fills with fat cells and becomes **yellow marrow**, a fat storage depot. By the time an individual reaches adulthood, only a few bones, such as the hip bones, sternum (breastbone), ribs, and vertebrae still produce RBCs. However, in times of need, yellow marrow can be converted back into active red marrow.

The number of RBCs in the blood remains more or less constant over long periods. Maintaining a constant concentration of RBCs is essential to homeostasis. This process is controlled by a hormone with a rather forbidding name, **erythropoietin** (eh-RITH-row-po-EAT-in) or, more commonly, **EPO**. As a side note, EPO is one of a dozen or more performance-enhancing chemical substances being used illegally by athletes. EPO increases the number of RBCs in the blood, enhancing oxygen supply to muscles.

Erythropoietin is produced by cells in the kidney when oxygen levels in the bloodstream decline—for example, when a person moves to high altitudes or loses a significant amount of blood in an accident. EPO circulates in the blood to the bone marrow where it stimulates the stem cells to divide, thus increasing RBC production. As the RBC concentration increases, oxygen levels in the blood increase. When oxygen levels return to normal, the production and release of erythropoietin decreases, thereby reducing the rate of RBC formation in a classical negative feedback mechanism so common among the body's homeostatic mechanisms.

> **KEY CONCEPTS**
>
> Red blood cells are replenished daily by specialized cells inside the red bone marrow of children and adults.

Oxygen Transport

Oxygen transport in the blood occurs in large part because of **hemoglobin** (HEME-oh-GLOBE-in), a large protein molecule found in abundance in the RBCs of your body. As shown in Figure 6-4, each hemoglobin molecule consists of four subunits.

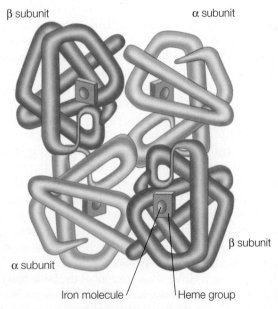

β subunit   α subunit

β subunit

α subunit

Iron molecule   Heme group

FIGURE 6-4 Porphyrin Ring The porphyrin ring of the hemoglobin molecule contains an iron ion that binds to oxygen and carbon monoxide.

Each of hemoglobin's four subunits contains a large, organic ring structure, known as a *heme group*. In the center of the heme group is iron.

Oxygen carried in the blood is largely attached to iron of hemoglobin. Oxygen is delivered to the blood stream via the respiratory system. Oxygen flows into the lungs in the air we breathe, then diffuses into the many capillary beds inside our lungs where it diffuses first into the blood plasma, then into the RBCs. Inside the RBC, oxygen binds to the iron in hemoglobin molecules where it is transported through the circulatory system. It is released inside capillary beds in the tissues and organs of the body.

Ninety-eight percent of all the oxygen in the blood is transported bound to iron in hemoglobin. Hemoglobin bound to oxygen is called *oxyhemoglobin*. The remaining 2% of the oxygen carried by blood is dissolved in the blood plasma.

Carbon dioxide, a waste product of cellular respiration, also binds to hemoglobin, but to a much lesser degree. Most carbon dioxide molecules are converted to bicarbonate and are transported in the plasma.

> **KEY CONCEPTS**
>
> Hemoglobin is a protein found inside red blood cells; it binds to oxygen, transporting it throughout the circulator system

Anemia

Homeostasis depends on the normal operation of the heart and blood vessels. It also depends on the blood's ability to absorb a sufficient amount of oxygen as it passes through the lungs. Unfortunately, the oxygen-carrying capacity of the blood can be impaired by a number of conditions. A reduction in the oxygen-carrying capacity of the blood is known as **anemia**. This condition may result from (1) a decrease in the number of circulating RBCs; (2) a reduction in the RBC hemoglobin content; or (3) the presence of abnormal hemoglobin in RBCs.

The number of RBCs in the blood may decline because of excessive bleeding or because of the presence of a tumor in the red marrow that reduces RBC production. Several infectious diseases (such as malaria) also decrease the RBC concentration in the blood. A reduction in the amount of hemoglobin in RBCs may be caused by iron deficiency or a deficiency in vitamin B_{12}, protein, or copper in the diet. Abnormal hemoglobin is produced in sickle-cell anemia and other genetic disorders.

Anemia generally results in weakness and fatigue. Individuals are often pale and tend to faint or become short of breath easily. People suffering from anemia often have an increased heart rate, because the heart beats faster to offset the reduction in the oxygen-carrying capacity of the blood. As a rule, anemia is not a life-threatening condition. However, it does weaken one's resistance to other diseases and also limits a person's productivity. Therefore, it should be treated quickly.

> **KEY CONCEPTS**
>
> Anemia is a condition in which oxygen transport in the blood is impaired; it can be caused by one of several different mechanisms.

6-4 White Blood Cells

White blood cells (WBCs), or **leukocytes**, are nucleated (they have nuclei) cells that play a huge role in maintaining homeostasis, primarily by eliminating harmful microorganisms such as bacteria and viruses. White blood cells are produced in the bone marrow alongside the red blood cells. They circulate in the bloodstream.

Function of WBCs

Although WBCs are officially blood cells, they do most of their work combating infections outside of the bloodstream, that is, in the body tissues. The blood and circulatory system merely transport the WBCs to sites of infection. When WBCs arrive at the "scene," they escape through the walls of the capillaries by "squeezing" between the endothelial cells (Figure 6-5).

Table 6-2 lists and describes the five types of WBCs found in the blood. The three most numerous, which are discussed here, are neutrophils, monocytes, and lymphocytes.

> **KEY CONCEPTS**
>
> White blood cells are a diverse group of cells that help protect the body from infections.

Neutrophils

Neutrophils (NEW-trow-fills) are the most abundant of the WBCs. Approximately twice the size of RBCs, these cells are distinguished by their multi-lobed nuclei (Table 6-2). Neutrophils circulate in the blood like a cellular police force awaiting microbial invasion. Attracted by chemicals released by infected tissue, neutrophils escape from the bloodstream, and then migrate to the site of infection by amoeboid movement.

Neutrophils are usually the first WBCs to arrive on the scene. When they arrive, they immediately begin to engulf (phagocytize) microorganisms, preventing the spread of bacteria and other organisms from the site of invasion. When a

neutrophil's lysosomes are used up, the cell dies and becomes part of the yellowish liquid, or pus, which oozes from wounds. Pus is a mixture of dead neutrophils, cellular debris, and bacteria, both living and dead.

> **KEY CONCEPTS**
>
> Neutrophils are a type of white blood cell that are the first to arrive at the site of an injury; they engulf microorganisms, helping prevent the spread of infection.

Monocytes

Monocytes (MON-oh-sites) are also phagocytic cells. Slightly larger than neutrophils, monocytes contain distinctive U-shaped or kidney-shaped nuclei. Like neutrophils, monocytes leave the bloodstream to do their "work" and migrate through body tissues via amoeboid motion. Once at the site of an infection, they begin phagocytizing microorganisms, dead cells, and dead neutrophils. Thus, while neutrophils are the "first-line" troops, monocytes constitute something of a mop-up crew.

Monocytes also take up residence in connective tissues of the body, where they are referred to as **macrophages** (MACK-row-FAY-jus). These cells remain more or less stationary, like watchful soldiers ever ready to attack and phagocytize invaders.

> **KEY CONCEPTS**
>
> Monocytes are another type of white blood cell that helps clean up injuries by phagocytizing dead cells and invading microorganisms.

Lymphocytes

The second most abundant WBC is the **lymphocyte**. These cells are not phagocytic but rather mount an immune response to bacteria and viruses. Although lymphocytes are found in the blood, most of them exist outside the circulatory system in lymphoid organs (LIM-foid). This includes organs such as the spleen, thymus, lymph nodes, and lymphoid tissue. Aggregations of lymphocytes also exist beneath the lining of the intestinal and respiratory tracts. It is in these locations that lymphocytes attack microbial intruders (Figure 6-6).

Two types of lymphocytes are found in the body. The first type is the **T lymphocyte**, or **T cell**. T cells attack foreign cells such as fungi, parasites, and tumor cells directly. They are thus said to provide *cellular immunity*.

The second type is called the **B lymphocyte**, or **B cell**. When activated, B cells transform into another kind of cell, known as the **plasma cells**. Plasma cells, in turn, synthesize and release *antibodies*, proteins that circulate in the blood and bind to foreign substances. The binding of antibodies to "foreigners" coats them and neutralizes them. Or, it marks microorganisms and tumor cells for destruction by macrophages.

Like the other formed elements of blood, the WBCs are involved in homeostasis. Their numbers can increase greatly during a microbial infection and other diseases. In fighting off a disease, they help return the body to normal function.

An increase in the number of WBCs, called *leukocytosis* (lew-co-sigh-TOE-siss), is a normal homeostatic response to

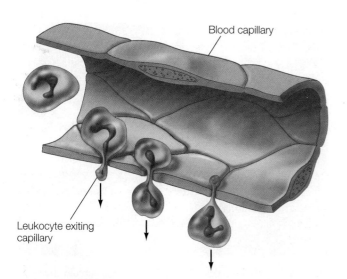

FIGURE 6-5 White Blood Cell Exit White blood cells (leukocytes) escape from capillaries by squeezing between endothelial cells. They then enter tissues of the body where they perform their protective functions.

Blood capillary

Leukocyte exiting capillary

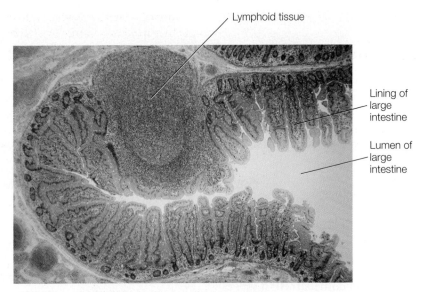

Lymphoid tissue

Lining of large intestine

Lumen of large intestine

FIGURE 6–6 Lymphoid Tissue The loose connective tissue beneath the lining of the large intestine and other sites is often packed with lymphocytes that have proliferated in response to invading bacteria. (© Gladden Willis, M.D./Visuals Unlimited.)

microbial intruders. It ends when the microbial invaders have been destroyed. Increases and decreases in various types of WBCs can be used to diagnose many medical disorders. For example, a dramatic increase in lymphocytes and intermittent abdominal pain, first felt in the upper abdomen or around the navel, are usually signs of **appendicitis**, an infection of the appendix. Because variations in the WBC count accompany many diseases, a blood test is a standard diagnostic procedure.

> **KEY CONCEPTS**
>
> Two types of lymphocyte are found in the body; they help provide immune protection, another means of protecting the body from foreign organisms like bacteria and viruses.

Leukemia

Like most other cells, WBCs can become cancerous, dividing uncontrollably in the bone marrow and then entering the bloodstream. A cancer of WBCs is called **leukemia** (lew-KEEM-ee-ah; literally, "white blood"). Leukemia may be either chronic or acute. *Chronic leukemia* unfolds slowly over a period of 3 to 10 years, depending on the type. The most serious type of leukemia is *acute leukemia*, so named because it comes on quickly and often kills patients within a few weeks of diagnosis. Children are the primary victims of this disease, although aged adults also fall victim.

In both forms of leukemia, WBCs fill the bone marrow, crowding out the cells that produce RBCs and platelets. Platelets are involved in blood clotting (discussed shortly). The proliferation of WBCs results in a decline in the production of RBCs, which leads to anemia. Anemia results in shortness of breath. The reduction in platelet production reduces blood clotting and thus increases internal bleeding or bleeding at surface wounds. Making matters worse, the cancerous WBCs produced in leukemia are often incapable of fighting infection. Leukemia patients typically succumb to infections and internal bleeding.

Acute leukemia patients may be treated by chemotherapy, drugs like vincristine (VIN-chris-teen) that stop the cancer-

ous white blood cells from dividing. Patients are also often given antibiotics to prevent infection and some patients may receive blood transfusions to boost the number of functional blood cells. Some receive radiation treatment to destroy cancerous WBCs in the brain.

Chronic leukemia patients also receive chemotherapeutic drugs and bone marrow transplants. To do this, a patient's own marrow is first destroyed by chemotherapy. This destroys the cancerous cells. Healthy bone marrow tissue is then injected into the bloodstream. It travels in the blood to the bone marrow, where it recolonizes the marrow.

In the mid-1970s, only one of every four children with leukemia survived. Today, thanks to vincristine, three of every four children with the disease survive! Vincristine was discovered in a plant known as the rosy periwinkle found in tropical rain forests (Figure 6-7). Thousands of other drugs have come from other tropical plants, underscoring the importance of protecting the rain forests and other ecosystems.

FIGURE 6–7 Rosy Periwinkle Many tropical plants contain chemicals that provide extraordinary medical benefits. One substance from the rosy periwinkle has helped physicians treat leukemia. (Courtesy of Forest & Kim Starr.)

Infectious Mononucleosis

Another common disorder of the WBCs is **infectious mononucleosis** (MON-oh-NUKE-clee-OH-siss), commonly called "mono" or "kissing disease." Mononucleosis is a viral infection that causes low-grade fever, sore throat, aches, and swollen lymph glands, especially in the neck. Swollen lymph glands are often painful.

Mono is caused by a virus transmitted through saliva and may be spread by kissing; by sharing silverware, plates, and drinking glasses; and possibly even through drinking fountains. The virus spreads through the body. In the blood, the virus infects lymphocytes in the bloodstream, which causes an increase in the number of monocytes and lymphocytes.

Physicians recommend that mono sufferers get plenty of rest and drink lots of liquids while the immune system eliminates the virus. Within a few weeks, symptoms generally disappear, although weakness may persist for two to three months.

6-5 Blood Clotting

The capillaries are rather delicate structures. Even minor bumps and scrapes can cause them to leak. Fortunately, leakage is quickly halted by blood clotting, one of the most intricate homeostatic systems to have evolved. A simplified version is discussed here.

Platelets

Among the most important agents of blood clotting are the **platelets**, tiny formed elements produced in the bone marrow by fragmentation of a huge cell known as the **megakaryocyte** (MEG-ah-CARE-ee-oh-site) (Figure 6-8). Like RBCs, platelets lack nuclei and organelles and therefore are not true cells. Also like RBCs, platelets are unable to divide. Carried passively in the bloodstream, platelets are coated by a layer of a sticky material, which causes them to adhere to irregular surfaces such as tears in blood vessels or atherosclerotic plaque or networks of the protein fibrin that form at the site of injury.

Clotting is a chain reaction stimulated by the release of a chemical called **thromboplastin** (THROM-bow-PLASS-tin) from injured cells lining damaged blood vessels. Be sure to view Figure 6-9a as you read this description. Thromboplastin is a lipoprotein. It acts on an inactive plasma enzyme in the blood known as **prothrombin** (produced by the liver). Thromboplastin causes prothrombin to be converted into its active form, **thrombin**. Thrombin, in turn, acts on another blood protein, fibrinogen, also produced by the liver. When activated, fibrinogen is converted into fibrin, long, branching fibers that produce a weblike network in the wall of the damaged blood vessel (Figure 6-9b). The fibrin web traps RBCs and platelets, forming a plug that stops the flow of blood to the tissue. Platelets captured by the fibrin web release additional thromboplastin, known as platelet thromboplastin, which causes more fibrin to be laid down, thus reinforcing the fibrin network. Blood clotting occurs fairly quickly. In most cases, a damaged blood vessel is sealed by a clot within 3 to 6 minutes of an injury; 30 to 60 minutes later, platelets in the clot begin to draw the clot inward, stitching the wound together. How do they perform this remarkable task?

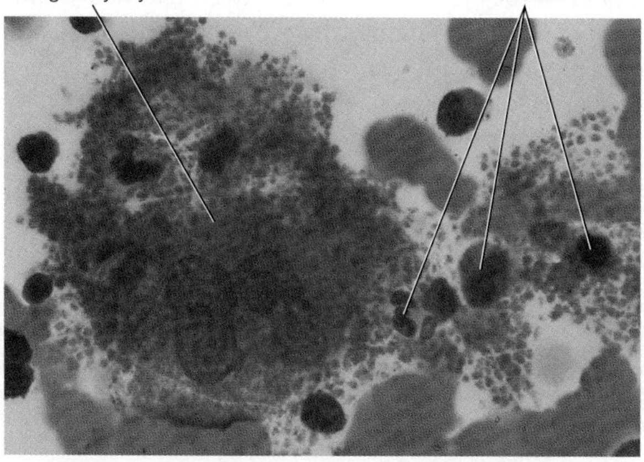

Platelets contain contractile proteins like those in muscle cells. Contraction of the protein fibers draws the fibrin network inward, pulling the edges of the cut or damaged blood vessel together. This closes the wound like a suture.

Blood clots do not stay in place indefinitely. If they did, the circulatory system would eventually become clogged, and blood flow would come to a halt. Instead, clots are dissolved by a bloodborne enzyme known as **plasmin** (PLAZ-min).

As important as blood clots are in protecting the body, clotting can also cause problems. For example, blood clots can break loose from their site of formation, circulate in the blood, and become lodged in arteries, especially those narrowed by plaque. In such cases, the blood clots restrict blood flow, causing considerable damage to tissues served by the vessel. Blood clots typically lodge in the narrow vessels of the heart, brain, and other vital organs. In the heart, they can lead to heart attacks. In the brain, they can lead to strokes.

As noted earlier, blood clotting may be severely impaired by a reduction in blood platelet production—for example, as a result of leukemia. A reduction in blood platelets may also result from exposure to excess radiation, which damages red bone marrow where the megakaryocytes reside. As you may recall, they produce platelets.

Reduced blood clotting can also result from liver damage, which decreases the production of blood-clotting factors. Liver

Megakaryocyte   Platelets

FIGURE 6-8 Megakaryocyte A light micrograph of a megakaryocyte, a large, multinucleated cell found in bone marrow; the megakaryocyte fragments, giving rise to platelets. (© John D. Cunningham/Visuals Unlimited.)

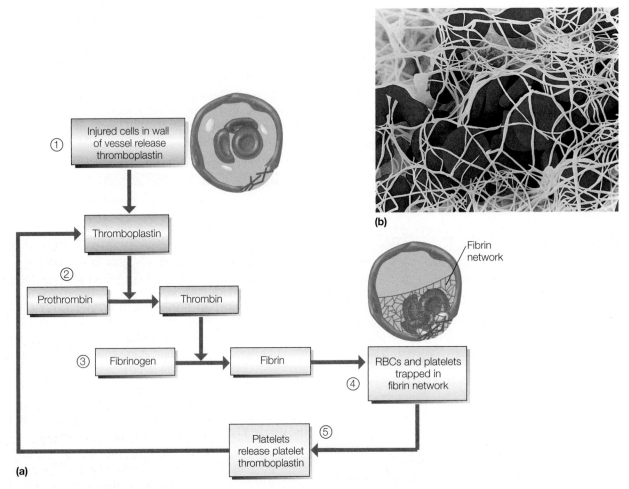

FIGURE 6-9   Blood Clotting Simplified   (a) Injured cells in the walls of blood vessels release the chemical thromboplastin (1). Thromboplastin stimulates the conversion of prothrombin, found in the plasma, into thrombin (2). Thrombin, in turn, stimulates the conversion of the plasma protein fibrinogen into fibrin (3). The fibrin network captures RBCs and platelets (4). Platelets in the blood clot release platelet thromboplastin (5), which converts additional plasma prothrombin into thrombin. Thrombin, in turn, stimulates the production of additional fibrin. (b) A scanning electron micrograph of a fibrin clot that has already trapped platelets and RBCs, plugging a leak in a vessel. The RBCs are red, and the fibrin network is turquoise. (© David M. Phillips/Visuals Unlimited.)

damage may result from infections of the liver (hepatitis), liver cancer, or excessive alcohol consumption. The most common cause of reduced blood clotting is a hereditary genetic defect known as *hemophilia* (he-moe-FEAL-ee-ah), discussed next.

> **KEY CONCEPTS**
>
> Platelets are elements of blood that help form blood clots needed to repair damaged blood vessels.

Hemophilia

Hemophilia is a rare genetic disease that in which the liver fails to produce enough blood clotting factors. Most cases of hemophilia are hereditary, that is, inherited from one's parents.

Problems often begin early in life, so that even tiny cuts or bruises can bleed uncontrollably, which can lead to death. Because of repeated bleeding into the joints, victims suffer great pain and often become disabled; they often die at a young age. In mild forms, the disease may go unnoticed until later in life. The disease is discovered after a traumatic accident or surgery. Hemophiliacs can be treated by transfusions of blood-clotting factors every few days. This therapy is expensive, however.

> **KEY CONCEPTS**
>
> Hemophilia is a potentially life-threatening clotting disorder caused by a genetic defect.

6-6   Blood Types

The surface of the red blood cell membrane, like that of other cells, contains many glycoproteins that form a unique cellular fingerprint (discussed elsewhere in the text). These glycoproteins are the basis of blood typing.

The Four Blood Types

In humans, four blood types exist: A, B, AB, and O. The letters refer to the type of glycoprotein on the plasma membrane of RBCs of an individual. As illustrated in Table 6-3, individuals with

TABLE 6-3 | Summary of Blood Types

Blood Type	Glycoproteins (antigens) on Plasma Membranes of RBCs	Antibodies in Blood	Safe to Transfuse To	Safe to Transfuse From
A	A	b*	A, AB	A, O
B	B	a	B, AB	B, O
AB	A+B	—	AB	A, B, AB, O
O	—	a+b	A, B, AB, O	O

*Lowercase b indicates antibody to B antigen.

type A blood have RBCs whose plasma membranes contain the A glycoprotein. The RBCs in individuals with type B blood contain type B glycoprotein. People with AB blood have both A and B glycoproteins, and people with type O have neither one.

Physicians learned a long time ago that blood could be successfully transfused from one person to another, but only if their blood types matched. In other words, an individual with type A blood could only receive type A blood, and individuals with type B blood can only receive type B blood.

Cross-matching blood is essential to prevent life-threatening immune reactions. These reactions result from antibodies in the blood of transfusion recipients. To understand this phenomenon, take a look at Table 6-3. As you can see, people with type A blood, for example, naturally contain antibodies to the B glycoprotein, indicated by a small "b" in the table. (This is why individuals with type A blood cannot receive type B blood.) People with type B blood contain antibodies to the A glycoprotein, indicated by small "a." For reasons not well understood, these antibodies appear in the blood during the first year of life.

What this means is that people with Type A blood will react to an infusion of Type B blood. Their antibodies will attack the Type B cells in the transfused blood. People with Type B blood react adversely to a transfusion containing Type A blood.

Serious health problems—even death—can occur when incompatible blood types are mixed. For example, consider what happens if an individual with type A blood is accidentally given type B blood (Figure 6-10). The antibodies to type B blood in the recipient quickly bind to the transfused RBCs (which contain type B glycoproteins). This causes the type B RBCs to clump together (agglutinate) and burst. This potentially deadly response is known as the *transfusion reaction* (Figure 6-10).

Health Tip 6-1

When seeking a way to maintain or lose weight, avoid the blood type diet.

Why?

According to the originator of this idea, Peter D'Amado, an individual's diet should be tailored to his or her blood type. Individuals with type O blood, he says, are told to eat plenty of red meat, while people with type A blood are advised to eat more grains. Despite what D'Amado says, nutritionists and others note that there is no scientific evidence supporting a different diet for people of different blood types.

RBC clumping restricts blood flow through capillaries, reducing oxygen and nutrient flow to cells and tissues. Massive breakdown of RBCs results in the release of large amounts of hemoglobin into the blood plasma. Hemoglobin precipitates in the kidney, blocking the tiny tubules that produce urine. This can result in acute kidney failure, which can result in death.

Because of the possibility of this potentially life-threatening reaction, successful transfusions require careful matching of the blood types of the donor and recipient. But there's more to blood typing and transfusion.

As shown in Table 6-3, individuals with type AB blood contain RBCs with both A and B glycoproteins. Their blood can be safely transfused into genetically similar people, that is, people with Type AB blood (because they are the only ones without A or B antigens). If Type AB blood is transfused into an individual with any other blood type, a deadly transfusion reaction will occur.

Also shown in Table 6-3, AB blood contains no antibodies related to the ABO system. As a result, people with AB blood can receive blood transfusion from all four blood types: A, B, AB, and O. Because of this, they are referred to as *universal recipients*.

Next on Table 6-3 you encounter individuals with type O blood. As shown, they have neither A nor B glycoproteins. The lack of these antigens means that type O blood can be transfused into individuals with all four blood types: A, B, AB, and O. As a result, type O individuals are said to be *universal donors*.

Also shown in Table 6-3, type O blood does contain antibodies to both A and B glycoproteins. What this means is that individuals with Type O blood cannot receive transfusions from individuals with Type A, B, or AB blood. Safe transfusions can be made only from people with Type O.

KEY CONCEPTS

Blood types are determined by the presence of specific glycoproteins found in the cell membranes of red blood cells.

Blood Transfusion

The terms universal donor and universal recipient are somewhat misleading because RBCs also contain nearly 200 other glycoproteins that can cause transfusion reactions. The most important of these is the **Rh factor**. This antigen was first identified in rhesus (REE-suss) monkeys; hence the designation Rh. People whose cells contain the Rh antigen, or Rh factor, are said to be *Rh-positive*. Those without it are *Rh-negative*.

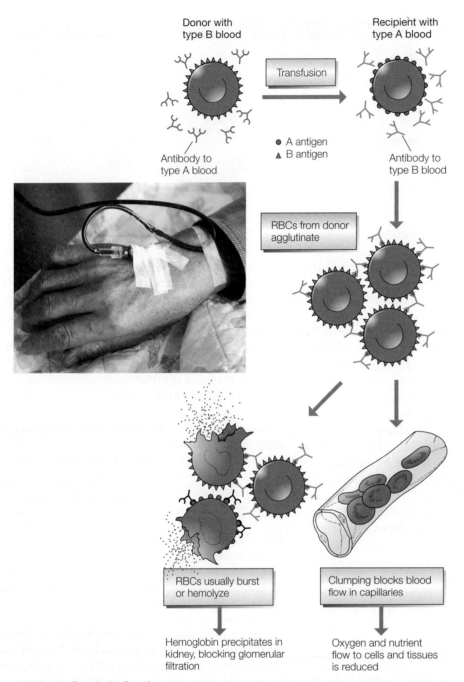

Donor with
type B blood

Recipient with
type A blood

Transfusion

● A antigen
▲ B antigen

Antibody to
type A blood

Antibody to
type B blood

RBCs from donor
agglutinate

RBCs usually burst
or hemolyze

Clumping blocks blood
flow in capillaries

Hemoglobin precipitates in
kidney, blocking glomerular
filtration

Oxygen and nutrient
flow to cells and tissues
is reduced

FIGURE 6-10 Transfusion Reaction Type B blood transfused into an individual with type A blood results in a transfusion reaction, characterized by agglutination and hemolysis. (Photo © Wa Li/Dreamstime.com.)

Unlike the ABO system, in which blood naturally contains antibodies to other blood types without prior exposure to them, in the Rh system, the antibodies are produced only when Rh-positive blood is transfused into the bloodstream of a person with Rh-negative blood. The first transfusion of Rh-positive blood into an Rh-negative person generally does not result in a transfusion reaction, but a second transfusion does. To reduce the likelihood of a transfusion reaction, Rh-negative people should receive only Rh-negative blood, and Rh-positive people should receive only Rh-positive blood.

The Rh factor becomes particularly important during pregnancy. Problems can arise if an Rh-negative mother has an Rh-positive baby (Figure 6-11). Even though the maternal and fetal bloodstreams are separate, small amounts of fetal blood usually leak into the maternal bloodstream at birth. Rh antibodies form in the maternal bloodstream, and the woman becomes sensitized to the Rh factor.

To prevent antibody production in Rh-negative women who give birth to Rh-positive babies, physicians routinely inject antibodies to fetal Rh-positive RBCs into the mother soon after she has given birth. (The antibody-containing serum is called RhoGAM.) These antibodies bind to Rh-positive RBCs from the fetus before a woman's immune system responds to them. This, in turn, prevents a woman from being sensitized. To be effective, however, the treatment must be given within 72 hours after the baby is born.

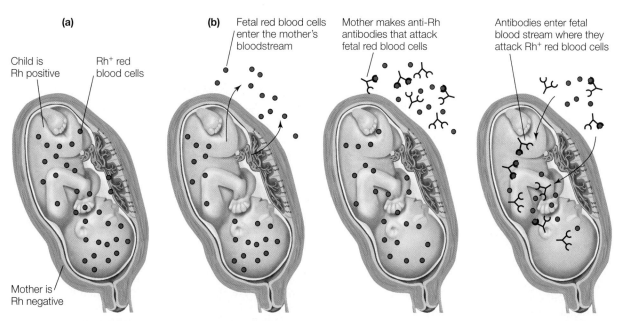

(a)

Child is Rh positive

Rh⁺ red blood cells

Mother is Rh negative

(b)

Fetal red blood cells enter the mother's bloodstream

Mother makes anti-Rh antibodies that attack fetal red blood cells

Antibodies enter fetal blood stream where they attack Rh⁺ red blood cells

FIGURE 6-11 The Rh Factor and Pregnancy (a) Rh-positive cells from the fetus enter the mother's blood at birth. If the mother is Rh-negative, her immune system responds, producing antibodies to the Rh-positive RBCs and destroying them. (b) Problems arise if the mother becomes pregnant again and has another Rh-positive baby. If the mother was not treated the first time, antibodies to Rh-positive RBCs cross the placenta and destroy fetal RBCs.

If the woman is not treated at the time and becomes pregnant again with an Rh-positive baby, maternal antibodies to the Rh factor will cross the placenta. In the fetal circulation, these antibodies cause fetal RBCs to clump together, then break down, resulting in anemia and lack of oxygen to tissues. Unless the baby receives a blood transfusion (of Rh-negative blood) before birth and several transfusions after birth, it is likely to have brain damage and may even die.

KEY CONCEPTS

Blood from one person can be used to supply another but the blood must match genetically with respect to blood type and Rh factors for a successful transfusion.

6-7 Health and Homeostasis

The blood is vitally important to homeostasis. It transports nutrients and waste to and from the cells, respectively. It protects against changes in pH. It transports excess heat to the body's surface, where it is eliminated. It plays a key role in the body's defense system, and it seals injuries in blood vessels through clots.

It is no surprise then to learn that cells depend mightily on the proper functioning of blood. Anything that upsets this function can upset cellular function throughout the body.

As you have seen in this chapter, many factors can harm the blood system, upsetting homeostasis. Genetic defects that result in a reduction in liver blood clotting agents lead to hemophilia. Excessive alcohol consumption can damage the liver and reduce the output of blood clotting factors. Leukemia, caused by exposure to cancer-causing agents, disrupts RBC and platelet production. This, in turn, decreases oxygen transport to the cells of the body and causes a marked decrease in blood clotting that leads to a dramatic increase in internal bleeding.

Another, even more common, disruptor of homeostasis is a common air pollutant, known as *carbon monoxide*. **Carbon monoxide (CO)** is a colorless, odorless gas produced by the incomplete combustion of organic fuels. It can be released by gas-powered cook stoves, furnaces, wood stoves, and water heaters in our homes. It is also produced by automobiles, trucks, and busses, causing increased levels along streets and highways, in parking garages, and in tunnels. It is released by power plants and factories, polluting the air in our cities. It is even a major pollutant in tobacco smoke. According to estimates by the Environmental Protection Agency, over 40 million Americans (one of every seven people) are exposed to levels of CO that are harmful to their health.

What makes CO so dangerous? Carbon monoxide, like oxygen, binds to hemoglobin. However, hemoglobin has a much greater affinity for CO than for oxygen (about 200 times greater). Consequently, CO "outcompetes" oxygen for the binding sites on the hemoglobin molecules and thus reduces the blood's ability to carry oxygen, which is essential for energy production.

For healthy people, low levels of CO in their blood do not create much of a problem. Their bodies simply produce more RBCs or increase the heart rate to augment the flow of oxygen to tissues. Carbon monoxide does create a problem, however, when levels are so high that the body cannot compensate. At very high levels, CO becomes deadly.

In levels commonly found in and around cities, CO is especially troublesome for the elderly. Carbon monoxide places additional strain on their hearts, making their already weakened hearts work even harder. Research shows that CO levels currently deemed acceptable by federal standards can trigger chest pain (angina) in adults with coronary artery disease. Chest pains result from a lack of oxygen in the heart muscle. High levels of CO can cause heart attacks among susceptible individuals. The elderly and people suffering from cardiovascular disease, therefore, are advised to stay inside on high-pollution days.

> **KEY CONCEPTS**
>
> Carbon monoxide is a common air pollutant found inside and outside our homes. It is produced by the combustion of fossil fuels. This pollutant binds to hemoglobin, reducing oxygen transport. At high levels, it can be lethal.

SUMMARY

Blood: An Overview

1. Blood is a watery tissue consisting of two basic components: (1) the plasma, and (2) the formed elements—red blood cells, white blood cells—and platelets, which are suspended in the plasma.

2. The functions of the formed elements are listed in Table 6-2. Take a moment to study it.

Blood Plasma

3. The plasma constitutes about 55% of the volume of a person's blood, and the formed elements make up the remainder.

4. Plasma is a water fluid that contains dissolved nutrients, proteins, gases, and wastes.

Red Blood Cells

5. Red blood cells (RBCs) are highly specialized cells that lack nuclei and organelles. Produced by the red bone marrow, RBCs transport oxygen in the blood.

6. Red blood cells contain the protein hemoglobin which contains iron atoms. Oxygen binds to the iron in the hemoglobin molecules and is transported through the bloodstream bound to iron.

7. The concentration of RBCs in the blood is controlled by the hormone erythropoietin, produced by the kidney in response to oxygen levels.

White Blood Cells

8. White blood cells (WBCs) are nucleated cells and are part of the body's protective mechanism to combat microorganisms. WBCs are produced in the bone marrow and circulate in the bloodstream, but do most of their work outside it, in the body tissues.

9. The most abundant WBCs are the neutrophils, which are attracted by chemicals released from infected tissue. Neutrophils leave the bloodstream and migrate to the site of infection by amoeboid movement.

10. Neutrophils are the first WBCs to arrive at the site of an infection, where they phagocytize microorganisms, preventing the spread of bacteria and other organisms.

11. The second group of cells to arrive is the monocytes, which phagocytize microorganisms, dead cells, cellular debris, and dead neutrophils.

12. Lymphocytes are the second most numerous WBCs and play a vital role in immune protection.

13. Two types of lymphocytes are T cells which provide cellular immunity and B cells which produce antibodies. Antibodies destroy foreign substances.

Blood Clotting

14. Platelets are fragments of large bone marrow cells and are involved in blood clotting.

15. Platelets are coated by a layer of a sticky material, which causes them to adhere to irregular surfaces such as tears in blood vessels. Blood clotting is summarized in Figure 6-9.

16. White blood cells can become cancerous. Two types of cancers are acute and chronic leukemia. In both diseases, cancerous white blood cells in the bone marrow crowd out RBCs and platelets, resulting in anemia and reduced blood clotting.

17. Cancerous white blood cells do a poor job of fighting infection, so patients typically succumb to infections and internal bleeding.

Blood Types

18. Blood types are determined by the types of glycoprotein found in RBC cell membranes.

19. Four blood types are encountered: A, B, AB, and O. A and B blood contain A and B antigens in the cell membranes of the RBCs. Each contains naturally occurring antibodies. Type A blood contains antibodies to Type B. Type B blood naturally contains antibodies to Type A. Because of this, blood transfusions require careful matching of donor and recipient blood types so antibodies in a recipient's blood do not react with the A or B glycoproteins.

20. Transfusions also require matching of the Rh factor.

Health and Homeostasis

21. Many factors influence the blood, upsetting homeostasis and leading to disease and even death.

22. Carbon monoxide is produced by the incomplete combustion of organic fuels in our homes, automobiles, factories, and power plants.

23. Carbon monoxide tightly binds to hemoglobin and reduces the oxygen-carrying capacity of the blood, which impairs cellular energy production, upsetting homeostasis. At high concentrations, carbon monoxide can be lethal. At lower concentrations, it is harmful to the elderly and to people with cardiovascular and lung disease.

THINKING CRITICALLY ANALYSIS

This analysis corresponds to the Thinking Critically scenario that was presented at the beginning of this chapter.

Although the results seem compelling, you haven't really proven anything yet. Why? Interestingly, if patients believe they are receiving a drug that will cure them, many will get better, even if the treatment is nothing more than a sugar pill. Recall that in this experiment, the control group received no treatment whatsoever. They were simply observed. The experimental group received medication. If they believed it would make them better, it just might.

So, how would you improve this experiment?

One way would be to give the control group a placebo—that is, a pill or injection that looks just like the one given to the experimental group. If the results were the same—that is, 60% of the experimental group and 10% of the control recovered—could you reliably report back to the drug company that their drug worked?

Not really.

Why?

If the researchers who doled out the drugs and placebos knew which was which—that is, which pill or injection was the real McCoy—they could give subtle clues to the patients that might bias the results. How can this potential problem be fixed?

Scientists like to solve this problem by performing *double blind studies*. That is, they set up the experiment so that patients don't know what they are getting and researchers don't know what they are giving (hence the name double blind). Only the person in charge of the experiment knows who gets what, and he or she is not involved in doling out the meds. He or she keeps it a secret until the end of the study. That way, any improvements in the health of the subjects can be assessed accurately. If the experiment is carried out this way, can one be sure of the results?

Not really. One would have to be sure that the groups were nearly identical. Did each of the groups consist of an equal number of men and women? Were the patients in each of the groups roughly the same age? Did they exercise similarly? Did they eat the same diet? How would a researcher control for all of these potentially complicating variables?

Researchers would randomly assign patients to each category. By using a large number of randomly assigned patients, they might be able to create an experimental group and a control group that were fairly identical. They could also monitor diet and exercise and study factors such as age, use of other medications, and smoking and alcohol use when analyzing the data at the end of the experiment to see if any of these factors influenced the outcome of the experiment.

KEY TERMS AND CONCEPTS

Anemia, p. 120
Antibodies, p. 117
Appendicitis, p. 122
B lymphocyte (B cell), p. 121
Bone marrow, p. 119
Carbon monoxide, p. 127
Carrier proteins, p. 117
Erythrocyte, p. 118
Erythropoietin (EPO), p. 120
Fibrin, p. 117
Fibrinogen, p. 117
Hematocrit, p. 116
Hemoglobin, p. 120

Hemophilia, p. 124
Infectious mononucleosis, p. 123
Leukemia, p. 122
Leukocytes, p. 121
Lymphocyte, p. 121
Macrophage, p. 121
Megakaryocyte, p. 123
Monocyte, p. 121
Neutrophil, p. 121
Plasma, p. 116
Plasma cell, p. 121
Plasmin, p. 123

Platelet, p. 123
Prothrombin, p. 123
Red blood cells, p. 118
Red marrow, p. 119
Rh factor, p. 125
Sickle-cell disease, p. 118
Stem cells, p. 119
T lymphocyte (T-cell), p. 121
Thrombin, p. 123
Thromboplastin, p. 123
White blood cells, p. 121
Yellow marrow, p. 119

CONCEPT REVIEW

1. To be successful in saving lives of accident victims, what functions must artificial blood perform? p. 116.
2. What is blood? What are its major components? About how many liters of blood do you have in your body? p. 116.
3. What is the hematocrit? What medical conditions does a lower hematocrit indicate? pp. 116–117.
4. What functions does blood plasma perform? Make a list and describe each one, including the component in the plasma that carries out this role. p. 117.

5. What role do carrier proteins play? Why are they so important for the transport of lipids in the blood? p. 117.
6. Why must RBCs be so flexible? What happens when RBCs lose their flexibility? p. 118.
7. Describe hemoglobin. What is it? What is it made of? What function does it perform? Where is the iron found in hemoglobin recycled? p. 120.
8. Describe the structure and function of each of the following: platelets, lym-

phocytes, monocytes, and neutrophils. pp. 121–124.
9. What is the difference between chronic and acute leukemia? Which one is more lethal? p. 122.
10. What are the symptoms of leukemia and how is this disease treated? p. 122.
11. List some causes of anemia. p. 120, 122.
12. What is hemophilia? How common is the disease? What are the symptoms? What are the causes of this disease? p. 124.

13. Explain how a blood clot forms and how it helps prevent bleeding. pp. 123–124.
14. Create a chart of the four blood types, including the type of antigens each carries, the antibodies that are found in each blood type and then list the blood types to which the blood can be safely transfused. Make a final column showing what blood types are safe to transfuse to each type. pp. 124–125.
15. What is the Rh factor? What should a woman who is Rh negative receive a shot of antibody if she gives birth to an Rh positive baby? pp. 125–127.
16. Describe the many ways blood participates in homeostasis. How can these homeostatic mechanisms be upset? pp. 127–128.

SELF-QUIZ: TESTING YOUR KNOWLEDGE

1. Plasma comprises _____ percent of the blood volume. p. 116.
2. Plasma proteins are the most abundant dissolved substance in blood and are responsible for maintaining _____ and the osmotic concentration of the blood, among other things. p. 117.
3. _____ proteins bind to smaller molecules and help transport them throughout the bloodstream. p. 117.
4. The _____ blood cell is the most abundant blood cell in the body and is involved in transporting oxygen throughout the body. p. 118.
5. Loss of flexibility of the RBC resulting from a genetic disorder is known as _____ disease. p. 118.
6. RBCs are produced in the _____ bone marrow. The hormone _____ is responsible for controlling the production of RBCs in the body. p. 119–120.
7. A reduction in the oxygen-carrying capacity of blood due to a decrease in RBCs or a decrease in the amount of _____ in RBCs is called _____. p. 120.
8. The first cell to arrive at the site of an infection is a _____. p. 121.
9. The second cell to arrive at a site of infection is the _____. These cells often reside in body tissues and are called _____. p. 121.
10. The _____ is a white blood cell involved in immune protection. p. 121.
11. Antibodies are produce by cells known as _____ cells that are formed from B lymphocytes after exposure to an antigen. p. 121.
12. Patients suffering from leukemia are vulnerable to _____ and suffer from internal bleeding. p. 122.
13. During blood clotting _____ becomes lodged in the _____ network formed at the wound site, stopping the flow of blood. p. 123.
14. The fibers found in a blood clot are formed from the protein _____ which is made from _____. p. 123.
15. Shortly after blood clot forms, platelets begin to pull the torn edges of blood vessels together because they contain _____ proteins like those found in muscle. p. 123.
16. People with type A blood contain the A _____ and B antibody (b). They can receive blood from type A and type O individuals. p. 125.
17. People with type AB blood contain both A and B glycoproteins but _____ antibodies. p. 125.
18. Type AB individuals are universal _____. p. 125.
19. Individuals with type O blood contain _____ glycoproteins. p. 125.
20. Type O individuals can receive blood from individuals with _____ blood types. p. 125.

biology.jbpub.com/chiras/8e/

The site features eLearning, an online review area that provides quizzes, chapter outlines, and other tools to help you study for your class. You can also follow useful links for in-depth information, research the differing views in the Point/Counterpoints, or keep up on the latest health news.

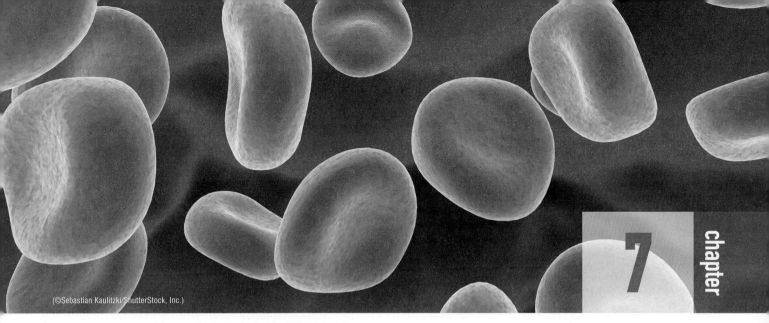
(©Sebastian Kaulitzki/ShutterStock, Inc.)

Nutrition and Digestion

It must be a law of human nature. Ask almost any couple, and they will tell you: She lies shivering under the covers on a cold winter night while he bakes. Out on a hike in winter, he stays warm in a light jacket while she bundles up in a down coat. What causes this difference between many men and women?

THINKING CRITICALLY

A group of researchers designed an experiment to test the widely held belief that sugar consumption leads to hyperactivity in children. They studied a group of 360 children on a school playground. The researchers divided the children into two groups, based on their observations. Of the 360 students, 300 were classified as hyperactive; the remaining 60 students were not. The researchers then asked the children if they had consumed candy prior to their assessment. In the hyperactive group, consisting of 300 children, 250 had consumed candy. In the nonhyperactive group, consisting of 60 children, 50 had consumed candy, and 10 had not. How would you interpret these data? Is the hypothesis that eating candy and other sweets leads to hyperactivity in children valid?

Part of the answer may lie in iron—not pumping iron, but dietary iron. Quite simply, many American women do not consume enough iron to offset losses that occur during menstruation, the monthly discharge of iron-rich blood from the lining of the uterus. Iron deficiencies in women may reduce internal heat production.

John Beard, a researcher at Pennsylvania State University, published a study that supports this conclusion. Beard compared two groups of women, one with low levels of iron in the blood and another with normal levels. Beard found that body temperature dropped more quickly in iron-deficient women exposed to cold than in those with normal iron levels. He also found that iron-deficient women generated 13% less body heat. Beard continued his study by asking the iron-deficient women to take iron supplements for 12 weeks, after which they responded normally to cold.

These findings illustrate how important one nutrient is to normal body function. Bear in mind that the human body needs dozens of nutrients, including vitamins, minerals, and a whole host of antioxidants to maintain health. In this chapter, we'll explore basic human nutrition and digestion.

7-1 An Introduction to Nutrition

The nutrients we need to survive and prosper physically and mentally can be divided into two broad categories: macronutrients and micronutrients. The term **macronutrients** refers to chemical substances in the food we eat that are required in fairly large quantity to maintain energy levels and health. Macronutrients include: proteins, carbohydrates, lipids, and water.

For many years, nutritionists concentrated most of their attention on only three macronutrients: carbohydrates, lipids, and proteins. A balanced diet, they asserted, consisted of proper proportions of each. What they found, however, was that to remain healthy, people also needed numerous micronutrients. As their name implies, **micronutrients** are substances required in rather small quantities such as the vitamins. Without them, a balanced diet, consisting of the proper ratio of proteins, lipids, and carbohydrates, actually resulted in serious deficiency diseases such as scurvy, rickets, and anemia, to name a few. In recent years, nutritionists and doctors have found that there are hundreds, if not thousands, of micronutrients that are required to maintain health and prevent disease. Table 7-1 lists the basic nutrients as well as some of the foods and beverages that you can consume to ensure you receive adequate quantities. Let's take a look at each group.

Macronutrients

Water is one of the most important of all the substances we ingest. Without it, you can survive only about 3 days. Despite its importance, water is not included on the food pyramid, a tool used to guide us in achieving proper nutrition (discussed shortly). In part, that is because it is supplied in so many different ways. For example, water constitutes the bulk of the beverages we drink and is present in virtually all of the solid foods we eat such as lettuce, celery, apples, and even meats and cheeses.

Maintaining the proper level of water in the body is important for several reasons. First, water participates in many chemical reactions in the body—for example, the breakdown of several large chemicals, such as starch. The breakdown of a chemical using water is called *hydrolysis* (high-DROL-ah-siss). A simplified version is shown in Figure 7-1. In these reactions, enzymes add water across covalent bonds, causing them to break apart.

Because of water's importance in the body, a decrease in internal levels can impair metabolism, including energy production. Athletic performance may drop significantly when the body's water level falls even slightly. When you are feeling tired, part of the reason may be that you are slightly dehydrated.

Maintaining an adequate water volume also helps to stabilize body temperature. A decline in the amount of water in your body decreases blood volume and extracellular fluid. Because the heat normally produced by body cells is being absorbed by a smaller volume of water, body temperature will

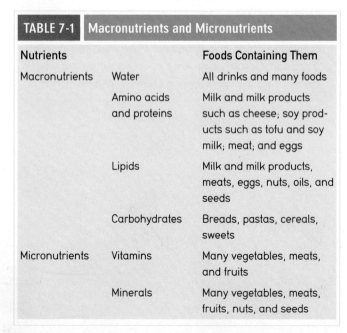

TABLE 7-1	Macronutrients and Micronutrients	
Nutrients		**Foods Containing Them**
Macronutrients	Water	All drinks and many foods
	Amino acids and proteins	Milk and milk products such as cheese; soy products such as tofu and soy milk; meat; and eggs
	Lipids	Milk and milk products, meats, eggs, nuts, oils, and seeds
	Carbohydrates	Breads, pastas, cereals, sweets
Micronutrients	Vitamins	Many vegetables, meats, and fruits
	Minerals	Many vegetables, meats, fruits, nuts, and seeds

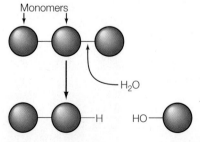

FIGURE 7-1 Hydrolysis
Hydrolysis is a reaction in which water is added across a covalent bond, causing the bond to split.

rise. A rise in body temperature can impair cellular function and this is the reason doctors advise patients suffering from fever to drink plenty of water. A high enough temperature may lead to death.

Water also helps maintain body temperature through perspiration, the release of water from sweat glands. The evaporation of water from the surface of the body, carries away heat, thereby cooling us off and preventing a rise in body temperature.

Maintaining proper water levels also helps individuals maintain normal concentrations of nutrients and toxic waste products in the blood and extracellular fluid. If you don't drink enough liquid, your urine will become more concentrated. An increase in the concentration of chemicals in the blood and urine increases your chance of developing kidney stones.

> **KEY CONCEPTS**
> Macronutrients are substances required in large quantity like water, carbohydrates, lipids, and proteins.

Carbohydrates

Carbohydrates are a group of organic compounds that includes such well-known examples as table sugar (sucrose), blood sugar (glucose), and starch. Carbohydrates consist primarily of carbon, oxygen, and hydrogen. Carbohydrates fit into one of three broad groups: monosaccharides, disaccharides, and polysaccharides.

Monosaccharides are the smallest carbohydrates These water-soluble molecules contain three to seven carbons. Many of these carbohydrates form ring structures. One of the most common and most important monosaccharides is **glucose** (GLUE-kose), a six-carbon sugar. Glucose is made by plants during photosynthesis. It is one of the main sources of energy in the human body.

Humans need a continuous supply of energy, even the most ardent couch potatoes! Interestingly, 70% to 80% of the total energy required by sedentary humans (individuals who get little exercise) is used to perform basic functions: metabolism, food digestion, absorption, and so on. The remaining energy is used to power body movements such as walking, talking, and turning on the television via the remote control. For more active people, these percentages shift considerably. That is, a higher percentage of energy goes to muscular activity. **Disaccharides** (dye-SACK-are-ides) are carbohydrates that consist of two monosaccharides covalently bonded to each other. Table sugar consists of the disaccharide sucrose, that is, two monosaccharides—glucose and fructose—covalently linked to each other.

In animals and plants, many monosaccharides combine to form much larger molecules known as *polymers*. A polymer is a general term used to describe any molecule consisting of numerous smaller molecules. Proteins, for instance, are polymers of amino acids. Starch is a polymer containing numerous glucose molecules covalently bonded to one another. We call these carbohydrate polymers **polysaccharides** (poly = many; saccharides = sugars).

The most common building block of polysaccharides is glucose. Plants synthesize two important polysaccharides from glucose: (1) **starch** and (2) **cellulose** (CELL-you-lose). Animals synthesize an equally important polysaccharide, known as glycogen.

Starch molecules are produced in the leaves of plants and are often stored in plant roots (for example, in potato plants) or seeds (as in wheat and rice). Starch is an important nutrient for plant-eating animals. In the digestive tract of humans, starch is broken down into glucose molecules. Glucose enters the bloodstream and is distributed to body cells, where it is broken down to release energy (Figure 7-2). Glucose can also be stored in cells of the liver and skeletal muscle for later use. It is not stored as glucose; however, it is recombined to form yet another polysaccharide, glycogen (GLYE-co-gen). **Glycogen** contains thousands of glucose molecules. Glycogen is often called "animal starch" because it serves a similar purpose and has a structure similar to starch found in plants.

Glycogen is an important player in the homeostatic mechanism that ensures constant blood sugar levels in humans and other animals. When blood glucose levels decline between meals, the liver breaks down its stored supplies of glycogen. The glucose molecules derived from the breakdown of glycogen are released into the bloodstream, maintaining blood glucose concentrations needed to keep us alive and well.

The release of glucose from the liver is stimulated by a hormone produced by the pancreas called *glucagon*. Glucagon secretion is triggered by declining levels of glucose. Glucagon circulates in the bloodstream and enters the liver where it stimulates the breakdown of glycogen into glucose molecules. Glucose restores blood sugar levels.

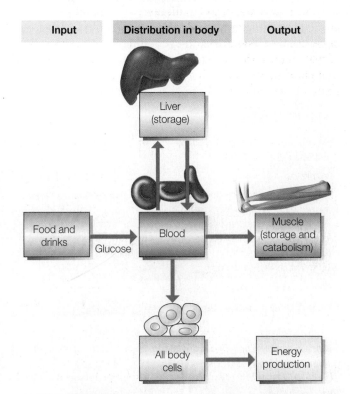

FIGURE 7–2 Glucose Balance Glucose levels in the blood result from a balance created by input and output. Input refers to glucose supplied from the digestion and absorption of food. Output results from storage in muscle and use in body cells for energy production.

Although muscle cells also store glucose in the form of glycogen, they do not participate in maintaining normal blood sugar levels. This glycogen is used only by muscle cells to provide energy.

Cellulose molecules are polysaccharides found only in plants. The covalent bonds that join the glucose molecules are slightly different than those in starch. Because of this, cellulose cannot be digested by humans. We lack the enzymes needed to break the covalent bonds joining its glucose subunits. Consequently, cellulose passes through the digestive systems of humans and most other animals relatively unchanged.

Even though it is not digested, cellulose ingestion is important to human health. In the human diet, cellulose is one of two types of dietary fiber. **Dietary fiber** is found in fruits, vegetables, and grains. Dietary fiber exists in two basic forms: water-soluble and water-insoluble.

Water-soluble fiber consists of gummy polysaccharides in fruits, vegetables, and some grains—including apples, bananas, carrots, barley, and oats—that dissolve in water. Water-soluble fiber helps lower blood cholesterol by acting as a sponge that absorbs cholesterol in food while inside the digestive tract. This prevents cholesterol from being absorbed into the bloodstream.

The second type of dietary fiber, **water-insoluble fiber**, is rigid cellulose molecules in foods such as celery, whole wheat pastas and bread, and brown rice. Some foods, such as green beans and green peas, contain a mixture of both types.

Water-insoluble fiber such as that in whole wheat bread increases the water content of the feces (FEE-seas), the semisolid waste produced by the large intestine. This makes the feces softer and facilitates their transport through the large intestine. Increasing the water content also reduces constipation. Water-insoluble fiber may also decrease cancer of the colon and rectum (parts of the large intestine). You can increase your intake of water-insoluble fiber by consuming celery, whole wheat bread and pasta, brown rice, green beans, and peas, among others.

Health Tip 7-1

When given the choice of an apple or a handful of pretzels, take the apple.

Why?

Both have the same number of calories, but apples provide many more nutrients, including fiber, vitamin C, and potassium, all of which are essential for a healthy lifestyle. Fruits such as apples are said to be "nutrient dense." Pretzels and other snack foods lack nutrients other than fat and carbohydrates, which most of us consume in excess already.

> **KEY CONCEPTS**
>
> Carbohydrates are a diverse group of biological chemicals that vary in complexity; they are one of two major sources of energy for the human body.

Lipids

Lipids are a source of energy but also play an important structural role in the human body.

Lipids are probably the most difficult nutrient to understand because they are a structurally diverse group of organic molecules. All lipids have two features in common: they are insoluble in water and soluble in nonpolar solvents such as ethyl alcohol. (Pour a tiny amount of vegetable oil in a glass of water, which is polar, and then pour some into a small amount of rubbing alcohol, which is nonpolar, to test this.)

Lipids are the waxy, greasy, or oily compounds found in plants and animals that serve a number of vital functions, discussed shortly. Let's look at two biologically important lipids: triglycerides and steroids.

> **KEY CONCEPTS**
>
> Lipids are a structurally diverse group of biological molecules; some are a major source of energy; others are important precursors to other important molecules; still others are an important structural component of cell membranes.

Triglycerides. **Triglycerides** are known to most of us as fats and oils. Cooking oil, for example, is a triglyceride, as is the fat in a steak or the butter on a piece of bread.

Triglycerides are composed of four organic subunits: one molecule of glycerol and three fatty acid molecules (Figure 7-3). As shown in the figure, **glycerol** is a three-carbon compound; **fatty acids** are long molecules containing many carbons and hydrogens and a COOH, or carboxyl group (car-BOX-ul), on one end.

Triglycerides are one of the main components of the cell membranes in your body. Triglycerides are also a main source of energy, easily rivaling glucose. That's because triglycerides contain many covalent bonds, each of which stores a small amount of energy. When cells break down triglycerides, these bonds are broken and energy is released. Energy liberated during the breakdown of triglycerides is used to drive a variety of cellular processes. Gram for gram, triglycerides yield more than twice as much energy as carbohydrates.

In adult humans, triglycerides provide about half of the cellular energy your body needs at rest. Glucose provides the rest. During moderate (aerobic) exercise such as riding a bicycle, triglycerides provide a larger proportion of your body's energy demand, explaining why aerobic exercise like jogging and bicycling can help you lose weight.

Triglycerides are stored in fat cells (Figure 7-4) under the skin in the abdomen around the intestinal track. These stores provide energy between meals. Storage depots around the intestines, heart, and kidneys also cushion them, protecting them from damage. Horseback riders, motor cyclists, runners, and other athletes can engage in activity for hours at a time without damaging internal organs in part because of these natural cushions. With a few exceptions, fats (from most animals) are solid at room temperature, and oils (from plants and fish) are liquid. The reason for this difference lies in their chemical structures. In fats, the carbon atoms of the fatty acids are joined by single covalent bonds (Figure 7-3). The remaining bond sites on the carbon atoms are taken up by hydrogens. Thus, the fatty acids are said to be *saturated* with hydrogens. When the carbon backbone of a fatty acid consists solely of single covalent bonds, the structure zigzags but remains fairly linear (Figure 7-5a). This allows the triglyceride

Formation of a triglyceride

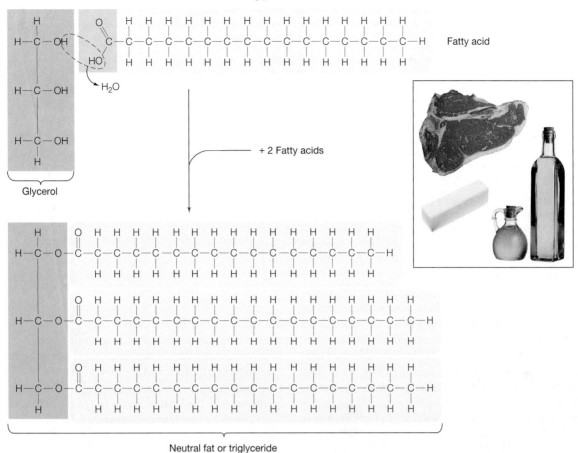

Glycerol

Fatty acid

H₂O

+ 2 Fatty acids

Neutral fat or triglyceride

FIGURE 7-3 Triglycerides The triglycerides are the fats and oils. Triglycerides consist of glycerol and three fatty acids, covalently bonded as shown. (Steak photo © Photodisc; butter photo, courtesy of Renee Comet/National Cancer Institute; oil photo © Photos.com.)

molecules of a fat to pack together, forming a solid at room temperature.

For reasons not well understood, saturated fatty acids increase cholesterol production by the liver, so a diet rich in saturated fat (animal fats) tends to increase an individual's chances of developing atherosclerosis (AH-ther-oh-skler-OH-siss), a disease that results from a buildup of cholesterol deposits, called atherosclerotic plaque, on the walls of arteries (Figure 7-6). Plaque restricts blood flow to the heart and brain that can lead to heart attacks and strokes.

Steak, hamburgers, cheese, chicken with its skin on, bacon, whole milk, and many common snack foods contain lots of saturated fat. In fact, fat is often added to snack foods as a flavoring.

Fat cells

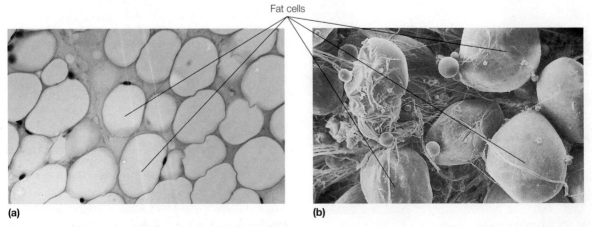

(a)　　　　　　　　　　　　　　　　　**(b)**

FIGURE 7-4 Fat Cells (a) A light micrograph of fat tissue. The clear areas are regions where the fat has dissolved during tissue preparation. Notice that the cytoplasm is reduced to a narrow region just beneath the plasma membrane. (© Donna Beer Stolz, Ph.D., Center for Biologic Imaging, University of Pittsburgh Medical School.) (b) A scanning electron micrograph of fat cells. (© Veronika Burmeister/Visuals Unlimited.)

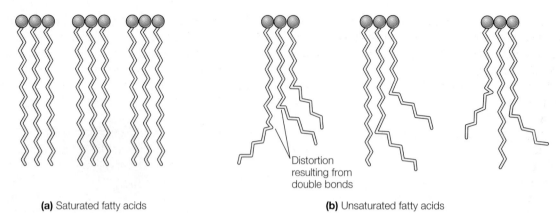

Distortion resulting from double bonds

(a) Saturated fatty acids   **(b)** Unsaturated fatty acids

FIGURE 7-5 **The Difference a Few Double Bonds Can Make** (a) Saturated fatty acids are principally derived from animal fats. The side chains are relatively straight and allow the molecules to pack tightly together, which explains why fats are solid at room temperature. (b) Double bonds in unsaturated fatty acids in oils cause the fatty acid chains to bend and thus prohibit tight packing. Oils, derived chiefly from plants, are therefore liquid at room temperature.

Unlike animal fats, fatty acids in oils from plants contain numerous double covalent bonds as shown in Figure 7-5b. These molecules are said to be *unsaturated*. Double covalent bonds create bends in the long hydrocarbon chains of fatty acid molecules that prevents tight packing. As a result, this somewhat "looser" arrangement of triglyceride molecules results in liquids at room temperature.

When numerous double bonds exist in a fatty acid molecule, the molecule is said to be *polyunsaturated*. Studies show that polyunsaturated fats reduce one's risk of developing atherosclerosis. As just noted, atherosclerosis is common in people who consume lots of fatty meat—hamburgers and steak, for instance. It is also more common in sedentary people and smokers. That said, some fairly lean individuals who eat fairly well can also develop atherosclerosis. These people are genetically predisposed to develop atherosclerosis. That is, their livers produce abnormally high levels of cholesterol.

Atherosclerosis was once thought to be a disease of adults, but studies show that children, teenagers, and young adults are starting to show a surprising amount of arterial plaque. Many researchers are concerned that the very high fat diets of many young children and the lack of exercise could dramatically increase heart disease in youngsters in more developed countries like the United States in the years to come.

Lowering the level of saturated fat in the diet can be accomplished by switching from whole milk to low-fat milk, dramatically reducing the consumption of red meat, trimming fat from chicken and other meats, cutting down on cheese, and cooking with unsaturated vegetable oil instead of animal fat (lard). Reducing snack foods in the diet can help, too. Manufacturers list the total fat and saturated fat content of all packaged foods. So, check out food labels very carefully. (For more on cholesterol, see Health Note 7-1.)

Vegetable oil can be converted to a solid (margarine) by chemically adding hydrogens to it. This process is called

Wall of artery

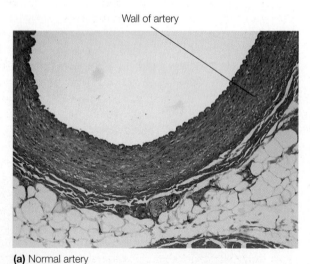

(a) Normal artery

Wall of artery   Plaque

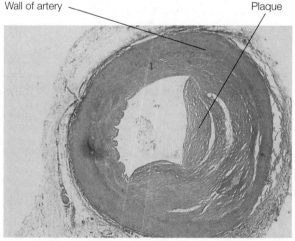

(b) Atherosclerotic artery

FIGURE 7-6 **Atherosclerosis** These cross sections of (a) a normal artery and (b) a diseased artery show how atherosclerotic plaque can obstruct blood flow. (Photo a © Donna Beer Stolz, Ph.D., Center for Biologic Imaging, University of Pittsburgh Medical School; photo b © William Ober/Visuals Unlimited.)

hydrogenation. It decreases the number of double bonds in the fatty acids of vegetable oils. This, in turn, tends to straighten the fatty acid chains, turning hydrogenated vegetable oils (liquids) into margarine (solid).

Although margarine has long been thought to be more healthful than butter, recent studies suggest this is not true. During hydrogenation, the polyunsaturated fatty acids of oils are converted to more saturated fatty acids known as *trans fatty acids*. Studies have shown that *trans* fatty acids, like saturated fatty acids of fatty meats and other foods, also increase one's risk of atherosclerosis. As a result, nutritionists advise us to minimize the intake of margarines and food products, such as crackers and baked goods, that contain *trans* fatty acids. Many fast food restaurants such as Taco Bell have recently vowed to eliminate *trans* fatty acids from their foods. New York City even banned the use of vegetable oils containing *trans* fatty acids in restaurants in 2006. Other cities are following their lead. But what about the advice that you should switch to butter or palm oil because *trans* fats are bad for you?

Don't fall victim to this bad advice. Scientific research doesn't back up such claims. *Trans* fat is no worse than saturated fats found in butter and palm oil. Your goal should be to minimize both types of fat. You can achieve this goal by minimizing your intake of fried foods and fatty meats.

Health Tip 7-2

To live a long, healthy life, avoid *trans* fats in your diet. *Trans* fats are found in foods made with hydrogenated vegetable oil, especially baked goods and snack foods. Check out labels on foods.

Why?

Trans fats increase the risk of coronary heart disease—the buildup of plaque in the arteries of the heart that can lead to heart attacks. Even when *trans* fats constitute 1% to 3% of a person's total daily caloric intake, they significantly increase one's risk of heart disease.

Steroids. As a group, the **steroids** (STEER-oids) are structurally quite different from their lipid cousins, the triglycerides and phospholipids. Steroids consist of three six-carbon rings and one five-carbon ring joined in one large structure resembling chicken wire. One of the best-known steroids is cholesterol. **Cholesterol** is a component of the plasma membrane in animal cells. It is also found in plant cells but typically in lower amounts. In humans, cholesterol is a raw material needed to synthesize other steroids, such as vitamin D, bile salts, and the sex hormones estrogen and testosterone. Cholesterol is also a major component of atherosclerotic plaque. In humans, cholesterol comes primarily from the liver. A lesser amount comes from the diet.

Amino Acids and Protein

Amino acids and proteins are important macronutrients. Their structure is discussed elsewhere in the text. **Protein** consumed in the diet is broken down in the small intestine into **amino acids** that are absorbed into the bloodstream. The amino acids are then used by cells to make a wide variety of body proteins. These include enzymes, hormones, and various structural proteins like the proteins that form the cytoskeleton or the proteins that make up your fingernails.

Proteins in the human body contain 20 different amino acids—all of which can be provided in the diet. The body, however, is capable of synthesizing 12 of the amino acids it needs from nitrogen and smaller molecules derived from carbohydrates and fats. Thus, if these 12 proteins are not present in the diet, they can be made. Two of the 12 are considered *semiessential amino acids*. They are produced in sufficient quantities in adults, but not in growing children. The remaining 10 amino acids cannot be synthesized by human cells under any circumstances. They must be provided by the diet. These amino acids are called **essential amino acids**. A deficiency of even one of the essential amino acids can cause severe physiological problems. As a result, nutritionists recommend a diet containing many different protein sources so individuals receive all of the amino acids they need.

Proteins are divided into two major groups by nutritionists: complete and incomplete. **Complete proteins** contain ample amounts of all of the essential amino acids. Complete proteins are found in milk, eggs, meat, fish, poultry, cheese, soy products, including soy milk and tofu, and at least one grain, known as quinoa (pronounced keen-wha). **Incomplete proteins** lack one or more essential amino acids and include those found in many plant products: nuts, seeds, grains, most legumes (peas and beans), and vegetables. Vegetarians can acquire the proteins they need by consuming milk and eggs as well as an assortment of vegetable protein sources such as whole grains, tofu, quinoa, seeds, nuts, and legumes like Navy beans and lentils. Those who avoid all animal products, including milk and eggs, are known as *vegans*. They acquire the essential amino acids they need by consuming tofu and quinoa, both complete protein sources, or combining two or more incomplete protein sources. Legumes such as black beans can be combined with grains or nuts and seeds, as shown in Figure 7-7, to provide all the amino acids you need. It's that simple. A meal consisting of a legume, for example, baked beans, combined with wheat bread will provide all of the amino acids your body needs. A bean burrito and a corn tortilla will do the same. Hard as it is for some to accept, you don't need to eat steak and chicken or fish to acquire the protein you need. As you will see in our study of MyPyramid in an upcoming section, we need very little protein to meet our daily requirements.

While vegans and vegetarians can acquire the protein they need by combining protein food sources their diet is often low in vitamin B_{12}. Over time, people who are deficient in the vitamin experience many adverse symptoms, such as fatigue, weight loss, upset stomach, diarrhea, or constipation, numbness, confusion, and memory loss, and others. As a result, The Vegan Society recommends: (1) eating B_{12} fortified foods (like fortified cereals) two or three times a day to get at least three micrograms per day, (2) taking B_{12} supplement daily providing at least 10 micrograms, or (3) taking a weekly B_{12} supplement that provides at least 2,000 micrograms. Although many people think of proteins as a source of energy, that's rarely the case. Proteins are only used for to make energy when dietary intake of carbohydrates and fats is severely restricted (when someone is starving, for instance) or when protein intake far exceeds demand.

Protein deficiencies occur in millions of children throughout the world. The arms and legs of these children

health**note**

7–1   Lowering Your Cholesterol

Diseases of the heart and arteries are leading causes of death in the United States. One of the most common diseases is atherosclerosis. *Atherosclerosis* is the accumulation of cholesterol in the linings of arteries. This material, called *plaque*, is responsible for nearly two of every five deaths in the United States each year. New research shows that atherosclerotic plaque is present even in children and teenagers who eat unhealthy diets. Thanks to improvements in medical care and diet, the death rate from atherosclerosis has been falling steadily in recent years, but it is still a major concern.

Researchers believe that atherosclerotic plaque begins to form after minor injuries to the lining of blood vessels, especially arteries. How does that occur? One source of damage is high blood pressure. Damage caused by high blood pressure and other causes allow cholesterol in the blood to seep through the lining into underlying cells of blood vessels. Certain cells (known as macrophages) in the walls of the blood vessels begin to absorb the cholesterol. When they do, they are called foam cells, so named because they appear foamy under the microscope. The blood vessel responds to the presence of foam cells by producing additional cells that grow over the fatty deposit, thickening the wall of the artery, reducing blood flow. Additional cholesterol is then deposited in the thickened wall, forming a larger and larger obstruction.

Cholesterol deposits impair the flow of blood in the heart and other organs, cutting off oxygen to tissues. Making matters worse, blood clots may form in the restricted sections of arteries, further reducing blood flow. When the oxygen supply to the heart is disrupted, cardiac muscle cells can die, resulting in heart attacks and death. Blood clots originating in other parts of the body may also lodge in diseased vessels, obstructing blood flow. Oxygen deprivation can weaken the heart, impairing its ability to pump blood. When the oxygen supply to the heart is restricted, the result is a type of heart attack known as a *myocardial infarction*. If the oxygen-deprived area is extensive, the heart may cease functioning altogether.

Atherosclerotic plaque also impairs the flow of blood to the brain. Blood clots circulating in the bloodstream may gather in the restricted areas, further blocking the flow of blood to vital regions of the brain. The loss of blood flow reduces the flow of oxygen and glucose to nerve cells, causing them to die. Whole regions of the brain may die. As a result, victims may lose the ability to speak or to move limbs. If the damage is severe enough, a patient may die. Those who survive often face long rehabilitation and usually recover lost functions as other parts of the brain take over for the damaged regions.

Atherosclerosis and cardiovascular disease are associated with nearly 40 risk factors. Of all the risk factors, three emerge as the primary contributors to cardiovascular disease: elevated blood cholesterol, smoking, and high blood pressure.

Consider cholesterol. Cholesterol is essential to normal body function, as noted in the text. In healthy individuals who consume a healthy diet, the majority of the cholesterol in the blood is produced by the liver. The liver regulates the concentration of cholesterol in the blood, keeping it fairly constant. Thus, if dietary input falls, the liver increases its output. If the amount of cholesterol in the diet rises, the liver reduces its production. So what's all the fuss about cholesterol in a person's diet?

Many people consume far more cholesterol or substances that contribute to high cholesterol levels than is healthy. Even though the liver regulates cholesterol levels, it cannot work fast enough to remove dietary excesses. That is, it may simply be unable to absorb, use, and dispose of

---

are often thin and wiry because their muscle cells break down protein in an effort to provide energy (Figure 7-8b). According to the United Nations, one of every six people in the world, over one billion people, suffer from protein malnutrition.

On the other end of the spectrum are the overfed populations of the world. The average American, for example, consumes twice the daily requirement of protein. Amino acids derived from the surplus dietary protein are broken down in the body to produce energy and fat.

Overnutrition

Residents of many countries, Americans included, ingest numerous nutrients in quantities far greater than is necessary. Excess food intake, called **overnutrition**, now rivals **undernutrition**, a lack of food. In the United States, nearly 70%

of all American adults are overweight, severely overweight, or obese (Figure 7-8a). The percentage of the population classified as severely overweight or obese has climbed from 15% to nearly 36% since 1980, according to the American

FIGURE 7–7 Complementary Protein Sources
By combining protein sources, a vegetarian who consumes no animal by-products can be assured of getting all of the amino acids needed. Legumes can be combined with foods made from grains or nuts and other seeds.

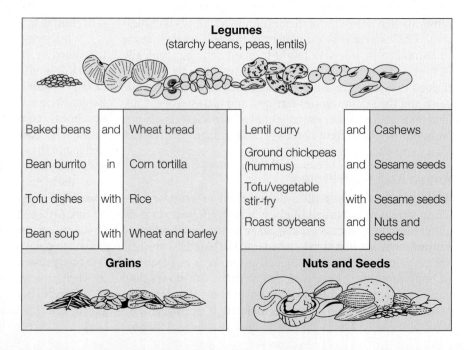

Legumes
(starchy beans, peas, lentils)

Baked beans	and	Wheat bread		Lentil curry	and	Cashews
Bean burrito	in	Corn tortilla		Ground chickpeas (hummus)	and	Sesame seeds
Tofu dishes	with	Rice		Tofu/vegetable stir-fry	with	Sesame seeds
Bean soup	with	Wheat and barley		Roast soybeans	and	Nuts and seeds

Grains   **Nuts and Seeds**

cholesterol quickly enough. Consequently, excess cholesterol circulates in the blood after a meal and is deposited in the arteries.

Cholesterol is carried in the bloodstream bound to protein. These complexes of protein and lipid fall into two groups: *high-density lipoproteins* (HDLs) and *low-density lipoproteins* (LDLs). HDLs and LDLs function very differently. LDLs, for example, transport cholesterol from the liver to blood vessels in body tissues. In contrast, HDLs are scavengers, picking up excess cholesterol and transporting it to the liver, where it is removed from the blood and excreted in the bile. Research shows that the ratio of HDL to LDL is an accurate predictor of cardiovascular disease. The higher the ratio, the lower the risk of cardiovascular disease.

High cholesterol (or hypercholesterolemia) tends to run in families. Thus, if a parent died of a heart attack or suffers from this genetic disease, his or her offspring are more likely to have high cholesterol levels.

To reduce your chances of atherosclerosis, the American Heart Association recommends that you (1) limit dietary fat to less than 30% of the total caloric intake; (2) limit dietary cholesterol to 300 milligrams per day (no more than one egg a day, for instance); and (3) acquire 50% or more of your calories from carbohydrates, especially polysaccharides (notably, starches found in potatoes, rice, vegetables, and other foods). Here's how you can achieve these goals: Cut back on saturated fats (animal fats). You can cut back on saturated fat by greatly reducing your consumption of red meat and trimming the fat off all meats before cooking. (Most Americans eat far more red meat than they require.) If you are a meat eater, consider buying grass-fed beef, especially low-fat varieties like Belted Galloways (Figure 1). It is now available at health food grocers and some mainstream supermarkets as well as directly from cattle growers. You should also

greatly increase your consumption of fruits, vegetables, and whole grains, letting these low-fat foods displace some of the fatty foods you might otherwise have eaten. These foods also provide powerful antioxidants and other chemicals that fight cancer and a whole host of diseases. In children under the age of 2, however, diets should not be restricted. A diet that is too restrictive may actually impair physical growth and development. What all children need is a well-balanced diet, low in fats, especially animal fat, with sufficient calories from other sources. Such a diet could encourage the eating habits necessary for good health throughout adult life.

Blood cholesterol levels can also be lowered with drugs and exercise. Research spanning several decades shows that a healthy diet, rich in fruits and vegetables, whole grain products, and low in eggs, meat, and milk products lower cholesterol levels in the blood and translates into a dramatic decline in cardiovascular disease as well as a large number of cancers.

FIGURE 1 (Courtesy of Daniel Chiras.)

 Visit Human Biology's Internet site for links to websites offering more information on this topic.

Obesity Association. Even children are afflicted, with 18% now considered overweight or obese. But the United States is not alone in this regard. Fifty percent of the adults in Russia, the United Kingdom, and Germany are overweight. Even in less developed nations, overnutrition is becoming more commonplace among the wealthier classes. The effects of overeating are many: heart attacks, stroke, late-onset diabetes, and arthritis. Why are so many people overweight?

Most Americans consume way too much food, especially too much carbohydrate, fat, and protein. Studies

show that Americans currently consume, on average, 27% more calories than they did in 1970.

Overconsumption results from many factors. For example, more and more of us are eating out these days, and are eating at fast-food restaurants that offer a plethora of unhealthy, fat-rich, high-calorie meals. Portion sizes are out of control

FIGURE 7-8 **Overnutrition and Undernutrition** Many of the world's people, especially those in more developed countries, suffer from overnutrition, resulting in (a) obesity, while a great many others suffer from undernutrition. (© Wendy Nero/ShutterStock, Inc.) (b) This child suffers from severe protein deficiency (kwashiorkor; kwash-EE-or-core). His arms and legs are emaciated because muscle protein has been broken down to supply energy. His belly is slightly swollen because of a buildup of fluid in the abdomen. (© AP Photos.)

(a)

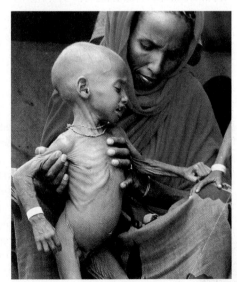
(b)

at many restaurants, and many meals offered by restaurants and fast-food establishments provide as many calories as a mature adult needs in an entire day. In addition, fatty meats dominate most meals and meat portions are way out of line with daily requirements. A 150-pound man, for instance, only needs about 70 ounces of protein a day for health—that's less than one sixth of a pound!

Many of us are also eating prepacked foods (frozen pizzas and the like) that we pop in the microwave and consume. Prepackaged foods often contain calorie-rich meats, sauces, and lots of juicy fat.

Another reason for overconsumption of calories is that many of us consume copious quantities of soft drinks that are packed with sugar. Many snack foods we munch on in between meals or in place of meals are also chock full of fats and sugars—providing many calories per ounce. To lower your chances of gaining weight and developing one of many diseases resulting from this condition, it's important to (1) cut back on serving sizes, (2) reduce consumption of soft drinks and other sweetened beverages, (3) trim back on snack foods rich in fats and/or sugars, and (4) follow the suggestions of the food pyramid discussed shortly. You'll also need to increase your level of physical activity. Read and heed the many health tips presented in this book.

Health Tip 7-3

To maintain or lose weight, watch your weight daily.

Why?

Daily monitoring allows you to adjust your food intake regularly and helps you catch small weight gains, preventing them from becoming large increases.

> **KEY CONCEPTS**
>
> Overnutrition, eating too much, has reached epidemic proportions in many countries.

Micronutrients

With this overview of the macronutrients, we now turn to the micronutrients.

> **KEY CONCEPTS**
>
> Micronutrients are nutrients that are required in minute quantities by the cells of our body but are vital for their function.

Vitamins

Although many Americans consume more than enough calories through meals, soft drinks, and snack foods every day to meet their needs, many diets are still deficient. Why is that?

Many foods provide little more than carbohydrate or carbohydrate and fat. Most soft drinks, for instance, are chock full of sugar, but provide very few, if any, essential micronutrients. For example, these and other foods like them lack vitamins, one type of micronutrient that the body needs to function normally and maintain our health. What is a vitamin?

Vitamins are a diverse group of organic compounds present in very small amounts in many foods; they play an important role in many metabolic reactions. These molecules are absorbed by the lining of the digestive tract without being broken down. Scientists have discovered 13 major vitamins.

Vitamins come from two sources. Most are ingested in the foods we eat—fruits, vegetables, meats, and grains. A few are synthesized by the body. Vitamin D, for instance, is manufactured by the skin when it is exposed to sunlight. However, most Americans spend so much time indoors that vitamin D should be supplied through the diet or by supplements to maintain good health. Unfortunately, vitamin D is naturally found in very few foods. Fish can be a good source of vitamin D as are fortified milk products and fortified orange juice. You may want to research some of your options on the Internet. Because vitamins are recycled many times during metabolic reactions, they are needed only in very small amounts. Table 7-2 lists the vitamins and their functions.

Vitamins fall into two broad categories: water-soluble and fat-soluble. **Water-soluble vitamins**—so named because they dissolve in water—include vitamin C and eight different forms of vitamin B. Water-soluble vitamins are transported in the blood plasma. Because they are water-soluble, they are readily eliminated by the kidneys and are not stored in the body in any appreciable amount.

Water-soluble vitamins generally work in conjunction with enzymes, promoting the cellular reactions that supply energy or synthesize cellular materials. Contrary to common myth, the vitamins themselves do not provide energy.

Many proponents of vitamin use believe that megadoses of water-soluble vitamins are harmless because these vitamins are excreted in the urine and do not accumulate in the body. Some water-soluble vitamins, such as vitamin B_6, however, can be toxic when ingested in excess (Table 7-2).

The **fat-soluble vitamins** are vitamins A, D, E, and K. They perform many different functions. Vitamin A, for example, is converted to light-sensitive pigments in receptor cells of the retina, the light-sensitive layer of the eye. These pigments play an important role in vision. Another member of the vitamin A group removes harmful chemicals (oxidants) from the body.

Unlike water-soluble vitamins, the fat-soluble vitamins are stored in body fat and accumulate in the fat reserves. The accumulation of fat-soluble vitamins can have many adverse effects (Table 7-2). An excess of vitamin D, for example, can cause weight loss, nausea, irritability, kidney stones, weakness, and other symptoms. Large doses of vitamin D taken during pregnancy can cause birth defects.

Vitamin excess is encountered largely in the more affluent developed countries and usually occurs only in people taking vitamin supplements. Each year, in fact, approximately 4,000 Americans are treated for vitamin supplement poisoning. To avoid problems from excess vitamins, nutritionists recommend eating a balanced diet that provides all of the vitamins the body needs, rather than taking vitamin pills. Megadoses should be avoided.

Vitamin Deficiencies. Because vitamins affect many cellular processes, deficiencies can lead to serious upsets of homeostasis and many adverse health effects. A deficiency of vitamin D, for example, can produce rickets (RICK-its), a disease that results in bone deformities. Vitamin K deficiencies can result in severe bleeding after injury. Vitamin C deficiency can result

TABLE 7-2	Important Information on Vitamins

Vitamin	Major Dietary Sources	Major Functions	Signs of Severe, Prolonged Deficiency	Signs of Extreme Excess
Fat-soluble				
A	Fat-containing and fortified dairy products; liver; provitamin carotene in orange and deep green fruits and vegetables	Vitamin A is a component of rhodopsin; carotenoids can serve as antioxidants; retinoic acid affects gene expression; still under intense study	Night blindness; keratinization of epithelial tissues including the cornea of the eye (xerophthalmia) causing permanent blindness; dry, scaling skin; increased susceptibility to infection	Preformed vitamin A: damage to liver, bone; headache, irritability, vomiting, hair loss, blurred vision 13-cis retinoic acid: some fetal defects; carotenoids: yellowed skin
D	Fortified and full-fat dairy products, egg yolk (diet often not as important as sunlight exposure)	Promotes absorption and use of calcium and phosphorus	Rickets (bone deformities) in children; osteomalacia (bone softening) in adults	Calcium deposition in tissues leading to cerebral, CV, and kidney damage
E	Vegetable oils and their products; nuts, seeds	Antioxidant to prevent cell membrane damage; still under intense study	Possible anemia and neurologic effects	Generally nontoxic, but at least one type of intravenous infusion led to some fatalities in premature infants; may worsen clotting defect in vitamin K deficiency
K	Green vegetables; tea	Aids in formation of certain proteins, especially those for blood clotting	Defective blood coagulation causing severe bleeding or injury	Liver damage and anemia from high doses of the synthetic form menadione
Water-soluble				
Thiamin (B-1)	Pork, legumes, peanuts, enriched or whole-grain products	Coenzyme used in energy metabolism	Nerve changes, sometimes edema, heart failure; beriberi	Generally nontoxic, but repeated injections may cause shock reaction
Riboflavin (B-2)	Dairy products, meats, eggs, enriched grain products, green leafy vegetables	Coenzyme used in energy metabolism	Skin lesions	Generally nontoxic
Niacin	Nuts, meats; provitamin tryptophan in most proteins	Coenzyme used in energy metabolism	Pellagra (multiple vitamin deficiencies including niacin)	Flushing of face, neck, hands; potential liver damage
B-6	High-protein foods in general	Coenzyme used in amino acid metabolism	Nervous, skin, and muscular disorders; anemia	Unstable gait, numb feet, poor coordination
Folic acid	Green vegetables, orange juice, nuts, legumes, grain products	Coenzyme used in DNA and RNA metabolism; single carbon utilization	Megaloblastic anemia (large, immature red blood cells); GI disturbances	Masks vitamin B-12 deficiency, interferes with drugs to control epilepsy
B-12	Animal products	Coenzyme used in DNA and RNA metabolism; single carbon utilization	Megaloblastic anemia; pernicious anemia when due to inadequate intrinsic factor; nervous system damage	Thought to be nontoxic
Pantothenic acid	Animal products and whole grains; widely distributed in foods	Coenzyme used in energy metabolism	Fatigue, numbness, and tingling of hands and feet	Generally nontoxic; occasionally causes diarrhea
Biotin	Widely distributed in foods	Coenzyme used in energy metabolism	Scaly dermatitis	Thought to be nontoxic
C (ascorbic acid)	Fruits and vegetables, especially broccoli, cabbage, cantaloupe, cauliflower, citrus fruits, green pepper, kiwi fruit, strawberries	Functions in synthesis of collagen; is an antioxidant; aids in detoxification; improves iron absorption; still under intense study	Scurvy; petechiae (minute hemorrhages around hair follicles); weakness; delayed wound healing; impaired immune response	GI upsets, confounds certain lab tests

Source: Adapted from *Nutrition for Living*, 4th ed., by J. L. Christian and L. L. Greger, Copyright © 1994 by The Benjamin/Cummings Publishing Company.

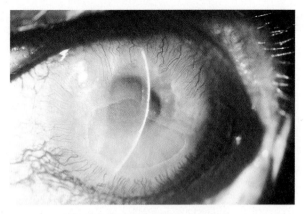

FIGURE 7-9　Vitamin A Deficiency　Vitamin A deficiency causes the cornea of the eye to dry and become irritated. If the deficiency is not corrected, corneal ulcers may form and rupture, resulting in permanent blindness. (© Tony Cubillas/ Visuals Unlimited.)

in delayed wound healing and reduced immunity, making people more susceptible to infectious disease.

One of the most common dietary illnesses is caused by a deficiency of vitamin A. This deficiency afflicts over 100,000 children worldwide each year. If not corrected, vitamin A deficiency causes the eyes to dry. Ulcers may form on the eyeball and can rupture, causing blindness (Figure 7-9).

Most people afflicted by vitamin deficiencies are those who fail to get enough to eat, although even well-fed individuals may suffer from a vitamin deficiency if they are not eating a well-rounded diet. Children with insufficient vitamin intake fail to grow. How do you know if you should take vitamins?

The American Dietetic Association recommends that you eat a variety of foods to achieve good health and reduce the risk of chronic disease. A diet rich in vegetables, fruits, and whole grain products like brown rice, whole grain tortillas and bread, and whole grain cereals can provide more than enough vitamins. Additional vitamins and minerals, if needed, can be obtained from fortified foods, such as vitamin D-enriched milk, and from supplements.

If you don't eat a well-balanced diet, consider making the switch. Ramp up your consumption of few fruits and vegetables, and lower your consumption of fat-rich meats (hamburgers) and desserts. While you make the shift, take a vitamin supplement.

> **KEY CONCEPTS**
> Vitamins are one of several key micronutrients; they play important roles in numerous chemical reactions occurring in the cells of our bodies.

Minerals

Humans require about two dozen inorganic substances known as **minerals**. These include calcium, sodium, iron, and potassium. On average, an adult contains about five pounds of minerals, naturally occurring inorganic substances vital to many life processes.

Minerals, like vitamins, are micronutrients and are derived from the food we eat and the beverages we drink. Minerals are divided into two groups: the major minerals and the trace minerals (Table 7-3). The major minerals are present in the body

in amounts larger than 5 grams. Calcium and phosphorus, for example, are major minerals, and make up three-fourths of all minerals in the human body. These two minerals form part of the dense extracellular matrix of bone. Trace minerals are required in a much smaller quantity. Zinc and copper are two trace minerals that are components of some enzymes.

The distinction between major minerals and trace minerals is not meant to imply that one group is more important than the other. They're all essential to health, just in different amounts. Table 7-3 lists some of the minerals, their function, and problems that arise when they are ingested in excess or deficient amounts.

> **KEY CONCEPTS**
> Minerals are key micronutrients required for many physiological processes.

Functional Foods

Nutritionists have found that a diet containing the micronutrients and macronutrients discussed so far is not necessarily a healthy diet. It lacks essential micronutrients. To make matters worse, most Americans don't follow dietary guidelines. They eat far too many proteins, much too much fat, and far too many carbohydrates. They consume way too few micronutrients. Study after study shows that the standard American diet is associated with much high rates of cancer, heart disease, late-onset diabetes, high blood pressure, and other often-fatal diseases. Those who eat the standard American diet—characterized by its abundance of calories but severe shortage of essential micronutrients—are actually chronically malnourished. Moreover, the food we eat could weaken our resistance to harmful bacteria and viruses. So what constitutes a healthy diet?

Nutritionists and medical researchers have found that humans need the macronutrients discussed so far, but in smaller quantity. We need the minerals and micronutrients just described as well. However, to be healthy we need lots more. To be healthy, live a long healthy life, and effectively combat common diseases—even bacterial and viral infections like those that cause the colds and flu—plant-based substances must be in our diet in much larger amounts.

The list of plant-based or **phytochemicals** chemicals is extensive—numbering in the thousands. Foods containing these and other beneficial micronutrients are referred to by nutritionists as functional foods. **Functional foods** consist of primarily of fruits, vegetables, seeds, nuts, and beans. Simply defined, functional foods provide a health benefit beyond basic nutrition. That is, they may provide proteins, fats, and lipids, but also contain hundreds of chemicals that promote health and combat disease.

Numerous research studies now clearly show that people that consume diets with a large portion of functional foods—fruits, vegetables, nuts, seeds, and beans—experience extremely low levels of cancer and other diseases that are prevalent in the United States. The reason for the high rates in the United States, studies strongly suggest, is that Americans consume only about 5% of their calories from fruits, vegetables, seeds, and nuts.

One group of beneficial chemicals are the antioxidants. **Antioxidants** are a large and diverse group of chemicals that react with naturally occurring harmful chemical substances

TABLE 7-3 | Information on Minerals

Minerals	Major Dietary Sources	Major Functions	Signs of Severe, Prolonged Deficiency	Signs of Extreme Excess
Major Minerals				
Calcium	Milk, cheese, dark green vegetables, legumes	Bone and tooth formation; blood clotting; nerve transmission	Stunted growth; perhaps less bone mass	Depressed absorption of some other minerals; perhaps kidney damage
Phosphorus	Milk, cheese, meat, poultry, whole grains	Bone and tooth formation; acid-base balance; component of coenzymes	Weakness; demineralization of bone	Depressed absorption of some minerals
Magnesium	Whole grains, green leafy vegetables	Component of enzymes	Neurologic disturbance	Neurologic disturbances
Sodium	Salt, soy sauce, cured meats, pickles, canned soups, processed cheese	Body water balance; nerve function	Muscle cramps; reduced appetite	High blood pressure in genetically predisposed individuals
Potassium	Meats, milk, many fruits and vegetables, whole grains	Body water balance; nerve function	Muscular weakness; paralysis	Muscular weakness; cardiac arrest
Chloride	Same as for sodium	Plays a role in acid-base balance; formation of gastric juice	Muscle cramps; reduced appetite; poor growth	High blood pressure in genetically predisposed individuals
Trace Minerals				
Iron	Meats, eggs, legumes, whole grains, green leafy vegetables	Component of hemoglobin, myoglobin, and enzymes	Iron-deficiency anemia, weakness, impaired immune function	Acute: shock, death Chronic: liver damage, cardiac failure
Iodine	Marine fish and shellfish, dairy products, iodized salt, some breads	Component of thyroid hormones	Goiter (enlarged thyroid)	Iodide goiter
Fluoride	Drinking water, tea, seafood	Maintenance of tooth (and maybe bone) structure	Higher frequency of tooth decay	Acute: GI distress Chronic: mottling of teeth; skeletal deformation

Source: Adapted from *Nutrition for Living*, 4th ed., by J. L. Christian and L. L. Greger. Copyright © 1994 by The Benjamin/Cummings Publishing Company.

called **free radicals**. Antioxidants react with and stabilize free radicals found in the blood and body tissues. Examples of some common antioxidants are beta-carotene, lycopene, and vitamins C and E. If you study this issue, you will find that there are hundreds more—and most of them are found in seeds, nuts, fruits, and vegetables.

Studies show that some free radicals are involved in cholesterol buildup in the walls of arteries. As a result, eating a diet rich in antioxidants may reduce heart and artery disease. A recent study, for instance, showed that consuming a serving of salad each day reduces the risk of stroke by 3% in women and 5% in men. A *stroke* occurs when an artery in the brain bursts or when cholesterol-choked arteries develop blood clots that restrict the flow of blood to brain cells.

Studies also show that antioxidants reduce the risk of many types of cancer, such as breast cancer in women and prostate cancer in men. Numerous, extensive studies of populations that consume large quantities of fruits and vegetables show that they have a much lower incidence of many forms of cancer than populations like ours where meat and dairy products dominate the diet and fruits and vegetables are con-

sumed in small quantity. Soy products also protect against cancer. For example, one group of naturally occurring antioxidants, isoflavones, which are found in soy products, have been shown to reduce prostate cancer in men.

Antioxidants can also help maintain vision. The antioxidants found in spinach and other dark, leafy vegetables, for example, may help prevent cataracts. Not all functional foods are created equal. Some foods provide a lot more protection than others. This includes: collard greens, Swiss chard, broccoli, cabbage, cauliflower, spinach, and vitamin C-rich foods such as oranges and orange juice. In what is good news for chocolate lovers, chocolate and cocoa powder are produced from beans that contain large quantities of natural antioxidants. They're called *flavonoids* (flay-vah-noids). Even commercially available chocolate products may contain significant amounts of these antioxidants. Researchers at Holland's National Institute of Public Health and Environment, for instance, found that dark chocolate had 53.5 mg of an antioxidant known as catechins per 100 g, while milk chocolate contains 15.9 mg per 100 g. In comparison, black tea contained 13.9 mg per 100 ml. Soy milk also contains flavonoids, as do tea and red wine.

Chocolate's flavonoids not only reduce plaque buildup, they help stimulate the relaxation of smooth muscle in arteries, which is important to maintaining healthy arteries. Flavonoid-rich foods may also benefit the heart by reducing the activity of blood platelets. Platelets are tiny cell-like structures that stimulate blood clotting. In arteries clogged by atherosclerotic plaque, blood clots can lead to heart attacks and strokes.

So does this mean you should load up on chocolate? Not really. Chocolates also contain a lot of saturated fat and calories (from sugars). Eat chocolate, but don't overindulge (as if that's possible!).

Not all micronutrients come from plant sources. One critical micronutrient that comes from animals that offers many health benefits are the **omega-3 fatty acids**. These **fatty acids** are vital for human health, but cannot be produced by the body. That is, they must be obtained through the food we eat. As a result, omega-3 fatty acids are classified as "essential fatty acids." The best sources are fatty fish, such as salmon, sardines, lake trout, tuna, and halibut, and other seafood including algae and krill. Grass-fed beef contains high amounts of omega-3 fatty acids. Omega-3s are also found in some plants products such as flax seeds and black raspberries. Eggs from free-range chickens, that is, chickens that are allowed to roam one's property where they eat vegetation and insects have higher levels of omega-3 fatty acids than penned or confined chickens feed a diet of corn and soy bean meal.

Omega-3 fatty acids are vital to human health for many reasons. For instance, they play a critical role in brain function, as well as normal growth and development. Research suggests that it may lower your risk of heart disease, cancer, and arthritis. Omega-3 fatty acids concentrate in the brain where they may help promote better memory, mental performance, and even behavior. Omega-3 fatty acid deficiencies may lead to fatigue, poor memory, dry skin, mood swings, depression, and poor blood circulation.

Interestingly, studies show that good health depends on achieving the correct ratio of omega-3 to omega-6 fatty acids. Omega-6 fatty acids are found in many unhealthy foods. Studies show that omega-3 fatty acids reduce inflammation, while most omega-6 fatty acids tend to increase inflammation. Unfortunately, according to the University of Maryland School of Medicine website, the "typical American diet tends to contain 14–25 times more omega-6 fatty acids than omega-3 fatty acids, which many nutritionally oriented physicians consider to be way too high on the omega-6 side."

To reduce your intake of omega-6 fatty acids and increase omega-3s, consider changing to a diet consisting of whole grains, such as whole grain bread, whole grain pasta, whole grain crackers, and brown rice. Increase your consumption of fresh fruits and vegetables as well as fish, olive oil, and garlic.

Clearly, micronutrients are a vital component of a healthy diet. In fact, "most health authorities today are in agreement that we should add more servings of healthy fruits and vegetables to our diet," notes Dr. Joel Fuhrman, a leading advocate of a diet rich in phytochemical micronutrients. But Fuhrman disagrees. "Thinking about our diet in this fashion," he claims, "doesn't adequately address the problem. Instead of considering adding protective fruits, vegetables, beans, seeds, and nuts to our disease-causing diet, *we must make these foods the main focus of the diet itself.*" (Italics his.) Fuhrman goes on to say, "There's no way around it; for superior health, we must eat more nutrient-rich foods and fewer calorie-rich foods." He and many others suggest strongly reducing caloric intake of animal products—milk, cheese, meats, and starchy foods, for instance—and decreasing the intake of foods that are completely empty of nutrients, such as white flour, sugar, processed foods, and fast foods. We can still obtain the lipids, carbohydrates and protein we need, but by eating nutrient-rich foods, we nourish our bodies *and* obtain all the protective phytochemicals needed to prevent or lessen the impact of common diseases that are causing the premature death of millions of people the world over.

Health Tip 7-4

Feeling clumsy? Having trouble remembering things? Blueberries may help.

Why?

Studies show that blueberries can boost weakened bioelectric signals in nerve cells, resulting in improved short-term memory, navigational skills, balance, coordination, and speed.

KEY CONCEPTS

Some chemicals in food (like plant-derived antioxidants) protect us against disease, including many types of cancer, heart attacks, and strokes.

A Healthy You

By now, it should be apparent that human health depends on a diet consisting of many different nutrients: protein, carbohydrate, fats, water, vitamins, minerals, and various beneficial chemicals. To acquire all of these nutrients it is important to consume a variety of foods. Nutritionists use the term *balanced diet* to describe the ideal food intake required to supply your cells, tissues, and organs with the nutrients needed to maintain proper function. What is a balanced diet?

In 2011, the U.S. Department of Agriculture (USDA) released a graphic representation of the proper diet, known as **MyPlate** (Figure 7-10). As you can see from Figure 7-10, useful guidelines on diet and exercise are provided. Let's start with diet.

As illustrated, a healthy diet consists of foods from five different groups: grains, vegetables, fruits, dairy, and protein. The size of each section indicates the approximate portion of our daily intake that should come from each group. Very specific recommendations are listed to make this information more useful. Let's take a look at some of the recommendations, using a 2,000-calorie-a-day diet shown in Figure 7-10. (You can customize your own needs by going online to www.choosemyplate.gov.)

According to the chart in Figure 7-10, a person eating a 2,000-calorie diet should eat 6 ounces of grains every day and most should be whole grains (like brown rice or whole-wheat bread). To help make the conversion, the MyPlate chart includes some common food servings. As shown, one ounce of grain is equal to a slice of whole-wheat bread or one cup of breakfast cereal or a half-cup of cooked rice, cereal, or pasta. Two slices of bread, a cup of cereal, and 1.5 cups of cooked rice meet the grain daily requirement.

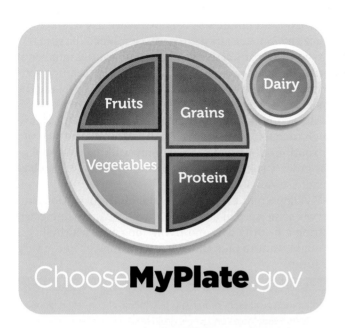

ChooseMyPlate.gov

It should be no surprise to you that vegetables and fruit are also a major component of a healthy diet. According to the USDA, you should consume 2.5 cups of vegetables and 2 cups of fruit every day. As in the case of grains, specific recommendations are also listed on the chart. For example, for vegetables, you should eat more dark-green vegetables such as spinach, Swiss chard, and broccoli. You should eat more orange vegetables, too, to obtain nutrients you need. And you should eat beans and lentils, too.

As for fruits, the USDA recommends eating a variety. Go easy on fruit juices, however, as they provide a lot of calories without other important dietary components like fiber (discussed shortly).

When it comes to fruit and vegetables, dieticians and nutrition-savvy doctors recommend that you "go for the color"—that is, that you eat a rainbow of fruits and vegetables. Why?

Eating a variety of colors is a great way to ensure that you are consuming vegetables from each of the five subgroups: dark green, orange, legumes, starchy vegetables, and others. This

Vegetables	Fruits	Grains	Dairy	Protein Foods
Eat more red, orange, and dark-green veggies like tomatoes, sweet potatoes, and broccoli in main dishes. Add beans or peas to salads (kidney or chickpeas), soups (split peas or lentils), and side dishes (pinto or baked beans), or serve as a main dish. Fresh, frozen, and canned vegetables all count. Choose "reduced sodium" or "no-salt-added" canned veggies.	Use fruits as snacks, salads, and desserts. At breakfast, top your cereal with bananas or strawberries; add blueberries to pancakes. Buy fruits that are dried, frozen, and canned (in water or 100% juice), as well as fresh fruits. Select 100% fruit juice when choosing juices.	Substitute whole-grain choices for refined-grain breads, bagels, rolls, break-fast cereals, crackers, rice, and pasta. Check the ingredients list on product labels for the words "whole" or "whole grain" before the grain ingredient name. Choose products that name a whole grain first on the ingredients list.	Choose skim (fat-free) or 1% (low-fat) milk. They have the same amount of calcium and other essential nutrients as whole milk, but less fat and calories. Top fruit salads and baked potatoes with low-fat yogurt. If you are lactose intolerant, try lactose-free milk or fortified soymilk (soy beverage).	Eat a variety of foods from the protein food group each week, such as seafood, beans and peas, and nuts as well as lean meats, poultry, and eggs. Twice a week, make seafood the protein on your plate. Choose lean meats and ground beef that are at least 90% lean. Trim or drain fat from meat and remove skin from poultry to cut fat and calories.
For a 2,000-calorie daily food plan, you need the amounts below from each food group. To find amounts personalized for you, go to ChooseMyPlate.gov.				
Eat 2½ cups every day **What counts as a cup?** 1 cup of raw or cooked vegetables or vegetable juice; 2 cups of leafy salad greens	**Eat 2 cups every day** **What counts as a cup?** 1 cup of raw or cooked fruit or 100% fruit juice; ½ cup dried fruit	**Eat 6 ounces every day** **What counts as an ounce?** 1 slice of bread; ½ cup of cooked rice, cereal, or pasta; 1 ounce of ready-to-eat cereal	**Get 3 cups every day** **What counts as a cup?** 1 cup of milk, yogurt, or fortified soymilk; 1½ ounces natural or 2 ounces processed cheese	**Eat 5½ ounces every day** **What counts as an ounce?** 1 ounce of lean meat, poultry, or fish; 1 egg; 1 Tbsp peanut butter; ½ ounce nuts or seeds; ¼ cup beans or peas

FIGURE 7-10 **My Plate** This graphic is designed to provide guidance on proper nutrition. (Courtesy of USDA.)

ensures that you ingest a variety of beneficial nutrients that each group provides. You'll learn more about these nutrients shortly.

Grains, fruits, and vegetables aren't all that you need to live a long, healthy life. You must also consume calcium-rich foods such as milk, soy milk, almond milk, and coconut milk and cheese and sources of protein such as meat and beans. As in previous categories, specific recommendations are given for each on MyPlate. For milk, you should consume three cups a day. Be sure to choose low-fat varieties (like low-fat milk and yogurt) to avoid saturated fatty acids that contribute to disease of the heart and arteries, discussed shortly. If you can't consume milk because you are lactose intolerant (lactose is a milk sugar that some people can't digest) or are allergic to milk protein, you will need to get calcium elsewhere. Green leafy vegetables like spinach contain a lot of calcium. Certain foods, such as IronKids bread, and drinks, like orange juice, are fortified with calcium.

As for protein, a person requiring 2,000 calories a day only needs 5.5 ounces a day. Choose lean (low-fat) protein sources whenever possible. Bake, broil, or grill meats like fish and steaks. Avoid deep fat fried meats. Don't forget that you can get plenty of protein from beans, peas, nuts, and seeds, too.

As noted earlier in the chapter, most people in more developed countries go wrong by eating too many fatty foods, especially meats, milk products like ice cream and whole milk, and snack foods like chips. They also typically consume too many carbohydrates. For example, they load up on pasta (rice and noodles) and bread and consume excessive amounts of sugar in sodas, fruit juices, cakes, and candy. We also consume way too few fruits and vegetables.

MyPlate also points out the importance of exercise. The USDA recommends at least 30 minutes of physical activity every day. However, to prevent weight gain, some people may need to engage in 60 minutes of physical activity every day of the week.

Health Tip 7-5

Try to include fruits and vegetables with every meal, even breakfast! Slice up an apple or an orange or eat a small bowl of raspberries or blueberries with your breakfast and include a small salad, raw or cooked vegetables, and fruit with every other meal. Fruit smoothies are a great way to consume a lot of fruit quickly—and enjoyably! Try blending various combinations of berries and fruit—such as blueberries, raspberries, apples, oranges, and strawberries—with some low-fat milk or soy or almond milk to create a delicious and healthy breakfast drink. Chocolate lovers can add a bit of chocolate for additional flavoring.

Why?

Scientific studies show that eating a diet rich in fruits and vegetables has many health benefits. It prevents many forms of cancer; dramatically reduces one's chances of heart disease and stroke; helps to prevent obesity and late-onset diabetes; results in healthier bones; and protects against loss of sight in older age, a disease known as macular degeneration. Switching to a diet rich in fruits and vegetables can lower blood pressure and reverse some diseases like late-onset diabetes, freeing patients from costly medication.

Although this goal sounds fairly impossible, remember that not all exercise occurs in the gym or on your bicycle or on the sports field. Even though the federal government recommends 60 minutes of exercise every day to stay trim and healthy, you don't have to spend your life in the gym trying to stay fit. Many activities burn a substantial number of calories, for example, walking, raking leaves, cleaning your apartment or dorm room, dancing, gardening, or taking the stairs instead of an elevator.

Although many people complain that they don't have enough time to exercise each day, if you work exercise into your routine—for example, by simply walking to school or to the grocery store or to a restaurant—you can get much of the exercise you need.

MyPlate contains a lot of valuable information and deserves some special attention, if you are truly interested in eating a healthy diet and living long and healthy lives, you might want to take time to study it now or in the very near future.

> **KEY CONCEPTS**
>
> Maintaining your health requires a healthful diet consisting of lots of fruits and vegetables and whole grain products, and exercise like walking, biking, and swimming. There's no way around it.

The Key to Successful Weight Loss

Maintaining optimum weight is the easiest way to stay healthy in the long term. It reduces one's risk of heart attack, stroke, and late-onset diabetes. It also reduces the risk of developing certain forms of cancer, arthritis, and a number of other illnesses. You can prevent costly and painful hip and knee replacements. Moreover, you'll live longer if you eat a healthy diet—the right foods in the right amount—and exercise regularly.

The same advice applies to those who have become overweight. They're often tempted to sign on to one of the latest diet fads—high protein diets or low-carb diets or even diets based on blood type!

Nutritionists and physicians often caution patients when it comes to fad diets. Many are based on very little, if any, science. In addition, some can be downright dangerous. A high-protein/low-carbohydrate diet like the Atkins diet may raise a person's risk for heart disease by increasing his or her intake of saturated fat in the form of meat products. Some individuals on this diet complain of feeling weak and nauseated. Low carbohydrate levels can lead to the buildup of dangerous chemicals (ketones) in your blood. In addition, adherence to this diet means you're eating too few fruits, vegetables, and whole grains, which provide a host of vitamins, minerals, and antioxidants, all of which are essential to good health.

The best way to lose weight is to reduce your total daily caloric intake gradually while eating a balanced diet. Combine these measures with an increase in the level of exercise and the pounds will come off.

Many health experts also caution against natural or herbal weight-loss products. Unfortunately, these products are rarely tested for safety and effectiveness. Moreover, just because a product is "natural," does not mean that it is safe. As an example, herbal products that contained the natural ingredient ephedra have been banned in the United States because they caused serious health problems and death in some individuals.

Health Tip 7-6

When you work out, rock out!

Why?

A recent study showed that tuning into music while exercising (in this study, walking) helped overweight individuals stick with their exercise programs and lose more weight!

> **KEY CONCEPTS**
>
> Although fad diets may promise weight loss, some can be potentially unhealthy; there's no substitute for increased physical activity and balanced consumption of nutrients, micronutrients, and various helpful chemicals like antioxidants.

7-2 The Digestive System

The human digestive system consists of a tube opened at both ends (**Figure 7-11**). Food enters the mouth and is chemically and physically altered as it travels along this tube, called the *gut*. During its journey through the digestive system, water, food molecules, vitamins, and minerals are absorbed through the wall of certain portions of the gut, where they enter the bloodstream. Waste is excreted at the opposite end, the *anus*.

The Mouth

The **mouth** is a complex structure in which food is broken down mechanically and, to a much lesser degree, chemically. Food taken into the mouth is sliced into smaller pieces by the sharp teeth in front; it is then ground into a pulpy mass by the flatter teeth toward the back of the mouth (**Figure 7-12a**). Chewing breaks food into smaller pieces and greatly increases the surface area. This makes it easier for digestive enzymes in the stomach and the small intestine, where most food digestion occurs, to break food down even further. The smaller the food particle, the more readily enzymes and food molecules come in contact.

As the food is pulverized in the mouth, it is liquefied by **saliva**, a watery secretion released by the salivary glands, three sets of exocrine glands located around the oral cavity (**Figure 7-13**). The release of saliva is triggered by the smell, feel, taste, and sometimes just the thought of food.

Saliva performs at least five functions. It liquefies the food, making it easier to swallow. It kills or neutralizes some bacteria via the enzymes and antibodies it contains. It dissolves substances so they can be tasted. It begins to break

down starch molecules with the aid of the enzyme amylase (AM-ah-lace). Saliva also cleanses the teeth, washing away bacteria and food particles.

> **KEY CONCEPTS**
>
> Digestion begins with the physical breakdown of food in the mouth.

The Teeth

Human teeth are hard structures embedded in the bones of the jaw. They aid in cutting and grinding food we eat. Take a moment to view Figure 7-12 to study the various types of teeth. As you have no doubt figured out by now, the teeth in front—the incisors and canines—are for slicing and tearing food. The teeth in back—the premolars and molars—are for grinding food.

As illustrated in Figure 7-12b, your teeth consist of a hard outer layer, called **enamel**. Enamel is the hardest substance in the human body. Beneath that is a softer layer of dentin, which forms the bulk of the tooth. In the inside of the tooth is a space filled with nerves and blood vessels, the pulp. Each tooth is embedded in the bone of the jaw.

Although enamel is hard, it is not impervious. It can be eroded by acids from the bacteria that live inside our mouths. These bacteria secrete a sticky material known as **plaque**. Plaque adheres to the surface of your teeth, trapping bacteria. They release small amounts of a weak acid that dissolves the enamel. The acid forms small pits, commonly referred to as cavities. Continued acid secretion may cause the cavities to deepen into the softer dentin layer beneath the enamel. If the cavity is left untreated, an entire tooth can be lost to decay.

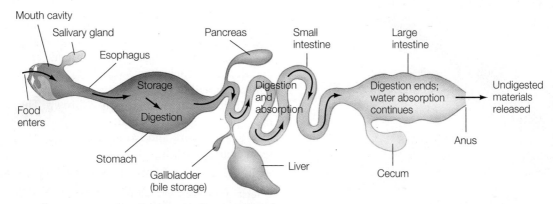

FIGURE 7-11 The Complete Gut This diagram illustrates the general anatomy of the tubular digestive tract present in humans and many other animals. Food enters the mouth and is excreted at the opposite end through the anus. Along the way, it is chemically and physically broken down, then transported into the bloodstream.

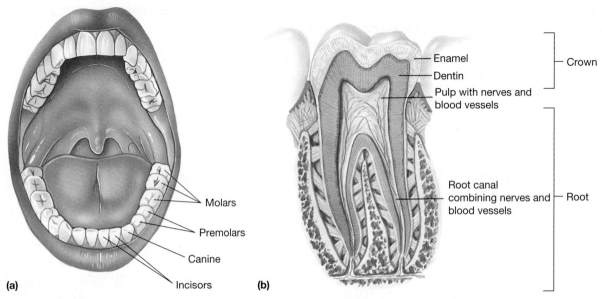

FIGURE 7–12 The Teeth (a) Adults contain 32 teeth. (b) Each tooth is embedded in the bone of the jaw.

Controlling the bacteria that live on the teeth by regular brushing is important. It removes plaque and thus helps reduce cavities. Regular cleaning at your dentist's office also helps remove plaque that brushing misses.

Besides causing decay, the bacteria in our mouths also give us bad breath. This is especially noticeable in the morning or sleep. That's because the release of saliva is greatly reduced during sleep. During the day saliva tends to keep your teeth clean. During sleep, however, bacteria can accumulate on the surface of your teeth where they break down microscopic food particles. This often produces foul-smelling chemicals that give us "dragon breath," or "morning mouth."

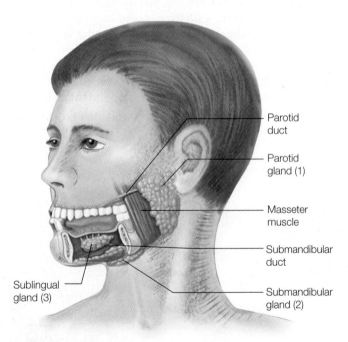

FIGURE 7–13 Salivary Glands Three salivary glands (parotid, submandibular, and sublingual) are located in and around the oral cavity and empty into the mouth via small ducts.

Brushing helps prevent the buildup of plaque and therefore helps to reduce the incidence of cavities. It also helps reduce bad breath. Most toothpastes also contain small amounts of fluoride, which hardens the enamel and helps reduce cavities. In most cities and towns, fluoride is added to drinking water as a preventive measure. Flossing your teeth also helps reduce cavities, by removing food particles that lodge between your teeth where toothbrushes can't reach. One study showed that chewing sugarless gum increases the flow of saliva and cleanses the teeth. By chewing sugarless gum within 5 minutes after a meal, and for at least 15 minutes, the researchers showed that individuals can reduce the incidence of cavities.

Gentle brushing of your gums when you brush your teeth is also essential to good dental health. Brushing the gums strengthens them and helps to reduce your chances of contracting **gingivitis** (gin-je-VIE-tiss), a potentially serious bacterial infection of the gums. This disease is caused by the buildup of plaque on one's teeth along the gum line. If not removed, plaque is converted to a hard deposit known as tartar. The presence of plaque and tartar irritate and inflame the gums. In addition, bacteria and the various chemical toxins they produce can cause the gums to become infected. Gums become swollen and tender.

What is more, inflammation (swelling and redness) and infection destroys the tissues that support the teeth, including the gums, small ligaments that hold the bones in place, and the boney sockets into which the teeth fit. If not treated, your teeth will likely fall out.

Healthy gums are pale pink or brown. They are firm and may appear speckled. If you have gingivitis, your gums will be red, soft, shiny, and swollen. They also bleed easily, even from gentle brushing.

KEY CONCEPTS

Teeth are exceedingly hard but can be damaged by the buildup of plaque that trap acid-producing bacteria that erode the hard outer layer.

The Tongue

After food is chewed, it must be swallowed. The **tongue**, a muscular organ in the mouth, plays an important role in this process by pushing food to the back of the oral cavity into the pharynx (FAIR-inks). The **pharynx** is a funnel-shaped chamber that connects the oral cavity with the esophagus (ee-SOFF-ah-gus), a long muscular tube that leads to the stomach (Figure 7-14).

The tongue aids in speech, that is, it helps us form sounds. It is also an important sensory organ because it contains numerous taste receptors, or taste buds, on its upper surface. **Taste buds** are cocktail-onion-shaped structures

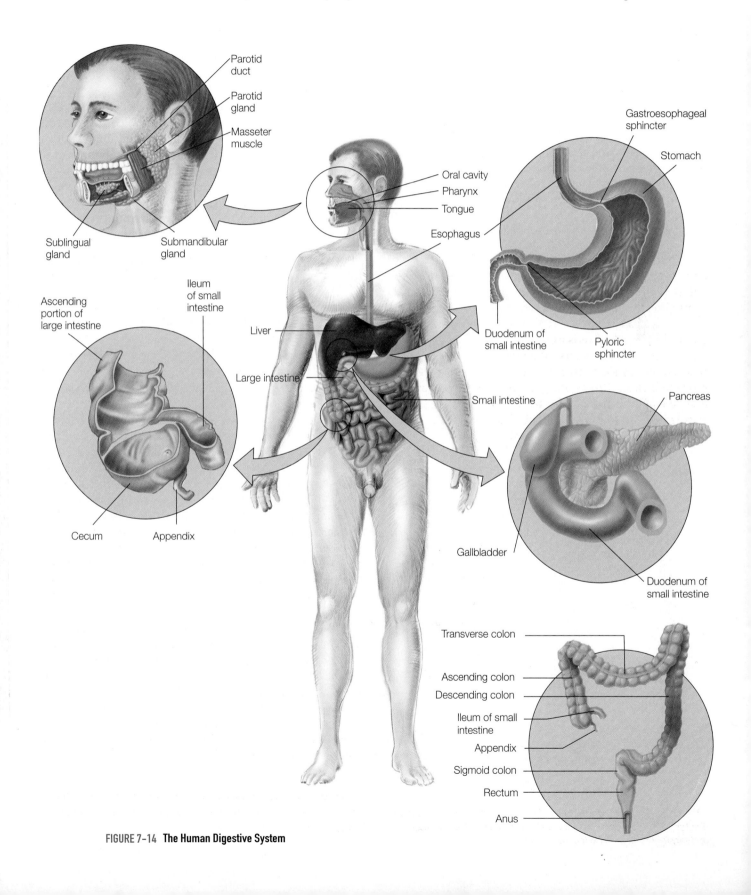

FIGURE 7-14 **The Human Digestive System**

located in protrusions called *papillae* (pah-pill-eye) on the upper surface of the tongue.

> **KEY CONCEPTS**
>
> The tongue is a muscular organ in the oral cavity that plays a key role in speech and swallowing.

Epiglottis

Food propelled from the pharynx into the esophagus is prevented from entering the trachea (TRAY-key-ah), or windpipe, which carries air to the lungs, by a flap of tissue known as the **epiglottis** (ep-ah-GLOT-tis). It is shown in Figure 7-15. The epiglottis is a door that closes off the trachea when you swallow food. Take a moment to study Figure 7-15 to see how.

Swallowing begins with a voluntary act—the tongue pushing food into the back of the oral cavity. Once food enters the pharynx, however, the process becomes automatic. Food in the pharynx stimulates stretch receptors in the wall of this organ. They then trigger the swallowing reflex, an involuntary contraction of the muscles in the wall of the pharynx. This forces the food into the esophagus.

> **KEY CONCEPTS**
>
> The epiglottis prevents food you swallow from entering the trachea and blocking airflow to the lungs.

The Esophagus

The esophagus is a long, muscular tube that transports food to the stomach by peristalsis.

Food travels from the oral cavity to the stomach via the **esophagus**. It is transported by involuntary contractions of the muscular wall of the esophagus. As Figure 7-16 shows, the muscles of the esophagus contract above the swallowed food mass, squeezing it along. This involuntary muscular action, called **peristalsis** (pear-eh-STALL-sis), occurs both in the esophagus and also in the stomach and intestine and propels food (and waste) along the digestive tract. It is so powerful that you can swallow when hanging upside down.

Scientists once thought that esophageal peristalsis could proceed in the opposite direction and called this process *reverse peristalsis*, or, more commonly, vomiting. Vomiting is a reflex action that occurs when irritants are present in the stomach. Now it is clear that vomiting is caused by an increase in abdominal pressure that forces food out. Vomiting is believed to be a protective adaptation that allows the body to rid the stomach of bad food and harmful viruses and bacteria, with obvious survival benefits to humans. Vomiting may also be caused by: (1) emotional factors, such as stress; (2) rotation or acceleration of the head, as in motion sickness; (3) stimulation of the back of the throat; and (4) elevated pressure inside the brain caused by injury.

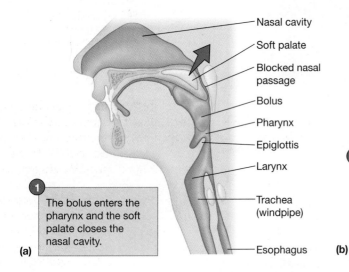

Nasal cavity
Soft palate
Blocked nasal passage
Bolus
Pharynx
Epiglottis
Larynx
Trachea (windpipe)
Esophagus

1 The bolus enters the pharynx and the soft palate closes the nasal cavity.

(a)

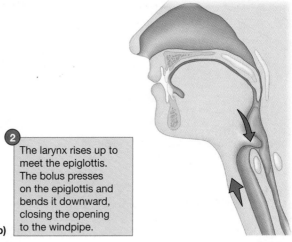

2 The larynx rises up to meet the epiglottis. The bolus presses on the epiglottis and bends it downward, closing the opening to the windpipe.

(b)

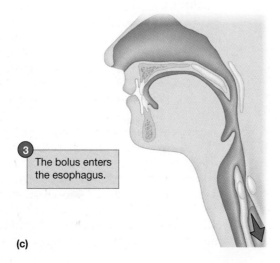

3 The bolus enters the esophagus.

(c)

FIGURE 7-15 **Swallowing** The epiglottis forms a trapdoor that prevents food from entering the trachea during swallowing. As illustrated, the trachea is lifted during swallowing, pushing against the epiglottis, which bends downward.

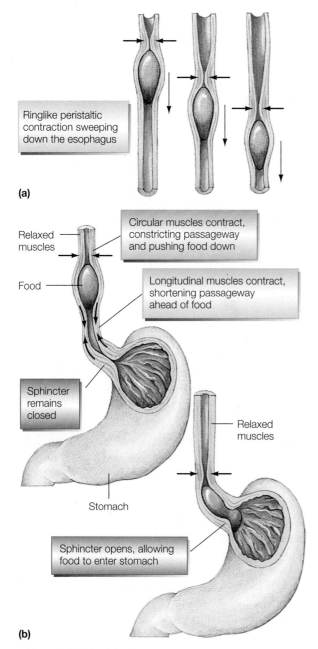

FIGURE 7-16 Peristalsis (a) Peristaltic contractions in the esophagus propel food into the stomach. (b) When food reaches the stomach, the gastroesophageal sphincter opens, allowing food to enter.

During vomiting, food is expelled from the stomach as a result of contractions of muscles in the abdomen and diaphragm (DIE-ah-fram). (The diaphragm is a large dome-shaped muscle that separates the thoracic [chest] and abdominal cavities and plays a key role in breathing.)

> **KEY CONCEPTS**
>
> The esophagus transports food by peristalsis to the stomach.

The Stomach

The **stomach** is a large temporary storage compartment for the food we eat. It also begins processing food for complete digestion in the small intestine. Food in the stomach is churned and liquefied and then released into the small intestine in spurts.

In humans, the stomach (Figures 7-14 and 7-17) lies on the left side of the abdominal cavity, partly under the protection of the rib cage. Food enters the stomach via the esophagus. The opening to the stomach, however, is closed off by a sphincter, a thickened layer of smooth muscle at the juncture of the esophagus and stomach (Figure 7-17).

As food enters the lower esophagus, the sphincter opens, allowing the food to enter the stomach. The sphincter then promptly closes, preventing food and stomach acid from percolating upward. Many people suffer from a disease known as GERD, which stands for gastroesophageal reflux disease. It is caused by defective gastroesophageal sphincters. When the sphincter fails to close, acid can "percolate" up into the esophagus and throat, causes irritation. In the esophagus, acid causes a common disease known as **heartburn**. It can also cause nausea after eating. In the throat, acid can cause soreness and tiny lesions (bumps) on the cords that lead to hoarseness. Difficulty swallowing, nasal drip, and other symptoms may also occur as a result of GERD. GERD may result from smoking, pregnancy, obesity, and several other causes.

> **Health Tip 7-7**
>
> Are you or a loved one suffering from heartburn caused by the stomach acids backing up into the esophagus? If so, dropping a little weight can help.
>
> *Why?*
>
> A recent study showed that even a modest increase in weight can double a person's chances of developing heartburn and acid reflux. In other words, you don't have to be considered overweight to develop heartburn. The study also showed that shedding extra weight can decrease a woman's risk of heartburn by as much as 40%.

Inside the stomach, food is "liquefied" by acidic secretions of tiny glands in the wall of the stomach, the **gastric glands**. These glands produce a watery secretion called *gastric juice*. Gastric juice contains hydrochloric acid (HCl) and an enzyme precursor called *pepsinogen* (pep-SIN-oh-gen; discussed below).

Inside the stomach, food is churned by peristalsis and mixed with gastric juice. The churning action of the stomach's muscular walls helps break down large pieces of food. Combined with the liquid from salivary glands and the gastric glands, the food becomes a rather thin, watery paste known as **chyme** (KIME). The stomach can hold 2–4 liters (2 quarts to 1 gallon) of chyme. It is gradually released into the small intestine at a rate suitable for efficient digestion and absorption.

Contrary to what many think, very little enzymatic digestion occurs in the stomach. The stomach's role is largely to prepare most food for the enzymatic digestion in the small intestine. There are some exceptions to this rule, however. Protein is one of them.

Proteins are denatured (coagulated) by HCl in the stomach. HCl also acts on pepsinogen, converting it to the active form pepsin. Pepsin is a protein-digesting enzyme that catalyzes the breakdown of proteins into large fragments. These fragments are further broken down in the small intestine, the next stop.

Besides denaturing protein and activating pepsinogen, HCl breaks down connective tissue fibers in meat and kills most bacteria, thereby protecting the body from infection.

FIGURE 7-17 **The Stomach** The stomach lies in the abdominal cavity. In its wall are three layers of smooth muscle that mix the food and force it into the small intestine, where most digestion occurs. The gastro-esophageal and pyloric sphincters control the inflow and outflow of food, respectively.

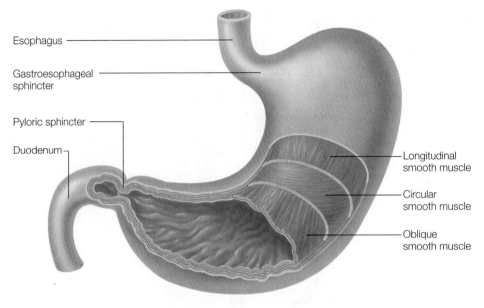

Esophagus

Gastroesophageal sphincter

Pyloric sphincter

Duodenum

Longitudinal smooth muscle

Circular smooth muscle

Oblique smooth muscle

Also contrary to popular belief, the stomach does not absorb many foodstuffs. Only a few substances, such as alcohol, can penetrate the lining of the stomach to enter the bloodstream. Alcohol consumed on an empty stomach therefore passes quickly into the blood, often producing an immediate dizzying effect. The presence of food in the stomach, however, slows alcohol absorption.

The stomach lining is protected from destruction by a basic secretion known as **mucus**. Produced by cells in the stomach lining, mucus coats the epithelium, protecting it from acid and pepsin. Acid is further prevented from penetrating the lining of the stomach and reaching underlying tissues by tight junctions in the cells of the epithelium. These structures join the epithelial cells very tightly, creating a leak-proof barrier. Like other protective epithelia, the lining of the stomach is continually renewed.

For years, scientists believed that stomach acid production could be increased by excess stress, coffee, aspirin, nicotine, and alcohol—or combinations of them. Excess acid, in turn, was thought to erode the stomach lining, forming **ulcers**—painful, open sores in the stomach lining that bleed into the stomach. Accordingly, most ulcers that were detected early were treated by reducing stress and changing a patient's diet. Patients were advised to eat smaller amounts of food and to reduce consumption of coffee, aspirin, and alcohol to lessen acid secretion in the stomach. When ulcers were not detected early and when damage was severe, parts of the stomach had to be removed surgically.

Today, researchers have found that at least 70% of all ulcers are caused by a fairly common bacterium, *Helicobacter pylori*, or *H. pylori* for short. Antibiotics can be used to rid the stomach of this agent, thus eliminating painful ulcers in many patients.

Chyme is ejected from the stomach into the small intestine by peristaltic muscle contractions. When the wave of contraction reaches the muscular sphincter at the juncture of the small intestine and stomach, it opens, and chyme squirts into the small intestine.

The stomach contents are emptied in 2 to 6 hours, depending on the size of the meal and the type of food. The larger the meal, the longer it takes the stomach to empty. Solid foods (meat) empty slower than liquid foods (milk shakes). Peristaltic contractions continue after the stomach is empty and are felt as hunger pangs.

As chyme and protein leave the stomach, gastric gland secretion declines and the stomach slowly shuts down its production of HCl until the next meal.

> **KEY CONCEPTS**
> The stomach is a muscular organ that stores and churns food, releasing it into the small intestine.

The Small Intestine

The **small intestine** is a coiled tube in the abdominal cavity that is about 6 meters (20 feet) long in adults (Figure 7-14). So named because of its small diameter, the small intestine consists of three parts in the following order: the duodenum (DUE-oh-DEEN-um), jejunum (jeh-JEW-num), and ileum (ILL-ee-um).

Inside the small intestine, large molecules are broken into smaller molecules (that is, digested) with the aid of enzymes. The smaller food molecules are then transported into the bloodstream and the lymphatic system. The lymphatic system, discussed elsewhere in the text, is a network of tiny vessels found throughout the body. They carry extracellular fluid from the tissues of the body to the circulatory system. Tiny lymphatic vessels in the small intestine are known as **lacteals** (LACK-teels). They absorb fats and transport them to the bloodstream.

> **KEY CONCEPTS**
> Food enters the small intestine from the stomach and is the site where most food digestion and absorption occur.

Digestive Enzymes

The digestion of food molecules inside the small intestine requires enzymes produced from two different sources: the pancreas and the lining of the small intestine.

The **pancreas** is an organ that lies beneath the stomach. It is nestled in a loop formed by the first portion of the small intestine (Figure 7-18). The pancreas is a dual-purpose organ—

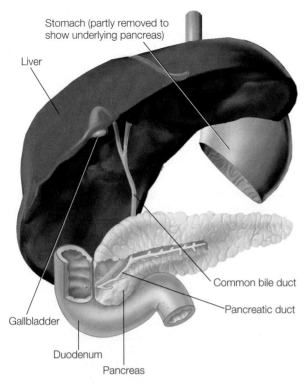

FIGURE 7-18 The Organs of Digestions The liver, gallbladder, and pancreas all play key roles in digestion. All empty by the common bile duct into the small intestine, in which digestion takes place.

TABLE 7-4	Digestive Enzymes	
Site of Production	**Enzyme**	**Action**
Salivary glands	Amylase	Digests polysaccharides in oral cavity
Stomach	Pepsin	Breaks proteins into peptides
Pancreas	Trypsin	Cleaves peptide bonds of polypeptides and proteins
	Chymotrypsin	Same as trypsin
	Carboxypeptidase	Cleaves peptide bonds on carboxyl end of polypeptides
	Amylase	Breaks starch molecules into smaller units (maltose)
	Phospholipase	Cleaves fatty acids from phosphoglycerides to form monoglycerides
	Lipase	Cleaves two fatty acids from triglycerides
	Ribonuclease	Breaks RNA into smaller nucleotide chains
	Deoxyribonuclease	Breaks DNA into smaller nucleotide chains
Epithelium of small intestine	Maltase	Breaks maltose into glucose subunits
	Sucrase	Breaks sucrose into glucose and fructose subunits
	Lactase	Breaks lactose into glucose and galactose subunits
	Aminopeptidase	Breaks peptides into amino acids

that is, it has both exocrine and endocrine functions. As an exocrine gland, it produces enzymes and sodium bicarbonate essential for the digestion of foodstuffs in the small intestine. Its endocrine function is fulfilled by special cells that produce hormones (insulin and glucagon) that help regulate blood glucose levels, thus assisting in maintaining homeostasis.

The digestive enzymes of the pancreas flow through the large pancreatic duct into the duodenum. Each day, approximately 1,200 to 1,500 milliliters (1.0–1.5 quarts) of pancreatic juice is produced and released into the small intestine. This liquid is composed of water, sodium bicarbonate, and several digestive enzymes (Table 7-4).

Sodium bicarbonate neutralizes the acidic chyme released by the stomach and thus protects the small intestine from stomach acid. It also gives the pancreatic juice a pH of about 8, creating an environment optimal for the function of the pancreatic enzymes.

The pancreatic enzymes released into the small intestine act on the large food molecules (proteins, starches, and so on). Table 7-4 lists each enzyme and the food molecules it breaks down. As a result of pancreatic enzymatic activity, fats, proteins, and carbohydrates are broken down into smaller molecules.

Like the stomach, the pancreas secretes its enzymes in an inactive form. This protects the gland from self-destruction. **Trypsinogen** (trip-SIN-oh-gin), for example, is the inactive form of the protein-digesting enzyme **trypsin** (TRIP-sin). Produced by the pancreas, trypsinogen is activated by a substance on the epithelial lining of the small intestine. Trypsin, in turn, activates other digestive enzymes.

The final stage of digestion occurs with the aid of enzymes produced by the epithelial cells that line the small intestine. These enzymes are embedded in the membranes of the epithelial cells. They are also listed in Table 7-4. These enzymes further break down the food molecules so they can be readily absorbed by the small intestine.

KEY CONCEPTS

Food is broken down in the digestive tract with the aid of enzymes produced by the pancreas and the lining of the small intestine.

The Liver

The **liver** is one of the largest and most versatile organs in the body. By various estimates, the liver performs as many as 500 different functions. Situated on the right side of the abdomen under the protection of the rib cage, the liver is one of the body's storage depots for glucose (see Figure 7-2). It also stores fats and many micronutrients such as iron, copper, and many vitamins. By storing glucose and lipids and releasing them as they are needed, the liver helps ensure a constant supply of energy-rich molecules for body cells. The liver also synthesizes some key blood proteins involved in blood clotting, and it is an efficient detoxifier of potentially harmful chemicals such as nicotine, barbiturates, and alcohol. These functions contribute significantly to homeostasis.

The liver also plays a key role in the digestion of fats through the production of a fluid called **bile**. Bile contains water, ions, and molecules such as cholesterol, fatty acids, and bile salts. **Bile salts** are steroids that emulsify fat—that is, they break the large fat globules into smaller ones. The fat becomes dispersed in the watery contents of the intestine. This process is essential for lipid digestion, because lipid-digesting enzymes in the small intestine do not work well on large fat globules. Unless large fat globules can be broken into smaller ones, fat digestion is incomplete.

Produced by the cells of the liver, bile is first transported to the **gallbladder**, a sac attached to the underside of the liver (Figure 7-18). The gallbladder removes water from the bile, concentrating it. Bile is stored in the gallbladder until needed. When chyme is present in the small intestine, the gallbladder contracts and bile flows out through the duct system into the small intestine (Figure 7-18).

Bile flow to the small intestine may be blocked by gallstones, deposits of cholesterol and other materials in the gallbladder of some individuals. Gallstones may lodge in the ducts draining the organ, thus reducing—even completely blocking—the flow of bile. The lack of bile salts greatly reduces lipid digestion. Because lipid digestion is reduced, fat is not digested and absorbed.

Gallstones occur more frequently in older, overweight individuals. The incidence in the elderly is about one in five. Gallstones are usually removed surgically, often through three or four tiny incisions in the abdominal wall. This procedure minimizes trauma and speeds up recovery compared to the older technique known as open surgery. It involved a much larger incision (six to eight inches) in the abdominal wall. Surgeons insert a tiny camera through one incision so he or she can locate and remove the organ using tiny instruments to secure and then remove the gall bladder. At the time of the operation, the surgeon also examines the bile ducts for blockages, removing them to ensure that bile still can flow into the small intestine.

After the gall bladder is removed, bile continues to be produced by the liver and it is stored in the common bile duct, which enlarges to accommodate the fluid.

Although minimally invasive treatments have been developed, medical researchers are working on a number of drugs that can dissolve gallstones. Several drugs taken orally appear to work well on small gallstones, but may take months or years to totally dissolve them. Researchers are also testing a drug (methyl tert-butyl ether) that is injected directly into the gallbladder to dissolve gallstones in short order—in one to three days.

KEY CONCEPTS

The liver is a vital organ that has hundreds of essential functions, including assisting in the process of digestion.

The Intestinal Epithelium

Virtually all food digestion occurs in the duodenum and jejunum. Once food molecules are digested, they are absorbed—that is, transported across the epithelial lining of the small intestine into the bloodstream or lymphatic system. In these regions, absorption is facilitated by three structural modifications of the small intestine. Figure 7-19a shows, for instance, that the lining of the small intestine is thrown into folds. This increases the surface area of the lining of the small intestine, which enhances absorption. On the surfaces of the folds are many fingerlike projections known as *villi* (VILL-eye; singular, **villus**). The villi also increase the surface area available for the absorption of food molecules. The intestinal surface area is further increased by **microvilli**, tiny protrusions of the plasma membranes of the epithelial cells lining the villi (Figure 7-19d). Each epithelial cell that functions in absorption contains an estimated 3,000 microvilli. Two hundred million microvilli occupy a single square millimeter of intestinal lining.

Together, the circular folds, villi, and microvilli result in a dramatic increase in the surface area of the lining of the small intestine, making it 600 times greater than if it were flat. As shown in Figure 7-19e, each villus of the small intestine is endowed with a rich supply of blood and lymph capillaries. These tiny vessels absorb nutrients that have passed through the lining of the small intestine.

Many mechanisms are involved in absorption—too numerous to be discussed here. Three of the most common are diffusion, osmosis, and active transport (discussed elsewhere in the text).

Although most nutrients diffuse from the epithelial cells into the blood capillaries, fatty acids and monoglycerides produced by the enzymatic digestion of triglycerides follow a different route. As noted earlier in the chapter, triglyceride molecules consists of three fatty acids attached to a single molecule of glycerin (Figure 7-5). When triglycerides are enzymatically digested in the small intestine they are broken down into two fatty acids and a monoglyceride—a molecule of glycerin still attached to one of its fatty acids. As shown in Figure 7-20, monoglycerides and fatty acids diffuse into the epithelial cells lining the small intestine. Inside these cells they react to reform triglycerides. The triglycerides accumulate to form fat droplets. The fat droplets are then coated with a layer of lipoprotein (lipid bound to protein) produced by the endoplasmic reticulum of the cells. This treatment renders the fat droplets, known as *chylomicrons* (KIE-low-MIE-krons), water soluble. The fat droplets are then released by the epithelial cells into the interstitial fluid—the fluid between the cells (Figure 7-20). Because blood capillaries are relatively impermeable to chylomicrons, most of the lipid globules enter the more porous lymph capillaries (known as *lacteals*) in the villi. As shown in Figure 7-20, small- or medium-chain fatty acids not incorporated in triglycerides pass directly into blood capillaries in the villi.

KEY CONCEPTS

The lining of the small intestine releases digestive enzymes that help break food molecules into much smaller and more readily absorbable molecules that easily pass into the blood vessels in the wall of the small intestine.

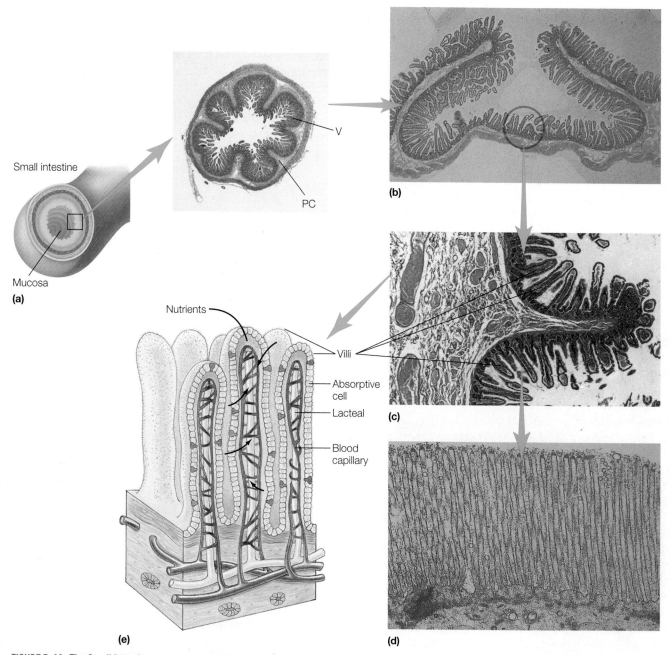

FIGURE 7-19 The Small Intestine The small intestine is uniquely "designed" to increase absorption. (a) A cross section showing the folds. V means villi, and PC means *plica circulares* (circular fold). (Courtesy of Douglas Burrin/USDA ARS.) (b) A light micrograph of folds and villi. (© Donna Beer Stolz, Ph.D., Center for Biologic Imaging, University of Pittsburgh Medical School.) (c) Higher magnification of villi. (© Donna Beer Stolz, Ph.D., Center for Biologic Imaging, University of Pittsburgh Medical School.) (d) An electron micrograph of the surface of the absorptive cells showing the microvilli. (© David M. Phillips/Visuals Unlimited.) (e) Each villus contains a loose core of connective tissue; a lacteal, or lymph, capillary; and a network of blood capillaries. Nutrients pass from the lumen of the small intestine through the epithelium and into the interior of the villi, where they are picked up by the lymph and blood capillaries.

The Large Intestine

The **large intestine** is about 1.5 meters (5 feet) long and consists of four regions. Starting at the small intestine, they are the cecum, appendix, colon, and rectum (Figure 7-21). The **cecum** (SEA-come) is a pouch that forms below the juncture of the large and small intestine. A small, wormlike structure, the **appendix**, attaches to the cecum. This organ measures a little over four inches long on average, The appendix is a common site for infection, resulting in a condition known as *appendicitis*. To

treat this potentially life threatening condition, the appendix is removed surgically.

The appendix was long thought to be a nonfunctional organ and was thus pronounced a *vestigial organ*—an organ that remains in the body but has lost its function. New research findings from Duke University Medical Center suggest that the appendix may not be vestigial. Their study suggests that the organ's location in the intestinal tract is where bacteria can hide during infection. The researchers

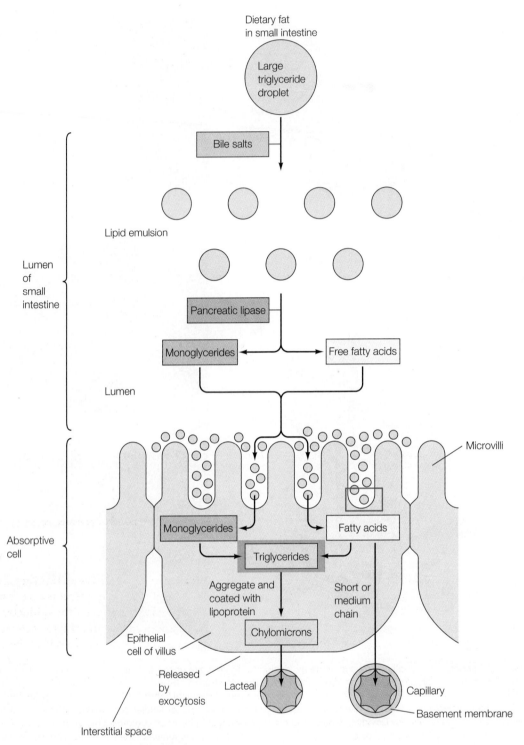

FIGURE 7–20 Fat Digestion and Absorption

found that the bacteria move out of the intestines and into the appendix during intestinal infections so that the immune system does not attack them while trying to rid itself of infection.

Research also suggests that the organ may play an important role in infants and adults. For example, it may manufacture certain hormones during fetal development and help to "train" the immune system, exposing the body to antibody producing chemicals, known as antigens, so that the immune system can produce antibodies.

Most of the large intestine consists of the **colon** (COE-lun). Unlike the small intestine, which is coiled and packed in a small volume, the colon consists of three relatively straight portions, the ascending colon, the transverse colon, and the descending colon. The colon empties into the **rectum** (WRECK-tum).

Although the large intestine is considerably shorter than the small intestine, it has a much greater diameter. So named because of its large diameter, the large intestine receives materials from the small intestine. The material entering the large intestine consists of a mixture of water, undigested or

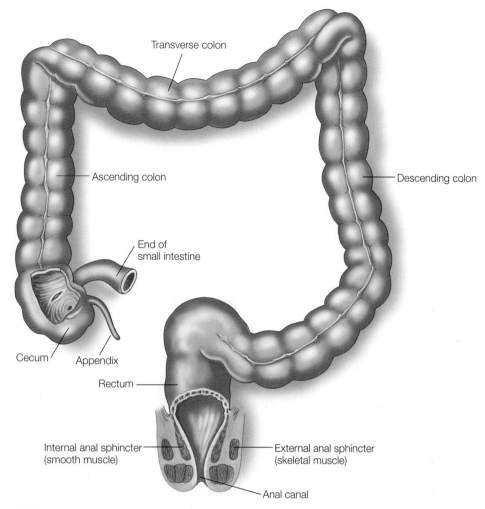

Transverse colon

Ascending colon

Descending colon

End of
small intestine

Cecum

Appendix

Rectum

Internal anal sphincter
(smooth muscle)

External anal sphincter
(skeletal muscle)

Anal canal

FIGURE 7-21 The Large Intestine This organ consists of four basic parts: the cecum, appendix, colon, and rectum.

unabsorbed food molecules, and indigestible food residues, such as cellulose. It also contains sodium and potassium ions.

The colon absorbs approximately 90% of the water and sodium and potassium ions that enter it. The undigested or unabsorbed nutrients feed a rather large population of *Escherichia coli* bacteria. These bacteria are vital to human health, for they synthesize several key vitamins: B$_{12}$, thiamine, riboflavin, and, most importantly, vitamin K, which is often deficient in the human diet. These vitamins are absorbed by the large intestine.

The contents of the large intestine (after water and salt have been removed) are known as the **feces**. The feces consist primarily of undigested food, indigestible materials, and bacteria. Bacteria, in fact, account for about one-third of the dry weight of the feces.

The feces are propelled by peristaltic contractions along the colon until they reach the rectum (Figure 7-21). As fecal matter accumulates in the rectum, it distends the organ. This action, in turn, stimulates stretch receptors in the wall of the rectum. These receptors stimulate the **defecation reflex** (DEAF-eh-KA-shun). In this reflex, nerve impulses from the stretch receptors in the rectum travel via nerves to the spinal cord. In the spinal cord, the nerve impulses stimulate neurons that supply the smooth muscle in the wall of the rectum, causing them to contract and expel the feces. Nerve impulses from the spinal

cord also travel to the internal anal sphincter, a ring of smooth muscle, which normally keeps fecal matter from entering the anal canal. These impulses cause the sphincter to relax.

Defecation does not occur until the external anal sphincter relaxes. This sphincter is composed of skeletal muscle and is under conscious control in all individuals except babies. If the time and place are appropriate, the external anal sphincter is relaxed, and defecation can occur. Defecation is usually assisted by voluntary contractions of the abdominal muscles and a forcible exhalation, both of which increase intra-abdominal pressure.

If circumstances are inappropriate, voluntary tightening of the external anal sphincter prevents defecation, despite prior activation of the reflex. When defecation is delayed, the muscle in the wall of the rectum relaxes, and the urge to defecate subsides—at least until the next movement of feces into the rectum occurs.

If defecation is delayed too long, additional water may be removed from the feces, making them hard and dry and difficult to pass. This condition is called **constipation**.

Besides being uncomfortable, constipation may result in a dull headache, loss of appetite, and depression. Constipation is generally caused by (1) ignoring the urge to defecate; (2) decreases in colonic contractions, which occur with age, emotional stress, or low-fiber diets; (3) tumors or colonic spasms that obstruct the movement of feces; (4) certain drugs like those containing codeine; and (5) nerve injury that impairs the defecation reflex.

Constipation is not only uncomfortable, it can result in serious problems. Hardened fecal material, for instance, may become lodged in the appendix. This, in turn, may lead to inflammation of the organ, a condition called **appendicitis** (ah-PEND-eh-SITE-iss). When this occurs, the appendix becomes swollen and filled with pus and must be removed surgically to prevent the organ from bursting and spilling its contents into the abdominal cavity. Fecal matter leaking into the abdominal cavity introduces billions of bacteria and can result in a deadly infection.

> **KEY CONCEPTS**
>
> The large intestine is the site of vitamin K production and water resorption.

7-3   Controlling Digestion

Digestion is an extremely complicated process that occurs at many levels and involves many different physiological processes. This intricate and important function is controlled largely by nerves and hormones. This section discusses some of the key events involved in the control of digestion.

Saliva and Salivation

As noted earlier, digestion begins in the oral cavity with the release of saliva. It helps us liquefy our food, which makes it easier to swallow. Saliva also contains some enzymes that begin breaking down long-chain carbohydrates. The excretion of saliva by the salivary glands is stimulated by the sight, smell, taste, and sometimes even the thought of food like a hot fudge sundae. Chewing also stimulates its release.

Secretion of saliva is controlled by the nervous system and is largely a reflex response. For a discussion of the discovery of this phenomenon, see Scientific Discoveries 7-1.

> **KEY CONCEPTS**
>
> Saliva is produced by the salivary glands; it helps lubricate food and contains some enzymes that start breaking the food down chemically. Its release is controlled by the nervous system.

The Gastric Glands

Besides activating salivary production, the stimuli listed in the previous paragraph also cause the brain to send nerve impulses along the *vagus nerve* (VAG-gus) to the stomach (Figure 7-22). This nerve starts at the brain and extends down into the abdominal cavity where it supplies the internal organs of the digestive system like the stomach. As shown in Figure 7-22, nerve impulses transmitted to the stomach by the vagus initiate the secretion of hydrochloric acid (HCl) and pepsinogen from cells of the gastric glands found in the lining of the stomach. Hydrochloric acid, you may recall, coagulates protein so it can be more easily digested by protein-digesting enzymes in the small intestine. Pepsinogen is a precursor to pepsin, a protein digesting hormone released by gastric glands. The most potent stimulus for gastric secretion, however, is the presence of protein in the stomach. Protein stimulates chemical receptors inside the stomach

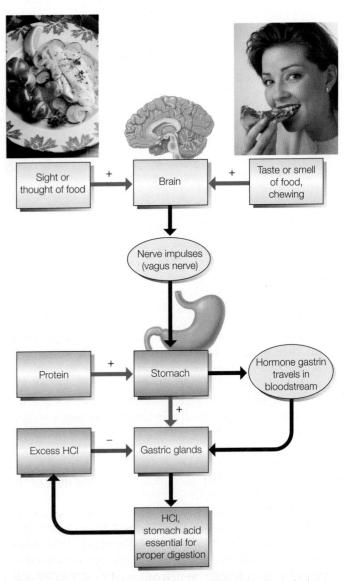

FIGURE 7-22 **Pathway Leading to the Release of HCl by the Stomach** (Plate photo courtesy of Len Rizzi/National Cancer Institute; pizza photo © Photodisc.)

Scientific Discoveries that Changed the World

7-1 Discovering the Nature of Digestion
Featuring the Work of van Helmont, Beaumont, and Pavlov

The foundation for the modern understanding of digestion was laid by a number of scientists. Two noteworthy contributors were a seventeenth-century Flemish physician, Jan Baptista van Helmont, and a nineteenth-century American, William Beaumont.

In van Helmont's time, most people thought that the digestion of food in the human was akin to cooking—that is, that food inside the stomach was broken down by body heat. Van Helmont, however, dismissed this theory with the simple observation that digestion occurs in cold-blooded animals such as fish. The lack of heat in such animals suggested to him that an alternative process was at work. But what was it?

The road to discovery began in a rather bizarre way. One day, a tame sparrow that often visited the scientist and would sit on his shoulder attempted to bite van Helmont's tongue. The physician detected a slight acidity in the bird's bite and hypothesized that acid must digest food.

To test this hypothesis, van Helmont tried to chemically break down meat with a vinegar solution, but he could not. Consequently, he hypothesized that the body must contain chemical substances in the stomach and small intestine that are specific for different types of food. Today, we know these chemicals as digestive enzymes.

Little insight into digestion was forthcoming in the period between van Helmont's work and the studies of the American scientist William Beaumont. Beaumont was a medical doctor. Soon after becoming a licensed physician, Beaumont joined the army, where he would make his important discoveries about digestion. He owed his great work to an accident. In 1822, a French-Canadian porter was wounded in the abdomen by a bullet from a musket that had accidentally discharged at close range. The porter, Alexis St. Martin, who was only 18, was brought to Beaumont for treatment. The doctor found that the bullet had ripped a hole in the man's stomach, out of which poured food he had eaten.

The patient survived, but the wound was slow to heal. For over 8 months, Beaumont tried in vain to close the hole in the stomach. During this time, it dawned on the young physician that he had stumbled across a golden opportunity to study digestion.

Beaumont's experiments consisted of feeding the patient various types of food, then studying what took place inside the man's stomach. In the next 9 years, Beaumont studied St. Martin's stomach contents with a wide assortment of foods, looking at the rate and temperature of digestion and the chemical conditions that favored different stages of the process. Through his studies, Beaumont demonstrated that digestion was a chemical phenomenon.

Beaumont did not address one important question: What caused gastric juice to be secreted? Was it the presence of food in the stomach or something else?

The answer came in 1889 from the work of Ivan Pavlov, a scientist whose experiments are legendary. In studies on dogs, Pavlov showed that the secretion of gastric juice was stimulated by the nervous system. How did he make this determination?

In one of many experiments, the Russian scientist surgically connected a fold of a dog's stomach to an opening in the animal's side; this permitted him to examine the production of gastric secretions. He next cut and tied off the esophagus of the dog, so that food the animal swallowed could not enter the stomach. Following the surgery, Pavlov gave the dog food and found that as soon as food entered the dog's mouth, gastric juice began to be secreted. These secretions continued as long as food was present, thus suggesting nervous system involvement.

Since that time, a great deal has been learned about digestion and its control. For example, as you learned in this chapter, hormones are also involved in the control of many digestive processes. But don't close the book on this subject; much more will inevitably be discovered.

---

that activate nerves in the stomach wall. These nerves, in turn, directly stimulate the gastric glands to secrete HCl and pepsinogen.

Nerve impulses from the brain also stimulate the secretion of a hormone by the stomach. This hormone is known as **gastrin**. As shown in Figure 7-22, gastrin is released into the blood. It acts on the gastric glands, stimulating additional HCl and pepsinogen secretion.

The concentration of HCl in the stomach is regulated by a negative feedback mechanism, illustrated in Figure 7-22. If the acid content rises too high, HCl inhibits gastrin secretion, shutting down the gastric glands.

KEY CONCEPTS

Gastric glands produce acid that coagulates proteins; its release is under the control of the nervous and the endocrine systems.

Control of Pancreatic Secretions

While the gastric glands of the stomach are stimulated by nerve impulses and the hormone gastrin, secretion by the pancreas is controlled by two hormones. The first hormone is **secretin**. Its release is stimulated by the presence of acidic chyme in the small intestine (Figure 7-23). Produced by the cells of the duodenum, secretin, like all hormones, travels

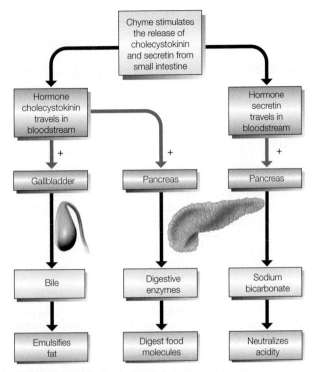

FIGURE 7-23 **Control of the Pancreas and Gallbladder**

in the bloodstream. In the pancreas, secretin stimulates the release of sodium bicarbonate. Sodium bicarbonate, in turn, is excreted by the pancreas into the small intestine, where it neutralizes the acidic chyme and creates an environment optimal for pancreatic enzymes.

The release of pancreatic enzymes is triggered by another intestinal hormone with the tongue-twisting name **cholecystokinin** (COAL-ee-sis-toe-KIE-nin) or CCK. This hormone is produced by cells of the duodenum in the presence of chyme (Figure 7-22). CCK also stimulates the gallbladder to contract, releasing bile into the small intestine.

Interestingly, recent evidence links abnormally low secretion of CCK to **bulimia** (buh-LEEM-ee-ah), a disorder characterized by recurrent binge eating followed by intentional vomiting. Some researchers believe that CCK may trigger the feeling of being full. Approximately 4% of America's young women, and a far smaller fraction of men, suffer from bulimia. This disorder is also thought to have psychological roots.

> **KEY CONCEPTS**
>
> The pancreas releases a substance that neutralizes stomach acids emptying into the small intestine and enzymes required for digestion. Pancreatic secretions are stimulated by two intestinal hormones.

7-4 Controlling Hunger

Hunger is an internal physiological signal that's vital to human survival. It ensures we replenish supplies of vital nutrients, including carbohydrates, proteins, and vitamins, from time to time so we can live.

Human hunger is controlled by a region in the lower part of the brain known as the *hypothalamus*, discussed elsewhere in the text. The hypothalamus, in turn, is influenced by several chemicals substances released by cells and organs of the body. One of them is a hormone known as leptin.

Leptin is a protein hormone consisting of 167 amino acids. It is synthesized primarily in white fat cells, but is also produced by brown fat and a number of other cells of the body. These include the ovaries, skeletal muscle, and stomach. Even breast tissue, the pituitary gland, and liver produce this hormone.

Leptin hormone travels through the bloodstream to the brain where it binds to receptors located in the plasma membranes of cells in the hypothalamus that control hunger. This group of cells is known as the *hunger center* (or more technically, the satiety center or even more technically, the arcuate nucleus). In the hunger center, leptin binds to nerve cells, decreasing their activity. This signals the brain that the body has had enough to eat.

Leptin operates by counteracting two very potent chemical feeding stimuli. One of these stimulants is a naturally produced chemical known as *neuropeptide Y* (NPY). This substance is produced by nerve cells in the hypothalamus. It has the opposite effect of leptin. It stimulates food intake. Another potent feeding stimulus is a neuropeptide called *anandamide*. It is also produced by cells of the brain.

The presence of leptin reduces signals that encourage us to eat. Unfortunately, a very small group of people become leptin-resistant. That is, their brains don't respond to leptin. As a result, leptin levels go higher and higher as they pile on more and more weight. Leptin production increases as the amount of fat in the body increases. In leptin resistance, a patient's leptin is high, which means they are overweight, but their brains can't see it.

Leptin is a multitalented hormone. It promotes heart and bone health, and also helps keep the immune system functioning well. For example, when leptin is working right, your bones are healthier—they have more calcium and are less likely to break. High levels of leptin may also promote the growth of melanoma, a type of skin cancer. It may even affect fertility in women. If a woman's brain doesn't properly react to leptin levels, her fertility drops. This is probably an evolutionary protection. If a person didn't have enough fat to survive a pregnancy, then it is to her benefit to not become pregnant.

Yet another hormone involved in hunger is ghrelin. This fast-acting hormone is produced primarily by cells lining part of the stomach, certain cells in the pancreas, and by cells in the hunger center as well. Ghrelin stimulates hunger. Studies show that ghrelin levels increase before meals and then decreases after meals. Ghrelin receptors are found in the hunger center.

> **KEY CONCEPTS**
>
> Hunger is controlled by the nervous and endocrine systems.

7-5 Health and Homeostasis

There are few places where the relationship between homeostasis and human health is as evident as in nutrition. Because cells and organs involved in homeostasis require an adequate supply of nutrients, nutritional deficiencies can have profound impacts on this vital physiological balancing act—and subsequently on your health. Unhealthy diets, in fact, can greatly increase your risk of contracting many diseases, including certain types of cancer, heart disease, late-onset diabetes, and hypertension.

Consider an example. Magnesium (a major mineral) is routinely ingested in insufficient amounts. New research suggests that such deficiencies may underlie a number of medical conditions, including diabetes, high blood pressure, problems during pregnancy, and cardiovascular disease.

Research shows that adding magnesium to the drinking water of rats with hypertension can eliminate the disease. Studies in rabbits show that magnesium decreases lipid levels in the blood and reduces plaque formation in blood vessels. Rabbits on a high-cholesterol, low-magnesium diet, for example, have 80% to 90% more atherosclerotic plaque than rabbits on a high-cholesterol, high-magnesium diet.

Studies have shown that in humans magnesium deficiencies during pregnancy result in migraine headaches, high blood pressure, miscarriages, stillbirths, and babies with low birth weight. Research suggests that magnesium deficiency causes spasms in the blood vessels of the placenta that reduce blood flow to the fetus. This can retard fetal growth and may even kill a fetus. Magnesium supplements greatly reduce the incidence of these problems.

Researchers believe that 80% to 90% of the American public may be magnesium deficient. One reason for this deficiency is that phosphates in many carbonated soft drinks bind to magnesium in the intestine, preventing it from being absorbed into the blood. Magnesium deficiencies can be reversed by eating more green leafy vegetables, seafood, and whole grain cereals and reducing consumption of carbonated soft drinks with high phosphate levels (colas such as Coke and Pepsi and similar dark carbonated soft drinks like Dr. Pepper—check the contents for phosphoric acid). Mineral supplements could help as well, but they should be used with extreme caution because excess magnesium can cause neurological disorders.

Over the years, numerous dietary recommendations have been issued to help Americans live healthier lives and reduce the risk of cancer and heart attack. Nutritionists recommend daily consumption of (1) fruits and vegetables, especially cabbage and greens, (2) high-fiber foods, such as whole wheat bread and celery, and (3) foods high in vitamins A and C. A healthy diet also minimizes the consumption of animal fat; red meat; and salt-cured, nitrate-cured, smoked, or pickled foods, including bacon and lunch meat.

A healthy diet results largely from habit and circumstance. In the hustle and bustle of modern society, many of us ignore proper nutrition, grabbing fat-rich fast foods because we haven't the time to sit down to a nutritionally balanced meal. Our fast-paced world places a high premium on saving time, often ignoring the importance of eating and living right. Fast food may allow us to hurry on to our next appointment, but unless the meal is nutritionally sound, it probably decreases the quality of our lives in the long run—and it may reduce our lifespan as well. Thus, some of the time we save by eating this way, will lessen the time many of us spend on planet Earth.

KEY CONCEPTS

Soft drinks contain phosphates that significantly lower the uptake of magnesium, a mineral required for human health.

SUMMARY

An Introduction to Nutrition

1. Humans acquire energy and nutrients from the food they eat. These nutrients fall into two categories: macronutrients, substances needed in large quantity, and micronutrients, substances required in much lower quantities.

2. The four major macronutrients are water, carbohydrates, lipids, and proteins.

3. Adequate water intake is important because water is involved in many chemical reactions in the body. It also helps to maintain body temperature and a constant level of nutrients and wastes in body fluids, all of which are vital to homeostasis. Water is contained in the liquids we drink and the foods we eat.

4. Glucose is a sugar that is broken down in the body to produce energy. In the human diet, glucose comes primarily from the polymer starch. Glucose molecules are stored in the liver and muscle as glycogen.

5. Another important dietary carbohydrate is cellulose. Although not digested, cellulose facilitates the passage of fecal matter through the intestinal tract and reduces the incidence of colon cancer.

6. Dietary protein is chiefly a source of amino acids for building proteins, enzymes, and protein hormones. Protein is broken down into amino acids and absorbed into the bloodstream. However, the body can synthesize some amino acids. Others cannot be synthesized in the body and must be supplied in the food we eat. These are known as essential amino acids.

7. To ensure an adequate supply of all amino acids, individuals should eat complete proteins, such as those found in milk, eggs, meat, tofu, and the grain quinoa, or combine incomplete protein sources such as legumes (starchy beans) and either grains or nuts and seeds.

8. Lipids are a diverse group of organic chemicals characterized by their lack of water solubility. The biologically important lipids serve many functions. Some lipids (triglycerides) provide energy. They also form layers of heat-conserving insulation. Other lipids (phospholipids and steroids) are part of the plasma membranes of cells.

9. Triglycerides are fats and oils. They consist of a single molecule of glycerol attached to three fatty acids. Triglycerides with many double bonds in their fatty acid side chains are known as unsaturated fatty acids lower your risk of a buildup of plaque on arterial walls. Triglycerides containing saturated fatty acids tend to increase this disease; they are commonly found in animal fats.

10. Micronutrients are needed in much smaller quantities and include two groups: vitamins and minerals.

11. Vitamins are a diverse group of organic compounds that are required in relatively small quantities for normal metabolism. A deficiency or surplus of one or more vitamins may alter homeostasis, with serious effects on human health.

12. Human vitamins fit into two categories: water-soluble and fat-soluble. The water-soluble vitamins include vitamin C and the B-complex vitamins. The fat-soluble vitamins include vitamins A, D, E, and K.

13. Minerals fit into one of two groups: trace minerals, those required in very small quantity, and major minerals, those required in greater quantity. Deficiencies and excesses of both types of minerals can lead to serious health problems.

14. In recent years, scientists have discovered numerous chemicals in the foods we eat that help prevent disease. Foods containing them are called functional foods because they provide a health benefit beyond basic nutrition. One important group is known as antioxidants, chemicals that react with naturally occurring chemical oxidants, substances in the blood that are involved in cholesterol buildup in the walls of arteries as well as cancer. Eating a diet rich in antioxidants may reduce heart and artery disease as well as cancer. Antioxidants are primarily found in fruits and vegetables.

The Digestive System

15. Food digestion begins in the mouth. The teeth mechanically break down the food. Saliva liquefies it, making it easier to swallow. An enzyme in saliva, salivary amylase, begins to break down starch molecules.

16. Food is pushed by the tongue to the pharynx, where it triggers the swallowing reflex. Peristaltic contractions propel the food down the esophagus to the stomach.

17. The stomach is an expandable organ that stores and further liquefies the food. The churning action of the stomach, created by peristaltic contractions, mixes the food, turning it into a paste known as chyme.

18. The stomach releases food into the small intestine in timed pulses, ensuring efficient digestion and absorption. Very limited chemical digestion and absorption occur in the stomach.

19. The stomach produces hydrochloric acid. This acid denatures (coagulates) protein, facilitating the enzymatic breakdown protein. The stomach also produces a protein-digesting enzyme called pepsin, which breaks proteins into peptides.

20. The functions of the stomach are regulated by the nervous and endocrine systems.

21. The small intestine is a long, coiled tubule in which most of the enzymatic digestion and absorption of food occur. Digestive enzymes are produced by the pancreas and cells lining the small intestine. Pancreatic enzymes break macromolecules into smaller fragments. Intestinal enzymes break these molecules into even smaller fragments that can be absorbed by the epithelial lining of the small intestine.

22. The pancreas also releases sodium bicarbonate, which neutralizes HCl entering the small intestine with the chyme and creates an environment suitable for pancreatic enzyme function.

23. The liver plays an important role in digestion. It produces a liquid called bile that contains bile salts. Bile is stored in the gallbladder and released into the small intestine when food is present. Bile salts break fats into small globules that can be acted on by fat-digesting enzymes.

24. Undigested food molecules pass from the small intestine into the large intestine, which absorbs water, sodium, and potassium, as well as vitamins produced by intestinal bacteria. It also transports the waste, or feces, to the outside of the body.

Controlling Digestion

25. Digestion begins in the mouth, continues in the stomach, and is completed in the small intestine.

26. Digestive processes are largely controlled by the nervous and endocrine systems.

27. Digestion begins with the release of saliva which is stimulated by the sight, smell, taste, thought of food, and chewing.

28. These stimuli also cause the brain to send nerve impulses to the gastric glands of the stomach, initiating the secretion of HCl and gastrin, a hormone that also stimulates additional HCl secretion. Protein in the stomach also stimulates the secretion of HCl and gastrin.

29. The release of pancreatic enzymes is controlled by two hormones, secretin and cholecystokinin. Both hormones are released by cells in the interior lining of the small intestine. Their release is stimulated by the presence of acidic chime in the small intestine. Secretin stimulates the release of sodium bicarbonate, which neutralizes stomach acid, creating an environment suitable for enzymatic destruction of food. Cholecystokinin stimulates the release of pancreatic enzymes and contraction of the gall bladder, which releases bile into the small intestine.

Health and Homeostasis

30. Human health is dependent on good nutrition. Numerous studies suggest that a healthy, balanced diet can maintain homeostasis and thus decrease the risk of certain forms of cancer, heart disease, hypertension, late-onset diabetes, and a host of other diseases.

31. Nutritionists recommend the daily consumption of (a) fruits and vegetables; (b) high-fiber foods, such as whole-wheat bread and celery; and (c) foods high in vitamins A and C.

32. In addition, nutritionists recommend reducing the consumption of animal fat; red meat; and salt-cured, nitrate-cured, smoked or pickled foods, including bacon and lunch meat.

THINKING CRITICALLY ANALYSIS

This analysis corresponds to the Thinking Critically scenario that was presented at the beginning of this chapter.

Most people who look at the data believe that it supports the hypothesis. Look at the large number of kids in the hyperactive group. Most of them consumed sweets. How can you interpret the data any other way? Do you agree?

If you do, think again. Note that the ratio of students who ate candy to those who did not in both groups is 5 to 1. In other words, in the hyperactive group there are 250 students who ate candy and 50 students who did not. In the nonhyperactive group there are 50 students who ate candy and 10 who did not. These data, and data from other similar studies, show that there is no correlation between eating sugar and hyperactivity. But why do so many who view these data misinterpret them?

We tend to ignore negative information. In this case, we ignore the nonhyperactive group and only focus on the apparent correlation in the hyperactive group. Keep this in mind when you analyze issues. Look carefully at the data.

KEY TERMS AND CONCEPTS

Amino acid, p. 137
Appendicitis, p. 158
Appendix, p. 155
Antioxidant, p. 142
Bile, p. 154
Bile salts, p. 154
Bulimia, p. 160
Carbohydrates, p. 133
Cecum, p. 155
Cellulose, p. 133
Cholecystokinin, p. 160
Cholesterol, p. 137
Chyme, p. 151
Colon, p. 156
Complete proteins, p. 137
Constipation, p. 157
Defecation reflex, p. 157
Dietary fiber, p. 134
Disaccharide, p. 133
Enamel, p. 147
Epiglottis, p. 150
Esophagus, p. 150
Essential amino acid, p. 137
Fat-soluble vitamins, p. 140
Fatty acids, p. 134
Feces, p. 157
Free radicals, p. 142

Functional food, p. 142
Gallbladder, p. 154
Gastric gland, p. 151
Gastrin, p. 159
Gingivitis, p. 148
Glucose, p. 133
Glycerol, p. 134
Glycogen, p. 133
Heartburn, p. 151
Incomplete proteins, p. 137
Lacteals, p. 152
Large intestine, p. 155
Leptin, p. 160
Lipid, p. 134
Liver, p. 153
Macronutrients, p. 132
Micronutrients, p. 132
Microvilli, p. 154
Mineral, p. 142
Monosaccharide, p. 133
Mouth, p. 147
Mucus, p. 152
MyPlate, p. 144
Omega-3 fatty acids, p. 144
Overnutrition, p. 138
Pancreas, p. 152
Peristalsis, p. 150

Pharynx, p. 149
Phytochemicals, p. 142
Plaque, p. 147
Polysaccharide, p. 133
Protein, p. 137
Rectum, p. 156
Saliva, p. 147
Secretin, p. 159
Small intestine, p. 152
Starch, p. 133
Steroid, p. 137
Stomach, p. 151
Taste buds, p. 149
Tongue, p. 149
Triglycerides, p. 134
Trypsin, p. 153
Trypsinogen, p. 153
Ulcer, p. 152
Undernutrition, p. 138
Villus, p. 154
Vitamin, p. 140
Water, p. 132
Water-soluble fiber, p. 134
Water-insoluble fiber, p. 134
Water-soluble vitamins, p. 140

CONCEPT REVIEW

1. The body requires proper nutrient input to maintain homeostasis. Give an example, and explain how nutrition affects homeostasis. p. 161.
2. What are the two major sources of cellular energy? p. 132.
3. Describe the conditions during which protein provides cellular energy. p. 137.
4. Describe how the different types of dietary fiber help protect human health. p. 134.
5. If you were considering becoming a vegan, a vegetarian that consumes no animal products whatsoever, even milk and eggs, how would you be assured of getting all of the amino acids your body needs? p. 137.
6. What are vitamins, and why are they needed in such small quantities? pp. 140–142.

7. A dietary deficiency of one vitamin can cause wide-ranging effects. Why? You might want to check vitamin B_{12} out on reliable sources of information on the Internet to determine how it is used in the body and the impacts of deficiencies of this vitamin. pp. 137, 140–142.
8. What organs physically break food down, and what organs participate in the chemical breakdown of food? pp. 147–153.
9. Describe the process of swallowing. pp. 149–150.
10. Describe the function of hydrochloric acid and pepsin in the stomach. How does the stomach protect itself from these substances? pp. 151–152.
11. How do ulcers form, and how can they be treated? p. 152.

12. The small intestine is the chief site of digestion and absorption. Where do the enzymes needed for this process come from, and how is the release of these enzymes stimulated? What other molecules (besides enzymes) are needed for proper digestion in the stomach and small intestine? pp. 152–153.
13. Describe the endocrine and nervous system control of the stomach function. pp. 158–159.
14. How does the smell or taste of food stimulate the digestive process? pp. 158–159.
15. Describe the functions of the large intestine. pp. 155–158.

SELF-QUIZ: TESTING YOUR KNOWLEDGE

1. Water is a type of nutrient known as a _____ because it is required in large quantities. p. 132.
2. Glucose is a monosaccharide that is combined to form a nutritive polysaccharide in plants known as _____. p. 133.
3. Dietary _____ may be either water soluble or water insoluble. p. 134.
4. The water-insoluble type is thought to lower _____ levels in the blood. p. 134.
5. _____ are a structurally diverse group including fats and oils and steroids that are all insoluble in water. p. 134.
6. A _____ is made of three fatty acid molecules and a molecule of glycerol. p. 134.
7. Cholesterol is produced in the _____ and ingested in some types of foods such as eggs. p. 135.

8. Amino acids that cannot be synthesized by the body are called _____ amino acids. p. 137.
9. A protein that contains all of the essential amino acids such as milk or tofu is known as a _____ protein. p. 137.
10. _____ is a condition in which a person is severely overweight. p. 138.
11. _____ are micronutrients needed in small quantity; most are not synthesized in the body and most play an important role in metabolic reactions. p. 140.
12. Copper is a _____ mineral required in very small quantity. p. 142.
13. A _____ food provides health benefits beyond nutrition. p. 142.
14. Salivary _____ is an enzyme released in the mouth which helps break down starch molecules. p. 147.

15. After food leaves the mouth, it enters the pharynx and is then passed into the _____, which transports it to the stomach. p. 149.
16. Muscular contraction of the stomach and intestines is called _____. p. 150.
17. Hydrochloric acid is released by the _____ glands in the stomach. p. 151.
18. Bile is produced by the _____ and stored in the _____ and released into the _____ intestine. p. 154.
19. The pancreas releases _____, which neutralizes the chyme and creates a pH conducive to enzyme digestion in the small intestine. p. 153.
20. _____ in the large intestine produce vitamin K, which is often deficient in the human diet. p. 157.

biology.jbpub.com/chiras/8e/

The site features eLearning, an online review area that provides quizzes, chapter outlines, and other tools to help you study for your class. You can also follow useful links for in-depth information, research the differing views in the Point/Counterpoints, or keep up on the latest health news.

(©Mircea BEZERGHEANU/Shutterstock, Inc.)

The Vital Exchange: Respiration

George F. Eaton was a robust and handsome Irishman who grew up in eastern Massachusetts, married, and raised three children. To support his family, Eaton worked as a firefighter for 30 years. Fire fighting is dangerous work, in part because it exposes men and women to smoke containing numerous potentially harmful air pollutants.

THINKING CRITICALLY

For years, representatives of the tobacco industry claimed that there was no proof that cigarette smoking caused cancer, despite the fact that over 40 studies on people and numerous studies on lab animals showed a strong correlation between smoking and lung cancer. Most of the studies on humans take into account factors that might cause lung cancer such as exposure to cancer-causing chemical pollutants in the workplace.

Why did representatives from the tobacco industry claim there was no proof that smoking caused lung cancer? How would you have responded to their assertion? Can a study be devised to prove the connection? What critical thinking rules does this exercise rely on?

Upon retirement, Eaton moved to Cape Cod, but his retirement years were cut short by emphysema—a debilitating respiratory disease resulting from the breakdown of the air sacs in the lungs, where oxygen and carbon dioxide are exchanged between the air and the blood. As the walls of the alveoli degenerate, the surface area for the diffusion of oxygen and carbon dioxide gradually decreases. Emphysema is irreversible and incurable. Patients suffer from shortness of breath; eventually, even mild exertion becomes difficult, prompting one patient to describe the disease as a kind of "living hell."

The degeneration of the lungs in patients with emphysema creates a negative domino effect. One of the dominoes is the heart. Because oxygen absorption in the lungs declines quite substantially over time, the heart of an emphysemic patient must work harder and harder. This puts additional strain on this already hard-working organ and can lead to heart failure.

George Eaton, my grandfather, died a slow death as his lungs grew increasingly more inefficient. Relatives mourning his death blamed it on the pollution to which he was exposed while working for the fire department. Fire fighting, however, was probably only part of the cause, for my grandfather had smoked most of his adult life. Cigarette smoking is the leading cause of emphysema.

This chapter describes the respiratory system—its structure and function and the diseases such as emphysema that affect it.

8-1 Structure of the Human Respiratory System

The human respiratory system functions automatically, drawing air into the lungs, then letting it out, in a cycle that repeats itself about 16 times per minute at rest—or about 23,000 times per day. If you got a dollar for each time you took a breath, you'd be a millionaire in a month and a half.

The process of breathing—drawing air in, then letting it out again—is vital to human health. It ensures a steady supply of oxygen to the cells of the body needed for cellular respiration, the complete breakdown of glucose by the cells of the body. This process generates energy. Breathing also helps us rid our bodies of waste, particularly carbon dioxide, a waste product of cellular energy production. The respiratory systems role as a provider of oxygen and a disposer of carbon dioxide waste helps us maintain the constant internal environment necessary for cellular metabolism. Thus, like many other systems, it plays an important role in homeostasis.

> **KEY CONCEPTS**
>
> The respiratory system supplies oxygen to cells of the body and gets rid of carbon dioxide, a waste product of cellular energy production.

Anatomy of the Respiratory System

The respiratory system consists of two basic parts: an air-conducting portion and a gas-exchange portion (Table 8-1). The air-conducting portion is an elaborate set of passageways that transports air to and from the **lungs**, two large, saclike organs in the chest (thoracic) cavity (Figure 8-1a). Like the arteries of the body, these passageways start out large. With each branching, they become progressively smaller and more numerous.

The lungs are the gas-exchange portion of the respiratory system. Each lung contains millions of tiny, thin-walled air sacs known as **alveoli** (singular, alveolus) (Figure 8-1b). The walls of the alveoli contain numerous capillaries that absorb oxygen from the inhaled air and release carbon dioxide (Figure 8-1b).

Air enters the respiratory system through the nose and mouth, then is drawn backward into the pharynx (FAIR-inks) (Figure 8-2). The **pharynx** is a funnel-shaped structure that opens into the nose and mouth in the front (anteriorly) and joins below with the larynx (LAIR-inks). The **larynx**, or voice box, is a rigid, hollow structure made of cartilage. To feel the larynx, gently place your fingers alongside your throat, and then swallow. The structure that moves up and down is the larynx.

The larynx plays an important role in swallowing. When food is swallowed, the larynx rises and presses against a flap of tissue called the **epiglottis**. As you may recall, the epiglottis is a flap of tissue that acts like a trapdoor that closes during swallowing. This closes off the larynx, prevents food from entering the trachea or windpipe, which could block airflow and result in asphyxiation, a complete and lethal blockage of air flow to the lungs. Food also pushes the epiglottis over the opening to the larynx when we swallow.

TABLE 8-1	Summary of the Respiratory System
Organ	**Function**
Air conducting	
Nasal cavity	Filters, warms, and moistens air; also transports air to pharynx
Oral cavity	Transports air to pharynx; warms and moistens air; helps produce sounds
Pharynx	Transports air to larynx
Epiglottis	Covers the opening to the trachea during swallowing
Larynx	Produces sounds; transports air to trachea; helps filter incoming air; warms and moistens incoming air
Trachea and bronchi	Warm and moisten air; transport air to lungs; filter incoming air
Bronchioles	Control air flow in the lungs; transport air to alveoli
Gas exchange	
Alveoli	Provide area for exchange of oxygen and carbon dioxide

FIGURE 8-1 The Human Respiratory System (a) This drawing shows the air-conducting portion and the gas-exchange portion of the human respiratory system. The insert shows a higher magnification of the alveoli, where oxygen and carbon dioxide exchange occurs. (b) A scanning electron micrograph of the alveoli, showing the rich capillary network surrounding them. (© David M. Phillips/Science Source, Inc.)

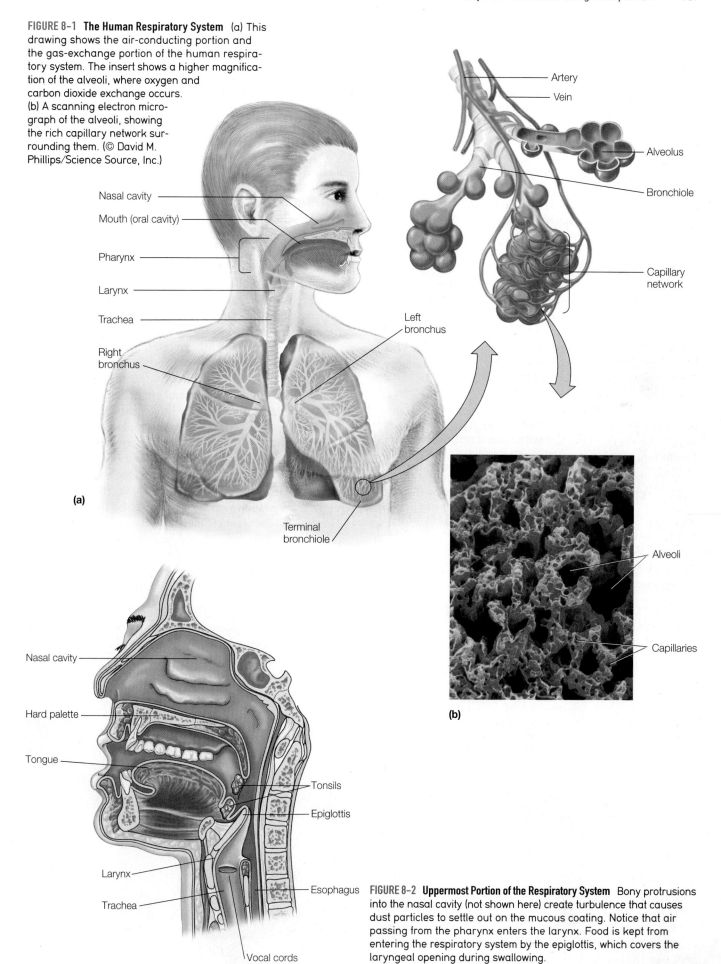

Artery

Vein

Alveolus

Bronchiole

Capillary network

Nasal cavity

Mouth (oral cavity)

Pharynx

Larynx

Trachea

Right bronchus

Left bronchus

(a)

Terminal bronchiole

Alveoli

Capillaries

(b)

Nasal cavity

Hard palette

Tongue

Tonsils

Epiglottis

Larynx

Esophagus

Trachea

Vocal cords

FIGURE 8-2 Uppermost Portion of the Respiratory System Bony protrusions into the nasal cavity (not shown here) create turbulence that causes dust particles to settle out on the mucous coating. Notice that air passing from the pharynx enters the larynx. Food is kept from entering the respiratory system by the epiglottis, which covers the laryngeal opening during swallowing.

The larynx also is involved in the production of sound. Sound is produced by the vocal cords. As shown in Figure 8-3c, the **vocal cords** consist of two folds of tissue. Inside each vocal cord is a band of elastic tissue that vibrates when we talk, sing, or hum. The sounds generated by the vocal cords are modified by changing the position of the tongue and the shape of the oral cavity.

The vocal cords vary in length and thickness from one person to the next. They also vary between men and women. In men, the vocal cords are longer and thicker than the vocal cords in women. This difference is due to the male sex hormone, testosterone, which is produced by the testes. Because their vocal cords are frequently thicker than those in women, men tend to have deeper voices than women.

The vocal cords are like the strings of a guitar or violin: They can be tightened or loosened, producing sounds of different pitch. The tighter the string on a guitar, the higher the note. In humans, muscles in the larynx that attach to the vocal cords make this adjustment possible. Relaxing the muscles lowers the tension on the cords, dropping the tone. Tightening the vocal cords has the opposite effect. This is what allows a singer to blast out notes of different pitch, even a male to sing the high-pitched notes of falsetto.

As just noted, the larynx opens into the **trachea** (TRAY-kee-ah) or windpipe below. You can feel the trachea beneath your Adam's apple, the protrusion of the laryngeal cartilage on your neck.

Although food is prevented from entering the larynx by the elevation of the larynx during swallowing and the epiglottis (ep-eh-GLOT-tis), food does occasionally enter the larynx. This unfortunate event leads to violent coughing, a reflex that helps eject the food from the trachea. If the food cannot be dislodged by coughing, steps must be taken to remove it—and fast—or the person will suffocate. Health Note 8-1 explains what to do when a person chokes.

The trachea, the next part of the air-conducting portion of the respiratory system, is a short, wide duct. Starting in the neck below the larynx, it enters the thoracic cavity, where it divides into two large branches, the right and left **bronchi** (BRON-kee). The bronchi (singular, bronchus) enter the lungs alongside the arteries and veins. Inside the lungs, the bronchi branch extensively, forming progressively smaller tubes that carry air to the alveoli.

The trachea and bronchi are reinforced by hyaline cartilage in the walls of these structures. This cartilage reinforcement prevents the structures from collapsing during breathing, thus ensuring a steady flow of air in and out of the lungs.

The smallest bronchi in the lungs branch to form **bronchioles** (BRON-kee-ols), small ducts that lead to the alveoli. Like the arterioles of the circulatory system, the walls of the bronchioles consist largely of smooth muscle. This permits them to open and close, and thus provides a means of controlling air flow in the lungs. During exercise or times of stress, the bronchioles open to increase the flow of air into the lungs. This homeostatic mechanism helps meet the body's need for oxygen during exercise, in much the same way that the arterioles of capillary beds dilate to let more blood into body tissues in times of need.

> **KEY CONCEPTS**
>
> The respiratory system consists of an air-conducting portion that delivers air to the lungs and a gas-exchange portion that allows oxygen to be transferred from inhaled air to the blood and carbon dioxide in the blood to be released into air within the alveoli, helping the body rid itself of a toxic waste byproduct of cellular respiration.

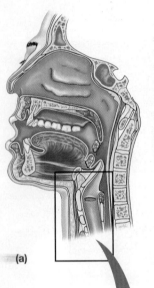

FIGURE 8–3 **Vocal Cords** (a) Uppermost portion of respiratory system showing location of the vocal cords. (b) Longitudinal section of the larynx showing the location of the vocal cords. Note the presence of the false vocal cord, so named because it does not function in phonation. (c) View into the larynx of a patient showing the true vocal cords from above. (Courtesy of James P. Thomas, M.D. [voicedoctor.net].)

(a)

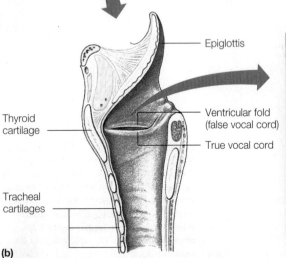

Epiglottis

Thyroid cartilage

Ventricular fold (false vocal cord)

True vocal cord

Tracheal cartilages

(b)

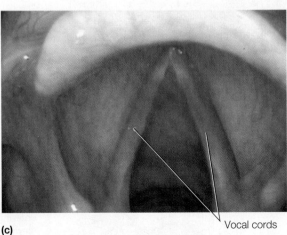

(c)

Vocal cords

Filtering Incoming Air

The respiratory system is in direct contact with the external environment and is therefore quite vulnerable to infectious organisms and air pollutants in the air we breathe. The respiratory systems of animals have evolved several protective mechanisms to enhance survival. In humans and other animals, for example, the air-conducting portion of the respiratory system filters airborne particles such as dust and pollen from the air we breathe.

Airborne particles or particulates come in many sizes. The large and medium-sized particles drop out of the air as they travel through the nasal passageways and the upper respiratory system. Smaller particles, known as fine particulates, however, are so tiny that they remain in the air breathed into our lungs. As a result, they can penetrate deeply into the lung. Some of these particles contain toxic metals such as mercury, which can cause lung cancer.

Particles that precipitate out of the inhaled air in the nose, trachea, and bronchi are trapped in a layer of mucus (MEW-kuss). **Mucus** is a thick, slimy secretion produced by the **mucous cells** in the epithelial lining of the nose, trachea, and bronchi (Figure 8-4).

The epithelium of the respiratory tract also contains numerous ciliated cells. **Cilia**, you may recall, are small organelles on the cells lining the respiratory tract and parts of the reproductive tract of women. The cilia in the respiratory system beat upward, propelling mucus containing bacteria and dust particles toward the mouth. Operating day and night, they sweep the mucus toward the oral cavity, where it can be swallowed or expectorated (spit out). This protects the respiratory tract and lungs from bacteria and potentially harmful particulates. It is another of the body's homeostatic mechanisms.

Like all homeostatic mechanisms, the respiratory mucous trap is not invincible. Bacteria and viruses do occasionally penetrate the lining, causing respiratory infections. Cigarette smoke and urban air pollution can make matters worse. That's because sulfur dioxide present in both tobacco smoke and urban air pollution can temporarily paralyze the cilia and, if in sufficient concentration, can even destroy them altogether. Sulfur dioxide gas in the smoke of a single cigarette, for instance, will paralyze the cilia for an hour or more, permitting bacteria and toxic particulates to be deposited on the lining of the respiratory tract and even enter the lungs. Ironically, the cilia of a smoker are put out of commission when they are needed the most!

Because smoking impairs a natural protective mechanism, it should come as no surprise that smokers suffer more frequent respiratory infections than nonsmokers. Interestingly, research also shows that alcohol paralyzes the respiratory system cilia, explaining why alcoholics are also prone to respiratory infections.

Health Tip 8-1

For healthy lungs, be sure to get plenty of vitamin D.

Why?

Studies show that lung function in adults is best in those with the highest vitamin D levels in their blood. Vitamin D is made naturally by the skin when exposed to sunlight. It is also found in certain foods or supplements.

A diet rich in vitamin D (1,000 IU per day) also seems to help protect us against several forms of cancer, including cancer of the colon, breast, and ovaries, and possibly the prostate gland in men. To obtain 1,000 IU of vitamin D daily, you need to eat vitamin D-rich foods such as enriched milk, eggs, salmon, pork, and ham. You may also need to take a vitamin D supplement. Supplements may not be necessary, however, if you eat a good diet rich in calcium and spend time outdoors every day, although that's not possible for many of us and not practical in the winter. The reason some people may not need to take vitamin D supplements, at least part of the year, is that exposure to ultraviolet radiation in sunlight stimulates vitamin D production by your skin. So, if you spend about 10 to 15 minutes a day in the sun with some of your skin exposed to the sun, your body should produce all the vitamin D you need. Individuals with darker skin tones, however, may need to spend two to three times longer to achieve the same goal. Remember, however, that sunscreen reduces UV exposure and, hence, decreases the synthesis of vitamin D. To tap into your body's natural ability to produce vitamin D on sunny days yet prevent sunburn, apply sunscreen after 15 minutes exposure to the sun.

KEY CONCEPTS

The air we breathe contains many tiny particles, including dust, particulate pollution, and bacteria, many of which are filtered out.

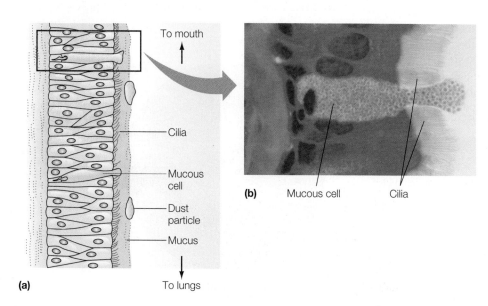

(a) To mouth / Cilia / Mucous cell / Dust particle / Mucus / To lungs

(b) Mucous cell / Cilia

FIGURE 8-4 **Mucous Trap** (a) Drawing of the lining of the trachea. Mucus produced by the mucous cells of the lining of much of the respiratory system traps bacteria, viruses, and other particulates in the air. The cilia transport the mucus toward the mouth. (b) Higher magnification of the lining showing a mucous cell and ciliated epithelial cells. (© David M. Phillips/Visuals Unlimited.)

health**note**

8-1   First Aid for Choking That May Save Someone's Life

What do you do if you encounter a person who is choking? If the person can cough, speak, or breathe, do not interfere. Encourage the individual to continue coughing. If he or she cannot cough, speak, or breathe, CALL OR HAVE SOMEONE CALL 911 OR YOUR LOCAL EMERGENCY NUMBER IMMEDIATELY.

If the person is conscious:
- Stand behind him and wrap your arms around his waist.
- Make a fist with one hand (thumb outside the fist), and place the fist just above the navel in the middle of the abdomen. Be sure that your fist is well below the lower tip of the sternum.
- Wrap the other hand around the fist and perform cycles of five quick inward and upward thrusts. After every five thrusts, check the person and repeat the cycles until the person begins to cough, speak, or breathe on his own.

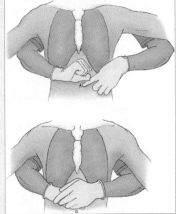

Adult Chest Thrust

If the person is unconscious:
- Carefully position her on her back.
- Open the airway by placing one hand on the forehead and one hand on the chin and gently tilt the head backward (head-tilt/chin-lift). Place your ear next to the person's mouth, listening for air exchange; feel for air exchange with your cheek; and look to see the chest rise and fall during air exchange.

Unconscious Victim Chest Thrust

- If you do not see, hear, or feel air exchange, pinch the person's nose closed and seal your mouth around the victim's mouth and give two slow breaths. The chest should rise and fall gently.
- If the first breath did not go in, reposition the victim's head and attempt another breath.
- If the breaths did not go in, straddle the victim's thighs and place the heel of one hand just above the navel and place the other hand on top, interlocking your fingers.
- Give five quick inward and upward thrusts.
- Return to the victim's head; perform a finger sweep of her mouth, searching for a dislodged object; perform a head-tilt/chin-lift; and give one slow breath.
- If the breath will not go in, give five more abdominal thrusts, followed by a finger sweep, and one breath.

Warming and Moistening Incoming Air

Beneath the epithelium of the respiratory tract is a rich network of capillaries. These capillaries release moisture that enters the incoming air. This prevents the lungs from drying out. Heat is also released by the capillaries, protecting the lungs from cold air. Except in extremely cold weather, by the time inhaled air reaches the lungs, it is nearly saturated with water and is warmed to body temperature.

On the way out of the respiratory system, much of the water that was added to the air condenses on the slightly cooler lining of the nasal cavity. This biological recycling mechanism conserves water and also accounts for the reason our noses tend to drip in cold weather.

> **KEY CONCEPTS**
>
> Incoming air is heated and humidified as it is inhaled to prevent damage to the lungs by the air-conducting portion of the respiratory system.

The Alveoli

The air we breathe consists principally of nitrogen and oxygen, with small amounts of carbon dioxide and other gases (Figure 8-5). Of these gases, oxygen is the most important to humans and other animals. Oxygen in the atmosphere is generated by the photosynthesis in plants and certain single-celled organisms, like algae. To survive, humans and other animals require a constant supply of oxygen. It is needed to sustain cellular energy production (remember that the complete breakdown of glucose in body cells requires oxygen).

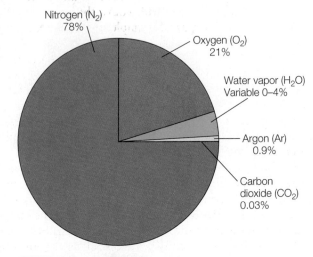

Nitrogen (N_2) 78%
Oxygen (O_2) 21%
Water vapor (H_2O) Variable 0–4%
Argon (Ar) 0.9%
Carbon dioxide (CO_2) 0.03%

FIGURE 8-5 Composition of Air

- Continue this cycle until the victim begins to cough, speak, or breathe on his or her own.
- Perform the same technique for children as you would for an adult. Children are considered those who are over 1 year of age and under the age of 8. The size of the child must also be a consideration.

For infants and very small children who cannot cough, speak, or breathe:
- Pick up the infant and place her face down on your forearm.
- Give the infant five back blows with the heel of your hand between the shoulder blades.

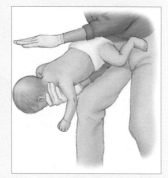

- Reposition the infant face up on the opposite forearm and give five chest thrusts using the pads of two or three fingers in the lower half of the sternum.
- Continue this cycle until the infant can breathe on her own or until the infant goes unconscious.

Infants and Small Children, Back Blows

If the infant is unconscious:
- Perform a head-tilt/chin-lift and look, listen, and feel for air exchange.

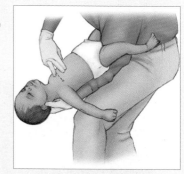

- If no exchange is present, seal your mouth around the mouth and nose of the infant and give two slow breaths. The chest should rise and fall gently.
- If the first breath did not go in, reposition the infant's head and attempt another breath.
- If the breaths did not go in, position the infant for five back blows, five chest thrusts; then, check her mouth for an object and give one breath.

Infants and Small Children, Chest Thrust

- Continue this cycle until the infant begins to breathe on her own.

In order to be skilled in these and other lifesaving skills, you should take a first aid and CPR class. Classes are offered in most communities through the National Safety Council. Call 1-800-621-7618 for information on classes in your area.

Source: Adapted by Jennifer Belcher, EMT-B from the guidelines of the American Red Cross, National Safety Council, and the American Heart Association.

 Visit Human Biology's Internet site for links to websites offering more information on this topic.

Oxygen is taken in by the respiratory system and delivered to the cells of our body by the circulatory system.

Oxygen is delivered to the alveoli of the lungs by the conducting portion of the respiratory system (Figure 8-6). The alveoli are the site of oxygen exchange with the bloodstream. In humans, each lung contains an estimated 150 million alveoli. Although they are extremely small, the surface area they create is about the size of a tennis court, that is, about 60 to 80 square meters. This provides ample area for the exchange of oxygen.

As shown in Figure 8-7, the alveoli are lined by a single layer of flattened cells (**Type I alveolar cells**). Aiding in the exchange of oxygen is an extensive network of capillaries, shown in Figure 8-1b. Each capillary is made of an equally thin layer of flattened cells. Together, the cells of the alveolar wall and the capillary present a fairly easy route for the passage of gases into and out of the alveoli.

The large surface area of the lungs created by the alveoli and the relatively thin barrier between the blood and the alveolar wall result in a rather rapid diffusion of gases across the alveolar wall. The alveoli therefore provide another example of the marriage of form and function produced by evolution.

Another important cell in the alveoli is the **alveolar macrophage**, sometimes known as the **dust cell**. Alveolar macrophages are the cellular equivalent of a janitorial staff. They wander freely through the alveoli, phagocytizing dust, bacteria, viruses, and other particulates that manage to escape filtration by the upper portions of the respiratory system (Figure 8-7). Once they are filled with particulates, they take up residence in the connective tissue surrounding the alveoli. Over time, numerous spent cells come to reside in this region. Because there are so many particulates in tobacco smoke, a smoker's lungs are blackened by dust cells packed with particulates. The lungs of urban residents and marijuana users also tend to be blackened by the accumulation of smoke and dust particles.

Health Tip 8-2

If you smoke, don't just cut back, quit.

Why?

Smoking leads to over 1,000 deaths from lung cancer, stroke, and heart attack *every day* in the United States. While reducing daily intake of tobacco smoke lessens one's chances of disease and death, several studies show that there is no safe level of tobacco smoke. Smoking even one or two cigarettes a day significantly increases one's risk of developing lung cancer.

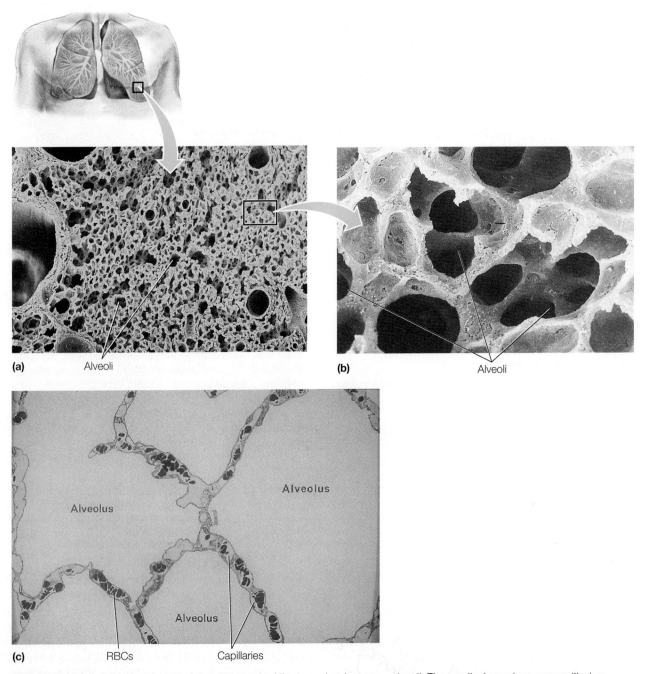

FIGURE 8–6   Alveoli   (a) A scanning electron micrograph of the lung showing many alveoli. The smallest openings are capillaries surrounding the alveoli. (© David M. Phillips/Visuals Unlimited.) (b) A higher magnification scanning electron micrograph of lung tissue showing alveoli. (© David M. Phillips/Visuals Unlimited.) (c) A transmission electron micrograph showing several alveoli and the close relationship of the capillaries containing RBCs (dark structures inside the capillaries). This close relationship between capillaries and alveoli ensures the rapid transport of oxygen and carbon dioxide. (© Don Fawcett/Visuals Unlimited.)

Yet another important cell in the alveoli is the **Type II alveolar cell** (Figure 8-7). Type II alveolar cells are large, round cells located in the alveolar epithelium. These prominent cells produce a vital chemical substance known as surfactant (sir-FACK-tant). **Surfactant** is a general term that refers to any of dozens of detergent-like substances. In human lung it contains a mixture of proteins and lipids. The lung's surfactant dissolves in the thin layer of water lining the alveoli, where it plays an important role in keeping the tiny alveoli open. How?

The water covering the inside surface of alveolar epithelium produces surface tension. Surface tension results from hydrogen bonds between water molecules. Hydrogen bonds are intermolecular bonds. They form between the slightly negatively charged oxygen atoms of water molecules and the slightly positively charged hydrogen atoms of other molecules. These hydrogen bonds draw water molecules together. At the surface of a watery fluid, the hydrogen bonds pull water molecules together more tightly than elsewhere, creating a slightly denser region referred to as *surface tension*. Surface tension on a pond permits some insects such as water striders to walk on water and is the reason a drop of water beads up on a car's windshield (Figure 8-8).

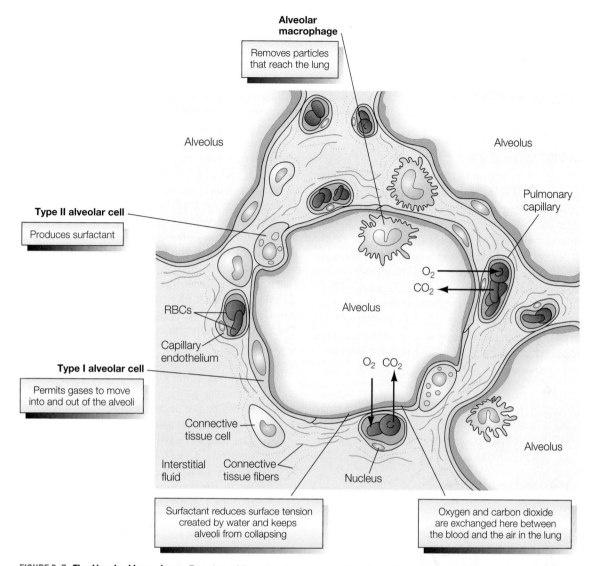

Alveolar macrophage
Removes particles that reach the lung

Alveolus

Alveolus

Pulmonary capillary

O₂
CO₂

Type II alveolar cell
Produces surfactant

Alveolus

RBCs

Capillary endothelium

Type I alveolar cell
Permits gases to move into and out of the alveoli

O₂ CO₂

Connective tissue cell

Interstitial fluid

Connective tissue fibers

Nucleus

Alveolus

Surfactant reduces surface tension created by water and keeps alveoli from collapsing

Oxygen and carbon dioxide are exchanged here between the blood and the air in the lung

FIGURE 8-7 The Alveolar Macrophage Drawing of the alveolus showing Type I and Type II alveolar cells and macrophages or dust cells.

FIGURE 8-8 Surface Tension Some insects can walk on water because of surface tension, the tight packing of water molecules along the surface of a pond. (© Vasiliy Koval/ShutterStock, Inc.)

In the alveoli, surface tension tends to draw the walls of the alveoli inward. In other words, it tends to make them collapse. Surfactant, however, reduces surface tension in the alveoli, decreasing forces that might otherwise cause the alveoli to collapse.

Some premature babies lack sufficient surfactant. High surface tension in the alveoli causes the larger alveoli to collapse. This, in turn, results in a dramatic, life-threatening condition known as **infant respiratory distress syndrome** or **hyaline membrane disease**. This occurs primarily in children born prematurely or children whose mothers are diabetic. Children born with this problem are admitted to the intensive care unit where they receive supplemental oxygen. In severe cases, they are put on ventilators. Within five or six days, most begin producing their own surfactant. In some cases, infants are administered an artificial surfactant that helps keeps the alveoli open until their lungs begin to produce enough of their own surfactant.

KEY CONCEPTS

The alveoli are tiny, thin-walled air sacs that are the site of gas exchange.

8-2   Breathing and the Control of Respiration

Air moves in and out of the lungs in much the same way that it moves in and out of the bellows that blacksmiths use to fan their fires. Breathing, however, is largely an involuntary action, controlled by the nervous system.

Inspiration and Expiration

During breathing, air must first be drawn into the lungs. This process is known as **inspiration**, or **inhalation**. Following inspiration, air must be expelled. This is known as **expiration**, or **exhalation**.

Inhalation is an active process controlled by the brain. Here's how it occurs. Nerve impulses traveling from the brain stimulate the **diaphragm**, a dome-shaped muscle that separates the abdominal and thoracic cavities (Figure 8-9a). These impulses cause the diaphragm to contract. When it contracts, the diaphragm flattens and lowers. The contraction of the diaphragm draws air into the lungs much the same way that pulling the plunger of a syringe draws air into the device.

Inhalation also involves the intercostal muscles, the short muscles that lie between the ribs (they are the meat on barbecued ribs). Nerve impulses traveling to these muscles cause them to contract as the diaphragm is lowered. When the intercostal muscles contract, the rib cage lifts up and out. Go ahead and try it right now. Place your hands on your chest, take a deep breath in, and feel the chest wall expand and rise. Together, the contractions of the intercostal muscles and the diaphragm increase the volume of the thoracic cavity (Figure 8-9b). This, in turn, decreases the pressure inside the lungs, known as **intrapulmonary pressure**. As a result, air naturally flows in through the mouth and nose.

In contrast to inhalation, exhalation is a passive process—that is, it does not require muscle contraction in a person at rest. Exhalation begins after the lungs have filled with air. At this point in the cycle, the diaphragm and intercostal muscles relax. The relaxed diaphragm rises and resumes its domed shape. The chest wall falls slightly inward. These changes reduce the volume of the chest cavity. This, in turn, raises the intrapulmonary pressure, forcing air out in much the same way that squeezing an inflated beach ball forces air out of an opening.

The lungs also contribute to passive exhalation. The lungs contain numerous elastic fibers. During inspiration, the lungs fill but when inhalation ceases, the elastic fibers inside the lungs cause the lungs to recoil, forcing air out. The lungs act a lot like balloons that are inflated, and then allowed to deflate.

Although exhalation is a passive process in an individual at rest, it can be made active by contracting the muscles of the wall of the chest and abdomen. The forceful expulsion of air is called *forced exhalation*.

Inhalation can also be consciously supplemented by a forceful contraction of the muscles of inspiration. (You can test this by taking a deep breath.) *Forced inhalation* increases the amount of air entering one's lungs. Athletes often actively inhale and exhale just before an event to increase oxygen levels in their blood. A competitive swimmer, for example, may take several deep breaths before diving into the pool for a race. Deep breathing, while effective, can be dangerous as it can lead to blackouts.

> **KEY CONCEPTS**
> Air moves in and out of the lungs as a result of changes in the pressure inside the chest cavity

Measuring Air Flow in the Lungs

Several measurements of lung capacity are routinely used to determine the health of a person's lungs. These measurements are taken under controlled conditions. As shown in Figure 8-10a, patients breathe into a machine that records the amount of air moving in and out of the lung at various times. Figure 8-10b shows a graph of some of the common measurements. The first is the **tidal volume**, the amount of air that moves in and

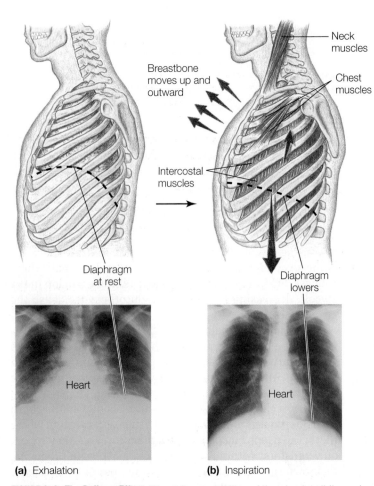

(a) Exhalation   **(b)** Inspiration

FIGURE 8-9 **The Bellows Effect** The rising and falling of the chest wall through the contraction of the intercostal muscles (muscles between the ribs) is shown in the diagram, illustrating the bellows effect. Inspiration is assisted by the diaphragm, which lowers. Like pulling a plunger out on a syringe, the rising of the chest wall and the lowering of the diaphragm draw air into the lungs. Illustrations and x-rays showing the size of the lungs in full exhalation (a) and full inspiration (b). (Photos a and b © SIU/Visuals Unlimited.)

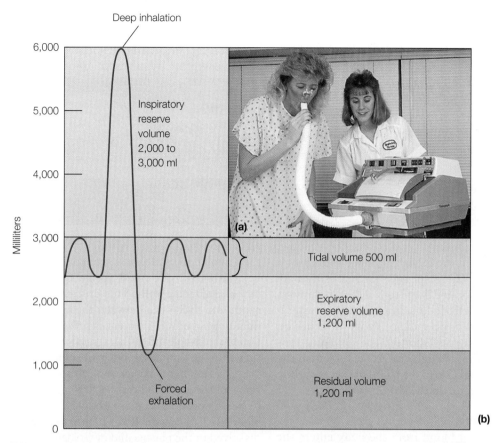

FIGURE 8-10 **Measuring Lung Capacity** (a) This machine allows healthcare workers to determine tidal volume, inspiratory reserve volume, and other lung-capacity measurements to determine the health of an individual's lung. (© SIU/Visuals Unlimited.) (b) This graph shows several common measurements.

out during normal (passive) breathing. At rest, each breath delivers about 500 milliliters of air to the lung.

After passive exhalation, however, the lungs still contain a considerable amount of air—about 2,400 milliliters. Forced exhalation will expel about half of that air. This is known as the **expiratory reserve volume**. The remaining 1,200 milliliters is known as the **residual volume**.

Another important measurement is the **inspiratory reserve volume (IRV)**. This is the amount of air that can be drawn into the lungs during a deep inhalation. Deep inhalation draws in four to six times more air than the tidal volume, or 2,000 to 3,000 milliliters, depending on the size of the individual.

Lung diseases often result in changes in these measurements. Asthma, for example, reduces the IRV by constricting the bronchioles and allowing mucus to build up, thus blocking air flowing into the lung. Children who are exposed to

tobacco smoke at home also experience a decrease in lung capacity—that is, a decrease in their ability to move air in and out of their lungs.

> **KEY CONCEPTS**
>
> Measurements of air flow into and out of the lungs can be used by physicians to assess the health of the lungs.

Control of Breathing

Breathing is controlled by a region of the brain known as the **breathing center**. It is located in the area called the **brain stem** (or medulla, [meh-DEW-lah]).

The breathing center contains nerve cells that generate periodic nerve impulses. These impulses stimulate inhalation by causing the periodic contraction of the intercostal muscles and the diaphragm. When the lungs fill, however, the nerve impulses cease and the muscles relax. Air is forced out of the lungs.

While the breathing center acts like a pacemaker, bringing about rhythmic contraction and relaxation of the muscles of inspiration, breathing can be increased or decreased by other factors. Chemical receptors inside the brain and arteries, for example, detect levels of carbon dioxide, and send nerve impulses to the breathing center to increase or decrease breathing as needed. When levels of carbon dioxide are high, for instance, during vigorous exercise, chemoreceptors in the aorta and carotid arteries (which deliver blood to the brain) are activated. They send impulses to the breathing center that

Health Tip 8-3

A diet rich in vegetables can help improve lung function as you age. *Why?*

Although you are probably not worrying about your declining years, it's a great idea to start developing good eating habits now. Studies show that nutrients from vegetables like beta-carotene from carrots, spinach, sweet potatoes, and winter squash help slow down the natural aging process in the lungs, which means better lung function as you get older.

cause it to increase the depth and rate of breathing to rid the body of excess carbon dioxide and replenish oxygen levels. Next time you run up a flight of stairs and start breathing hard, you'll know why.

Stretch receptors in the lung also help to control breathing. They detect when the lungs are full and send impulses back to the breathing center, causing it to cease firing. Stretch receptors probably function only during exercise, when large volumes of air are moved in and out of the lungs.

The body also contains oxygen receptors. These receptors are not as sensitive as the other receptors, however. In fact, oxygen levels must fall considerably before these receptors begin generating impulses that increase breathing.

> **KEY CONCEPTS**
>
> Breathing is controlled by the breathing center in the brain; it contains nerve cells that periodically fire, sending impulses to the muscles that control breathing.

8-3   Gas Exchange in the Lungs

The respiratory system serves many functions. For example, the vocal cords, located in the larynx, produce sounds that allow us to communicate. The respiratory system also houses the olfactory membrane located in the roof of the nasal cavity. It contains sensors that detect odors. In addition, the respiratory system helps maintain pH balance by its influence on carbon dioxide levels (discussed shortly). The main functions of the respiratory system, though, are to (1) replenish the blood's oxygen supply and (2) rid the blood of excess carbon dioxide. Much of this occurs in the alveoli of the lungs.

Function of the Alveoli

Recall that deoxygenated blood from the body enters the lungs via the pulmonary arteries. This blood is laden with carbon dioxide picked up as it circulates through body tissues. In the capillary beds of the lungs, oxygen in the inhaled air is added to the blood while carbon dioxide is released into the alveolar air, and then expelled during exhalation.

Carbon dioxide and oxygen diffuse across the thin capillary and alveolar walls. This process is "driven" by the concentration difference between the alveoli and capillaries. As illustrated in Figure 8-11, oxygen is present in a higher concentration in the alveoli and diffuses through the alveolar epithelium, then into the capillaries, where it enters the blood plasma (Figure 8-12). From here, oxygen molecules cross the plasma membrane of the red blood cells (RBCs). Inside the RBCs, it binds to hemoglobin molecules. About 98% of the oxygen in the blood is carried in the RBCs bound to hemoglobin; the rest is dissolved in the plasma and cytoplasm of the RBCs.

In order to understand the details of carbon dioxide diffusion in the lung, we must go back to the body cells, where CO_2 is formed. As is shown in Figure 8-13, cells produce CO_2 during

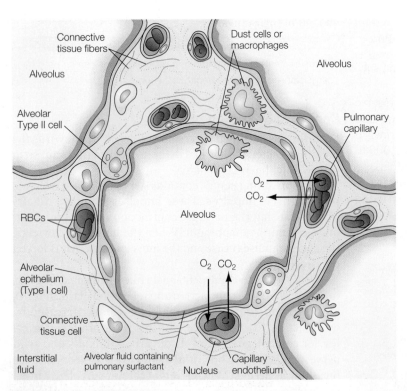

FIGURE 8-11 **Close-Up of the Alveolus** Oxygen diffuses out of the alveolus into the capillary. Carbon dioxide diffuses in the opposite direction, entering the alveolar air that is expelled during exhalation.

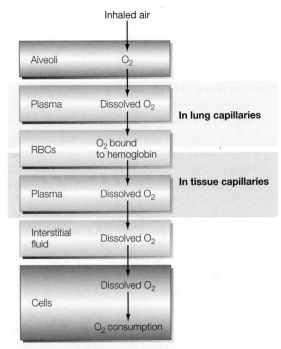

FIGURE 8-12 **Oxygen Diffusion** Oxygen travels from the alveoli into the blood plasma, then into the RBCs, where much of it binds to hemoglobin. When the oxygenated blood reaches the tissues, oxygen is released from the RBCs and diffuses into the plasma, then into the interstitial fluid and body cells.

cellular respiration, the chemical breakdown of glucose. In body tissues, carbon dioxide diffuses out of the cells and enters the blood plasma. Some carbon dioxide (7%) is dissolved in the plasma. The rest diffuses into the RBCs. Here, some of it (15% to 25%) binds to hemoglobin. The rest reacts with water to form carbonic acid, H_2CO_3 (Figure 8-13). This reaction is catalyzed by the enzyme *carbonic anhydrase* found inside RBCs.

As shown at the top of Figure 8-13, carbonic acid molecules readily dissociate to form bicarbonate ions and hydrogen ions. Many of the bicarbonate ions then diffuse out of the RBCs into the plasma, where they are carried with the blood. Hydrogen ions stay behind.

When blood rich in carbon dioxide reaches the lungs, this process reverses. First, bicarbonate ions in the blood plasma diffuse into the RBCs, as shown in Figure 8-14. They then combine with hydrogen ions to form carbonic acid. Carbonic acid, in turn, reforms carbon dioxide. The CO_2 diffuses out of the RBCs into the blood plasma and then diffuses into the alveoli. Carbon dioxide is then expelled from the lungs during exhalation.

The uptake of oxygen and the discharge of carbon dioxide in the lungs "replenish" the blood in the alveolar capillaries. The oxygenated blood then flows back to the left atrium of the heart via the pulmonary veins. From here it empties into the left ventricle and is pumped to the body tissues via the aorta and its many branches where it releases its oxygen and picks up more carbon dioxide.

KEY CONCEPTS

Oxygen flows into the capillaries surrounding the alveoli and carbon dioxide flows out.

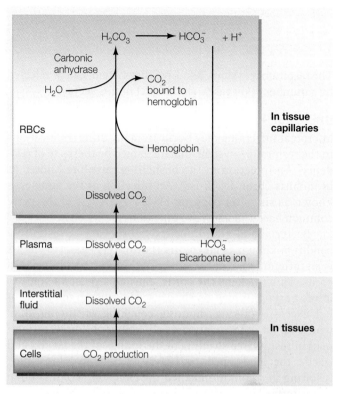

FIGURE 8–13 Bicarbonate Ion Production Carbon dioxide (CO_2) diffuses out of body cells where it is produced and into the tissue fluid, then into the plasma. Although some carbon dioxide binds to hemoglobin and some is dissolved in the plasma, most is converted to carbonic acid (H_2CO_3) in the RBCs. Carbonic acid dissociates and forms hydrogen ions and bicarbonate ions. Hydrogen ions remain inside the RBCs, but most bicarbonate diffuses into the plasma where it is transported.

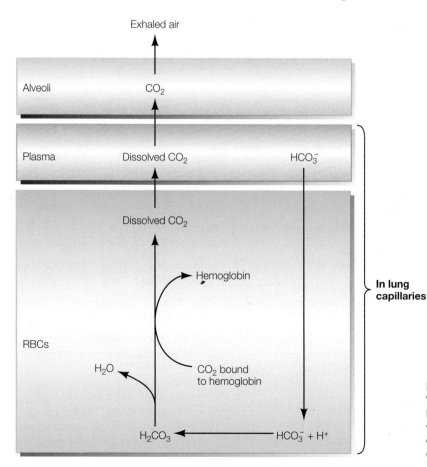

FIGURE 8–14 Carbon Dioxide Production from Bicarbonate When the carbon dioxide-laden blood reaches the lungs, bicarbonate ions diffuse back into the RBCs and combine with hydrogen ions in RBCs forming carbonic acid, which dissociates, forming carbon dioxide gas. CO_2 diffuses out of the RBCs into the plasma, then into the alveoli.

8-4   Diseases of the Respiratory System

The respiratory system, like any other system, can be affected by a number of diseases, the subject of this section.

Bacterial and Viral Infections

Microbial infections may occur in many different locations in the respiratory system and are named by their site of residence. An infection in the bronchi is therefore known as **bronchitis** (bron-KITE-iss). An infection of the sinuses is known as **sinusitis** (sigh-nu-SITE-iss). (A few of the more common bacterial and viral infections are listed in Table 8-2.)

Bacterial and viral infections of the larynx result in a condition known as **laryngitis** (lair-in-JITE-iss). Laryngitis is an inflammation of the lining of the larynx and the vocal cords. This thickens the cords, causing a person's voice to lower. Laryngitis may also be caused by tobacco smoke, alcohol, excessive talking, shouting, coughing, or singing, all of which irritate the vocal cords. In young children, inflammation results in a swelling of the linings of the larynx and trachea that may impede the flow of air and impair breathing, resulting in a condition called the *croup* (crewp).It is often characterized by a barking cough that sounds a bit like a seal bark or hoarseness, especially when a child cries. Most cases of croup are caused by infections from various viruses, usually the parainfluenza virus and sometimes others.

Once inside the respiratory tract, bacteria, viruses, and other microorganisms can spread to other organ systems. For example, **meningitis** (MEN-in-JITE-iss) is a bacterial or viral infection of the meninges (meh-NIN-jees), the fibrous layers surrounding the brain and spinal cord. This potentially fatal disease usually starts out as an infection of the sinuses or

> ### Health Tip 8-4
>
> Worried about catching colds or the flu? Don't count on antibacterial soaps to help you.
>
> *Why?*
>
> A year-long study of 224 households showed that residents in homes using antibacterial soaps were no healthier than those who used ordinary soap. Bacteria are all around us and, more important, very few of them cause disease. Washing with ordinary soap is sufficient to remove those that do.

the lungs. That's why you're advised to see a doctor if these infections last more than a few weeks.

The most common causes of meningitis are viral infections. Patients typically recover without treatment. Bacterial meningitis, on the other hand, is extremely serious, and can result in brain damage or even death, even if treated.

The lungs are also susceptible to airborne materials, among them asbestos fibers. Asbestos is a naturally occurring fiber that has been used in thousands of products from pipe and sound insulation to car brake pads. Helpful as it is, inhaled asbestos can be extremely dangerous. Workers exposed to asbestos, for example, develop two types of lung cancer, one of which (mesothelioma) is highly lethal. The death rate in asbestos insulation workers is four times greater than expected. Interestingly, though, those who were exposed to asbestos and smoked were 92 times more likely to develop lung cancer.

Other individuals exposed to asbestos develop a debilitating disease known as asbestosis (ass-bes-TOE-sis). **Asbestosis**

TABLE 8-2	Common Respiratory Diseases		
Disease	**Symptoms**	**Cause**	**Treatment**
Emphysema	Breakdown of alveoli; shortness of breath	Smoking and air pollution	Administer oxygen to relieve symptoms; quit smoking; avoid polluted air. No known cure.
Chronic bronchitis	Coughing, shortness of breath	Smoking and air pollution	Quit smoking; move out of polluted area; if possible, move to warmer, drier climate.
Acute bronchitis	Inflammation of the bronchi; yellowy mucus coughed up; shortness of breath	Many viruses and bacteria	If bacterial, take antibiotics and decongestant tablets; use vaporizer.
Sinusitis	Inflammation of the sinuses; mucus discharge; blockage of nasal passageways; headache	Many viruses and bacteria	If bacterial, take antibiotics, cough medicine; use vaporizer.
Laryngitis	Inflammation of larynx and vocal cords; sore throat; hoarseness; mucus buildup and cough	Many viruses and bacteria	If bacterial, take antibiotics, cough medicine; avoid irritants like smoke; avoid talking.
Pneumonia	Inflammation of the lungs ranging from mild to severe; cough and fever; shortness of breath at rest; chills; sweating; chest pains; blood in mucus	Bacteria, viruses, or inhalation of irritating gases	Consult physician immediately; go to bed; take antibiotics, cough medicine; stay warm.
Asthma	Constriction of bronchioles; mucus buildup in bronchioles; periodic wheezing; difficulty breathing	Allergy to pollen, some foods, food additives; dandruff from dogs and cats; exercise	Use inhalants to open passageways; avoid irritants.

is a buildup of scar tissue. Scarred lung tissue does not expand and contract normally, so lung capacity is reduced. The more asbestos one was exposed to the greater the reduction in lung capacity. Symptoms do not usually appear until 20 years after exposure. Patients suffer from shortness of breath, chest pain, and cough. Unfortunately, there is no cure.

Because asbestos is so dangerous, virtually all of its uses have been banned in the United States, and asbestos used for insulation of pipes, ducts, and ceilings is either being removed by specialists from buildings or stabilized so it won't flake off.

KEY CONCEPTS

Bacteria and viruses can affect many parts of the respiratory system.

Asthma

Another common disease of the respiratory system is **asthma**. Characterized by periodic episodes of wheezing and difficult breathing, asthma is a chronic disorder—that is, it's a disease that persists for many years.

Most cases of asthma are caused by allergic reactions (abnormal immune reactions) to common stimulants such as dust, pollen, and skin cells (dander) from pets. In some individuals, it is brought on by certain foods, such as eggs, milk, chocolate, and food preservatives. Still other cases are triggered by drugs, such as antibiotics. Vigorous exercise and physiological stress are known to trigger asthma attacks, too.

In asthmatics, irritants such as pollen and dander cause a rapid increase in the production of mucus by the bronchi and bronchioles. Irritants also stimulate the constriction of the muscles in the walls of the bronchioles. Mucous production and constriction of the bronchioles make breathing difficult. Asthmatics also suffer a chronic inflammation of the lining of the respiratory tract.

Although asthma is fairly common in school children, it often disappears as they grow older. But that may be changing. Studies show that the incidence of asthma in adults is on the rise. Many more children suffer from the disease, too. In fact, the number of Americans now suffering from asthma is estimated to be about 25 million in 2009, the latest year for which data are available. This is up from 20 million in 2001. Public health officials are concerned about the growing epidemic of asthma because periodic attacks can be quite disabling. Some even lead to death. According to the Centers for Disease Control and Prevention (CDC), nearly 3,400 Americans die each year from severe asthma attacks. Victims are generally elderly individuals who suffer from other diseases as well.

As in many diseases, prevention is always the best strategy. A person with asthma, for example, should learn to recognize conditions that result in an attack, including exposure to allergens, respiratory infections, and cold weather. Screening tests can help a patient find out what substances trigger an asthmatic attack so they can be avoided. They then must learn to avoid their triggers if at all possible.

If asthma attacks are severe, unpredictable, and frequent—occurring more than twice a week—patients generally receive preventive medications. Taken daily, these meds can control asthma symptoms.

Among the most effective preventive treatments are corticosteroids, anti-inflammatory drugs that reduce inflammation occurring in the linings of the respiratory tract. When treatment begins early in life, corticosteroids can "normalize" lung function and can even prevent irreversible damage to one's airways. A single dose taken every day is often sufficient to control asthma.

In conjunction with anti-inflammatories, many patients take bronchodilators, which are drugs that cause the bronchioles to expand, allowing air to flow more easily into the lungs. One of the most common is an oral spray (inhalant) containing the drug albuterol. This drug relaxes the muscles in the bronchioles.

Vaccines can also be administered to patients with allergic asthma. These are especially helpful for asthma triggers that cannot be avoided like pollen and animal dander. This immunological treatment increases one's tolerance to substances that trigger asthma.

A more recent treatment is anti-IgE (immunoglobulin E). This antibody stops the allergic reaction before it can begin. This treatment is currently approved in the United States for patients age 12 and older who suffer from moderate-to-severe allergic asthma. IgE and other pharmaceutical treatments also are discussed elsewhere in the text.

KEY CONCEPTS

Asthma is an abnormal allergic reaction to food and substances in the air we breathe such as skin cells, dust, and pollen.

Chronic Bronchitis and Emphysema

Another common lung disease is **chronic bronchitis**, a persistent irritation of the bronchi. Characterized by excess mucous production, coughing, and difficulty in breathing, chronic bronchitis afflicts over 15 million American men and women. The disease is most prevalent, however, in men, affecting one out of every five American men between the ages of 40 and 60. Cigarette smoking is the single most important cause of chronic bronchitis. Workers exposed to dusts and fumes in the workplace—such as coal miners, grain handlers, and metal workers—are also at high risk of developing chronic bronchitis. Certain pollutants in urban air pollution irritate the lung and bronchial passages. Air pollution therefore can aggravate the disease, worsening symptoms.

Unfortunately, people who suffer from chronic bronchitis often ignore the symptoms of the disease until it is too late, believing that the disease is merely a nuisance and not life-threatening. Unfortunately, waiting too long may result in serious damage to one's lungs. Patients who wait too long can develop respiratory problems and heart failure.

Fortunately, if diagnosed early on chronic bronchitis can be treated and managed. Quitting smoking, bronchodilators and anti-inflammatory drugs are common.

Another lung disease of great significance is emphysema. **Emphysema**, mentioned in the introduction, is the progressive deterioration of the alveoli caused by a breakdown of their walls. Emphysema is one of the fastest growing causes of death in the United States. Resulting principally from smoking and urban air pollution, emphysema afflicts over 4.3 million Americans, mostly men. Not surprisingly, smokers who live in polluted urban settings have the highest incidence of the disease. This condition is a progressive and incurable disease.

health**note**

8-2   Smoking and Health: The Deadly Connection

Urban air pollution worries many Americans, and with good reason. However, city air that many of us breathe is benign compared with the "air" that more than 45 million Americans over the age of 18 voluntarily inhale from cigarettes. Loaded with dangerous pollutants in concentrations far greater than those in the air of our cities, cigarette smoke takes a huge toll on citizens of the world. In the United States, for example, an estimated 443,000 people (2011 CDC data)—over 1,200 every day—die from the many adverse health effects of tobacco smoke, including heart attacks, lung cancer, and emphysema. Making matters worse, smoking is addictive. Nicotine, one of the many components of tobacco smoke, hooks many people on a dangerous activity.

Smoking costs society a great deal in medical bills and lost productivity. According to the CDC, smokers cost the country $96 billion a year in direct healthcare costs and an additional $97 billion a year in lost productivity.

Smoking is a principal cause of lung cancer, claiming the lives of nearly 159,000 men and women in the United States each year. Smokers are 11 to 25 times more likely to develop lung cancer than nonsmokers. The more one smokes, the more risk one suffers.

Unfortunately, nonsmokers are also affected by the smoke of others. Non-smokers inhale tobacco smoke in meetings, in restaurants, at work, at home or while traveling in vehicles with smokers. Research has shown that non-smokers—sometimes referred to as "passive smokers" because they "smoke" involuntarily—are more likely to develop lung cancer than those nonsmokers who manage to steer clear of smokers. In a study of Japanese women married to men who smoked, researchers found that the wives were as likely to develop lung cancer as people who smoked half a pack of cigarettes a day!

New, more carefully designed and extensive studies suggest that while second-hand smoke does cause cancer in nonsmokers, the risk is much lower than the Japanese study suggested. As a case in point, an extensive ten-year study in Europe showed that the risk was modestly elevated among nonsmokers exposed as adults at home or in the workplace. The average increase among passive smokers is around 20%. Risk did increase with the amount of exposure (although not consistently with all measures used). The study noted that there was no increase in risk associated with exposures to environmental tobacco smoke (ETS) in childhood.

According to the National Cancer Institute, at least 69 chemicals in second-hand smoke can cause cancer. A report by the U.S. Environmental Protection Agency estimates that passive smoking causes 500 to 5,000 cases of lung cancer a year in the United States. Passive smokers who are exposed to tobacco smoke for long periods also suffer from impaired lung function equal to that seen in light smokers (people who smoke less than a pack a day).

Cigarette smoke in closed quarters can cause angina (chest pains) in smokers and nonsmokers afflicted by atherosclerosis of the coronary arteries. Carbon monoxide in cigarette smoke is responsible for the attacks.

Smokers are also more susceptible to colds and other respiratory infections. And smoking affects children. One study showed that children from families in which both parents smoked suffered twice as many upper respiratory infections as children from nonsmoking families. They also suffer from more lower respiratory infections such as bronchitis and pneumonia. Recently, researchers from the Harvard Medical School reported finding a 7% decrease in lung capacity in children raised by mothers who smoked. The researchers believe that this change may lead to other pulmonary problems later in life. According to the Environmental Protection Agency (EPA), exposure to secondhand smoke can cause asthma in children who have not previously exhibited symptoms.

The effects of smoking extend far beyond respiratory disease. Studies, for instance, show that smoking even decreases fertility in women. For example, women who smoke more than a pack of cigarettes a day are half as fertile as nonsmokers. Smoking may also affect the outcome of pregnancy. Studies show that women who smoke several packs a day during pregnancy are much more likely to miscarry and are also more likely to give birth to smaller children. On average, the children of these women are 200 grams (nearly 0.5 pound) lighter than children born to nonsmoking mothers. Second-hand smoke also increases the risk of sudden infant death syndrome. Finally, children of women who smoke heavily

As it worsens, lung function deteriorates, and victims eventually require supplemental oxygen to perform even routine functions, such as walking or speaking.

Chronic bronchitis is often referred to as **chronic obstructive pulmonary disease (COPD)** This term also applies to people who have emphysema or both bronchitis and emphysema. Most patients have symptoms of both.

One of the most prevalent respiratory diseases is lung cancer caused by smoking, a topic discussed in Health Note 8-2.

KEY CONCEPTS

Chronic bronchitis and emphysema are known as chronic obstructive pulmonary disease and result from exposure to chemicals present in tobacco smoke and, to a lesser extent, air pollution.

8-5   Health and Homeostasis

The respiratory system plays a vital role in homeostasis. It helps regulate oxygen and carbon dioxide levels in the blood and in body tissues. It helps maintain the pH (acidity) of extracellular fluid by controlling the rate at which acid-forming carbon dioxide is removed from the blood. Furthermore, the respiratory system helps to protect us from

during pregnancy generally score lower on mental aptitude tests during early childhood than children whose mothers do not smoke.

Numerous studies also show that smoking is a major contributor to impotence in men. In fact, smokers are 50% more likely to be impotent than nonsmokers. Smoking also causes arterial changes that restrict blood flow to the penis. Changes begin to occur early in life. Teenage smokers could have problems in their thirties if they continue to smoke.

Tobacco smoke contains numerous hazardous substances that damage the lining of the respiratory system. Sulfur dioxide, for example, paralyzes the cilia lining the respiratory tract. Tobacco smoke is also laden with microscopic carbon particles. These carbon particles penetrate deeply into the lungs, where they accumulate in the alveolar walls, turning healthy tissue into a blackened mass that often becomes cancerous (Figure 1).

Tobacco smoke may also paralyze the alveolar macrophages, cells that removed particles from the lung, making a bad situation even worse. Why? Toxic chemicals in cigarette smoke, many of which are known carcinogens, attach to carbon particles. Toxin-carrying particles adhere to the lungs, larynx, trachea, and bronchi. Virtually any place they stick, they can cause cancer, explaining why smokers are five times more likely than nonsmokers to develop laryngeal cancer and four times more likely to develop cancer of the oral cavity.

Nitrogen dioxide and sulfur dioxide in tobacco smoke penetrate deep into the lungs, where they dissolve in the watery layer inside the alveoli. Nitrogen dioxide is converted to nitric acid; sulfur dioxide is converted to sulfuric acid. Both acids erode the alveolar walls, leading to emphysema.

The dangers of smoking are becoming well known. As a result of widespread publicity and public pressure, smoking has dropped substantially in the United States. In 2011, for example, about 19.3% of Americans smoke, down from 25% in 2004 and 34% in 1985 and down substantially from the 1950s and 1960s, when well over half of all men and over one-third of all women engaged in this potentially lethal habit.

Despite the downturn in smoking, as noted earlier, more than 45 million American adults (over the age of 18) still smoke. An estimated three million

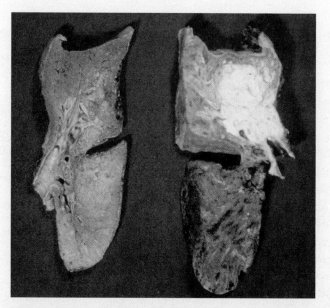

FIGURE 1 **The Normal and Cancerous Lung** (left) The normal lung appears spongy. (right) The cancerous lung from a smoker is filled with particulates and a large tumor. (© St. Bartholomew's Hospital/Science Source.)

teenagers and 600,000 middle schoolers smoke. Smoking occurs in greatest proportion in certain minority groups—particularly African-American men, Native Americans, and Native Alaskans.

Smoking is a personal choice with public and private hazards. In the U.S., there's been a definite trend downward in cigarette smoking, which will have many social, economic, and environmental benefits.

Visit Human Biology's Internet site for links to websites offering more information on this topic.

infectious disease. The respiratory system is also vital to the proper functioning of other organ systems that play key roles in homeostasis.

Like other body systems, the respiratory system can be damaged by pathogenic organisms and chemical contaminants in the environment. Changes in respiratory function alter homeostasis and can lead to disease.

The air of the industrialized world contains a multitude of chemical pollutants that damage human health by upsetting homeostasis. Air pollution affects millions of us on a daily basis. Unfortunately, most people are unaware of the dangers of air pollution because the line between cause and effect is not always clear. Consider, for instance, the headache you experienced in traffic going home from school or work. Was it caused by tension, or could it have been caused by carbon monoxide emissions from cars, buses, and trucks? And what about the

runny nose and sinus condition you suffered from last winter? Was that caused by a virus or bacterium or by pollution?

One classic study of air pollution in New York City in the 1960s clearly illustrated the relationship between air pollution and respiratory problems—and shocked many physicians. Researchers found that an increase in the level of sulfur dioxide, a pollutant produced by automobiles, power plants, and factories, increased the incidence of colds, coughs, nasal irritation, and other symptoms fivefold in just a few days. Soon after the pollution levels returned to normal, the symptoms subsided. Few people knew they'd been affected by the air they were breathing. Most thought they had been stricken by some infectious agent.

KEY CONCEPTS

Air pollutants in urban air can result in respiratory diseases like colds.

SUMMARY

Structure of the Human Respiratory System

1. The respiratory system consists of an air-conducting portion and a gas-exchange portion.
2. The air-conducting portion transports air from outside the body to the alveoli in the lungs, the site of gaseous exchange.
3. The alveoli are tiny, thin-walled sacs. Surrounding the alveoli are capillary beds that pick up oxygen and expel carbon dioxide.
4. The lining of the alveoli is kept moist by water. Surfactant, a chemical substance produced in the lung, reduces the surface tension inside the alveoli and prevents their collapse.

Breathing and the Control of Respiration

5. Breathing is an involuntary action controlled by the breathing center in the brain stem. Nerve cells in the breathing center send impulses to the diaphragm and intercostal muscles, causing them to contract. This increases the volume of the thoracic cavity, which draws air into the lungs through the nose or mouth.
6. When the lungs are full, the muscles relax and the lungs recoil, expelling air.
7. The breathing center is regulated by factors such as the levels of carbon dioxide in the blood.
8. Expiration can be supplemented by enlisting the aid of abdominal and chest muscles, as can inspiration.

Gas Exchange in the Lungs

9. The respiratory system conducts air to and from the lungs, exchanges gases, and helps produce sounds. Sound is generated as air rushes past the vocal cords, causing them to vibrate. The sounds are modified by movements of the tongue and changes in the shape of the oral cavity.
10. Oxygen and carbon dioxide diffuse across the alveolar wall of the lungs, driven by concentration differences between the blood and alveolar air. Oxygen diffuses into the blood plasma, and then into the RBCs, where most of it binds to hemoglobin. Carbon dioxide diffuses in the opposite direction.
11. Carbon dioxide, a waste product of cellular respiration, is picked up by the blood flowing through capillaries. Some carbon dioxide is dissolved in the blood. Most of it, however, enters the RBCs in the bloodstream, where it is converted into carbonic acid, which diffuses out of the RBCs and is transported in the plasma.
12. The rate of respiration can be increased by rising blood carbon dioxide levels and by an increase in physical exercise.

Diseases of the Respiratory System

13. Bacterial and viral infections of the respiratory tract can cause considerable discomfort, and some can be fatal.
14. The lungs are also susceptible to airborne materials, among them asbestos fibers, which can cause two types of lung cancer and a debilitating disease called asbestosis.
15. Another common disease of the respiratory system is asthma, an allergy (abnormal immune reaction) to dust, pollen, and other common substances. It is characterized by periodic episodes of wheezing and difficult breathing.
16. Chronic bronchitis is a persistent irritation of the bronchi and is caused by several common air pollutants primarily in tobacco smoke, and also in urban air pollution.
17. Emphysema is a breakdown of the alveoli that gradually destroys the lung's ability to absorb oxygen. Despite the role of air pollution in causing emphysema, smoking remains the number one cause of this disease.

Health and Homeostasis

18. The respiratory system contributes mightily to homeostasis by regulating concentrations of oxygen and carbon dioxide in the blood and body tissues. It also helps control the pH of the blood and tissues. In addition, it supports other homeostatic organs. Proper functioning of the respiratory system is therefore essential for health.
19. Respiratory function can be dramatically upset by microorganisms as well as by pollution from factories, automobiles, power plants, and even our own homes.

THINKING CRITICALLY ANALYSIS

This analysis corresponds to the Thinking Critically scenario that was presented at the beginning of this chapter.

For years, the tobacco industry insisted that there was no link between smoking and lung cancer because, they said, there was no positive proof. Semantically, they were quite correct. Science does not prove anything. Sure, scientists have performed many studies on people that show smokers are much more likely to develop lung cancer than nonsmokers and that indicate the more an individual smokes, the higher his or her chances are of developing this disease. Technically, the scientists haven't proved anything, they've shown a correlation.

The tobacco industry was being a bit deceptive, though, for when a correlation like that between smoking and lung cancer shows up again and again or when the correlation is high, you can be fairly confident in the cause-and-effect relationship under study.

Can a study be devised to prove the connection? It's unlikely. As I mentioned earlier, science is in the business of supporting hypothesis and generating theory, not providing proof positive.

This exercise requires one to look at definitions (what constitutes proof and the difference between proving something and showing a strong correlation). It also shows that one usually needs to dig deeper to understand controversies. In this instance, the tobacco industry has, for years, hung its argument on semantics. This exercise also suggests the importance of looking at who's talking—and uncovering any hidden biases. The tobacco industry clearly has a vested interest in assuring people that smoking is not harmful.

KEY TERMS AND CONCEPTS

Alveolar macrophage, p. 171
Alveoli, p. 166
Asbestosis, p. 178
Asthma, p. 179
Brainstem (medulla), p. 175
Breathing center, p. 175
Bronchi, p. 168
Bronchioles, p. 168
Bronchitis, p. 178
Chronic bronchitis, p. 179
Chronic obstructive pulmonary disease (COPD), p. 180
Cilia, p. 169
Diaphragm, p. 174

Dust cell, p. 171
Emphysema, p. 179
Epiglottis, p. 166
Exhalation, p. 174
Expiration, p. 174
Expiratory reserve volume, p. 175
Hyaline membrane disease, p. 173
Infant respiratory distress syndrome, p. 173
Inhalation, p. 174
Inspiration, p. 174
Inspiratory reserve volume (IRV), p. 175
Intrapulmonary pressure, p. 174
Laryngitis, p. 178
Larynx, p. 166

Lung, p. 166
Meningitis, p. 178
Mucous cell, p. 169
Mucus, p. 169
Pharynx, p. 166
Residual volume, p. 175
Sinusitis, p. 178
Stretch receptors, p. 176
Surfactant, p. 172
Tidal volume, p. 174
Trachea, p. 168
Type I alveolar cells, p. 171
Type II alveolar cell, p. 172
Vocal cord, p. 168

CONCEPT REVIEW

1. Trace the flow of air from the mouth and nose to the alveoli, and describe what happens to the air as it travels along the various passageways. pp. 166–168.
2. Draw an alveolus, including all cell types found there. What does each cell do? Be sure to show the relationship of the surrounding capillaries. Show the path that oxygen and carbon dioxide take. pp. 166–168.
3. Trace the movement of oxygen from alveolar air to the blood in alveolar capillaries. Describe the forces that cause oxygen to move in this direction. Do the same for the reverse flow of carbon dioxide. pp. 170–171.
4. Why would a breakdown of alveoli in emphysemic patients make it difficult

for them to receive enough oxygen? pp. 179–180.
5. Describe how sounds are generated and refined. p. 168.
6. A baby is born prematurely and is having difficulty breathing. As the attending physician, explain to the parents what the problem is and how it could be corrected. p. 173.
7. Smoking irritates the trachea and bronchi, causing mucus to build up and paralyzing cilia and alveolar macrophages. How do these changes affect the lung? p. 169.
8. Describe inspiration and expiration, being sure to include discussions of what triggers them and the role of muscles in effecting these actions. p. 174.

9. How does the breathing center regulate itself to control the frequency of breathing? pp. 175–176.
10. Exercise increases the rate of breathing. How? pp. 175–176.
11. Turn to Figure 8-10. Explain each of the terms. p. 174–175.
12. What is asthma? What are its symptoms? How does it affect one's inspiratory reserve volume? p. 179.
13. Debate this statement: Urban air pollution has very little overall impact on human health. pp. 179–181.
14. Do you agree or disagree that the single most effective way of reducing deaths in the United States would be to ban smoking? Explain. pp. 180–181.

SELF-QUIZ: TESTING YOUR KNOWLEDGE

1. The _____ are located in the larynx and are responsible for the production of sound. p. 168.
2. The larynx opens below into the _____, which then splits to form the right and left bronchi. p. 168.
3. _____ force mucus in the lining of the upper respiratory tract upward toward the mouth where it can be swallowed or expectorated. p. 169.
4. The air that moves in and out with each breath is known as the _____ volume. p. 174.

5. When the lungs are full, _____ receptors in the lung send signals to the breathing center, causing them to cease firing. p. 176.
6. The tiny air sacs in the lung that permit oxygen to be transported into the bloodstream are known as the _____. p. 166.
7. Most oxygen is transported in the blood bound to _____ in the RBCs. p. 176.
8. Most carbon dioxide is transported in the blood as _____ in the blood plasma. p. 177.

9. Inhaled particles are deposited into a layer of _____. p. 169.
10. Particles that make it into the lung are engulfed by _____ cells, also known as alveolar macrophages. p. 171.
11. Large round cells in the lining of the alveoli produce a substance consisting of lipid and protein known as _____ that helps reduce surface tension in the lungs, keeping the alveoli open. p. 172.
12. The muscles of inspiration are the _____ and intercostal muscles between the ribs. p. 174.

13. Breathing is controlled by the _____ center in the brain. p. 174.
14. Inflammation of the vocal cords is known as _____. p. 178.
15. Persistent irritation of the bronchi, which causes mucus production and coughing, is known as chronic _____. p. 179.
16. _____ is a common disease caused by an allergic reaction to common environmental components such as pollen, dust, dander, and certain foods. p. 179.
17. Chronic _____ pulmonary disease includes chronic bronchitis and _____. pp. 179–180.
18. _____ _____ in cigarette smoke paralyzes the cilia, reducing natural cleansing of the respiratory system, so essential to human health. p. 169.
19. Smoke from tobacco consumed by others that is present in our homes, vehicles, and at work is known as _____ smoke and may cause cancer and a host of other symptoms in nonsmokers. p. 180.
20. Label the diagram.

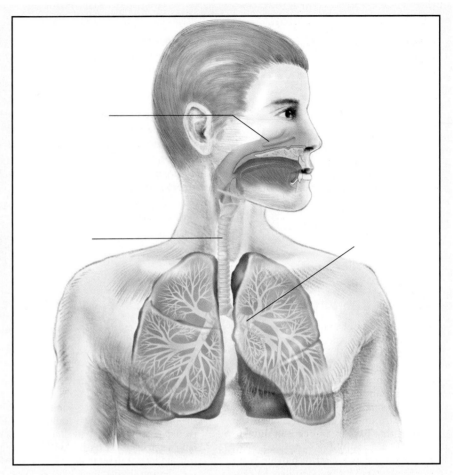

 biology.jbpub.com/chiras/8e/

The site features eLearning, an online review area that provides quizzes, chapter outlines, and other tools to help you study for your class. You can also follow useful links for in-depth information, research the differing views in the Point/Counterpoints, or keep up on the latest health news.

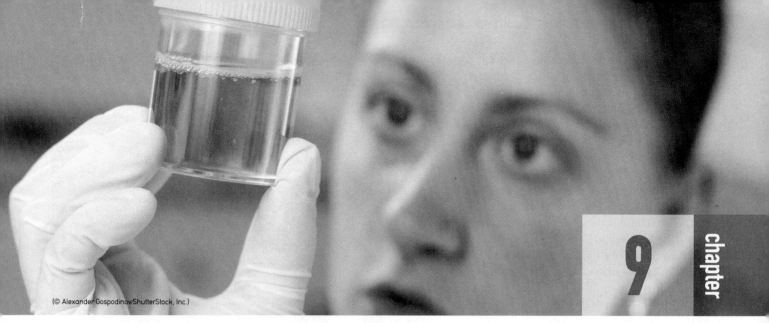
(© Alexander Gospodinov/ShutterStock, Inc.)

9 chapter

The Urinary System: Ridding the Body of Wastes and Maintaining Homeostasis

Eileen Markowitz lies on a bed in a special room in the hospital. Physicians position two large, cylindrical devices in over her kidneys, the organs that filter the blood. Over the next few hours, as Markowitz listens to tapes of her favorite music, ultrasound waves, undetectable by the human ear, will smash the kidney stones that are obstructing the flow of urine and causing excruciating pain in this young teacher.

THINKING CRITICALLY

In 2012, the National Cancer Institute estimated that there were nearly 65,000 new cases of kidney cancer diagnosed in the United States. Over 13,500 people will die from this disease. Chemotherapy, the only treatment available for patients whose cancers have spread to other locations, helps fewer than 10% of them.

Recently, a promising new treatment was introduced. The treatment, known as ALT (autolymphocyte therapy), costs about $22,000 per patient. ALT is designed to strengthen the immune system. Here's how it works: Blood is withdrawn from patients diagnosed with kidney cancer. Cancer-fighting cells called *lymphocytes* are then extracted from each patient's blood. These cells are exposed to antibodies in the kidney tumor. (Antibodies are blood proteins that are produced by immune system cells.) Exposure of the lymphocytes to antibodies activates the cells, causing them to produce a chemical known as a *cytokine*. Cytokines boost the activity of lymphocytes.

The cytokines produced by the lymphocytes are extracted from the samples, divided into six batches, and stored. Once a month, for the next 6 months, patients donate more of their own lymphocytes. The cells are treated with cytokines, activated, and then reinjected, where they apparently go to work on the tumors.

Results of one study on the effectiveness of ALT in 90 individuals showed that patients who had undergone the procedure survived an average of 22 months after treatment, two and a half times longer than patients treated by chemotherapy. Less than a year after the publication of this clinical trial, a treatment center opened in Boston. If you were considering investing in the company that owns and operates the treatment center, what questions would you ask before investing?

A decade earlier, surgeons would have had to cut an incision 15 to 20 centimeters (6 to 8 inches) long in Markowitz's side to access the kidney to remove the stone. She would have spent 7 to 10 days recovering in the hospital and another 8 weeks at home recuperating before returning to work. With this technique, known as *ultrasound lithotripsy* (LITH-oh-TRIP-see) ("stone crushing"), patients usually return home within a day (Figure 9-1).

In this chapter, we examine the kidneys and other portions of the urinary system, which function to rid our bodies of potentially harmful wastes. We briefly examine other organs, too, such as the skin and liver that help eliminate waste. They are collectively referred to as *excretory organs*. Throughout the chapter, you will see how these organs contribute to homeostasis and thus human health. We end with a discussion of common diseases, such as kidney stones, that disturb the function of this important organ.

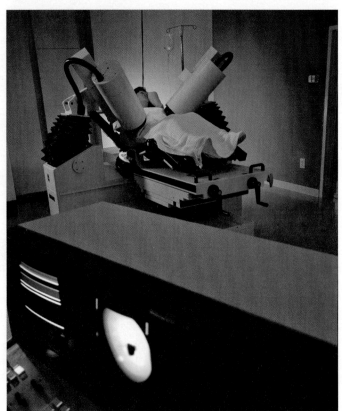

FIGURE 9-1 Lithotripsy Ultrasound waves, undetectable to the human ear, bombard the kidney stones in this man, smashing them into sandlike particles that can be passed relatively painlessly in the urine. (© Phototake, Inc./Alamy Images.)

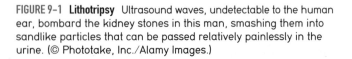

9-1 Organs of Excretion

Ralph Waldo Emerson once said that as soon as there is life, there is risk. A biologist might say as soon as there is life, there is waste. All living things produce waste, and humans are no exception. In this chapter, we're not concerned with the mountains of waste we humans produce in our homes and businesses, however. We're concerned about wastes produced by the cells of the body. These wastes like the waste produced by human society must be eliminated—either recycled or dumped. They cannot accumulate inside the human body. If they do, they will cause serious damage, even death. So, like all other animals, humans must get rid of internally produced wastes, quickly and efficiently.

Table 9-1 lists the major metabolic wastes—that is, wastes produced by the cells during metabolism—and other chemicals excreted from the body. The first three entries are nitrogen-containing wastes—ammonia (ah-MOAN-ee-ah), urea (yur-EE-ah), and uric acid (YUR-ick).

Ammonia (NH_3), is a product of the breakdown of amino acids in the liver. Amino acid breakdown generally occurs when there's excess protein in the diet. It also occurs when there's a shortage of carbohydrate in a diet—a very rare occurrence in more developed countries. During shortages, the body breaks down protein in muscle to acquire amino acids. The amino acids are then broken down to produce energy. During the breakdown of amino acids, the amino groups (NH_2) are stripped from the molecules. The reaction is called *deamination* (dee-AM-in-A-shun). The amino groups are converted into ammonia. Ammonia is a highly toxic chemical.

TABLE 9-1	Important Metabolic Wastes and Substances Excreted from the Body	
Chemical	**Source**	**Organ of Excretion**
Ammonia	Deamination (removal of amine group) of amino acids in liver	Kidneys
Urea	Derived from ammonia	Kidneys, skin
Uric acid	Nucleotide breakdown in liver	Kidneys
Bile pigments	Hemoglobin breakdown in liver	Liver (into small intestine)
Urochrome	Hemoglobin breakdown in liver	Kidneys
Carbon dioxide	Breakdown of glucose in cells	Lungs
Water	Food and water; breakdown of glucose	Kidneys, skin, and lungs
Inorganic ions*	Food and water	Kidneys and sweat glands

*Ions are not a metabolic waste product like the other substances shown in this table. Nonetheless, ions are excreted to maintain constant levels in the body.

Most ammonia is converted to **urea** in the liver. It then enters the bloodstream and is excreted by the kidneys. A small amount is excreted in perspiration, too.

Health Tip 9-1

To lose or maintain weight, take larger portions of vegetables and whole grains and smaller meat portions.

Why?

Vegetables and whole grains provide less fat and fewer calories than meats. Vegetables also provide many nutrients the body needs for long-term health—especially vitamins and minerals. Vegetables also contain many helpful chemicals such as antioxidants that help prevent serious disease like cancer. As you may recall from elsewhere in the text, most of us consume much more meat protein than we need. Plant foods, especially whole grain products supply protein in the diet, too, without the fat of meat.

Another byproduct of metabolism in humans is **uric acid**. Produced by the liver, uric acid derives not from the breakdown of amino acids but rather from the breakdown of nucleotides. Nucleotides are the building blocks of DNA and RNA. Uric acid, like ammonia, is excreted in the urine.

In adults, excess uric acid production may result in the deposition of uric acid crystals in the bloodstream and in joints, where they cause considerable pain. This condition is known as *gout* (gowt). Gout is considered a type of arthritis.

It may be chronic, that is, long term. Chronic gout results in repeated episodes of pain and inflammation in multiple joints. Some cases are acute, that is, it is short term and affects one joint. The exact cause is unknown, but it results when the body either produces too much uric acid or can't excrete it fast enough. **Bile pigments** are another important metabolic waste. They are produced during the breakdown of RBC hemoglobin in the liver. Bile pigments are transferred from the liver to the gallbladder, which is attached to the underside of the liver. The gallbladder empties by a duct into the small intestine. Bile pigments are then excreted with the feces.

Next on the list is **urochrome** (YUR-oh-chrome). Urochrome is a water-soluble pigment also produced in the liver during the breakdown of hemoglobin. This yellow pigment is dissolved in the blood and passes to the kidneys, where it is excreted with the urine. It gives urine its yellowish color.

Table 9-1 also lists inorganic ions excreted from the body. Even though these ions are not end products of metabolism, they are removed from the body by various excretory organs such as the skin and kidneys. Excretion is essential to maintain optimal levels in the blood, tissue fluids, and the cytoplasm of cells—so vital for homeostasis and health and survival.

KEY CONCEPTS

Cells of the body produce many wastes that must be removed by organs of excretion to maintain homeostasis and health.

9-2 The Urinary System

As the previous section demonstrates, evolution has "provided" several avenues by which animals get rid of, or excrete, wastes. In humans, excretion of wastes occurs in the lungs, the skin, the liver, the kidneys, and the intestines. These organs are consequently classified as excretory organs.

Of all the organs that participate in removing waste, however, the kidneys rank as one of the most important. That's because they rid the body of the greatest variety of dissolved wastes. In the process, the kidneys also play a key role in regulating the chemical constancy of the blood.

As Figure 9-2a shows, humans come equipped with two kidneys. They are part of the urinary system, one of the body's most important excretory systems. The urinary system consists of three additional organs: (1) the ureters (YUR-eh-ters); (2) the urinary bladder; and (3) the urethra (you-REETH-rah). The functions of these organs are described in more detail below and are listed in Table 9-2.

KEY CONCEPTS

The urinary system is one of several body systems that remove wastes.

The Anatomy of the Urinary System

The **kidneys** lie on either side of the backbone, the vertebral column. About the size of a person's fist, each kidney is surrounded by a layer of protective fat. The kidneys are located high in the posterior (back) abdominal wall beneath the diaphragm.

The human kidneys are oval structures, slightly indented on one side, much like kidney beans (Figure 9-2b). (Kidney beans get their name because they're shaped like this organ.) As illustrated, each kidney is served by a **renal artery**, which enters at the indented region. The renal arteries are major branches of the abdominal aorta, a large blood vessel that delivers blood to the abdominal organs and lower limbs.

The kidneys filter wastes and excess chemicals from the blood entering via the renal arteries. After the blood has been filtered, it leaves the kidneys via the **renal veins**, which drain into the inferior vena cava. It transports venous blood from the legs, abdomen, and chest to the heart.

Wastes extracted from the blood are eliminated in the **urine** (YUR-in), a yellowish fluid produced by the kidneys. It contains water, nitrogenous wastes (ammonia, urea, and uric acid), small amounts of hormones, ions, and other substances, including various medications people take like blood-pressure meds and antidepressants.

Dissolved wastes are removed from the blood by numerous microscopic filtering units in the kidney, the **nephrons** (NEFF-rons). The urine produced by the nephrons drains from the kidney via the **ureters**. These tubules transport urine to the urinary bladder with the aid of peristaltic contractions—smooth muscle contractions. The **urinary bladder** lies in the pelvic cavity just behind the pubic bone. The walls of the bladder contain lots of smooth muscle that stretches as the organ fills. As you know, the bladder serves as a temporary receptacle

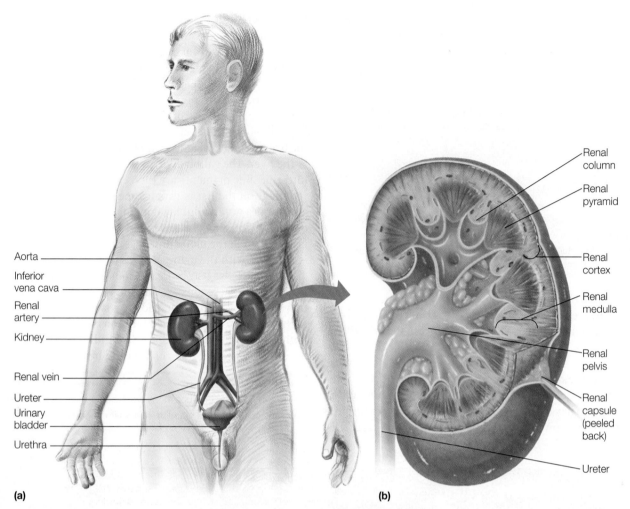

Aorta
Inferior vena cava
Renal artery
Kidney
Renal vein
Ureter
Urinary bladder
Urethra

Renal column
Renal pyramid
Renal cortex
Renal medulla
Renal pelvis
Renal capsule (peeled back)
Ureter

(a)
(b)

FIGURE 9–2 The Urinary System (a) Anterior view showing the relationship of the kidneys, ureters, urinary bladder, and urethra. (b) A cross section of the human kidney showing the cortex, medulla, and renal pelvis.

for urine produced by the kidneys. When the bladder is full, its walls contract, which forces the urine out through the urethra.

The **urethra** is a narrow tube, measuring approximately 4 centimeters (1.5 inches) in women and 15 to 20 centimeters (6 to 8 inches) in men. The additional length in men largely results from the fact that the urethra travels through the penis (Figure 9-3).

The difference in the length of the urethra between men and women has important medical implications. Doctors speculate that the shorter urethra in women makes women more susceptible to bacterial infections of the bladder. Bladder infections may also be more common in women because the

urethral opening lies close to the vagina and the anus. As a result, it may also be easier for bacteria to enter the urethra and make their way into the bladder.

Bladder infections may result in an itching or burning sensation and an increase in the frequency of urination. They may also cause blood to appear in the urine. Urinary tract infections can be treated with antibiotics. Untreated infections may spread up the ureters to the kidneys, where they can seriously damage the blood-filtering tubules, the nephrons, and impair kidney function.

> **KEY CONCEPTS**
>
> The urinary system consists of the kidneys, ureters, bladder, and urethra; waste is removed from the blood in the kidneys, transported to the bladder where it is stored, then released through the urethra.

The Anatomy of the Kidney

Each kidney is surrounded by a connective tissue capsule, appropriately called the *renal capsule*. Immediately beneath the capsule is a region known as the *renal cortex*. This region contains many nephrons, tiny biological filters. The inner zone is the *renal medulla*. As shown in Figure 9-2b, the medulla consists of cone-shaped structures (renal pyramids)

TABLE 9-2	Components of the Urinary System and Their Functions
Component	**Function**
Kidneys	Eliminate wastes from the blood; help regulate body water concentration; help regulate blood pressure; help maintain a constant blood pH
Ureters	Transport urine to the urinary bladder
Urinary bladder	Stores urine; contracts to eliminate stored urine
Urethra	Transports urine to the outside of the body

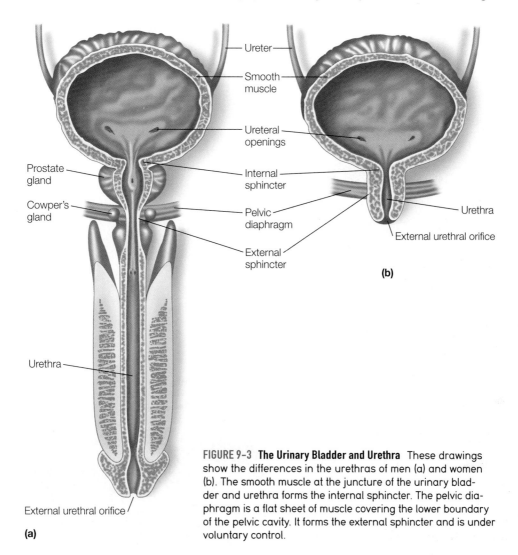

Ureter

Smooth
muscle

Ureteral
openings

Prostate
gland

Cowper's
gland

Internal
sphincter

Pelvic
diaphragm

External
sphincter

Urethra

External urethral orifice

(a)

Urethra

External urethral orifice

(b)

FIGURE 9-3 The Urinary Bladder and Urethra These drawings show the differences in the urethras of men (a) and women (b). The smooth muscle at the juncture of the urinary bladder and urethra forms the internal sphincter. The pelvic diaphragm is a flat sheet of muscle covering the lower boundary of the pelvic cavity. It forms the external sphincter and is under voluntary control.

and intervening tissue (renal columns). The cone-shaped sections contain small ducts that transport urine from the nephrons into a central receiving chamber, the *renal pelvis* (Figure 9-2b). It joins with the ureters.

> **KEY CONCEPTS**
>
> Kidneys are large organs that contain the blood filtering nephrons.

The Nephron

Each kidney contains 1 to 2 million nephrons, visible only through a microscope. One of the marvels of evolution, each nephron consists of a tuft of capillaries, the **glomerulus** (glom-ERR-you-luss), and a long, twisted tube, the **renal tubule** (Figure 9-4b). The renal tubule consists of four segments: (1) Bowman's capsule, (2) the proximal convoluted tubule, (3) the loop of Henle, and (4) the distal convoluted tubule (Table 9-3). You may want to take a moment to study Figure 9-4b to acquaint yourself with these structures.

As illustrated in Figure 9-4b, the glomerulus is surrounded by a saclike portion of the renal tubule, called **Bowman's capsule** (after the scientist who first described it). It is a double-walled structure; the inner wall fits closely over the glomerular capillaries and is separated from the outer wall by a small space, Bowman's space.

To understand the relationship between the glomerulus and Bowman's capsule, imagine that your fist is a glomerulus. Then imagine that you are holding a balloon in your other hand. If you push your fist (glomerulus) into the balloon, the layer immediately surrounding your fist would be the inner layer of Bowman's capsule. It is separated from the outer layer of the capsule by Bowman's space.

The outer wall of Bowman's capsule is continuous with the **proximal convoluted tubule** (PCT), a contorted section of the renal tubule. The PCT soon straightens, then descends a bit, makes a sharp hairpin turn, and ascends. In the process, it forms a thin, U-shaped structure known as the **loop of Henle** (HEN-lee). The loops of Henle of some nephrons extend into the medulla.

The loop of Henle drains into the fourth and final portion of the renal tubule, the **distal convoluted tubule** (DCT), another contorted segment. This structure, shown in Figure 9-4b, empties into a straight duct called a **collecting tubule**. The collecting tubules, in turn, merge to form larger ducts. They course through the cone-shaped renal pyramids and drain into the renal pelvis. From here, urine exits the body via the urethra.

> **KEY CONCEPTS**
>
> Nephrons consist of two parts, a glomerulus and renal tubule.

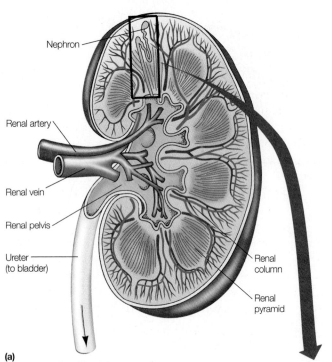

(a)

FIGURE 9-4 **The Nephron** (a) A cross section of the kidney showing the location of the nephrons. (b) A drawing of a nephron.

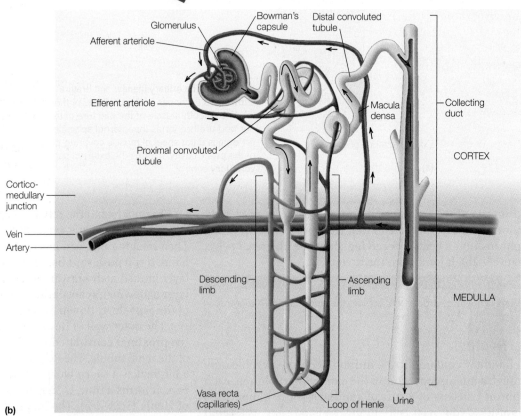

(b)

TABLE 9-3	Components of the Nephron and Their Function
Component	**Function**
Glomerulus	Mechanically filters the blood
Bowman's capsule	Mechanically filters the blood
Proximal convoluted tubule	Reabsorbs 75% of the water, salts, glucose, and amino acids
Loop of Henle	Participates in countercurrent exchange, which maintains the concentration gradient
Distal convoluted tubule	Site of tubular secretion of H^+, potassium, and certain drugs

9-3 Function of the Urinary System

With the basic anatomy of the kidney and urinary system in mind, let us turn our attention to the function of the urinary system, focusing first on the nephron.

Blood Filtration in the Kidney

The nephrons filter enormous amounts of blood and produce from 1 to 3 liters of urine per day, depending on how much fluid a person ingests. Table 9-3 summarizes the functions of each segment of the nephron.

> **KEY CONCEPTS**
>
> Blood filtration in the nephrons involves three processes.

Glomerular Filtration

The first step in the purification of the blood is **glomerular filtration** (Figure 9-5). It's a rather straightforward physical filtration process not unlike that which occurs when you pour a fluid containing particles into a filter. Here's how it works. As blood flows through the glomerular capillaries, water and dissolved materials are forced through the walls of these tiny vessels. This liquid, called the **glomerular filtrate**, travels through the inner layer of Bowman's capsule. Red blood cells and large proteins in the blood can't escape because of the barrier formed by the inner layer of Bowman's capsule. From here, the filtrate passes into the proximal convoluted tubule, then into the loop of Henle, and finally out the distal convoluted tubule.

> **KEY CONCEPTS**
>
> Blood is physically filtered first in the glomeruli, a process that removes much more from the blood than is necessary; subsequent steps reabsorb valuable water and nutrients so as not to waste them.

Tubular Reabsorption

Glomerular filtration is a rather crude filtering process. In fact, it allows quite a lot of valuable nutrients like blood sugar and valuable ions like sodium to pass into the nephron, but prevents formed elements like blood cells and platelets from entering. In the remaining steps, most of the valuable ions and nutrients reenter the blood stream, leaving behind the wastes like urea, ammonia, uric acid, and urochrome. To give you an idea of the scale of this operation: Each day, approximately 180 liters (45 gallons) of filtrate is produced in the glomeruli. However, the kidneys produce only about 1 to 3 liters of urine per day. Thus, 99% of the filtrate is reabsorbed back into the bloodstream and only about 1% of the filtrate will leave the kidneys as urine. What happens to the rest of the fluid filtered by the glomerulus?

The movement of valuable materials such as water, ions, and molecules from the filtrate in the nephron back into to the bloodstream is referred to as **tubular reabsorption**. Most tubular reabsorption occurs in the proximal convoluted tubule.

During tubular reabsorption, water containing valuable nutrients and ions—that need to be preserved—passes from the renal tubule and enters the networks of capillaries that surround the nephrons. As shown in Figure 9-4, these capillaries are branches of the efferent arterioles (arterioles leaving the glomerulus). They empty into the renal veins, returning nutrients to the body.

To reiterate, tubular reabsorption is valuable because it conserves both water and important ions and nutrients that are indiscriminately filtered out in the glomerulus. Waste products like urea and uric acid, however, remain in the glomerular filtrate and will eventually be excreted in the urine. Tubular reabsorption also helps to control blood pH. Table 9-4 shows the reabsorption rates of various molecules.

> **KEY CONCEPTS**
>
> Tubular reabsorption is the second phase of blood filtration; it is responsible for the selective reabsorption of water and other materials filtered indiscriminately from the blood by the glomeruli.

Tubular Secretion

Waste disposal is supplemented by a third process, **tubular secretion** (Figure 9-5). Peritubular secretion removes wastes from the blood—specifically wastes that escaped filtration in the glomerulus.

As shown in Figure 9-4b, blood leaves the glomerulus via the efferent arteriole. It flows into the peritubular (pear-ee-TUBE-you-ler) capillaries that surround the distal convoluted tubules of the nephrons. Wastes that escaped filtration are pumped through the capillary walls into the nephron. This is referred to as tubular secretion.

Tubular secretion not only rids the blood of wastes, it helps regulate the H^+ concentration of the blood. If the blood is too acidic, for example, H^+ ions are transported from the blood into the urine to reduce its acidity.

Figure 9-5 summarizes the somewhat confusing process of blood filtration in the kidney. You might want to take a

TABLE 9-4	Fate of Various Substances Filtered by Kidneys	
	Filtered Substance (Average %)	
Substance	Reabsorbed	Excreted
Water	99	1
Sodium	99.5	0.5
Glucose	100	0
Waste products		
Urea	50	50
Phenol	0	100

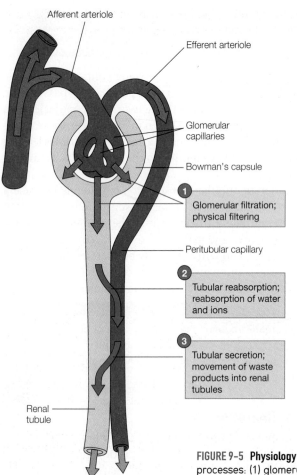

Afferent arteriole

Efferent arteriole

Glomerular capillaries

Bowman's capsule

1 Glomerular filtration; physical filtering

Peritubular capillary

2 Tubular reabsorption; reabsorption of water and ions

3 Tubular secretion; movement of waste products into renal tubules

Renal tubule

Urine   Filtered blood

moment to study it. Bottom line: Blood draining from the capillaries surrounding the nephrons is fairly well cleansed by the time it empties into the nearby veins, thanks to glomerular filtration, tubular reabsorption, and tubular secretion. Veins that drain the blood from the interior of the kidney converge to form the renal vein. It transports the highly filtered blood out of the kidney and into the inferior vena cava, which carries it to the heart.

Urine produced by the nephrons flows out of the nephron into the collecting tubules. The urine leaving the nephron consists mostly of water and a variety of dissolved waste products. The collecting tubules descend through the medulla and converge to form larger ducts, the collecting ducts. They empty their contents into the renal pelvis. As the collecting tubules descend through the medulla, much of the remaining water escapes by osmosis, further concentrating the urine and conserving body water.

KEY CONCEPTS

Tubular secretion is the third phase of blood filtration; it involves the transport of waste products from the peritubular capillaries back into the renal tubule.

FIGURE 9–5 **Physiology of the Nephron** The nephron carries out three processes: (1) glomerular filtration, (2) tubular reabsorption, and (3) tubular secretion. All contribute to the filtering of the blood.

9-4   Controlling Urination

Urine emptying into the renal pelvis consists mostly of water and dissolved wastes. Urine is produced continuously by the kidneys and flows down the ureters to the urinary bladder, where it is stored. Leakage out of the bladder and into the urethra is prevented by two sphincters—muscular "valves" similar to those found in the stomach (Figure 9-3).

The first sphincter, the *internal sphincter*, is formed by a smooth muscle in the neck of the urinary bladder at its junction with the urethra. The second valve, the *external sphincter*, is a flat band of skeletal muscle that forms the floor of the pelvic cavity. When both sphincters are relaxed, urine is propelled into the urethra and out of the body.

Urination is stimulated by the accumulation of urine inside the bladder—about 200 to 300 milliliters start the process. When the bladder fills, its walls stretch. Special sensors in the walls, called *stretch receptors*, detect the distension of the bladder and begin sending impulses via sensory nerves to the spinal cord. In the spinal cord, the incoming nerve impulses stimulate other nerve cells. Nerve impulses generated in these cells leave the spinal cord and travel along nerves that terminate on the muscle cells in the wall of the bladder, causing these cells to contract. Nerve impulses also cause

the relaxation of the muscle cells in the internal sphincter to relax, allowing urine to enter the urethra. Urination cannot occur in children over 2 to 3, teenagers, and adults, however, until the external sphincter is relaxed. That's because the external sphincter is under conscious control. It will not relax until you consciously permit it.

In babies and very young children, urination is entirely reflexive. Once the bladder expands to a certain size, muscles of the wall contract and the sphincters relax, allowing the bladder to enter. Not until children grow older (2 to 3 years) can they begin to control urination.

Unfortunately, adults sometimes lose control over urination, resulting in a condition referred to as *urinary incontinence* (in-KAN-teh-nance). Urinary incontinence may be caused by a traumatic injury to the spinal cord that disrupts neural pathways from the brain that allow us to suppress urination. In such instances, the bladder empties as soon as it reaches a certain size, much as it does in a baby or young child.

Mild urinary incontinence is fairly common in older adults. In this disorder, urine escapes when a person sneezes or coughs. Mild urinary incontinence is most common in women and usually results from damage to the external sphincter during childbirth. Childbirth stretches the skeletal muscles of the external sphincter, reducing their effectiveness. To avoid this, many women perform exercises (called Kegel exercises) to strengthen these muscles before and after childbirth. Urinary incontinence may also occur in men whose external sphincters have been injured in surgery on the prostate gland, a gland that surrounds the neck of the urinary bladder. Prostate surgery is performed to remove tumors.

> **KEY CONCEPTS**
>
> Urination is a reflex response in babies but is controlled consciously in older children and adults.

9-5   Controlling Kidney Function and Maintaining Homeostasis

The kidneys help the body control the chemical composition of the blood and thus help to maintain homeostasis—ensuring good health. Kidney function is controlled by two hormones, discussed in this section. As you shall soon see, the concentration of ions and nutrients in the blood are regulated by controls on water concentration in the blood.

As noted earlier, much of the water filtered out of the blood by the glomeruli is reabsorbed by the renal tubules and returned to the bloodstream. The rate of water reabsorption can be increased or decreased. When dehydrated, water reabsorption is increased and urine output declines. When overhydrated, water reabsorption is decreased and urine output increases. Two hormones play a major role in this important homeostatic process: ADH and aldosterone.

> **KEY CONCEPTS**
>
> Concentrations of important ions and other substances in the blood are controlled by regulating water levels.

ADH

Water reabsorption is controlled in part by **antidiuretic hormone** (ANN-tie-DIE-yur-eh-tick) or **ADH**. ADH is released by an endocrine gland at the base of the brain known as the *pituitary* (peh-TWO-eh-TARE-ee; discussed elsewhere in the text). As illustrated in Figure 9-6, ADH secretion is regulated by two receptors, one in the brain and one in the heart. Take a moment to check this out.

The brain sensor is a group of nerve cells in a region of the brain called the *hypothalamus* (HIGH-poe-THAL-ah-muss). It is located just above the pituitary gland. Cells that regulate ADH monitor the osmotic concentration (the concentration of dissolved substances) of the blood, as you shall soon see. The second sensor, located in the heart, detects changes in blood volume, which reflect water levels.

When the water concentration of the blood falls, ADH is released. To understand this process, imagine you're out in the hot desert sun on a hike or playing. Your body's sweating and you haven't had a drink of water for hours. As water losses continue, the blood volume decreases and the osmotic concentration of the blood increases. The decrease in blood volume is detected by the heart receptors. The rise in osmotic

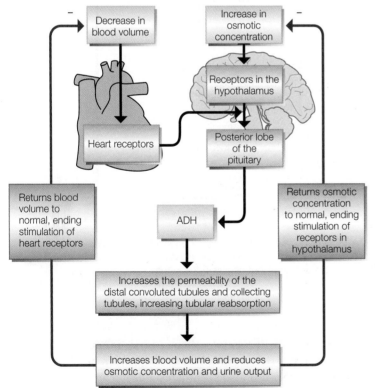

FIGURE 9-6 ADH Secretion ADH secretion is under the control of the hypothalamus. When the osmotic concentration of the blood rises, receptors in the hypothalamus detect the change and trigger the release of ADH from the posterior lobe of the pituitary. Detectors in the heart also respond to changes in blood volume. When it drops, they send signals to the brain, causing the release of ADH. (Photo © Photos.com.)

concentration is detected by sensors in the brain. As shown in Figure 9-6, both of these stimuli trigger the release of ADH from the pituitary gland.

ADH circulates in the blood. In the kidney, ADH stimulates tubular reabsorption of water in the distal convoluted tubules and the collecting tubules. This increases the flow of water from the nephron and collecting tubules into the peritubular capillaries. This, in turn, reduces urinary output and restores the volume and osmotic concentration of the blood—if you are not too dehydrated.

Excess water intake, on the other hand, has the opposite effect. Drinking excess water increases the blood volume and decreases its osmotic concentration. Sensors in the brain and heart note these changes and cause a reduction in ADH secretion. As ADH levels in the blood fall, tubular reabsorption decreases and more water is lost in the urine. As a consequence, blood volume and osmotic concentration are restored. Urine output increases.

> **KEY CONCEPTS**
>
> The pituitary hormone ADH increases water reabsorption, helping to reduce water loss during times of reduced intake or excess output (sweating).

Aldosterone

Water balance is also regulated by the hormone **aldosterone** (al-DOS-ter-own). Aldosterone is a steroid hormone produced by the adrenal glands, endocrine glands that sit atop the kidneys (discussed elsewhere in the text). Aldosterone levels in the blood are controlled by two factors: (1) blood pressure and (2) the volume of the fluid in the nephrons.

This gets confusing, so consult Figure 9-7 as you read the text. To understand this process, begin at the top of Figure 9-7. Let's imagine that you are dehydrated due to intense physical activity on a hot summer day. Dehydration leads to a decline in blood pressure. It also decreases the volume of filtrate in the nephrons. In response to these stimuli, certain cells in the kidney produce an enzyme called **renin** (REE-nin).

Renin is released into the bloodstream where it encounters a large plasma protein known as **angiotensinogen** produced by the liver (AN-gee-oh-TEN-SIN-oh-gin). Renin slices a segment off angiotensinogen, producing a small peptide molecule known as **angiotensin I**. Angiotensin I is inactive. It must be converted into the active form, **angiotensin II**, by enzymes it encounters as blood flows through the lungs.

Health Tip 9-2

Some people are sensitive to salt intake. Excess salt intake raises blood pressure. If you are one of them, you may want to lower your salt intake by using low-sodium salt.

Why?

A recent study in elderly men showed that reducing sodium intake resulted in a marked decrease in deaths due to cardiovascular disease. The low-sodium salt (which is now commercially available) consisted of half sodium chloride and half potassium chloride. The researchers believe that the beneficial effect resulted not so much from lower sodium intake, but from increased potassium intake. This can also be achieved by eating potassium-rich fruits and vegetables. Potassium is believed to help lower blood pressure.

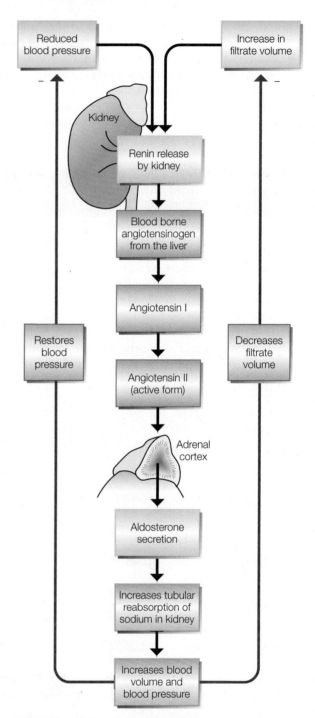

FIGURE 9–7 Aldosterone Secretion Aldosterone is released by the adrenal cortex. Its release, however, is stimulated by a chain of events that begins in the kidney.

Angiotensin II has two effects. First, it causes blood vessels to constrict, resulting in an increase in blood pressure. Angiotensin II also stimulates aldosterone secretion by the adrenal glands. Aldosterone released from the adrenal glands circulates in the blood. In the kidney, it stimulates an increase in tubular reabsorption of sodium in the distal convoluted tubules and collecting ducts. Water follows sodium ions out of the nephron into the peritubular capillaries. This, in turn, increases blood volume, which increases blood pressure. The increase in blood volume reduces renin release in a classic negative feedback mechanism, as shown in Figure 9-7.

As water leaves the filtrate and enters the blood, urine output decreases.

Aldosterone also stimulates sodium ion and water reabsorption in other sites, including the intestinal tract, salivary glands, and sweat glands, helping to restore blood volume.

Water balance is also affected by several chemicals present in the daily diet of many people, two of the most influential being caffeine and alcohol.

> **KEY CONCEPTS**
>
> Aldosterone is a hormone produced by the adrenal cortex; it also stimulates water reabsorption by the kidneys, helping to conserve body water.

Caffeine

Have you ever noticed that you have to urinate more frequently when drinking lots of coffee or Coke or Pepsi? The reasons for this are twofold. First, you are increasing your intake of liquids. Secondly, caffeine is a *diuretic* (DIE-yur-ET-ick), that is, a chemical that increases urination.

As you probably know, caffeine is found in coffee, most non-herbal teas, most soft drinks, and a whole host of energy drinks now flooding the market. Believe it or not, you'll even find caffeine in decaffeinated coffee, but in much smaller amounts. (A recent study of decaffeinated coffee showed that caffeine concentrations were about 5% to 8% of the levels found in caffeinated coffees.)

Caffeine increases urine production by increasing glomerular blood pressure. Higher pressure inside the glomerular capillaries increases the amount of fluid passing into Bowman's space, that is, increases glomerular filtration. As a result, more filtrate is formed.

Caffeine also decreases the tubular reabsorption of sodium ions in the distal convoluted tubule. As noted earlier, water follows sodium ions out of the distal convoluted tubule during tubular reabsorption. A decrease in sodium reabsorption thus results in a decline in the amount of water leaving the renal tubule and an increase in urine output.

> **KEY CONCEPTS**
>
> Caffeine increases blood pressure in the glomeruli, increasing the movement of water out of the blood, thus increasing urine production.

Ethanol

Like caffeine, ethyl alcohol (or ethanol, for short) is a diuretic. Ethanol is found in a variety of beverages, including beer, wine, wine coolers, and hard liquor like bourbon and gin. Ethanol passes into the bloodstream in the stomach and small intestine and circulates throughout the body. When it reaches the brain, ethanol inhibits the cells that secrete antidiuretic hormone. When ADH levels fall, water reabsorption declines. The result is a marked increase in water loss by the kidneys, giving credence to the quip that you don't buy wine or beer, you rent it!

> **KEY CONCEPTS**
>
> Ethanol increases urine output by decreasing ADH secretion from the pituitary.

9-6   Diseases of the Urinary System

As you now know, urine contains dissolved wastes and various ions. Ninety percent of the dissolved waste consists of three substances: urea, sodium ions, and chloride ions. Varying amounts of other chemical substances are also present, but in minute amounts.

In many ways, the urine provides a window to the chemical workings of the body. Increases or decreases in the level of substances in the urine thus may signal underlying problems—upsets in homeostasis and possible health problems. Consequently, physicians routinely analyze the urine of their patients to test for metabolic disorders. One of the most common is **diabetes mellitus** (DIE-ah-BEE-tees mell-EYE-tus), or sugar diabetes. This disease results in a defect in glucose uptake by cells in the body. Because body cells cannot absorb glucose, blood and urine glucose levels are elevated.

Like other body systems, the urinary system can malfunction. Tumors sometimes develop in the kidneys and urinary bladder. These and other disorders alter the function of this vital body system, creating a homeostatic nightmare.

This section discusses three of the most common disorders: kidney stones, kidney failure, and diabetes insipidus.

Health Tip 9-3

When you have a choice of seafood, select cold-water fish such as salmon, mackerel, herring, or sardines.

Why?

Studies show that cold-water fish species such as salmon have 20 to 30 times more omega-3 fatty acids and three to five times more vitamin D than other species, such as cod, tuna, freshwater fish, shrimp, lobster, and crayfish. Swedish researchers showed that women who consumed fatty cold-water fish at least once a week were 44% less likely to develop kidney cancer than women who ate no fish whatsoever. Other studies have found similar results. Omega-3 fatty acids also help lower the risk of heart disease and arthritis. Because of this, the American Heart Association recommends eating fish, especially particularly fatty fish such as mackerel, lake trout, herring, sardines, albacore tuna, and salmon at least twice a week. You can also increase your intake of omega-3 fatty acids by eating grass-fed beef and vegetables, beans, and tree nuts like pistachios.

Kidney Stones

Excess chemicals and ions in the urine not only signal problems elsewhere, they can damage the kidneys. For instance, higher than normal concentrations of calcium, magnesium, and uric acid in the urine, often caused by inadequate fluid intake, may crystallize inside the kidney, most often in the renal pelvis. Small deposits enlarge by a process called *accretion* (ah-CREE-shun)—the deposition of materials on the outside of the stones, causing them to grow in much the same way that a pearl grows in an oyster (Figure 9–8). These small crystals, or **kidney stones**, eventually grow into fairly large deposits.

Calcium stones are common and occur most often in young men between the ages of 20 and 30. Calcium often combines with other substances in the urine, such as oxalate.

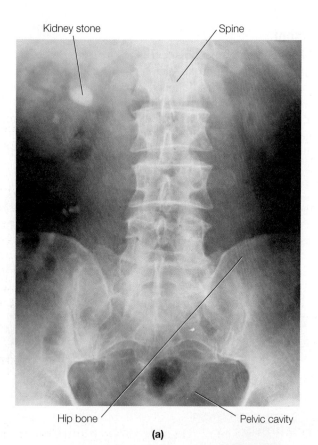

FIGURE 9–8 **Kidney Stones** (a) An X-ray of a kidney stone. (© NMSB/Custom Medical Stock Photo.) (b) Kidney stones removed by surgery. (© Eskimo71/Dreamstime.com.)

It is found in certain foods such as spinach and vitamin C supplements.

Kidney stones produce intense pain that starts suddenly and then often disappears just as quickly. Pain may be felt in the abdomen or the side and may move to the groin or testicles (in men).

Kidney stones are a health threat and an economic liability. Fortunately, many kidney stones are flushed out of the kidney into the urinary bladder and out the urethra on their own. Although small stones that enter the urinary bladder are often excreted in the urine, the sharp edges of the stone can dig into the walls of the ureters and urethra, causing pain.

Problems arise when stones become lodged inside the kidney or the ureters and obstruct the flow of urine. This causes an increase in internal pressure that may result in considerable damage to the nephrons if left untreated.

For years, kidney stones were removed surgically. Since the 1980s, however, the medical technique of *ultrasound lithotripsy* is generally used. In this technique, mentioned in the opening paragraph of this chapter, physicians bombard kidneys with ultrasound waves, which shatter the stones, producing fine, sandlike grains that are passed in the urine without incident. The procedure is nearly painless and much safer than surgery. Table 9–5 lists additional urinary system disorders.

> **KEY CONCEPTS**
>
> Kidney stones are deposits inside the kidneys that can block the flow of urine and cause serious damage if not removed.

Renal Failure

The importance of the kidneys is most obvious when they stop working, a condition known as **renal failure**. It results in a decrease in glomerular filtration and thus impairment of the kidney's ability to filter blood.

Renal failure may be acute—that is, it comes on quickly—or chronic—the kidneys deteriorate slowly over time. Acute renal failure has many causes. It may, for instance, be caused by toxic chemicals in the blood—for example, an overdose of drugs, including antibiotics and chemotherapeutic agents, or as a result of an immune reaction to certain antibiotics. In other cases, it is caused by severe kidney infections. Accidents in which huge amounts of blood are lost also cause renal failure due to the sudden decrease in blood flow to the kidneys. Surgical procedures such as coronary artery bypass transplants may also result in a decrease in blood flow to the kidneys, causing them to shut down. In acute renal failure, the kidneys lose function over a few hours or a few days.

The most common cause of chronic renal failure is diabetes mellitus, often referred to as sugar diabetes. It damages blood vessels in vital organs such as the kidneys. The second most common is chronic, uncontrolled high blood pressure. In chronic renal failure, kidney function deteriorates slowly over many years, resulting in chronic renal impairment. Chronic renal impairment may lead to a complete or nearly complete shutdown, known as end-stage failure.

Kidney failure (whether acute or end stage) is life-threatening, for when the kidneys stop working, water and toxic wastes begin to accumulate in the body. This disrupts

TABLE 9-5	Common Urinary Disorders	
Disease	**Symptoms**	**Cause**
Bladder infections	Especially prevalent in women; pain in lower abdomen; frequent urge to urinate; blood in urine; strong smell to urine	Nearly always bacteria
Kidney stones	Large stones lodged in the kidney often create no symptoms at all; pain occurs if stones are being passed to the bladder; pains come in waves a few minutes apart	Deposition of calcium, phosphate, magnesium, and uric acid crystals in the kidney, possibly resulting from inadequate water intake
Kidney failure	Symptoms often occur gradually; more frequent urination, lethargy, and fatigue; should the kidney fail completely, patient may develop nausea, headaches, vomiting, diarrhea, water buildup, especially in the lungs and skin, and pain in the chest and bones	Immune reaction to some drugs, especially antibiotics; toxic chemicals; kidney infections; sudden decreases in blood flow to the kidney resulting, for example, from trauma
Pyelonephritis	Infection of the kidney's nephrons; sudden, intense pain in the lower back immediately above the waist, high temperature, and chills	Bacterial infection

homeostasis. If untreated, a patient will die in 2 to 3 days. Patients usually die from an increase in the concentration of potassium ions in the blood and tissue fluids. Why?

Although potassium is essential for the function of heart muscle, excess potassium destroys the rhythmic contraction of the heart, causing fibrillation. As you may recall from elsewhere in the text, fibrillation is the uncoordinated contraction of heart muscle that may render the heart ineffective as a pumping organ. Patients often die.

The treatment of renal failure varies, depending on the underlying cause. If the problem is caused by an acute loss of blood, transfusions may be required. Patients whose kidneys have shut down, even temporarily, may require renal dialysis (DIE-AL-eh-siss). In this procedure, blood is drawn out of a vein and passed through a piece of tubing that transports the blood to an artificial filter that removes harmful wastes. After filtration, the blood is pumped back into the patient's bloodstream. Dialysis requires several hours and must be repeated every 2 or 3 days. Some patients have dialysis units at home and simply hook themselves up each night before they go to bed.

Another method of treating renal failure is *continuous ambulatory peritoneal dialysis (CAPD)*. In this procedure, 2 liters of dialysis fluid are injected into a person's abdomen through a permanently implanted tube or catheter (CATH-eh-ter). Waste products diffuse out of blood vessels into the abdominal cavity across the peritoneum (PEAR-eh-tah-KNEE-um), a thin membrane that lines the organs and wall of the abdominal cavity. The fluid, containing waste products, is drained from the abdominal cavity a couple of times a day.

This form of dialysis is much simpler. Patients can take care of it themselves. In addition, it allows for more frequent filtering of the blood, although there is a slightly higher risk of infection than dialysis performed in a facility by professionals.

Complete kidney failure can be treated by kidney transplantation. Transplants are generally most successful when they come from closely related family members. Such transplants are not rejected by the immune system.

Until the 1960s the complete or nearly complete destruction of kidney function was almost always fatal. Thanks to the technological advances in renal dialysis and kidney transplantation, many patients today can live fairly normal, healthy lives. These procedures, especially transplants, are costly, however. The public currently picks up the tab for individuals without insurance—or they go without needed treatments and die. For a debate on prioritizing medical expenditures, see this chapter's Point/Counterpoint.

> **KEY CONCEPTS**
>
> Renal failure may occur suddenly or gradually and is treated by periodic renal dialysis, artificially cleansing the blood.

Diabetes Insipidus

Hormonal imbalances can also lead to disruptions in the urinary system. Severe head injuries, for example, may result in a reduction or a complete cessation of ADH secretion. This results in a disease known as **diabetes insipidus** (DIE-ah-BE-teas in-SIP-eh-duss). This condition is not to be confused with diabetes mellitus (commonly called sugar diabetes), a disorder involving the hormone insulin.

Insufficient ADH output results in excess water loss—up to 20 liters (5 gallons) per day. (Normal urine output is 1 to 3 liters per day!) Because so much water is lost, the urine is dilute and colorless. To replace this water, a person must consume large quantities of liquids.

Diabetes insipidus can also result from an insensitivity of the kidneys to ADH. This disorder is called *nephrogenic diabetes insipidus*. Two other types are also encountered.

Diabetes insipidus can be treated in a variety of ways, depending on the severity of the disorder and the cause.

Point/Counterpoint 9-1 Prioritizing Medical Expenditures

We Need to Learn to Prioritize Medical Expenditures *by Richard D. Lamm*

Heath care in the United States finds itself in (to paraphrase Dickens)—the best of times and the worst of times. It is the epoch of medical miracles, yet at the same time millions go without even basic health care.

The basic dilemma of U.S. medicine is that we have invented more health care than we can afford to deliver. Healthcare costs are rising at three times the rate of inflation. They are absorbing funds desperately needed elsewhere in our country to educate our kids, rebuild our infrastructure, and revitalize our industries.

Healthcare costs are rising for many reasons. One major reason is that no one prioritizes the myriad procedures that health care can deliver. No one asks, "How do we buy the most health for the most people with our limited funds?" Policy decisions in Oregon and California, relating to the public funding of transplant operations, illustrate how two states attempted to deal with a crisis in health care.

Richard D. Lamm is the director of the Center for Public Policy and Contemporary Issues at the University of Denver, where his primary research and teaching interests have been in the area of health policy. Before assuming his position, he served three terms as governor of Colorado, from 1975 until 1987. (© Richard D. Lamm.)

Oregon received adverse publicity for its decision not to publicly fund soft tissue transplant operations. As is often the case, much of the focus has been on the handful of individuals who have been adversely affected by the policy rather than the numerous, but anonymous, people who will benefit. The Oregon policy sparked a public outcry over society's lack of compassion for individuals who desperately need transplants. Yet, when Oregon policymakers weighed the needs of a few transplant patients against the basic healthcare needs of the medically indigent, they decided against the former in favor of the latter. They have now set up a process that sets priorities throughout the healthcare system.

California took a contrary approach. Policymakers in California decided to publicly fund transplant operations. Then, 1 week later, they removed 270,000 low-income people from their state's medical assistance program.

Which was the wiser decision? The answer seems clear. Oregon bought more health for more of its citizens for its limited funds. Some say Oregon should have raised its taxes and funded transplants. That is very easy for nonpoliticians to say. Polls show people believe all Americans should have access to quality healthcare. But the polls also show that most are unwilling to accept even modest tax increases to provide the care they say they favor.

Illinois passed legislation giving "universal access to major organ transplants," but appropriated less than one-third of the funds needed. Politically, of course, Illinois played it far safer than Oregon but, in doing so, failed to confront one of the most pressing social issues of our time. Avoidance may be a politically expedient tactic, but sooner or later our society will be forced to allocate scarce health resources.

Clearly, medical science is inventing faster than the public is willing or able to pay. Realistically, no system of health care can avoid rationing medicine. In fact, rationing already exists. But instead of rationing medicine according to need or a patient's prospects for recovery, we ration by seniority and ability to pay. Those who pretend that we do not ration medicine forget that 43 million Americans do not have full access to health care. Most states have staggering numbers of medically indigent, yet they provide a small percentage with a full program of coverage. Oregon is the first state to cover 100% of the people living under the federal poverty line with basic healthcare.

I suggest this yardstick: Which state best asked itself, "How do we buy the most health for the most people?" Which state tried to maximize limited resources? Which state benefited the greatest number of people and which state harmed the greatest number of people?

Oregon attempted to weigh the basic health needs of the medically indigent against other programs for which the state paid. It recognized that the money the state paid for transplants and other high-cost procedures for a few people could buy more health care elsewhere, providing basic low-technology services to people not covered at that time.

Prioritizing health care does not abandon the poor, it seeks to serve the largest number with the more effective procedures.

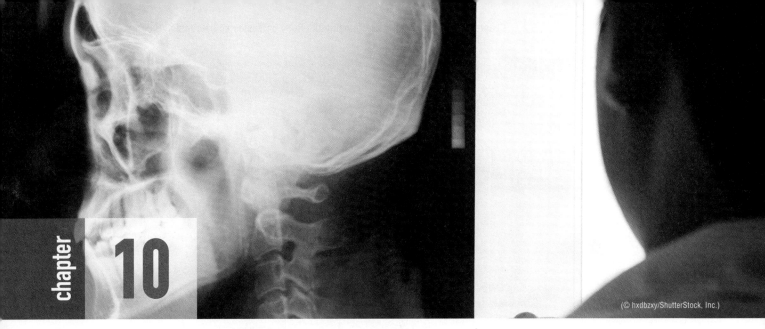

(© hxdbzxy/ShutterStock, Inc.)

The Skeleton and Muscles

Thomas Powers is a 40-year-old college professor. Like many of his friends and colleagues at the University where he teaches, he likes to ski. But lately, his skiing has been curtailed by chronic knee pain from an old skiing injury. After visiting his doctor, he found that, years ago, a spill on the slopes tore some cartilage in his knee. Although it never really healed, the tear caused few problems until he recently took up mountain biking. The additional strain combined with some degeneration of the knee joint conspired to create persistent pain that became so severe it disturbed Professor Powers' sleep.

THINKING CRITICALLY

A two-year study on osteoporosis in Australian women involving 120 nonsmokers, ranging in age from 50 to 60, concluded that the best prevention against bone fractures is estrogen-replacement therapy. (Estrogen is a female sex hormone that stimulates bone formation.) During the study, the researchers examined the density of bone in the forearms of the women at three-month intervals. Initial measurements showed very low levels of calcium, which indicates a high risk of fractures.

The women were then divided into three groups. The first group received a placebo, the second group received 1 gram of calcium per day, and the third group received daily doses of estrogen and another hormone, progesterone. All the women were encouraged to take two brisk 30-minute walks every day and to engage in one low-impact aerobics class each week.

The study showed that bone loss continued in the control group (group 1, exercise only) at a rate of 2.5% per year. Bone loss also continued in group 2, which received a daily dose of calcium, but was slower than in group 1 (0.5% to 1.3% per year). In group 3, whose participants exercised and received hormone therapy, bone density increased from 0.8% to 2.7% per year.

Given the potential side effects of hormone therapy, the researchers recommended that only women with the lowest bone densities should receive hormone therapy and that others should exercise and take calcium to reduce their risk of osteoporosis.

Can you pinpoint any potential flaws in the experiment's design or execution? How could they be corrected?

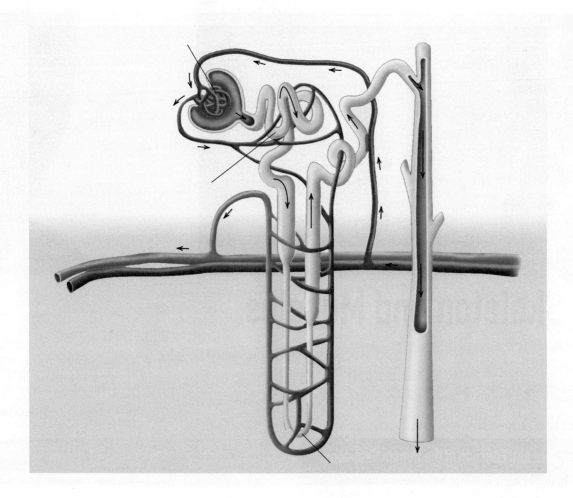

biology.jbpub.com/chiras/8e/

The site features eLearning, an online review area that provides quizzes, chapter outlines, and other tools to help you study for your class. You can also follow useful links for in-depth information, research the differing views in the Point/Counterpoints, or keep up on the latest health news.

Like the professor many of us will develop problems with our skeletal system, particularly our joints. They are prone to wear and tear. Carrying extra weight also speeds up their degeneration. This chapter discusses the human skeleton, including the joints, like those troubling Professor Powers and many readers, if not now, sometime in the future. It also examines the skeletal muscles. Together, the bones, joints, and muscles constitute the **musculoskeletal system**. In adults, bones and muscles make up a remarkable 50% to 60% of the body weight.

10-1 Structure and Function of the Human Skeleton

The human skeleton consists of bone and cartilage. The word **skeleton** is derived from a Greek word that means "dried-up body." Your first impression of bone might not be much different (Figure 10-1). The bones you've probably encountered in biology classes or on hikes through the woods after all, appear to be a dry, dead structure. But appearances can be deceiving. The truth be known, in live animals bone is a metabolically active tissue. Bone tissue contains numerous cells, known as **osteocytes** (OSS-tee-oh-SITES). These cells are embedded in an extracellular material, known as the **matrix** (MAY-tricks). The matrix consists of calcium phosphate crystals, which impart hardness and strength. It is deposited on an organic component, **collagen fibers** (COLL-ah-gin), which are made of protein. The collagen imparts flexibility.

The human skeleton consists of 206 bones, discrete structures made of bone tissue (Figure 10-2). Bones provide internal structural support, giving shape to our bodies and enabling us to maintain an upright posture. Some bones protect internal organs. The rib cage, for instance, protects the lungs and heart, and the skull protects the brain. Bones serve as the site of attachment for the tendons of many skeletal muscles whose contraction results in purposeful movements. Inside many bones are cells that give rise to red blood cells, white blood cells, and platelets. Bones are also a storage depot for fat, needed for cellular energy production both at work and at rest. Finally, bones are a reservoir of calcium and other minerals. Bones release and absorb calcium as needed, thus helping to maintain normal blood levels. That's important, in part, because calcium is required for muscle contraction. Disturbances in blood calcium levels can impair muscle contraction.

FIGURE 10-1 **Bone** When most people think about bones, they picture a dried-out, lifeless structure. In reality, bones are living tissues that perform many functions essential to homeostasis. (© Robynrg/ ShutterStock, Inc.)

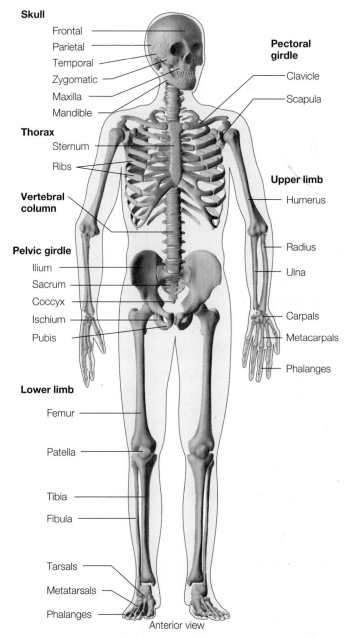

FIGURE 10-2 **The Human Skeleton** Over 200 bones of all shapes and sizes make up the human skeleton. The cartilage is shaded gray.

The Skeleton

The human skeleton consists of the axial and appendicular skeletons (Figure 10-2). The **axial skeleton** (AXE-ee-al) forms the long axis of the body. It consists of the skull, the vertebral column, and the rib cage. The **appendicular skeleton** (ah-pen-DICK-you-lar) consists of the bones of the arms and legs as well as the bones of the shoulders and pelvis, to which the upper and lower extremities are attached.

> **KEY CONCEPTS**
>
> The human skeleton is made of bones and consists of two parts, the axial and the appendicular.

Compact and Spongy Bone

Take a moment to study the skeleton in Figure 10-2. As you examine it, you may notice that bones come in a variety of shapes and sizes. Some are long, some are short, and some are flat and irregularly shaped. Nevertheless, all bones share some characteristics. Consider one of the long bones of the body, the humerus (HUGH-mer-us), which is located in the upper arm.

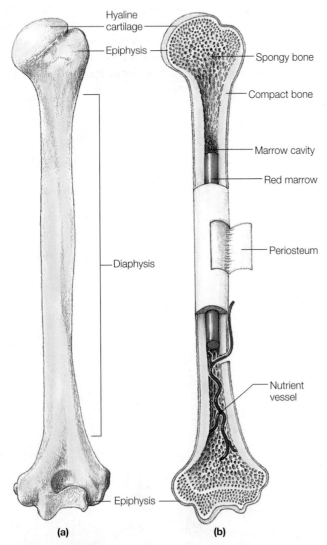

Hyaline cartilage

Epiphysis

Spongy bone

Compact bone

Marrow cavity

Red marrow

Periosteum

Diaphysis

Nutrient vessel

Epiphysis

(a) (b)

FIGURE 10-3 Anatomy of Long Bones (a) Drawing of the humerus. Notice the long shaft and dilated ends. (b) Longitudinal section of the humerus showing compact bone, spongy bone, and marrow.

Figure 10-3 illustrates the anatomy of the humerus. As shown, this bone, like other long bones of the body, consists of a long, narrow shaft, the *diaphysis* (die-AF-eh-siss), and two expanded ends, the *epiphyses* (ee-PIF-eh-seas). The ends of the epiphyses are covered with a thin layer of hyaline cartilage (discussed elsewhere in the text), which reduces friction in joints. Like other bones, the humerus consists of an outer shell of dense material called **compact bone**. Inside the humerus is a mass of **spongy bone**, so named because of its spongy appearance. Spongy bone forms an interlacing network that contains many small cavities. It is therefore less dense than compact bone.

On the surfaces of many bony spicules of spongy bone are many large, multinucleated cells called **osteoclasts** (OSS-tee-oh-KLASTS; "bone breakers"). As you may recall from elsewhere in the text, osteoclasts digest bone with the aid of enzymes. These cells are part of a homeostatic system that ensures proper blood calcium levels. When calcium levels in the blood fall, osteoclasts are activated by a hormone from the thyroid gland, known as parathyroid hormone (PTH). Once activated, the osteoclasts digest small portions of the spongy bone. This, in turn, releases calcium into the bloodstream. Calcium is necessary for a number of body functions, including muscle contraction.

On the outer surface of the compact bone is a layer of connective tissue, the **periosteum** (PEAR-ee-oss-TEA-um; "around the bone"). The periosteum serves as the site of attachment for skeletal muscles (via tendons). It also contains cells that make new bone during remodeling or repair.

The periosteum is richly supplied with blood vessels, which enter the bone at numerous locations. Blood vessels then travel through small canals in the compact bone and course through the inner spongy bone, providing nutrients and oxygen and carrying off cellular wastes such as carbon dioxide. The periosteum is also richly supplied with nerve fibers. The majority of the pain felt after a person bruises or fractures a bone results from stimulation of pain fibers in the periosteum.

Inside the shaft of long bones is a large cavity, the **marrow cavity** (MARE-row). The marrow cavities in most of the bones of the fetus and newborn contain **red marrow**, so named because of its color. Red marrow is tissue that serves as a blood-cell factory. It produces red blood cells, white blood cells, and platelets to replace those routinely lost each day. As an individual ages, most red marrow is slowly "retired" and becomes filled with fat, creating **yellow marrow**. Yellow marrow begins to form during adolescence and, by adulthood, is present in all but a few bones. Red blood cell formation, however, continues in the bodies of the vertebrae, the hip bones, and a few others. Yellow marrow can be reactivated to produce blood cells under certain circumstances—for example, after an injury.

> **KEY CONCEPTS**
>
> Bone consists of a dense outer layer known as compact bone, and a spongy inside known as spongy bone.

Meeting New Stresses

Bone is a dynamic tissue. As just noted, bones of your body store and release calcium, helping to maintain concentrations needed for normal body function. Bone also undergoes considerable remodeling to adjust for changes in activity. Spongy bone, for example, is "remodeled" to increase bone strength when a person's level of activity increases after periods of inactivity. The leg bones of a previously sedentary student who begins backpacking or hiking, for instance, are remodeled to accommodate the additional demands of her newfound hobbies. During remodeling, osteoclasts break down some of the spongy bone in the leg bones while another type of bone cell, **osteoblasts** (OSS-tea-oh-BLASTS), lay down new spongy bone to accommodate the stresses and strains of exercise. When an osteoblast becomes surrounded by calcified extracellular material, it is referred to as an osteocyte.

Compact bone is also remodeled by osteoclasts and osteoblasts when activity levels change, thus helping the entire bone structure withstand greater stress. Within a few weeks of starting an exercise program, our sedentary college student's bones will have been considerably remodeled and able to meet the demands of hiking. If she goes back to her sedentary ways at a later time, however, her bones will revert to their previous, weaker state. The compact bone will decrease in thickness and the spongy bone will change.

> **KEY CONCEPTS**
>
> Bones are remodeled for strength when exposed to new stresses.

The Joints

Mobility in humans and other animals is made possible by bone and muscles acting together. However, mobility also requires joints. **Joints** are the structures that connect the

Health Tip 10-1

For healthy bones and a longer, healthier life, exercise regularly and incorporate it into your daily activities. Take the stairs instead of an elevator or escalator. Walk or ride your bike to work or to school. Do sit-ups or ride a stationary bike when you are watching television.

Why?

Exercise may be hard to work into a busy schedule, but if you make it part of your daily routine, you'll get plenty. Regular exercise adds years to one's life and life to one's years—that means you'll be in better health as you age. You won't be suffering as much as your sedentary cohorts as you grow older. If you are overweight or out of shape, it is a good idea to seek a doctor's approval before you begin an exercise program. Start slowly. If you're terribly out of shape, start with 10 minutes of exercise every other day for a week or two. Then gradually increase the activity to 20 minutes. Starting out too quickly and working out too hard can cause soreness and pain and may discourage you from continuing.

bones of the skeleton. They can be classified by the degree of movement they permit. Those that permit no movement are *immovable joints*, those that allow little movement are *slightly movable*, and those that allow free movement are simply referred to as *movable joints*. Consider some examples.

The bones of the skull are shown in **Figure 10-4a**. As illustrated, these bones interlock to form immovable joints,

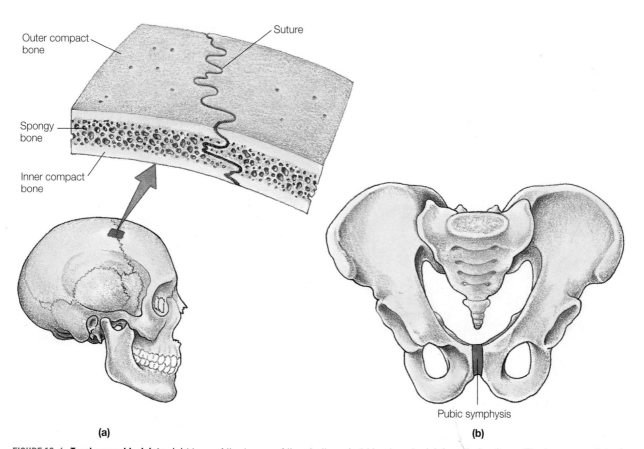

(a)

(b)

FIGURE 10-4 Two Immovable Joints (a) Many of the bones of the skull are held in place by joints called *sutures*. The bones are linked by fibrous tissue, and the joints are immovable. (b) The pubic symphysis is another immovable joint. During childbirth it softens and expands to permit delivery.

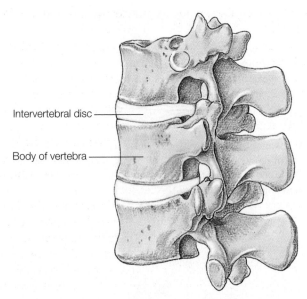

FIGURE 10–5 **A Slightly Movable Joint** The intervertebral discs allow for some movement, giving the vertebral column flexibility.

known as *sutures*. Fibrous connective tissue spans the space between the interlocking bones, holding them together. Another immovable joint is the pubic symphysis (PEW-bick SIM-fa-siss), formed by the two pubic bones (Figure 10-4b). These bones are held in place by fibrocartilage. Near the end of pregnancy, however, hormones loosen the fibrocartilage of the pubic symphysis, allowing the pelvic outlet to widen enough to permit a baby to pass through—an adaptation of enormous value to our species!

The bodies of the vertebrae are united by slightly movable joints (**Figure 10-5**). Each vertebra is separated from its nearest neighbors by an **intervertebral disc**. The inner portion of the disc acts as a cushion, softening the impact of walking and running. The outer, fibrous portion holds the disc in place and joins one vertebra to its nearest neighbor. The joints between

the vertebrae offer some degree of movement, resulting in a fair amount of flexibility. If they did not, we would be unable to bend over to tie our shoes or to curl up on the couch for an afternoon nap.

The most common type of joint is the freely **movable joint**, or **synovial joint** (sin-OH-vee-al). The synovial joints are more complex than other types and permit varying degrees of movement. Although synovial joints differ considerably, they share several features. The first commonality is the hyaline cartilage located on the articular (joint) surfaces of the bones (**Figure 10-6a**). This thin cap of hyaline cartilage reduces friction, which facilitates movement. The second commonality is the **joint capsule**, a structure that joins one bone to another (Figure 10-6a). The outer layer of the joint capsule consists of dense connective tissue that attaches to the periosteum of adjoining bones. In many joints, parallel bundles of dense connective tissue fibers in the outer layer of the capsule form distinct ligaments. A **ligament** is a structure that joins bones to bones together in joints. They provide additional support to the joint. (You can remember this by the initials of the American president LBJ—ligaments connect bones to bones at joints.)

As a rule, ligaments are fairly inflexible. Some individuals have remarkably flexible ligaments and tendons, however. (**Tendons** are connective tissue that joins muscles to bones at their points of origin and insertion.) The increased flexibility of ligaments and tendons results in extraordinary flexibility. For example, these individuals are often able to extend their fingers and thumbs so much that they can touch the back of their hand. People with this ability are said to be "double-jointed," a misnomer that suggests the flexibility is permitted by the joint when in fact it is due to the tendons and ligaments.

The inner layer of the joint capsule is the **synovial membrane**. It produces a fairly thick, slippery substance called **synovial fluid**. Synovial fluid provides nutrients to the articular cartilage (the hyaline cartilage on the articular surfaces of the bone) and also acts as a lubricant. Normally, the synovial membrane produces only enough fluid to create a thin film on

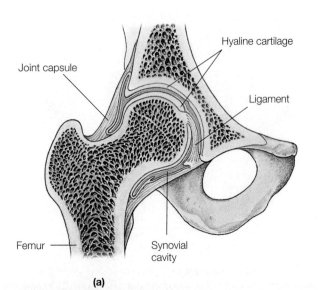

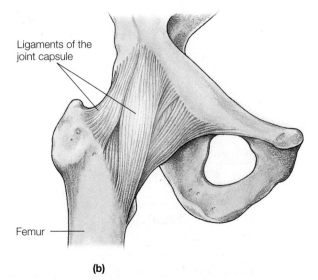

(a) (b)

FIGURE 10–6 **A Synovial Joint** (a) A cross section through the hip joint (a ball-and-socket joint) showing the structures of the synovial joint. (b) Ligaments in the outer portion of the joint capsule help support the joint.

the articular cartilage. Injuries to a joint, however, can result in a dramatic increase in synovial fluid production, causing swelling and pain in joints.

Even though synovial joints allow for movement, they require some stability. Stability is provided by the ligaments of the joint capsule, tendons, and, in some cases, nearby muscles. In the shoulder, for example, muscles help hold the head of the arm bone or humerus in place. Because muscles strengthen joints, individuals who are in poor physical shape are much more likely to suffer a dislocation during exercise or when engaged in sports than someone who is in good shape. A **dislocation** is an injury in which a bone is displaced from its proper position in a joint because of a fall or some other unusual body movement. In some instances, bones slip back in without assistance. In others, the bone can be put back in place usually with the assistance of a trained healthcare worker.

Synovial joints come in many shapes and sizes and are classified on the basis of structure. Two of the most common are the hinge joint and the ball-and-socket joint. The knee joint is a hinge joint, as are the joints in the fingers. **Hinge joints** open and close like hinges on a door. Such joints provide for movement in only one plane. The hip and shoulder joints are **ball-and-socket joints**. They allow a wider range of motion. (Compare the movements permitted by the shoulder joint to those permitted by the knee joint.)

> **KEY CONCEPTS**
>
> Joints connect bones in the body; some allow no movement while others allow varying degrees of flexibility.

Repairing Damaged Joints

Joint injuries are common among physically active people, especially athletes (Table 10-1). A hard blow to the knee of a football player, for instance, can tear the ligaments in the knee joint.

Torn ligaments, tendons, and cartilage in joints heal very slowly because they have few blood vessels to bring nutrients needed for repair. For years, joint repair required major surgery that was so traumatic it put patients out of commission for several months. Today, however, **arthroscopic surgery** allows physicians to repair joints with a minimum of trauma (Figure 10-7). Through small incisions in the skin over the joint, surgeons insert a device called an **arthroscope** (ARE-throw-scope), which allows them to view the damage inside the joint. Through another tiny incision, the surgeon inserts another instrument that is used to remove the damaged cartilage. The surgeon can thus repair damaged cartilage without making a large incision to open the joint, as was common previously. Athletes are usually back on their feet and on the playing field in a matter of weeks. New surgical techniques can rebuild torn ligaments and return joints to nearly their original state. During this procedure, small

TABLE 10-1	Common Injuries of the Joints	
Injury	**Description**	**Common Site**
Sprain	Partially or completely torn ligament; heals slowly; must be repaired surgically if the ligament is completely torn.	Ankle, knee, lower back, and finger joints
Dislocation	Occurs when bones are forced out of a joint; often accompanied by sprains, inflammation, and joint immobilization. Bones must be returned to normal positions.	Shoulder, knee, and finger joints
Cartilage tears	Cartilage may tear when joints are twisted or when pressure is applied to them. Torn cartilage does not repair well because of poor vascularization. It is generally removed surgically; this operation sometimes makes the joint less stable.	Knee joint

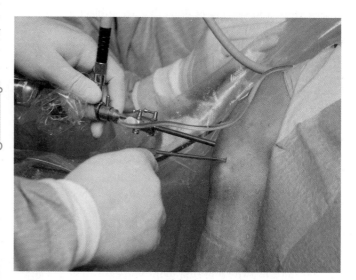

(a)

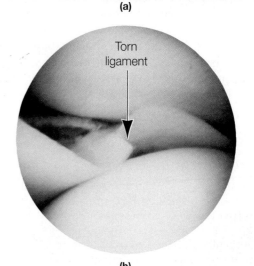

Torn ligament

(b)

FIGURE 10-7 Arthroscopic Surgery (a) A physician performing arthroscopic surgery. (© piotrwzk/ShutterStock, Inc.) (b) Inside view of knee joint through an arthroscope showing torn ligament. (© CNRI/Science Source, Inc.)

sections of ligaments from other parts of the body are removed and sutured into the damaged joint.

Arthritis

Doctors recognize over 100 different types of arthritis. One of the most common is **osteoarthritis (OA)**. Osteoarthritis or degenerative joint disease results from wear and tear on a joint or from an injury. Excess wear, for instance, causes the articular cartilage on the ends of bones to flake and crack. Hyaline cartilage is the firm yet rubbery tissue that performs two functions: it cushions bones at joints and also reduces friction as bones move over one another. As the cartilage degenerates, the bones come in contact and grind against each other during movement. This causes considerable swelling, pain, and discomfort.

OA occurs most often in the weight-bearing joints of the knees, hips, and spine—regions that are subject to the most wear. Osteoarthritis may also develop in the finger joints.

OA is extremely common in adults and is due to aging. In fact, X-ray studies of people over 40 years of age show that most people have some degree of degeneration in one or more joints. Fortunately, many people do not even notice this condition, and the disease rarely becomes a serious medical problem.

OA tends to run in families—that is, it is genetic. It may also occur as a result of previous injury, such as bone fractures or injuries to joints. Wear and tear on joints is often seriously worsened by obesity. The extra pressure of the weight wears away the cartilage more quickly, causing intense pain and suffering. Because it hurts so much to walk, many people become less mobile, which only piles on more weight, aggravating an already painful situation

Eating a healthy diet that helps you control your weight early in life can reduce one's chances of developing severe, debilitating arthritis. For those who are already overweight, weight loss can reduce the rate of degeneration and ease the pain. Painkillers such as aspirin and other anti-inflammatory drugs are commonly used to treat the pain and swelling. Injec-

tions of steroids may help reduce inflammation, although repeated injections can damage the joint.

Rheumatoid Arthritis

Another common disorder of the synovial joint is rheumatoid arthritis (RA). **Rheumatoid arthritis (RA)** (RUE-mah-toid) is the most painful and crippling form of arthritis. It is caused by an inflammation of the synovial membrane. The inflammation often spreads to the articular cartilage. If the condition persists, RA can wear through the cartilage and cause degeneration of the underlying bone. Thickening of the synovial membrane and degeneration of the bone often disfigure the joints, reduce mobility, and cause considerable pain (Figure 10-8). Afflicted joints can become completely immobilized. In severe cases, the bones become dislocated, causing the joints to collapse.

RA generally occurs in the joints of the wrist, fingers, and feet. It can also affect the hips, knees, ankles, and neck. RA afflicts an estimated 1.3 million Americans. Although it usually afflicts middle aged adults, it can appear in one's 20s and 30s. RA is much more common in women than men—at least 2.5 times more prevalent in women.

Research suggests that rheumatoid arthritis results from an autoimmune reaction—that is, an immune response to the cells of one's own synovial. It also tends to run in families, so it may be inherited from one's parents. Rheumatoid arthritis is usually a permanent condition, although the degree of severity varies widely. Because damage to the joints occurs within the first two years, treatment should commence early. Treatments include rest, joint protection (e.g., weight control), heat, physical therapy, painkillers, and medication. Medications include anti-inflammatory drugs (like aspirin), low doses of

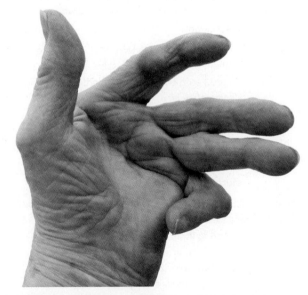

FIGURE 10-8 Hand Disfigured by Rheumatoid Arthritis Degeneration of the finger joints causes mild to extreme disfiguration. (© Peterfactors/Dreamstime.com.)

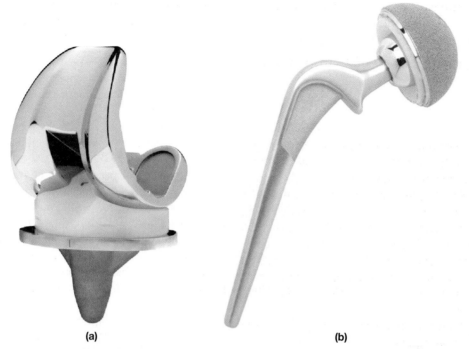

(a)   **(b)**

FIGURE 10-9 **Artificial Joints** An artificial knee joint (a) and hip joint (b). (Photos courtesy of DePuy Orthopaedics, Inc. Used with permission.)

corticosteroids like prednisone, and antirheumatic drugs. Severe damage can be addressed by joint replacement surgery.

Diseased joints can also be replaced by *prostheses* (pross-THEE-seas), artificial joints that restore mobility and reduce pain. Plastic joints are used to replace the finger joints, greatly improving the appearance of the hands by eliminating the gnarled, swollen joints. Moreover, patients regain the use of previously crippled fingers. Day-to-day chores (such as buttoning a shirt) that often required assistance become noticeably easier. Once impossible tasks—such as opening screw-top jars—are again feasible. Severely damaged knee and hip joints are replaced with special steel or Teflon substitutes, which, if fitted properly, may last 10 to 15 years (Figure 10-9).

In knee-replacement surgery, the ends of the bones are first sawed off. Artificial joints are made of plastic, metal, or both. The joint is fitting into the existing bone and may be cemented into place or not cemented so that the existing bone will grow into it. In some cases, surgeons use both methods to ensure that the new joint is firmly attached. In a hip replacement, the end of the leg bone, the femur, is sawed off and replaced with an artificial ball and socket joint. The socket into which the head of the femur once fit is replaced with a Teflon-coated socket.

> **KEY CONCEPTS**
>
> Rheumatoid arthritis is an autoimmune disease that results in an inflammation of the synovial membrane of movable joints.

Bone Formation

During embryonic development, hyaline cartilage forms in the arms, legs, head, and torso where bone will eventually be. The cartilage is then converted to bone. This process is known as **endochondral ossification** (en-doh-CON-drall). It continues after birth.

In a newborn baby, the bones of the leg (the tibia and fibula) are bowed. Cramped inside the mother's uterus, the baby's bones do not grow very straight. During the first 2 years of life, however, as the child begins to walk and run, the leg bones generally straighten (Figure 10-10). The bones are remodeled

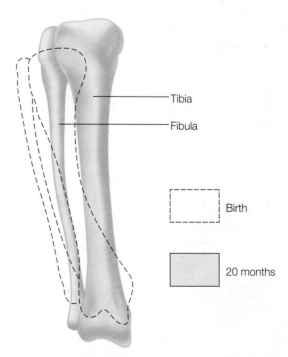

FIGURE 10-10 **Remodeling of a Baby's Leg Bones** Notice how dramatically the bones change to meet the changing needs of the toddler.

by osteoclasts and osteoblasts to meet the markedly different stresses placed on them by upright posture.

> **KEY CONCEPTS**
>
> Most of the bones of the human skeleton start out as hyaline cartilage.

Maintaining Calcium Levels

In addition to their role in remodeling bone, osteoblasts and osteoclasts help control blood calcium levels and are therefore part of a homeostatic mechanism. They are controlled by two hormones: parathormone and calcitonin.

When blood calcium levels fall, the parathyroid glands (small glands embedded in the thyroid glands in the neck) release a hormone known as **parathormone** (PAIR-ah-THOR-moan; **PTH**) into the bloodstream. PTH travels throughout the body. When it reaches the bone, it stimulates the osteoclasts, causing them to digest bone in their vicinity. The calcium released by the activity of the osteoclasts replenishes blood calcium levels.

When calcium levels rise—for example, after a meal—the thyroid gland (a large endocrine gland in the neck) releases **calcitonin** (CAL-seh-TONE-in). This hormone inhibits osteoclasts, stopping bone destruction. It also stimulates osteoblasts, causing them to deposit new bone. The inhibition of

osteoclasts and the formation of new bone help lower blood calcium levels, returning them to normal. This process is described in more detail elsewhere in the text.

> **KEY CONCEPTS**
>
> Blood levels of calcium are controlled in large part by the release and absorption of calcium from the bones, a process that is under hormonal control.

Repairing Bone Fractures

If you're like most people, chances are that some time in your life you'll break a bone. Bone fractures can vary considerably in their severity. Some only involve hairline cracks, which mend fairly quickly. Others involve considerably more damage—a complete break, for instance—that takes much longer to repair. In some instances, the broken ends must be realigned. A plaster cast is then applied to hold the limb in position and keep the ends together to optimize healing. Severe fractures may require that surgeons insert steel pins or screws to hold the bones together.

Bones, like other tissues, are capable of self-repair. Figure 10-11 illustrates the process. As shown in the first drawing on the left, after a fracture, blood flows into the fracture. It comes from broken blood vessels in the periosteum

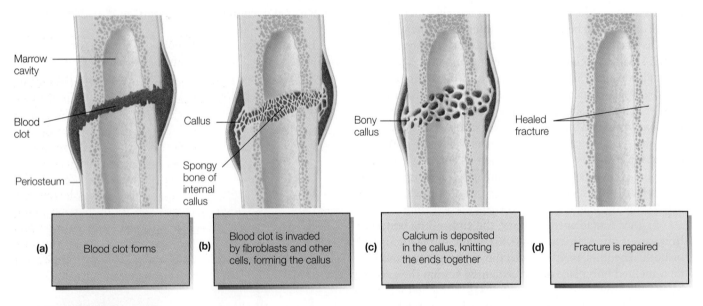

Marrow cavity

Blood clot

Periosteum

Callus

Spongy bone of internal callus

Bony callus

Healed fracture

(a) Blood clot forms

(b) Blood clot is invaded by fibroblasts and other cells, forming the callus

(c) Calcium is deposited in the callus, knitting the ends together

(d) Fracture is repaired

FIGURE 10-11 **Stages in Fracture Repair** (Photo © Robert O. Brown Photography/ShutterStock, Inc.)

and marrow cavity pours into the fracture. A blood clot forms. Within a few days, the clot is invaded by **fibroblasts**, connective tissue cells from the periosteum. Fibroblasts secrete collagen fibers, thus forming a mass of cells and fibers, the **callus** (CAL-us), which bridges the broken ends internally and externally. This is shown in the second drawing from the left.

The callus is next invaded by osteoblasts from the periosteum. Osteoblasts slowly convert the callus to bone, thus "knitting" the broken ends together. As Figure 10-11 shows, the callus is initially much larger than the bone itself. But the excess bone is gradually removed by osteoclasts, leaving little, if any, evidence that a break has occurred.

> **KEY CONCEPTS**
>
> Bone fractures repair naturally as a result of two cells, fibroblasts and osteoblasts.

Osteoporosis

One of the most common problems adults face is **osteoporosis** (OSS-tea-oh-puh-ROE-siss; "porous bone"), a condition characterized by a progressive loss of bone calcium (Figure 10-12). Bone loss begins rather early in life, especially in women. By the time women are in their thirties, many have experienced a 33% decrease in bone density. As they age, further losses occur. In some individuals, calcium loss is so severe that bones become extremely brittle—so much so that even normal activities such as getting out of bed in the morning or doing house-work cause fractures.

Osteoporosis occurs in nearly one of every 10 American adults over the age of 50. It is more prevalent in women. Sixteen percent of women and 4% of men ages 50 and older have osteoporosis, according to a report from the Centers for Disease Control and Prevention (CDC). Osteoporosis is common in postmenopausal women. Menopause (MEN-oh-pause) usually occurs between 45 and 55 years of age. It results from the loss of estrogen production by a woman's ovaries. Although estrogen is primarily a reproductive hormone, it also helps to maintain bone.

Osteoporosis also occurs in people who are immobilized for long periods. Hospital patients who are restricted to bed for 2 or 3 months, for example, show signs of osteoporosis.

Osteoporosis may also result from environmental factors, too. In the 1960s, women living along the Jintsu River

in Japan developed a painful bone disease known as itai-itai (which literally means "ouch, ouch"). The women lived downstream from zinc and lead mines that released large amounts of the heavy metal cadmium into the river. They used river water for drinking and for irrigating rice paddies. Even though men, young women, and children were exposed to cadmium, 95% of the cases of itai-itai occurred in postmenopausal women.

> **KEY CONCEPTS**
>
> Osteoporosis is a thinning of the bone that results from a loss of calcium; bones become brittle and easily fractured.

Researchers hypothesized that the accelerated bone loss in the postmenopausal Japanese women was caused by an overdose of cadmium. They tested this hypothesis by feeding mice diets containing various levels of cadmium chloride. The group receiving the highest dose showed significant reductions in bone calcium when their ovaries had been removed.

These findings may also explain why older women who smoke are more likely to develop osteoporosis and why they experience more bone fractures and tooth loss than non-smokers. Cadmium is one of several harmful substances present in cigarette smoke. Smoking, however, may also decrease estrogen levels, making it doubly dangerous to a woman's health. For a discussion of ways to prevent osteoporosis, see Health Note 10-1.

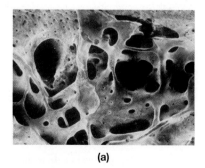
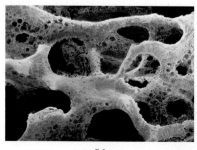

(a)   (b)

FIGURE 10-12 Osteoporosis The loss of estrogen or prolonged immobilization weakens bone. In these situations, bone is dissolved and becomes brittle and easily breakable. (a) Normal bone; (b) bone weakened by osteoporosis. (Photo a, © Photo Insolite Realite/Science Source, Inc. Photo b, © Professor Pietro M. Motta/Science Source, Inc.)

health**note**

10-1 Preventing Osteoporosis: A Prescription for Healthy Bones

Twenty-eight million Americans, mostly women, suffer from osteoporosis (Figure 1). Caused by a gradual deterioration of the bone, osteoporosis results in nagging pain, discomfort, and loss of function.

Each year, nearly 60,000 Americans—again, most of them women—will die from complications resulting from this common malady. Unfortunately, most women do not realize they have the disease until it has progressed quite far.

Recent research shows that osteoporosis begins much earlier than researchers once thought—by the time a woman reaches her mid-twenties. Bone demineralization occurs very rapidly. In fact, by age 30, many women have lost one-third of their bone calcium! Between the ages of 30 and 50, many women's bones continue to deteriorate, becoming extremely brittle.

Calcium loss begins so early because many weight-conscious women in their mid-twenties avoid foods such as whole milk, cheese, and ice cream in an effort to control their weight. Although these foods are rich in calories, including fat, they are also a major source of calcium. Milk products are also shunned by many adults who develop an intolerance to lactose (a sugar) in milk and other dairy products. Lactose intolerance results from a sharp reduction in the production and secretion of the intestinal enzyme lactase as one ages, which makes it difficult to digest lactose. Because of these and other factors, adult women often consume only about one-half of the 1,000–1,500 milligrams of calcium they need every day.

Osteoporosis can be prevented and even reversed by eating calcium-rich foods, such as spinach, milk, cheese, oysters, and tofu (soybean curd).

Calcium supplements can also help retard or even stop bone deterioration and restore calcium levels. Because vitamin D increases the absorption of calcium in the intestines, vitamin supplements can help. Table 1 shows the recommended intake of calcium and vitamin D based on age.

Studies also suggest that osteoporosis can be prevented by exercise. To increase bone strength you can walk, hike, jog, play tennis, dance, climb stairs, and lift weights. If you smoke, consider cutting back or quitting altogether, for reasons that will become clear shortly.

Osteoporosis can be reversed by exercise even after the disease has reached the dangerous stage. Only 45 minutes of moderate exercise (walking) 3 days a week, for example, decreases the rate of calcium loss in older individuals and can stimulate the rebuilding process, replacing calcium lost in previous years.

Another effective treatment for postmenopausal women is estrogen. Low doses of estrogen promote bone formation. Because women who are given estrogen suffer an increased risk of cancer of the uterine lining, physicians often prescribe a mixed dose of estrogen and progesterone, which reduces the likelihood of this cancer.

Studies have also shown that high doses of calcium fluoride stimulate bone development. Calcium fluoride treatment increases bone mass approximately 3% to 6% per year and decreases bone fractures. The average patient in one study experienced one fracture every 8 months before treatment. After treatment, that figure dropped to one fracture every 4.5 years.

Unfortunately, large doses of fluoride may erode the stomach lining, causing internal bleeding. They may also stimulate abnormal bone development and cause pain and swelling in joints. To offset these problems, researchers have developed a pill that releases the fluoride gradually.

For millions of young women, early detection and sound preventive measures, including exercise, vitamin D supplements, dietary improvements, and fluoride treatments, can prevent this unnecessary disease.

(© Val Thoermer/ShutterStock, Inc.)

TABLE 1	Recommended Calcium and Vitamin D	
Life-stage group	Calcium mg/day	Vitamin D (IU/day)
Infants 0 to 6 months	200	400
Infants 6 to 12 months	260	400
1 to 3 years old	700	600
4 to 8 years old	1,000	600
9 to 13 years old	1,300	600
14 to 18 years old	1,300	600
19 to 30 years old	1,000	600
31 to 50 years old	1,000	600
51- to 70-year-old males	1,000	600
51- to 70-year-old females	1,200	600
> 70 years old	1,200	800
14 to 18 years old, pregnant/lactating	1,300	600
19 to 50 years old, pregnant/lactating	1,000	600

Definitions: mg = milligrams; IU = International Units
Source: Food and Nutrition Board, Institute of Medicine, National Academy of Sciences, 2010.

10-2 The Skeletal Muscles

The muscles in your body enable you to get around, play music, engage in sports, send text messages, and much more. Most muscles are attached to the skeleton, so it is the combination of bone, joints, and muscles that permit us to move about and perform all the tasks we need to work, play, drink, and eat. Muscles that attach to bone are appropriately referred to as **skeletal muscles**. They are virtually all

are under conscious control thanks to the nervous system. **Figure 10-13** shows many of the skeletal muscles of the body. Take a moment to study it and identify some of the major muscles of your body.

As you can see from the illustration, most skeletal muscles cross one or more joints. As a result, when they contract, they produce movement.

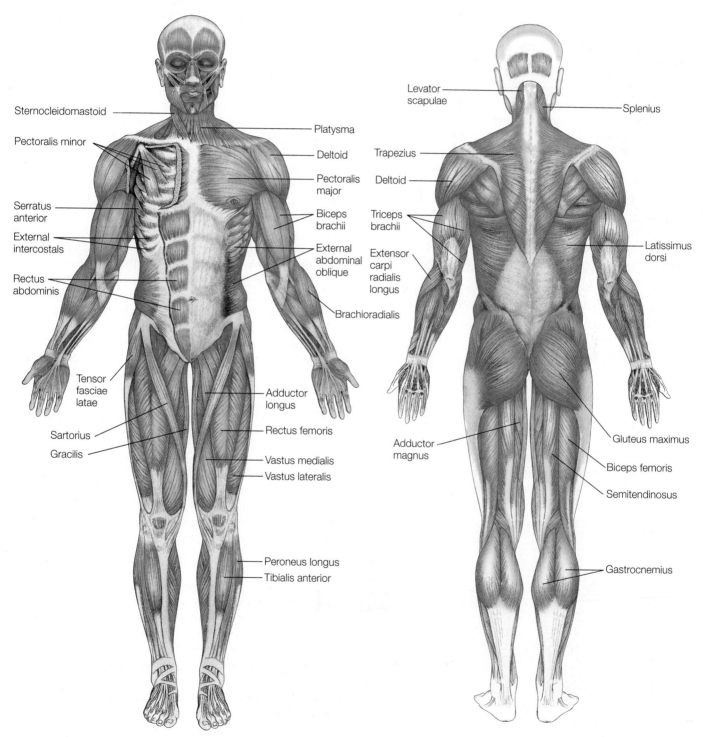

FIGURE 10-13 The Skeletal Muscles

As a rule, muscles generally work in groups to bring about various body movements. Groups of muscles are often arranged in such a way that one set produces one movement and another set on the opposite side of the joint causes an opposing movement. For example, the biceps (BUY-seps) is a muscle in the front of the upper arm. Locate it now. When it and other members of its group contract, they cause the arm to bend, a movement called *flexion* (FLECK-shun). On the back side of the upper arm is another muscle, the *triceps* (TRY-seps). Locate it as well on the drawing and on yourself. When the triceps contracts, it causes the arm to straighten, or extend. Such opposing muscles are called antagonists. When one muscle contracts to produce a movement, the antagonistic muscles relax.

Not all skeletal muscles make bones move. Some muscles simply steady joints, allowing other muscles to act. These muscles help to maintain our posture, permitting us to stand or sit upright despite the constant pull of gravity.

Another group of muscles that doesn't move bones is the muscles of the face. Anchored to the bones of the skull and to the skin of the face, these muscles allow us to wrinkle our skin, open and close our eyes, and move our lips. They're the muscles that help us eat and give us a wide variety of facial expressions, so crucial to human communication. They allow us to show anger, happiness, stress, frustration, and more.

Muscles also produce enormous amounts of heat as a byproduct of metabolism. Working muscles produce even more heat—so much, in fact, that you can cross-country ski or jog in freezing weather wearing only light clothing.

> **KEY CONCEPTS**
>
> Skeletal muscles are for the most part under conscious control; they connect to bones and when contracted permit a wide range of motions.

Anatomy of a Skeletal Muscle

Skeletal muscles consist of long, unbranched **muscle fibers**. Each fiber contains many nuclei derived from individual cells that fuse during embryonic development. Viewed with the light microscope, skeletal muscle fibers appear striped or striated (STRY-a-tid; Figure 10-14).

Like nerve cells, the muscle fiber is an excitable cell. Muscle fibers are stimulated by nerve impulses arriving form the nerves that terminate on them. The ends of these

nerve cells contain many small swellings known as *terminal boutons*. Nerve impulses stimulate the release of a chemical substance from the terminal bouton. This neurotransmitter substance stimulates an impulse in the muscle fiber. It travels along the membrane of the muscle.

Muscle fibers are also contractile. When stimulated, the contractile proteins inside the fibers cause the cells to shorten. Muscle fibers are also elastic and therefore capable of returning to normal length after a contraction has ended.

> **KEY CONCEPTS**
>
> Skeletal muscle cells actually consist of many cells so are called muscle fibers; they are packed with contractile proteins that allow the muscles to contract when stimulated, producing movement.

Muscle Fibers

Understanding how a muscle contracts first requires an understanding of its structure. To begin, imagine that you could tease a single skeletal muscle fiber (cell) free from a skeletal muscle (Figure 10-15a). Under the microscope, you would find that each muscle fiber is a long cylinder encased by plasma membrane and containing many nuclei. Each muscle fiber is further characterized by a series of dark and light bands (Figure 10-14).

When the muscle fiber is viewed through an electron microscope, you will find that inside each muscle fiber are numerous threadlike filaments. They exist in tiny bundles called **myofibrils** (MY-oh-FIE-brills). Take a moment to study Figure 10-15a to locate the muscle, bundles of muscle fibers, individual muscle fibers, and myofibrils. Myofibrils contain contractile filaments (**myofilaments**) and are striated. As shown in Figure 10-15c, myofibrils contain light and dark bands arranged in a uniform pattern, which gives the myofibril and muscle fiber their striated appearance.

Figure 10-15d shows a close up of a myofibril. As you can see, the myofibrils contain thick and thin filaments. The thick filaments consist of the protein **myosin** (MY-oh-sin). The thin filaments are composed primarily of the protein **actin** (ACK-tin).

> **KEY CONCEPTS**
>
> Muscle fibers contain small bundles of contractile proteins.

How Muscle Fibers Contract

When a muscle contracts, it shortens. This is achieved by the sliding of actin filaments over the myosin filaments. However, the actin filaments do not passively slide to the middle; they are pulled inward by the myosin molecules. How is this accomplished?

As Figure 10-16 shows, each myosin filament consists of numerous golf-club–shaped myosin molecules arranged with their "club ends," or heads, projecting toward the actin filaments. During muscle contraction, the heads of the myosin molecules attach to actin filaments, forming cross bridges, which pull the filaments toward the center of the sarcomere inward.

The attachment of the myosin molecules to the actin molecules is stimulated by the release of calcium inside muscle cells. Calcium is stored in the smooth endoplasmic reticulum, which forms an extensive network inside the muscle cells. The release of calcium, in turn, is stimulated by impulses from the nerves that supply our muscles.

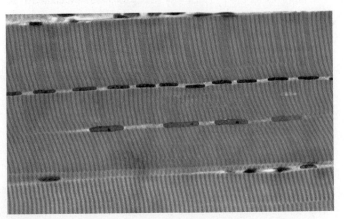

FIGURE 10–14 Light Micrograph of Skeletal Muscle Notice the banding pattern on these muscle fibers. (© Innerspace Imaging/Science Source.)

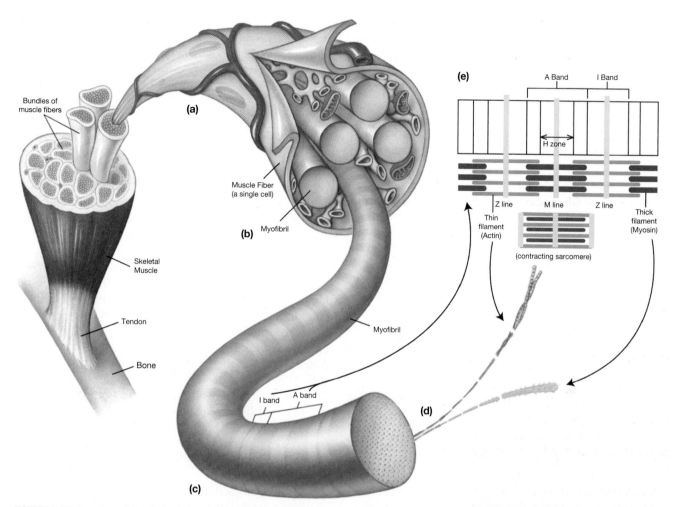

FIGURE 10-15 **Structure of the Skeletal Muscle Fiber, Myofibril, and Sarcomere** (a) A single muscle fiber teased out of the muscle. (b) Each muscle fiber consists of many myofibrils. (c) Note the banded pattern of the myofibril. (d) Sarcomeres consist of thick and thin filaments, as shown here. (e) Molecular structure of the thick (myosin) and thin (actin) filaments.

Neurotransmitters released from the terminal boutons of the motor nerves cause impulses to be generated in muscle cells. These impulses travel along the membrane of muscle cells and enter the interior via small infoldings of the plasma membrane. These infoldings come in close contact with the smooth endoplasmic reticulum. When an impulse penetrates into the interior, it stimulates calcium release. Calcium release causes the heads of the myosin molecules to attach to the actin filaments. ATP in the muscle provides the energy needed to pull the actin filaments inward.

> **KEY CONCEPTS**
>
> During muscle contraction the actin filaments are pulled toward the center of the sarcomere, causing them to shorten and the muscle to contract.

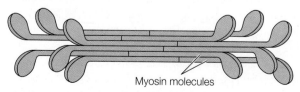

FIGURE 10-16 **Structure of Myosin Filaments** Myosin molecules join to form a myosin filament. Note the presence and orientation of the heads of the myosin molecules.

ATP and Muscle Contraction

Adenosine diphosphate (ADP) is a molecule that, when converted to ATP, gives off energy. This energy is used to power many cellular processes. Interestingly, ATP is in relatively short supply in cells, but is required in large quantity. To meet the cell's demands, ATP is recycled over and over again in rapid succession. That is, ATP is converted to ADP and then back again. How is ATP regenerated?

In muscles, some of the ATP required for muscle contraction is regenerated by a naturally occurring chemical substance known as creatine phosphate (CREE-ah-teh-NEEN). This molecule contains a high-energy phosphate bond, indicated by the squiggly line (below). Stored in muscle in high concentrations, creatine phosphate reacts with ADP to regenerated ATP:

$$\text{creatine} \sim P_i + ADP \rightarrow ATP + \text{creatine}$$

ATP is also regenerated in muscle by the breakdown of glucose and fatty acids. Cellular respiration—the complete breakdown of glucose involves several steps: glycolysis, the transition reaction, the citric acid cycle, and the electron transport system. As you may recall, the main source of ATP during the complete breakdown of glucose is the electron transport system. As you may also recall, cellular respiration requires oxygen.

During vigorous muscle contraction, however, oxygen supplies inside muscle cells may fall precipitously. If the circulatory system cannot replace oxygen as quickly as it is being used, the citric acid cycle and electron transport system shut down. How does the cell meet its demands for energy?

To generate ATP, the cell must resort to fermentation—the metabolic breakdown of glucose in the absence of oxygen. Besides being inefficient, this process results in the buildup of lactic acid in muscle fibers.

Exercise also depletes the muscle of glycogen. Glycogen is a long-chain molecule in muscle (and the liver). It is used to store glucose. Glucose is broken down in cellular respiration to produce ATP. Finally, oxygen levels are also depleted during exercise, creating an oxygen debt.

The shortage of ATP, the buildup of lactic acid, and the depletion of muscle glycogen that occur during vigorous exercise result in muscle fatigue. Shortages of glycogen and oxygen must be remedied for the muscle to continue to function, just as gasoline in your car must be added for you to continue driving. Glycogen is replaced during rest by glucose from the bloodstream. Oxygen is also replenished from the bloodstream by the diffusion of oxygen from blood capillaries that surround the muscle fibers. This process occurs rather quickly, often right after you exercise, which explains why you keep breathing hard for a while after you stop exercising. It may take one to three days for lactic acid to be cleared from muscles and the soreness to go away.

> **KEY CONCEPTS**
>
> ATP provides energy to muscle required for contraction, which results when actin filaments are pulled toward the center of each sarcomere by the myosin filaments.

Recruitment

When stimulated by a nerve impulse, individual skeletal muscle fibers contract. A single contraction, followed by relaxation, is called a **twitch**. Generating the force needed to move arms and legs or eyelids, however, requires the action of many muscle fibers. The engagement of a large number of muscle fibers during muscle contraction is called **recruitment**.

To understand how recruitment occurs in muscle, take a look at Figure 10-17. The drawing shows the nerve cells that end on muscle fibers. They branch and end on many small swellings, the terminal boutons. A single nerve cell, typically ends on dozens of individual muscle fibers in a skeletal muscle. A nerve and the muscle fibers it supplies constitute a **motor unit**. When the neuron supplying a motor unit is activated, all of the muscle fibers in the unit are stimulated.

Motor units account for the degree of control that various muscles exert. The fewer muscle fibers in a motor unit, the finer the control. Such is the case in the muscles of the fingers, which are capable of very fine movements. In contrast, the more muscle fibers in a motor unit, the cruder the control. Thus, the neurons that supply the muscles of the leg, which produce a crude but strong propulsive force, may end on as many as 2,000 muscle fibers. Each time an impulse is delivered to the motor unit, it stimulates all 2,000 muscle fibers.

To increase the force of contraction in a skeletal muscle, more motor neurons are stimulated. More muscle fibers thus are called into action. Therefore, if you were to lift a piece of foam, only a small fraction of the motor neurons going to the

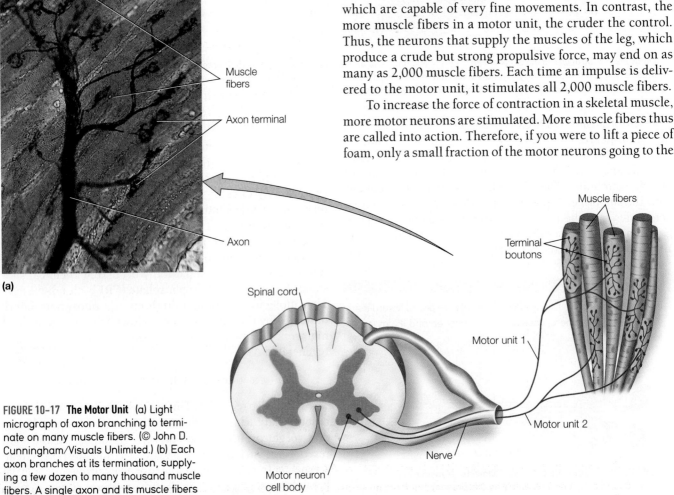

FIGURE 10-17 The Motor Unit (a) Light micrograph of axon branching to terminate on many muscle fibers. (© John D. Cunningham/Visuals Unlimited.) (b) Each axon branches at its termination, supplying a few dozen to many thousand muscle fibers. A single axon and its muscle fibers constitute a motor unit.

Labels: Muscle fibers · Axon terminal · Axon · (a) · Spinal cord · Muscle fibers · Terminal boutons · Motor unit 1 · Motor unit 2 · Nerve · Motor neuron cell body · (b)

arm muscles would be used. If you were to lift a 40-pound bag of dog food, your brain would recruit many more motor neurons and muscle fibers.

The strength of muscle contraction can also be increased by a process called **wave summation**. Let me explain what this means. Each time a muscle fiber is stimulated, it contracts to produce a twitch. Figure 10-18a shows the strength of twitch (contractions) resulting from two separate nerve impulses, the second arriving after the muscle fiber has contracted and then relaxed. Panel (b) shows what happens if a second nerve impulse reaches a muscle fiber before it has had time to relax. As illustrated, this creates additional tension and the muscle fiber contracts more forcefully. In such instances, the second contraction "piggybacks" on the first. The waves of contraction are added up; hence the term *wave summation*.

If the nerve impulses arrive frequently at enough skeletal muscle fibers, a smooth, sustained contraction occurs in the muscle (Figure 10-18c). This is called *tetanus* (not to be confused with the serious, often fatal bacterial infection of the same name). These contractions occur in your arm muscles when you carry your textbooks to class or a heavy bag of groceries from your car to your home. In such instances, the muscles contract to support the weight and remain contracted throughout the activity. Tetanic contractions eventually cause muscle fatigue and muscles stop contracting, even though the neural stimuli may continue.

> **KEY CONCEPTS**
> The strength of muscle contraction is increased by stimulating many muscle fibers; the more fibers that are called into duty, the stronger the contraction.

Muscle Tone

Touch one of your muscles. Even if you are not in peak physical condition, you will notice that the muscle is relatively firm. This firmness is called **muscle tone**. Muscle tone is essential for maintaining posture. Without it, you would fall into a heap on the floor. Muscle tone also generates heat in warm-blooded animals like us.

Muscle tone results from the contraction of muscle fibers during periods of inactivity. But not all fibers contract—just enough of them to keep the muscles slightly tense. Muscle tone is maintained, in part, by the muscle spindle, a receptor that monitors muscle stretching. The spindle "alerts" the brain and spinal cord to the degree of stretching. When muscles relax, signals travel to the spinal cord and then back out to motor axons, which stimulate a low level of muscle contraction to maintain muscle tone.

> **KEY CONCEPTS**
> Muscle tone, the firmness you feel even when touching a relaxed muscle, results from the continuous contraction of a small number of muscle fibers.

Slow- and Fast-Twitch Fibers

Physiological studies have revealed the presence of two types of skeletal muscle fibers: fast-twitch and slow-twitch fibers. **Slow-twitch muscle fibers** contract relatively slowly but have incredible endurance. It is not surprising, then, that the muscles of endurance athletes (for example, long-distance runners) contain a high proportion of slow-twitch muscle fibers (Figure 10-19). These permit such athletes to perform for long periods without tiring.

Fast-twitch muscle fibers contract swiftly. The muscles of sprinters and other athletes whose performance depends on quick bursts of activity contain a high proportion of fast-twitch fibers.

Skeletal muscles generally contain a mixture of slow- and fast-twitch fibers, giving each muscle a wide range of performance abilities. A muscle that performs one type of

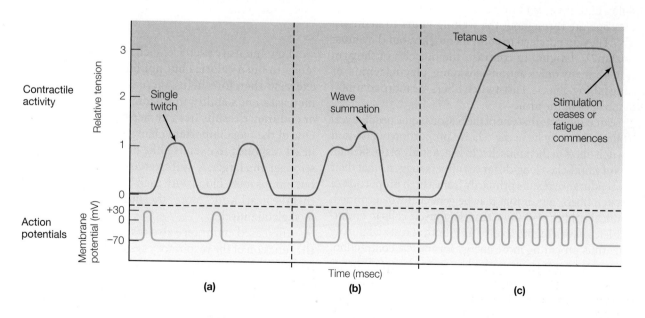

The duration of the action potentials is not drawn to scale but is exaggerated.

FIGURE 10-18 **Graph of Muscle Contraction** (a) The force generated by two separate stimuli (action potentials from motor neuron). (b) The force generated by two closely timed stimuli. (c) The force generated by many closely spaced stimuli.

FIGURE 10-19 Built to Last An abundance of slow-twitch fibers gives the cross-country runners endurance. (© Val Thoermer/ShutterStock, Inc.)

Increasing Muscle Size and Endurance

Milo of Croton was a champion wrestler in ancient Greece who stumbled across a revolutionary way to build muscle. His method was simple. Milo borrowed a newborn calf, and then proceeded to carry the animal around every day. Day after day, he faithfully followed this routine so that by the time the calf had become a bull, so had he. Milo's program worked, for he went on to win six Olympic championships in wrestling.

The Greek wrestler discovered a principle of muscle physiology familiar to weight lifters today: When muscles are made to work hard, they respond by becoming larger and stronger. The increase in size and strength results from an increase in the amount of contractile protein inside muscle cells.

Unfortunately, muscle protein is quickly made and quickly destroyed. In fact, about half of the muscle you gain in a weight-lifting program is broken down 2 weeks after you stop exercising. In order to build muscle, you must keep up with your exercise program.

High-intensity exercise such as weight lifting builds muscle, and it takes surprisingly little exercise to have an effect. Working out every other day for only a few minutes will result in noticeable changes in muscle mass. According to some sources, 18 contractions of a muscle (in three sets of six contractions each) are enough to increase muscle mass, if the contractions force your muscles to exert over 75% of their maximum capacity.

Health Tip 10-4

Exercise in context—for example, by playing volleyball or softball.
Why?
For many people, exercise in a gym or weight room is pure drudgery. Time passes very slowly and the whole thing feels like an uphill battle. When playing games like soccer or softball or volleyball, however, the thrill of competition often makes us forget we're working out.

function more often than another, however, tends to have a disproportionately higher number of fibers corresponding to the type of activity it performs. The muscles of the back, for example, contain a larger number of slow-twitch fibers. These muscles operate throughout the waking hours to maintain posture. They do not need to contract quickly, but they must be resistant to fatigue. In contrast, the muscles of the arm are used for many quick actions—waving, playing tennis, or grasping falling objects. Fast-twitch fibers are more common in the muscles of the arm.

Research suggests that one of the reasons some people excel in certain sports may lie in the relative proportion of fast- and slow-twitch fibers in their muscles. In fact, a study of the skeletal muscles of world-class long-distance runners suggests that their physical endurance results primarily from a high proportion of slow-twitch fibers, a trait that may be genetically determined. Biochemical studies also show that the muscle cells in endurance athletes have a higher level of ATP, both at rest and during exercise, thus providing more energy for muscle contraction. Endurance athletes start out with a larger storehouse of energy and maintain a larger supply throughout exercise.

KEY CONCEPTS

Two types of muscle fibers are found in the body, slow-twitch and fast-twitch; slow twitch are found in muscles that require endurance; fast-twitch are found in muscles that contract quickly.

Low-intensity exercise, such as aerobics and swimming, tends to burn calories but not build muscle bulk. Aerobic exercise therefore helps increase endurance—that is, it increases one's ability to sustain muscular effort. Stamina or endurance results from numerous physiological changes. One of the most important changes occurs in the heart. The heart responds to exercise like any other muscle—it grows stronger and larger. A well-exercised heart beats more slowly but pumps more blood with each beat. The net result, then, is that the heart works more efficiently, delivering more oxygen to skeletal muscles.

Increased endurance also results from improvements in the function of the respiratory system. For example, exercise increases the strength of the muscles involved in breathing. These muscles become stronger and can operate longer without tiring. Breathing during exercise becomes more efficient.

Increased endurance may also be attributed to an increase in the amount of blood, another benefit of aerobic exercise. An increase in blood volume results in an increase in the number of RBCs, which increases the amount of oxygen available to cells. This improvement, combined with others, allows an individual to work out longer without growing tired.

When you set out on an exercise program, it is important to establish your goals first. If you are interested in increasing your endurance, you should pursue an exercise regime that works the heart and muscles at a lower intensity over longer periods such as riding a stationary bicycle. If you are after bulk, the answer lies in high-intensity exercises such as weight lifting.

Many health clubs and university gyms offer advice and exercise programs to help you achieve your goals. And many offer a variety of machines to help you build muscles or simply tone them up. Most of the exercise machines work one particular muscle group—for example, the muscles of the upper arm. In most gyms, a dozen or more machines are usually placed in a line so you can go down the line, working one set of muscles after another until you have exercised your entire body.

Health Tip 10-5

Get in the habit of exercise now and exercise as often as you can to maintain muscle mass.

Why?

Starting in midlife, adults lose as much as one-third to one-fourth of a pound of muscle mass per year if they don't do something to prevent it. Adopting a healthy lifestyle now, which includes plenty of exercise, makes it easier to stay active as you get older.

Health Tip 10-6

Chart a weekly exercise plan with goals and accomplishments. Set reasonable goals each week (for example, the number of sit-ups you'll perform each day or miles you will ride or walk) and keep track of your daily accomplishments.

Why?

Keeping track of the exercise one performs each week helps many people stick to an exercise program as it provides tangible evidence of progress.

Exercise machines are popular because they are safer than free weights (barbells and dumbbells). These machines eliminate the chances of your dropping a weight on your toes—or someone else's! Moreover, it is almost impossible to strain your back if you make a mistake using one. In contrast, lifting free weights requires caution and training.

Exercise machines also reduce the amount of time a person needs to exercise by about half because they've been designed to require work when a joint is both flexed and extended. The biceps machine, for example, requires you to pull the weights up, then return them slowly, forcing your biceps muscles to work in both directions.

> **KEY CONCEPTS**
>
> High intensity exercise causes an increase in muscle mass, while low-intensity exercise, aerobic exercise, burns calories but does not build mass.

10-3　Health and Homeostasis

The skeleton and muscles provide many benefits, the most evident being mobility. These two systems also contribute to homeostasis. The skeleton, for instance, serves as a calcium reservoir, essential for controlling blood levels of calcium, which are crucial for normal muscle contraction. Muscle helps provide the heat necessary for the normal operation of cells. Calcium is also essential for nerve cell transmission.

Efforts to enhance the performance of the muscular system can upset homeostasis and have serious effects on other systems. Consider anabolic steroids.

Anabolic steroids are synthetic hormones that resemble the male sex hormone, testosterone. When taken in large doses, anabolic steroids stimulate muscle formation. They increase muscle size and strength by stimulating protein synthesis in muscle fibers. High doses may also reduce the inflammation that frequently results from heavy exercise, allowing athletes to work out harder and longer.

When it comes to building muscle, steroids and exercise are an unbeatable combination. Some users claim that steroids even increase aggression, which may be helpful to competitive athletes.

Despite the benefits of steroids, most physicians view them as a dangerous proposition. For example, steroids can result in mental and behavioral problems. One set of researchers interviewed 41 athletes who used steroids in doses 10 to 100 times greater than those used in medical treatment. The athletes also reported using as many as five or six steroids simultaneously in cycles lasting 4 to 12 weeks, a practice known as "stacking." In this study, researchers found that one-third of the athletes developed severe psychiatric complications.

Athletes in the study reported episodes of severe depression during and after steroid use. Some reported feelings of invincibility. One man, in fact, deliberately drove a car into a tree at 40 miles per hour while a friend videotaped him. Some subjects reported psychotic symptoms in association with steroid use, including hearing voices. Withdrawal from steroids resulted not only in depression but also in suicidal tendencies.

Other studies have shown that steroids used in excess may damage the heart and kidneys and reduce testicular size in men. In women, steroids deepen the voice and may cause enlargement of the clitoris. Steroids also cause severe acne and liver cancer. Unfortunately, there are no scientific studies on the long-term health effects of steroids.

Despite the fact that anabolic steroids are banned by professional sports organizations like the National Football League, the International Olympic Committee, and college athletic programs, athletes continue to use them. Although periodic drug tests are often given, these steroids, along with a host of other performance-enhancing drugs, often escape detection. Although U.S. athletes are subjected to numerous unannounced drug tests, other countries are more lenient.

A recent survey of 46 public and private high schools across the United States involving over 3,000 teenagers suggests that steroid use is especially prevalent in high school seniors. About 1 of every 15 senior boys reported taking anabolic steroids. The study also showed that the use of anabolic steroids begins in junior high school. Two-thirds of the students surveyed said that they had used steroids by age 16. Nearly half the users said they took the drugs to boost athletic performance, and 27% said their primary motive was to improve their appearance. Researchers say that adolescents who use steroids may be putting themselves at risk of stunted growth, infertility, and psychological problems.

Steroid use in the United States illustrates our dependence on quick fixes. It is a tragic result of the almost obsessive focus on performance and achievement in our highly competitive society, a social environment that may be endangering the health of our children and our athletes.

SUMMARY

Structure and Function of the Human Skeleton

1. Bones serve many functions. They provide internal support, allow for movement, and help protect internal body parts. They also produce blood cells and platelets, store fat, and serve as a calcium reservoir that is used to help maintain blood calcium levels.

2. Most bones have an outer layer of compact bone and an inner layer of spongy bone. Inside the bone is the marrow cavity, filled with either fat cells (yellow marrow) or blood cells and blood-producing cells (red marrow) or with combinations of the two.

3. The joints unite bones. Some are immovable, but most provide some degree of movement. The movable joints allow for flexion, extension, and other important movements.

4. Joints are vulnerable to injury and disease. Torn ligaments and ripped cartilage are common injuries among athletes.

5. Wear and tear on some joints can cause the articular cartilage to crack and flake off, resulting in degenerative joint disease, or osteoarthritis. This occurs most commonly in the weight-bearing joints and is aggravated by excess weight.

6. Rheumatoid arthritis results from an autoimmune reaction that produces a painful inflammation and thickening of the synovial membrane, which leads to an erosion of the articular cartilage as well as disfigurement and stiffening of the joints. It most often occurs in the joints of the hand and wrist as well as the feet.

7. Most of the bones form from hyaline cartilage that forms inside the body during fetal development.

8. Bone is constantly remodeled after birth to accommodate changing stresses. During bone remodeling, osteoclasts destroy bone. Osteoclasts are stimulated by parathormone, produced by the parathyroid glands. Osteoclasts also participate in the homeostatic control of blood calcium levels. When activated, these cells free calcium from the bone, raising blood calcium levels.

9. Osteoblasts are bone-forming cells stimulated by calcitonin from the thyroid gland. Calcitonin secretion decreases blood calcium levels and increases the amount of calcium in bones.

10. Osteoporosis is a disease of the bone caused by progressive loss of calcium. The bones become brittle and easily broken. Osteoporosis occurs with age and is most common in postmenopausal women and results from the loss of the ovarian hormone estrogen. It also occurs in people who are immobilized for long periods.

11. Osteoporosis can be prevented and reversed by exercise, calcium supplements, consuming calcium-rich foods, vitamin D supplements, and estrogen therapy.

The Skeletal Muscles

12. Skeletal muscles are involved in body movements; they help maintain our posture and produce body heat both at rest and while we are working or exercising.

13. Each skeletal muscle consists of many muscle fibers. Muscle fibers are both excitable and contractile.

14. Inside each muscle fiber are numerous myofibrils, bundles of the contractile filaments actin and myosin.

15. Muscle contraction is stimulated by nerve impulses from nerve cells that terminate on the muscle fibers, called motor neurons. The energy for muscle contraction comes from ATP.

16. Individual muscle fibers contract when stimulated, producing a twitch. Contractions of varying strength can be generated in whole muscles by recruitment and wave summation.

17. Recruitment results from the involvement of many motor units. A motor unit is termination of a nerve cell (a motor axon) and all of the muscle cells it innervates.

18. Wave summation is a piggybacking of muscle fiber contractions that occurs when stimuli arrive before the fiber relaxes.

19. The body contains two types of skeletal muscle fibers: slow-twitch and fast-twitch fibers. Skeletal muscles generally contain a mixture of the two types, but fast-twitch fibers are found in greatest numbers in muscles that perform rapid movement. Slow-twitch fibers are found in muscles such as those of the back that perform slower motions or are involved in maintaining posture.

20. Muscle mass can be increased by certain forms of exercise, such as weight lifting. An increase in muscle mass results from an increase in the amount of contractile protein in muscle fibers.

21. Endurance can be increased by other forms of exercise such as aerobics.

22. Endurance is a function of at least three factors: the condition of the heart, the condition of the muscles of inspiration, and the blood volume. Improvement in all three factors increases the efficiency of oxygen delivery to muscles.

Health and Homeostasis

23. Many athletes are using synthetic anabolic steroids to improve performance and build muscle. Unfortunately, the massive doses of steroids they use can have many serious side effects, including an increase in aggression and mental imbalance such as severe depression. Steroid use may also result in damage to the heart and kidneys.

THINKING CRITICALLY ANALYSIS

This analysis corresponds to the Thinking Critically scenario that was presented at the beginning of this chapter.

If you noted that no mention was made of dietary calcium intake by the various groups, you're correct. Unfortunately, the diets of the women in this study were not monitored to determine how much calcium was being ingested through food and beverages. Second, this study focused on the bones of the forearm, not the load-bearing bones that are at greatest risk of fractures. Walking would probably influence the bones of the legs, hips, and back. To accurately assess the effects of the exercise regime prescribed in this study, the researchers probably should have looked at bone density in all three sites, not the arms. Given these inadequacies, do you think this study's conclusions are reliable?

KEY TERMS AND CONCEPTS

Actin, p. 216
Appendicular skeleton, p. 206
Arthroscope, p. 209
Arthroscopic surgery, p. 209
Axial skeleton, p. 206
Ball-and-socket joint, p. 209
Calcitonin, p. 212
Callus, p. 213
Collagen fibers, p. 205
Compact bone, p. 206
Dislocation, p. 209
Endochondral ossification, p. 211
Fast-twitch muscle fiber, p. 219
Fibroblast, p. 213
Hinge joint, p. 209
Intervertebral disk, p. 208
Joint, p. 207

Joint capsule, p. 208
Ligament, p. 208
Marrow cavity, p. 206
Matrix, p. 205
Motor unit, p. 218
Movable joints, p. 208
Muscle fiber, p. 216
Muscle tone, p. 219
Musculoskeletal system, p. 205
Myofibril, p. 216
Myofilament, p. 216
Myosin, p. 216
Osteoarthritis (OA), p. 210
Osteoblasts, p. 207
Osteoclasts, p. 206
Osteocytes, p. 205
Osteoporosis, p. 213

Parathormone (PTH), p. 212
Periosteum, p. 206
Recruitment, p. 218
Red marrow, p. 206
Rheumatoid arthritis (RA), p. 210
Skeletal muscle, p. 215
Skeleton, p. 205
Slow-twitch muscle fibers, p. 219
Spongy bone, p. 206
Synovial fluid, p. 208
Synovial joint, p. 208
Synovial membrane, p. 208
Tendon, p. 208
Twitch, p. 218
Wave summation, p. 219
Yellow marrow, p. 206

CONCEPT REVIEW

1. Describe the functions of bone. In what ways does bone participate in homeostasis? p. 205.
2. The synovial joints move relatively freely. Why? What structures support the joint, helping to keep the bones in place? p. 208.
3. A young patient comes to your office with swollen joints and complains about pain and stiffness in the joints of his hands and wrists. Friends have suggested that the boy has arthritis, but his parents argue that he is too young. Only old people get arthritis, they say. How would you respond to them? pp. 210–211.
4. Bone is constantly remodeled, from infancy through adulthood. Explain when bone remodeling occurs and what cells and hormones participate in the process. pp. 211–212.
5. Using what you know about bone, explain why an office worker who exercises very little is more likely to break a bone on a skiing trip than a counterpart who works out every night after work. pp. 206–207.
6. Describe bone healing that occurs after a fracture. pp. 212–213.
7. A 30-year-old friend of yours who smokes, exercises very little, and avoids milk products because she's concerned that the small amounts of fat in milk will cause her to gain weight says: "Why should I worry about osteoporosis? That's a disease of old women." Based on what you know about bone, how would you respond to her? pp. 213–214.
8. Describe the major functions of skeletal muscle. pp. 215–216.
9. What is a muscle fiber? Why does it have so many nuclei? Describe the structure of a skeletal muscle fiber. p. 216.
10. Describe the molecular events involved in muscle contraction. How is calcium involved? Where does the calcium come from? pp. 216–218.
11. Describe the two types of skeletal muscle fibers, and explain how they differ. Where would you expect to find them? pp. 219–220.
12. A friend comes to you complaining that he can't seem to lose weight. He works out on barbells three times a week for an hour or so each time. His diet hasn't changed much, but with the increase in exercise, he thinks he should be losing weight, rather than staying even. Why do you think he isn't losing weight? pp. 220–221.
13. A friend is thinking about using anabolic steroids to improve his performance in gymnastics. He argues that he is young and will be taking the drug only for a year or so. He points out that other young men are using steroids and they really help build muscle and endurance. What advice would you give him? pp. 221–222.

SELF-QUIZ: TESTING YOUR KNOWLEDGE

1. Bone cells, known as _____, are embedded in a hard matrix. p. 205.
2. These cells (in question 1) form from another type of cell known as an _____. p. 205.
3. Bone consists of calcium phosphate which provides rigidity and _____ which makes bones a bit more flexible. p. 205.
4. The dense bone on the outside of the arm bone is known as _____ bone. p. 206.
5. The connective tissue layer surrounding bone is called the _____. p. 206.
6. When a bone is broken, a person often feels intense pain. The pain fibers associated with this pain are found in the _____. p. 206.
7. The spongy bone inside a bone may be filled with blood-forming tissue called the _____ bone marrow. p. 206.
8. Inactive bone marrow—that is, bone marrow not involved in making blood cells—is known as _____ bone marrow. p. 206.
9. The bone cells that help raise blood calcium levels are known as _____. p. 206.
10. The dense connective tissue structures that join bones in joints are known as _____. They are often thickenings of the _____ _____. p. 208.
11. The fluid inside joints that helps lubricate the bones is called _____ fluid. p. 208.
12. _____ or degenerative joint disease results from excess wear and tear on the bones in joints. p. 210.
13. _____ arthritis is a degenerative joint disease thought to be caused by an autoimmune reaction. p. 210.
14. Joint surgery often involves use of an instrument known as an _____, which allows doctors to peer inside the joint to fix damage without having to create large incisions. p. 209.
15. The process of bone formation is known as _____ ossification. p. 211.
16. A muscle cell consists of many cells fused together and is known as a muscle _____. p. 216.
17. Inside the muscle cells are smaller units known as _____; they contain many smaller myofilaments. p. 216.
18. Two myofilaments are involved in muscle contraction, actin, and _____. p. 216.
19. Contraction results from the release of _____ ions by the smooth endoplasmic reticulum of muscle cells. p. 216.
20. Energy required for muscle contraction is provided by _____. p. 217.

biology.jbpub.com/chiras/8e/

The site features eLearning, an online review area that provides quizzes, chapter outlines, and other tools to help you study for your class. You can also follow useful links for in-depth information, research the differing views in the Point/Counterpoints, or keep up on the latest health news.

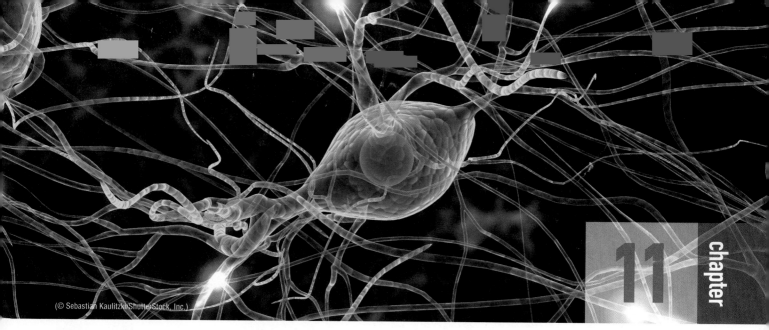

(© Sebastian Kaulitzki/ShutterStock, Inc.)

The Nervous System

Ernest Hemingway is one of America's most celebrated writers. His novels, including *The Old Man and the Sea*, and short stories won him the Nobel Prize in literature. Despite success and widespread popularity, Hemingway was a troubled man who eventually committed suicide. What ultimately caused him to take his life no one can know, but some believe that one contributing factor was a rare and painful nervous system disorder known as *trigeminal neuralgia* (try-GEM-in-al ner-AL-gee-ah). People suffering from this disease complain

THINKING CRITICALLY

Many people believe that subliminal tapes can enhance a person's memory, self-esteem, concentration, and vocabulary. If you search the Internet, in fact, you'll find what appears to be valid research that supports these beliefs. Psychologists decided to test this hypothesis. They gave one group of subjects a subliminal tape designed to improve memory. The other group received a tape that supposedly improved self-esteem. Each group was told the purpose of the tape.

A month later, the researchers polled each group, and, sure enough, those who received the memory tape reported enhanced memory. Those who received the self-esteem tape reported higher self-esteem. Pretty convincing?

What's wrong with the conclusion?

of periodic, unexplained flashes of intense pain along the course of the sensory nerve that supplies the face (trigeminal nerve). A slight breeze or the pressure of a razor can set off this pain, which lasts a minute or more.

Some people believe that the pain Hemingway felt may have become unbearable. Combined with other personal conflicts, this neurological disorder may have driven him to commit suicide.

This chapter describes the structure and function of the human nervous system. You will study the nerve that caused Hemingway so much trouble, and you will also examine ways in which the nervous system contributes to homeostasis.

11-1   An Overview of the Nervous System

The human nervous system governs many functions in the body. It controls muscles, glands, and organs. It also controls heartbeat and breathing. Even digestion and urination are under its control. The nervous system also helps regulate blood flow through the body as well as maintain the concentration of chemicals in the blood. As such, the nervous system plays a major role in maintaining homeostasis.

The nervous system doesn't act in isolation. Like the various branches of government, it receives input from a large number of sources. This input helps it "manage" body functions. The human nervous system also provides functions not seen in other animal species. For example, the brain is the site of ideation—the formation of ideas. Our brain allows us to think about and plan for the future. It enables us to reason—that is, to judge right from wrong and logical from illogical. (In other words, to think critically.)

The nervous system also allows us to change our environment to better serve our needs. Although actions such as draining swamps to create farmland help humankind survive and prosper, they can be detrimental to the many species that share this planet with us—for example, the ducks and muskrats that once inhabited the swamplands converted to farmland. In the long run, some human actions that appear beneficial to us now may end up being detrimental to our own kind. Fortunately, our own hope of survival also depends on our brain—an evolutionary by-product unrivaled in the biological world.

> **KEY CONCEPTS**
>
> The nervous system plays a key role in regulating body activities, including many functions related to homeostasis and hence human health.

The Central and Peripheral Nervous Systems

The human nervous system consists of three components: the brain, the spinal cord, and nerves (Figure 11-1). The brain and spinal cord constitute the **central nervous system** (CNS). They are housed in the skull and vertebral canal, respectively. Three layers of connective tissue, known as the **meninges** (men-IN-gees), surround the brain and spinal cord. The space between the middle and inner layers is filled with a liquid

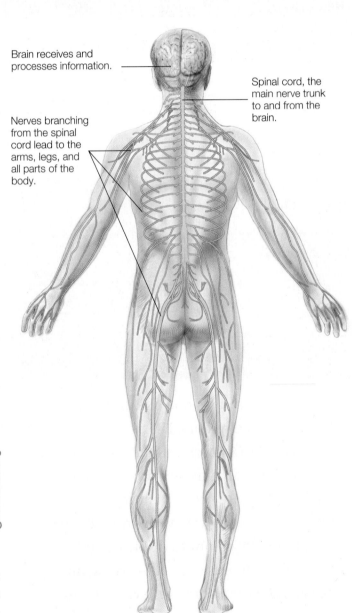

Brain receives and processes information.

Spinal cord, the main nerve trunk to and from the brain.

Nerves branching from the spinal cord lead to the arms, legs, and all parts of the body.

FIGURE 11-1 The Nervous System The human nervous system is a network of nerves connected to the brain and spinal cord. Nerves comprise the peripheral nervous system. The spinal cord and brain make up the central nervous system.

called cerebrospinal fluid (CSF; sir-REE-bro-SPIE-nal), covered in more detail shortly.

While the brain and spinal cord constitute the central nervous system, the nerves form the peripheral nervous system. **Nerves** consist of bundles of nerve fibers. Each fiber is part of a nerve cell, or **neuron** (NER-on). Nerves transport messages to and from the CNS. Those nerves that transport messages to the CNS end on receptors. Receptors, in turn, respond to a variety of internal and external stimuli. They help you remain aware of your environment. The brain and spinal cord receive an enormous amount of sensory information from the body. Right now, for instance, your brain and spinal cord are being sent information from various receptors in your body that alert you to the room temperature, traffic sounds, people talking, music playing, and the touch of the page. The information is transmitted along sensory nerves. This massive inflow of information is managed by the CNS. Some information may be stored in memory. Some incoming information is ignored. Other information may elicit a response. Responses are produced when the brain and spinal cord send nerve impulses to muscles and glands (effectors) via the nerves. In summary, then, nerves carry two types of information: (1) sensory impulses traveling to the CNS from sensory receptors in the body and (2) motor impulses traveling away from the CNS to muscles and glands.

Somatic and Autonomic Divisions of the PNS

As **Figure 11-2** shows, the **peripheral nervous system** (PNS) consists of the somatic nervous system and the autonomic nervous system. The **somatic nervous system** (so-MAA-tick) is that portion of the PNS that controls voluntary functions of the body, such as the muscle contractions that cause the limbs to move. It also controls certain involuntary reflex actions, such as the knee-jerk response.

The part of the PNS that controls the majority of the involuntary functions such as heart rate and breathing is the **autonomic nervous system** (ANS; au-toe-NOM-ick; Figure 11-2). Many other functions are under the control of the autonomic nervous system, including digestion and body temperature regulation. Right now, your autonomic nervous system is busily controlling numerous functions like heartbeat and breathing so you can concentrate on reading.

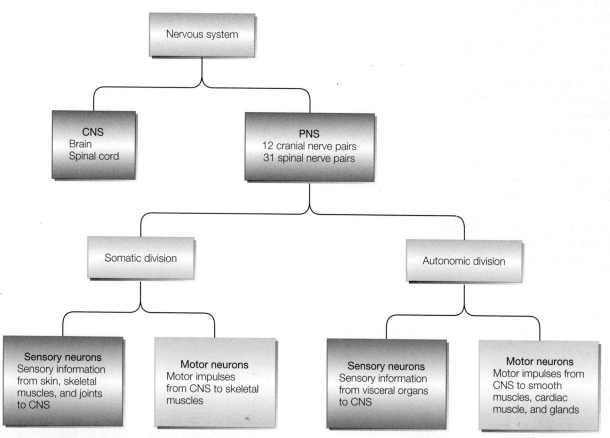

FIGURE 11-2 **Subdivisions of the Nervous System** The nervous system is divided into two parts, the central nervous system (CNS) and the peripheral nervous system (PNS). The PNS consists of autonomic and somatic divisions. The activities of the autonomic and somatic divisions often overlap. The PNS consists of the somatic and autonomic subdivisions.

11-2 Structure and Function of the Neuron

The **neuron** is the fundamental structural unit of the nervous system. This highly specialized cell generates bioelectric impulses and transmits them from one part of the body to another. Such signals alert us to a variety of internal and external stimuli and permit us to respond to them.

> **KEY CONCEPTS**
> Neurons are conductive cells that transmit impulses throughout the central and peripheral nervous systems.

The Structure of the Neuron

Neurons come in several shapes and sizes, but they all share several characteristics. All neurons, for example, consist of a more or less spherical central portion, the cell body

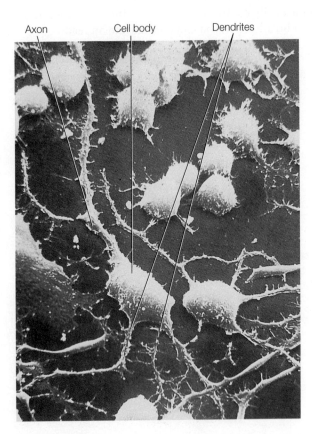

Axon   Cell body   Dendrites

(Figure 11-3). It houses the nucleus, most of the cell's cytoplasm, and numerous organelles. Metabolic activities provide energy and synthesize materials necessary for proper cell function, thus sustaining the entire neuron. All nerve cells contain processes called **dendrites** (DEN-drights) that transmit bioelectric impulses toward the cell body.

Nerve cells also contain processes called **axons** (AXE-ons) that transmit impulses away from the cell body. When an axon reaches its destination, it often branches, giving off many small fibers. These fibers terminate in tiny swellings called **terminal boutons** (boo-TAWNS; boutons is the French word for buttons). Terminal boutons serve as a communication link with other neurons, muscle fibers, or glands. That is, they transmit impulses from one nerve cell to another or from a nerve cell to an effector such as a muscle fiber.

> **KEY CONCEPTS**
> Neurons contain dendrites that transmit impulses to the cell body, the cell body where the nuclei are located, and axons that transmit impulses away from the cell body.

The Myelin Sheath

The axons of many multipolar neurons in both the central and peripheral nervous systems are coated with a protective layer called the **myelin sheath** (MY-eh-lin; Figure 11-4). The myelin sheath consists of many layers, and is mostly made of lipid. It appears glistening white when viewed with the naked eye. As shown in Figure 11-4, segments of myelin are separated by small unmyelinated segments known as the **nodes of Ranvier** (RON-vee-A). For reasons explained later, the myelin sheath permits nerve impulses to travel with great speed along axons, "jumping" from node to node like a stone skipping along the surface of water (Figure 11-4). Although most axons are covered with myelin, some are unmyelinated. The unmyelinated axons conduct impulses much more slowly than their

FIGURE 11-3 A Neuron The multipolar neuron resides within the central nervous system. Its multiangular cell body has several highly branched dendrites and one long axon. Collateral branches may occur along the length of the axon. When the axon terminates, it branches many times, ending on individual muscle fibers. (Photo © David M. Phillips/Visuals Unlimited.)

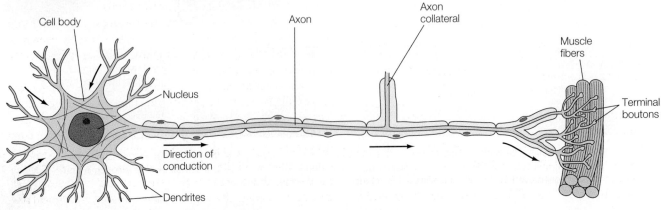

Cell body   Axon   Axon collateral   Muscle fibers

Nucleus   Terminal boutons

Direction of conduction   Dendrites

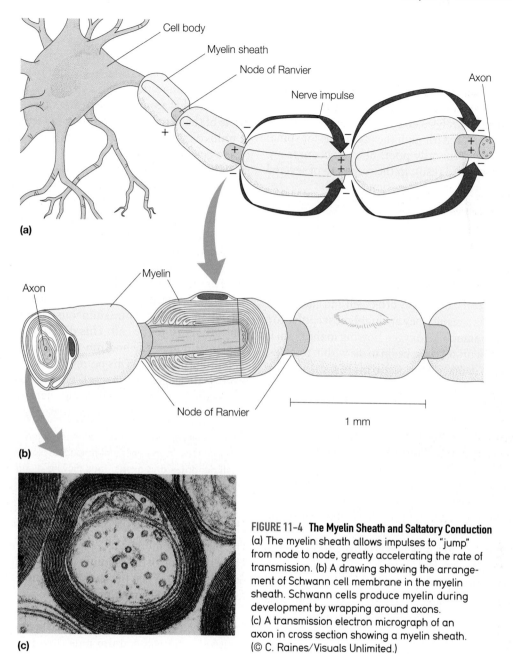

FIGURE 11-4 The Myelin Sheath and Saltatory Conduction
(a) The myelin sheath allows impulses to "jump" from node to node, greatly accelerating the rate of transmission. (b) A drawing showing the arrangement of Schwann cell membrane in the myelin sheath. Schwann cells produce myelin during development by wrapping around axons. (c) A transmission electron micrograph of an axon in cross section showing a myelin sheath. (© C. Raines/Visuals Unlimited.)

myelinated counterparts. As a rule, the most urgent types of information are transmitted via myelinated fibers; less urgent information is transmitted via unmyelinated fibers.

> **KEY CONCEPTS**
>
> The myelin sheath is a layer of lipid that surrounds the axons of many nerve cells; it helps to greatly accelerate the transmission of nerve impulses along myelinate nerves.

Nerve Cell Repair

Unlike many other cells of the body, nerve cells cannot divide. Thus, when nerve cells die as a result of injury or old age, they cannot be replaced by cell division. But not all cell death results in irreversible loss of function.

Consider what happens in the case of a stroke. A **stroke** is the loss of blood to a part of the brain. It can be caused by breaks in arteries of the brain, for example, an arterial blowout called an aneurysm. It can also be caused by blood clots that become lodged in arteries of the brain narrowed by atherosclerosis, or blood clots that form in narrowed portions of arteries. Although blood flow to important parts of the brain may cause nerve cells to die, undamaged neurons elsewhere in the brain can take over the function of damaged brain cells, permitting partial to complete recovery.

Nerve cells can also be damaged without permanent loss of function, but it depends on where the damage occurs. Nerves in the arm (part of the PNS), for instance, can be severed in accidents. Severed axons normally degenerate from the point of injury to the muscle or gland they supplied. Over time, segments of the axons still attached to the cell body of the damaged neuron may elongate and replace the degenerated sections. During regeneration, the axons extend along the hollow tunnel in the myelin sheath that was left by the degenerated part. Eventually, the axons reestablish connections with the muscles or glands they once supplied, making partial or nearly complete recovery of control possible.

In the brain and spinal cord, however, severed axons cannot be repaired. Consequently, spinal cord injuries that sever nerves result in paralysis and loss of sensation in the region below the damage. Although the prognosis for spinal cord injury victims is generally poor, new procedures are being developed that could result in the regrowth of severed axons, allowing recovery of lost function.

> **KEY CONCEPTS**
> Nerve cells are highly specialized cells that cannot divide and therefore generally cannot be regenerated when damaged.

Oxygen Sensitivity of Nerve Cells

Nerve cells require a constant supply of oxygen and glucose for energy production. As a result, they are highly susceptible to a loss of oxygen. If the amount of oxygen flowing to the brain is drastically reduced, neurons, thus, begin to die within minutes.

To prevent brain damage from occurring in someone who has collapsed with a severe heart attack or who has drowned, rescuers must start resuscitating the victim within 4 to 5 minutes. Although victims may be revived after this crucial period, the lack of oxygen in the brain often results in permanent brain damage. Generally, the longer the deprivation, the greater the damage. The only exception is when someone has drowned and has been submerged in very cold water. In such instances, resuscitation may be successful if begun within an hour. In most cases, victims recover without any detectable brain damage. Recovery is possible in such instances because cold water greatly slows brain metabolism, which greatly reduces oxygen demand. As a result, brain cells are preserved, and brain damage is minimized or prevented.

> **KEY CONCEPTS**
> Nerve cells are highly sensitive to oxygen levels in the blood and will die within minutes without oxygen.

Nerve Cell Impulses

Nerve impulses are not like the electric current that powers computers or light bulbs, which is formed by the flow of electrons. Rather, nerve impulses are small ionic changes in the membrane of the neuron. They move along the plasma membrane of a nerve cell.

To understand the **nerve impulse**, we begin by examining the plasma membrane of a neuron. We first place tiny electrodes on the outside and inside of the plasma membrane of a neuron and hook them up to a voltmeter, a device that measures voltage. Voltage is a measure commonly used when studying electricity. You buy batteries with different voltages that you can measure with a simple voltmeter. Voltage is rather difficult to explain. In a battery, voltage is a measure of the tendency of charged particles (electrons) to flow from one pole of the battery to the other. The higher the voltage, the greater the tendency for electrons to flow through a wire connected to the poles. In the nerve cell, however, electrons do not flow from one side of the membrane to another; sodium and potassium ions do.

In neurons, the potential difference, or voltage, is a measure of the force that can drive sodium ions from one side of the membrane to the other. For now, it is important just to remember that a small voltage exists across the plasma membrane of the neuron. It is so small, in fact, that it is measured in millivolts (MILL-ee-volts). A millivolt is 1/1,000 of a volt. Just to give you a point of reference, the voltage of the electricity running through the wires in your home is about 120 volts.

The potential difference in a typical neuron is about −70 millivolts. This is known as the **resting potential** because it is the voltage of a nerve cell at rest (a nerve not transmitting an impulse). The minus sign is added because the plasma membrane is positively charged on the outside and negatively charged on the inside.

To understand the bioelectric impulse, it is important to know that sodium ions are found in greater concentration outside the neuron. Potassium ions are found in greater concentration inside the cell. This difference is due to an active transport pump, the **sodium–potassium pump**, in the cell membrane of nerves. This pump consists of many membrane proteins that actively transport sodium ions out of the cell and potassium ions into the cell, using energy in the form of ATP.

> **KEY CONCEPTS**
> Nerve cell impulses involve movement of ions across the plasma membrane of nerve cells that propagate along the length of nerve cell dendrites and axons, thus moving along the length of the nerve cell.

The Action Potential

The plasma membrane of a nerve cell has a built-up charge. This charge consists of sodium ions concentrated on the outside of the cell. When the neuron is stimulated, its membrane suddenly becomes more permeable to sodium ions.

Neurophysiologists believe that stimulating the nerve cell causes protein pores in the plasma membrane to open. Positively charged sodium ions flow into the cell through these pores.

Electrodes implanted in a nerve cell detect the sudden inflow of positively charged sodium ions. The electrodes then register a shift in the resting potential from −70 millivolts to +30 millivolts. The change in voltage occurs at the site of stimulation and is called **depolarization**. Immediately after depolarization, the membrane returns to its previous state, which is referred to as **repolarization**.

The depolarization/repolarization of the membrane is shown graphically in Figure 11-5. This tracing is an **action potential**. It appears on a screen hooked up to the voltmeter. The action potential consists of (1) a brief upswing, depolarization, as the voltage goes from −70 millivolts to +30 millivolts and (2) a rapid downswing, repolarization, the return to the resting potential. The graph also shows changes in sodium and potassium ion permeability—how well these ions can move through the membrane during the action potential.

The depolarization occurs so rapidly and the membrane returns to the resting state so quickly (in about 3/1,000 of a

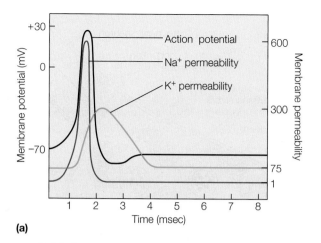

(a)

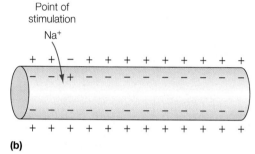

Point of
stimulation

(b)

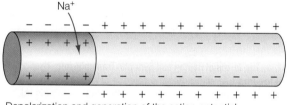

Depolarization and generation of the action potential

(c)

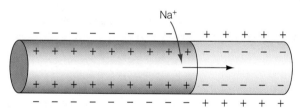

Propagation of the action potential

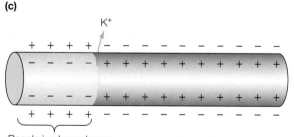

Repolarized membrane

(d)

FIGURE 11–5 **Action Potential** (a) Stimulating the neuron creates a bio-electric impulse, which is recorded as an action potential. The resting potential shifts from 270 millivolts to 130 millivolts. The membrane is said to be depolarized. This graph shows the shift in potential and the change in the permeability of sodium (Na⁺) and potassium (K⁺) ions, which is largely responsible for the action potential. (b) The influx of sodium ions and the depolarization that occur at the point of stimulation. (c) The impulse travels along the membrane as a wave of depolarization. (d) The efflux of potassium ions restores the resting potential, allowing the neuron to transmit additional impulses almost immediately.

second) that neurons can be stimulated in rapid succession. The quick recovery allows us to perform rapid muscle movements. Nerve cells can also transmit many impulses one after another because only a small number of sodium ions are exchanged with each nerve impulse.

While depolarization results in the rapid inflow of sodium ions, repolarization occurs as a result of two factors: (1) a sudden decrease in the membrane's permeability to sodium ions that stops the influx of sodium ions and (2) a rapid outflow of positively charged potassium ions (Figure 11-5). The efflux of potassium ions reestablishes the resting potential.

When a neuron is no longer stimulated, the sodium-potassium pump quickly reestablishes the sodium and potassium ion concentrations inside and outside the cell. It does this by pumping sodium ions that flowed into the cell during activation out of the axon and by pumping potassium ions that flowed out of the neuron during repolarization back in.

> **KEY CONCEPTS**
> Stimulating a nerve cell results in a change in the permeability to ions, resulting in a change in the voltage measured across the plasma membrane; this change is called the action potential.

Nerve Cell Transmission

Researchers have found that depolarization in one area of a neuron stimulates depolarization in adjacent regions. This occurs because depolarization in one region increases membrane permeability in adjacent regions, resulting in an influx of sodium ions and subsequent depolarization of the new region. This process continues along the length of axons and dendrites.

In unmyelinated fibers in the human nervous system, nerve impulses travel like waves in water from one region to the next. In myelinated fibers, however, the depolarization "jumps" from one node of Ranvier to another. Shown in Figure 11-4, impulse jumping greatly increases the rate of transmission. In fact, a nerve impulse travels along an unmyelinated fiber at a rate of about 0.5 meter per second (1.5 feet per second). That's a little over one mile per hour. In a myelinated neuron, the impulse travels 400 times faster or about 200 meters per second. That's about 450 miles per hour!

> **KEY CONCEPTS**
> Nerve impulses are self-propagating waves of depolarization that move along the length of nerve cells.

Synaptic Transmission

Nerve impulses travel from one neuron to another across a small space that separates them (Figure 11-6b). This juncture is called a **synapse** (SIN-apse). A synapse consists of (1) a terminal bouton (or some other kind of axon terminus), (2) a gap between the adjoining neurons, the **synaptic cleft**, and (3) the membrane of the dendrite or postsynaptic cell (Figure 11-6c). The neuron that transmits the impulse is the presynaptic neuron; the one that receives the impulse is the postsynaptic neuron. Does the impulse simply jump the space like an electric spark?

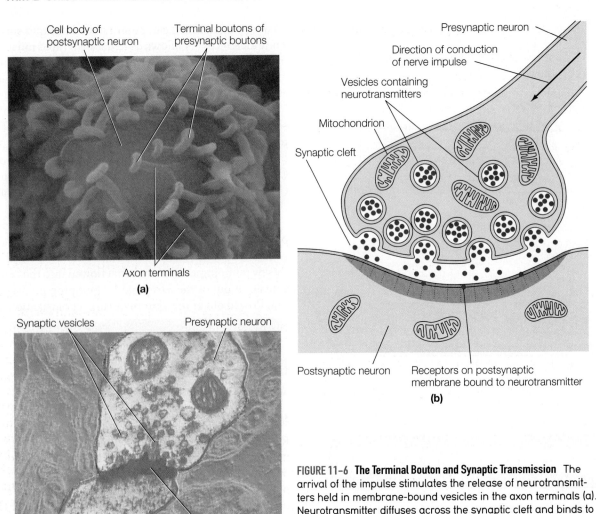

Cell body of postsynaptic neuron

Terminal boutons of presynaptic boutons

Axon terminals

(a)

Synaptic vesicles

Presynaptic neuron

Postsynaptic neuron

Synaptic cleft

(c)

Presynaptic neuron

Direction of conduction of nerve impulse

Vesicles containing neurotransmitters

Mitochondrion

Synaptic cleft

Postsynaptic neuron

Receptors on postsynaptic membrane bound to neurotransmitter

(b)

FIGURE 11–6 The Terminal Bouton and Synaptic Transmission The arrival of the impulse stimulates the release of neurotransmitters held in membrane-bound vesicles in the axon terminals (a). Neurotransmitter diffuses across the synaptic cleft and binds to the postsynaptic membrane (b and c), where it elicits another action potential that travels down the dendrite to the cell body. (Photo a, © E. R. Lewis/Science VU/Visuals Unlimited. Photo c, © T. Reese-D. W. Fawcett/Visuals Unlimited.)

No. The nerve impulse is transmitted chemically. Here's how synaptic transmission occurs. When a nerve impulse reaches a terminal bouton, depolarization of the plasma membrane of the bouton stimulates a rapid inflow of calcium ions. Calcium ions from the extracellular fluid surrounding the bouton flow into the structure. Calcium ions, in turn, stimulate the release (by exocytosis) of a chemical substance stored in small vesicles in the terminal bouton. Take a look at Figure 11-6b and 11-6c. These chemicals are known as **neurotransmitters**. To date, researchers have identified approximately 30 neurotransmitters. One of the most common neurotransmitters is known as **acetylcholine** (a-SEA-tol-KO-leen). It is released by the terminal boutons of neurons that supply the skeletal muscle cells of your body.

Neurotransmitters are released into the synaptic cleft. They migrate to the membrane of the plasma membrane of the postsynaptic neuron where they bind to receptors embedded in the membrane. Binding of a neurotransmitter to receptors of the postsynaptic neurons will either stimulate a rapid increase in the permeability of the membrane of the post-

synaptic cell to sodium ions or a decrease. Whether a nerve impulse travels from one cell to another depends on the sum of the excitatory and inhibitory impulses.

KEY CONCEPTS

Synapses are gaps between nerve cells. Nerve impulses are transmitted across these synapses by neurotransmitters.

Excitation and Inhibition

In the brain, a single nerve cell may form as many as 50,000 synapses with other nerve cells. In some synapses, the neurotransmitters released by the presynaptic neurons stimulate the uptake of sodium ions. This depolarizes the postsynaptic neuron. That is, it makes it more likely to trigger a new action potential. In other synapses, the neurotransmitters make the membrane less excitable. If the number of excitatory impulses exceeds the number of inhibitory impulses, a nerve impulse will be generated in the post synaptic cell. If not, the neuron will not fire. This phenomenon provides the nervous system with a way of integrating incoming

lack of initiative also occur early on. As mental functions decline, patients lose the ability to remember recent events. Over time, losses may be so great that a parent may be unable to recognize his or her own children.

Scientists do not know the cause of Alzheimer's, although there is some evidence to suggest that previous severe brain injuries could be the cause in some instances. In a very small number of cases, the cause is genetic, that is, the disease is inherited from one's parents. Lifestyle may also play a role. Researchers are studying the link between cognitive decline (loss of brain function) and between several "lifestyle" diseases running rampant in developed countries, including heart disease, stroke, high blood pressure, diabetes, and obesity. As you have learned, these are diseases caused largely by an unhealthy lifestyle—a lack of exercise and exceedingly poor diets rich in fatty foods (including snack foods) and low in vegetables, fruits, whole grain products.

The brains of patients with Alzheimer's contain fibrous tangles and deposits of a material called plaque. The fibrous clumps are protein from degenerated nerve cells. The plaque consists of a protein known as beta-amyloid. Research suggests that beta-amyloid binds to nerve cells in the brain and then causes cell death. This leads to the loss of memory and other symptoms. The tangles of fibers are believed to be remnants of the cells' proteins. They're formed when beta-amyloid binds to the nerve cells.

Health Tip 11-6

To reduce your chances of developing Alzheimer's, eat a healthy diet with lots of fruits and vegetables. Limit your intake of saturated fats found in snack foods, higher-fat content dairy products, and meats. And be sure to get plenty of exercise.

Why?

All the steps you take to reduce weight gain and risk of heart disease also appear to lower your chances of developing Alzheimer's disease.

Although there is no cure, there are several drugs that can be given to patients with Alzheimer's to slow down the deterioration of the brain. Experiments with mice suggest that antibodies to beta-amyloid could someday be given to humans to prevent or slow or even stop the development of Alzheimer's disease. When mice that had the disease were given these antibodies, beta-amyloid accumulation in the brain was prevented. Existing deposits were removed. Results of experiments on eight Alzheimer's patients released in 2005 suggest that antibodies to beta-amyloid work similarly in humans. If further research bears out this result, those who suffer from this devastating disease or who will fall victim to it could be greatly benefited. Currently, there are numerous clinical trials of various Alzheimer's treatments meant to slow or halt the disease. Many researchers believe that treatment may call on a combination of drugs as is common in certain forms of cancer and AIDS.

KEY CONCEPTS

Loss of memory and confusion in older individuals is often the result of Alzheimer's disease.

Parkinson's Disease

Another important nervous system disorder is **Parkinson's disease**. Parkinson's disease is caused by a progressive deterioration of certain brain centers that control movement, especially semiautomatic movements such as swinging the arms when walking. It afflicts about 1 million Americans, most over the age of 60. In fact, approximately 1 in 100 Americans in this age group has the disease.

Parkinson's disease is characterized by tremors. In many people with the disease, the hands or the head (sometimes both) shake involuntarily. Tremors decrease or disappear, however, when one moves the afflicted part of the body intentionally—for example, when one reaches for an object. If the disease worsens, symptoms worsen and it becomes difficult to write or walk. Speaking may deteriorate too, as it becomes difficult to move one's mouth and tongue. Falls become more frequent. It may be difficult to get out of a chair. In later stages of the disease, memory and thinking may deteriorate.

Like Alzheimer's, the cause or causes of Parkinson's are unknown. Environmental chemical pollutants may contribute to the disease. Amalgam fillings in teeth and chemicals used in wood stains may play a role in the onset of the disease. In some cases, Parkinson's may result from an inherited genetic mutation in a chromosome.

Parkinson's is characterized by a lack of a chemical substance in the brain called *dopamine*. Because of this, patients can be treated with a drug known as levodopa. Its benefits may diminish over years of use, however. Some patients suffer serious side effects, including severe nausea and vomiting. Several other drugs are also used.

Some studies show that stem cell transplants into the affected brain area may eliminate the symptoms of the disease. In one study, a year after receiving a stem cell transplant, 13 of 19 people showed elevated levels dopamine production. Newer studies reveal conflicting results. Some show significant improvement after transplanting fetal stem cells; others don't. Surgeons can also destroy a small region of the brain that appears to be hyperactive in Parkinson's patients. This improves a patient's score on standardized movement tests, although several patients had serious side effects including facial paralysis. Electrodes can also be implanted in the brain to stop uncontrollable movements. Patients are equipped with tiny electronic stimulators, much like pacemakers, that can be switched on or off as needed. Researchers in England have discovered a drug that may be helpful in reversing Alzheimer's, though it does not pass through the blood-brain barrier. They are currently searching for related compounds that might work.

KEY CONCEPTS

Parkinson's disease is a neurological disorder caused by a lack of dopamine in a region of the brain that helps control movement.

Multiple Sclerosis

Destruction of the myelin sheath of nerve cells in the central nervous system results in a condition known as **multiple sclerosis** (skler-OH-siss; MS). The damaged myelin results in nerve cell death that leads to numbness, slurred speech, and paralysis. Thought to be an autoimmune disease, multiple sclerosis can affect any part of the CNS. Early symptoms are generally mild weakness or a tingling or numb feeling in one part of the body. Temporary weakness may cause a person to stumble and fall. Some people report blurred vision, slurred speech, and difficulty controlling urination. In many cases, these symptoms disappear, never to return. Other individuals suffer repeated attacks. Because recovery after each attack is incomplete, patients gradually deteriorate, losing vision and becoming progressively weaker. Fortunately, many treatments are available, and only a small number of multiple sclerosis patients are crippled by the disease.

> **KEY CONCEPTS**
> Multiple sclerosis (MS) is believed to be an autoimmune disease caused by a destruction of the myelin sheath of nerve cells in the CNS leading to numbness, slurred speech, and paralysis.

Brain Tumors

Two types of tumors develop in the brain tissue, benign and malignant. Benign tumors are cellular growths in the brain that do not grow uncontrollably and do not spread to other parts of the body. Even as such, benign tumors can cause problems. They may, for instance, place pressure on crucial areas of the brain, resulting in problems. Benign tumors have distinct borders and can be removed surgically; unlike malignant tumors, they do not recur.

Malignant tumors grow rapidly, especially late in the disease. They are likely to place pressure on neighboring structures and may invade into the tissue around them or to other parts of the body.

Brain tumors occur at any age, but the most common affected groups are children in the 3- to 12-year-old age bracket and adults in the 40- to 70-year-old bracket. Although researchers still do not know what causes brain tumors, there are some risk factors that increase one's likelihood for developing a cancerous tumor. Workers in certain industries, for example, oil refining, rubber manufacturing, and drug manufacturing, are at a higher risk of developing a brain tumor. Chemists and embalmers have a higher incidence of brain tumors as well. These tumors may result from exposure to chemicals in the workplace.

Some researchers believe that exposure to certain viruses may be responsible for brain tumors. And some brain tumors seem to run in families, so there may be a genetic cause that is passed from parents to offspring. In most cases, however, patients with brain tumors have no clear risk factors, so researchers believe that the disease is probably the result of several factors acting together.

There's a great deal of concern these days over the potential of cell phones to cause brain cancer. Although several recent studies, including one study that tracked subjects for 10 years, show no link between cell phone use and brain cancer, remember that most cancers take 20 to 30 years to develop. Another fact to keep in mind is that cell phones emit radio waves. Unlike ionizing radiation, for example, from nuclear power plants, radio waves do not impart much energy in body tissues and may have little, if any, effect on cells of the body. Only time will tell.

> **KEY CONCEPTS**
> Brain tumors afflict the young and the old and may have many causes.

11-8 Health and Homeostasis

In healthy individuals, a decline in blood glucose is normally prevented by homeostatic mechanisms. Serious problems can result in people whose glucose homeostasis is not operating properly. Many changes may occur in the brain and the behaviors controlled by this mysterious organ. In diabetics, for instance, blood glucose levels can fall dangerously low if too much insulin is taken or if not enough glucose is ingested. Deprived of glucose, brain cells begin to falter. Individuals may become dizzy and weak. Vision may blur. Speech may become awkward. Diabetics are sometimes mistaken for drunks. Chronic low blood glucose often results in severe headaches.

Some diabetics become aggressive when blood sugar levels fall. In extreme cases, low blood sugar triggers convulsions and death. Interestingly, other body cells are not adversely affected by a decline in blood glucose because, unlike the brain, they can switch to alternative fuels, notably fats and amino acids, to provide cellular energy needed to function.

> **KEY CONCEPTS**
> The brain requires blood glucose for energy. A lack of glucose, caused by disorders occurring in other organs, notably the pancreas, can result in numerous adverse effects on the brain.

SUMMARY

An Overview of the Nervous System

1. The nervous system controls a wide range of functions and plays a key role in ensuring homeostasis. The human nervous system also performs many other functions, such as controlling muscle movement and higher mental functions such as planning for the future, learning, and remembering.

2. The nervous system consists of the central nervous system (CNS), made up of the brain and spinal cord, and the peripheral nervous system (PNS), which consists of the spinal and cranial nerves.

3. Receptors in the skin, skeletal muscles, joints, and organs transmit sensory input to the CNS via sensory neurons. The CNS integrates all sensory input and generates appropriate responses. Impulses travel to the effector organs via motor neurons.

4. The PNS has two functional divisions. The autonomic division controls involuntary actions such as heart rate and breathing. The somatic division largely controls voluntary actions such as skeletal muscle contractions. It also provides the neural connections needed for many reflex arcs.

Structure and Function of the Neuron

5. The fundamental unit of the nervous system is the neuron. This highly specialized cell generates and transmits bioelectric impulses from one part of the body to another.

6. All neurons have more or less spherical cell bodies. Extending from the cell body are two types of processes: dendrites, which conduct impulses to the cell body, and axons, which conduct impulses away from the cell body.

7. Many axons are covered by a layer of myelin, which greatly increases the rate of impulse transmission.

8. The terminal ends of axons branch profusely, forming numerous fibers that end in small knobs called terminal boutons.

9. During cellular differentiation, nerve cells lose their ability to divide. Because of this, neurons that die cannot be replaced by existing cells.

10. The small electrical potential across the membrane of nerve cells is known as the membrane potential or resting potential.

11. When a nerve cell is stimulated, its plasma membrane increases its permeability to sodium ions. Sodium ions rush in, causing depolarization, which spreads down the membrane.

12. Depolarization is followed by repolarization, a recovery of the resting potential stemming largely from the outflow of potassium ions. The depolarization and repolarization of the neuron's plasma membrane constitute an action potential.

13. When a nerve impulse reaches the terminal bouton, it stimulates the release of neurotransmitters contained in membrane-bound vesicles. They bind to receptors in the postsynaptic membrane.

14. Neurotransmitters may excite or inhibit the postsynaptic membrane. Whether a neuron fires depends on the sum of excitatory and inhibitory impulses it receives.

The Spinal Cord and Nerves

15. The spinal cord descends from the brain through the vertebral canal to the lower back. It carries information to and from the brain, and its neurons participate in many reflexes.

16. Two types of nerves emanate from the CNS: spinal and cranial. Spinal nerves arise from the spinal cord and may be sensory, motor, or mixed. Cranial nerves attach to the brain and supply the structures of the head and several key body parts.

The Brain

17. The brain is housed in the skull. The cerebrum with its two cerebral hemispheres is the largest part of the brain.

18. The cerebral cortex consists of many discrete functional regions, including motor, sensory, and association areas.

19. Consciousness resides in the cerebral cortex, but a great many functions occur at the unconscious level in parts of the brain beneath the cortex.

20. The cerebellum coordinates muscle movement and controls posture. The hypothalamus regulates many homeostatic functions. The limbic system houses instincts and emotions. The brain stem, like the hypothalamus, regulates basic body functions.

21. A watery fluid known as cerebrospinal fluid surrounds the brain and spinal cord. It cushions the brain and spinal cord, protecting them from traumatic injury.

22. Electrodes applied to different parts of the scalp detect electrical activity in the brain and produce a tracing known as an electroencephalogram (EEG). The type of brain wave recorded depends on one's level of cortical activity. EEGs are used to diagnose some brain dysfunctions.

23. Headaches are the most common form of pain and generally result from tension, swelling of the membranes lining the sinuses, eyestrain, or dilation of the cerebral blood vessels. Very serious headaches may result from increased intracranial pressure caused by a brain tumor, intracranial bleeding, or inflammation caused by an infection of the meninges (meningitis) or the brain itself (encephalitis). Less threatening, but still extraordinary, are migraines, a form of severe, recurrent headache.

The Autonomic Nervous System

24. The autonomic nervous system or ANS is a division of the peripheral nervous system. It supplies all internal organs and has two subdivisions: the sympathetic and the parasympathetic.

25. The sympathetic division of the ANS functions in emergencies and is responsible in large part for the fight-or-flight response. The parasympathetic division of the ANS brings about a relaxed state.

Learning and Memory

26. Learning is a process in which an individual acquires knowledge and skills.

27. Newly acquired knowledge is first stored in short-term memory. Short-term memories may be transferred to long-term memory, which holds information for periods of days to years.

28. Memory is stored in multiple regions of the brain.

29. Short- and long-term memories appear to involve structural and functional changes of the neurons.

Diseases of the Brain

30. Alzheimer's disease is a progressive loss of mental function, which usually begins later in life.

31. Parkinson's disease is a neurological disorder caused by a progressive deterioration of certain brain centers that control movement. It is characterized by tremors that typically worsen as one ages. Like Alzheimer's, the cause or causes of Parkinson's are unknown.

32. Multiple sclerosis is a disease caused by degeneration of the myelin sheath of axons. Damage to the myelin results in nerve cell death that leads to numbness, slurred speech, and paralysis.

33. Two types of tumors develop in the brain tissue, benign and malignant. Although benign tumors do not grow uncontrollably or spread to other areas, they can cause problems by placing pressure on areas of the brain.

34. Malignant tumors grow rapidly, and often place pressure on neighboring structures. They may also invade the tissue around them or may spread to other parts of the body.

Health and Homeostasis

35. Abnormal brain activity can result from a disruption of homeostasis, for example, in the mechanisms that control blood glucose levels.

THINKING CRITICALLY ANALYSIS

This analysis corresponds to the Thinking Critically scenario that was presented at the beginning of this chapter.

Although this study may sound scientific, it is not, and the results are meaningless. They are based solely on personal testimonials—what people believe has happened to them. Furthermore complicating matters, the patients were told in advance what to expect. Knowing the intent of a treatment can fool people into believing that they've been changed.

As it turns out, the researchers were well aware of the pitfalls of their study. They did it to prove a point, notably that advance knowledge of the expected outcome can bias the results.

The researchers conducted a second experiment. They analyzed two additional groups. The first group was given the memory tape, but subjects were told it was a self-esteem tape. The second group was given the self-esteem tape, but subjects were told it would improve their memory.

A month later, subjects who received the fake self-esteem tape reported improved self-esteem. Those who received the fake memory tape reported better memory. These results illustrate once again that advance knowledge of the outcome of an experimental treatment can convince people that they've been altered. Put another way, people thought their memory or self-esteem improved because that's what they were expecting.

The researchers didn't stop here, however. They actually measured memory and self-esteem before and after the experiment and found no improvement in either group. In other words, the tapes were meaningless.

What's the lesson in all of this?

Beware of studies that rely on testimonials. To test for changes, patients must be unaware of the anticipated outcome. That requires a tightly controlled experiment.

KEY TERMS AND CONCEPTS

Acetylcholine, p. 232
Acetylcholinesterase, p. 233
Action potential, p. 230
Anesthetics, p. 233
Association cortex, p. 240
Autonomic nervous system, p. 227
Axon, p. 228
Brain, p. 237
Central nervous system, p. 226
Cerebellum, p. 240
Cerebral cortex, p. 238
Cerebral hemispheres, p. 238
Cerebrospinal fluid (CSF), p. 242
Cerebrum, p. 238
Cranial nerve, p. 237
Dendrite, p. 228
Depolarization, p. 230
Electroencephalogram, p. 243
Fight-or-flight response, p. 245

Headaches, p. 243
Hypothalamus, p. 240
Learning, p. 245
Limbic system, p. 240
Long-term memory, p. 245
Memory, p. 245
Meninges, p. 226
Meningitis, p. 242
Migraine, p. 243
Multiple sclerosis, p. 247
Myelin sheath, p. 228
Nerve, p. 227
Nerve impulse, p. 230
Neuron, p. 228
Neurotransmitter, p. 232
Node of Ranvier, p. 228
Parkinson's disease, p. 247
Peripheral nervous system, p. 227
Premotor cortex, p. 239

Primary motor cortex, p. 239
Primary sensory cortex, p. 239
Reflex arc, p. 235
Remembering, p. 245
Repolarization, p. 230
Resting potential, p. 230
Reticular activating system, p. 241
Reticular formation, p. 241
Short-term memory, p. 245
Sodium-potassium pump, p. 230
Somatic nervous system, p. 227
Spinal cord, p. 235
Spinal nerve, p. 235
Stroke, p. 229
Synapse, p. 231
Synaptic cleft, p. 231
Terminal boutons, p. 228
Thalamus, p. 240
Vertebral canal, p. 235

CONCEPT REVIEW

1. The nervous system performs many functions. What functions do you think are unique to humans? p. 226.
2. Draw a typical multipolar neuron, and label its parts. Show the direction in which an action potential travels. p. 228.
3. Draw a typical synapse, label the parts, and explain how a nerve impulse is transmitted from one nerve cell to another. pp. 231–232.
4. Describe how the resting and action potentials are generated. Describe how the plasma membrane of a neuron is repolarized. pp. 230–231.

5. How do common insecticides affect synaptic transmission? p. 233.
6. How do antidepressants increase serotonin levels in the brain? p. 233.
7. Name and describe the various divisions of the nervous system. pp. 226–233, 235–243.
8. Draw a cross section through the spinal cord showing the spinal nerves. Label the parts, and explain a reflex arc and how nerve impulses entering a spinal nerve also travel to the brain. pp. 235–237.

9. A physician can stimulate various parts of the brain and get different responses. What effects would you expect if the electrodes were placed in the primary motor cortex? The primary sensory cortex? pp. 239–240.
10. What is association cortex? Some association cortex carries out specific functions. What are these areas and what functions do they perform? p. 240.

SELF-QUIZ: TESTING YOUR STRENGTH

1. The brain and spinal cord form the _____ nervous system. p. 226.
2. Three layers of connective tissue surround the brain; they're known as the _____. p. 226.
3. Nerves carry _____ impulses from receptors in the body to the brain and spinal cord. p. 227.
4. The spinal and cranial nerves are part of the _____ nervous system. p. 227.
5. Nerve cells or neurons contain a cell body and two processes. The one that carries impulses away from the cell body is known as the _____. p. 228.
6. Axons branch at their ends, terminating in small swellings known as _____. p. 228.

7. A nerve impulse is caused by a change in the permeability of the plasma membrane of nerve cells that results in an influx of _____ ions. p. 230.
8. An electrical tracing of the change in the membrane potential of a nerve cell is called an _____. p. 243.
9. Nerve impulses travel from one nerve cell to another, thanks to the release of a chemical from the presynaptic neuron known as _____. p. 232.
10. Located in the spinal canal, the _____ extends from the brain to the lower back. p. 235.
11. The largest portion of the brain is known as the _____. p. 238.

12. The part of the cerebral cortex that receives sensory information is known as the _____ cortex. p. 239.
13. The _____ ensures muscle synergy—the coordination of muscular activity. p. 240.
14. The _____ consists of many aggregations of nerve cells that control a variety of autonomic functions such as appetite and body temperature. p. 240.
15. The brain is kept alert, thanks to nerve impulses arising from the _____ activating system. p. 241.

 biology.jbpub.com/chiras/8e/

The site features eLearning, an online review area that provides quizzes, chapter outlines, and other tools to help you study for your class. You can also follow useful links for in-depth information, research the differing views in the Point/Counterpoints, or keep up on the latest health news.

(© Sunny studio-Igor Yaruta/ShutterStock, Inc.)

The Senses

Debra Cartwright noticed something strange. Over a period of weeks she had slowly lost her senses of taste and smell. Doctors were puzzled at first, as Debra seemed to be in fine health. She was not suffering from a cold, sinus infection, or even any allergies that might have blocked her nasal passages and impaired her senses of taste or smell. Her doctors ordered a blood test. One physician noticed that blood levels of the micronutrient zinc were quite low. She prescribed a zinc supplement, and in short order Debra's senses of taste and smell returned.

THINKING CRITICALLY

Many people believe in extrasensory perception (ESP), the ability to perceive things by a sense supposedly outside of the conventional senses—sight and hearing, for instance. One scientist (Michael Shermer) who wanted to study ESP went to an organization that conducts experiments on ESP to see what it was all about. When he arrived, an experiment was under way. Individuals were asked to discern the shapes that another person was looking at. The shapes were printed on cards and consisted of a plus sign, a square, a star, a circle, or a wavy line. The person looking at the cards—the sender—was asked to concentrate on the symbol. A second individual—the receiver—was told to concentrate on the sender's forehead and attempt to sense what he was seeing. All told, 35 people participated in the experiment.

The instructor told the group that when asked to identify 25 cards, individuals should correctly identify 5 out of 25 cards by chance. That's because there is a one in five chance of guessing each card correctly. He went on to say that anyone who identifies seven or more cards correctly has ESP.

In the present group, three people correctly identified eight cards in the first trial. Did these people really have ESP?

This story illustrates the importance of a nutritionally balanced diet and the impact of a dietary deficiency on two important body functions—the senses of taste and smell. It is just one of many examples presented in this book that illustrates how body functions can be altered by factors such as stress, pollution, and diet. This chapter examines the senses, dividing the discussion into two broad categories: the general senses and the special senses.

12-1 The General and Special Senses

Your health and survival depend on a sophisticated surveillance system that keeps track of changes inside and outside your body. This surveillance system consists of numerous receptors that monitor internal and external conditions. In humans, receptors are located in the skin, internal organs, bones, joints, and muscles. They detect stimuli that give rise to the **general senses**: pain, temperature, light touch, pressure, and a sense of body and limb position (Table 12-1). All of these senses allow us to function in an ever-changing and sometimes dangerous world.

The human body is also endowed with five additional senses, known as the **special senses**. They are taste, smell, vision, hearing, and balance. They are made possible by sophisticated sensory organs such as the eye. Receptors in humans involved in the general and special senses fit into five categories: (1) mechanoreceptors, (2) chemoreceptors, (3) thermoreceptors, (4) photoreceptors, and (4) nociceptors (pain receptors). **Mechanoreceptors** are those activated by mechanical stimulation—for example, touch or pressure. **Chemoreceptors** are activated by chemicals in the food we eat, in the air we breathe, or in our blood. **Thermoreceptors** are activated by heat and cold, and **photoreceptors** are sensitive to light. **Nociceptors** (no-see-SEP-tors) are pain receptors stimulated by pinching, tearing, or burning.

> **KEY CONCEPTS**
>
> The human body contains receptors that monitor numerous internal and external stimuli essential for homeostasis and our well-being.

TABLE 12-1	Summary of General and Special Senses	
Sense	**Stimulus**	**Receptor**
General senses	Pain	Naked nerve endings
	Light touch	Merkel's discs; naked nerve endings around hair follicles; Meissner's corpuscles; Ruffini's corpuscles; Krause's end-bulbs
	Pressure	Pacinian corpuscles
	Temperature	Naked nerve endings
	Proprioception	Golgi tendon organs; muscle spindles; receptors similar to Meissner's corpuscles in joints
Special senses	Taste	Taste buds
	Smell	Olfactory epithelium
	Sight	Retina
	Hearing	Organ of Corti
	Balance	Crista ampularis in the semicircular canals; maculae in utricle and saccule

12-2 The General Senses

Sit back in your chair for a moment, close your eyes, and concentrate on what you feel. You may detect the pressure of the chair on your back and heat from a reading lamp. You may feel your cat brushing against the hairs on your arm. You may detect pain from gas pressure in your intestines from the burritos you ate yesterday. Now move your arm. Even though your eyes are closed, you can feel it moving.

The various sensations you have just experienced fall into the group of general senses. Receptors for the general senses detect internal and external stimuli and relay messages to the spinal cord and brain via sensory nerves.

Sensory input to the central nervous system may elicit a conscious response—for example, the touch of a cat may cause you to pet your furry friend. Some stimuli will cause unconscious responses—for example, heat may cause you to perspire. Others may simply be registered in the cerebral cortex, making you aware of the stimulus, but do not elicit a response. Still other stimuli may be blocked so that they are not perceived at all.

Receptors for the general body senses come in many shapes and sizes, but they generally fit into two groups based on structure: naked nerve endings and encapsulated receptors (Figure 12-1).

health**note**

12-1 Old and New Treatments for Pain

Millions of people suffer from persistent pain. In the United States, for example, an estimated 70 to 80 million people are tormented by back pain. Another 36 million suffer from migraine headaches.

Minor aches and pains can be treated with painkillers such as acetaminophen and ibuprofen. More intense pain is treated with stronger painkillers, such as codeine. Codeine is often mixed with ibuprofen and acetaminophen. Although codeine and other strong painkillers are used today, codeine is addictive, so doctors are sometime reluctant to prescribe it for long-term pain control. Pain can also be treated by stress relief and exercise, as noted in the accompanying Health Tip. Being that I spend long hours at the computer working on books, I've suffered excruciating upper back and neck pain. I found that exercises that strengthened the muscles of my upper back dramatically reduced pain. I take a break every 50 to 60 minutes to engage in physical activity. This helps too. Careful attention to the height of my computer screen and use of an ergonomic keyboard also help me prevent pain. A good desk chair has helped too. You may find some similar changes helpful if you suffer from pain. Even purchasing a more comfortable mattress or altering the way you sit at your desk can dramatically reduce pain.

Another proven pain reliever is acupuncture. Used by the Chinese for thousands of years, acupuncture relies on thread-thin needles inserted in the skin near nerves (Figure 1). No one knows how acupuncture works. The Chinese assert that it encourages the free flow of energy through various pathways in the body whose purpose is to irrigate and nourish the tissues. They contend that blood and nervous impulses follow these pathways and that they can become blocks. Any obstruction in the movement of the energy flow, they contend, results in deficiencies of energy and blood, as well as pain. The acupuncture needles supposedly unblock obstructions and reestablish normal flow.

The scientific explanation for the way acupuncture works is that this technique blocks pain by overloading the neuronal circuitry. They note that two types of nerve fibers transmit sensory information from the body to the central nervous system: small- and large-diameter fibers. Small-diameter fibers carry pain messages. Large-diameter fibers carry many other forms of sensory information from receptors in the skin—for example, pressure and light touch. The dendrites of both small- and large-diameter sensory nerve cells often terminate in the same location in the spinal cord. From here, they send impulses to the brain, signaling pain or some other sense.

Scientists believe that acupuncture needles stimulate the large-diameter nerve fibers. This stimulation blocks nerve impulses carried by the smaller nerve fibers, blocking pain messages to the brain.

Research on the mechanism of action of acupuncture points to several other possibilities as well. Acupuncture needles, for instance, may stimulate the release of endorphins, the body's natural painkillers. Endorphins are produced in the brain.

Many U.S. physicians trained to use drugs and surgery to solve most pain remain skeptical about the usefulness of acupuncture. Over the past 30 years, however, a small but steady stream of research has confirmed the painkilling effect of this treatment. Today, most insurance companies will pay the cost of acupuncture to relieve pain caused, for example, by a car crash.

Joseph Helms, a physician with the American Academy of Acupuncture in Berkeley, California, performed acupuncture on 40 women with menstrual pain. Some women received real acupuncture treatment. Others received placebo treatments (shallow needle treatments that did not reach the acupuncture points). In the group of women receiving acupuncture, 10 out of 11 showed a marked decrease in pain. Patients reported an approximately 50% decrease in pain. In the placebo group, only 4 out of 11 reported a lessening of pain. Only 1 of 10 people given no treatment showed improvement. Acupuncture also reduced the need for painkilling drugs during treatment by more than 50%.

Muscle relaxants can also help reduce pain. I have found that they can work wonders when combined with other techniques such as exercise, stretching, massage, acupuncture, a change of mattress, and other adjustments. Some healthcare practitioners are using another procedure called *transcutaneous electrical nerve stimulation* (TENS). Patients are fitted with electrodes attached to the skin above the nerves that transmit pain signals to the central nervous system (CNS). When the pain begins, patients press a button on a battery pack that sends a tiny current to the electrode. Conducted through the skin, the current blocks the pain impulses. TENS can be used to reduce pain after surgery.

Severe pain can also be treated by surgery. Doctors may cut nerves, sever nerve tracks in the spinal cord that carry pain, or even destroy small parts of the brain to rid a patient of chronic pain. Nerves can also be destroyed by applying certain chemical substances.

Although all of these methods are effective, there are dangers in these procedures. Cutting nerve tracks in the spinal cord, a structure that's approximately the size of your little finger, can be risky. One slip and a surgeon could destroy vital motor neurons. Nerves carrying pain signals also transmit sensory information. Some doctors think that patients should not risk loss of other senses or motor functions for the sake of pain relief. In addition, pain recurs in nine of ten patients who have undergone pain-relieving surgery, usually within a year or so, as a result of the partial regrowth of axons. Even after another operation, the pain frequently returns, often with much greater intensity.

Another promising measure is deep brain stimulation. Electrodes can be implanted in parts of the brain and stimulated to block pain impulses before they reach the sensory cortex, where pain is perceived. The electrodes are connected to a portable battery worn on the belt or implanted under the skin. When the pain begins, the patient turns on the current, blocking the pain impulses. Research shows that deep brain stimulation is a very effective blocker of even the most powerful pain stimuli.

Chronic pain is one of medicine's greatest enigmas. Over the years, I have found that a holistic approach to pain works extremely well. It greatly reduces or eliminates pain and can prevent its recurrence. If you are suffering from pain, take stock of your life. Where's the pain located? What can you do to alleviate or eliminate it? What exercises will help strengthen the muscles in the area? Take breaks while you study or work that allow muscles to relax. Try stretching a little before and after exercise or during the day. Try changes in body position during sleep, work, and study. You'll be amazed at how small things can add up. But be patient. It took years to develop chronic pain. It could take a while to achieve relief!

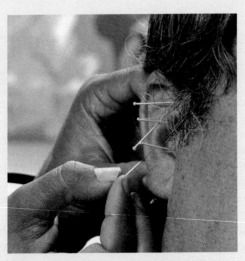

FIGURE 1 Patient Undergoing Acupuncture Acupuncture is generally used to alleviate pain, but it has other applications as well. (© Photodisc.)

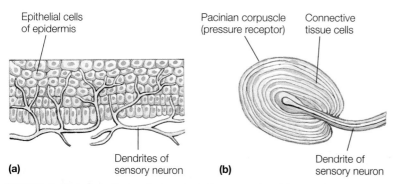

Epithelial cells
of epidermis

Pacinian corpuscle
(pressure receptor)

Connective
tissue cells

(a)

Dendrites of
sensory neuron

(b)

Dendrite of
sensory neuron

FIGURE 12–1 Receptors General sense receptors are either (a) naked nerve endings or (b) encapsulated nerve endings.

Health Tip 12-1

Suffering from back pain? Try stress relief measures like meditation and yoga to help ease the pain.

Why?

Studies show that emotional stress often worsens back pain. It can even trigger back pain. Subjects suffering from psychological stress are three times more likely to develop back pain than those who are not stressed. Relieving stress through exercise, psychotherapy, relaxation, yoga, Pilates, and other measures can reduce back pain.

Naked Nerve Endings

Naked nerve endings are the ends of dendrites of sensory neurons. They are found in abundance in your skin, bones, and internal organs and in and around joints. Naked nerve endings are responsible for at least three sensations: pain, light touch, and temperature.

Pain. Your body perceives two basic types of pain: somatic and visceral. Somatic pain results from injuries in the skin, joints, muscles, and tendons. You can think of it as external pain, for reasons that will be clear very soon. Somatic pain receptors alert us to several types of stimuli. Some pain receptors, for instance, send impulses when we are cut. Others respond to crushing or pinching. Still others respond to temperature. Some pain fibers respond to irritating chemicals released from injured tissues. All these sensations alert the brain to danger.

Visceral pain is internal pain. It results from the stimulation of naked nerve endings inside the body, specifically in the organs (the viscera) of the body. Organs are also referred to as the viscera. In some organs, pain receptors are stimulated by internal expansion. For example, the intestinal pain you feel after eating a bean burrito, for instance, results from gas released by bacteria in your intestine. Gas, in turns, stretches naked nerve fibers in the wall of the intestine, causing pain. These nerve endings are mechanoreceptors. In many organs, pain receptors are stimulated by a lack of oxygen, or anoxia (ah-NOCKS-see-ah). For example, the pain one feels during a heart attack results from the lack of oxygen in the heart muscle. The nerve fibers send pain impulses to the brain, altering it to the danger.

As you can see, somatic pain and visceral pain are caused by very different stimuli, they are also perceived differently. Somatic pain is easily pinpointed. A sore joint or bruised muscle is pretty easy to locate. The origin of visceral pain, however, is often difficult to localize. Making matters worse, it is often felt on the body surface at a site some distance from its origin. For example, pain caused by a lack of oxygen to the heart muscle appears in the neck and jaw, and along the inside of the left arm (Figure 12-2).

Visceral pain that appears on the body surface away from the location of the source of the pain is called **referred pain**. Physiologists do not know the cause of this phenomenon, but most think it is because pain fibers from internal organs enter the spinal cord at the same location that the sensory fibers from the skin enter. The brain, they hypothesize, interprets the impulses from pain fibers supplying internal organs as pain from a

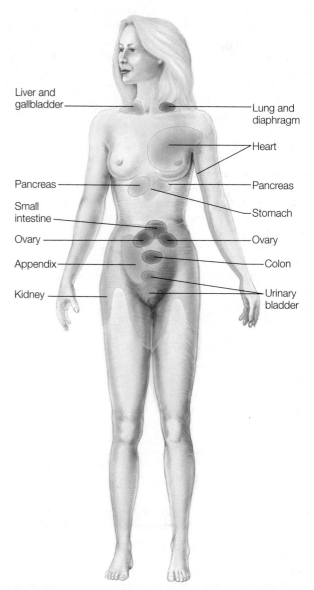

Liver and
gallblader

Lung and
diaphragm

Heart

Pancreas

Pancreas

Small
intestine

Stomach

Ovary

Ovary

Appendix

Colon

Kidney

Urinary
bladder

FIGURE 12–2 Referred Pain Visceral pain is often felt on the body surface at the points indicated by the colored areas.

somatic source. (**Health Note 12-1** describes techniques that relieve pain.)

Light Touch. Gently brush the hair on your arm or touch the skin on your forearm very lightly. The sensation you felt is called light touch. Light touch is perceived by two anatomically distinct mechanoreceptors. As shown in Figure 12-3, the first type consists of naked nerve endings (dendrites) that wrap around the base of the hair follicles. When a hair is moved—for example, when you brush the hair on the back of your arm—these nerve fibers are stimulated.

The second light-touch mechanoreceptor, Merkel's disc, is also shown in Figure 12-3. This structure consists of small cup-shaped cells and naked nerve endings that make contact with them. Located in the outer layer of the epidermis of the skin, these receptors are activated by gentle pressure applied to the skin. They were activated when you gently touched your skin. Go ahead. Try it again.

Temperature. Certain naked nerve endings in the skin detect heat and cold within specific temperature ranges. If the temperatures are hotter or colder, pain receptors are activated. Temperature receptors are important for many reasons. Can you think of any?

> **KEY CONCEPTS**
>
> Naked nerve endings in tissues detect pain, temperature, and light touch.

Encapsulated Receptors

Shown in Figure 12-3, the largest encapsulated nerve ending is the Pacinian corpuscle (pah-SIN-ee-an CORE-puss-l). It consists of a naked nerve ending surrounded by numerous concentric cell layers. The Pacinian corpuscle resembles a small onion pierced by a thin wire. Located in the deeper layers of the skin, in the loose connective tissue of the body, and elsewhere, Pacinian corpuscles are stimulated by pressure such as the pressure you feel sitting in your chair.

Another common encapsulated sensory receptor is the Meissner's corpuscle (MEIZ-ners). Smaller than Pacinian corpuscles, these oval receptors contain two or three spiraling dendritic ends surrounded by a thin cellular capsule (Figure 12-3). Like the naked nerve endings wrapped around the base of hair follicles and Merkel's discs, Meissner's corpuscles are thought to respond to light touch. Located in the outer-most layer of the dermis, Meissner's corpuscles are most abundant in the sensitive parts of the body, such as the lips and the tips of the fingers.

Two additional encapsulated receptors are Krause's end-bulbs and Ruffini's corpuscles (Figure 12-3). These receptors are probably variations of the Meissner's corpuscle and are stimulated by light touch.

Proprioception (PRO-pree-oh-CEP-shun) is the sense of position. It is detected by encapsulated receptors located in the joints of the body. Resembling Meissner's corpuscles, these receptors inform us of the position of our limbs and alert us to movements of the body.

Position is also detected by sensors, the muscle spindles, located in skeletal muscles. Muscle spindles consist of several modified muscle fibers with sensory nerve endings wrapped around them. The nerve endings are stimulated when muscles are stretched. Nerve impulses generated by the spindle are then transmitted to the spinal cord, where they may ascend to the cerebral cortex, helping us remain aware of body position. Other nerve impulses may stimulate motor neurons in the spinal cord, eliciting a reflex—a contraction of the muscle being stretched. This is what happens when you fall asleep sitting upright in a chair and your head falls forward—and quickly snaps back.

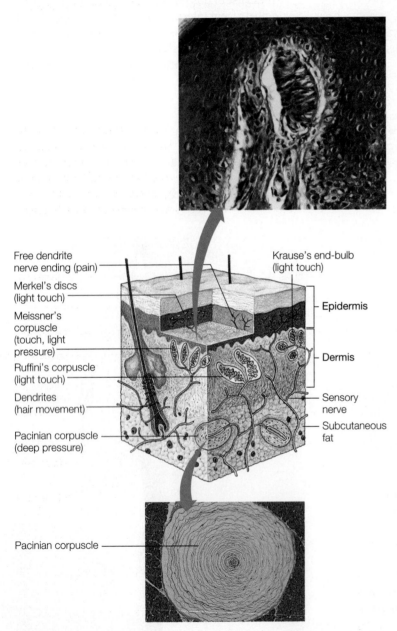

Free dendrite nerve ending (pain)

Merkel's discs (light touch)

Meissner's corpuscle (touch, light pressure)

Ruffini's corpuscle (light touch)

Dendrites (hair movement)

Pacinian corpuscle (deep pressure)

Krause's end-bulb (light touch)

Epidermis

Dermis

Sensory nerve

Subcutaneous fat

Pacinian corpuscle

FIGURE 12-3   General Sense Receptors The skin houses many of the receptors for general senses. Receptors fall into two categories: naked nerve endings and encapsulated receptors. The Pacinian corpuscle, often located in the dermis of the skin, detects pressure. Meissner's corpuscle, found just beneath the epidermis, detects light touch. (Top photo © Astrid & Hans-Frieder Michler/Science Source, Inc. Bottom photo © Cabisco/Visuals Unlimited.)

Another proprioceptor is the Golgi tendon organ (GOAL-jee). These mechanoreceptors, named after the Italian scientist who discovered them, are located in tendons—the structures that connect muscles to bones. Golgi tendon organs are composed of connective tissue fibers surrounded by dendrites and encased in a capsule. When a muscle contracts, the tendon stretches and stimulates the receptor, alerting the brain of movement and body position.

> **KEY CONCEPTS**
> Encapsulated receptors consist of nerve endings surrounded by one or more layers of cells.

Adaptation

Pain, temperature, and pressure receptors are all subject to a phenomenon called **adaptation**. This occurs when sensory receptors stop generating impulses even though the stimulus is still present. You have probably experienced the phenomenon many times. Recall, for example, the first time you wore a ring or contact lenses. At first, the sensation of pressure from the ring or lenses may have nearly driven you mad, but after a short period, the stimulus seemed to have disappeared. What happened was that pressure receptors that were originally alerting the brain stopped generating impulses, releasing you from what could have been unrelenting discomfort.

Not all receptors adapt. Muscle stretch receptors and joint proprioceptors are two examples. Because the central nervous system must be continuously informed about muscle length and joint position to maintain posture, adaptation in these receptors could be dangerous.

> **KEY CONCEPTS**
> Many receptors stop generating impulses after extended exposure to stimuli.

12-3 Taste and Smell

The special senses are taste, smell, sight, hearing, and balance. In this section, we'll examine taste and smell.

Taste Buds

In humans and other mammals, the tongue contains receptors for taste. Known as **taste buds**, these microscopic, onion-shaped structures are located in the surface of the tongue and on small protrusions, known as *papillae* (pah-PILL-ee) (Figure 12-4a). Taste buds are also found on the roof of the oral cavity, the pharynx, and the larynx, but in smaller numbers.

Taste buds are stimulated by chemicals in the food we eat and are therefore referred to as *chemoreceptors*. Chemical

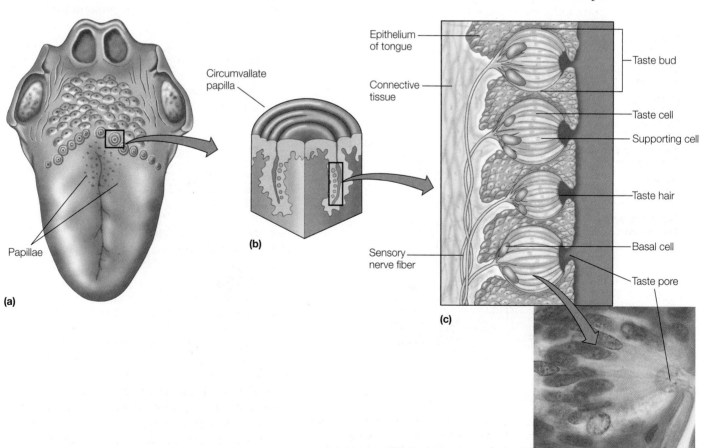

FIGURE 12–4 Taste Receptors (a) Taste buds are located on the upper surface of the tongue and are concentrated on small protrusions called (b) papillae. (c) The structure of the taste bud. (d) Taste pore. (© John D. Cunningham/Visuals Unlimited.)

substances in food and drinks dissolve in the saliva and enter the taste pores, small openings that lead to the interior of the taste bud (Figure 12-4c). Taste buds contain receptor cells, the ends of which possess large microvilli, known as *taste hairs*.

Taste hairs project into the taste pore. The plasma membranes of taste hairs contain clusters of protein molecules that serve as receptors. These receptors bind to food molecules dissolved in water in saliva and the foods we eat and beverages we drink. When food molecules bind to protein receptors, they trigger nerve impulses in the dendrites of the sensory nerves that are wrapped around the receptor cells. These impulses are then transmitted to the brain.

In the introduction to this chapter, you learned about a student who lost her sense of taste because of a dietary zinc deficiency. Why's that?

Zinc stimulates division of the cells in taste buds, helping to replace cells of the taste buds that are lost from normal wear and tear. Zinc deficiencies reduce cellular replacement, so the taste buds eventually cease operation.

> **KEY CONCEPTS**
> Taste buds are located in the tongue and respond to chemicals dissolved in foods and drinks.

Primary Flavors

Humans can discriminate among thousands of taste sensations. The taste sensations are a combination of five basic flavors: sweet, sour, bitter, salty, and umami. Sweet flavors result from sugars and some amino acids. Sour flavors result from acidic substances. Salty tastes result from metal ions (like sodium in table salt). Bitter flavors result from chemical substances belonging to a group called *alkaloids*, among them caffeine, and some nonalkaloid substances such as aspirin. Umami is the meaty taste associated with the flavorant monosodium glutamate (MSG) once commonly added to Chinese food. All taste buds respond, in varying degree, to all five taste sensations, but each responds preferentially to one taste.

Taste buds are distributed unevenly on the surface of the tongue. The tip of the tongue, for instance, is most sensitive to sweet flavors as it contains a higher proportion of taste buds that respond to sugars. The sides of the tongue are most sensitive to sour flavors, and the back of the tongue is most sensitive to bitter flavors. Salty taste is more evenly distributed, with slightly increased sensitivity on the sides of the tongue near the front. Food contains many different flavors. What we taste, therefore, depends on the relative proportion of the five basic flavors.

> **KEY CONCEPTS**
> Taste buds respond preferentially to one of five primary flavors.

The Olfactory Epithelium

Smell is a chemical sense like taste. The chemoreceptors responsible for the detection of odors are located in the roof of each nasal cavity in a patch of cells called the **olfactory epithelium** (ole-FACK-tore-ee; Figure 12-5). Odors are perceived by receptor cells. They are neurons whose dendrites extend to the surface of the olfactory membrane as shown in

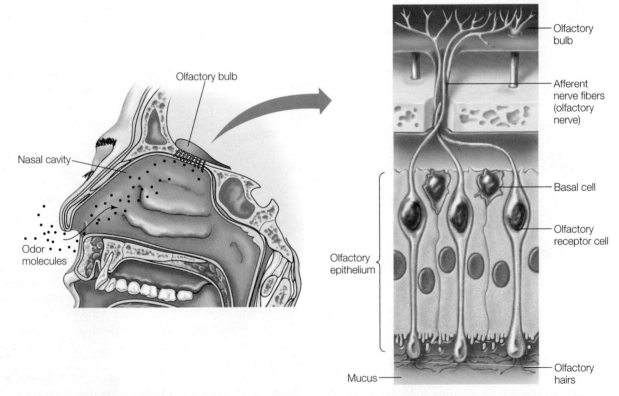

FIGURE 12-5 Location and Structure of the Olfactory Epithelium Olfactory receptors are located in the olfactory epithelium in the roof of the nasal cavity. Chemicals in the air dissolve in the watery fluid bathing the surface of the cells, then bind to receptors on the plasma membranes of the olfactory hairs. The olfactory receptors terminate in the olfactory bulb. From here, nerve fibers travel to the brain.

Figure 12-5. In humans, there are an estimated 50 million olfactory receptor cells in the olfactory membrane. These cells terminate in six to eight long projections, called *olfactory hairs*. The membranes of the olfactory hairs shown in Figure 12-5 contain protein receptors to which airborne molecules bind. To do so, airborne molecules must first be trapped on the thin, watery layer on the surface of the cell olfactory hairs. Binding of odor molecules to the receptors activates the neurons, causing them to generate impulses. They are transmitted to the olfactory bulb, a part of the brain that lies just above the olfactory membrane. Here, dendrites of the receptor cells synapse with nerve cells. Axons of the neurons in the olfactory bulb cells then travel to the brain via the *olfactory nerve*.

Like taste, odor discrimination is thought to depend on combinations of primary odors. Unfortunately, neuroscientists do not agree on what the primary odors are. One system of classification recognizes seven primary odors ranging from peppermint to floral to putrid. It is thought that molecules that produce a similar odor are similarly shaped and that they bind to one type of receptor on the olfactory hairs. Various combinations of the primary odors give rise to the many odors we perceive. Some researchers hypothesize that there may be thousands of kinds of smell receptors providing odor discrimination. Research shows that there are at least 1,000 different protein receptor molecules. New studies suggest that these molecules may be combined in different ways to provide the wide range of odor detection. Receptors for smell adapt within a short period—about a minute, which is a good thing, for example, if you live or work on a dairy or pig farm.

> **KEY CONCEPTS**
>
> The olfactory epithelium is a patch of receptor cells in the roof of the nasal cavity that detects odors.

Smell and Taste

Hold a piece of hot apple pie near your nose and inhale. It smells so good you can almost taste it. In fact, you are tasting it. Molecules given off by the pie enter the nose, where they stimulate receptors in the olfactory epithelium. They also reach the mouth through the pharynx and dissolve in the saliva, where they stimulate taste receptors on the tongue. That's what's meant when we say that small influence taste.

Just as odors stimulate taste receptors, food in our mouths also stimulates olfactory receptors. Molecules released by food enter the nasal cavities, dissolving in the water on the surface of the olfactory membrane and stimulating the receptor cells.

The complementary nature of taste and smell is not often evident. You may notice it most when you suffer from nasal congestion. In fact, cold sufferers often complain that they cannot taste their food. This phenomenon results from the buildup of mucus in the nasal cavities, which blocks the flow of air. Mucus may also block the olfactory hairs. Therefore, food loses its "taste" when you have a stuffy nose because your sense of smell is impaired. (You can test this by holding your nose while you eat.)

> **KEY CONCEPTS**
>
> Smell influences taste and vice versa.

12-4   The Visual Sense: The Eye

The human eye is one of the most extraordinary products of evolution. It contains a patch of photoreceptors that permits us to perceive the remarkably diverse and colorful environment we live in.

Anatomy of the Eye

Human eyes are roughly spherical organs located in the eye sockets, or orbits, cavities formed by the bones of the skull. The eye is attached to the orbit by six small muscles that control eye movement.

The Sclera and Cornea. As Figure 12-6 shows, the wall of the human eye consists of three layers (Table 12-2). The outermost is a durable fibrous layer, which consists of the **sclera**, the white of the eye, and the **cornea**, the clear part in front, which lets light into the interior of the eye (Figure 12-6). Tendons of the extrinsic eye muscles attach to the sclera.

The Choroid, Ciliary Body, and Iris. The middle layer consists of three parts: the choroid, the ciliary body, and the iris. The **choroid** is the largest portion of the middle layer. It contains a large amount of melanin (MELL-ah-nin), a pigment that absorbs stray light the way the black interior of a camera does. The blood vessels of the choroid supply nutrients to the eye. Anteriorly, the choroid forms the ciliary body. The **ciliary body** contains smooth muscle fibers that control the shape of the lens. Like the lens of a camera, it permits us to focus incoming light. This helps us view objects at different distances.

The **iris** is the colored portion of the eye visible through the cornea. Looking in a mirror, you can see a dark opening in the iris called the pupil. The **pupil** allows light to penetrate the eye. The blackness you see through the pupil is the choroid layer and the pigmented section of the retina, discussed below.

Like the ciliary body, the iris contains smooth muscle cells. The smooth muscle of the iris regulates the diameter of the pupil. Opening the pupil lets more light in; narrowing it reduces the amount of light that enters. You have experienced this phenomenon when you entered a dark theater. At first you were practically blind. In a minute or so, however, you began to see. This change was due to your pupils opening, allowing more light in.

The pupils open and close reflexively in response to light intensity. This reflex is an adaptation that protects the light-sensitive inner layer, the retina. It also allows more light to enter to improve vision in dark conditions and restricts light when too much is present, protecting the eye from damage.

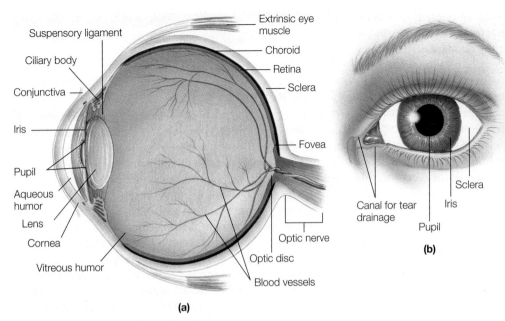

FIGURE 12-6 Anatomy of the Human Eye

The Retina, Rods, and Cones. The innermost layer of the eye is the retina. The **retina** consists of an outer, pigmented layer and an inner layer consisting of photoreceptors (modified nerve cells that detect light) and associated nerve cells. The retina is weakly attached to the choroid and can become separated from it as a result of trauma to the head. A detached retina can lead to blindness if not quickly repaired by laser surgery.

The photoreceptors of the retina are highly modified nerve cells. Two types of photoreceptors are present in the retina: rods and cones (Table 12-3). The **rods**, so named because of their shape, are sensitive to low light (Figure 12-7b and 7c). Rods function at night and produce grayish, somewhat vague, black-and-white images. The **cones**, also named because of their shape, operate only in brighter light. They are responsible for visual acuity—sharp vision—and color vision.

As Figure 12-7b shows, the rods and cones synapse with the bipolar neurons in the retina. These, in turn, synapse with ganglion cells, also located in the retina. The axons of the ganglion cells unite at the back of the eye in a central location to form the **optic nerve**. Because this area contains no photoreceptors, it is insensitive to light. Eye doctors call this the *blind spot*. The blood vessels that enter and leave the eye do so with the optic nerve. An ophthalmologist (eye doctor) can easily see these blood vessels and their branches by shining a light through the pupil onto the back wall of the eye.

Rods and cones are found throughout the retina, but the rods are most abundant. There are an estimated 150 million rods in each eye and only about 6 million cones. However, the cones are concentrated in a tiny region of each eye. In the center of this region is a small depression, about the size of the head of a pin, known as the **fovea centralis** (FOE-vee-ah), which contains only cones. The sharpest vision occurs at the fovea because cones are responsible for visual acuity. The number of cones in the retina decreases progressively from the fovea outward, whereas the number of rods increases. The greatest concentration of rods thus is found in the periphery of the retina.

Images from our visual field are cast onto the retina, creating bioelectric impulses that are transmitted to the visual cortex of the brain. Some processing of the image occurs in the retina. The rest takes place in the brain.

A New Light Sensor in the Eye. Although the rods and cones are sensitive to light, recent research suggests that a very small number of the ganglion cells are also sensitive to light. These cells, researchers believe, send signals to the suprachiasmatic

TABLE 12-2 Structures and Functions of the Eye		
Structure		**Function**
Wall		
Outer layer	Sclera	Provides insertion for extrinsic eye muscles
	Cornea	Allows light to enter; bends incoming light
Middle layer	Choroid	Absorbs stray light; provides nutrients to eye structures
	Ciliary body	Regulates lens, allowing it to focus images
	Iris	Regulates amount of light entering the eye
Inner layer	Retina	Responds to light, converting light to nerve impulses
Accessory structures and components	Lens	Focuses images on the retina
	Vitreous humor	Holds retina and lens in place
	Aqueous humor	Supplies nutrients to structures in contact with the anterior cavity of the eye
	Optic nerve	Transmits impulses from the retina to the brain

TABLE 12-3 Summary of Rods and Cones			
Photoreceptor	**Day or Night**	**Color Vision**	**Location**
Rods	Night vision	No	Highest concentration in the periphery of the retina
Cones	Day vision	Yes	Highest concentration in the macula and fovea

nucleus (SCN), a group of nerve cells in the hypothalamus that is often referred to as the master clock. Some people liken the SCN to the clock chip in a computer. It keeps track of time and controls biological rhythms like sleep and daily hormonal cycles.

Research further indicates that these cells are most sensitive to blue light. Blue light can be used to treat patients who suffer from depression during the short days of winter, a condition known as **seasonal affective disorder (SAD)**. Blue light also stimulates alertness, a finding that some companies are using to boost on-the-job performance. These companies adjust the color of the lighting in their offices in the morning and in the afternoon after lunch to combat sleepiness.

The Lens. Light is focused on the retina by the lens. The **lens** is a transparent, flexible structure that lies behind the iris, as shown in Figure 12-6. The lens is attached to the ciliary body by thin fibers. This connection allows the smooth muscle of the ciliary body to alter the shape of the lens, an action necessary for focusing the eye (discussed shortly).

> ## Health Tip 12-2
>
> Having trouble getting to sleep at night? Try avoiding late-night computer work.
>
> *Why?*
>
> Interestingly, studies have found that exposure to blue light from television and computer screens may throw off the sleep cycle of teens by creating greater alertness late in the evening. This, in turn, results in later bedtimes. If you have to get up early to attend class or go to work, you'll naturally be more tired.

As we age, the lens may develop cloudy spots, or **cataracts**, that obstruct vision. This condition is most commonly encountered in people who have been exposed to excessive ultraviolet radiation in sunlight, although other factors may also be responsible for the disease, including genetics.

Currently, an estimated 22 million Americans over the age of 40 have cataracts. Scientists believe that the risk of developing cataracts may increase as the Earth's ozone layer, which blocks dangerous ultraviolet radiation from the sun, is eroded by chlorofluorocarbons (CFCs) from refrigerators, air-conditioning units, and other sources. Although these chemicals have been banned and replaced with ozone friendly substances, CFCs are persistent. They remain in the atmosphere for a very long time, eating away at the ozone layer. Some scientists estimate that it could take 50 to 100 years for the ozone layer to fully recover.

Eye surgeons correct cataracts by replacing the cloudy lens with a clear plastic lens. Studies suggest that vitamin E and

FIGURE 12–7 The Retina (a) Cross section through the wall of the eye, showing (b) arrangement of the cellular components of the retina. (c) The structure of the rods and cones.

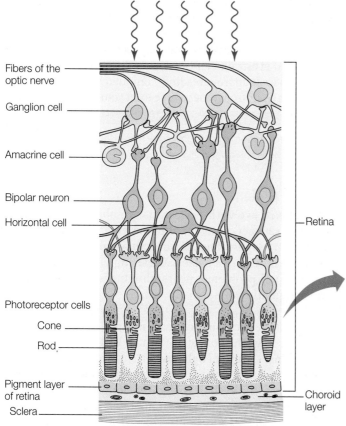

(a)

Direction of light

Fibers of the optic nerve

Ganglion cell

Amacrine cell

Bipolar neuron

Horizontal cell

Retina

Photoreceptor cells

Cone

Rod

Pigment layer of retina

Sclera

Choroid layer

(b)

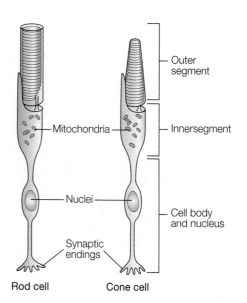

Outer segment

Mitochondria

Innersegment

Nuclei

Cell body and nucleus

Synaptic endings

Rod cell   Cone cell

(c)

antioxidants found in fruits and vegetables like spinach and other foods can protect us against cataracts (**Health Tip 3**).

Health Tip 12-3

Eat spinach and other dark leafy green vegetables like kale and collard greens to prevent cataracts.

Why?

Laboratory studies of the cells of the lens show that certain anti-oxidants in dark, leafy green vegetables, such as spinach, protect the cells from ultraviolet radiation, a leading cause of cataracts. The antioxidants (lutein and zeaxanthin) were nearly 10 times more powerful than vitamin E, another antioxidant found in many foods and thought to protect the eyes from cataracts.

Studies of adult women also show that those who have taken vitamin E supplements for 10 years or more were much less likely to develop cataracts than those who took no supplements. This research study also showed that two B vitamins, riboflavin and thiamin, help reduce the incidence of cataracts. Previous work had shown that vitamin C also protects against cataracts. These vitamins can be acquired either by taking supplements or by consuming a healthy diet consisting of many fruits and vegetables.

The lens separates the interior of the eye into two cavities of unequal size. Everything in front of the lens is the **anterior chamber**; everything behind it is the **posterior chamber**. The posterior cavity is filled with a clear, gelatinous material, the **vitreous humor** (VEH-tree-ous HUE-mur; "glassy liquid"). The anterior cavity contains a thin liquid, chemically similar to blood plasma, called the **aqueous humor** ("watery fluid"). The aqueous humor provides nutrients to the cornea and lens and carries away cellular wastes.

In normal, healthy individuals, aqueous humor production is balanced by absorption. If the outflow is blocked, however, the aqueous humor builds up inside the anterior chamber, creating internal pressure. This disease, called **glaucoma** (glaw-COE-mah), progresses gradually and imperceptibly. If untreated, the pressure inside the eye can damage the retina and optic nerve, causing blindness.

Health Tip 12-4

While eating fruits and vegetables is good for one's long-term health, few people follow this advice.

Why?

One reason is that many misunderstand what is meant by a serving. A serving is a fairly small amount—only half a cup. Take a look at a measuring cup and you'll see that half a cup isn't much! In 2005, the U.S. government issued guidelines recommending 13 half-cup servings of fruits and vegetables a day. Translated, that's about one-half of each meal should be fruits and vegetables, which is what the 2011 MyPlate guidelines recommend. To be sure you eat correctly, you can eat a salad at every meal and sometimes make it your main meal. Also consume lots of dishes featuring vegetables like vegetable stews and soups. Tiny amounts of meat can be added for flavor. You can pour spaghetti sauce with a few meatballs, if you like, over cooked vegetables like carrots, beans, summer squash, peppers that have been steamed or boiled. That makes a very delicious and healthy meal.

KEY CONCEPTS

The eye consists of three distinct layers.

Focusing Light on the Retina

Light entering the eye is focused on the photoreceptors of the retina by the cornea and the lens. The lens is a flexible structure whose shape is controlled by the muscles in the ciliary body. These muscles are attached to the lens by the suspensory ligament, tiny filamentous strands. As Figure 12-8a shows, light from distant objects falls on the retina. As shown in Figure 12-8b, light rays from nearby objects are divergent—spread out. To focus on nearby objects, the lens thickens and shortens, and thus becomes more curved, as shown in Figure 12-8c. This occurs when the muscles of the ciliary body relax. The effect is much the same as a rubber ball flattened between your hands returning to normal when you reduce the pressure on it. This automatic adjustment in the curvature of the lens as it focuses on a nearby object is called **accommodation** (Figure 12-9).

Synchronized Eye Movement and Convergence. Six muscles located outside the eye are responsible for eye movement (Figure 12-10). As noted earlier, these muscles attach to the bony eye socket and to the sclera (the white of the eye).

The eyes generally move in unison like a pair of synchronized swimmers. Synchronized movement is an evolutionary adaptation that ensures that images are focused on the foveas of both eyes at the same time. The fovea, you may recall, is the region of the eye with the highest concentration of cones. When a nearby object is viewed, the eyes turn inward—that is, they converge. Convergence ensures that near images are focused on each fovea. Because convergence occurs during all near-point work—reading, writing, sewing, knitting—it tends to strain the extrinsic eye muscles, creating eyestrain and headaches. Continued eyestrain can cause your vision to deteriorate, so be sure to look up from reading, computer work, or any other near point tasks from time to time to let your eye muscles relax.

KEY CONCEPTS

The cornea and lens focus light on the retina; the lens is adjustable, thus provides fine control of focusing.

Visual Problems

In a relaxed eye with perfect vision, objects farther than 6 meters (20 feet) away fall into perfect focus on the retina (Figure 12-11a). Many of us, however, have imperfectly shaped eyeballs or defective lenses. These imperfections result in three visual problems: nearsightedness, farsightedness, and astigmatism.

Nearsightedness. Nearsightedness (myopia) results when the eyeball is slightly elongated (Figure 12-11b). Without corrective lenses, parallel light rays arising from distant images come into focus in front of the retina, creating a fuzzy image. (Remember, the eyeball is slightly elongated, so light rays naturally come into focus in front of the retina, as shown in Figure 12-11b.) Because light from nearby images is divergent light, light rays come into focus on the retina. People with

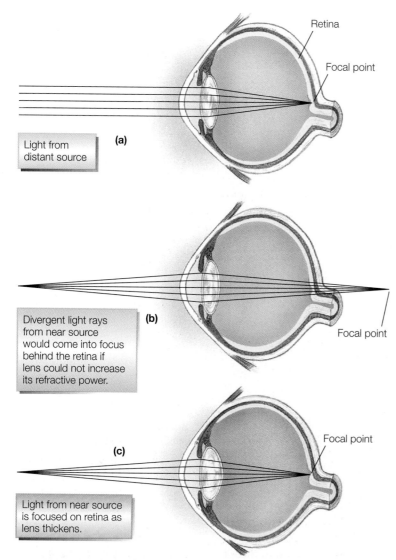

Retina

Focal point

Light from
distant source

(a)

Divergent light rays
from near source
would come into focus
behind the retina if
lens could not increase
its refractive power.

(b)

Focal point

(c)

Focal point

Light from near source
is focused on retina as
lens thickens.

FIGURE 12-8 Refraction of Light by the Cornea and Lens (a) Light rays from distant objects are parallel when they strike the eye. The refractive power of the cornea and resting lens are sufficient to bring them into focus on the retina. (b) Light rays from nearby objects are divergent. The cornea has a fixed refractive power and cannot change. The rays would focus behind the retina if the lens could not alter its refractive power. (c) To focus the image, the ciliary muscles contract. This lessens pressure on the suspensory ligaments, which allows the lens to thicken and shorten, becoming more curved and more refractive.

nearsightedness, therefore, can see near objects without corrective lenses—hence the name.

Nearsightedness may also result when the lens is too strong—that is, too concave. This lens bends the light coming from distant objects too much, causing the image to come into focus in front of the retina.

Nearsightedness is caused by many factors and tends to run in families; it generally appears around age 12, often worsening until a person reaches 20. Nearsightedness can be corrected by contact lenses or prescription glasses that cause incoming light rays to diverge a little so that parallel light rays from distant images come in focus on the retina (Figure 12-11b). It can also be corrected by laser surgery, discussed shortly.

Farsightedness. **Farsightedness** (hyperopia) is the opposite of nearsightedness. It results when the eyeballs are too short or the lens is too weak (too convex). In the eyes of farsighted individuals, parallel light rays from distant objects usually fall into focus on the retina. However, divergent rays from nearby objects cannot be bent inward enough to be sufficiently focused. Without corrective lenses, farsighted individuals see distant objects well, but nearby objects are fuzzy. Farsightedness is generally present from birth and tends to run in families. Glasses or contact lenses that bend the light inward bring near objects into sharp focus on the retina (Figure 12-11c). It can also be corrected by laser surgery.

Astigmatism. The cornea and lens are generally uniformly curved, but in some individuals they are slightly disfigured. An eye doctor explained it to me this way. He said that the cornea of a normal eye is round like a basketball. In an astigmatic eye, the cornea is shaped more like the surface of a football. This unequal curvature is called an **astigmatism** (a-STIG-mah-tiz-em). Astigmatism creates fuzzy images because light rays are bent differently by the different parts of the cornea. Astigmatism is usually present from birth and does not grow worse with age. It can also be corrected with glasses or contact lenses and laser surgery.

Laser Surgery. Visual defects can also be corrected by laser surgery. One of the most popular types of surgery is LASIK (which stands for laser in-situ keratomileusis). In this procedure, a computer is used to map the surface of the cornea and calculate the amount of corneal tissue that needs to be removed to ensure proper vision. A thin flap of cornea is then partially shaved off using an extremely sharp blade. This creates a partial flap (Figure 12-12a). A laser is then used to remove underlying tissue and reshape the cornea (Figure 12-12b). This laser machine delivers a cool pulsing beam of ultraviolet light. The flap is then folded back into its original position (Figure 12-12c). It heals quickly.

Laser eye surgery is quick and accurate, with 95% of patients achieving 20/40 visual acuity or better. Worldwide, many hundreds of thousands of patients have been successfully treated with this procedure. Problems arise in under 1% of the cases.

Another less invasive type of laser surgery is photorefractive keratectomy (PRK). It is used to correct mild to moderate nearsightedness, farsightedness, and/or astigmatism. During this technique, an eye surgeon uses a laser to reshape the surface of the cornea, not underneath a flap as in LASIK. PRK may also be done with computer imaging of the cornea.

A third variant is known as LASEK (laser epithelial keratomileusis). In this procedure, an epithelial flap of the cornea is created. The epithelial cells are then loosened by applying

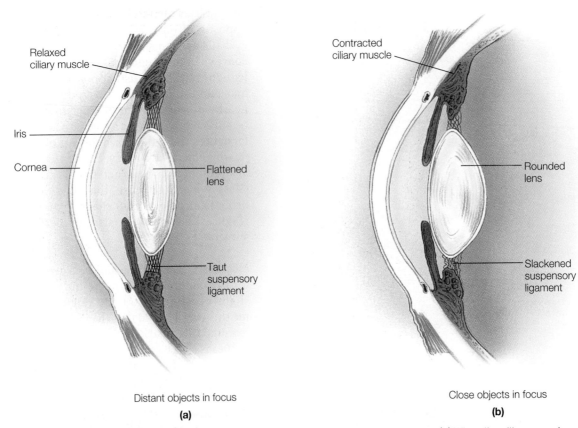

FIGURE 12-9   **Accommodation**   (a) The lens is flattened when the ciliary muscles are relaxed. (b) When the ciliary muscles contract, tension on the suspensory ligaments is reduced and the lens shortens and thickens.

alcohol. A laser is then used to reshape the cornea. The flap is repositioned and secured by a soft contact lens that is worn until the flap has healed.

Presbyopia.   The aging process brings with it many changes: hair loss, hearing loss, and arthritis. It also results in a decline in vision, known as **presbyopia** (PREZ-bee-OPE-ee-ah). It occurs between the ages of 40 and 50. While most people think of presbyopia as the inability of the eye to focus on nearby objects, it really is an inability to focus on objects at

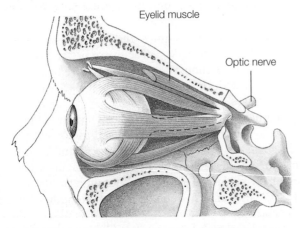

FIGURE 12-10   **Extrinsic Eye Muscles**   These muscles move the eye in all directions. They attach to the bony orbit and the sclera.

all distances. It is usually noticed when fine print starts to become blurry.

Eye doctors are not 100% certain about what causes presbyopia. Most of them believe it is caused by a stiffening of the lenses which makes it difficult for the eyes to focus. Others hypothesize that it could be caused by continued growth of the lens or atrophy (shrinkage) of the muscles that control the lens.

Presbyopia was typically remedied by wearing reading glasses or multifocal lenses—that is, bifocal or progressive lenses—that allow one to use different parts of the lens to focus on nearby and more distant images. In recent years, eye surgeons have been able to correct this condition with surgery.

One of the first effective surgical procedures was a process known as "monovision." This can be achieved by wearing contact lenses or via LASIK surgery. To understand, remember that both of your eyes work together—and equally—when you look at something. The result is binocular vision.

Most of us, however, have a dominant eye—one that your brain tends to be favor when "sighting." (For most right-handed individuals, the right eye is dominant.) To correct for presbyopia, eye doctors will fit a patient's dominant eye with a contact lens designed for optimum distance vision. The nondominant eye is fitted with a contact lens that permits good near vision.

Surgeons can do the same think with LASIK. That is, they leave the nondominant eye slightly nearsighted so patients can see nearby images without glasses. The other eye is corrected so that it focuses on more distant objects. Not all

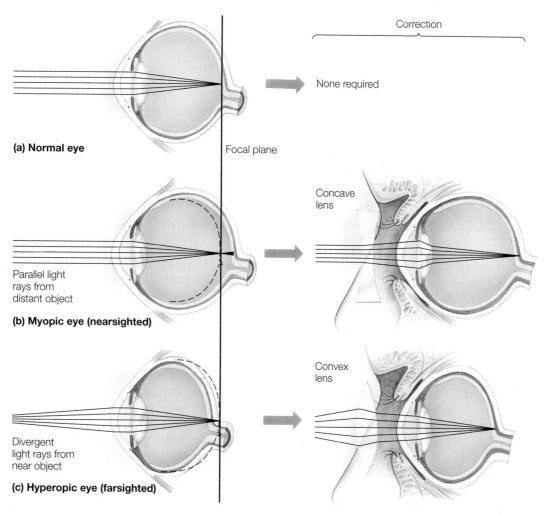

Correction

None required

(a) Normal eye

Focal plane

Parallel light rays from distant object

(b) Myopic eye (nearsighted)

Concave lens

Divergent light rays from near object

(c) Hyperopic eye (farsighted)

Convex lens

FIGURE 12–11 Common Visual Problems

patients can become accustomed to monovision, so it is wise to try this first with contact lenses.

A second procedure is conductive keratoplasty (CK). It is used to achieve monovision. Rather than lasers, surgeons use low-level, radio-frequency waves to shrink collagen fibers in the periphery of the cornea. CK is performed on only one eye—and usually the nondominant eye. What this procedure does is lengthen the eyeball, allowing that eye to focus on nearby objects.

Color Blindness. About 5% of the human population suffers from **color blindness**, a hereditary disorder. Color blindness is more prevalent in men than in women and ranges from an inability to distinguish certain shades of color to a complete inability to perceive color. The most common form of this disorder is red-green color blindness. In individuals with red-green color blindness, the red or green cones may either be missing altogether or be present in reduced numbers. If the red cones are missing, red objects

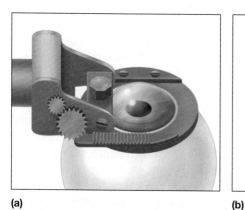

(a)

(b)

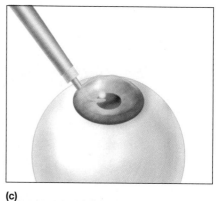

(c)

FIGURE 12–12 LASIK Surgery (a) Microkeratome slices off a thin layer of corneal tissue. (b) Laser burns away corneal tissue to correct eyesight. (c) Flap is restored, as is vision.

appear green. If the green cones are missing, however, green objects appear red.

Many color-blind people are either unaware of their condition or untroubled by it. They rely on a variety of visual cues such as differences in intensity to distinguish red and green objects. They also rely on position cues. In vertical traffic lights, for example, the red light is always at the top of the signal; green is on the bottom. Although the colors may appear more or less the same, the position of the light helps color-blind drivers determine whether to hit the brakes or step on the gas.

> **KEY CONCEPTS**
>
> Alterations in the shape of the lens and the eyeball cause the most common visual problems: nearsightedness, farsightedness, and astigmatism.

12-5   Hearing and Balance: The Cochlea and Middle Ear

The human ear is an organ of special sense. It serves two functions: it detects sound, and it detects body position, enabling us to maintain balance (Table 12-4).

The Anatomy of the Ear

The Outer Ear. The outer ear consists of an irregularly shaped piece of cartilage covered by skin, the *auricle* (OR-eh-kul), and the earlobe, a flap of skin that hangs down from the auricle (Figure 12-13). The outer ear also consists of a short tube, the **external auditory canal**, which transmits airborne sound waves to the middle ear (Figure 12-13a). The external auditory canal is lined by skin containing modified sweat glands that produce earwax. Earwax traps foreign particles, such as bacteria, and contains a natural antibiotic substance that may reduce ear infections.

The Middle Ear. The middle ear lies entirely within the skull (Figure 12-13b). The eardrum, or **tympanic membrane** (tim-PAN-ick), separates the middle ear cavity from the external auditory canal. The tympanic membrane vibrates when struck by sound waves in much the same way that a guitar string vibrates when a note is sounded by another nearby instrument.

Inside the middle ear are three minuscule bones, the **ossicles** (OSS-eh-kuls). Starting from the outside, they are the **malleus** (pronounced MAL-ee-us; hammer), **incus** (IN-cuss; anvil), and **stapes** (STAY-pees; stirrup). As illustrated in Figure 12-13b, the hammer-shaped malleus lies next to the tympanic membrane. When the membrane is struck by sound waves, it vibrates. This causes the malleus to rock back and forth. The malleus, in turn, causes the incus to vibrate, which causes the stapes, the stirrup-shaped bone, to move in and out against the **oval window**, an opening to the inner ear covered with a membrane like the head of a drum. Thus, vibrations created in the eardrum are amplified as they are transmitted to the inner ear, where the sound receptors are located.

> **KEY CONCEPTS**
>
> The ear consists of three anatomically separate portions: the outer ear, the middle ear, and the inner ear.

The Eustachian Tube

As Figure 12-13 illustrates, the middle ear cavity opens to the pharynx via the **auditory**, or **eustachian**, **tube** (you-STAY-shun). The eustachian tube serves as a pressure valve. Normally, the eustachian tube is closed. Yawning and swallowing, however, cause it to open, allowing air to flow into or out of the middle ear cavity. This equalizes the internal and external pressure on the eardrum. You have noticed your ears "pop" when you are taking off in an airplane. This is the result of pressure being released by the eustachian tubes.

The Inner Ear. The inner ear occupies a large cavity in the temporal bone of the skull and contains two sensory organs, the cochlea and the vestibular apparatus. The **cochlea** (COE-klee-ah) is shaped like a snail shell and houses the receptors for hearing.

The **vestibular apparatus** consists of two parts: the semicircular canals and the vestibule (Figure 12-13b). The **semicircular canals** are three ringlike structures set at right angles to one another. They house receptors for body position and movement. The **vestibule** is a bony chamber lying between the cochlea and semicircular canals. It houses receptors that respond to body position and movement.

> **KEY CONCEPTS**
>
> The Eustachian tube helps to equalize pressure in the middle ear.

TABLE 12-4	Structures and Functions of the Eye	
Part	Structure	Function
Outer ear	Auricle	Funnels sound waves into external auditory canal
	Earlobe	
	External auditory canal	Directs sound waves to the eardrum
Middle ear	Tympanic membrane or eardrum	Vibrates when struck by sound waves
	Ossicles	Transmit sound to the cochlea in the inner ear
Inner ear	Cochlea	Converts fluid waves to nerve impulses
	Semicircular canals	Detect head movement
	Saccule and utricle	Detect head movement and linear acceleration

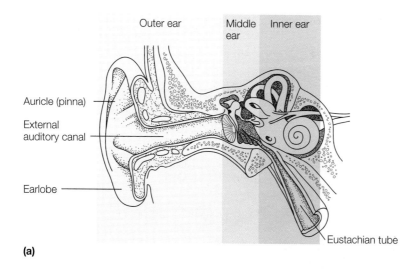

(a)

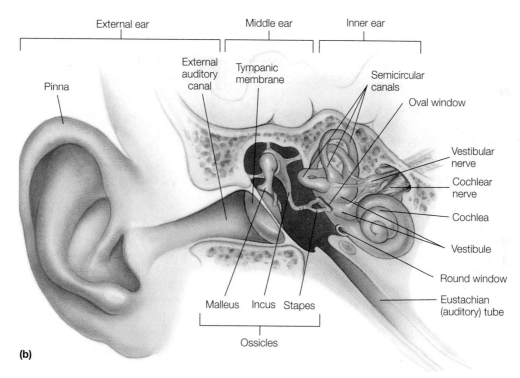

(b)

FIGURE 12-13 The Structures of the Ear (a) Cross section showing the structures of the outer, middle, and inner ears. (b) The receptors for balance and sound are located in the inner ear.

Structure and Function of the Cochlea

As noted previously, sound waves enter the external auditory canal, where they strike the eardrum. The eardrum then vibrates back and forth, causing the bones of the middle ear cavity to vibrate. This transmits sound waves to the cochlea, which houses the receptor for sound.

The cochlea is a hollow, spiral structure. A cross section through this structure reveals three fluid-filled canals (Figure 12-14a). Inside the middle cavity is the receptor for sound, the **organ of Corti**. It rests on a flexible membrane called the **basilar membrane** (BAZ-eh-ler).

The organ of Corti contains receptor cells, the **hair cells** (Figure 12-14b). When sound waves are transmitted into the inner ear, they create pressure waves in the fluid in the upper-

most canal of the cochlea. Fluid pressure waves created in the vestibular canal then travel through the vestibular membrane into the middle canal. From here, the pressure waves pass through the basilar membrane into the lowermost canal. Pressure is relieved by the outward bulging of the **round window** (Figure 12-15a), an opening in the bony cochlea, which, like the oval window, is spanned by a flexible membrane.

In their course through the cochlea, the pressure waves cause the basilar membrane to vibrate. This vibration stimulates the hair cells, which contact the overlying tectorial membrane (teck-TORE-ee-al). The hair cells stimulate the dendrites wrapped around their bases and they create nerve impulses that travel to the auditory area of the brain via the **vestibulocochlear nerve** (vess-TIB-you-low-COE-klee-er), one of the 12 cranial nerves.

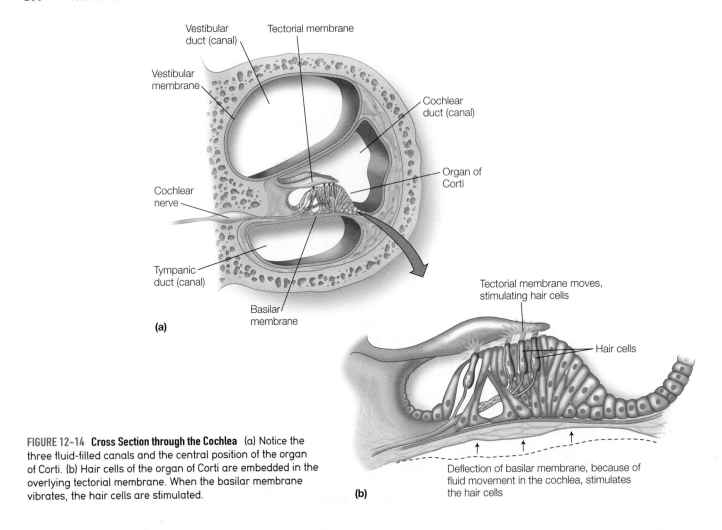

FIGURE 12-14 Cross Section through the Cochlea (a) Notice the three fluid-filled canals and the central position of the organ of Corti. (b) Hair cells of the organ of Corti are embedded in the overlying tectorial membrane. When the basilar membrane vibrates, the hair cells are stimulated.

Distinguishing Pitch and Intensity. A pitch pipe helps a singer find the note to begin a song. It has a range of notes from high to low. The ear can distinguish between various pitches, or frequencies, produced by a pitch pipe and other sources of sound in large part because of the structure of the basilar membrane. The basilar membrane underlying the organ of Corti is stiff and narrow at the oval window where fluid pressure waves are first established inside the cochlea (Figure 12-15b). As the basilar membrane proceeds to the tip of the spiral, however, it becomes wider and more flexible. The change in width and stiffness results in marked differences in its ability to vibrate. The narrow, stiff end, for example, vibrates maximally when pressure waves from high-frequency sounds ("high notes") are present (Figure 12-15c). The far end of the membrane vibrates maximally with low-frequency sounds ("low notes"). In between, the membrane responds to a wide range of intermediate frequencies.

Pressure waves caused by the sound of a certain pitch stimulate one specific region of the organ of Corti. The hair cells stimulated in that region send impulses to the brain, which it interprets as a specific pitch. Each region of the organ of Corti sends impulses to a specific region of the auditory cortex in the temporal lobe of the brain.

The intensity of a sound, or its loudness, depends on how much vibration occurs in a portion of the basilar membrane. The louder the sound, the more vigorous the vibration. The more vigorous the vibration of the eardrum, the greater the deflection of the basilar membrane in the area of peak responsiveness. The greater the deflection of the basilar membrane, the more hair cells are stimulated. Loud rock music or sirens, however, cause extreme vibrations in the basilar membrane, destroying hair cells over time and causing partial deafness.

Hearing Loss. As people grow older, many lose their hearing gradually—so slowly, in fact, that most people are unaware until the losses are significant. In some cases, though, people lose their hearing suddenly. A loud explosion, for example, can damage the hair cells or even break the tiny ossicles.

Hearing losses may be temporary or permanent, partial or complete. Hearing loss falls into one of two categories, depending on the part of the system that is affected. The first is *conduction deafness*, which occurs when the conduction of sound waves to the inner ear is impaired. Conduction deafness may result from a rupture of the eardrum or damage to the ossicles. Conduction deafness most often results from infections in the middle ear. Bacterial infections may result in the buildup of scar tissue that causes the ossicles to fuse and lose their ability to transmit sound. Infections of the middle ear usually enter through the eustachian tube. Thus, an infection in the throat can easily spread to the ear, where it requires prompt treatment. Ear infections are especially common in babies and young children. If undetected, middle ear infections in children can slow down the development of

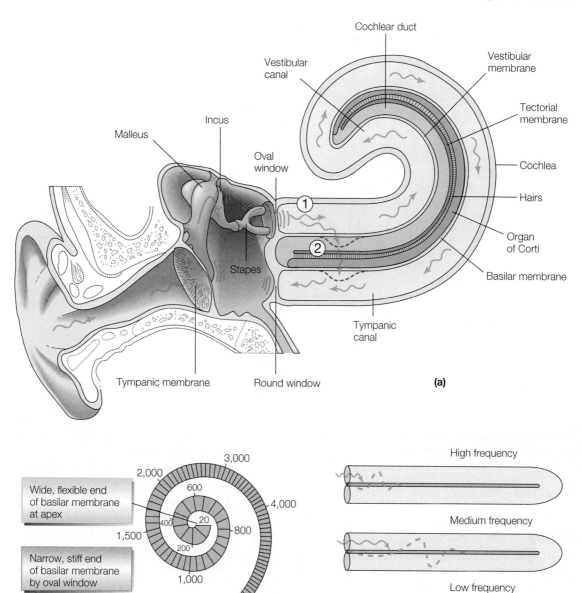

The numbers indicate the frequencies with which different regions of the basilar membrane maximally vibrate.

FIGURE 12–15 The Transmission of Sound Waves through the Cochlea The cochlea is unwound here to simplify matters. (a) Vibrations are transmitted from the stirrup (stapes) to the oval window. Fluid pressure waves are established in the vestibular canal and pass to the tympanic canal, causing the basilar membrane to vibrate. (b) A representation of the basilar membrane, showing the points along its length where the various wavelengths of sound are perceived. Notice that the basilar membrane is narrowest at the base of the cochlea at the oval window end and widest at the apex. (c) High-frequency sounds set the basilar membrane near the base of the cochlea into motion. Hair cells send impulses to the brain, which interprets the signals as a high-pitch sound. Low-frequency sounds stimulate the basilar membrane where it is widest and most flexible.

speech. If untreated, such infections can lead to permanent deafness. Conduction deafness is treated by hearing aids that fit in the ear or just behind it.

The second type of hearing loss is neurological and is called *nerve deafness*. It results from physical damage to the hair cells, the vestibulocochlear nerve, or the auditory cortex. Explosions, extremely loud noises, and some antibiotics can damage the hair cells in the organ of Corti, creating partial to complete deafness. The auditory nerve, which conducts

impulses from the organ of Corti to the cortex, may degenerate, thus ending the flow of information to the cortex. Tumors in the brain or strokes also can destroy the cells of the auditory cortex.

KEY CONCEPTS

Sounds are detected by vibrations in the basilar membrane of the organ of Corti located in the cochlea; vibrations stimulates hair cells that send bioelectric impulses to the brain.

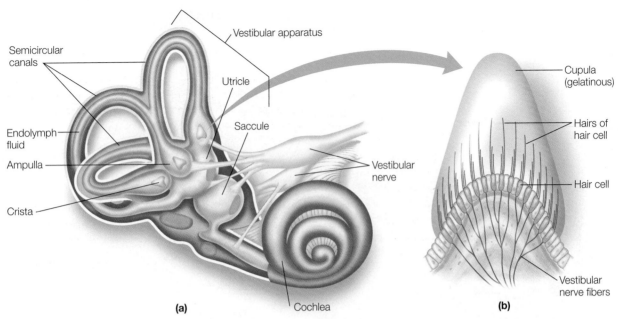

FIGURE 12-16 Vestibular Apparatus and Semicircular Canals (a) This illustration shows the location of the cristae in the ampullae of the semicircular canals. The semicircular canals are filled with endolymph. (b) When the head spins, the endolymph is set into motion, deflecting the gelatinous cupula of the crista, thus stimulating the receptor cells.

The Vestibular Apparatus

The vestibular apparatus consists of the two parts, the semicircular canals and the vestibule (Figure 12-16). Both are involved in proprioception.

The Semicircular Canals. The three semicircular canals are arranged at right angles to one another. Each canal is filled with a fluid. As Figure 12-16a shows, the base of each semicircular canal expands. On the inside wall of the enlarged area is a small ridge of tissue, or crista (CHRIS-tah). Each crista consists of a patch of receptor cells. Each receptor cell contains numerous microvilli and a single cilium embedded in a cap of gelatinous material, the cupula (CUE-pew-lah), shown in Figure 12-16b.

Rotation of the head causes the fluid in the semicircular canals to move. The movement of the fluid deflects the cupula, which stimulates the hair cells. They, in turn, stimulate the dendrites of sensory nerves that wrap around the base of the receptor cells. The sensory nerves send impulses to the brain, alerting it to the rotational movement of the head and body.

Because the semicircular canals are set in all three planes of space, movement in any direction can be detected. By alerting the brain to rotation and movement, the semicircular canals contribute to our sense of balance. They are, therefore, important to *dynamic balance*—helping us stay balanced when in movement.

The Utricle and Saccule. Two additional receptors play a role in balance and the detection of movement, the utricle (YOU-treh-kul) and saccule. These membranous compartments inside the vestibule provide input under two distinctly different conditions: at rest and during acceleration in a straight line. These receptors therefore contribute not just to dynamic balance but also static balance—that is, staying balanced when we are not moving.

The utricle and saccule contain small receptor organs similar to the cristae of the semicircular canals. The nerve impulses generated by the receptors in the semicircular canals and the utricle and saccule are sent to a cluster of nerve cell bodies in the brain stem. Here all of the information the receptors generate on position and movement is integrated with input from both the eyes and receptors in the skin, joints, and muscles (discussed earlier in this chapter). From this center, information flows in many directions. One major pathway leads to the cerebral cortex. This path makes us conscious of our position and movement. Another path leads to the muscles of the limbs and torso. Signals to the muscles maintain our balance and, if necessary, correct body position.

Health Tip 12-5

Eat right and exercise regularly for greater flexibility and balance.

Why?

Studies show that exercise and diet together are much more effective in helping to maintain or lose weight than diet alone—and they're healthier. Diet alone, for instance, reduces body fat, but also results in a loss of muscle. Exercise helps maintain muscle mass. Exercise also contributes to greater flexibility and greater balance.

In some people, activation of the vestibular apparatus results in motion sickness, characterized by dizziness and nausea. The exact cause of motion sickness is not known.

KEY CONCEPTS

The vestibular apparatus houses receptors that detect body position and movement.

12-6 Health and Homeostasis

Many general sense receptors play an important role in homeostasis. Mechanoreceptors that detect changes in blood pressure and chemoreceptors that respond to the ionic concentration of the blood, for example, help maintain blood pressure and proper water balance through mechanisms involving the kidney. Chemoreceptors that detect levels of carbon dioxide and hydrogen ions in the blood and cerebrospinal fluid help to regulate respiration. Additional chemoreceptors that detect levels of various hormones, nutrients, and ions in the blood and body fluids are discussed elsewhere in the text. These detectors may stimulate hormonal responses that correct potentially disruptive chemical imbalances.

The special senses help us—and other animals—acquire food and other necessities essential for normal health and homeostasis. They also stimulate behavioral responses that are essential to homeostasis.

Our environment, however, can alter homeostasis. Consider noise. Noise damages hearing over time. It may also disturb rest and sleep, both important in maintaining homeostasis and health. Noise also raises our level of stress, which, in turn, shortens our lives. Reducing noise in our environment can therefore help us live less stressful and much healthier lives.

KEY CONCEPTS

Homeostasis requires the nervous system and its sensors.

SUMMARY

The General and Special Senses

1. The body contains two types of receptors—those that provide general senses and those that provide special senses.

The General Senses

2. The general senses are pain, light touch, pressure, temperature, and position sense. Receptors for these senses may be naked or encapsulated nerve endings.
3. Receptors fit into five functional categories: (1) mechanoreceptors; (2) thermoreceptors; (3) photoreceptors; (4) chemoreceptors; and (5) nociceptors (pain receptors).
4. Sensory stimuli may be internal or external. Some stimuli cause reflex actions. Still others may stimulate physiological changes.
5. Many sensory receptors cease responding to prolonged stimuli, a phenomenon called sensory adaptation. Pain, temperature, pressure, and olfactory receptors all adapt.

Taste and Smell

6. The special senses include taste, smell, vision, hearing, and balance.
7. Taste receptors called taste buds are located principally on the upper surface of the tongue. Food molecules dissolve in the saliva and bind to the membranes of the microvilli of the receptor cells inside taste buds.
8. Taste buds respond to five flavors: salty, bitter, sweet, sour, and umami.
9. The receptors for smell are located in the olfactory membrane in the roof of the nasal cavities. The receptor cells in the olfactory membrane respond to thousands of different molecules.

The Visual Sense: The Eye

10. The eye is the receptor for visual stimuli.
11. The wall of the human eye consists of three layers. The outermost layer consists of the sclera (the white of the eye) and the cornea (the clear anterior structure that lets light shine in).
12. The middle layer consists of the choroid (the pigmented and vascularized region), the ciliary body (whose muscles control the lens), and the iris (which controls the amount of light entering the eye).
13. The innermost layer is the retina, the light-sensitive layer, which contains two types of photoreceptors: rods and cones. Rods function in dim light and provide black-and-white vision, and cones operate in bright light and provide color vision. Cones are also responsible for visual acuity.
14. Light is focused on the retina by the cornea and lens.
15. The eye is divided into two cavities. The posterior cavity lies behind the lens and is filled with a gelatinous material, the vitreous humor. The anterior cavity is filled with a plasmalike fluid called the aqueous humor, which nourishes the lens and other eye structures in the vicinity. If the rate of absorption of aqueous humor decreases, however, pressure can build inside the anterior cavity, resulting in glaucoma.
16. As people age, the lens becomes less resilient and less able to focus on nearby objects.
17. Abnormalities in shape of the lens or eyeball can result in visual defects. The two most common are nearsightedness and farsightedness. Another abnormality in the cornea is astigmatism. All three conditions can be corrected by eyeglasses, contact lenses, or surgery.
18. Color blindness is a genetic disorder, more common in men, and results from a deficiency or an absence of one or more types of cones. Red-green color blindness is the most common type.

Hearing and Balance: The Cochlea and Middle Ear

19. The human ear consists of three portions: the outer, middle, and inner ear.
20. The outer ear consists of the auricle and external auditory canal, both of which direct sound to the eardrum.
21. The middle ear consists of the eardrum and three small bones that transmit vibrations to the inner ear.
22. The inner ear contains the cochlea, where the receptors for sound are located. The inner ear also houses receptors for movement and head position: the semicircular canals, utricle, and saccule.
23. Sound waves create vibrations in the eardrum and ossicles, which are transmitted to fluid in the cochlea. Pressure waves in the cochlea cause the basilar membrane to vibrate, which stimulates the hair cells of the organ of Corti. Pressure waves resulting from any given sound cause one part of the membrane to vibrate maximally. The hair cells stimulated in that region send signals to the brain, which it interprets as a specific frequency.
24. The semicircular canals are three hollow rings filled with a fluid. The receptors for head movement are located in an enlarged portion at the base of each canal. Fluid movement inside the semicircular canals deflects the gelatinous cap, stimulating the underlying receptor cells and alerting the brain to head movements.
25. Two membranous sacs in the vestibule, the utricle and saccule, contain receptors that respond to linear acceleration and tilting of the head.

Health and Homeostasis

26. Receptors that provide for the general and special senses are extremely important to homeostasis. They help us adjust internally to respond to external changes.
27. Some external stimuli, such as noise, can alter homeostasis. Noise, for instance, damages the structures of the internal ear. It can also disturb sleep and can lead to stress, both of which impair human health.

28. Noise damages the ears. Extremely loud noises can rupture the eardrum or break the ossicles. Less intense noises, however, generally destroy hearing gradually by damaging hair cells. In most people, hearing loss occurs so gradually as to be undetected.

29. Besides deafening us and cutting us off from the important sounds of modern life, noise disturbs communications, rest, and sleep—the last two of which are particularly important to homeostasis and health. Noise also raises our level of stress, which, in turn, shortens our lives.

THINKING CRITICALLY ANALYSIS

This analysis corresponds to the Thinking Critically scenario that was presented at the beginning of this chapter.

While it is tempting to believe that this test actually revealed three people with ESP, the results can be easily explained by chance and probability, two factors which we often forget when observing our world and drawing conclusions from our observations.

By pure chance alone, some people will guess correctly more often than you'd expect. That is, some people will "identify" more than five correct cards. It is bound to happen. They don't have ESP, they just guessed correctly seven times. It was pure luck.

Probability and chance are two important factors to remember when observing the world. If you toss a coin 10 times, you'd expect that half the time it would turn up heads and half the time it would turn up tails. That's what probability theory tells us. However, there will be some times when the coin toss turns up more heads than tails. That's just the way of the world. The same can be said about the ESP experiment.

If the results of the ESP experiment could be plotted on graph paper, they'd very likely resemble a bell-shaped curve. The bell-shaped curve, which you have probably heard about, represents the distribution of the results. Some people will guess correctly five times, some will guess correctly only two or three times, and a few will guess correctly seven, eight, or even nine times.

Can you think of a way to determine whether those who were thought to have ESP actually have it?

One way would be to test the same people again. If you did, it's doubtful that those who were thought to have ESP would do any better than the average. They might even do worse. Why don't you try this experiment with some classmates and see what happens?

KEY TERMS AND CONCEPTS

Adaptation, p. 257
Anterior chamber, p. 262
Aqueous humor, p. 262
Astigmatism, p. 263
Auditory (eustachian) tube, p. 266
Basilar membrane, p. 267
Cataracts, p. 261
Chemoreceptors, p. 253
Choroid, p. 259
Ciliary body, p. 259
Cochlea, p. 266
Color blindness, p. 265
Cones, p. 260
Cornea, p. 259
External auditory canal, p. 266
Farsightedness, p. 263
Fovea centralis, p. 260
General senses, p. 253

Glaucoma, p. 262
Hair cells, p. 267
Incus, p. 266
Iris, p. 259
Lens, p. 261
Malleus, p. 266
Mechanoreceptors, p. 253
Nearsightedness, p. 262
Nociceptors, p. 253
Olfactory epithelium, p. 258
Optic nerve, p. 260
Organ of Corti, p. 267
Ossicles, p. 266
Oval window, p. 266
Photoreceptors, p. 253
Posterior chamber, p. 262
Presbyopia, p. 264
Proprioception, p. 256

Pupil, p. 259
Referred pain, p. 255
Retina, p. 260
Rods, p. 260
Round window, p. 267
Sclera, p. 259
Seasonal affective disorder (SAD), p. 261
Semicircular canals, p. 266
Special senses, p. 253
Stapes, p. 266
Taste bud, p. 257
Thermoreceptors, p. 253
Tympanic membrane, p. 266
Vestibular apparatus, p. 266
Vestibule, p. 266
Vestibulocochlear nerve, p. 267
Vitreous humor, p. 262

CONCEPT REVIEW

1. Define the terms general senses and special senses. p. 253.
2. Using your knowledge of the senses and of other organ systems gained from previous chapters, describe the role that sensory receptors play in homeostasis. Give specific examples to illustrate your points. p. 271.
3. Make a list of both the encapsulated and the nonencapsulated general sense receptors. Note where each is located and what it does. pp. 253, 255–257.
4. Define the term adaptation. What advantages does it confer? Can you think of any disadvantages? p. 257.
5. Describe the receptors for taste, explaining where they are located, what they look like, and how they operate. pp. 257–258.
6. Taste buds detect five basic flavors. What are they? How do you account for the thousands of different flavors that you can detect? p. 258.

7. Describe the olfactory membrane. How do the receptor cells operate? In what ways are taste receptors and olfactory receptors similar? In what ways are they different? pp. 258–259.

8. Explain the following statement: Taste and smell are complementary. p. 259.

9. Draw a cross section of the human eye and label its parts. pp. 259–260.

10. Define the following terms: retina, rods, cones, fovea centralis, and optic nerve. p. 260.

11. Compare and contrast rods and cones. p. 260.

12. Define the following terms: nearsightedness, farsightedness, presbyopia, and astigmatism. pp. 262–264.

13. What is color blindness? What is the most common type? Explain what your world would look like if you were afflicted by red-green color blindness. pp. 265–266.

14. What is a cataract, and how is it treated? pp. 261–262.

15. Describe the disease known as glaucoma. p. 262.

16. Describe the anatomy of the ear and the role of the outer, middle, and inner ear in hearing. pp. 266–267.

17. How do the semicircular canals operate? How do the utricle and saccule operate? p. 270.

SELF-QUIZ: TESTING YOUR KNOWLEDGE

1. Pain, temperature, and light touch are considered _____ senses, while vision and hearing are considered _____ senses. p. 253.

2. The rods and cones of the retina are a type of receptor known as _____. p. 260.

3. Pain felt on the surface of the body away from its internal source is known as _____ pain. p. 255.

4. Naked nerve endings wrapped around the base of hair follicles is responsible for the sensation known as _____. p. 256.

5. The sense of position is also known as _____. p. 256.

6. _____ are structures that house receptors that respond to chemicals dissolved in the food we eat. p. 257.

7. Smells are detected by sensors found in the _____ epithelium in the roof of the nasal cavities. p. 258.

8. The white of the eye is known as the _____; it joins the clear portion of the eye in front known as the _____. p. 259.

9. Light is focused on the retina by the _____ and the _____. p. 260.

10. _____ are receptors that permit color vision. p. 260.

11. Clouding of the lens is a condition known as _____. p. 261.

12. Fluid buildup in the anterior chamber of the eye is called _____. p. 262.

13. The gelatinous material in the posterior cavity is known as _____ humor. p. 262.

14. An irregularly shaped cornea (football shaped) is responsible for a visual defect known as _____. p. 263.

15. As one ages, the _____ becomes less elastic and unable to focus light from nearby objects resulting in a condition known as presbyopia. p. 264.

16. The tiny bones in the middle ear are known as _____. They transmit sound waves from the _____ to the oval window. p. 266.

17. Fluid waves in the cochlea stimulate hair cells in the organ of _____. p. 267.

18. The _____ canals are located in the inner ear; they are important in dynamic balance. p. 270.

19. Label the diagram.

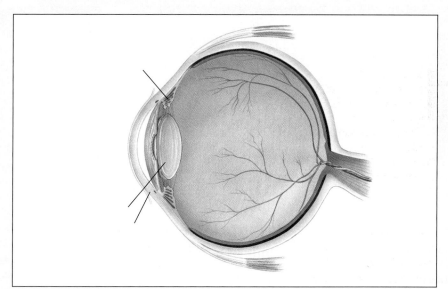

biology.jbpub.com/chiras/8e/

The site features eLearning, an online review area that provides quizzes, chapter outlines, and other tools to help you study for your class. You can also follow useful links for in-depth information, research the differing views in the Point/Counterpoints, or keep up on the latest health news.

(© Louie Schoeman/ShutterStock, Inc.)

The Endocrine System

One day while on his morning jog, President George Bush (the elder) began to feel weak. His heartbeat became irregular. He felt breathless. Fearing a heart attack, his security entourage rushed him to the hospital. There a blood test revealed elevated levels of the hormone thyroxine produced by the thyroid gland in the neck. This substance accelerates cellular energy production and affects a number of mental and physical processes. Elevated levels are indicative of a condition

THINKING CRITICALLY

Plastics like those used to make soft drink bottles and car interiors contain compounds known as *plasticizers*. These chemicals are designed to make them a bit more flexible and durable. Most are used for a very common plastic known as PVC. It's used to make beach balls, pipe, furniture, fencing, home siding, and toys. Researchers have long been concerned about the potential health effects of these plasticizers on humans. Numerous studies on fish exposed to plasticizers, for example, show that these compounds can dramatically alter reproduction. That's because plasticizers act like hormones inside the bodies of fish. Many of them mimic natural estrogen, the female sex hormone. In the wild, these hormones have been shown to have a feminizing effect on male fish—that is, they turn males into females.

A recent study in mice showed that one plasticizer, bisphenol-A, which is found in polycarbonate plastics used to line beverage cans, had the opposite effect on female mice. That is, it had a masculinizing effect. The females acted more like males. In addition, it had a masculinizing effect on the brains of females.

Researchers injected female mice with specific amounts of bisphenol-A so that concentrations found in the pregnant and later nursing mothers would be similar to those found in the blood of humans as a result of drinking beverages in cans lined with polycarbonate plastic. The babies were exposed *in utero* through the bloodstream and also after birth by consuming milk from their mothers that was contaminated with this plasticizer. The researchers did not conclude that similar results might occur in humans, but did suggest that it was a possibility. What are your thoughts on this? Can these results be extrapolated to humans? Why or why not?

called *hyperthyroidism* (HIGH-per-THIGH-roid-izm), or Graves disease (after its discoverer, not the outcome). This rare disease can occur at any age. It is one of many diseases caused by a hormonal imbalance. This chapter discusses hormones and the endocrine system and how they help ensure homeostasis and other functions such as growth and reproduction. It also presents some examples of homeostatic imbalance resulting from endocrine disorders.

13-1   Principles of Endocrinology

The human **endocrine system** consists of numerous small glands scattered throughout the body (Figure 13-1). These glands produce and secrete chemical substances known as **hormones**. (The word hormone comes from the Greek *hormon*, which means "to stimulate" or "excite.") A hormone is a chemical produced and released by cells or groups of cells that form the endocrine (ductless) glands.

All hormones are transported via the bloodstream to distant sites where they exert some effect. The hormone insulin, for example, is produced by the pancreas and travels in the blood to skeletal muscle. Here the hormone stimulates glucose uptake and glycogen synthesis. (Glycogen is a polysaccharide that serves as a storage form of glucose in muscle.) The cells affected by a hormone are called its **target cells**. For a discussion of some of the early research that contributed to our understanding of hormones, see Scientific Discoveries 13-1.

Hormones function in five areas: (1) homeostasis; (2) growth and development; (3) reproduction; (4) energy production, storage, and use; and (5) behavior. At any one moment, the blood carries dozens of different hormones. The cells of the body are, therefore, exposed to many different chemical stimuli.

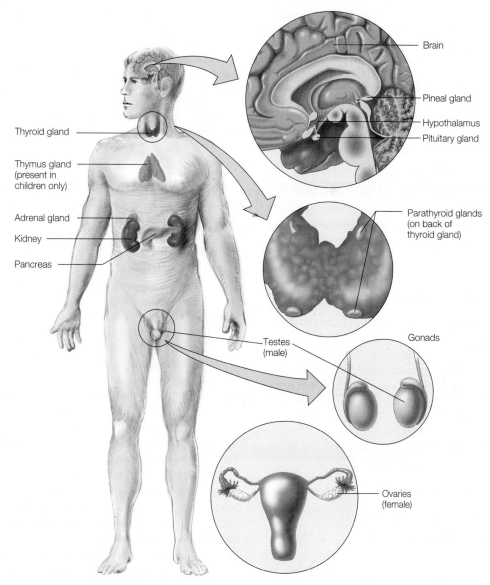

FIGURE 13-1 **The Human Endocrine System** The endocrine system consists of a scattered group of glands that produce hormones, chemicals that regulate growth and development, homeostasis, reproduction, energy metabolism, and behavior.

Scientific Discoveries that Changed the World

13-1   Pancreatic Function: Is It Controlled by Nerves or Hormones?

Featuring the Work of Bayliss and Starling

The concept of chemical control of body functions, like many other important biological ideas, is rooted in numerous experiments.

Of the many experiments crucial to our understanding of endocrinology, one by two British physiologists, W. M. Bayliss and E. H. Starling, stands out. Published in 1902, this experiment set to rest a debate about the mechanism controlling the release of pancreatic secretions.

Several experiments before the publication of this work suggested that hydrochloric acid in chyme (produced in the stomach) stimulated nerve endings in the small intestine. This, in turn, triggered a reflex that caused the release of pancreatic juices containing sodium bicarbonate, which neutralizes the acid. In these studies, researchers injected substances into the small intestine of anesthetized dogs to see what stimulated the release of pancreatic secretions. To determine the nature of the reflex, the researchers cut the nerves supplying the small intestine. Even though severing the nerves did not halt the secretion of pancreatic juices, these researchers still argued that a nervous system reflex was involved. They hypothesized that clumps of nerve cells (ganglia) and their network of interconnected fibers in the wall of the intestine participated in local reflex arcs.

Bayliss and Starling repeated these experiments but were careful to cut all nerve connections to the small intestine. They then tied off a piece of the small intestine and injected small amounts of hydrochloric acid into it. This treatment resulted in the production of pancreatic juices. Because it was previously known that acid introduced into the bloodstream had no effect on the release of pancreatic secretions, the researchers concluded that "the effect was produced by some chemical substance finding its way into the veins of the loop of jejunum in question . . ." This substance, they wrote, was "carried in the bloodstream to the pancreatic cells."

To verify their hypothesis, the researchers tried another experiment. In this one, they scraped off the cells lining the section of the small intestine, exposed them to acid, ground them up, and then extracted the fluid, which they injected into the bloodstream. After a brief latent period, the pancreas began secreting, a clear indication that these cells lining the first part of the small intestine were producing a chemical substance that stimulated the pancreas.

Bayliss and Starling referred to the mystery substance as "secretin," because it stimulated pancreatic secretion. This name remains in use today.

Besides settling the debate over the control of pancreatic function, Bayliss and Starling helped clarify our understanding of endocrinology, furnishing definitions for some of the most important terms in the field. One good example is the term endocrine gland, which they defined as an organ that secretes into the blood specific substances that affect some other organ or process located at a distance from the gland.

KEY CONCEPTS

The endocrine system produces chemicals called hormones that are transported through the bloodstream to distant sites where they affect many functions including growth, metabolism, and reproduction.

Target Cells

Despite the fact that the cells of the body are flooded with hormonal signals 24 hours a day, each cell responds only to very specific hormones. Target cells are therefore said to be selective. Selectivity occurs because of protein receptors. That is to say, each cell contains receptors that bind to specific hormones to which the cell is genetically programmed to respond. In some target cells, the hormone receptors are embedded in the plasma membrane; in others, they're located in the cytoplasm. (More on this later.)

KEY CONCEPTS

Target cells are genetically programmed to respond to specific hormones; they contain receptors that bind to those hormones.

Tropic and Nontropic Hormones

Hormones fall into two broad categories. The first are the **tropic hormones** (TROW-pick; "to nourish"). Tropic hormones stimulate other endocrine glands to produce and secrete hormones. An example is thyroid-stimulating hormone (TSH). Produced by the pituitary gland, TSH travels in the blood to the thyroid gland located in the neck on either side of the larynx. In the thyroid, TSH stimulates the release of a hormone, thyroxine (thigh-ROX-in). Thyroxine, in turn, circulates in the blood and stimulates metabolism in many types of body cells. Thyroxine is a nontropic hormone. Growth hormone, once thought to stimulate growth directly, stimulates the release of hormones by the liver. They, in turn, stimulate growth of bone and cartilage.

Nontropic hormones stimulate vital cellular processes, including metabolism, but do not stimulate the release of other hormones. Prolactin, a hormone from the anterior pituitary, for instance, stimulates the production of milk in a woman's breast tissue. Testosterone released by the testes of men stimulates bone growth directly. It, too, is a nontropic hormone.

Hormones can also be classified into three groups according to their chemical composition: (1) steroids; (2) proteins and polypeptides; and (3) amines.

Steroid hormones are derivatives of cholesterol. Figure 13-2 illustrates two common steroid hormones. Only minor differences in the chemical structure of these molecules exist, but they produce profound functional differences. Protein and polypeptide hormones are the largest group of hormones. Proteins and polypeptides are long chains (polymers) of amino acids. Growth hormone and insulin are two examples. Amine hormones are much smaller molecules derived from the amino acid tyrosine. Figure 13-3 shows the structure of two

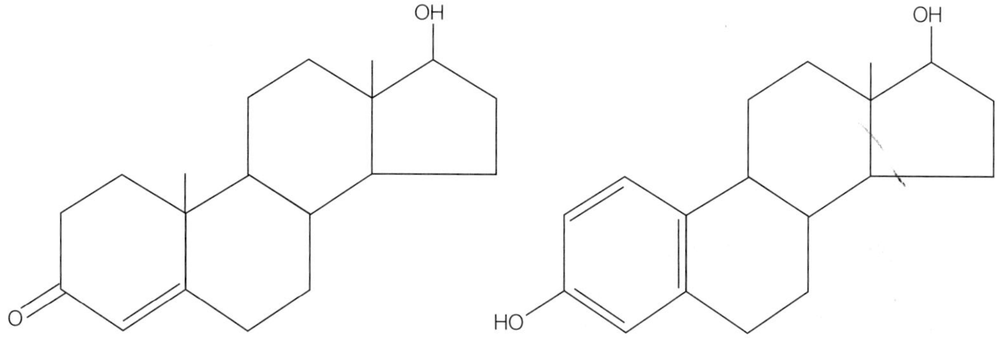

Testosterone,
a masculinizing
hormone

Estradiol,
a feminizing
hormone

FIGURE 13-2 **Two Common Steroids**

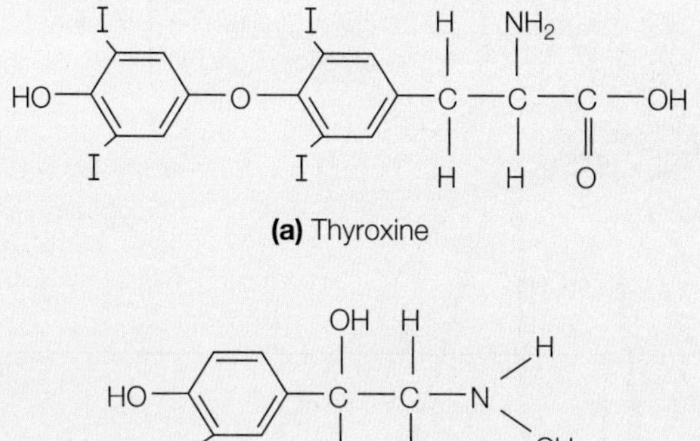

(a) Thyroxine

(b) Epinephrine

FIGURE 13-3 **Representative Amine Hormones** (a) Thyroxine from the thyroid and (b) adrenalin (epinephrine) from the adrenal medulla.

of the four amine hormones produced in the body, thyroxine and epinephrine (adrenalin).

> **KEY CONCEPTS**
>
> Tropic hormones stimulate the synthesis and release of hormones by other endocrine glands; nontropic hormones affect a wide assortment of cellular processes in other cells.

Negative Feedback Control

Hormones help to control many homeostatic mechanisms. Not surprisingly, their production and release are generally controlled by negative feedback loops, described elsewhere in the text. As you may recall, in **negative feedback loops**, the end product of a biochemical process inhibits its own production. Consider an example, the hormone glucagon (GLUE-ka-gone) and glucose. Glucagon is a hormone secreted by cells in the pancreas. It is released when glucose concentrations fall—for example, between meals. Glucagon, in turn, stimulates the breakdown of glycogen stored inside liver cells into glucose molecules. Glucose is released into the bloodstream, restoring normal levels. When normal glucose levels are achieved, glucose molecules cause the glucagon-producing cells in the pancreas to cease secretion.

Positive feedback loops are also encountered in the endocrine system, but only rarely. A positive feedback loop results when the hormonal product of a cell or organ stimulates the production of another hormone. This, in turn, causes more secretion, which stimulates the hormone production and so on and so on. Noise in a restaurant is a good example of a positive feedback loop. As the noise level increases, people talk louder. This increases the overall noise, forcing people to talk even louder. The noise can become intolerable.

Positive feedback loops perform specialized functions. Ovulation, the release of the ova from the ovary, for example, is stimulated by a positive feedback. Fortunately, each positive feedback loop has a built-in mechanism that ends the escalating cycle, preventing the response from getting completely out of hand. With this background information in mind, let's turn our attention to the pituitary hormones.

> **KEY CONCEPTS**
>
> Hormones control many physiological processes through negative and positive feedback loops that help maintain homeostasis.

13-2 The Pituitary and Hypothalamus

Attached to the underside of the brain by a thin stalk is the **pituitary gland** (peh-TWO-eh-TARE-ee; Figure 13-4). About the size of a pea, the pituitary lies in a depression in the base of the skull.

The pituitary gland is divided into parts: the **anterior pituitary** and the **posterior pituitary**. Together, they secrete a large number and variety of hormones, affecting many different body functions. The anterior pituitary, for instance, produces seven protein and polypeptide hormones, six of which are discussed in this chapter (Table 13-1).

The anterior pituitary is controlled by a region of the brain known as the **hypothalamus**. Lying just above the pituitary gland, the hypothalamus contains receptors that

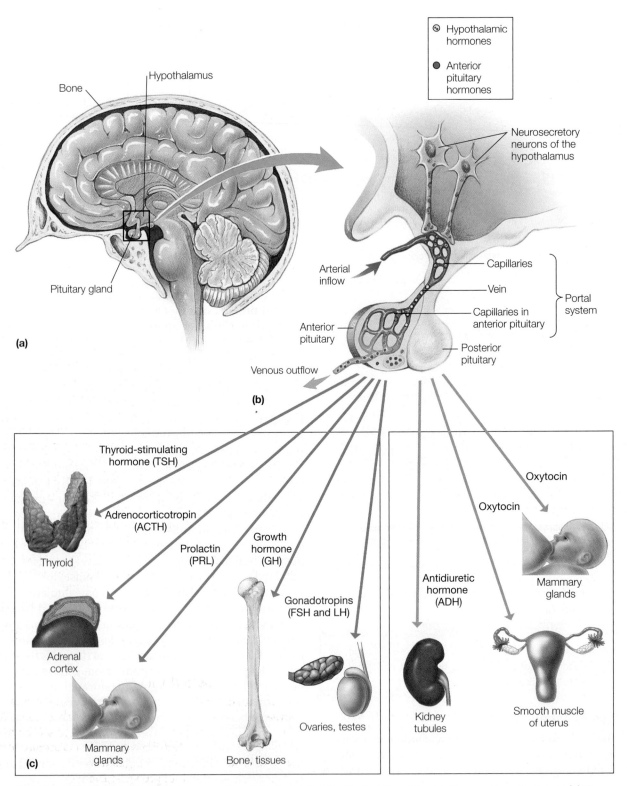

FIGURE 13-4 The Pituitary Gland (a) A cross section of the brain showing the location of the pituitary and hypothalamus. (b) The structure of the pituitary gland. (c) Releasing and inhibiting hormones travel via the portal system from the hypothalamus to the anterior pituitary, where they affect hormone secretion.

monitor blood levels of hormones, nutrients, and ions. When activated, the receptors stimulate specialized nerve cells in the hypothalamus. These cells are a special brand of neurons called **neurosecretory neurons**—so named because they synthesize and secrete hormones. The hormones they produce control the production and release of hormones by the anterior pituitary located nearby (Figure 13-4).

The hypothalamus produces two types of hormones—those that stimulate the release of anterior pituitary hormones, called **releasing hormones**, designated **RH**, and those that inhibit the release of hormones from the anterior pituitary. They're called **inhibiting hormones**, designated **IH**.

The releasing and inhibiting hormones are produced in the cell bodies of the neurosecretory neurons. They travel down

TABLE 13-1	Hormones Secreted by the Pituitary Gland
Hormone	**Function**
Anterior pituitary	
Growth hormone (GH)	Stimulates cell growth. Primary targets are muscle and bone, where GH stimulates amino acid uptake and protein synthesis. It also stimulates fat breakdown in the body.
Thyroid-stimulating hormone (TSH)	Stimulates release of thyroxine and triiodothyronine.
Adrenocorticotropic hormone (ACTH)	Stimulates secretion of hormones by the adrenal cortex, especially glucocorticoids.
Gonadotropins (FSH and LH)	Stimulate gamete production and hormone production by the gonads.
Prolactin	Stimulates milk production by the breast.
Melanocyte-stimulating hormone (MSH)	Function in humans is unknown.
Posterior pituitary	
Antidiuretic hormone (ADH)	Stimulates water reabsorption by nephrons of the kidney.
Oxytocin	Stimulates ejection of milk from breasts and uterine contractions during birth.

the axons of the neurosecretory cells, which terminate in the lower part of the hypothalamus, just above the pituitary gland. The hormones are stored in the ends of the axons until needed.

When released, the hormones diffuse into nearby capillaries. As shown in Figure 13-4, these capillaries drain into a series of veins in the stalk of the pituitary. These veins drain into a capillary network located in the anterior pituitary. From here, the hormones diffuse into the anterior pituitary. They then bind to their target cells, causing them to release hormones. This unusual arrangement of blood vessels in which a capillary bed drains to a vein, which then drains into another capillary bed, is called a **portal system**.

> **KEY CONCEPTS**
> The pituitary gland consist of two parts, the anterior and posterior pituitary; these glands produce numerous hormones vital to human health and reproduction.

The Anterior Pituitary

The anterior pituitary produces a number of hormones that play extremely important roles in human physiology. This section examines these hormones.

> **KEY CONCEPTS**
> The anterior pituitary secretes seven hormones with widely different functions.

Growth Hormone

People differ in height and body build. These differences are largely attributable to nutrition and **growth hormone (GH)**, a protein hormone produced by the anterior pituitary. Growth hormone stimulates growth in the body by promoting cellular enlargement (hypertrophy; HIGH-per-TROW-fee) and an increase in the number of cells through division (hyperplasia; HIGH-per-PLAY-za).

Although growth hormone affects virtually all body cells, it acts primarily on bone and muscle. In muscle, it stimulates the uptake of amino acids and protein synthesis, which results in an increase in muscle mass. In bone, growth hormone

stimulates cell division and protein synthesis in bone cells, resulting in an increase in the length and width of bones.

Recent evidence shows that growth hormone does not affect bone and muscle directly. It actually operates by stimulating the production of several polypeptide hormones produced primarily by the liver. Known as **somatomedins** (so-MAT-oh-ME-dins), these hormones stimulate bone and cartilage cells.

As a rule, the more growth hormone produced during the growth phase of an individual's life cycle, the taller and heftier he or she will be. In men, body growth is also stimulated by testosterone, a sex steroid produced by the testes. It stimulates bone and muscle growth.

Growth hormone secretion undergoes a diurnal (daily) cycle. As Figure 13-5 shows, the highest blood levels are present during sleep and during strenuous exercise. It is no wonder that sleep is so important to a growing child. Growth hormone secretion declines gradually as we age.

> **KEY CONCEPTS**
> Growth hormone is a protein released by the anterior pituitary; it stimulates the growth of muscles and bones.

Control of Growth Hormone Section

The release of growth hormone by the pituitary is controlled by the hypothalamus in a classic negative feedback loop. When levels are low, the hypothalamus secretes growth hormone-releasing hormone or GH-RH into the blood stream. It flows down the stalk of the pituitary gland where it stimulates the production and release of growth hormone.

Growth hormone release is also stimulated directly through the nervous system. Studies show that stress and vigorous exercise both stimulate the hypothalamus via neural pathways to release GH-RH.

Deficiencies in growth hormone can result in dramatic changes in body shape and size, depending on when the deficiency occurs. Undersecretion, or hyposecretion, that occurs during the growth phase of a child is one cause of stunted growth (dwarfism) (Figure 13-6a). Oversecretion, or hypersecretion, results in giantism if the excess occurs during the growth phase (Figure 13-6b). If the pituitary begins producing

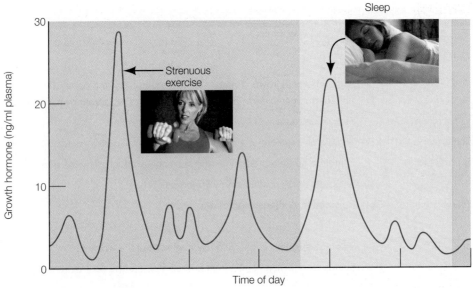

ng/ml = nanograms per mililiter

FIGURE 13-5 **Growth Hormone Secretion in an Adult** Growth hormone stimulates muscle and bone growth and is released during exercise and sleep. (Left photo, © LiquidLibrary; right photo © Photos.com.)

excess growth hormone after growth is complete, the result is a relatively rare disease called **acromegaly** (AH-crow-MEG-al-ee). Facial features become coarse, and hands and feet continue to grow throughout adulthood (Figure 13-7). Growth of the vertebrae results in a hunched back. The tongue, kidneys, and liver often become quite enlarged.

> **KEY CONCEPTS**
> The secretion of growth hormone is controlled by growth hormone levels, stress, and exercise.

Thyroid–Stimulating Hormone

Thyroid-stimulating hormone (TSH) is a protein hormone produced by the anterior pituitary and is controlled by TSH-releasing hormone (TSH-RH). As shown in Figure 13-8,

(a)

(b)

FIGURE 13-6 **Disorders of Growth Hormone Secretion** (a) Pituitary dwarf. (© NYPL/Science Source, Inc.) (b) Pituitary giant. (© AP Photos.)

(a) (b)

(c) (d)

FIGURE 13-7 **Acromegaly** Hypersecretion of growth hormone in adults results in a gradual thickening of the bone, which is especially noticeable in the face, hands, and feet. There is no sign of the disorder at age 9 (a) or age 16 (b). Symptoms are evident at age 33 (c) and age 52 (d). (Reproduced from *Am. J. Med.*, vol. 20, Mendelhoff, A. I., and Smith, D. E., Acromegaly, diabetes, hypermetabolism . . . , pp. 133–144. Copyright 1956, with permission from Elsevier.)

TSH-RH is produced in the hypothalamus. TSH-RH release is regulated primarily by the concentration of thyroxine (T_4; thigh-ROX-in) and a chemically similar hormone, triiodothyronine (T_3; try-eye-OH-doe-THIGH-row-neen) in the blood in a negative feedback loop, also shown in Figure 13-8. As you may recall, thyroxine is a hormone produced by the thyroid glands, which are located on the kidneys. Receptors in the hypothalamus monitor the level of circulating thyroxine. When blood levels are low, these receptors signal the hypothalamus to release TSH-RH. This causes a TSH-release that stimulates the production and release of the two thyroid hormones, T_4 and T_3. When the levels of T_4 and T_3 increase, TSH-RH secretion declines. As shown in Figure 13-8, TSH-RH secretion is also stimulated by cold and stress. Thus, when you're under a lot of stress, TSH-RH secretion is higher than when you are calm. Thyroxine increases metabolic rate.

T_3 and T_4 circulate in the bloodstream and influence many body cells. One of their chief functions is to stimulate the breakdown of glucose by body cells. Because this process produces

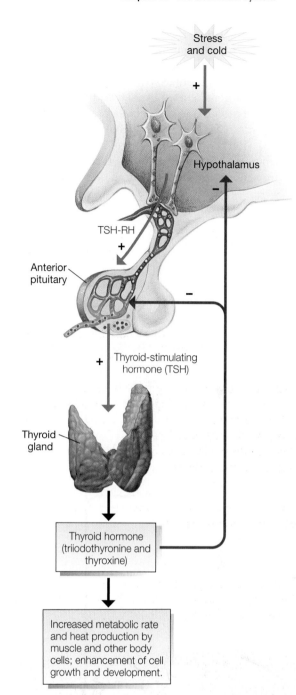

FIGURE 13-8 **Negative Feedback Control of TSH Secretion** Triiodothyronine (T_3) and thyroxine levels are detected by receptors in the hypothalamus. They also feed back on the anterior pituitary. When levels of T_3 and thyroxine are low, the hypothalamus releases TSH-RH, which stimulates the pituitary to release TSH. TSH, in turn, stimulates the thyroid gland to secrete T_3 and thyroxine. Other factors such as stress and cold influence the release of TSH via the hypothalamus. (+ denotes stimulation; − denotes inhibition.)

energy and heat, increased levels of these hormones raise body temperature. Secretion of T_3 and T_4 is therefore typically higher during cold winter months than during hot summer months.

KEY CONCEPTS

TSH is a hormone produced by the anterior pituitary that stimulates the thyroid gland to produce thyroxine, a metabolic hormone.

Adrenocorticotropic Hormone

Another important hormone of the anterior pituitary is **adrenocorticotropic hormone** (**ACTH**; ad-REE-no-core-tick-oh-TRO-pick). Its target organ is the outer regions of the adrenal gland, the adrenal cortex. As shown in Figure 13-9, ACTH stimulates the production and release of a group of steroid hormones known as the **glucocorticoids** (GLUE-co-CORE-teh-KOIDS). Glucocorticoids are steroids. Their main function is to increase blood glucose levels, through mechanisms beyond the scope of this book. Bottom line: glucocorticoids

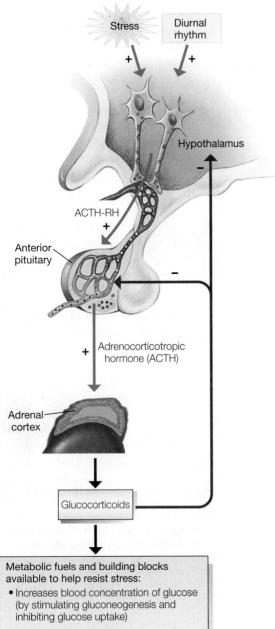

FIGURE 13-9 Feedback Control of ACTH Cortisol regulates hypothalamic and pituitary activity, but stress and the biological clock also influence the release of ACTH-RH.

help maintain proper blood glucose levels that are essential for the function of brain cells and all other body cells. It is therefore vital to homeostasis. Like other end products of hormonally controlled processes, glucocorticoids participate in a negative feedback loop.

As shown in Figure 13-9, the production and release of ACTH are controlled by a small peptide known as ACTH-RH from the hypothalamus. Like other releasing hormones, it travels to the pituitary via the portal system. ACTH secretion is also controlled by stress, acting through the nervous system. The release of ACTH-RH and ACTH during stressful situations ensures that you have the additional energy you need to operate body systems, especially muscles, when under stress.

> **KEY CONCEPTS**
>
> ACTH is a hormone released by the anterior pituitary; it controls the production and release of glucocorticoids from the adrenal cortex; glucocorticoids help control blood sugar levels.

The Gonadotropins

Reproduction in men and women is under the control of the anterior pituitary. It produces two hormones that affect the gonads, appropriately known as **gonadotropins** (go-NAD-oh-TROW-pins).

> **KEY CONCEPTS**
>
> The gonadotropins are hormones produced in and released by the anterior pituitary; they help control reproduction in males and females.

Prolactin

Prolactin (pro-LACK-tin) is a protein hormone produced by the anterior pituitary. In women, prolactin is secreted at the end of pregnancy and acts on the mammary glands (breasts) where it stimulates milk production.

Prolactin secretion is controlled by a **neuroendocrine reflex**, a reflex involving both the nervous and endocrine systems. As Figure 13-10 shows, suckling stimulates sensory fibers in the breast, which send nerve impulses to the hypothalamus. In the hypothalamus, these impulses stimulate the release of prolactin releasing hormone (PRH). PRH, in turn, travels to the anterior pituitary in the bloodstream, where it stimulates the secretion of prolactin.

Prolactin production persists for many months, as long as suckling continues. As babies begin to eat solid food, they reduce their dependence on their mother's milk. Reduced suckling reduces prolactin secretion, causing milk production to decline.

> **KEY CONCEPTS**
>
> Prolactin is an anterior pituitary hormone that stimulates milk production in the breasts of women; its release is controlled by suckling.

The Posterior Pituitary

Have you ever wondered why alcoholic beverages make a person urinate so much or what actually causes milk to flow from a woman's breast? Have you ever wondered why a woman's uterus contracts when she delivers a baby?

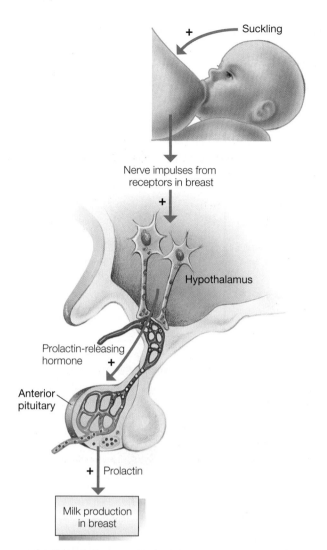

FIGURE 13-10 **Neuroendocrine Reflex and Prolactin Secretion**
Suckling stimulates prolactin release by the anterior pituitary.
Prolactin stimulates milk production by the breast. Milk release
requires another hormone, oxytocin, from the posterior pituitary.

These phenomena can be attributed to two hormones from the posterior pituitary. Like the hypothalamus, the posterior pituitary is a neuroendocrine gland—that is, a gland made of neural tissue that produces hormones. Derived from brain tissue during embryonic development, the posterior pituitary remains connected to the brain throughout life.

The posterior pituitary secretes two hormones: oxytocin (OX-ee-TOE-sin) and antidiuretic hormone (ADH; antie-DIE-yur-ET-ick). Both hormones consist of nine amino acids. As shown in Figure 13-11, ADH and oxytocin are produced in the cell bodies of the neurosecretory cells located in the hypothalamus. They then travel down the axons into the posterior pituitary. The posterior pituitary itself consists of the axons and terminal ends of neurosecretory cells in which these hormones are stored until they are released into the surrounding capillaries.

> **KEY CONCEPTS**
> The posterior pituitary is a neuroendocrine gland, a gland consisting of modified nerve cells that produce hormones; it produces and releases two hormones, ADH and oxytocin.

Antidiuretic Hormone

Antidiuretic hormone (ADH) helps regulate water balance in humans. As you may recall, ADH is released when water levels drop, for example, when you become dehydrated. ADH does this by increasing water absorption in the nephrons located in the kidneys (Figure 13-12). As a result, water reenters the bloodstream, counteracting dehydration by increasing blood volume and maintaining the normal osmotic concentration of the blood.

ADH secretion is controlled by a negative feedback loop (Figure 13-12). Receptors in the hypothalamus monitor the concentration of dissolved substances (especially sodium ions) in the blood. When the concentration exceeds the normal level—for example, when a person becomes dehydrated—ADH is released into the bloodstream. ADH stimulates the reabsorption of water by the nephrons. When the osmotic concentration of the blood approaches homeostatic levels, ADH secretion ceases.

As noted elsewhere in the text, ADH is inhibited by ethanol in alcoholic beverages. The reduction in ADH decreases water reabsorption in the kidneys and increases urine output. This, in turn, can lead to dehydration—the dry mouth and intense thirst a person may experience as part of a hangover.

Inadequate ADH secretion, for example, caused by traumatic head injuries that damage the hypothalamus, result in a condition known as *diabetes insipidus*. Patients with this disease produce several gallons of dilute urine a day because of the reduction in ADH secretion. To counteract the extreme water loss, they must drink enormous quantities of water each day.

> **KEY CONCEPTS**
> ADH released by the posterior pituitary increases water reabsorption in the kidneys, thus preventing dehydration.

Oxytocin

Prolactin from the anterior pituitary stimulates milk production in the breasts. But it's **oxytocin** from the posterior pituitary that stimulates the contraction of the smooth musclelike cells around the glands in the breast that "pump" the milk out. Oxytocin release is stimulated by suckling. Oxytocin travels in the blood to the breast, where it stimulates milk let-down—the ejection of milk from the glands soon after suckling begins.

Oxytocin also plays a role in childbirth. At the end of pregnancy, in the hours prior to delivery, the walls of the uterus and cervix stretch. Impulses from the stretch receptors in the uterus travel to the cells of the hypothalamus that produce oxytocin. The impulses cause the neurosecretory cells to release oxytocin into the blood. The hormone then travels in the blood to the uterus. There it stimulates smooth muscle contraction, aiding in the expulsion of the baby and, later, the placenta.

> **KEY CONCEPTS**
> Oxytocin is a hormone produced by the posterior pituitary. It stimulates the ejection of milk from breasts during suckling and expulsion of the baby during childbirth.

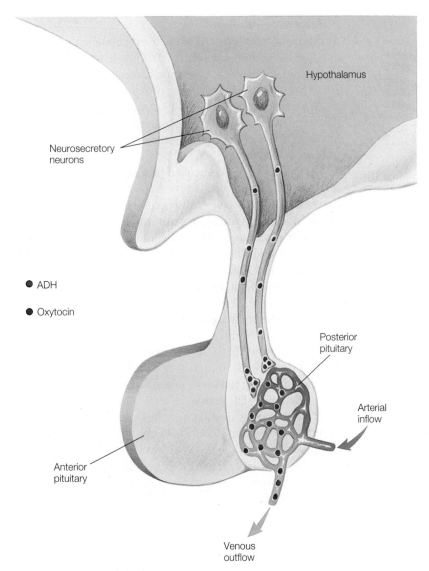

FIGURE 13-11 **The Posterior Pituitary** Neurosecretory neurons that produce oxytocin and ADH originate in the hypothalamus and terminate in the posterior pituitary. Hormones are produced in the cell bodies of the neurons and are stored and released into the bloodstream in the posterior pituitary.

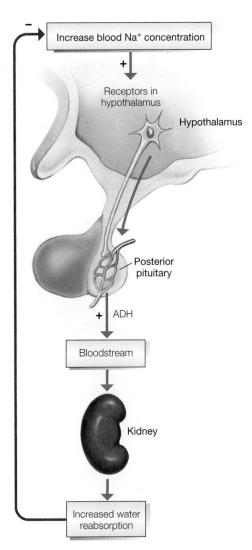

FIGURE 13-12 **Role of ADH in Regulating Fluid Levels** ADH secretion is stimulated by an increase in the blood concentration of sodium caused by dehydration. ADH increases water reabsorption in the kidney, thus, eliminating the stimulus for ADH secretion.

13-3 The Thyroid Gland

On a 50°F (10°C) day in the fall, you bundle up in a sweater to feel warm. However, 50°F (10°C) in the spring feels warm. You may not even need a sweater. Why?

The answer lies in the thyroid gland and two of its hormones. The **thyroid gland** is a U- or H-shaped gland (it varies from one person to the next) located in the neck (Figure 13-13a). The thyroid gland produces three hormones: (1) thyroxine (tetraiodothyronine, or T_4); (2) a chemically similar compound, triiodothyronine (T_3); and (3) calcitonin. The first two are involved in controlling metabolism and heat production; the last helps regulate blood levels of calcium.

As Figure 13-13b shows, the thyroid gland consists of large, spherical structures known as follicles. Each follicle consists of a central region containing a gellike glycoprotein called *thyroglobulin* and a surrounding layer of cuboidal follicle cells that produce it. **Thyroxine** and **triiodothyronine** are relatively small molecules (amines) cleaved off the thyroglobulin molecules in the follicles.

T_3 and T_4, shown in Figure 13-3a, contain iodide ions, indicated by I. To make these hormones, the thyroid requires large quantities of iodide. It comes from the diet through vegetables and iodized (iodine-fortified) salt. A shortage of iodide in the diet results in a decline in the levels of thyroxine and triiodothyronine in the blood. The hypothalamus responds by producing more thyroid-stimulating hormone (TSH). TSH causes the thyroid to increase thyroglobulin production. Unfortunately, thyroglobulin produced under these conditions yields very little T_3 and T_4. TSH secretion rises and the thyroid gland continues to produce thyroglobulin. This causes the gland to enlarge, sometimes forming softball-sized

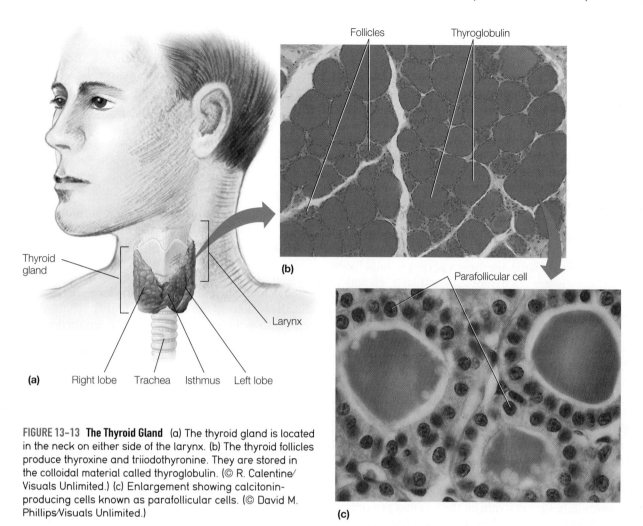

FIGURE 13–13 **The Thyroid Gland** (a) The thyroid gland is located in the neck on either side of the larynx. (b) The thyroid follicles produce thyroxine and triiodothyronine. They are stored in the colloidal material called thyroglobulin. (© R. Calentine/Visuals Unlimited.) (c) Enlargement showing calcitonin-producing cells known as parafollicular cells. (© David M. Phillips/Visuals Unlimited.)

enlargements on the neck. This condition is known as *goiter* (GOY-ter; Figure 13-14).

Goiter was once common in areas of Europe and the United States where iodine had been leached out of the soil by rain—for example, near the Great Lakes. Crops grown in iodine-poor soils failed to provide sufficient quantities. To counteract this deficiency, iodine was added to table salt. Today, few people in industrialized nations develop goiter because they eat iodized salt or ingest sufficient quantities of iodine in their food and through mineral supplements.

KEY CONCEPTS

The thyroid gland located in the neck secretes hormones that accelerate metabolism, increase body heat, and lower calcium levels.

Actions of Thyroxine and Triiodothyronine

Thyroxine and triiodothyronine are virtually identical and, as a result, have nearly identical functions. Both hormones, for example, accelerate the rate of glucose breakdown in most body cells. That is, they boost metabolism. These hormones also stimulate cellular growth and development. Bones and muscles are especially dependent on them during the growth phase. Even normal reproduction requires these hormones. A deficiency of T_3 and T_4, for example, delays sexual maturation in both sexes.

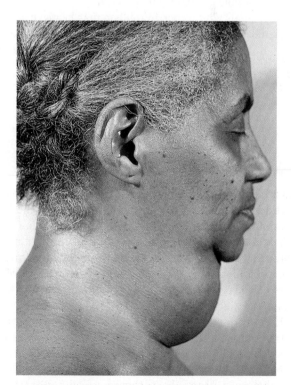

FIGURE 13–14 **Goiter** An enlargement of the thyroid gland most often results from a lack of iodine in the diet. (© Ken Greer/Visuals Unlimited.)

In children, depressed thyroid output stunts mental as well as physical growth. If the deficiency is not detected and treated, the effects are irreversible. In adults, reduced thyroid activity, or hypothyroidism, has less severe results. Hypothyroidism decreases the metabolic rate, making a person feel cold much of the time. People suffering from hypothyroidism also feel tired and worn out. Even simple mental tasks become difficult. Their heart rate may slow to 50 beats per minute. They may have difficulty thinking clearly and suffer from memory problems. If left untreated, hypothyroidism can raise blood cholesterol levels and render a patient more likely to suffer from a heart attack or stroke. Untreated hypothyroidism in pregnant women can harm a developing baby.

Hypothyroidism can affect people of all ages, but is more common in older adults—especially women age 60 and older. This disease is also genetic, meaning a person is more likely to contract the disease if a parent suffers from it. Hypothyroidism is treated by pills containing artificially produced thyroid hormone.

Excess thyroid activity—**hyperthyroidism**—in adults results in elevated metabolism, excessive sweating (due to overheating), and weight loss despite increased food intake. The increase in thyroid hormone levels results in increased mental activity, resulting in nervousness and anxiety. People suffering from hyperthyroidism often find it difficult to sleep. Their heart rate may accelerate, as in the case of former President Bush, and they may lose their sensitivity to cold. Some people suffering from hyperthyroidism exhibit a condition called *exophthalmos* (EX-oh-THAL-mus), or bulging eyes. This is brought about by a swelling of the soft tissues of the eye socket, which causes the eyes to protrude. In such individuals, the eyes may protrude so much that the eyelids cannot close completely. Exophthalmos can cause double vision or blurred vision.

Individuals with hyperthyroidism may be given antithyroid medications, drugs that block the effects of thyroid hormones. If the gland is cancerous, surgeons may remove part or all of it. The most common treatment for hyperthyroidism, however, is radioactive iodine. Taken orally, radioactive iodine concentrates in the thyroid gland, where it damages the cells that produce thyroglobulin, reducing the output of thyroid hormones. This procedure, while effective, may lead to other problems (notably cancer) later in life.

> **KEY CONCEPTS**
>
> Thyroxine and triiodothyronine raise the metabolic rate by increasing the breakdown of glucose in cells and stimulate growth and development.

Calcitonin

The thyroid gland also produces the hormone **calcitonin**. This polypeptide hormone is also known as **thyrocalcitonin**. It is produced by large, round cells in the perimeter of the thyroid follicles known as *parafollicular cells*. They are visible in Figure 13-13c.

Calcitonin helps control calcium levels in the blood. Specifically, it lowers the blood calcium concentrations, thus helping to maintain homeostasis. Calcitonin does so in two ways. The first is by inhibiting bone cells known as *osteoclasts*. As you may recall, osteoclasts are bone-resorbing cells. They digest small amounts of bone when calcium levels decline. The calcium then enters the bloodstream. By inhibiting osteoclasts, helps reduce calcium concentrations in the blood. Calcitonin also stimulates bone-forming cells, known as *osteoblasts*, to deposit calcium in bone, again helping to lower blood levels of this important ion. Finally, calcitonin increases the excretion of calcium (and phosphate) ions by the kidneys.

Calcitonin is involved in a negative feedback loop with calcium ions in the blood. When the calcium-ion concentration increases, calcitonin secretion increases. As calcium concentrations fall, calcitonin secretion falls.

> **KEY CONCEPTS**
>
> Calcitonin produced by cells in the thyroid lowers blood calcium levels, thus helping to maintain homeostasis.

13-4 The Parathyroid Glands

The **parathyroid glands** are four small nodules of tissue embedded in the back side of the thyroid gland. They produce a polypeptide hormone known as **parathyroid hormone**, or **parathormone (PTH)**. PTH works in conjunction with calcitonin to control calcium levels in the blood. However, while calcitonin lowers blood calcium levels, PTH increases them. PTH secretion is therefore stimulated when calcium levels in the blood drop.

PTH restores blood calcium levels by increasing intestinal absorption of calcium. It also stimulates bone resorption by osteoclasts. Finally, PTH increases calcium reabsorption in the kidney.

Calcium levels are also influenced by vitamin D, available in some of the foods we eat, such as milk. Vitamin D is also produced in the skin under the influence of sunlight. Vitamin D increases calcium absorption in the intestine and increases the responsiveness of bone to parathyroid hormone.

Like other endocrine glands, the parathyroid glands occasionally malfunction, producing too much or too little hormone. Excess secretion of parathyroid hormone is the most common condition. This condition (known as **hyperparathyroidism**) may occur as a result of a tumor in one of the four glands. Excess PTH from tumors or other causes results in a loss of calcium from the bones and teeth. An excess of this hormone also upsets several metabolic processes, resulting in indigestion and depression. High serum calcium levels may also result in irritability and inability to concentrate. Because bones contain enormous reserves of calcium, most symptoms of excess PTH other than high blood calcium levels do not appear until 2 to 3 years after the onset of the disease. By the time the disease is discovered, kidney stones may already have formed from calcium, cholesterol, and other substances, and bones may have become more fragile and susceptible to breakage. To prevent further complications, surgeons typically remove the gland con-

taining the tumor. The remaining glands provide sufficient amounts of PTH.

> **KEY CONCEPTS**
>
> The parathyroid gland produces the hormone PTH that increases calcium levels.

13-5 The Pancreas

A young girl's parents complain to her doctor about their daughter's frequent urination, bladder infections, fatigue, and weakness. Tests show that she is suffering from **diabetes mellitus**, a disorder of the pancreas (discussed below). Diabetes mellitus has several causes, but in young people it is generally caused by a lack of insulin production.

> **KEY CONCEPTS**
>
> The pancreas is a dual purpose organ. It produces digestive enzymes and two hormones that regulate blood sugar, insulin and glucagon.

Insulin

Insulin is a protein hormone produced by the pancreas, an organ located in the abdominal cavity (Figure 13-15a). The head of the pancreas lies in the curve of the duodenum, and its tail stretches to the left kidney. Most of the pancreas consists of tiny clumps of cells that produce digestive enzymes and

sodium bicarbonate. Scattered throughout these enzyme-producing cells, however, are small islands of hormone-producing cells, known as the islets of Langerhans (EYE-lits of LANG-er-hahns) after the scientist who described them (Figure 13-15b).

Insulin is released by the **pancreas** within minutes after glucose levels in the blood begin to rise (Figure 13-16). Insulin has many functions, but one of the most important is that it stimulates the uptake of glucose in liver and muscle cells. Inside muscle and liver cells, glucose molecules are combined to form the polysaccharide glycogen. Glucose is said to be stored as glycogen. In healthy individuals, the amount of glycogen in the liver and muscle begins to increase immediately after meals.

> **KEY CONCEPTS**
>
> Insulin causes blood sugar levels to decline rapidly after meals by increasing the uptake and storage of glucose in the liver and muscle.

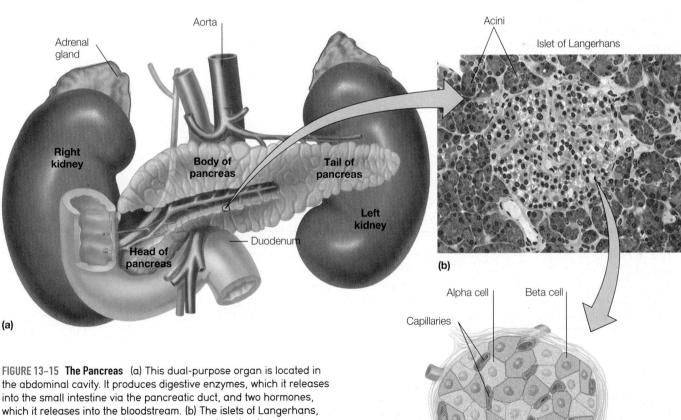

(a)

(b)

Alpha cell   Beta cell

Capillaries

(c)

FIGURE 13-15 The Pancreas (a) This dual-purpose organ is located in the abdominal cavity. It produces digestive enzymes, which it releases into the small intestine via the pancreatic duct, and two hormones, which it releases into the bloodstream. (b) The islets of Langerhans, located among the acini of the pancreas, produce two hormones: glucagon, which increases blood glucose, and insulin, which lowers blood glucose. (© Donna Beer Stolz, Ph.D., Center for Biologic Imaging, University of Pittsburgh Medical School.) (c) Hormones are produced by small clumps of cells called islets of Langerhans.

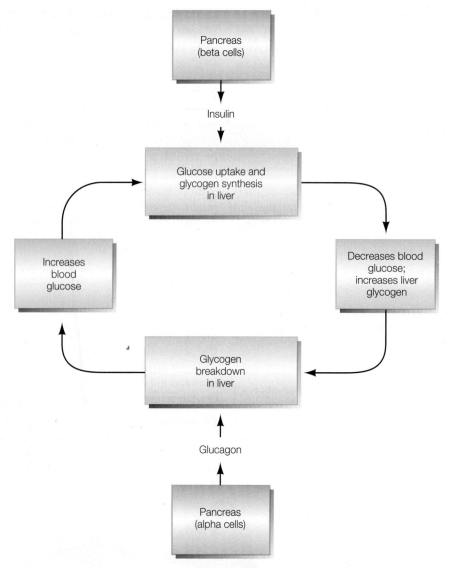

FIGURE 13–16 The Role of the Liver and Pancreas in Controlling Blood Glucose Levels Glucagon and insulin are antagonistic hormones that regulate blood glucose levels through different mechanisms.

Glucagon

The islets of Langerhans also produce a polypeptide hormone known as **glucagon**. Described earlier, this hormone has the opposite effect of insulin—that is, it increases blood levels of glucose. It does so by stimulating the breakdown of glycogen in liver cells (Figure 13-16). This process helps maintain proper glucose levels in the blood between meals. One molecule of glucagon causes 100 million molecules of glucose to be released.

Glucagon also elevates blood glucose levels by stimulating the synthesis of glucose from amino acids and glycerol molecules derived from the breakdown of one type of fat, the triglycerides. This process occurs in the liver.

Glucagon secretion, like that of insulin, is regulated by a negative feedback mechanism stimulated by glucose concentrations in the blood. When glucose levels fall, the hormone is released into the bloodstream. When glucose levels rise, glucagon secretion declines. Glucagon release is also stimulated by an increase in the concentration of amino acids in the

blood. Because glucagon stimulates the conversion of amino acids to glucose, this process ensures that excess amino acids in high-protein meals are not wasted.

> **KEY CONCEPTS**
> Glucagon is a hormone produced by the pancreas; it has the opposite effect of insulin, that is, it increases blood sugar.

Diabetes Mellitus

Millions of people the world over suffer from sugar diabetes Physicians recognize two types of diabetes (Table 13-2). **Type 1 diabetes** can occur at any age, but appears usually in children, teens, or young adults. It is also called **juvenile diabetes** or **early-onset diabetes**.

Type 1 diabetes typically results from a deficiency in insulin production and is also known as insulin-dependent diabetes. Surprisingly, the cause of early onset diabetes is unknown. Many doctors believe that it is caused by damage to the insulin-producing cells of the pancreas. A leading hypothesis suggests

TABLE 13-2	Type 1 and Type 2 Diabetes	
Characteristic	Type 1 Diabetes	Type 2 Diabetes
Level of insulin secretion	None or almost none	May be normal or exceed normal
Typical age of onset	Childhood	Adulthood
Percent of diabetics	10%–20%	80%–90%
Basic defect	Destruction of beta cells	Reduced sensitivity of insulin's target cells
Associated with obesity?	No	Usually
Speed of development of symptoms	Rapid	Slow
Development of ketosis	Common if untreated	Rare
Treatment	Insulin injections; dietary management	Dietary control and weight reduction; occasionally oral hypoglycemic drugs

that it is the result of an autoimmune reaction (in which an individual's immune system attacks parts of his or her own body). In other cases, researchers speculate that Type 1 diabetes may be caused by a viral infection or an environmental pollutant that damages the insulin-producing cells.

Insulin production varies in patients with Type 1 diabetes. In some, it is only slightly reduced; in others, it is completely suppressed. In the latter, the lack of insulin starves body cells for glucose. Energy must be provided by the breakdown of fats.

A reduction in insulin secretion results in elevated blood glucose levels, especially right after meals.

Figure 13-17 illustrates this point. This graph shows blood glucose levels in nondiabetic and diabetic people following a procedure known as a **glucose tolerance test**. During this procedure, physicians give their patients an oral dose of glucose and then check blood levels at regular intervals. The blue line on the graph in Figure 13-17 represents the blood glucose levels in an individual with a healthy pancreas (a nondiabetic person). As you can see, glucose levels increase slightly after the administration of glucose. Within 2 hours, glucose levels decrease to normal, thanks to insulin secretion.

The pink line illustrates the response in a diabetic person, who produces insufficient amounts of insulin. As shown, glucose levels nearly double after the ingestion of glucose and they remain elevated much longer than in a nondiabetic—for 4 to 5 hours. A similar response occurs in untreated diabetics after they eat. Blood glucose levels remain high for such a long period because virtually no glucose can enter skeletal muscle cells or liver cells due to the lack of insulin. You'll see why shortly.

Glucose levels eventually decline as this nutrient is used by brain cells, red blood cells, and others. The kidneys also excrete excess glucose. Unfortunately, this wastes a valuable nutrient and causes the kidneys to lose large quantities of water. Excessive urination and constant thirst are the chief symptoms of untreated diabetes. Diabetics may have to urinate every hour, day and night.

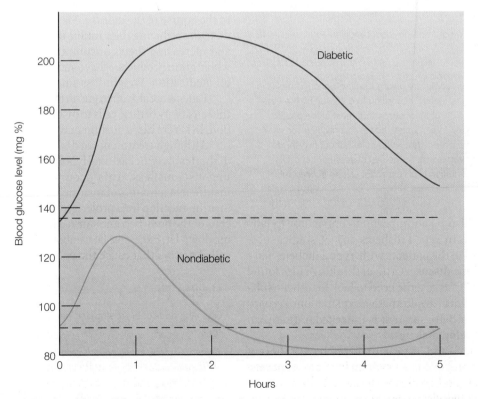

FIGURE 13-17 Glucose Tolerance Test In a diabetic (pink line), blood glucose levels increase rapidly and remain high after ingestion of glucose. In a normal person (blue line), blood glucose levels rise but are quickly reduced by insulin.

The second type of diabetes known as **Type 2 diabetes** usually occurs in people over 40 and is also called **late-onset diabetes** (Table 13-2). In this disease, the pancreas produces normal or above-normal levels of insulin. However, the cells of the body are unable to absorb glucose because of a decline in the number of insulin receptors in target cells. The cells are said to be unresponsive to insulin.

Type 2 diabetes is commonly associated with obesity, a problem growing rapidly in developed nations like the United States where people dine on high-fat, high-calorie diets. In the United States, diabetes has increased dramatically in adults—and children—almost entirely as a result of a rise in Type 2 diabetes caused by a corresponding rise in obesity. Lack of exercise and calorie-rich diets are the main culprits. According to the US Department of Health and Human Services, approximately 26 million Americans, age 20 and above, suffer from Type 1 and Type 2 diabetes—that's 11% of all adults in that age group. About 2 million people ages 20 and above are diagnosed with diabetes each year.

Heredity may also be an important contributor to late-onset diabetes, as it is in Type 1 diabetes. In fact, studies show that about one-third of the patients with Type 2 diabetes have a family history of the disease. Critical thinking rules found elsewhere in this text suggest the need to look for other explanations. Can you think of an alternative explanation? Perhaps it is a tendency toward obesity that is inherited, rather than the tendency to develop diabetes.

Although Type 1 and Type 2 diabetes have different causes, both forms of this disease exhibit similar symptoms. Excessive urination and thirst are generally the first signs of trouble. Patients often feel tired, weak, and apathetic. Weight loss and blurred vision are also common. Excess glucose in the urine may result in frequent bacterial infections in the bladder.

Treating Diabetes

Type 1 (early-onset) diabetes is treated with insulin injections. Patients receive injections of insulin, usually two to three times per day. Patients are also required to eat meals and snacks at regular intervals to maintain constant glucose levels in the blood and to ensure that regular insulin injections always act on approximately the same amount of blood glucose.

Insulin injections are tailored to an individual's lifestyle and body demands. To mimic the body's natural release, some diabetics must use a device called an *insulin pump*. This device delivers predetermined amounts of insulin. Because the pump cannot actually measure blood glucose levels, it may deliver inappropriate amounts at certain times. Medical researchers are also experimenting with ways to transplant healthy insulin-producing cells into the pancreas.

Researchers have also developed a nasal spray that dispenses insulin. In addition, researchers have developed new ways to monitor blood glucose in diabetics. Currently, diabetics must prick their skin three or more times a day to gather a drop of blood that allows them to test blood glucose levels. Newer devices measure blood glucose in much less painful ways. Researchers are also working on ways to prevent early-onset diabetes by eliminating the autoimmune reaction that attacks the insulin-producing cells of the pancreas.

In patients with Type 2 diabetes, doctors typically recommend weight loss via diet and exercise as the first line of attack. This can be extremely effective if undertaken early in the course of the disease. Many patients who change their diet—by reducing their intake of fatty meats and snack foods and increasing their consumption of fruits, vegetables, and whole grain products—have successfully weaned themselves off medication. Weight-loss surgery can be effective as well.

Patients are also often put on medications that either increase the sensitivity of their cells to insulin or stimulate insulin secretion. In severe cases, insulin may need to be administered.

Although treatments of diabetes have greatly improved, diabetics may occasionally suffer from *diabetic comas* or unconsciousness. This occurs when a diabetic receives insufficient amounts of insulin or if he or she goes without an insulin injection for a prolonged period. Without insulin, the cells of the body become starved for glucose and begin breaking down fat. Excessive fat catabolism releases toxic chemicals called *ketones* that cause the patient to lose consciousness.

Diabetics may also suffer from *insulin shock* as a result of an overdose of insulin. This reduces blood glucose levels. The severity of the symptoms is dose related. In mild overdoses, symptoms include tremor, fatigue, sleepiness, and the inability to concentrate. These symptoms result from a lack of glucose in the brain. (Remember, brain cells can only use glucose to make energy.) In severe cases, unconsciousness and death may occur.

Elevated levels of glucose common in all diabetic patients can cause serious damage to nerve cells and blood vessels. Over the long term, early- and late-onset diabetics frequently suffer from loss of vision, nerve damage, and kidney failure, symptoms that appear 20 to 30 years after the onset of the disease, even if the diabetics are being treated. Loss of vision results from recurring breakages in the capillaries in the retinas. Damage to the circulatory system can cause gangrene (GANG-green) in the fingers and toes, requiring amputation of limbs, especially the lower extremities. These serious complications result from the inevitable elevations in blood glucose levels that occur periodically over the years. Careful control over blood glucose, cholesterol, and blood pressure help reduce the risk of damage to the kidneys, eyes, nervous system, heart, reducing the incidence of heart attack and stroke, among others.

> **KEY CONCEPTS**
>
> Type 1 (early-onset) diabetes is typically treated with insulin injections while Type 2 (late-onset) diabetes, caused principally by obesity, is treated through diet and medications.

13-6   The Adrenal Glands

You stand on the banks of a raging river. As you watch kayakers head into the rapids, your heart starts to race, and your intestines churn in excitement and fear. You're next.

This natural response results from the secretion of two hormones by the adrenal glands (ah-DREE-nal). As Figure 13-18 shows, the **adrenal glands** perch atop the kidneys, and each consists of two zones. The central region, or adrenal medulla, produces the hormones that increase the heart rate and accelerate breathing when a person is excited or frightened. The outer zone, the adrenal cortex, produces a number of steroid hormones whose functions are discussed shortly.

> **KEY CONCEPTS**
>
> The adrenal glands are endocrine glands that are perched on top of the kidneys; they consist of an inner medulla and outer cortex, each of which releases several important hormones.

Adrenal Medulla

The **adrenal medulla** produces two hormones: **adrenalin (epinephrine)** and **noradrenalin (norepinephrine)**. In humans, about 80% of the adrenal medulla's output is adrenalin.

Helping us meet the stresses of life, adrenalin and noradrenalin are instrumental in the **fight-or-flight response**—the physiological reactions that take place when an animal is threatened. As suggested by its name, this response enhances our ability to fight or flee in a stressful or dangerous situation. Adrenalin and noradrenalin are secreted under all kinds of stress—for example, when a careless driver cuts in front of you in heavy traffic or as you wait outside a lecture hall before an exam. Nerve impulses traveling from the brain to the adrenal medulla trigger the release of adrenalin and noradrenalin. The response is part of the autonomic nervous system (discussed elsewhere in the text).

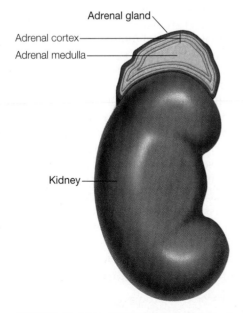

Adrenal gland
Adrenal cortex
Adrenal medulla

Kidney

FIGURE 13-18 Adrenal Gland The adrenal glands sit atop the kidney and consist of an outer zone of cells, the adrenal cortex, which produces a variety of steroid hormones, and an inner zone, the adrenal medulla. The adrenal medulla produces adrenalin and noradrenalin, the secretion of which is controlled by the autonomic nervous system. (Photo © Photos.com.)

The physiological changes these hormones cause are many. For example, adrenalin and noradrenalin elevate blood glucose levels, making more energy available to cells, particularly skeletal muscle cells. They also increase the breathing rate, which provides additional oxygen to skeletal muscles and brain cells. These hormones cause the heart rate to accelerate as well. This increases circulation and ensures adequate glucose and oxygen for cells that might be called into action. In addition, adrenalin and noradrenalin cause the bronchioles in the lungs to dilate, permitting greater movement of air in and out of them. Furthermore, these hormones cause blood vessels in the intestinal tract to constrict, putting digestion on temporary hold, while the blood vessels in skeletal muscles dilate, increasing flow through them. Mental alertness increases as a result of increased blood flow and hormonal stimulation. You are ready to fight or flee.

> **KEY CONCEPTS**
>
> The adrenal medulla produces stress hormones adrenalin and noradrenalin, both of which are involved in the fight-or-flight response.

Adrenal Cortex

Surrounding the adrenal medulla is the **adrenal cortex** (Figure 13-18). The adrenal cortex produces three types of steroid hormones, each of which has a different function. The first group, the **glucocorticoids**, affect glucose metabolism and maintain blood glucose levels. The second group, the **mineralocorticoids** (MIN-er-al-oh-CORE-teh-KOIDS), regulates the ionic concentration of the blood and tissue fluids. The final group, the **sex steroids**, is identical to the hormones produced by the ovaries and testes. In healthy adults, the secretion of adrenal sex steroids is insignificant compared to the amounts produced by the gonads.

Glucocorticoids. The prefix *gluco* reflects the fact that these steroid hormones affect carbohydrate metabolism. (Glucose, of course, is one of many carbohydrates in living things.) The secretion of glucocorticoid is governed by ACTH from the anterior pituitary, as noted earlier. Several chemically distinct glucocorticoids are secreted, the most important being cortisol. **Cortisol** increases blood glucose levels by stimulating the synthesis of glucose from amino acids and glycerol. This makes more glucose available for energy production in times of stress. Cortisol also stimulates the breakdown of proteins in muscle and bone, freeing amino acids that can then be converted to glucose in the liver.

In pharmacological doses (levels beyond those normally seen in the body), cortisol inhibits inflammation, the body's response to tissue damage. Such doses also depress the allergic reaction. Because they reduce inflammation, cortisol and cortisone, a synthetic glucocorticoid-like hormone, can be used to treat inflammation resulting from physical injuries or diseases such as rheumatoid arthritis. Physicians will often inject these steroids directly into inflamed joints like the knee joint to provide relief from swelling and pain.

Mineralocorticoids. The mineralocorticoids are involved in electrolyte or mineral salt balance. The most important mineralocorticoid is **aldosterone**. It is the most potent mineralocorticoid and constitutes 95% of the adrenal cortex's hormonal output.

Although mineralocorticoids regulate the level of several ions, their main function is to control sodium and potassium ion concentrations. Their chief target organ is the kidney. Aldosterone increases the movement of sodium ions out of the nephron and into the blood, a process called tubular reabsorption. Aldosterone also stimulates sodium ion reabsorption in sweat glands and saliva and potassium excretion by the kidney. When sodium is shunted back into the blood in the kidney and the skin, water follows. Aldosterone, therefore, helps conserve body water. Aldosterone secretion is controlled by the sodium ion concentration in a negative feedback loop.

Diseases of the Adrenal Cortex. One of the most common diseases involving hormones from the adrenal cortex is **Cushing's syndrome**, named after the researcher who first described it. Cushing's syndrome generally results from pharmacological doses of cortisone—that is, from injections of large amounts of cortisone.

Cortisone, as noted earlier, is a synthetic glucocorticoid. It is used to treat rheumatoid arthritis and asthma. In a few instances, the disease may be caused by a pituitary tumor that produces excess ACTH or a tumor of the adrenal cortex that secretes excess glucocorticoid.

Patients with Cushing's disease often suffer persistent hyperglycemia—high blood sugar levels—because of the presence of high levels of glucocorticoid. Bone and muscle protein may also decline sharply because glucocorticoids stimulate the breakdown of protein. Individuals complain of weakness and fatigue. Loss of bone protein increases the ease with which bones fracture.

Water and salt retention are also common in Cushing's patients, resulting in tissue edema (swelling). Because of this, Cushing's patients typically have a rounded "moon face." Swelling occurs because in high concentrations glucocorticoids have mineralocorticoid effects—that is, they stimulate the reabsorption of sodium and potassium, increasing the amount of water in the body. Because most cases of Cushing's syndrome result from steroids taken for health reasons, treatment is a simple matter: gradual reduction of the dose. Tumors in the pituitary and the adrenal cortex may be treated with radiation or may be surgically removed.

Another disease of the adrenal cortex is **Addison's disease**, which is extremely rare. Most cases of Addison's disease are thought to be autoimmune reactions in which the cells of the adrenal cortex are destroyed by the body's own immune system. Long-lasting infections, for example tuberculosis and HIV, can cause the disease.

Addison's disease is characterized by a variety of symptoms including low blood pressure, loss of appetite, weight loss, fatigue, and weakness. Although insulin and glucagon are present, control of blood glucose levels is impaired due to the absence of cortisol. The body's reaction to stress is also impaired. The lack of aldosterone results in electrolyte imbalance. Because aldosterone maintains sodium levels and blood pressure, patients with Addison's disease have low blood sodium levels and low blood pressure. Addison's disease can be treated with steroid tablets that replace the missing hormones. Treatment allows patients to lead fairly normal, healthy lives.

> **KEY CONCEPTS**
>
> The adrenal cortex produces three types of hormones with markedly different functions: glucocorticoids, mineralocorticoids, and sex steroids.

13-7 Health and Homeostasis

Hormones orchestrate an incredible number of body functions and help to create a dynamic balance that's vital for human health. Hormones influence homeostasis primarily by controlling the rate of various metabolic reactions and by regulating ionic balance. When this balance is altered, our health suffers (Table 13-3).

The endocrine system, like other systems, is sensitive to outside factors. Stress, for example, can lead to an imbalance in adrenal hormones, resulting in high blood pressure and other complications. High blood pressure, in turn, puts strain on the heart and can result in heart attacks.

Hormonal balance can also be upset by toxic pollutants. One example is **dioxin** (die-OX-in), a contaminant found in some herbicides and in paper products. Scientists believe that dioxins bind to plasma membranes and cytoplasmic receptors to which hormones normally attach, upsetting body functions. Thus, dioxin is referred to as an "environmental hormone." One form of dioxin, TCDD, suppresses the immune system in mice at least 100 times more effectively than corticosterone, one of the body's glucocorticoids.

Studies show that a number of common herbicides (the thiocarbamates) may upset the thyroid's function. These herbicides are chemically similar to thyroid hormone and therefore block the secretion of TSH, resulting in goiter. At higher levels, they may cause thyroid cancer.

Numerous studies also show that chemicals released from plastics—like the plastic lining of steel cans and plastic bottles—are environmental hormones. Some of these chemicals, for instance, act like the hormone estrogen. Released into the aquatic environment, these chemicals alter the reproductive ability of male fish. These chemicals are also ingested by humans and could impact human reproduction. Studies in mice, for instance, show that one such chemical, known as bisphenol-A (BPA), has a masculinizing effect on female mice that were exposed to the chemical while still *in utero* and during breast feeding. What their effect on humans who ingest many foods and beverages from plastic containers or steel cans lined by plastic is the subject of ongoing research.

TABLE 13-3	Summary of Some Endocrine Disorders	
Disease	**Cause**	**Symptoms**
Giantism	Hypersecretion of GH starting in infancy or early life	Excessive growth of long bones
Dwarfism	Hyposecretion of GH in infancy or early life	Failure to grow
Acromegaly	Hypersecretion of GH after bone growth has stopped	Facial features become coarse; hands and feet enlarge; skin and tongue thicken
Hyperthyroidism	Overactivity of the thyroid gland	Nervousness; inability to relax; weight loss; excess body heat and sweating; palpitations of the heart
Hypothyroidism	Underactivity of the thyroid gland	Fatigue, reduced heart rate; constipation; weight gain; feel cold; dry skin
Hyperparathyroidism	Excess parathyroid hormone secretion, usually resulting from a benign tumor in the parathyroid gland	Kidney stones; indigestion; depression; loss of calcium from bones
Hypoparathyroidism	Hyposecretion of the parathyroid glands	Spasms in muscles; numbness in hands and feet; dry skin
Diabetes insipidus	Hyposecretion of ADH	Excessive drinking and urination; constipation
Diabetes mellitus	Insufficient insulin production or inability of target cells to respond to insulin	Excessive urination and thirst; poor wound healing; urinary tract infections; excess glucose in urine; fatigue and apathy
Cushing's syndrome	Hypersecretion of hormones from adrenal cortex or, more commonly, from cortisone treatment	Face and body become fatter; loss of muscle mass; weakness; fatigue; osteoporosis
Addison's disease	Gradual decrease in production of hormones from adrenal gland; most common cause is autoimmune reaction	Loss of appetite and weight; fatigue and weakness; complete adrenal failure

SUMMARY

Principles of Endocrinology

1. The endocrine system consists of a widely dispersed set of glands that produce hormones. Hormones are chemicals released into the bloodstream that affect five aspects of our lives: (1) homeostasis; (2) growth and development; (3) reproduction; (4) energy production and storage; and (5) behavior.
2. Hormones act on specific cells known as target cells. Whether a cell responds to a hormone depends on the type of hormone receptors located in its cell membrane or in the cytoplasm.
3. Three types of hormones are produced in the body: (1) steroids; (2) proteins and polypeptides; and (3) amines.
4. Hormone secretion is controlled primarily by negative feedback loops, many of which involve the hypothalamus.

The Pituitary and Hypothalamus

5. The pituitary is a pea-sized gland suspended from the hypothalamus by a thin stalk. It consists of two parts: the anterior pituitary and the posterior pituitary.
6. The anterior pituitary produces seven protein and polypeptide hormones. Their release is controlled by releasing and inhibiting hormones produced by the hypothalamus.
7. Hormones from the hypothalamus are transported to the anterior pituitary via a portal system, specialized blood vessels that run down the stalk of the pituitary gland.
8. The release of the hypothalamus's hormones is controlled by various chemical stimuli as well as nerve impulses.
9. The hormones of the anterior pituitary and their functions are summarized in Table 13-1.
10. The posterior pituitary produces two hormones, antidiuretic hormone (ADH) and oxytocin, whose functions are summarized in Table 13-1.

The Thyroid Gland

11. The thyroid gland, located in the neck, produces three hormones: thyroxine (T_4), triiodothyronine (T_3), and calcitonin.
12. Thyroxine and triiodothyronine accelerate the rate of glucose breakdown in most cells, increasing body heat. These hormones also stimulate cellular growth and development.
13. Calcitonin lowers blood calcium levels.

The Parathyroid Glands

14. The parathyroid glands are located on the back of the thyroid gland and produce a polypeptide hormone called parathyroid hormone (PTH) or parathormone.
15. PTH increases blood calcium levels.

The Pancreas

16. The pancreas produces two hormones, insulin and glucagon, from the islets of Langerhans.
17. Insulin is the glucose-storage hormone. It stimulates the uptake of glucose by many body cells. Insulin also increases the uptake of amino acids and stimulates protein synthesis in muscle cells.
18. Glucagon raises glucose levels in the blood between meals by stimulating glycogen breakdown and the synthesis of glucose from amino acids and fats.
19. Diabetes mellitus is a disease involving insulin and blood glucose. It has two principal forms: Type 1 and Type 2.
20. Type 1 diabetes, also called early-onset diabetes, usually occurs early in life and may be caused by an autoimmune reaction. It can be treated by insulin injections.
21. Type 2, or late-onset, diabetes, results from a reduction in the number of insulin receptors on target cells. It is caused by obesity and genetic factors and can often be treated successfully by dietary management and/or medication.

The Adrenal Glands

22. The adrenal glands lie atop the kidneys and consist of two separate portions: the adrenal medulla, at the center, and the adrenal cortex, a surrounding band of tissue.
23. The adrenal medulla produces two hormones under stress: adrenalin and noradrenalin. These hormones stimulate heart rate and breathing, elevate blood glucose levels, constrict blood vessels in the intestine, and dilate blood vessels in the skeletal muscles.
24. The adrenal cortex produces three classes of hormones: glucocorticoids, mineralocorticoids, and sex steroids.
25. The glucocorticoids affect carbohydrate metabolism and tend to raise blood glucose levels. The principal glucocorticoid is cortisol.
26. The chief mineralocorticoid is aldosterone. It acts on the kidneys, sweat glands, and salivary glands, causing sodium and water retention and potassium excretion.

Health and Homeostasis

27. Hormones influence an incredible number of body functions, fostering homeostasis, a dynamic balance necessary for good health. When homeostasis is altered, our health suffers.
28. The endocrine system, like others, is sensitive to upset from stress and environmental factors such as pollutants. Dioxin, a contaminant in some herbicides and paper products, may exert most of its effects through the endocrine system. A number of common herbicides (the thiocarbamates) upset the thyroid's function and may even cause thyroid tumors.

THINKING CRITICALLY ANALYSIS

This analysis corresponds to the Thinking Critically scenario that was presented at the beginning of this chapter.

A representative from the plastics industry argues that the results cannot be extrapolated to humans with certainty. His argument is that human exposure is oral. In the experiment, female mice were given injections so that the chemical was incorporated in the blood and then circulated in the blood to the fetus and to the breasts after birth. Humans ingest plasticizers in drinks and, he added, when ingested, these chemicals are chemically modified by the liver so they no longer act like estrogen. In order for the results to be relevant for people, he argues, the chemical would have to be given orally to the mice. Does this sound reasonable to you?

Because you are not versed in toxicology, it might be hard for you to make an assessment. The fact is that the route of exposure does make a huge difference in toxicology. The representative from the plastics industry does have a good point. The results of the mouse study may or may not pertain to humans. To know for sure, the researchers would have to orally expose the pregnant and nursing female mice to bisphenol-A.

KEY TERMS AND CONCEPTS

Acromegaly, p. 280
Addison's disease, p. 292
Adrenal cortex, p. 292
Adrenal gland, p. 291
Adrenal medulla, p. 291
Adrenalin (or epinephrine), p. 291
Adrenocorticotropic hormone (ACTH), p. 282
Aldosterone, p. 292
Anterior pituitary, p. 277
Antidiuretic hormone (ADH), p. 283
Calcitonin (or thyrocalcitonin), p. 286
Cortisol, p. 292
Cushing's syndrome, p. 292
Diabetes mellitus, p. 287
Dioxin, p. 293
Early-onset diabetes (Type 1), p. 288
Endocrine system, p. 275
Fight-or-flight response, p. 291
Glucagon, p. 288
Glucocorticoid, p. 292

Glucose tolerance test, p. 289
Gonadotropins, p. 282
Growth hormone (GH), p. 279
Hormone, p. 275
Hyperparathyroidism, p. 286
Hyperthyroidism, p. 286
Hypothalamus, p. 277
Inhibiting hormone (IH), p. 278
Insulin, p. 287
Juvenile diabetes (Type 1), p. 288
Late-onset diabetes (Type 2), p. 290
Mineralocorticoids, p. 292
Negative feedback loop, p. 277
Neuroendocrine reflex, p. 282
Neurosecretory neuron, p. 278
Nontropic hormone, p. 276
Noradrenalin (norepinephrine), p. 291
Oxytocin, p. 283
Pancreas, p. 287
Parathyroid glands, p. 286

Parathyroid hormone or parathormone (PTH), p. 286
Pituitary gland, p. 277
Portal system, p. 279
Positive feedback loop, p. 277
Posterior pituitary, p. 277
Prolactin, p. 282
Releasing hormone (RH), p. 278
Sex steroids, p. 292
Somatomedin, p. 279
Target cells, p. 275
Thyroid gland, p. 284
Thyroid-stimulating hormone (TSH), p. 280
Thyroxine, p. 284
Triiodothyronine, p. 284
Tropic hormone, p. 276
Type 1 diabetes, p. 288
Type 2 diabetes, p. 290

CONCEPT REVIEW

1. Define the following terms: endocrine system, hormone, and target cell. p. 275.
2. Hormones function in five principal areas. What are they? Give some examples of each. p. 275.
3. Define the terms tropic and nontropic hormones, and give several examples of each. p. 276.
4. Give two examples of negative feedback loops in the endocrine system: a simple feedback mechanism and a more complex one that operates through the nervous system. p. 277.
5. List the hormone(s) involved in maintaining blood glucose levels. Where does each one come from? What does each one do? p. 277.
6. What hormones are vital for normal growth? p. 279.
7. What is the hypothalamus? Why is it so important for homeostasis? pp. 277–278.
8. Describe the role of the hypothalamus in controlling the production and release of anterior pituitary hormones. pp. 277–278.

9. Describe the ways in which the posterior pituitary differs from the anterior pituitary. pp. 282–283.
10. Acromegaly and giantism are both caused by the same problem. Why are these conditions so different? pp. 279–280.
11. Describe the neuroendocrine reflex involved in prolactin secretion. pp. 282–283.
12. Offer some possible explanations for the following experimental observation: Milk production occurs late in pregnancy and is thought to be stimulated by the hormone prolactin. A nonpregnant rat is injected with prolactin but does not produce milk. p. 282.
13. Where is ADH produced? Where is it released? Describe how ADH secretion is controlled. What effects does this hormone have? pp. 283–284.
14. Where is oxytocin produced? Where is it released? Describe how oxytocin secretion is controlled. What effects does this hormone have? p. 283.

15. A patient comes into your office. She is thin and wasted and complains of excessive sweating and nervousness. What tests would you run? p. 286.
16. A patient comes into your office. He is suffering from indigestion, depression, and bone pain. An x-ray of his bone shows some signs of osteoporosis. You think that the disorder might be the result of an endocrine problem. What test would you order? p. 292.
17. How are the two basic types of diabetes mellitus different? How are they similar? How are they treated, and why are these treatments chosen? pp. 288–291.
18. Describe the physiological changes that occur under stress. What hormones are responsible for them? pp. 291–292.
19. Aldosterone is a mineralocorticoid. Describe its chief functions. How does it help retain body fluid? Under what conditions is aldosterone secreted? p. 292.
20. Explain how hormones ensure homeostasis and give some examples. p. 293.

SELF-QUIZ: TESTING YOUR KNOWLEDGE

1. The cell a hormone affects is called a _____ cell. p. 275.
2. A hormone that stimulates the production and release of another hormone is known as a _____ hormone. p. 276.
3. The production and release of most hormones are controlled by _____ feedback. p. 277.

4. The _____ gland is attached to the underside of the brain by a thin stalk. p. 277.
5. A _____ produced by the _____ travels in the bloodstream to the anterior pituitary, causing the production and release of its hormones. p. 278.

6. _____ hormone released by the anterior pituitary stimulates growth but primarily affects muscle and _____. p. 279.
7. _____ is a disease caused by overproduction of growth hormone in adults and results in coarse facial features and continued growth of the hands and feet. p. 280.

8. _____ and triiodothyronine are hormones produced by the thyroid gland; their production and release are controlled by _____, a hormone from the anterior pituitary. p. 281.

9. ACTH stimulates the production and release of _____, a group of steroid hormones that increase blood glucose. p. 282.

10. Milk production in the breasts of women is stimulated by the hormone _____, which is produced by the anterior pituitary. Its release is stimulated by _____. pp. 282–283.

11. Milk ejection from glands in the breast is controlled by the hormone _____ from the _____ pituitary. p. 283.

12. _____ is a hormone of the posterior pituitary gland that stimulates water reabsorption by the kidney. p. 283.

13. An insufficiency of iodine results in enlargement of the thyroid gland, a condition known as _____. p. 285.

14. The pancreas produces two hormones, insulin and _____. pp. 287–288.

15. Insulin increases the uptake of glucose by liver and _____ cells, and lowers blood glucose levels. p. 287.

16. Late-onset diabetes is caused primarily by _____. p. 290.

17. Early-onset diabetes is treated by _____ injections. p. 290.

18. The hormone _____ is released by the adrenal medulla and is responsible for physiological changes associated with the fight-or-flight response. p. 291.

19. The mineralocorticoid aldosterone is produced by the adrenal _____. p. 292.

20. The hormone _____ produced by the thyroid gland lowers blood calcium levels. p. 286.

biology.jbpub.com/chiras/8e/

The site features eLearning, an online review area that provides quizzes, chapter outlines, and other tools to help you study for your class. You can also follow useful links for in-depth information, research the differing views in the Point/Counterpoints, or keep up on the latest health news.

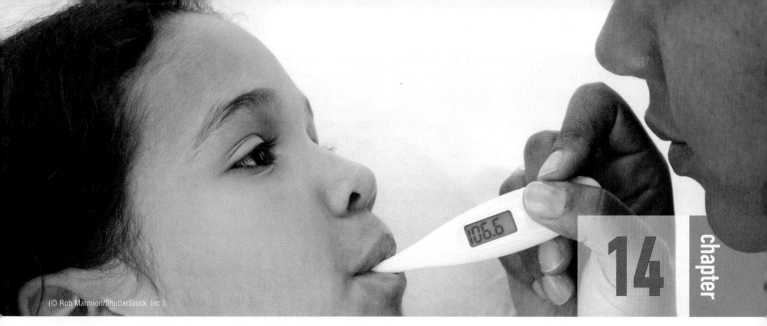

(© Rob Marmion/ShutterStock, Inc.)

The Immune System

Floating in the air and circulating in the water are billions upon billions of microorganisms, mostly bacteria and viruses but also single-celled fungi. Although the vast majority of these microbes in our environment are harmless, occasionally you'll encounter a few microorganisms that are not. If you are traveling in a busy airport, for instance, you may encounter a bacterium or virus that could cause illness such as the flu.

THINKING CRITICALLY

You're the parent of a young child and are debating whether to have your child vaccinated. Some friends tell you that it's dangerous—in fact, they say, some children have been paralyzed by vaccinations. They also tell you that vaccinations can damage the immune system and may cause autism in children, a disease that severely impairs a child's ability to communicate and form normal social relationships. They contend that there's no need to subject a child to this treatment because dangerous infectious diseases have been eliminated.

After reading this chapter, how would you respond? Is there merit to the claim that a vaccination damages the immune system? Have the diseases been eliminated? Do you have enough information to make an informed decision?

Why don't we succumb to these potentially harmful microbes? The answer is that although some microbes cause illness, many of the illnesses are not life threatening. In addition, we humans come equipped with a system of armor (the skin and other protective layers) that protects us from potential intruders. If microorganisms break through these barriers, the immune system is ready to combat them. Even then, however, some microorganisms can wreak havoc, causing a range of diseases from the flu and colds to AIDS and even cancer.

In this chapter, we'll examine the body's defense against potentially harmful microbes. We will also see how the body's defense mechanisms fight cancer and how they combat tissue and organ transplants. We'll begin with a brief overview of infectious agents.

14-1 Viruses and Bacteria: An Introduction

Infectious agents come in many forms. In this section we'll look primarily at viruses and bacteria. A closer look at a host of other infectious agents is covered elsewhere in the text.

Viruses

A **virus** consists of a nucleic acid core, containing either DNA or RNA (Figure 14-1). Surrounding the nucleic acid core is a protein coat known as the *capsid*. Some viruses have an additional protective layer that lies outside the capsid. This outer layer, the *envelope*, is structurally similar to the plasma membranes of eukaryotic cells.

Viruses are something of an enigma in the biological world. Although some can be quite deadly, viruses are not considered to be living organisms. Why? Viruses do not possess the features we attribute to living organisms, discussed elsewhere in the text. For example, viruses cannot multiply and divide on their own like bacteria. They also have no metabolism. So what are they?

Viruses have been likened to pirates because they invade cells and then commandeer their metabolic machinery in much the same way that pirates take over ships. After invading a cell, viruses convert their so-called host cells into miniature virus factories. That is, they use each host cell's organelles and biochemical reserves to make new viral protein and nucleic acid. Inside the host cell, hundreds of thousands of new viruses are formed. They are released from infected cell by the thousands. So completely spent is the host cell that it dies. Once released, viruses can then spread to other cells via the bloodstream and the lymphatic vessels, repeating this cycle over and over again until a host dies or its immune system eliminates the virus.

Viruses most often enter the body through the respiratory and digestive systems. However, other avenues of entry are also possible—for example, through the skin during sexual contact.

Fortunately for us, the immune system kills many harmful viruses, but usually not until the unwitting host (you and I) has experienced considerable discomfort. Influenza viruses, for example, result in the flu, characterized by fevers, chills, headaches, and muscle pains that can last up to 2 weeks (Table 14-1). During this time, the immune system

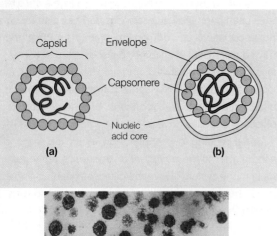

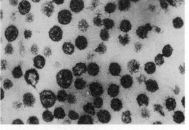

(c)

FIGURE 14-1 **General Structure of a Virus** (a) The virus consists of a nucleic acid core of either RNA or DNA. Surrounding the viral core is a layer of protein known as the capsid. Each protein molecule in the capsid is known as a capsomere. (b) Some viruses have an additional protective coat known as the envelope. (c) Electron micrograph of the human immunodeficiency virus (HIV). (© G. Musil/Visuals Unlimited.)

TABLE 14-1	How Do You Know If You Have A Cold or the Flu?	
Symptoms	**Cold**	**Flu**
Onset	Gradual	Sudden
Fever	Rare	Common, may reach 101°F (38°C); may last 3 to 4 days
Headache	Rare	Common and can be severe
Cough	Hacking	Dry cough
Muscle aches and pains	Slight	Typical, often severe
Tiredness and weakness	Mild	Common, often severe
Chest discomfort	Mild to moderate	Common
Stuffy nose	Common	Sometimes
Sneezing	Usual	Sometimes
Sore throat	Common	Sometimes; may last 3 to 4 days
Caused by	Any of 200 viruses	Influenza virus

mounts an attack, but it usually takes 10 to 14 days before it can eliminate the virus. Colds are also caused by approximately 200 different viruses. Like the flu, colds cannot be treated with antibiotics.

Some viruses, like those that cause fever blisters and genital herpes, evade the immune system. They take refuge in body cells, reemerging from time to time, usually when an individual is stressed, to cause problems. The virus responsible for genital herpes, for example, causes tiny sores that emerge periodically on the genitals, thighs, and buttocks of infected men and women.

> **KEY CONCEPTS**
>
> Viruses are simple, nonliving, biological entities that can invade cells, causing disease.

Bacteria

Bacteria are unlike viruses in many ways. For instance, **bacteria** (singular, **bacterium**) are classified as living organisms. As a result, they are capable of reproducing without taking over host cells. In fact, bacteria do not enter cells at all. They proliferate outside of the cells of the body.

Bacteria belong to a group of single-celled organisms known as *prokaryotes*, cells without nuclei. (Human cells are eukaryotes—meaning true nucleus.) The fact that they don't have nuclei doesn't mean that bacteria lack DNA. They do. It is just that their DNA is not enclosed in a nuclear membrane. Unlike the cells in your body, their DNA forms circular strands. While prokaryotes have cytoplasm which is enveloped by a plasma membrane—just like the cells in your body—they have no cellular organelles (Figure 14-2). Outside the plasma membrane

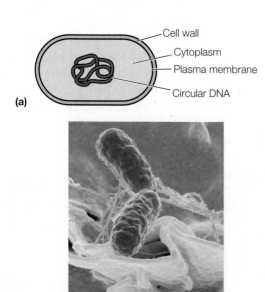

(a)

Cell wall
Cytoplasm
Plasma membrane
Circular DNA

(b)

FIGURE 14-2 General Structure of a Bacterium
(a) Bacteria come in many shapes and sizes, but all have a circular strand of DNA, cytoplasm, and a plasma membrane. Surrounding the membrane of many bacteria is a cell wall. (b) Electron micrograph of salmonella bacteria. (Courtesy of Rocky Mountain Laboratories, NIAID, NIH.)

of bacteria is a thick, protective rigid cell wall. Moreover, many bacteria also contain tiny circular pieces of extrachromosomal DNA, called **plasmids**. Plasmids carry functional genes that help bacteria survive—for example, they can confer bacterial resistance to antibiotics (discussed shortly).

Like viruses, bacteria enter the body through the respiratory and gastrointestinal (GI) tracts. They may also enter through the epithelium of the urinary system or via cuts and abrasions in the skin. Inside the body, bacteria proliferate, using nutrients to make more of their kind. Some bacteria produce toxins, substances that make us ill.

Although best known for their role in causing sickness and death, it is important to remember that most bacteria are quite useful to our survival and well-being. Soil bacteria, for example, help recycle nutrients in rotting plants and animals, a process that is essential to the continuation of life on Earth. Without these bacteria, biological systems would quickly grind to a halt. Some of the biologically useful bacteria are discussed elsewhere in the text.

> **KEY CONCEPTS**
>
> Bacteria are living organisms that can invade the body, producing toxins that can cause illness and sometimes lethal diseases.

Treating Bacterial and Viral Infections

One of the greatest discoveries in the field of medicine is the discovery of antibiotics. *Antibiotics* are chemical substances that kill bacteria. They do so by selectively inhibiting protein synthesizing bacteria—that is, they shut down bacterial protein synthesis but do not affect protein synthesis of the cells of your body.

Antibiotics are given to patients with bacterial pneumonia, bronchitis, sinus infections, and sore throats. Because all of these diseases can be caused by viruses, too, a doctor should be consulted. He or she can usually tell whether the infection is viral or bacterial. If it is bacterial, the doctor will prescribe antibiotics.

Important as antibiotics have been in fighting disease, there's been a tendency to overprescribe them. For years, many doctors have prescribed antibiotics even when a patient was suffering from a viral infection. Unfortunately, this overuse of antibiotics has resulted in a significant problem: antibiotic resistance in bacteria. Nationwide, one out of every four bacterial infections today involves a resistant strain. Most of these infections occur in children under the age of five. According to the Centers for Disease Control and Prevention (CDC), nearly 100,000 people die each year just from infections caused by antibiotic-resistant bacteria they are exposed to in hospitals. To treat antibiotic resistant strains of bacteria, doctors often prescribe higher doses or newer antibiotics. Although these strategies work, bacteria frequently develop resistance to the higher doses as well as to new antibiotics.

Bacteria become resistant to drugs in one of several ways. They can, for example, chemically alter antibiotics, rendering them ineffective. Bacteria also undergo chemical changes so that antibiotics no longer bind to their cell membranes. Some resistance occurs when bacteria develop mechanisms that enable them to pump antibiotics out of their interior, lowering internal concentrations so that the drugs are no longer effective.

health**note**

14-1 Maggots: The Latest in Medical Care?

Elainor Brocknor is a diabetic who for 30 years has been injecting the hormone insulin to regulate her blood sugar. Careful as she has been, she's developed serious medical problems. The blood vessels in her feet have been damaged by periodic high concentrations of glucose. The tissues in two of her toes have begun to die and rot. She's developed gangrene and doctors fear they may have to amputate these two toes and perhaps even her foot if the infection does not stop. Making matters worse, the bacteria infecting the dying tissues in her feet don't seem to respond to antibiotics.

One day, her doctor makes a bizarre recommendation. He says he may be able to save her foot by using maggots. Maggots are the immature form of common ordinary flies. They're called larvae. They develop from eggs laid by flies. They're voracious eaters.

Elainor and her husband are mortified, at first, by the doctor's seemingly crazy suggestion. What a preposterous idea! But the doctor explains that maggots raised in a sterile environment have been used very successfully to remove dead tissue in people with gangrene. They not only feed on dead tissue, they gobble up harmful bacteria, so the infection doesn't spread.

Reluctant, Elainor agrees to try the procedure. The next day, her doctor visits her in her hospital room. He's carrying a glass jar full of writhing maggots. He removes the bandages from her feet, and then prepares a new one. On the surface of the gauze, the physician sprinkles 100 tiny maggots. He then places the bandage over her wound and tapes it shut.

Elainor watches in amazement and is surprised that she feels nothing at all. A day later, the doctor returns, removes the bandage, and applies another. In a few days, he announces that the wound is clean and free of infection. They won't have to amputate.

Elainor is just one of thousands of people each year treated with maggots. In this day of high-tech medicine, such an approach may seem ridiculously primitive. Absurd as it may sound, this treatment works when high-tech solutions don't. And it is saving many people the agony of amputation.

The idea of using maggots to clean wounds is an old one. During the Civil War, doctors who treated wounded soldiers noticed that wounds in men returning from the battlefields several days away were often infested with maggots. They'd come from eggs flies laid in the open wounds. Surprisingly, the maggot-filled wounds were remarkably clean.

In Africa, rural doctors noticed a similar phenomenon. In modern times, doctors have found that when they removed maggots from wounds and put a patient on antibiotics, the patients died—apparently from antibiotic-resistant bacteria that had been under control. Maggots aren't the only creepy cure. Some doctors are using leeches to restore blood flow in severed limbs or fingers that have been sewn back in place.

In 2004, the U.S. Food and Drug Administration (FDA) approved the production and sale of maggots for use in humans. Maggots are also being used to treat certain wounds in horses. Leeches are turning out to be valuable tools in modern medicine.

Because of concern for antibiotic resistance, many doctors now prescribe antibiotics more judiciously, refusing to give them to patients suffering from viral infections, even though patients might insist they get some kind of medication. For the flu, a viral disease, doctors prescribe bed rest and medications that reduce some of the symptoms, such as acetaminophen, which reduces fever, muscle aches, and headaches. They also recommend consumption of warm fluids such as tea to reduce the feeling of congestion while the immune system battles the virus. To help reduce the spread of disease, it is wise to stay at home when sick. (Bear in mind, however, that some diseases are contagious several days before an individual exhibits symptoms.)

Pharmaceutical companies now produce several antiviral drugs such as Tamiflu. To be effective, however, antiviral drugs must be taken very early in the course of the disease, that is, within a day or two of the onset of symptoms. To prevent the flu, many doctors recommend flu shots. Flu shots contain a vaccine that prevents the influenza virus from gaining a foothold and are discussed in more detail shortly.

The common cold is also a viral disease. In fact, over 200 viruses can cause colds. Cold viruses infect the upper respiratory tract, primarily the nose. Symptoms include coughing, sore throat, runny nose, and fever. Symptoms usually last seven to ten days, although some symptoms may persist for up to three weeks.

For colds, doctors recommend bed rest in a warm setting. Doctors also recommend that their patients drink fluids and over-the-counter medications to reduce symptoms such as coughing. Mild pain relievers can reduce muscle pain and headaches, if you have them. Decongestants can help relieve a stuffy nose and will help you breathe better. Be sure to read the **Health Tip 14-1**, as new evidence suggests that many of the over-the-counter remedies are not effective. As a side note, some doctors in the United Kingdom and the United States are now using maggots to clean wounds infected by bacteria, including antibiotic-resistant strains (Health Note 14–1).

Health Tip 14-1

Got a cough? You may want to skip the over-the-counter cough medicines.

Why?

There is no clinical evidence to show that over-the-counter cough medicines actually work. Does anything work?

Yes. Both Benadryl (an antihistamine) and Sudafed (a decongestant) appear to provide relief, according to authors of the study published in the January 2006 issue of *Chest*, a medical journal.

KEY CONCEPTS

Bacteria can be treated with antibiotics; viruses are not responsive to these drugs and are generally naturally removed from the body by the immune system.

14-2 The First and Second Lines of Defense

The First Line of Defense

The human body is like a fort with an outer barrier that wards off potential invaders. One component of this outer wall is the skin, a protective layer that repels many potentially harmful microorganisms.

The skin consists of a relatively thick and impermeable layer of epidermal cells, the **epidermis**, overlying the rich vascular layer, the **dermis**. Epidermal cells are produced by cell division in the base of the epidermis. As the basal cells proliferate, they move outward, become flattened, and die. The dead cells contain a protein called *keratin* (CARE-ahtin). Keratin forms a fairly waterproof protective layer that not only reduces moisture loss but also protects underlying tissues from microorganisms. The cells of the epidermis are joined by special structures known as *tight junctions*, which impede water loss and microbial penetration.

Although the skin protects us from infections, the human body contains three passageways that penetrate into its interior, providing a fairly direct route for microbes to enter the body. The passages into the interior are the respiratory, digestive, and urinary tracts. All three of them are protected by epithelial linings, although they are not as thick and impenetrable as the skin. As a result, they are more easily penetrated by microorganisms.

The body's first line of defense also consists of chemical barriers. The skin, for example, which serves as a physical barrier, also produces slightly acidic secretions that impair bacterial growth. The stomach lining produces hydrochloric acid, which destroys many ingested bacteria. Tears and saliva contain an enzyme called *lysozyme* (LIE-so-zime) that dissolves the cell walls of bacteria, killing them. Cells in the lining of the respiratory tract produce mucus, which has antimicrobial properties.

The skin, the epithelial linings, and chemical barriers are all nonspecific. Like a castle wall, they operate indiscriminately against all invaders.

> **KEY CONCEPTS**
>
> The first line of defense against microorganisms is a nonspecific physical and chemical barrier created by epithelia.

The Second Line of Defense

The first line of defense is not impenetrable. Even tiny breaks in the skin or in the lining of the respiratory, digestive, or urinary tracts may permit potentially harmful microorganisms to enter the body. Fortunately, a second line of defense exists. It involves a host of chemicals and cellular agents. Like the first line, these mechanisms are nonspecific. This section discusses four components of the second line of defense: the inflammatory response, pyrogens, interferons, and complement.

> **KEY CONCEPTS**
>
> The second line of defense nonspecifically combats microorganisms that manage to penetrate the first line of defense; it involves numerous chemical and cellular agents.

The Inflammatory Response

Damage to body tissues triggers a series of reactions collectively referred to as an inflammatory response. The **inflammatory response** is characterized by redness, swelling, and pain. (The word *inflammatory* refers to the heat given off by a wound.)

The inflammatory response is a kind of chemical and biological warfare waged against bacteria, viruses, and other microorganisms. Because it is so complicated, be certain to refer to Figure 14-3 as you read this material. The inflammatory response begins with the release of a variety of chemical substances by the injured tissue. Some chemicals attract macrophages that reside in body tissues and neutrophils in the blood. These cells engulf (phagocytize) bacteria that enter a wound. Soon after these cells begin to work, a yellowish fluid begins to exude from the wound. Called **pus**, it contains dead white blood cells (mostly neutrophils), microorganisms, and cellular debris that accumulate at the wound site.

Another chemical substance released by injured tissues is histamine (HISS-tah-mean). **Histamine** stimulates the arterioles in the injured tissue to dilate, causing the capillary networks to swell with blood. The increase in the flow of blood through an injured tissue is responsible for the heat and redness around a cut or abrasion. Heat, in turn, increases the metabolic rate of cells in the injured area and accelerates healing.

Still other substances released by injured tissues increase the permeability of capillaries, which increases the flow of plasma into a wounded region. Plasma carries oxygen and nutrients that aid in healing. It also carries the molecules necessary for blood clotting. The clotting mechanism walls off injured vessels and reduces blood loss.

Plasma leaking into injured tissues causes swelling, which stimulates pain receptors in the area. Pain receptors send nerve impulses to the brain. Pain also results from chemical toxins released by bacteria and from chemicals released by injured cells. One important pain-causing chemical is **prostaglandin** (PROSS-tah-GLAN-din). Aspirin and other mild painkillers work by inhibiting the synthesis and release of prostaglandins.

Inflammation even comes with its own cleanup crew in the form of late-arriving monocytes. These cells phagocytize dead cells, cell fragments, dead bacteria, and viruses.

> **KEY CONCEPTS**
>
> The inflammatory response is a complex series of changes that represents a kind of chemical and biological warfare against microorganisms; it is characterized by pain, swelling, and redness.

Pyrogens, Interferons, and Complement

Pyrogens (PIE-rah-gins) are molecules released primarily by macrophages that have been exposed to bacteria. Pyrogens travel to the hypothalamus (a region of the brain). In the hypothalamus is a group of nerve cells that controls the body's temperature, in much the same way that a thermostat

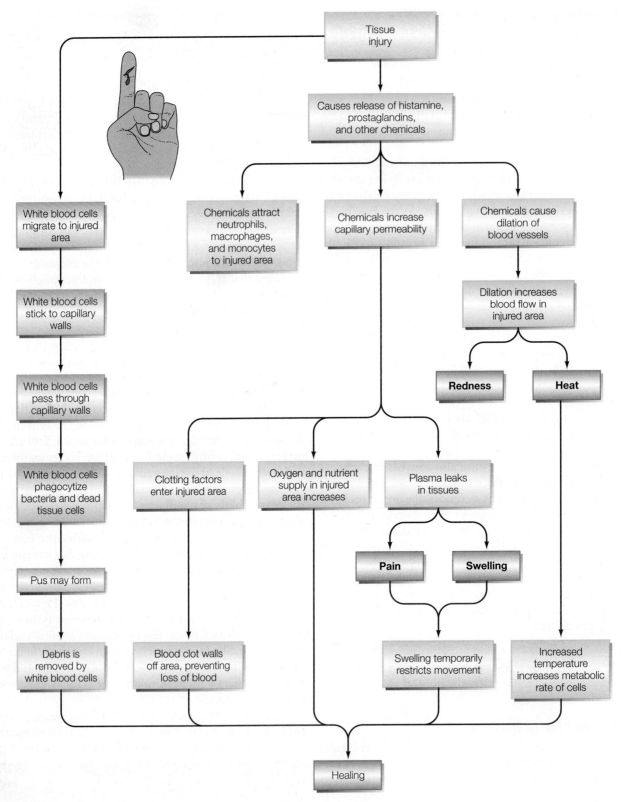

FIGURE 14–3 The Inflammatory Response

regulates the temperature of a room. Pyrogens turn the thermostat up, increasing body temperature and producing a fever. Mild fevers cause the spleen and liver to remove additional iron from the blood. Because many disease-causing bacteria require iron to reproduce, fever helps reduce their replication. Fever also increases metabolism, which facilitates healing and accelerates cellular defense mechanisms such as phagocytosis. Important as it is, fever can also be debilitating, and a severe fever (over 105°F [41°C]) is potentially life-threatening because it begins to denature (coagulate) vital body proteins, especially enzymes needed for biochemical reactions in body cells.

Another chemical safeguard is a group of small proteins known as the interferons (in-ter-FEAR-ons). **Interferons** are released from cells infected by viruses. They diffuse away from the site of production through the interstitial tissue and bind to receptors on the plasma membranes of non-infected body cells (Figure 14-4). The binding of interferon to uninfected cells, in turn, triggers the synthesis of cellular enzymes that inhibit viral replication. Thus, when viruses enter the previously uninfected cells, they cannot replicate and spread.

Interferons do not protect cells already infected by a virus; they help stop the spread of viruses from one cell to another. In essence, the production and release of interferon are the dying cell's last act to protect other cells of the body. Interferons help to slow the spread of viruses, while the immune response (discussed shortly) attacks and destroys the viruses outside the cells.

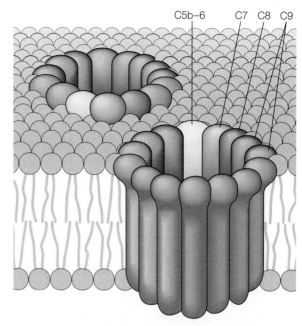

FIGURE 14-5 **The Membrane–Attack Complex** Five complement proteins combine and embed in a cell's membrane, causing it to leak, swell, and burst. (Reproduced from Cohn, Z. A. and Young, J. D.-E., *Sci. Am.* 258 [January 1988]: 38–44. Illustration courtesy of Dana Burns.)

Another group of chemical agents that fight infection are the **complement proteins**. These blood proteins form the **complement system**, so named because it complements the action of antibodies.

The details of the complement system are very complex. A few points will demonstrate how this remarkable system works. Complement proteins circulate in the blood in an inactive state. When foreign cells such as bacteria invade the body, the complement protein is activated. This triggers a chain reaction in which one complement protein activates the next.

Five proteins in the complement system join to form a large protein complex, known as the **membrane-attack complex** (Figure 14-5). The membrane-attack complex embeds in the plasma membrane of bacteria, creating an opening into which water flows. The influx of water causes bacterial cells to swell, burst, and die.

Several of the activated complement proteins also function on their own and are part of the inflammatory response. Some of them, for example, stimulate the dilation of blood vessels in an infected area, described earlier. Others increase the permeability of the blood vessels, allowing white blood cells and nutrient-rich plasma to pass more readily into an infected zone. Certain complement proteins may also act as chemical attractants, drawing macrophages, monocytes, and neutrophils to the site of infection, where they phagocytize foreign cells. Yet another complement protein (C3b) binds to microorganisms, forming a rough coat on the intruders that facilitates their phagocytosis.

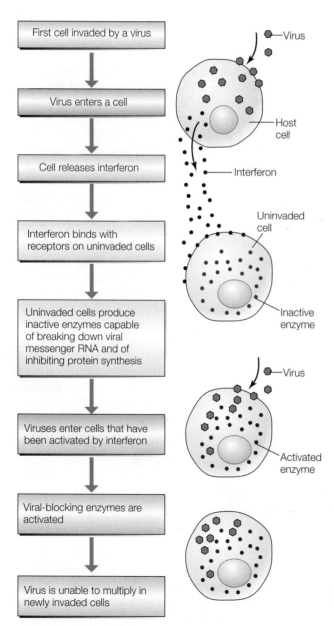

FIGURE 14-4 **How Interferon Works** Interferon protects cells from viral infection.

KEY CONCEPTS

Additional protection is provided during the inflammatory response by pyrogens, interferons, and complement.

14-3 The Third Line of Defense: The Immune System

The **immune system** is the third line of defense. Unlike the respiratory or digestive systems, the immune system is rather diffuse—spread out and indistinct. Lymphocytes, for example, are part of the immune system. They circulate in the blood and lymph and also take up residence in the lymphoid organs such as the spleen, thymus, lymph nodes, tonsils, and body tissues. All of these are part of the immune system.

The immune system, like the first and second lines of defense, is an important homeostatic mechanism that eliminates foreign organisms—including bacteria, viruses, single-celled fungi, and many parasites. However, the cells of the immune system selectively target foreign substances and foreign organisms. As a result, the immune system is said to be *specific*.

The immune system comes into play when foreign organisms penetrate the outer defenses of the body—the first and second lines of defense. The immune system also helps prevent the emergence and development of cancer.

As in other body systems, optimal immune system function depends on optimum health—excellent nutrition, exercise, and sufficient sleep. For example, studies show that eating foods rich in micronutrients, especially phytochemicals, dramatically reduces one's chances of developing common illnesses like colds and flus, or the duration of these illnesses. It also reduces severe illnesses such as autoimmune disease, heart disease, and cancer. As noted in the chapter on nutrition, this includes fruits, vegetables, nuts, seeds, and beans (see **Health Tip 14-2**).

One of the chief functions of the immune system is to identify what belongs in the body and what does not. Once a foreign substance has been detected, the immune system mounts an attack to eliminate it. Therefore, like all homeostatic systems, the immune system requires receptors to detect changes and effectors to bring about appropriate responses. In the immune system, the lymphocytes serve both functions.

Health Tip 14-2

A balanced diet, with plenty of fruits and vegetables, is not only good for the heart, it is good for the immune system.

Why?

Many nutrients are essential to achieve optimal immune system function. Adequate protein intake, for example, is essential for antibody production. Most people consume plenty of protein. Other nutrients, such as zinc, iron, vitamin C, vitamin D, and carotenoids, have been shown to play roles in the immune system—with some being more important than others. Many of these vitamins and minerals come from fruits and vegetables. Zinc comes from many food sources, among them wheat bran flake cereals, chicken and turkey (dark meat), refried beans, yogurt, and ground beef. Eat a varied diet with proportionately more fruits and vegetables to be sure you are getting all of the essential nutrients your body needs to prevent or combat illness and disease.

KEY CONCEPTS

Lymphocytes selectively target foreign substances, viruses, and foreign cells, including bacteria; they may attack the cells directly or via antibodies.

Antigens

The immune response is triggered by a variety of foreign cells, among them, bacteria, viruses, fungi, parasites, and cells derived from skin grafts or blood transfusion from genetically dissimilar individuals. Why?

The common denominator in all of these cases is that all these cells—and even viruses—to which our immune systems respond contain large molecules in their outermost coats or membranes, proteins and polysaccharides. These molecules are called **antigens** (AN-tah-gins), which is an abbreviation for antibody-generating substances.

When a bacterium or virus enters your body, your immune system recognizes that it doesn't belong there—that is, it is a foreigner. It's the unique assemblage of proteins and carbohydrates in the outer membranes of the foreign organisms that your immune system recognizes as not belonging. These include viruses, bacteria, single-celled fungi, and parasites such as the protozoan that causes malaria. Cells transplanted from one person to another also elicit an immune response. That's because each individual's cells contain a unique "cellular fingerprint" resulting from the unique array of plasma membrane glycoproteins. The immune system is activated by these antigens on the foreign cells.

Cancer cells that arise within your body also present a slightly different chemical fingerprint, making them essentially foreign cells within our own bodies to which the immune system responds. Although cancer cells evoke an immune response, it is often not sufficient to stop the disease.

As a rule, small molecules generally do not elicit an immune reaction. In some individuals, however, small, nonantigenic molecules such as formaldehyde, penicillin, and the poison ivy toxin bind to naturally occurring proteins in the body, forming complexes. These large complexes are unique compounds that are foreign to the body and can cause an immune response.

KEY CONCEPTS

Antigens are substances that trigger an immune response; they consist of large molecular weight carbohydrates (polysaccharides) and proteins located in the membranes that enclose cancer cells and foreign organisms.

Immunocompetence

The immune response is largely due to the stimulation of two types of types of white blood cells: **T lymphocytes**, commonly called **T cells**, and **B lymphocytes**, also called **B cells**. The "T" and "B" in their names will be explained shortly. B cells and T cells respond to different antigens and function quite differently.

As a rule, B cells recognize and react to microorganisms such as bacteria and **bacterial toxins**, chemical substances released by bacteria. B cells also respond to a few viruses. When activated, B cells produce antibodies to these antigens.

In contrast, T cells recognize and respond to our own body cells that have gone awry. This includes cancer cells as well as body cells that have been invaded by viruses. T cells also respond to transplanted tissue cells or transfused blood cells. In addition, they respond to larger disease-causing organisms, such as single-celled fungi and parasites. Unlike B cells, T cells attack their targets directly.

Lymphocytes are produced in the red bone marrow and released into the bloodstream. These immature cells circulate in the blood and lymph but are not able to function until they become immunologically competent. This process occurs in specific organs in the body called **lymphoid organs**. Let's start with the T cell.

T Cells. T cells get their name from the place where they mature, the thymus. The **thymus** (THIGH-muss) is a lymphoid organ located above the heart. After they're formed in the bone marrow, immature lymphocytes enter the bloodstream. Some of them end up in the thymus. Over the next few days, these lymphocytes mature and become functional T cells. Immunologists, scientists who study the immune system, say that the lymphocytes have developed **immunocompetence** (IM-you-know-COM-pah-tense)—that is, they've developed the capacity to respond to antigens. But not just any antigens. In ways not completely understood, T lymphocytes leave the organ transformed to respond to very specific antigens.

When the T cells leave the thymus, each of their membranes contains a unique type of membrane receptor that will bind to one—and only one—type of antigen. Over an individual's lifetime, thousands upon thousands of antigens will be encountered. Thanks to the immunocompetence developed during fetal development, each of us is equipped with an army of uniquely preprogrammed T cells that respond to the onslaught of antigens.

B Cells. In humans, B cells mature and differentiate in the bone marrow. The "B" in B cell, however, did not come originally from the bone marrow. It came from the bursa, an organ found in chickens. Early researchers studied lymphocytes in chickens and came up with the name then.

After they mature, immunologically competent B cells circulate in the blood and take up residence in connective and lymphoid tissues. They therefore become part of the body's vast cellular reserve, stationed at distant outposts, awaiting the arrival of the microbial invaders.

By various estimates, several million immunologically distinct B and T cells are produced in the body early in life. Over a lifetime, only a relatively small fraction of these cells will be called into duty.

> **KEY CONCEPTS**
>
> Two types of lymphocyte play key roles in immunity, B cells and T cells. These cells are incapable of responding to antigens until they become immunologically competent.

B Cells and Antibodies

The immune response consists of two separate but related reactions: humoral immunity, provided by the B cells, and cell-mediated immunity, involving T cells (Table 14-2). Let's consider humoral immunity first.

When a bacterium first enters the body, B cells programmed to respond to the unique proteins found in the microbe's cell membrane bind to it (Figure 14-6). The B cells soon begin to divide, producing a population of immunologically similar cells known as **lymphoblasts**. As shown in Figure 14-6, some of the lymphoblasts differentiate to form another kind of cell, the **plasma cell**. Plasma cells produce antibodies. **Antibodies** are proteins that help eliminate antigens, that is, bacteria or bacterial toxins.

Released from plasma cells, antibodies circulate in the blood and lymph, where they bind to the antigens that triggered the response. Because the blood and lymph were once referred to as body "humors," this arm of the protective immune response is called **humoral immunity**. Once the invaders have been removed, no new plasma cells are formed, and existing plasma cells die.

> **KEY CONCEPTS**
>
> B cells provide humoral immunity through the production of antibodies.

Primary Response

The first time an antigen enters the body, it elicits an immune response, but the initial reaction—or **primary response**—is relatively slow (Figure 14-7). During the primary response, antibody levels in the blood do not begin to rise until approximately the beginning of the second week after the intruder was detected, partly explaining why it takes most people about 7 to 10 days to combat a cold or the flu. This delay occurs because it takes time for B cells to multiply and form a sufficient number of plasma cells. Antibody levels usually peak about the end of the second week, then decline over the next 3 weeks.

> **KEY CONCEPTS**
>
> The initial B cell response to an antigen is known as the primary response. It is slower and weaker than subsequent responses.

TABLE 14-2	Comparison of Humoral and Cell-Mediated Immunity

Humoral

Principal cellular agent is the B cell.

B cell responds to bacteria, bacterial toxins, and some viruses.

When activated, B cells form memory cells and plasma cells, which produce antibodies to these antigens.

Cell-Mediated

Principal cellular agent is the T cell.

T cell responds to cancer cells, virus-infected cells, single-cell fungi, parasites, and foreign cells in an organ transplant.

When activated, T cells differentiate into memory cells, cytotoxic cells, suppressor cells, and helper cells; cytotoxic T cells attack the antigen directly.

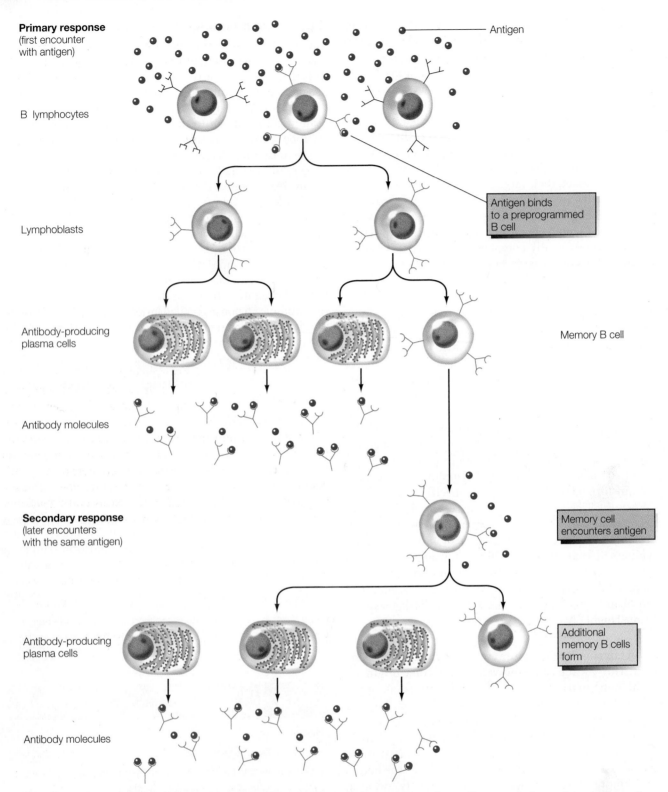

FIGURE 14-6 B-Cell Activation Immunocompetent B cells are stimulated by the presence of an antigen, producing an intermediate cell, the lymphoblast. The lymphoblasts divide, producing plasma cells and some memory cells. Memory cells respond to subsequent antigen encroachment, yielding a rapid, secondary response.

Secondary Response

The secondary response of a B cell to an antigen is much quicker, due to the production of memory cells during the primary response. This ensures that subsequent exposures to the same antigen do not typically result in illness.

If the same antigen enters the body at a later date, however, the immune system acts much more quickly and more forcefully (Figure 14-7). This stronger reaction constitutes the **secondary response**. During a secondary response, antibody levels increase quickly—a few days after the antigen has entered the body. The amount of antibody produced also

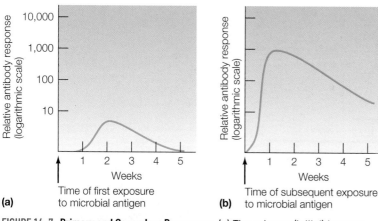

(a)

Time of first exposure to microbial antigen

(b)

Time of subsequent exposure to microbial antigen

FIGURE 14-7 Primary and Secondary Responses (a) The primary (initial) immune response is slow. It takes about 10 days for antibody levels to peak. Almost no antibody is produced during the first week as plasma cells are being formed. (b) The secondary response is much more rapid. Antibody levels rise almost immediately after the antigen invades. T cells show a similar response pattern.

produce copious amounts of antibody to combat the foreign invaders. Memory cells also generate additional memory cells during the secondary response. These remain in the body in case the antigen should enter the body at some later date. Immune protection afforded by memory cells can last 20 years or longer, which explains why a person who has had a childhood disease such as the mumps or chicken pox is unlikely to contract it again.

> **KEY CONCEPTS**
>
> The secondary response of a B cell to an antigen is much quicker, due to the production of memory cells during the primary response. This ensures that subsequent exposures to the same antigen do not typically result in illness.

How Antibodies Work

Antibodies belong to a class of blood proteins called the globulins (GLOB-you-lynns), introduced in Chapter 7. Antibodies are specifically called **immunoglobulins** (im-YOU-know-GLOB-you-lynns).

Each antibody is a T-shaped molecule consisting of four peptide chains (Figure 14-8). The arms of the T bind to antigens.

Antibodies destroy foreign organisms and antigens via one of four mechanisms. One process is *neutralization*. During neutralization, antibodies bind to viruses and bacterial toxins and form a complete coating around them. This prevents viruses from binding to the plasma membrane receptors of body cells and entering them (Figure 14-9, far right). They are effectively neutralized. Bacteria and bacterial toxins are also coated and rendered ineffective. In time, bacteria, bacterial toxins, and viruses

greatly exceeds quantities generated during the primary response. Consequently, the antigen is quickly destroyed, and a recurrence of the illness is prevented.

Why is the secondary response so different? Figure 14-6 shows that during the primary response, some lymphocytes divide to produce memory cells. Memory cells are immunologically competent B cells. Although they do not transform into plasma cells, they remain in the body awaiting the antigen's reentry. These cells are produced in large quantity and thus create a relatively large reserve of antigen-specific B cells. When the antigen reappears, memory cells proliferate rapidly, producing numerous plasma cells that quickly begin to

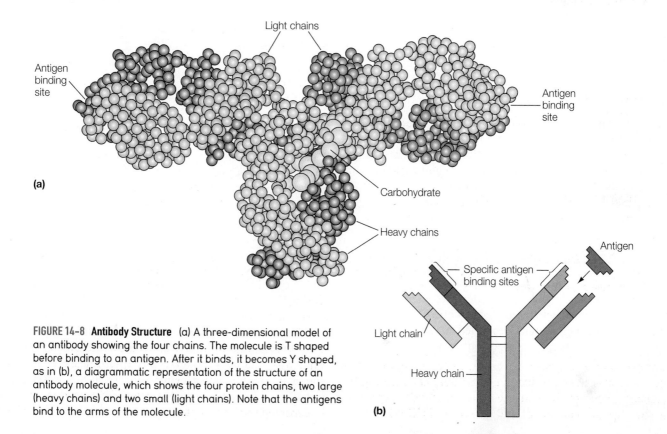

FIGURE 14-8 Antibody Structure (a) A three-dimensional model of an antibody showing the four chains. The molecule is T shaped before binding to an antigen. After it binds, it becomes Y shaped, as in (b), a diagrammatic representation of the structure of an antibody molecule, which shows the four protein chains, two large (heavy chains) and two small (light chains). Note that the antigens bind to the arms of the molecule.

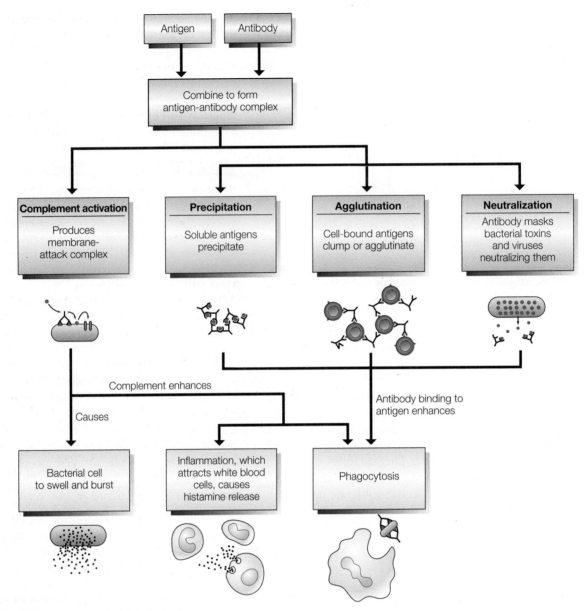

FIGURE 14-9 How Antibodies Work

neutralized by antibody coatings are engulfed by macrophages and other phagocytic cells (Figure 14-9).

Other antibodies eliminate antigens by a process called *agglutination* (ah-GLUTE-tin-A-shun). Agglutination occurs when antibodies bind to numerous antigens, causing them to clump together, again rendering them ineffective. The antigen-antibody complexes are then engulfed by macrophages and other phagocytic cells.

Antibodies can also bind to soluble antigens (for example, proteins), forming much larger, water-insoluble complexes. These precipitate out of solution, where they are engulfed by phagocytic cells. The process is called *precipitation*.

Antibodies also help to rid the body of bacteria by activating the complement system, described above.

KEY CONCEPTS

Antibodies are a type of proteins known as immunoglobulins that are produced by plasma cells; they eliminate antigens in one of four ways.

T Cells

T cells provide a much more complex form of protection than B cells. Like B cells, they respond to the presence of antigens by undergoing rapid proliferation. T cells, however, differentiate into numerous types including: (1) cytotoxic T cells, (2) natural killer cells, (3) memory T cells, (4) helper T cells, and (5) suppressor T cells (Table 14-3).

Cytotoxic T cells play a key role in cell-mediated immunity. Some cytotoxic T cells attack and kill body cells infected by viruses. Other cytotoxic T cells attack and kill bacteria, parasites, single-celled fungi, cancer cells, and foreign cells introduced during blood transfusions or tissue or organ transplants. These organisms or cells have proteins in their membranes that identify them as foreign. How do cytotoxic T cells kill other cells?

Cytotoxic T cells bind to antigenic molecules in the membranes of cells they attack. They then release a chemical substance known as **perforin-1** (purr-FOR-in). As shown in Figure 14-10, perforin-1 molecules are housed in membranous sacs in the cytoplasm of the cytotoxic T cells. These structures bind

TABLE 14-3	Summary of T Cells
Cell Type	**Action**
Cytotoxic T cells	Destroy body cells infected by viruses, and attack and kill bacteria, fungi, parasites, and cancer cells
Helper	Produce a growth factor that stimulates B-cell proliferation and differentiation and also stimulates antibody production by plasma cells; enhance activity of cytotoxic T cells
Suppressor	May inhibit immune reaction by T cells decreasing B- and T-cell activity and B- and T-cell division
Memory	Remain in body awaiting reintroduction T cells of antigen, at which time they proliferate and differentiate into cytotoxic T cells, helper T cells, suppressor T cells, and additional memory cells

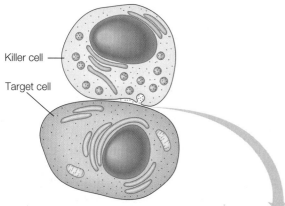

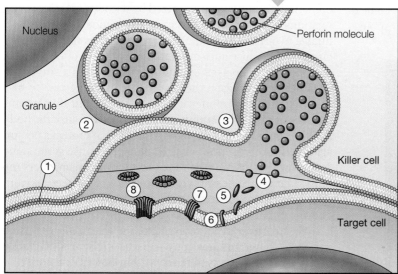

FIGURE 14-10 How Cytotoxic T Cells Work Cytotoxic T cells, containing perforin-1 granules, bind to their target and release perforin-1, then detach in search of other invaders. Perforin-1 molecules congregate in the target plasma membrane, forming a pore that disrupts the plasma membrane, causing the cell to die. (Reproduced from Cohn, Z. A. and Young, J. D.-E., *Sci. Am.* 258 [January 1988]: 38–44. Illustration courtesy of Dana Burns.)

to the membrane of the cytotoxic T cell. The membranes at the point of contact break down and the perforin-1 molecules are released into the space between the killer cell and its prey.

Perforin molecules then embed in the plasma membrane of the target cell. There, they join to form small pores, similar to those produced by the membrane-attack complex of the complement system. These pores cause the plasma membrane to leak, destroying the target cell within a few hours. After it has delivered its lethal payload, the cytotoxic cell detaches and is free to hunt down other antigens.

Natural killer (NK) cells are like cytotoxic T cells. They respond rapidly to viral infection of cells and to newly formed tumors, acting at around 3 days after infection. NK cells recognize stressed cells in the absence of typical immune system signals, like the presence of antibodies or the presences of large molecular weight proteins and carbohydrates on the plasma membranes of foreign organisms. This permits a very rapid response to tumors and infection.

Memory T cells are also produced when antigens are present. As in B cells, the memory T cells form an immunologically competent cellular reserve. They exist to protect the body in the event of a further incursion at some other time.

Helper T cells are also produced from T cells. **Helper T cells (T_h cells)** enhance the immune response—both the humoral and the cell-mediated immune response. Helper T cells release a chemical known as *interleukin 2*. It increases the activity of cytotoxic T cells. Mature T_h cells contain a surface glycoprotein referred to as CD4 and are, therefore, often referred to as **CD4+ T cells** or **CD4 cells**.

Helper T cells are the most abundant of all the T cells (comprising 60%–70% of the circulating T cells). Some immunologists liken helper T cells to the immune system's master switch. Without them, antibody production and T-cell activity are greatly reduced. This phenomenon occurs in individuals infected with human immunodeficiency virus (HIV), an RNA that causes acquired immunodeficiency syndrome (AIDS). As you will see in this chapter's supplement, **A Closer Look**, helper T cells are targeted by HIV. Infection of helper T cells by HIV essentially disables a person's immune system. Patients with AIDS therefore contract several infectious diseases to which they are unable to mount an effective immune response. Unless a cure can be found, people infected with the AIDS virus will eventually die from these diseases.

Another type of T cells is the **regulatory T cell**, formerly known as the **suppressor T cell**. Several types have been discovered. The role of regulatory T cells is several-fold. Research shows that they "turn off" the immune reaction as the antigen begins to disappear, that is, as the antigen is phagocytized. The activity of suppressor T cells, therefore, increases as the immune system finishes its job. Suppressor cells release chemicals that reduce T-cell division. Regulatory T cells also suppress autoimmunity, that is, they prevent the immune system from attacking cells of the body.

KEY CONCEPTS

T cells attack and destroy foreign cells directly.

Macrophages, Dendritic Cells, and Immunity

As noted earlier, macrophages play a key role in the immune system by stimulating T and B cells. This process begins when macrophages phagocytize antigens such as bacteria. Macrophages then transfer the antigens from the surface of the bacteria to their own cell membranes. Macrophages then cluster around B cells, presenting the bacterial antigens to them. B cells programmed to respond to that antigen are converted to plasma cells.

Macrophages also present antigen to helper T cells. Helper T cells, in turn, produce a growth factor that stimulates the formation of additional plasma cells as well as antibody production by plasma cells.

In recent years, immunologists have also discovered an immune system cell, known as **dendritic cell**—so named because its extensive branching fibers make it look like a tree. (Dendrology is the study of trees.) Figure 14-11 shows a typical dendritic cell. Like macrophages, they are antigen-presenting cells. Dendritic cells are found in tissues in close contact with the external environment, for example, under the skin and beneath the linings of the respiratory and digestive systems. They are also found in the blood and bone marrow, although in an immature state.

Immature dendritic cells are formed in blood marrow, like other white blood cells. These cells constantly sample their environment for potential pathogens including viruses and bacteria. This function is made possible by cell membrane receptors, known as pattern recognition receptors (PRRs). They recognize the large molecular weight carbohydrates and proteins on the outer surface of foreign biological agents like viruses and bacteria. Once they come into contact with a foreign agent, dendritic cells phagocytize them, and break their proteins into smaller pieces. These proteins migrate to the plasma membrane of the dendritic cell, as in macrophages. Immature dendritic cells are now referred to as mature dendritic cells. Mature dendritic cells migrate to the lymph nodes of the body through the lymphatic system and to the spleen via the bloodstream. Here they come in contact with T cells and B cells to initiate an immune response.

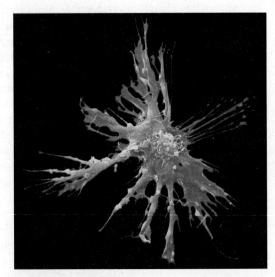

FIGURE 14-11 **Scanning Electron Micrograph of a Dendritic Cell** This cell plays a key role in activating the immune system. (© RGB Ventures, LLC, dba Superstock Alamy.)

They activate helper T-cells, killer T-cells, and B-cells by presenting them with antigens from the pathogen and through other avenues as well.

> **KEY CONCEPTS**
>
> Macrophages and dendritic cells play a key role in humoral and cell-mediated immunity by concentrating and presenting antigens to B and helper T cells.

Active and Passive Immunity

One of the major medical advances of the 1800s was the discovery of vaccines (vac-SEENS). A **vaccine** is a substance that provides immunity to a disease. Vaccines typically contain inactivated or weakened viruses, bacteria, or bacterial toxins. When injected into the body, the "disabled" antigens in vaccines elicit an immune response. The immune system responds to the vaccine by making antibody-producing plasma cells or T cells, depending on the antigen. In essence, the vaccine stimulates the primary response to a dummy infectious agent. Now the immune system is primed and ready for the real McCoy. Should a real pathogen arrive, the immune system can respond quickly. But won't an individual get sick from the vaccine?

Vaccines stimulate the immune reaction because the weakened or deactivated organisms (or toxins) they contain still possess the antigenic proteins or carbohydrates that trigger B- and T-cell activation. However, because the infectious agents have been seriously weakened or deactivated, the viruses, bacteria, and bacterial toxins in vaccines do not cause sickness. Some individuals may develop minor symptoms, but they're not life-threatening.

Problems may arise if the vaccines are given when a person is suffering from a cold, flu, or other sicknesses. About one-third of all parents in the United States are concerned that exposure to vaccines during these times may be connected to chronic illnesses, such as autism. Autism is a severe developmental disability characterized by withdrawal from social relations. Autistic children have reduced motor skills and are unable to or refuse to talk. Numerous studies show that vaccines do not cause autism. Moreover, the risk of developing autism is not increased by multiple vaccines in a single day or multiple vaccines in the first two years of life.

Vaccination provides a form of protection that immunologists call **active immunity**—so named because the body actively produces memory T and B cells that protect a person against future infections. The immune response is the same as the one that occurs when a pathogenic organism enters the body. Many vaccines provide immunity or protection from microorganisms for long periods, sometimes for life. Others give only short-term protection.

Vaccinations are vital in controlling deadly diseases such as polio, typhus, and smallpox—diseases that can kill people before their immune system mounts an effective response. In fact, in the wealthier nations of the world, such as the United States, vaccines have nearly eliminated many infectious diseases such as polio and smallpox.

The second type of immunity, called **passive immunity**, is a temporary form of protection, resulting from the injection of immunoglobulins (antibodies to specific antigens).

As you may recall, immunoglobulins are antibodies. They are produced by injecting antigens in animals such as sheep. The antibodies are then extracted from the animals' blood for use in humans.

Passive immunity is so named because the cells of the immune system are not activated. T cells and B cells are not called into duty. Protective immunoglobulins remain in the blood for a few weeks, protecting an individual from infection. Because the liver slowly removes immunoglobulins from the blood, a person gradually loses this protection. Why would anyone want to be passively immunized?

Immunoglobulins are administered to protect people from illnesses during travel to less developed nations where diseases are less under control. For instance, travelers are often given immunoglobulins to viral hepatitis (liver infection) as a preventive measure. Immunoglobulins are also used to treat individuals who have been bitten by poisonous snakes.

Passive immunity can also be conferred naturally—for instance, from a mother to her fetus. A fetus receives antibodies from the mother via the placenta (plah-SEN-tah), the organ that transfers nutrients from the mother's bloodstream to the fetal blood. Maternal antibodies transferred via the placenta remain in the blood of a newborn infant for several months, protecting the youngster from bacteria and viruses while its immune system is developing. Mothers also transfer antibodies to their babies in breast milk. The maternal antibodies in milk attack bacteria and viruses in the intestine, protecting the infant from infection. (For more on this topic, see Health Note 14-2.)

Vaccination Fears in the United States. In the United States and other relatively affluent industrialized nations, vaccines have lowered the incidence of many infectious diseases by 99% or more. Vaccines for smallpox, diphtheria, tetanus, whooping cough, polio, measles, mumps, and congenital rubella (German measles) have all but eliminated these deadly or crippling diseases.

Despite the success of vaccines, publicity concerning their very rare side effects has convinced some parents not to have their children vaccinated. Another cause for the decline in vaccination stems from the success of previous immunization programs. Parents who were reared in an environment free from diseases that have been nearly eradicated by vaccines may be unaware of how dangerous these diseases can be and the dangers they pose. Having their children immunized seems unimportant. Some people, notably Christian Scientists, refuse to vaccinate children on religious grounds.

Although avoiding vaccinations may seem logical, public health officials are quick to point out that pathogenic organisms that once took a huge toll on humans have not been eradicated. Many exist in less developed countries and could be spread to more developed countries. Without continuing widespread vaccination, epidemics could occur again. In a world with rapidly increasing population, in which transportation links greater numbers of people, infectious disease could spread very rapidly, with devastating results. Consider polio. In the United States, prior to the advent of the polio vaccine, there were an estimated 50,000 cases of polio reported each year. Striking children,

this virus attacked part of the spinal cord, damaging nerve tracts that controlled the muscles to the legs and sometimes the diaphragm, which is essential to breathing. The virus left thousands of victims—mostly children—permanently paralyzed. They were often relegated to a life dependent on braces, crutches, or wheelchairs and those whose diaphragms became paralyzed lived out their lives in iron lungs (artificial breathing machines).

Fortunately for those of us living in the United States and other developed countries, the polio vaccine has eradicated the polio virus. So why continue to vaccinate against it? Because the virus still remains a threat in some less developed countries and could easily be transported back into the developed countries by an unwary traveler.

Another highly contagious and extremely common but vaccine-preventable disease is pertussis or whooping cough. Caused by a bacterium, this disease is characterized by coughing spells that can last for many weeks or even months. It was so named because children make a "whooping" sound as they try to catch their breath between coughs. Pertussis is easily transmitted from person to person through personal contact, coughing, and sneezing. It is especially serious for babies as it can be lethal. It can even cause them to stop breathing. Each year, hundreds of young children—most of them under the age of one—are hospitalized for whooping cough. Some die from it.

Polio and pertussis are just two of about 16 diseases that can be treated by vaccination. To many health experts, vaccination is more important today than ever before. But what about harmful side effects of vaccines?

Harmful side effects from vaccines were often caused by reactions to certain "nonessential" antigens on the injected microorganism. These antigens frequently played little or no role in immunity. By eliminating them from vaccines, researchers have developed safer alternatives to the vaccines in use today.

Another concern of those who oppose vaccines is the use of the mercury-containing preservative, thimerosol, which has been used in some vaccines since the 1930s. Thimerosol causes redness and swelling at the site of injection, but there is no evidence that low doses present in vaccines causes any other, according to the Centers for Disease Control. Even so, in 1999 the Public Health Service agencies, the American Academy of Pediatrics, and vaccine manufacturers agreed to reduce or eliminate thimerosal from vaccines as a precautionary measure.

> **KEY CONCEPTS**
>
> Two types of immunity are possible, active immunity, which is acquired when one is exposed to an antigen, and passive immunity, acquired when one is administered a serum containing antibodies.

Developing a Strong Immune System

Because bacteria and viruses can cause illness, many people assume that the best way to keep their children healthy is to protect them from exposure early in life. We've all witnessed the outpouring of soaps, cleaning agents, and other products that contain antibacterial agents. Unfortunately, research

health**note**

14–2 Bringing Baby Up Right: The Immunological and Nutritional Benefits of Breast Milk

A baby is born into a dangerous world in which bacteria and viruses abound. Complicating matters, the immune system of a newborn child is poorly developed. Fortunately, the newborn is protected by passive immunity—antibodies that have traveled from its mother's blood. Antibodies also travel to the infant in breast milk (Figure 1).

Several immunoglobulins are present in breast milk. One of these is called secretory IgA. It is present in very high quantities in colostrum—a thick fluid produced by the breast immediately after delivery—before the breast begins full-scale milk production. Colostrum is so important, in fact, that some hospitals give "colostrum cocktails" to newborns who are not going to be breast-fed by their mothers. Nurses remove the colostrum from the mother's breast with a breast pump and feed it to the baby in a bottle.

"Colostrum," said the late Sarah McCamman, a nutritionist at the University of Kansas Medical School, "coats the lining of the intestines. The IgA antibodies in colostrum prevent bacteria ingested by the infant from adhering to the epithelium and gaining entrance. In addition, colostrum creates an environment that allows for the growth of beneficial intestinal bacteria." Breast milk also contains lysozyme, an enzyme that breaks down the cell walls of bacteria, destroying them.

In general, breast-fed babies are healthier than bottle-fed babies. The incidence of gastroenteritis (inflammation of the intestine), otitis (ear infections), and upper respiratory infections is lower in breast-fed babies. As a result, breast-fed babies require fewer hospitalizations and trips to the doctor than bottle-fed babies.

FIGURE 1 Mother breast-feeding her newborn infant. (© Photodisc/Getty Images.)

Studies also show that children breast-fed for at least 6 months contract fewer childhood cancers than their bottle-fed counterparts. Breastfeeding reduces your baby's risk of developing asthma or allergies. The incidence of childhood lymphoma, a cancer of the lymph glands, in bottle-fed babies is nearly double the rate in breast-fed children for reasons not yet understood.

Research also suggests that certain proteins in breast milk may stimulate the development of a newborn's immune system. In laboratory experiments, the proteins speed up the maturation of B cells and prime them for antibody production. These soluble proteins may also activate macrophages, which play a key role in the immune system.

Studies have shown a link between IQ scores in later childhood and breastfeeding, too.

Breast milk is also more digestible and more easily absorbed by infants than formula. Formula is a mixture of cow's milk, proteins, vegetable oils, and carbohydrates. It is only an approximation of mother's milk and is not broken down and absorbed as completely as breast milk.

Breast feeding also has benefits to mothers. For example, breastfeeding burns extra calories, helping a woman lose "pregnancy weight" faster. Suckling releases the hormone oxytocin. This causes contraction of the uterus, allowing it to return to its pre-pregnancy size. Breastfeeding may also reduce bleeding from the uterus after birth. Studies even show that breastfeeding can lower a woman's risk of developing cancer of the breast and ovaries and may lower a woman's risk of osteoporosis.

Because you don't have to buy and measure formula, sterilize nipples, or warm bottles, breastfeeding saves you time and money. Deciding to breastfeed provides you with regular time for relaxing quietly with your newborn as you grow close and emotionally bond.

Due to a growing awareness of the benefits of breastfeeding, virtually every national and international organization involved with maternal and child health supports breastfeeding, says McCamman. Breastfeeding can have a major health impact in this country. Unfortunately, the benefits are not as widely known as many people would like, even among healthcare professionals. Fortunately, more and more healthcare workers are beginning to understand the benefits of breastfeeding and are promoting this option. As a result, many middle-class American women are now choosing to breastfeed.

Unfortunately, says McCamman, there is "a huge population of low-income, poorly educated . . . women who choose not to nurse." The federal government may be playing an unwitting role in their decision. A national program aimed at improving child nutrition provides free infant formula to needy women, perhaps discouraging mothers from breastfeeding.

Another reason is economic. Low-income women often work at jobs that do not provide maternity leave. Thus, these women must return to work soon after giving birth, making breastfeeding difficult.

Still another reason for the low rate of breast-feeding among low-income women is that women need a lot of support to nurse. "People think nursing is innate, natural, and easy," says McCamman, "but this is not always the case." Getting started often requires guidance and education. Without such support, breastfeeding can be difficult and painful. Breastfeeding among all women, rich and poor, may also be discouraged by cultural attitudes and fear of embarrassment. In fact, many otherwise open-minded people find breastfeeding in public or even semipublic settings embarrassing.

Given the many benefits of breastfeeding, McCamman recommends it to all mothers who are able. Assistance is also available in every state to help women learn to breastfeed.

suggests that the best way to develop a healthy immune system is for children to use them when they're young. That is, the best way to ensure maximum health is not to protect a child from exposure to non-life-threatening microbes.

Although this idea seemed preposterous when first proposed, more and more evidence is showing that children who have been sheltered from normal childhood illnesses develop more asthma and allergies—a lot more! In fact, a study released in 2000 showed that infants who go to day care centers or who have older siblings—both of which increase their risk of catching colds or the flu—are less likely to develop asthma later in childhood than those who stay at home. The reason for this is not known, but researchers believe that early exposure to common microbes, especially nonpathogenic ones, may somehow start the immune system off on the right path of development. A study published in 2013, in fact, showed that exposure to harmless microorganisms in the soil and our food as well as all surfaces in a home or our environment help the immune system prepare for future infections of harmful microorganisms. They do so by programming CD4 cells, discussed earlier, to respond to harmful microorganisms to which we will be exposed later in life.

Small families, good sanitation, and widespread use of antibiotics, say researchers, reduce childhood exposure to bacteria and viruses. This may be responsible for the dramatic increase in allergies and asthma in the United States and other developed countries. What is more, studies show that children in "sterilized" homes are not less likely to develop respiratory infections.

The advice some healthcare professionals now give is not to sterilize the environments in which we live. Avoid the many new products like soaps and sprays with germ-fighting chemicals. They're not needed, and they may have the opposite effect of making children less healthy in the long run. There are some concerns that the widespread use of antibiotics in cleansing agents could lead to super strains of resistant microorganisms.

> **KEY CONCEPTS**
> The immune system may operate optimally when exposed to antigens early in life.

Tissue Transplantation

Although the immune system is important in protecting us from microorganisms, it presents something of a challenge to surgeons and patients during tissue and organ transplants and blood transfusions.

Organ and tissue transplantations are fairly common medical practices these days. Diseased hearts, kidneys, and livers, for instance, can be removed and replaced by healthy organs from suitable donors. However, only three conditions exist in which a person can receive a transplant and not reject it.

One is if the tissue comes from an individual's own body. For burn victims, surgeons might use healthy skin from one part of the body to cover a badly damaged region elsewhere. Hair transplants also qualify. In hair transplants, hair from the back of the head is often moved to balding spots up front or on top.

The second instance is when a tissue is transplanted between identical twins—individuals derived from a single fertilized ovum that splits early in embryonic development to form two embryos. These individuals are genetically identical and have identical cellular antigens.

A third instance occurs when tissue rejection is inhibited by drugs that suppress the cell-mediated response mounted by the immune system. This treatment must be continued throughout the life of the patient. If not, the body will reject the transplant. Unfortunately, most immune suppressants have numerous side effects and often leave the patient vulnerable to bacterial and viral infections.

> **KEY CONCEPTS**
> Tissue transplants from one individual can evoke a cell-mediated immune response that results in tissue rejection unless anti-rejection drugs are administered.

14-4 Diseases of the Immune System

The immune system, like all other body systems, can malfunction. This section looks at two disorders: allergies and autoimmune diseases. AIDS, a sexually transmitted disease that affects the immune system, is discussed in the chapter supplement.

Allergies

An **allergy** is an overreaction to some environmental substance, such as pollen or certain foods. Antigens that stimulate an allergic reaction are called **allergens** (AL-er-gens). The process is outlined in **Figure 14-12**. As illustrated, allergens cause plasma cells to produce one type of immunoglobulin known as the immunoglobulin E (IgE) antibodies. IgE antibodies are one of five types of antibodies.

Unlike most other antibodies that bind to antigens such as bacteria, effectively annihilating them, IgE antibodies bind to mast cells, as shown in Figure 14-12. Mast cells are found in many tissues of the body, but especially in the connective tissue surrounding blood vessels. Mast cells contain large cytoplasmic vesicles containing the chemical histamine.

As shown in the right side of Figure 14-12, allergens not only stimulate the production of IgE antibodies, they bind to the IgE antibodies after they have attached to the mast cells. This binding triggers the release of histamine from the vesicles via exocytosis (Figure 14-12). Histamine, in turn, causes nearby arterioles to dilate. In the lungs, histamine released from mast cells causes the bronchioles to constrict, reducing airflow and making breathing difficult. This condition is called **asthma** (AS-ma). Asthma is discussed in more detail elsewhere in the text.

Allergic reactions usually occur in specific body tissues, where they create local symptoms that, while irritating, are not life-threatening. For example, an allergic response may occur

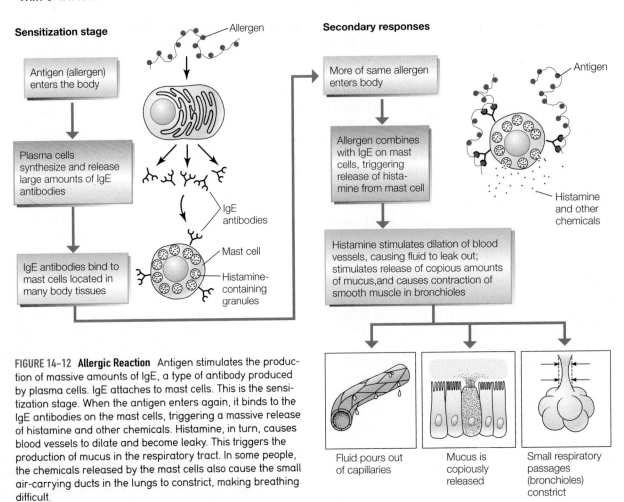

Sensitization stage

Antigen (allergen) enters the body

Plasma cells synthesize and release large amounts of IgE antibodies

IgE antibodies bind to mast cells located in many body tissues

Allergen

IgE antibodies

Mast cell

Histamine-containing granules

Secondary responses

More of same allergen enters body

Allergen combines with IgE on mast cells, triggering release of histamine from mast cell

Histamine stimulates dilation of blood vessels, causing fluid to leak out; stimulates release of copious amounts of mucus, and causes contraction of smooth muscle in bronchioles

Antigen

Histamine and other chemicals

Fluid pours out of capillaries

Mucus is copiously released

Small respiratory passages (bronchioles) constrict

FIGURE 14-12 Allergic Reaction Antigen stimulates the production of massive amounts of IgE, a type of antibody produced by plasma cells. IgE attaches to mast cells. This is the sensitization stage. When the antigen enters again, it binds to the IgE antibodies on the mast cells, triggering a massive release of histamine and other chemicals. Histamine, in turn, causes blood vessels to dilate and become leaky. This triggers the production of mucus in the respiratory tract. In some people, the chemicals released by the mast cells also cause the small air-carrying ducts in the lungs to constrict, making breathing difficult.

in the eyes, causing redness and itching. Or, it may occur in the nasal passageway, causing stuffiness.

The allergic response can also occur in the bloodstream, where it can be fatal if not treated quickly. For example, the presence of penicillin or bee venom in the bloodstreams of certain people causes a massive release of histamine and other chemicals. This, in turn, causes extensive dilation of blood vessels in the skin and other tissues. The blood pressure then falls quickly, shutting down the circulatory system. Histamine released by mast cells also causes severe constriction of the bronchioles (the ducts in the lungs that open onto the alveoli), making breathing difficult. The decline in blood pressure and constriction of the bronchioles result in **anaphylactic shock** (ANN-ah-fah-LACK-tic). Death may follow if measures are not taken to reverse this catastrophic event. One such measure is an injection of the hormone epinephrine (commonly known as adrenalin), which rapidly reverses the constriction of the bronchioles.

Allergies are treated in three ways. First, patients are advised to avoid allergens—for instance, to avoid milk and milk products, or stay clear of dogs and cats. Second, patients may also be given antihistamines (an-tea-HISS-tah-means), drugs that counteract the effects of histamine. Third, patients may also be given allergy shots, injections of increasing quantities of the offending allergen. In many cases, this treatment makes an individual less and less sensitive to the allergen.

Desensitization results from the production of another class of antibodies, the IgG antibodies, which bind to allergens. This, in turn, prevents the allergen from binding to the plasma cells, which reduces IgE antibody release.

KEY CONCEPTS

Allergies are an immune system overreaction to common environmental substances such as pollen and to certain foods.

Autoimmune Diseases

Autoimmune diseases are an abnormal immune response of the body in which the immune system mounts an attack on its own cells. In other words, the immune system mistakes some part of the body—a substance or a cell—as a foreign object, like a pathogen, and attacks its. To date, there are more than 100 diseases classified as autoimmune diseases and many more that are suspected.

Autoimmune diseases result from many causes. For example, in some instances, normal body proteins can be modified by certain air pollutants like formaldehyde, which is present in furniture and other modern building materials. The binding of formaldehyde to native body proteins renders them foreign. The immune system attacks these foreign proteins. Viruses and genetic mutations can also modify native body proteins, making them appear foreign to the immune system.

In some cases, normal body proteins that are usually isolated from the immune system enter the bloodstream and evoke an immune response. For example, a protein called *thyroglobulin* is produced by the thyroid gland in the neck. Thyroglobulin is stored inside the gland and not exposed to cells of the immune system. If the gland is injured, however, thyroglobulin may enter the bloodstream. Lymphocytes encountering this protein may then mount an immune response to it.

Yet another cause of autoimmune reaction is exposure to antigens that are nearly identical to body proteins. The bacterium that causes strep throat, for example, contains an antigen that's structurally similar to one of the proteins found in the plasma membranes of the cells lining the heart valves of some individuals. The body mounts an attack on the bacterium, but antibodies also bind to the lining of the heart valve, causing a local inflammation and scar tissue to develop. This can damage the valve, resulting in valvular incompetence.

> **KEY CONCEPTS**
>
> Autoimmune diseases occur when a person's immune system attacks his or her own cells.

14-5 Health and Homeostasis

Human beings would not survive past early infancy without some means of protection against the potentially harmful viruses and microorganisms that abound in our world. But what does protection have to do with homeostasis?

The answer is, plenty. Without the protective mechanisms you've studied in this chapter, homeostasis could not be maintained. Bacteria and viruses would take over, sapping us of our strength and damaging vital organs that play a key role in homeostasis. The immune system and other protective mechanisms of our bodies, therefore, play a vital role in maintaining internal constancy.

Despite its prowess, the immune system is not impenetrable. Stress can suppress the immune system, making us more prone to infectious disease. Poor nutrition, especially a lack of micronutrient phytochemicals, and some chemical pollutants in the environment—for example, certain pesticides—can affect our immune systems, making us more susceptible to infectious agents and more vulnerable to diseases.

> **KEY CONCEPTS**
>
> By fighting infectious microorganisms and cancer, the immune system helps maintain homeostasis; our environment, in turn, profoundly affects the function of the immune system.

SUMMARY

Viruses and Bacteria: An Introduction

1. Two of the most important common infectious agents are viruses and bacteria. Viruses consist of a nucleic acid core, made up of either DNA or RNA, and an outer protein coat. Viruses must invade cells to reproduce.

2. Viruses most often enter the body through the respiratory and digestive systems and spread from cell to cell in the bloodstream and lymphatic system. However, other avenues of entry are also possible—for example, sexual contact.

3. Bacteria are single-celled microorganisms that consist of a circular strand of DNA and cytoplasm enclosed by a plasma membrane.

4. Unlike viruses, bacteria can be killed with antibiotics. Unfortunately, overuse of antibiotics in medicine has created a growing problem: antibiotic resistance in bacteria. Many thousands of people die each year of infections caused by antibiotic-resistant bacteria that they contract in hospitals or in other places.

5. Although they are best known for their role in causing sickness and death, most bacteria perform useful functions.

The First and Second Lines of Defense

6. The first line of defense against infectious agents is the skin and the epithelia of the respiratory, digestive, and urinary systems. Some epithelia also produce protective chemical substances that kill microorganisms.

7. The second line of defense consists of cells and chemicals that the body produces to combat infectious agents that penetrate the epithelia. One of the chief agents in the second line of defense is the macrophage. Located in connective tissue beneath epithelia, macrophages phagocytize infectious agents, preventing their spread. Neutrophils and monocytes also invade infected areas from the bloodstream and destroy bacteria and viruses.

8. Another combatant in the second line of defense consists of the chemicals released by damaged tissue, which stimulate arterioles in the infected tissue to dilate. The increase in blood flow raises the temperature of the wound. Heat stimulates macrophage metabolism, accelerating the rate of the destruction of infectious agents. Heat also speeds up the healing process.

9. Still other chemicals increase the permeability of the capillaries, causing plasma to flow into the wound and increasing the supply of nutrients for macrophages and other protective cells.

10. Another part of the secondary line of defense are the pyrogens, chemicals released primarily by macrophages exposed to bacteria, which raise body temperature and lower iron availability, thus decreasing bacterial replication.

11. Interferons, a group of proteins released by cells infected by viruses, are also part of the second line of defense. Interferons travel to other non–virus-infected cells, where they inhibit viral replication.

12. The blood also contains the complement proteins, which circulate in an inactive state, becoming activated only when the

body is invaded by bacteria. Some of the complement proteins stimulate the inflammatory response. Others embed in the plasma membrane of bacteria, creating holes and killing these pathogens.

The Third Line of Defense: The Immune System

13. The immune system consists of billions of lymphocytes that circulate in the blood and lymph and take up residence in the lymphoid organs and lymphoid tissues.

14. Lymphocytes recognize antigens foreign cells like bacteria and parasites because their membranes contain a unique set of proteins and large molecular-weight polysaccharides unlike that found in the cells of the body. This allows lymphocytes to distinguish between self and not-self.

15. T and B cells are two types of lymphocytes produced in red bone marrow. Immunocompetent B cells encounter antigens (often presented to them by macrophages) to which they are programmed to respond. They then begin to divide, forming plasma cells and memory cells.

16. Plasma cells produce antibodies.

17. The memory cells enable the body to respond more quickly to future invasions by the same antigen.

18. Antibodies are small protein molecules that bind to specific antigens, destroying them either by precipitation, agglutination, or neutralization, or by activation of the complement system.

19. When T and B cells first encounter an antigen, they react slowly. The initial response is called the primary response. Because the body responds slowly at first, there is often a period of illness before the pathogen is removed by the immune system.

20. Numerous memory cells produced during the first assault ensure that a reappearance of the antigen will elicit a much faster and more powerful reaction, the secondary response. The resistance created by a response to an antigen is called immunity.

21. When activated by an antigen, T cells multiply and differentiate, forming memory T cells, cytotoxic T cells, helper T cells, and regulatory T cells, whose functions are summarized in Table 14-3.

22. A solution containing a dead or weakened virus, bacterium, or bacterial toxin that is injected into people to create active immunity is called a vaccine.

23. Vaccines have helped to virtually eliminate many fatal or crippling infectious diseases like polio and German measles (rubella) in more developed nations, prompting some individuals to not vaccinate their children.

24. Unfortunately, the microorganisms that cause these diseases are still found in less developed nations and could spread quickly thanks in large part to international travel, suggesting that vaccines should be continued even in more developed nations.

25. Passive immunity can be achieved by injecting antibodies into a patient or by the transfer of antibodies from a mother to her baby through the bloodstream or breast milk. Passive immunity is short-lived, lasting at most only a few months, compared to active immunity, which lasts for years.

26. Tissue transplantation requires careful matching of donor and recipient. In such instances, only cells from the same individual or an identical twin will be accepted. All others are rejected by the T cells, unless the system is suppressed with drugs.

Diseases of the Immune System

27. The most common malfunctions of the immune system are allergies, extreme reactions to some antigens referred to as allergens.

28. Allergies are caused by IgE antibodies, produced by plasma cells.

29. IgE antibodies first bind to histamine-containing mast cells. The allergens then bind to the antibodies. This causes the mast cells to release massive amounts of histamine and other chemical substances that induce the symptoms of an allergy—production of mucus, sneezing, and itching.

30. Autoimmune diseases, another immune system disorder, result from an immune attack on the body's own cells.

31. Autoimmune diseases may occur when normal proteins are modified by chemicals or genetic mutations so that they are no longer recognizable by the body. Other causes are the sudden presence of proteins that are normally isolated from the immune system and exposure to antigens that are nearly identical to body proteins.

32. To date, around 50 diseases have been classified as autoimmune disorders and numerous others are suspected as being caused by autoimmunity.

Health and Homeostasis

33. The immune system and protective mechanisms that constitute the first and second lines of defense protect the body from harmful viruses and microbes. They help maintain a constant internal state either directly, by warding off infectious agents, or secondarily, by protecting other body cells that are essential to homeostasis.

34. Chemicals in the environment can damage the immune system, upsetting homeostasis. Stress and poor nutrition can also lower our immunity.

THINKING CRITICALLY ANALYSIS

This analysis corresponds to the Thinking Critically scenario that was presented at the beginning of this chapter.

This chapter offers a brief overview of the topic of vaccination. The first conclusion that you might draw is that the text does not provide enough information to analyze the claims made by your friends about vaccination. You'd be advised to study an immunology book as well as the scientific literature. Once you've read some more, you will be better able to discern whether vaccination is worth the relatively small risk.

Despite the brief coverage, this chapter does point out that infectious diseases have not been eliminated. Vaccination has only kept them under control by reducing the potential population of hosts. Thus, if large numbers of people are unvaccinated against infectious disease, outbreaks could occur. The results could be quite dramatic because we're living on a crowded planet, and infectious organisms spread quickly in crowded environments.

KEY TERMS AND CONCEPTS

Active immunity, p. 310
Allergens, p. 313
Allergy, p. 313
Anaphylactic shock, p. 314
Antibodies, p. 305
Antigens, p. 304
Asthma, p. 313
Autoimmune disease, p. 314
Bacteria, p. 299
Bacterial toxin, p. 305
Bacterium, p. 299
B lymphocyte (B cell), p. 304
CD4 cells, p. 309
CD4+ T cells, p. 309
Complement protein, p. 303
Complement system, p. 303
Cytotoxic T cell, p. 308

Dendritic cell, p. 310
Dermis, p. 301
Epidermis, p. 301
Helper T cell (T_h cell), p. 309
Histamine, p. 301
Humoral immunity, p. 305
Immune system, p. 304
Immunocompetence, p. 305
Immunoglobulin, p. 307
Inflammatory response, p. 301
Interferon, p. 303
Lymphoblast, p. 305
Lymphocyte, p. 305
Lymphoid organs, p. 305
Membrane-attack complex, p. 303
Memory T cell, p. 309
Natural killer (NK) cell, p. 309

Passive immunity, p. 310
Perforin-1, p. 308
Plasma cell, p. 305
Plasmid, p. 299
Primary response, p. 305
Prostaglandin, p. 301
Pus, p. 301
Pyrogen, p. 301
Regulatory T cell, p. 309
Secondary response, p. 306
Suppressor T cell, p. 309
Thymus, p. 305
T lymphocyte (T cells), p. 304
Vaccine, p. 310
Virus, p. 298

CONCEPT REVIEW

1. Describe the structure of a typical virus. p. 298.
2. Based on your previous studies, in what ways are bacteria similar to human cells, and in what ways are they different? p. 299.
3. The human body consists of three lines of defense. Briefly describe what they are and how they operate. pp. 301–310.
4. Describe the inflammatory response, and explain how it protects the body. You may want to draw a diagram showing all aspects of the inflammatory response. pp. 301–303.
5. Define each of the following terms, and explain how they help protect the body: pyrogen, interferon, and complement proteins. pp. 301–303.
6. The first and second lines of defense differ substantially from the third line of defense. Describe the major differences. pp. 301–310.
7. How does the immune system detect foreign substances? What role do macrophages play? p. 304.

8. Define the term immunocompetence. Why is this process necessary for normal immunity? p. 305.
9. Describe how the B cell operates. Be sure to include the following terms in your discussion: bone marrow, plasma membrane receptors, primary response, plasma cell, antigen, antibody, and secondary response. pp. 305–307.
10. Describe the four mechanisms by which antibodies "destroy" antigens. pp. 307–308.
11. Describe the events that occur after a T cell encounters its antigen. Be sure to include the terms helper T cell, cytotoxic T cell, and regulatory T cell in your answer. pp. 308–309.
12. What role does the macrophage play in immunity? p. 310.
13. What is the difference between active and passive immunity? Give some examples of each. How are these two processes different? Which one lasts the longest? pp. 310–311.
14. What is a vaccine? Why are vaccines given? Why do vaccines not cause

disease when injected into a person? p. 310.
15. Spend some time on the Internet reading information on web sites of those who favor vaccination and those who oppose vaccination. Make a list of the key points made by each group? Which viewpoint do you find most scientifically accurate? p. 311.
16. A child is stung by a bee, swells up, and collapses, having great difficulty breathing. What has happened? What can be done to save the child's life? p. 314.
17. What are some of the main advantages of breastfeeding? p. 312.
18. What is an allergy? Is it a normal immune system reaction? Why or why not? pp. 313–314.
19. Explain how histamines are released during allergic reactions and the effects they have on the body. pp. 313–314.
20. What is an autoimmune disease? Explain the reasons it forms. pp. 314–315.

SELF-QUIZ: TESTING YOUR KNOWLEDGE

1. Viruses contain a nucleic acid core consisting of either _____ or _____. p. 298.
2. Bacteria are living cells, but lack organelles and true _____. p. 299.
3. _____ respond to antibiotics, but viruses do not. p. 299.
4. Viral infections may lead to secondary _____ infections that require antibiotics to treat. p. 299.

5. The _____ and epithelial linings of the respiratory, urinary, and digestive tracts are part of the first line of defense against potentially harmful microbes. p. 301.
6. The acidic secretions of the skin and stomach help destroy bacteria and other microorganisms and are therefore part of the _____ line of defense. p. 301.

7. The _____ response is a series of changes characterized by redness, swelling, and pain that protect us from potentially harmful microbes after a cut or abrasion in the skin. p. 301.
8. _____ is a pain-causing chemical released by injured cells. p. 301.
9. _____ is a chemical released by virus-infected cells. It prevents viruses

from successfully replicating in other cells. p. 301.

10. The membrane-attack complex is formed by proteins of the _____ system. This structure creates pores in the membranes of bacteria, causing them to leak, swell, and burst. p. 303.

11. _____ lymphocytes are responsible for cell-mediated immunity, while _____ lymphocytes are responsible for humoral immunity. p. 305.

12. Lymphocytes cannot respond to antigens until they become _____. p. 305.

13. B lymphocytes are converted to _____ cells, which produce antibodies. p. 305.

14. The _____ response is more powerful and swifter than the first response to an antigen. p. 306.

15. Antibodies belong to a group of plasma proteins known as _____. p. 307.

16. Helper T cells _____ the immune response. p. 309.

17. Once an infection is beginning to subside, _____ T cells help to reduce the immune response. p. 309.

18. Vaccines provide _____ immunity, while injections of antibodies provide _____ immunity. p. 310.

19. Allergies are an abnormal immune reaction caused by the production of_____ antibodies from plasma cells. p. 313.

20. One of the advantages of breast feeding is that it provides _____ to the newborn that helps protect it from microorganisms. p. 311.

biology.jbpub.com/chiras/8e/

The site features eLearning, an online review area that provides quizzes, chapter outlines, and other tools to help you study for your class. You can also follow useful links for in-depth information, research the differing views in the Point/Counterpoints, or keep up on the latest health news.

Acquired Immunodeficiency Syndrome

In 1987, Damion Knight, a bright, young cabinetmaker, began to lose weight and experience bouts of unexplained fever. His lymph nodes became swollen. He complained of feeling weak and drowsy. A doctor found that Damion had **acquired immunodeficiency syndrome**, commonly known as **AIDS**. Like millions of other people worldwide, Knight died several years after his diagnosis.

The Human Immunodeficiency Virus

AIDS is a deadly sexually transmitted disease. In fact, in the United States AIDS is currently the sixth leading cause of death among individuals ages 25 to 44. In 1995, it was the number one cause.

AIDS is caused by a virus (Figure 1a), known as the **human immunodeficiency virus (HIV)**. HIV is an RNA virus that attacks helper T cells, described elsewhere in the text. Because the helper T cells dramatically boost immunity, HIV seriously weakens the immune system. HIV infection, therefore, leaves its victims highly vulnerable to a number of life-threatening infections such as pneumonia from common bacteria, yeast, parasites, and viruses that do not normally cause serious disease in individuals with healthy immune systems. AIDS progresses through stages, as you shall soon see. In the last stage, untreated patients grow progressively weaker and generally fall victim to other infectious agents. For example, many die from an otherwise rare form of pneumonia. HIV also leaves its victims vulnerable to certain types of cancer (discussed shortly).

HIV comes in several varieties. In the United States, HIV-1 is the prevalent form. In Africa, HIV-2 predominates. What is more, HIV is a highly mutable virus. That is, it mutates inside the body. This makes it more difficult to develop a preventive vaccine. (More on this shortly.)

KEY CONCEPTS

HIV is a sexually transmitted disease that causes AIDS, a lethal immune system disorder.

Where Did HIV Come From?

AIDS rose to infamy in the early 1980s. But the first documented case was that of a young Missouri boy who died at age 15 in 1969. The first known case of AIDS, however, appeared

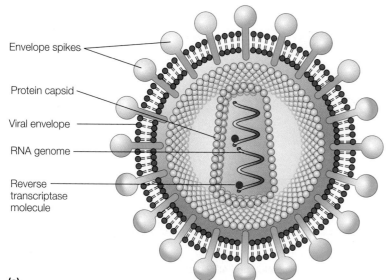

Envelope spikes
Protein capsid
Viral envelope
RNA genome
Reverse transcriptase molecule

(a)

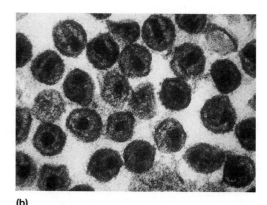

(b)

FIGURE 1 **The Human Immunodeficiency Virus** (a) The virus consists of two molecules of RNA and molecules of reverse transcriptase. A protein capsid surrounds the genome. (b) A photomicrograph of HIV. (© Science VU/Visuals Unlimited.)

10 years earlier. Blood samples taken in 1959 from a man from the Congo showed that he was infected by HIV. Studies suggest that AIDS may have actually begun spreading among humans as long ago as 1930.

Research suggests that HIV-1 came from wild chimpanzees in southern Cameroon, Africa. Chimpanzees are infected with a strain of simian immunodeficiency virus (SIVcpz) that is very similar to HIV-1. Researchers speculate that the virus was transferred to humans either by exposure to the blood of chimps killed by humans living in the area or by consumption of the meat.

KEY CONCEPTS

HIV very likely entered the human population from eating the meat or being exposed to the blood of chimpanzees in west-central Africa that carry a similar virus.

The Rising Incidence of AIDS

In the United States, the first cases of AIDS were reported in 1981. According to the Centers for Disease Control and Prevention's (CDC) latest statistics, in 2010 approximately 1.2 million people were living with HIV in the United States. Unfortunately, one in five of these individuals do not know they have contracted the virus, creating a potential for further spread of the disease. According to the CDC, every 9.5 minutes someone becomes infected with HIV.

AIDS has taken a huge toll. Since 1981, for instance, nearly 620,000 American men, women, and children have died from AIDS.

As shown in Figure 2, however, there is some encouraging news. Although the number of people living with HIV continues to increase, the number of newly diagnosed cases of AIDS has remained pretty much the same since the 1990s thanks to prevention.

Additional encouraging news is shown in Figure 3. As you can see, the top line in this graph shows the number of individuals diagnosed with AIDS each year. As illustrated, this number has decreased dramatically since the late 1980s and early 1990s. That said, there are approximately 40,000 new

- People living with HIV
- New HIV infections using back-calculation methodology
- New HIV infections using incidence surveillance methodology
- New HIV infections using updated incidence surveillance methodology

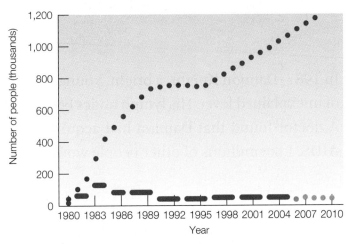

FIGURE 2 HIV Prevalence and Incidence (1980–2010) (Reproduced from Estimates of New HIV Infections in the United States, 2007–2010/CDC.)

cases diagnosed each year. In 2010, for instance, the CDC reported 47,500 new cases. As shown in Table 1, most new cases appear between the ages of 20 and 49.

Figure 3 also shows that 17,000 to 20,000 people die of AIDS each year.

Among men, the hardest hit groups are African-Americans, Hispanics, and caucasians (white), as shown in Table 2. Nearly two-thirds of all new cases occur in these groups. AIDS is now a leading killer of African-American males. In 2010, for instance, African-Americans, who make up approximately 14% of the U.S. population, accounted for slightly more than half of the estimated number of HIV cases diagnosed.

HIV infections and AIDS are on the rise worldwide, prompting one researcher to liken AIDS to the bubonic plague that spread through Europe in the fourteenth and fifteenth

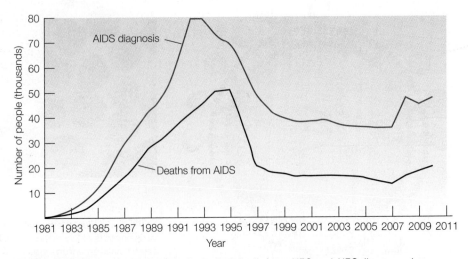

FIGURE 3 AIDS Diagnoses This graph shows that death from AIDS and AIDS diagnoses have dropped substantially. Even so, people still contract and die from the disease each year.

TABLE 1	Estimated Number of New Cases of HIV Infection in the United States in 2010
Age (Years)	Estimated Number of Diagnoses of HIV Infection
< 13	217
13–14	34
15–19	2,200
20–24	7,565
25–29	6,823
30–34	5,954
35–39	5,523
40–44	5,720
45–49	5,296
50–54	3,671
55–59	2,154
60–64	1,119
≥ 65	853

Source: Centers for Disease Control and Prevention

TABLE 2	Estimated Number of Diagnoses of HIV Infection, 2010
Race or Ethnicity	Estimated Number of Diagnoses of HIV Infection, 2010
American Indian/Alaska Native	225
Asian	814
Black/African American	21,854
Hispanic/Latino[a]	9,653
Native Hawaiian/Other Pacific Islander	64
White	13,878
Multiple Races	642

[a]Hispanics/Latinos can be of any race.
Source: Centers for Disease Control and Prevention

centuries. In the 1300s, the plague killed one-quarter of the adult population and untold children. By comparison, as many as 50% of the residents in some African villages today are testing HIV-positive. More than one-third of the population of Botswana and Swaziland has AIDS.

HIV infection is a global epidemic. To date, approximately 30 million people have died from AIDS, worldwide, according to the United Nations. As of July 2012, an estimated 34.2 million people worldwide were infected with HIV, according to the United Nations Programme on AIDS (UNAIDS)—up 18% from a decade earlier (2001) at which time 28.9 million were living with HIV. Of those 34.2 million living with HIV, an estimated 3.3 million are under the age of 15, according to Amfar, the Foundation for AIDS Research. According to the UN, there were 2.5 million new HIV infections in 2011, as well, including an estimated 330,000 among children. That's nearly 300 people every hour.

Women account for about half of all people living with HIV. Children represent nearly 7% of the total.

Some areas have been hit harder by AIDS than others. For example, more than two-thirds (69%) of the people living with HIV worldwide are in sub-Saharan Africa. That's 23.5 million people and 91% of the world's HIV positive children. Other areas of the world are also witnessing a rise in AIDS cases. Myanmar, Viet Nam, Cambodia, and India, for example, are experiencing a rapid increase in AIDS. Eastern Europe has witnessed a dramatic increase as well.

KEY CONCEPTS

The number of AIDS cases increased dramatically in the United States in the 1980s but has slowed significantly in recent years.

Stages of AIDS

As shown in Figure 4, AIDS progresses through three phases: acute infection, clinical latency, and full-blown AIDS.

Acute Phase. Within two to four weeks of being infected with HIV, some HIV-infected individuals experience symptoms of the flu—that is, fever, chills, aches, and swollen lymph nodes. In fact, it is often described as the "worse flu ever." It is the body's response to the virus. In some people, flu-like symptoms appear up to three months after infection,

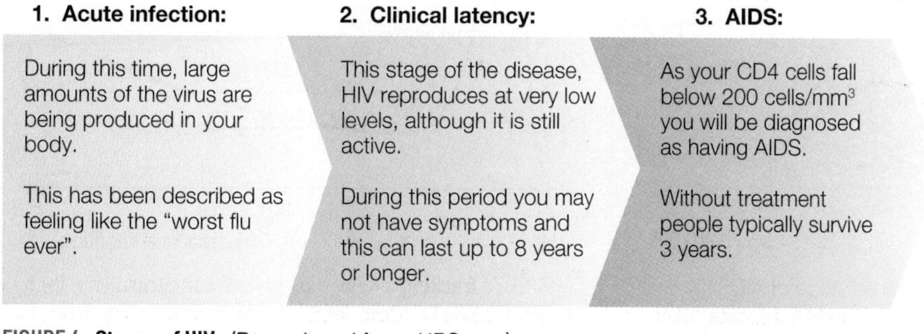

1. Acute infection:

During this time, large amounts of the virus are being produced in your body.

This has been described as feeling like the "worst flu ever".

2. Clinical latency:

This stage of the disease, HIV reproduces at very low levels, although it is still active.

During this period you may not have symptoms and this can last up to 8 years or longer.

3. AIDS:

As your CD4 cells fall below 200 cells/mm^3 you will be diagnosed as having AIDS.

Without treatment people typically survive 3 years.

FIGURE 4 **Stages of HIV** (Reproduced from AIDS.gov.)

but many HIV-positive individuals develop no symptoms. As a result, the acute phase is also sometimes referred to as the asymptomatic phase, meaning the individual exhibits no symptoms (Figure 5a). The acute phase may last for 2 to 15 years.

To study the progression of the disease, take a look at Figure 6. As you can see, this figure consists of three panels. The top panel plots the number of helper T cells (also called T4 cells) during the three stages of AIDs. As you may recall, T4 cells (also known as CD4 cells) enhance the immune response. They are the most abundant of all the T cells. Some immunologists liken them to the immune system's master switch. Without them, antibody production and T-cell activity are greatly reduced. Symptoms patients experience are listed in the middle panel. The bottom panel shows the concentration of HIV and the HIV antibody levels.

As illustrated in Figure 6, during the acute phase, the T4-cell count is initially quite high (Figure 6a)—as you would expect in a normal, healthy individual. Late during this phase, however, the T4 cell count begins to decline.

(a)

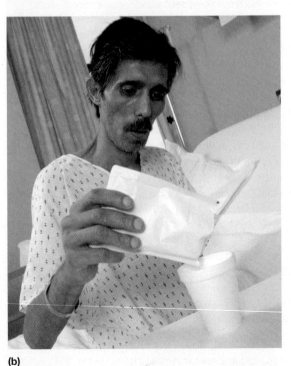

(b)

FIGURE 5 **Phases of AIDS** (a) Acute stage of AIDS. (© Gary Conner/PhotoEdit, Inc.) (b) Patient with full-blown AIDS. (© Yoav Levy/Phototake/Alamy Images.)

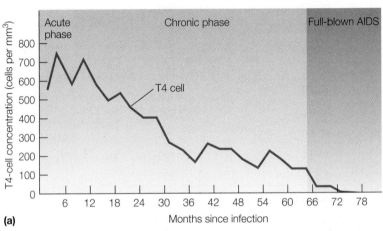

(a)

Symptoms

Asymptomatic Phase	AIDS-Related Complex (ARC)	Full-Blown AIDS
Lasts 9 months or more	10 to 63 months	64 to 83 months
Typically no symptoms	Fatigue	Swollen lymph nodes
	Persistent and recurrent fever	Persistent infections
	Persistent cough	Extreme loss of weight
	Diarrhea	Severe weakness
	Loss of memory	Persistent diarrhea
	Difficulty thinking	Persistent cough
	Depression	Kaposi's sarcoma
	Impaired judgement	

(b)

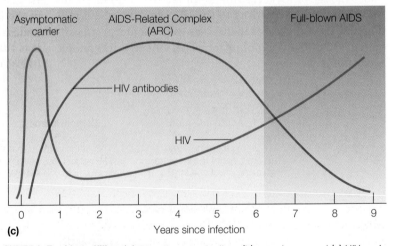

(c)

FIGURE 6 **Tracking a Killer** (a) T4-cell concentration, (b) symptoms, and (c) HIV and antibody levels. (Data for parts a and c from Redfield, R. R., and Burke, D. S., "HIV Infection: The Clinical Picture," *Sci. Amer.* 259 [October 1988]: 90–98.)

As illustrated in Figure 6b, most patients show no symptoms during the first few months after infection, as just noted. They appear quite healthy (Figure 5a). Even if a patient experiences chills and fever, these symptoms vanish shortly thereafter, and individuals go on about their business, often unaware that they have contracted a deadly disease—and in such cases unaware that they're transmitting it to others.

As Figure 6c shows, the number of viruses (blue line) rises rapidly during the chronic phase. Antibodies to HIV begin to increase as a result. Antibodies, as you may know, are proteins produced by certain cells of the immune system, notably plasma cells. They destroy infectious agents, or at least attempt to do so.

Clinical Latency. During the second phase of this disease, clinical latency, the T4-cell count begins to plummet as shown in Figure 6a. This decrease is a result of the slow but steady proliferation of HIV in the body, as shown in Figure 6c.

While HIV proliferates, the body produces increasing amounts of antibody. However, as shown in Figure 6c, antibodies levels rise for some time, then plateau and begin to decline. That's the result of the continued decline in helper T-cell population. As they die off, the immune system begins to falter.

During the second phase, patients begin to show outward signs of the disease caused by the decline in the number of T4 cells. As shown in Figure 6b, patients complain of severe fatigue and unexplained, persistent fever. Some patients experience a persistent cough and loss of memory, difficulty thinking, and depression. When recurring infections set in, the third and final phase of AIDS is about to begin.

Full-Blown AIDS. The final phase, known as **full-blown AIDS**. As illustrated in Figure 6c, T4 cells have been annihilated, and HIV antibodies have decreased accordingly. HIV is thriving in this new nearly totally immune-suppressed environment.

Patients suffer from persistent infections, extreme loss of weight, and weakness (Figure 5b). Most patients succumb to one of a handful of noninfectious organisms—microbes not ordinarily capable of producing serious infections. When a patient is in a state of extreme immune compromise, though, these organisms take hold and become life-threatening.

Without treatment, people who are diagnosed with AIDS typically survive about 3 years, according to the CDC. Once someone develops a dangerous opportunistic infection, however, life-expectancy typically decreases to about 1 year.

> **KEY CONCEPTS**
> AIDS is a disease that progresses through three stages.

Cancer, Brain Deterioration, and HIV

HIV affects more than a person's immune system. AIDS patients, for example, often contract a rare form of cancer called *Kaposi's sarcoma* (kah-PO-sees sar-KOME-ah; Figure 7). Kaposi's sarcoma (KS) primarily appears in the skin, mouth, and lymph nodes, but can appear in the lungs and GI tract. HIV is now the most common cause of this cancer.

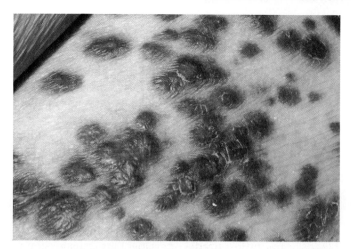

FIGURE 7 Kaposi's Sarcoma This cancer is a serious opportunistic disease of the HIV-infected person. (Courtesy of National Cancer Institute.)

The most visible signs of Kaposi's sarcoma are lesions on the skin. These non–life-threatening tumors appear as red or purple spots on Caucasian skin or bluish, brown, or black lesions on dark skin. Flat and painless, these lesions may enlarge, forming raised bumps. Some grow together. In some patients, the lesions change very slowly; in others, new lesions may appear weekly.

Researchers have found that Kaposi's sarcoma is caused by a strain of the herpes virus known as *human herpes virus 8* (HHV8). It is believed to be transmitted by deep kissing, kissing that involves a lot of contact with saliva. When the immune system is impaired, this virus proliferates and eventually leads to Kaposi's sarcoma. Most severe cases of Kaposi appear late in the progression of AIDS, although some KS skin lesions can appear rather early.

Kaposi's sarcoma appearing in the skin can be treated locally like other forms of skin cancer. Tumors appearing inside the body are treated systemically, that is, by administering tumor-fighting medications—usually combinations of these chemotherapeutic drugs—that circulate through the bloodstream.

AIDS patients also experience a number of neurological disorders, beginning in the second phase, as noted earlier. These include memory loss and progressive mental deterioration. Why? Researchers have found that inside helper T cells, HIV produces several proteins that are incorporated into the viral capsid. One of those proteins is known as gp120. This protein, researchers believe, travels in the blood to the brains of some patients, where it kills neurons, thus producing neurological defects.

Loss of mental function may also be caused by a single-celled parasite that is normally found in cats but sets up residence in people whose immune systems are compromised by HIV. This parasite causes a brain infection (encephalitis) that leads to a loss of brain cells, seizures, and weakness.

> **KEY CONCEPTS**
> Besides attacking and weakening the immune system, HIV causes a rare form of cancer and neurological disorders resulting from deterioration of the brain.

HIV Transmission

Research has shown that HIV is passed from one person to the next primarily by two routes: sexual contact and contaminated needles shared by intravenous drug abusers.

The AIDS virus can be detected in a number of body fluids and tissue of HIV-positive individuals. "It is important to understand, however, that finding a small amount of HIV in a body fluid or tissue does not mean that HIV is transmitted by that body fluid or tissue," says the Centers for Disease control. They go on to point out that only specific fluids—blood, semen, vaginal secretions, and breast milk—containing the HIV virus can transmit it to another. Moreover, they point out, these fluids must come in direct contact with a mucous membrane or damaged tissue or be directly injected into the bloodstream (from a needle or syringe) for transmission to possibly occur.

In the United States, HIV is most often transmitted by sexual behaviors, notably anal or vaginal sex, or by sharing needles with an HIV-positive individual, according to the CDC. HIV is less commonly transmitted via oral sex. It is also rare for HIV to pass from an HIV-positive woman to her baby before or during childbirth or after birth through breastfeeding or by prechewing food for her infant.

The CDC notes that it is also possible to acquire HIV through exposure to infected blood, transfusions of infected blood, blood products, or organ transplants, though they point out that "this risk is extremely remote due to rigorous testing of the U.S. blood supply and donated organs."

As shown in Figure 8a, in the United States, male-to-male sexual contact is the leading cause of transmission among men. This occurs between gay and bisexual men. While gay, bisexual, and other men who have sex with men constitute only about 2% of the U.S. population, they accounted for 61% of all new HIV infections in 2009. Among this group, African Americans bear the greatest disproportionate burden of HIV, according to the CDC. Between 2006 to 2009, HIV infections in this group increased 48%.

Surprisingly, 15% of all cases in men result from heterosexual contact—that is, infection of men from an HIV-positive female partner. A nearly equal number of cases result from the use of contaminated needles (injection drug use).

As shown in Figure 8b, 80% of all cases in women occur as a result of heterosexual contact—that is, sexual relations with an infected male partner. The vast majority of the remaining cases result from exposure via contaminated needles. According to data gathered by the CDC, women constituted 24% of all new cases of HIV infection in the United States in 2009. In 2008, the CDC reports, an estimated 25% of all adults and adolescents living with HIV infection were female. Among women, black and Latina women are disproportionately affected, as shown in Figure 9.

HIV can also be transmitted from an infected mother to her baby through the placenta and breast milk, though this is now rare thanks to better medical treatment. In 1990, 2,000 babies were born in the United States with an HIV infection they acquired from their mothers. In 2003, the number had dropped to 200. Two years later, the number had dropped to about 50. This dramatic decline was largely attributed to the treatment of infected mothers with a drug known as azidothymidine (AZT) during pregnancy.

According to researchers, about 25% of the babies of HIV-infected women would be born infected if they receive no treatment. However, the risk of transmission decreases to approximately 4% if a woman receives the drug AZT during pregnancy and delivery, and her baby is given AZT as well. The risk drops to 2% or less if the mother receives combination antiretroviral therapy (ART)—that is, a combination of anti-HIV drugs. There's no evidence that HIV can be transmitted in saliva or by sharing utensils, plates, or cups. In fact, laboratory studies suggest that even though HIV can be found in the saliva of infected individuals, saliva has natural components that reduce HIV's ability to infect cells. Researchers have found that it is possible to become infected through oral sex, although the risk is much lower than that posed by

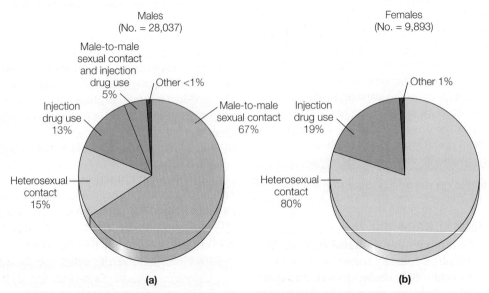

(a)   (b)

FIGURE 8 Routes of Infection (a) In men, the AIDS virus, HIV, is primarily transmitted via male-to-male sex. (b) In women, the main route of transmission is heterosexual sex, having sex with an infected male partner. (Data for a and b courtesy of CDC.)

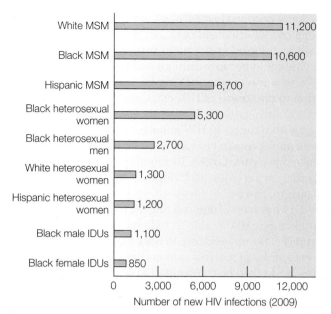

FIGURE 9 New HIV Infections (2009) MSM = men who have sex with men; IDUs = intravenous drug users. (Reproduced from *Estimates of New HIV Infections in the United States, 2007–2010*/CDC.)

unprotected sexual intercourse with a man or woman. Deep kissing is also an extremely low risk activity, with only one confirmed case of HIV infection.

Individuals with the genetic disorder hemophilia were once at risk for AIDS. Hemophiliacs are given clotting factors from pooled human plasma. Before 1984, blood donors were not screened for HIV. Consequently, many of the blood clotting preparations were contaminated with this deadly virus. As a result, a majority of the estimated 15,000 hemophiliacs in the United States who received clotting factors between 1975 and 1984 contracted AIDS and died.

To prevent the spread of AIDS through blood transfusions, blood is now routinely tested for HIV. Tissues and organs for transplantation are also tested. Improvements in screening have dramatically reduced the risk of HIV infection. However, despite improvements in screening, blood transfusion is not a fail-safe proposition. Individuals who will need blood for an operation are therefore encouraged to donate some of their own blood for transfusion ahead of time.

> **KEY CONCEPTS**
>
> HIV is transmitted in a number of ways, though, sexual contact and contaminated needles are the two leading causes.

Preventing the Spread of AIDS

People recently infected with HIV, that is, people exposed within 2 or 3 months, are much more likely to transmit HIV than others. Why?

As you examine Figure 6c, you will see that their viral load—the amount of HIV in their body—is the higher during this period than any other, except very late in the progression of AIDS when patients are severely ill. As a general rule, then, the higher the viral load, the higher the risk of transmission.

HIV, although lethal, does not spread as readily as the flu virus or cold viruses; individuals can protect themselves by practicing sexual abstinence before marriage, by engaging in safe sex (using condoms, for example), and by avoiding multiple sexual partners. According to AIDS.org, a leading educational group, having sex in a monogamous relationship is safe if both partners are uninfected (that is, HIV-negative), sex is confined to the relationship, and neither partner gets exposed to HIV through drug use or other routes. This pertains to both homosexual and heterosexual relationships.

The spread of HIV among intravenous drug users can be reduced by seeking help for addiction so an individual can cease using them. Users can prevent infection by not injecting drugs and not sharing needles and syringes and other "equipment." If they must share, they should clean needles and syringes with bleach and water before every use. To prevent the spread of HIV among intravenous drug users, some countries and some U.S. cities distribute clean hypodermic needles to addicts.

As noted previously, pregnant women infected with HIV can be treated with the drug AZT (described shortly) and other drug combinations during pregnancy to reduce the likelihood of infecting the fetus. If the baby is delivered by cesarean section (through an incision made in the lower abdomen), its chances of contracting the disease can be reduced dramatically. It is important for an HIV-positive woman who becomes pregnant to begin prenatal care immediately. The sooner she does, the lower the chance that her offspring will become infected.

A recent report suggests that universal circumcision of men and boys in sub-Saharan Africa could dramatically reduce HIV infections. This practice, which is rare in sub-Saharan populations, could prevent 2 million new infections and avert 300,000 deaths over the next 10 years. Over the following decade, such efforts could prevent 3.7 million additional HIV infections and 2.7 million deaths. According to the researchers, circumcision reduces the risk of a sexually active man acquiring HIV by more than 50%. Why? Removing the foreskin eliminates cells that are easily infected.

> **KEY CONCEPTS**
>
> AIDS can be prevented by a number of measures, including sexual abstinence and the use of condoms.

Promising Developments

For years, healthcare workers have determined the presence of HIV via an immunologic test, which detects antibodies to HIV in the blood. The results, however, were not available for up to 2 weeks. In 2002, the Food and Drug Administration approved a new and much quicker test for HIV. This test, known as OraQuick Rapid HIV-1 Antibody Test, is 99.6% accurate. It requires a single drop of blood and produces results within 20 minutes.

Researchers have also developed a new, more sensitive test that detects antigens—proteins in the HIV coat—that has greatly improved the screening of blood and tissue. There are even tests individuals can use at home which require a pin prick to draw blood or a mouth swab to sop up antibodies to HIV.

Unfortunately, there's no cure for AIDS. However, since the 1980s, numerous drugs have been developed that, when used in combination, can help to control the virus. These drugs fall into one of five categories, based on how they work.

Each of them blocks the virus at some stage in its life cycles, but in different manners. Best results occur when doctors combine at least three drugs from two different classes. This helps prevent the creation of drug-resistant strains of HIV—viruses that are resistant to a single medication.

The first group consists of drugs that disable a specific enzyme needed by HIV to make copies of itself. The second group consists of drugs that are building blocks needed by HIV to make copies of itself inside cells. However, these are faulty or incomplete building blocks so the resultant HIV is faulty. A third group consists of medications that inhibit protease, another enzyme needed to make copies of itself. The fourth group blocks HIV's entry into T4 helper cells. The fifth group inhibits the enzyme integrase, which is required by HIV to insert its genetic material into helper cells.

Currently, more than three dozen drugs are used to combat HIV and to treat infections, Kaposi's sarcoma, and other complications including weight loss. The first on the market and the most widely known drug is AZT. AZT works by blocking one of the key enzymes (reverse transcriptase) needed by HIV to make copies of itself. Reverse transcriptase allows the HIV to make DNA on its RNA. The DNA is then inserted into a host cell's chromosomes, and there successfully converts the cell to a virus-producing factory. Protease inhibitors block another enzyme (protease) that is also vital in the replication of HIV in infected cells.

AZT and other similar drugs reduce, even halt, the proliferation of HIV in the body. However, as noted above, HIV can develop a resistance to AZT and other drugs when used alone. That's why a combination treatment is required to effectively suppress the virus.

In the 1990s, an American diagnosed with the AIDS virus could expect to live less than 10 years and easily spend $15,000 a year just on medication. In 1993, for instance, life expectancy was a mere 7 years. Today, thanks to new drugs and other advances in AIDS treatment, patients diagnosed with AIDS can expect to live, on average, approximately 24 years. The cost of health care, including medicine, during this period is more than $600,000. Researchers estimated that the monthly cost of care was $2,100, two-thirds of which was spent on medications.

> **KEY CONCEPTS**
>
> The battle against AIDS has been facilitated by new, accurate, and fast screening tests and by drugs that slow the replication and spread of HIV.

HIV Vaccine

Researchers hope eventually to be able to cure AIDS and prevent HIV infections by administering vaccines. As you may recall, vaccines for viral infections are inactivated or altered viruses (or parts of them) that stimulate an immune response. They are used to prevent a host of diseases.

Currently, researchers at numerous universities, drug and biotech companies, and government agencies are working hard to create an HIV vaccine. In fact, according to one source, researchers have spent more money on developing an HIV vaccine than on any other vaccine in human history. Unfortunately, they currently have no vaccine that works. Why is it taking so long?

Unbeknownst to most of us, developing a vaccine is often a long and laborious process. For example, it took scientists 47 years to develop the polio vaccine. Developing an HIV vaccine could be more difficult and time-consuming. Why?

One reason is that HIV replicates very rapidly. Another is that many strains of HIV exist, and new ones continue to arise. Unfortunately, HIV is notorious for its ability to mutate. Even within the body, HIV mutates fairly freely. In one study, for example, researchers analyzed viruses isolated from two infected patients. Over a 16-month period, they found 9 to 17 different varieties, all thought to have been formed from the original virus.

HIV has ways of outsmarting the human immune system. For example, HIV is able to hide in the body. HIV may, for instance, take up residence in bone marrow stem cells—that is, cells in bone marrow that give rise to lymphocytes. The virus could then be transmitted to new white blood cells by cell division. Thus, once the virus is in the body, it may be there forever. Eliminating the virus from the body may be virtually impossible.

Finally, immunologists are still trying to understand how the immune system must behave to prevent HIV infections.

Despite these hurdles, medical researchers remain determined to create a vaccine. Researchers are working on two types of HIV vaccines: preventive and therapeutic.

Preventive HIV vaccines, immunologists hope, will some day be administered to HIV-negative men and women, perhaps even children, and will enable them to prevent an infection. The vaccine, when developed, will produce the proper immune response that allows the body to recognize and destroy HIV before it can establish an infection. To date, there are currently more than 35 trials in 25 countries underway.

A few medical researchers are also working on therapeutic HIV vaccines that could be administered to HIV-positive patients whose immune systems are still healthy. It is hoped that those vaccines will help control the infection, perhaps slowing the progression of the disease by killing HIV-infected cells or by preventing or limiting HIV from replicating.

Researchers are also looking into the DNA of chimpanzees for clues for a cure. As mentioned at the beginning of this section, many wild chimps carry a virus called SIV that is very similar to HIV. Studies suggest that the virus has been in the chimp population for over 10,000 years. Interestingly, however, chimps are not affected by SIV. Because chimps contain 98% of the same genes as humans, studying the genetic differences of the two species could lead to a better understanding of why humans are so devastatingly affected by this virus. This, too, could help researchers develop a cure.

To date, success in treating or curing AIDS via vaccines has been limited. Despite discouraging results, several recent studies in lab animals suggest that a human vaccine may be forthcoming, although it could still take a long time to test the vaccines on humans and, if successful, bring the vaccine to the marketplace.

> **KEY CONCEPTS**
>
> Researchers are primarily working on vaccines that could prevent an individual from contracting AIDS but also on vaccines that could slow the disease in people already exposed to HIV.

SUMMARY

1. AIDS is a disease of the immune system caused by HIV, an RNA virus that attacks helper T cells (T4 cells), severely impairing a person's immune system.

2. HIV is a global disease, affecting the rich and the poor. The highest incidence is in sub-Saharan Africa.

3. AIDS progresses through three stages: acute, clinical latency, and full-blown AIDS.

4. During the first phase, an individual is highly infectious—able to transmit the disease to others. Symptoms may or may not appear.

5. During the first phase, T4 cells decline in number and viruses replicate.

6. During the second phase, patients grow progressively weaker as their immune system falters. Lymph nodes swell and patients report recurrent fevers and a persistent cough. Mental deterioration may also occur.

7. During the second phase, T4 cells decline further and HIV begins a second, long-duration increase in number.

8. During the last phase, patients suffer from severe weight loss and weakness. Many develop cancer and bacterial infections because of their diminished immune response.

9. AIDS is spread primarily through body fluids during sexual contact and through needles shared by drug users.

10. AIDS is a fatal disease, though one's life expectancy after exposure to the virus has more than doubled in the past 20 years thanks to the discovery of numerous HIV-fighting drugs.

11. To be effective, antiHIV drugs are administered in combinations. This prevents the virus from becoming resistant to the medications.

12. Stopping the virus has proved difficult, in large part because symptoms of AIDS do not appear until several months to several years after the initial HIV infection.

13. Numerous researchers are developing vaccines that they hope will protect people and eventually eradicate the virus, but so far, efforts to develop a vaccine have met with limited success.

14. One reason is that creating an HIV vaccine has proven difficult is that HIV is highly mutable. It also seeks shelter in the body.

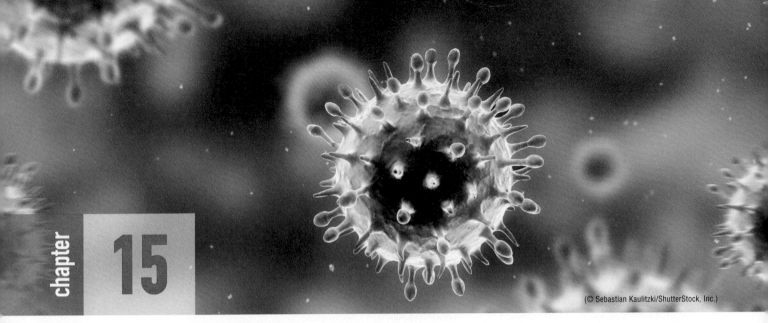

(© Sebastian Kaulitzki/ShutterStock, Inc.)

Microorganisms (commonly called microbes) are found in all environments on Earth. They are found in the air, on food and plants, in and on animals, in the soil and water, and on just about every surface. Microbes range in size from microscopic single-celled organisms to parasitic worms that grow to several feet in length. Although microorganisms are everywhere, few of them cause disease in humans. Those that do usually are dealt with by our immune systems, discussed in the previous chapter. Even though the vast majority of microbes are harmless, some of those that cause disease can be quite deadly. Disease-causing microorganisms make us

THINKING CRITICALLY

You work out regularly at a local gym and jog three times a week with a friend. One day you and your friend see a story on the local TV news about an outbreak in your city of a bacterial disease called listeriosis. Discovered in some packages of a brand of luncheon meat sold in your local supermarket, the bacterium, if ingested, may cause serious complications including stillbirth in pregnant women late in pregnancy. It may also cause meningitis (infection of the connective tissue layers surrounding the brain) in the elderly. After hearing the story, your friend announces that he has a package of this luncheon meat in his refrigerator and, in fact, ate some of it two days ago.

Because he is a healthy 20-year-old and is not sick, he dismisses the story. He says the package must be fine, because he hasn't gotten sick. He announces that he is going to have some of the meat in a sandwich for lunch and invites you to join him for lunch. What advice would you give your friend?

ill because they cause symptoms before our immune systems can respond. Making matters worse, some microorganisms undergo mutations, creating new forms that are able to breach our immune system's defenses, causing disease, even death. Disease may occur in normal healthy individuals, but those with weakened immune systems are particularly vulnerable. As you learned elsewhere in the text, poor nutrition may lead to a weakened immune system.

In this chapter, you will learn about the agents that cause infectious disease in humans. You will also learn about the stages of human diseases and discover how microorganisms cause disease. You will study the ways microbes are transmitted from one person to the next. In addition, you will find out about newly emerging infectious diseases and bioterrorism. Before we examine these topics, however, let's review what you learned about infectious agents—the viruses and bacteria—and expand your understanding of these organisms.

Health Tip 15-1

Don't skip breakfast. Your mother's right, it is the most important meal of the day!

Why?

Breakfast breaks a huge fast caused by sleep, a time when no food enters the body. Body stores of nutrients like liver glycogen, a source of blood sugar, run low during sleep. When morning comes, it's best to replenish these supplies.

Breakfast supplies about 17% of one's daily caloric input at a crucial time. It also typically supplies about 25% of the recommended daily intake of dairy products. If you're eating right, that is, including fruits in your diet, breakfast will supply up to 28% of the recommended intake of fruit. It will provide 20% of the daily recommended grains and 28% of most vitamins and minerals. It's a quick meal with a big impact.

15-1 The Infectious Agents of Human Disease

When most of us hear the words *bacteria* and *virus*, we usually think of infections and the diseases they cause such as the flu or the common cold. In reality, most of these microorganisms are harmless to humans. Only a few actually cause disease. Microbes that cause disease are called **pathogens**. Microbes are considered **infectious** if they invade and grow in the tissues of the body. They are considered to be **contagious** if they can be readily transmitted from one organism to another—for example, from one person to another.

Disease-Causing Agents

History is full of accounts of infectious diseases that periodically flared up, creating huge epidemics. Infectious diseases, such as the plague, cholera, and smallpox, were once known as "slate-wipers" because they killed millions of people in Europe and other countries. Knowledge of impending epidemics often generated shock and terror in populations.

Until the 1850s, the fear of such diseases was complicated by ignorance because no one knew the cause of these diseases. To combat these diseases, people often dressed in strange costumes that were thought to ward off disease (**Figure 15-1**). Many strange potions and procedures, such as bloodletting, were used to cure people who had become ill. Unfortunately, few of these "cures" were effective, and some methods such as bloodletting often made a bad situation worse.

Before the turn of the twentieth century, the work of scientists such as Louis Pasteur, Robert Koch, and others had shown that certain microbes caused many of the devastating diseases of the time. In fact, the research of Pasteur and Koch culminated in the formulation of the **germ theory of disease** (as described in Scientific Discoveries that Changed the World 15-1). Thanks to their pioneering work, we now know the cause of most infectious diseases.

FIGURE 15-1 **Beak Doctors** Some physicians during the time of plague believed that wearing a beak-like mask would protect them from the disease. Such physicians were called "Beak Doctors." (© National Library of Medicine.)

Scientific Discoveries that Changed the World

15-1 The Germ Theory of Infectious Disease
Featuring the Work of Louis Pasteur and Robert Koch

Throughout history, infectious diseases such as plague, tuberculosis, typhoid fever, and diphtheria ravaged peoples and populations around the world. Neither royalty nor common folk were spared from the devastation. Fear was rampant during epidemics because no one knew how to cure people afflicted with the disease. Ways to eradicate such diseases were equally elusive.

Although cures for infectious diseases were not available in the nineteenth century, scientists were beginning to suspect infectious agents ("germs") as the cause. The link between these diseases and infectious agents, however, was an idea that many in the 1800s found hard to believe because it did not fit with the perceptions at the time. Most scientists believed that infectious diseases, such as the plague and malaria, were spread by an altered chemical quality of the atmosphere or a poisoning of the air (the word malaria comes from mala aria, meaning "bad air"). These qualities of the air might arise from decaying or diseased bodies.

In the mid-1850s, evidence for germs (microorganisms) as the cause of infectious disease gained stronger footing thanks to the work of the French scientist Louis Pasteur. Pasteur was a chemist by training. In 1854, at the age of 32, he was appointed Professor of Chemistry at the University of Lille in northern France. In 1857, he was asked to unravel the mystery of why local French wines were turning sour. Pasteur observed that although both good and soured wines contained tiny yeast cells, only the soured wines contained populations of barely visible rod-like organisms, known then and now as bacteria. His studies showed that yeast cells and bacteria are tiny, living factories in which important chemical changes take place. Pasteur's work also drew attention to microorganisms as agents of change because bacteria appeared to make the wine "sick." He surmised that, if microorganisms could sour wine, perhaps others could make people ill. In 1857, Pasteur published a short paper on wine souring by bacteria. In the paper, he implied that germs were related to human illness.

Other studies furthered Pasteur's understanding of microbes. He came to view microorganisms as entities that were everywhere. Some of these ubiquitous organisms, he concluded, might be agents of human disease. So sure was he of this fact that he formulated the germ theory of infectious disease. The germ theory holds that microorganisms are responsible for infectious diseases. Between 1865 and 1868, Pasteur showed that a mysterious disease of silkworms that was sweeping through France was caused by a protozoan that infected both the silkworms and the mulberry leaves fed to them. Pasteur showed that the disease could be controlled by separating the healthy silkworms from the diseased silkworms and their food. In quelling the spread of the silkworm disease, he strengthened the germ theory of infectious disease.

Pasteur made yet another important contribution to science and medicine in the 1870s. During this period, huge numbers of sheep and cattle in his home country were dying of anthrax, a deadly blood disease. Pasteur's research team showed that the disease was passed from animal to animal by infectious organisms, and he was confident that they could find a way of stopping the spread of anthrax by producing a vaccine.

Unfortunately, the work proved more difficult than he and his coworkers thought. Anthrax is highly infectious, and healthy animals have a strong chance of catching the disease simply by grazing over the burial site of anthrax victims. Still, he used the bacterium to produce an untested vaccine. Things came to a head in 1881 when a French veterinarian challenged Pasteur to test out his vaccine. Rather than appear unsure of his work, Pasteur accepted the challenge. His vaccine worked—his vaccinated sheep survived the injection of live anthrax spores. Again, Pasteur saved another French industry.

In 1885, Pasteur reached the zenith of his career when he successfully immunized a young boy against the dreaded disease rabies. Although he never saw the causative agent of rabies, Pasteur grew it in the brains of animals and injected the boy with bits of the brain tissue. Once again, success presented itself. Many monetary rewards followed, including a generous gift from the Russian government after Pasteur immunized 20 peasants against rabies. The funds helped establish the Pasteur Institute in Paris, one of the world's foremost scientific institutions. Pasteur presided over the institute until his death in 1895.

Pasteur's work stimulated others to investigate the nature of microorganisms and their association with disease. One of the most significant investigators was Robert Koch, a country doctor from East Prussia (now part of Germany). Koch's work provided the crucial evidence for complete acceptance of the germ theory.

In 1876, Koch observed the anthrax-causing bacteria with a microscope and identified the bacterium as that found in animals that had died from anthrax. Koch watched for hours as the rod-shaped bacteria multiplied, formed tangled threads, and finally reverted to highly resistant spores. He then took several spores on a sliver of wood and injected them into healthy mice. The symptoms of anthrax appeared within hours. Koch autopsied the animals, found their blood swarming with the same rod-shaped bacteria, and then re-isolated the bacteria. The cycle was now complete. The rod-shaped bacteria definitely caused anthrax. Koch established a set of criteria (called Koch's postulates) for proving that a specific microorganism caused a specific disease.

Koch's work on anthrax verified the germ theory of infectious disease. In 1881, he outlined his methods at an international medical congress, and several days later, Koch received a personal letter of congratulations from Pasteur.

Koch also reached the height of his influence in the 1880s. In 1883, he interrupted his work on tuberculosis to lead groups studying cholera in Egypt and India. In both countries, Koch isolated the infectious bacterium by following his previous methods. In 1891, he became Director of Berlin's Institute for Infectious Diseases. At various times, he studied malaria, plague, and sleeping sickness, but his work with tuberculosis ultimately gained him the 1905 Nobel Prize in Physiology or Medicine. He died of a stroke in 1910 at the age of 66.

By the end of the nineteenth century, the pioneering discoveries of Pasteur and Koch had created almost universal acceptance of the germ theory of infectious disease. With the passing of Pasteur and Koch, a new generation of international microbiologists stepped in to expand our understanding of microorganisms and infectious diseases.

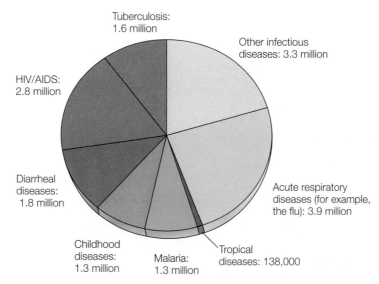

FIGURE 15–2 **Infectious Disease Deaths Worldwide** This pie chart depicts the leading causes of infectious disease and the number of worldwide deaths, as reported by the World Health Organization. Tropical diseases include schistosomiasis and filariasis.

Once researchers and public officials understood the cause of infectious diseases, they set out to find ways to control or eliminate them. Over the years, better sanitation, improvements in water treatment, and the development of antibiotics and vaccines helped to control, even eliminate many infectious diseases. Despite these advances, however, infectious disease remains a major threat to human health worldwide. Moreover, some diseases still create anxiety and alarm in people. Just think about the fear that AIDS, the H1N1 virus, severe acute respiratory syndrome (SARS), and West Nile virus have caused in the United States and other countries. Figure 15-2 shows the human impact of some of the most prolific deadly diseases that exist worldwide today.

> **KEY CONCEPTS**
>
> Throughout history many infectious diseases are caused by bacteria and viruses have decimated human populations; advances in waste treatment and drinking water purification as well as the advent of vaccines and antibiotics have brought many of these diseases under control.

Viruses and Bacteria

Many of the most common infections are caused by viruses and bacteria. As their name suggests, most microorganisms are tiny and cannot be seen without the aid of a microscope. To put their size into perspective, consider this analogy: If a virus were the size of a baseball, a typical bacterium would be the size of the pitcher's mound. On this scale, one of your body's cells that might be infected by the virus or bacterium would be the size of the entire ballpark.

Viruses are the simplest of all traditional infectious agents (Figure 15-3). As you may recall, viruses contain genetic material—either DNA or RNA. The genetic information, containing genes, is surrounded by a protein coat, called the

capsid. In some viruses, the capsid is covered by a membrane-like structure known as the **viral envelope**.

Viruses are not cells, and they are not even considered living organisms because they lack the enzymes and cellular structures required for metabolism and growth. Because of this, viruses in the air or soil cannot replicate (make copies of themselves). To reproduce, they must invade living cells. Within these **host** cells, they find the chemicals and energy required to multiply.

All viruses go through similar life cycles. Consider, for example, the virus that causes chickenpox, one type of herpes

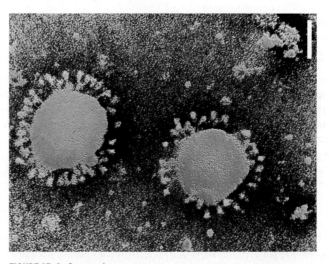

FIGURE 15–3 **Coronaviruses** False color transmission electron micrograph of two human coronaviruses. In humans, coronavirus typically causes colds. The spikes can be seen clearly extending from the viral envelope. Viruses similar to these are responsible for severe acute respiratory syndrome (SARS). (Bar = 60 nm.) (© Dr. Steve Patterson/Science Source, Inc.)

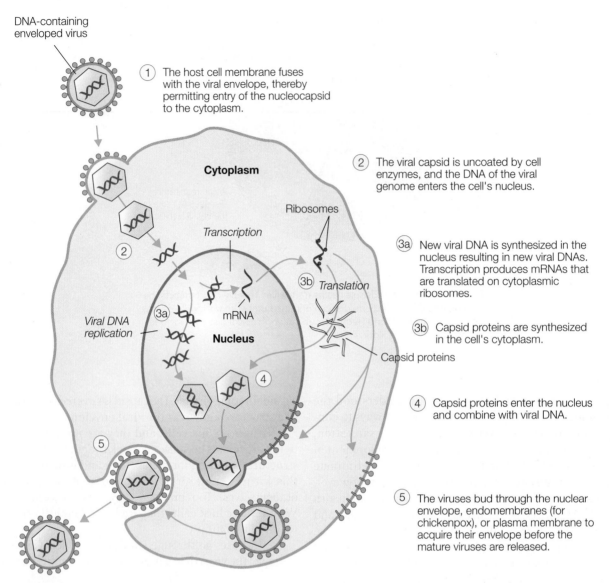

DNA-containing enveloped virus

① The host cell membrane fuses with the viral envelope, thereby permitting entry of the nucleocapsid to the cytoplasm.

Cytoplasm

② The viral capsid is uncoated by cell enzymes, and the DNA of the viral genome enters the cell's nucleus.

Ribosomes

Transcription

③a New viral DNA is synthesized in the nucleus resulting in new viral DNAs. Transcription produces mRNAs that are translated on cytoplasmic ribosomes.

③b *Translation*

Viral DNA replication

mRNA

Nucleus

③b Capsid proteins are synthesized in the cell's cytoplasm.

Capsid proteins

④ Capsid proteins enter the nucleus and combine with viral DNA.

⑤ The viruses bud through the nuclear envelope, endomembranes (for chickenpox), or plasma membrane to acquire their envelope before the mature viruses are released.

FIGURE 15–4 Replication of a DNA Animal Virus The virus illustrated here is a herpesvirus (such as one that might cause chickenpox), and the host cell is from human skin.

virus, which is shown in Figure 15-4. To gain entry into a cell, the virus first attaches to protein receptors in the plasma membrane of target cells; in this case, it attaches to cells of the skin. The virus is then brought into the cell, often by phagocytosis, and its nucleic acid (DNA for the herpesvirus) is released into the cytoplasm. The virus' nucleic acid then takes control of the host cell's metabolic machinery, causing it to produce numerous copies of the virus' proteins. Once the components of new viruses are made, they are assembled into new viruses. In this process, an infected cell may produce hundreds of new viruses. After the viruses are assembled, they are released, usually when the cell bursts open. The host cell is typically destroyed in the process. Table 15-1 includes several examples of common diseases caused by a variety of different viruses.

Bacteria are single-celled organisms that usually are visible with a light microscope. Even so, they are so small

that it would take a thousand of them laid end to end to span the diameter of a pencil eraser. Bacteria may be shaped like short rods, spheres, or spirals, depending on the species (Figure 15-5). Because bacteria are prokaryotes, they lack most of the organelles found in human cells. Nonetheless, most bacteria can reproduce and grow on their own—that is, they can reproduce without invading and taking over the metabolic functions of host cells as viruses do. During reproduction, bacteria replicate their DNA, then split into two identical cells, each with one copy of the DNA (Figure 15-6a). This process often occurs very rapidly, producing large populations of living cells in a very short time (Figure 15-6b).

As noted earlier, very few bacteria are harmful to humans. In fact, less than 1% of the bacteria known to science cause disease. Some bacteria that live in our bodies are beneficial. For example, a number of bacterial species live on our skin surface and protect us from potentially infectious pathogens.

TABLE 15-1	Human Diseases and Their Infectious Agent

	Infectious Agent					
Disease	Virus	Bacterium	Fungus	Protozoan	Helminth	Other Agent
AIDS	✓					
Anthrax		✓				
Athlete's foot			✓			
Botulism		✓				
Chickenpox	✓					
Cholera		✓				
Common cold	✓					
Creutzfeldt-Jakob disease (variant)						✓
Cryptosporidiosis				✓		
Diarrhea	✓	✓		✓		
Diphtheria		✓				
Dengue fever	✓					
Ebola hemorrhagic fever	✓					
Filariasis					✓	
Genital herpes	✓					
Gonorrhea		✓				
Hantavirus pulmonary syndrome	✓					
Hepatitis	✓					
Histoplasmosis			✓			
Hookworm					✓	
Infectious mononucleosis	✓					
Influenza	✓					
Leprosy		✓				
Listeriosis		✓				
Lyme disease		✓				
Malaria				✓		
Measles	✓					
Monkeypox	✓					
Mumps	✓					
Oral thrush			✓			
Pinworm					✓	
Plague		✓				
Pneumonia	✓	✓	✓			
Rabies	✓					
Ringworm			✓			

(continued)

TABLE 15-1 | Human Diseases and Their Infectious Agent (Continued)

Disease	Virus	Bacterium	Fungus	Protozoan	Helminth	Other Agent
Rubella	✓					
Salmonellosis		✓				
SARS	✓					
Schistosomiasis					✓	
Smallpox	✓					
Staph infections		✓				
Strep throat		✓				
Tetanus		✓				
Toxoplasmosis				✓		
Trichinosis					✓	
Tuberculosis		✓				
Typhoid fever		✓				
Valley fever			✓			
West Nile fever	✓					
Whooping cough (pertussis)		✓				

Likewise, many species of harmless bacteria live in our intestines and help us digest food, provide us with nutrients, and protect us against pathogens.

Pathogenic bacteria, on the other hand, can infect the body, causing illness, even death. Illness results from some combination of their rapid growth, their production of poisonous chemicals known as *toxins*, and our immune system's response to the infection. *Listeria*, the bacterium that causes human listeriosis, mentioned in the critical thinking exercise at the beginning of the chapter, is such an example (Figure 15-6). Table 15-1 lists several other common diseases caused by bacteria.

KEY CONCEPTS

Some species of bacteria and viruses cause disease; viruses invade cells of the body, taking over their metabolic machinery to reproduce; bacteria replicate outside of cells and produce toxins that can make us ill.

Eukaryotic Pathogens and Parasites

Not all infectious diseases are caused by bacteria and viruses. Several eukaryotic microbes such as fungi, certain protozoa, and helminths can cause infections and some very serious diseases in humans (Table 15-1).

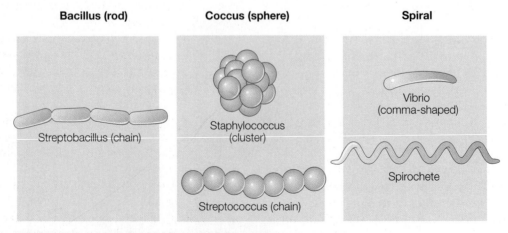

Bacillus (rod)	Coccus (sphere)	Spiral

Streptobacillus (chain)

Staphylococcus (cluster)

Streptococcus (chain)

Vibrio (comma-shaped)

Spirochete

FIGURE 15–5 Variations in Bacterial Shape and Cell Arrangements

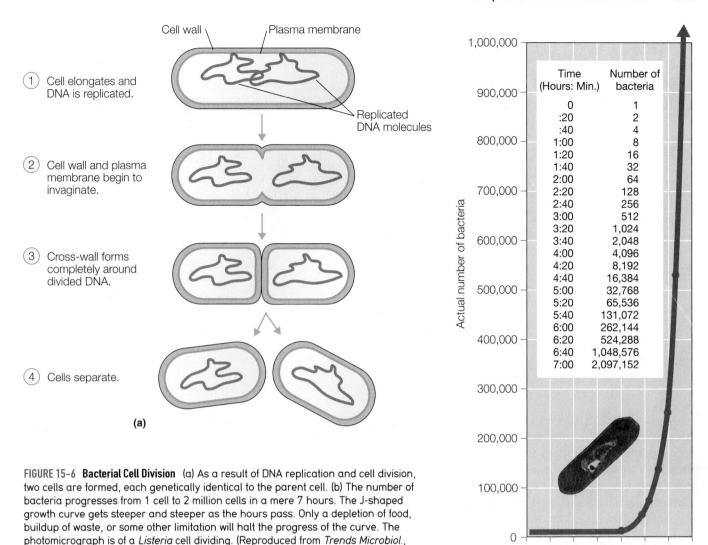

Time (Hours: Min.)	Number of bacteria
0	1
:20	2
:40	4
1:00	8
1:20	16
1:40	32
2:00	64
2:20	128
2:40	256
3:00	512
3:20	1,024
3:40	2,048
4:00	4,096
4:20	8,192
4:40	16,384
5:00	32,768
5:20	65,536
5:40	131,072
6:00	262,144
6:20	524,288
6:40	1,048,576
7:00	2,097,152

FIGURE 15-6 **Bacterial Cell Division** (a) As a result of DNA replication and cell division, two cells are formed, each genetically identical to the parent cell. (b) The number of bacteria progresses from 1 cell to 2 million cells in a mere 7 hours. The J-shaped growth curve gets steeper and steeper as the hours pass. Only a depletion of food, buildup of waste, or some other limitation will halt the progress of the curve. The photomicrograph is of a *Listeria* cell dividing. (Reproduced from *Trends Microbiol.*, vol. 1, Tilney, L. G., and Tilney, M. S., The wily ways of a parasite. . . , pp. 25–31. Copyright 1993, with permission from Elsevier.)

Fungi consist of the yeasts, molds, and a group you are probably most familiar with—mushrooms. Yeasts are single-celled organisms. Molds are multicellular and often are visible on spoiled foods, such as cheese or bread, or on overripe fruit. Mushrooms also are multicellular. Although they are not infectious organisms, some species of mushroom can produce toxic chemicals that can be deadly if consumed.

Fungi live in the air, water, and soil, and on plants. Some can live in our bodies, usually without causing illness. Some species of fungi are beneficial. For example, the antibiotic penicillin is produced by a fungus. This drug kills many species of infectious bacteria. Several species of yeast are also important in making foods such as bread and cheese. Yeasts also are used in the production of beer and wine.

A few fungi cause human illness and disease. For example, athlete's foot is caused by one species of fungi that grows in the skin. A yeast-like fungus known as *Candida* (Figure 15-7) can infect the mouth, causing oral thrush in infants, in people taking antibiotics, and in people with weakened immune systems. *Candida* is also responsible for most types of infection-induced diaper rash and is the organism responsible for "yeast infections" in the vaginas of adult women. Certain species of *Candida* can also grow inside the blood and tissues of the body. Fungal pathogens can produce respiratory diseases, such as valley fever in the American Southwest and histoplasmosis in the Ohio and Mississippi River valleys.

Protozoa also are single-celled eukaryotic organisms. Many protozoans inhabit the intestinal tracts of humans. Although most protozoans are harmless, some are pathogenic. Pathogenic protozoans spend part of their life outside of humans, living in soil, water, or insects, or infecting other animals. If they infect the human body, they often live off body fluids, surviving as parasites. A **parasite** is an organism that lives within the body of another, often causing some level of damage.

Protozoa invade the human body through contaminated food or water. One such parasite, *Cryptosporidium parvum*, invaded the Milwaukee city water supply in 1993 and was responsible for the largest waterborne disease

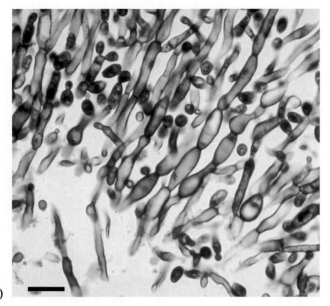

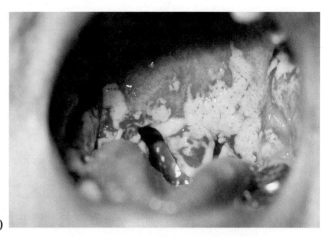

FIGURE 15-7 **The Agent of Oral Thrush** (a) A photomicrograph of stained *Candida albicans* cells (bar = 40 μm). (Copyright © 1987 American Society of Clinical Pathologists. Reprinted with permission. Photo courtesy of CDC.) (b) Oral thrush. (Courtesy of CDC.)

outbreak in U.S. history, causing illness in more than 400,000 people. Other protozoa are transmitted through sexual contact. Still others are carried by vectors, organisms such as mosquitoes and ticks that transmit the protozoa from one person to the next. Malaria is an example. Malaria is perhaps the most prevalent and deadly of all protozoan diseases in the world, killing more than 1 million people (mostly children in tropical countries) each year. The malaria-causing protozoan, called *Plasmodium*, is transmitted from person to person by mosquitoes (Figure 15-8). For example, a mosquito sucks blood from one individual infected with *Plasmodium*. When that same mosquito attacks another person, it may transfer some of the protozoa to the uninfected person.

Helminths are parasitic worms. (The term *helminth* comes from the Greek for "worm.") Most helminths are visible to the naked eye. Nonetheless, they are designated as "microorganisms" because they can cause infectious disease.

The most common helminths are flatworms and roundworms. When adult helminths or their microscopic eggs enter the body, they live parasitically off the body's nutrients within the intestinal tract, lungs, liver, skin, or brain, depending on the species. **Flatworms** range in length from 6 inches to more than 25 feet! One of the most prevalent flatworms, *Schistosoma*, is responsible for a disease known as schistosomiasis, which affects more than 250 million people worldwide. Victims suffer from fever, muscle pain, diarrhea, coughing, vomiting, and a burning sensation during urination. **Tapeworms** are one type of flatworms made up of hundreds of segments, each of which is capable of breaking off and developing into a new tapeworm (Figure 15-9a).

Another type of helminth is the roundworm. **Roundworms** damage their hosts by forming large masses of worms in blood vessels, lymphatic vessels, or the intestines. The roundworm responsible for the disease filariasis, for example, blocks lymphatic vessels after many years of infection. The lymphatic vessels swell and become distorted with fluid. The condition is called elephantiasis (Figure 15-9b).

The most common roundworm disease in the U.S. is pinworm disease. Surprisingly, an estimated 30% of children and 16% of adults serve as hosts. Among the other roundworm diseases are trichinosis in pork and hookworm disease. Hookworm infects hundreds of thousands worldwide, causing a dry cough, mild fever due to the presence of worms in the lung, and abdominal pain due to the presence of worms in the intestine.

Besides these "traditional" microbial pathogens, other infectious agents exist. The most newsworthy nontraditional infectious agent is the prion. **Prions** are proteins that are responsible for several diseases, including mad cow disease (bovine spongiform encephalopathy). The word prion comes from the words *protein* and *infection*. Prions are misfolded proteins—that is, proteins that have not folded properly. You will recall from elsewhere in the text that proteins are strings of amino acids; the amino acid chains bend and fold to give proteins their three-dimensional structure. Prions are incorrectly folded proteins. In the body, they induce other proteins to become misfolded as well, leading to disease. Prions are described in Health Note 15-1.

> **KEY CONCEPTS**
>
> Diseases can also be caused by single celled fungi and protozoa as well as multicellular parasitic worms.

by few obvious symptoms. Many people, for example, have experienced subclinical cases of mumps or infectious mononucleosis. A **clinical disease** is one in which the symptoms are apparent (such as occurs with a common cold). Symptoms may be anywhere from mild to severe, depending on the pathogen.

The **acme period** or climax is the third stage in disease progression. This is the critical stage of the disease. Very specific signs and symptoms appear during this phase. During a bad case of the flu, for instance, signs and symptoms include a severe cough and muscle pain in the chest, back, and legs. Many patients experience high fever and shaking chills (chills are a result of the difference in temperature between the superficial and deep areas of the body). Individuals suffering from the flu are most contagious during this period.

Complications can arise during this phase, depending on the specific form of the disease and the state of the host. While many people contract the flu each year in the United States, approximately 200,000 of them require hospitalization because of complications—for example, because of a fever lasting more than five days (which is usually a sign of bacterial or viral pneumonia that has developed because the individual is in a weakened state). In many elderly individuals or people who are in poor health, complications such as this can lead to death if treatment is not sought early on. In fact, each year, influenza kills more than 35,000 Americans. Most people who are in good health at the time of a flu infection survive.

As the signs and symptoms of a disease subside, victims enter a **period of decline**. Sweating is common during this phase of the disease, as the body releases excessive amounts of heat. The recovery period varies in length. For individuals ill with the flu, it is often quite rapid. For other infectious diseases, however, recovery may take a long time.

The disease sequence concludes after the body passes through a **period of convalescence**, during which time the body's systems return to normal. Interestingly, during the period of decline—and even convalescence—individuals can transmit the disease to others. During this period, pathogens exit the body. Bacteria responsible for gastrointestinal illness may pass through the feces, while flu viruses can be released by coughing and sneezing.

> **KEY CONCEPTS**
>
> An infectious disease typically follows a series of five stages: (1) incubation, (2) prodromal, (3) acme, (4) decline, and (5) convalescence.

15-3 How Pathogens Cause Disease

Here we examine some of the factors that determine whether a microorganism can cause disease in an individual. We focus most of our attention on the properties of microorganisms that enable them to overcome a host's immune defense.

Infection is dependent on the pathogen's ability to adhere to cells in specific tissues. The bacterium that causes the sexually transmitted disease known as *gonorrhea*, for example, only adheres to the lining of the urogenital tract. Many human viruses, including the flu virus, have protein structures called spikes that protrude from their capsids or envelopes. The spikes attach to specific host cells. Once a pathogen binds to a cell membrane, it is usually engulfed by the cell by phagocytosis (Figure 15-11).

Enzymes and Toxins

The ability of many pathogenic bacteria to penetrate tissues and cause adverse health effects depends on their release of enzymes or toxins. Certain enzymes, for instance, help pathogenic bacteria resist body defenses and thus increase the virulence of a microbe. These enzymes may interfere with certain cellular functions or alter barriers that are meant to thwart invasion. Microbial enzymes, such as coagulase, for example, produce blood clots that encompass clusters of pathogens, protecting them from immune system attack. Other bacterial enzymes target and destroy cells of the immune system. The bacterium *Staphylococcus* produces enzymes that digest the "cellular cement" that holds cells of tissues together. This enables the bacteria to penetrate deep into body tissues. These bacteria are responsible for a variety of illnesses including pneumonia, food poisoning, and skin infections.

Several human diseases involve the formation of biofilms (impenetrable colonies) in tissues that are very difficult for the immune system or even medical intervention to eliminate. Health Note 15-2 describes biofilms.

Some bacteria cause disease by producing toxins. **Toxins** are microbial poisons. They fall into two categories: exotoxins and endotoxins.

Exotoxins are protein molecules, often enzymes, which are manufactured by bacteria in body tissues. When released from the bacteria, the toxins may be transported throughout the body.

Usually, only minute amounts of an exotoxin are needed to cause disease. The exotoxin that causes botulism is among

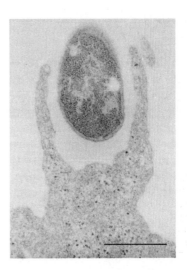

FIGURE 15-11 **Tissue Invasion** A bacterium is being engulfed by an epithelial cell (bar = 1 mm). (Reprinted with permission from the American Society of Microbiology from Finlay, B. B., et al., *ASM News* 58 (1992): pp. 486–490. Photo courtesy of B. Brett Finlay, The University of British Columbia.)

health**note**

15-2   Infectious Disease and Biofilms

In your reading of this chapter, you may have the impression that pathogens act as independent agents to cause disease. In a few cases, that might be true. However, for the most part, pathogens do not act as individuals; rather, they survive in complex communities called **biofilms**. A biofilm consists of an immobilized population of bacteria (or other microorganisms) caught in a sticky web of tangled polysaccharide fibers that adheres to various surfaces.

Biofilms develop on virtually all surfaces in contact with a watery environment. This includes the surfaces of aquatic plants or animals, water pipes, and stones. Contact with a fluid environment ensures a plentiful supply of nutrients, and as the bacteria grow, they secrete sticky polysaccharides.

Researchers have found that biofilms are highly organized structures. They contain water channels that serve to deliver nutrients to the bacteria and remove wastes. The bacteria share these passageways, interact metabolically, and benefit from each other's metabolic byproducts. They also form communities that store nutrients and resist predators such as protozoa and viruses. Some scientists estimate that in nature, 99% of all microbial activities occur in biofilms.

Bacteria in biofilms act very differently than individual cells and are extremely difficult to treat. In biofilm, for example, the bacteria often are impervious to drugs such as antibiotics, disinfectants, and antiseptics, which are designed to attack individual cells (Figure 1). Pathogens in biofilms may resist the immune system's attempts to eliminate them and, therefore, can be extremely virulent. For example, white blood cells have difficulty reaching the pathogens in this slimy conglomeration of armor-like material.

Health officials at the Centers for Disease Control and Prevention (CDC) estimate that more than 80% of human infections involve biofilms. If you or someone you know has had a middle ear infection, the cause was a bacterial biofilm. Dental plaque is a type of biofilm as well. Brushing and flossing teeth and seeing a dentist regularly are essential to reduce and control this biofilm. Eye doctors are also concerned about the formation of biofilms on contact lenses, which can produce serious eye infections. Proper cleaning and storage of contact lenses is important to prevent biofilm formation.

Biofilms can also form on catheters and medical devices such as artificial hearts. Biofilms in urinary catheters often provide starting points for urinary tract infections, as bacteria creep up the catheters. Once in body tissue, the bacteria sequestered in the slimy conglomerates are shielded from attack by the body's immune system, and they are difficult to kill with antibiotics. In males, biofilms also have been implicated in prostate gland infections, which are accompanied by chronic pain and sexual dysfunction. Finding a way to penetrate this polysaccharide barrier is important in stemming the tide of urinary tract infections.

Besides being involved in human disease, biofilms can cause problems in industry. Biofilms, for example, can corrode water pipes. When they contaminate computer chips, biofilms act as conductors and thereby interfere with electronic signals. Indeed, one researcher has called biofilms the "venereal disease of industry."

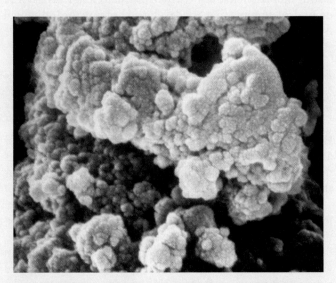

FIGURE 1   Biofilms Biofilms are communities of microorganisms. Some biofilms can cause infectious diseases, such as these *Pseudomonas* bacteria that have formed a biofilm on the lungs of a cystic fibrosis patient. (Courtesy of Hiroyuki Kobayashi/Kyorin University School of Medicine.)

---

the most lethal toxins known. One pint of the pure toxin would be sufficient to destroy the entire human population of around 6.8 billion people.

Exotoxins destroy cellular structures or inhibit essential metabolic functions. Thus, the disease symptoms vary depending on the exotoxin involved. Table 15-3 highlights several examples.

Endotoxins are a lipid portion of the cell wall of many bacteria. They usually are released only upon disintegration (death) of the bacterial cell. When released, endotoxins cause chills, fever, weakness and aches, and general malaise. Endotoxins may also damage the circulatory system, causing a massive increase in the permeability of the blood vessels. This, in turn, causes blood to leak into the intercellular spaces, where it is useless. Tissues swell, the blood pressure drops, and the patient may lapse into a coma.

Like exotoxins, endotoxins add to the virulence of pathogens and enhance their ability to cause disease. However, they are required in larger doses than are exotoxins. Typhoid fever and some urinary tract infections are the result of endotoxin production.

KEY CONCEPTS

Pathogenicity and virulence depend on key metabolic characteristics of a pathogen.

TABLE 15-3	Diseases Caused by Exotoxins	
Disease	**Bacterium**	**Signs or Symptoms; Mechanism of Action**
Botulism	*Clostridium botulinum*	Muscle paralysis; inhibits the release of acetylcholine at the synaptic junction
Whooping cough (pertussis)	*Bordetella pertussis*	Spasmodic and recurrent coughing, ending with a loud inspiratory whoop with choking on mucus; paralyzes ciliated cells and impairs mucus movement
Cholera	*Vibrio cholerae*	Severe diarrhea; causes massive loss of water from the intestines, which can produce a rapid drop in blood pressure
Diphtheria	*Corynebacterium diphtheriae*	Thick membrane coating the upper respiratory mucous membranes. Fever, sore throat, cough; interferes with protein synthesis in the cytoplasm of epithelial cells of the upper respiratory tract; respiratory blockage
Anthrax	*Bacillus anthracis*	Inflammation, hemorrhage, shock (inhalational anthrax); three exotoxins generate an accumulation of fluid and killing of host cells

15-4   How Infectious Diseases Are Transmitted

For a disease to spread, pathogens must be transmitted to other hosts. In some cases, an infectious disease spreads only within a given region. If this occurs and the level of infection is relatively low, microbiologists and health officials consider it an **endemic disease**. Plague in the American Southwest is an example of an endemic disease. If, on the other hand, the disease breaks out in explosive proportions within a population, it is considered an **epidemic**. Influenza often causes epidemics. A more contained occurrence is considered an **outbreak**. An abnormally high number of measles cases in one U.S. city would be classified as an outbreak, whereas it would be classified as an epidemic if it occurred in several states. A **pandemic disease** (or **pandemic**) occurs worldwide. The most recent example of a pandemic would be AIDS, although other pandemics have occurred (Table 15-4).

So, how are the microbial agents that can cause outbreaks, epidemics, and pandemics spread? Although diseases can be transmitted in many ways, all modes of transmission fall into two broad categories: direct and indirect transmission methods.

> **KEY CONCEPTS**
>
> Infectious agents can be transmitted from one person to the next directly or indirectly.

Direct Transmission Methods

Direct transmission occurs in a variety of ways. The most common involves person-to-person contact, during which time bacteria, viruses, or other infectious agents are transferred from people who have the disease to uninfected individuals. Direct physical contact usually occurs during hand-shaking, kissing, or exchanging body fluids—for example, from sexual intercourse or blood transfusions (Figure 15-12). Infectious mononucleosis, a disease common among college students, typically is transmitted by kissing. Gonorrhea and genital herpes are transmitted directly during sexual contact. Hepatitis B is transmitted by blood transfusions.

Infectious diseases can also spread from a mother to her unborn child through the placenta or via the vagina during childbirth. The AIDS virus is an example of an infectious agent that can be passed from mother to her fetus during childbirth. Disease can also be transmitted after birth through breast milk.

Diseases can also spread from animals to people. A pet cat or dog, for example, can carry disease-causing microbes such as rabies. Rabies can be transmitted to a pet by a bite from another animal (raccoons, bats, cattle, rabbits, skunks, and foxes) that has rabies. A bite from the infected cat or dog can then spread the viral disease to a human.

Toxoplasmosis, which is sometimes called *litter box disease*, results from contact with a protozoan parasite in cat

TABLE 15-4	Some of the Major Historical Pandemics and Resulting Human Deaths	
Pandemic	**Estimated Deaths**	
Bubonic plague/"Black death" (6th, 14th, 17th centuries)	137 million	
Smallpox (1900–1977)	300–500 million	
Influenza		
"Spanish flu" (1918–1919)	20–50 million	
"Asian flu" (1957–1958)	1–4 million	
"Hong Kong flu" (1968–1969)	1–4 million	
"Russian flu" (1977–1978)	No accurate account	
"Swine flu" (2009–2010)	18,209+	
AIDS (1981–2011)	35 million	

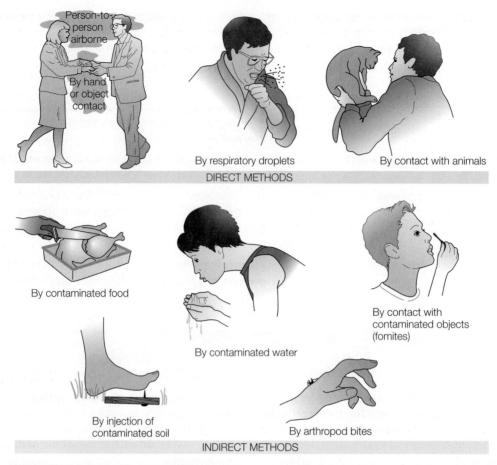

DIRECT METHODS

By respiratory droplets

By contact with animals

Person-to-person airborne

By hand or object contact

By contaminated food

By contaminated water

By contact with contaminated objects (fomites)

By injection of contaminated soil

By arthropod bites

INDIRECT METHODS

FIGURE 15–12 **Methods of Transmitting Disease**

feces. Although the infected animal lacks signs and symptoms, initial symptoms in a human after contact are similar to those of the flu (swollen lymph glands, fatigue, fever, and headache). To be safe, women who are pregnant and have a cat should allow someone else to clean the litter box because the disease can cause miscarriage, premature births, and mental retardation.

Infectious diseases also are transmitted by less common pets such as reptiles and birds. All reptiles, especially turtles and iguanas, can carry the bacterium *Salmonella*. If transmitted to humans, the bacterium can cause gastroenteritis, an inflammation of the stomach and intestine, which results in diarrhea and vomiting.

Infectious disease also can be spread directly by respiratory droplets and airborne particles. For example, when you cough or sneeze, you expel droplets into the air around you. If you have a cold, the flu, or another contagious respiratory illness, these droplets contain the infectious agent. Respiratory droplets travel only about 3 feet because they are too large to stay suspended in the air for a long period. However, if a droplet contacts the eyes, nose, or mouth of someone else, that individual may become infected and may experience signs and symptoms of the disease. Crowded places, especially indoor environments—including classrooms, airports, and airplanes—can increase the chances of respiratory droplet contact, which may explain the increase in such respiratory infections in the winter months when more people congregate in enclosed places.

KEY CONCEPTS

Direct transmission of disease organisms involves person-to-person contact such as hand-shaking, kissing, or sexual contact.

Indirect Transmission

Disease-causing organisms may also be transmitted indirectly (see Figure 15-12). **Indirect transmission** occurs when a person comes in contact with pathogens on inanimate objects, such as handkerchiefs, doorknobs, or faucet handles. For example, if you touch a doorknob that has been touched by someone who had the flu, you may pick up the viruses he or she left behind. If you then touch your eyes, mouth, or nose before washing your hands, you may become infected. Skin punctures by contaminated objects also can spread a disease-causing organism, as mentioned earlier for tetanus.

Some pathogens travel through the air on much smaller airborne particles, known as *aerosols*. **Aerosols** consist of moisture droplets and fine dust particles. They can remain suspended in the air for extended periods and can travel in air currents. Aerosols containing pathogens such as viruses or bacteria may be inhaled, causing disease in the host. Tuberculosis and SARS are examples of diseases that usually spread through the air as respiratory droplets and aerosols.

Arthropods, such as mosquitoes, flies, fleas, lice, and ticks, are also responsible for the indirect transmission of disease. Such carriers are called **vectors**. Some mosquitoes, for example, may carry the malaria parasite or the West Nile virus. A bite from a mosquito carrying West Nile virus can transfer the virus into the blood and lead to West Nile fever. Deer ticks can carry the bacterium responsible for Lyme disease. Fleas may carry the plague bacterium. Even common houseflies may carry diseases. Thus, when a housefly lands on your dinner plate, it may transfer pathogens to the food. If a sufficient dose is transmitted, eating the contaminated food may make you sick.

Finally, some infectious agents can spread through the food we eat and the water we drink. Poor food processing or food preparation can introduce pathogens into meats and other foods. The food then serves as the vehicle by which the pathogens are spread. Toxin-producing strains of *E. coli* often make the news because improper food-processing procedures have accidentally introduced the bacterium into a food product, often hamburger. In 1998, the U.S. Department of Agriculture recalled 25 million pounds of raw hamburger contaminated with a toxin-producing strain called *E. coli*

O157:H7. If contaminated raw foods are not cooked to the proper temperature needed to kill the bacteria, chances are people will become ill and, in extreme cases, even die. In 2006, two outbreaks of *E. coli* occurred in the United States, resulting in numerous deaths. The bacteria were found on spinach in the first outbreak, and lettuce in the second, and were present as a result of contaminated irrigation water. In June 2009, two people died and 28 became ill on the East Coast from *E. coli*–contaminated beef, and in November 2009, 65 people fell sick in 29 states from eating chocolate-chip cookie dough tainted with the bacterium. *Salmonella* in peanut products killed at least six people and sickened hundreds in January 2009.

KEY CONCEPTS

Pathogens can also be transmitted by contact with infected inanimate objects, by inhalation of airborne particles, by vectors insects or other organisms that carry the disease, and through food and water.

15-5 Emerging Infectious Diseases and Bioterrorism

In the latter half of the twentieth century, many infectious diseases that once ravaged human populations seemed to be under control. After millions upon millions of deaths, humans had finally triumphed over infectious disease. Our success stemmed from the use of antibiotics, the development of vaccines, a vigilant public health system, better sanitation, and water purification. In fact, in 1980, the virus that causes smallpox was eradicated from the face of the Earth through a massive global vaccination effort.

Over the last several decades, however, a number of new infectious diseases, such as AIDS and SARS, have emerged (Figure 15-13). We have also witnessed a resurgence of some infectious diseases, such as tuberculosis and dengue fever,

which health officials thought were under control. Why are new infectious diseases emerging and other infectious diseases reemerging?

Emerging and Reemerging Infectious Diseases

Emerging infectious diseases are those that have recently surfaced in a population. Among the more newsworthy have been AIDS, hantavirus pulmonary syndrome, Lyme disease, Ebola hemorrhagic fever, mad cow disease, SARS, West Nile fever (in the Americas), and the H1N1 virus.

The most recent, the H1N1 virus, is a type of influenza A virus and is the most common cause of influenza (the flu) in human beings. In 2010, the virus spread quickly throughout the

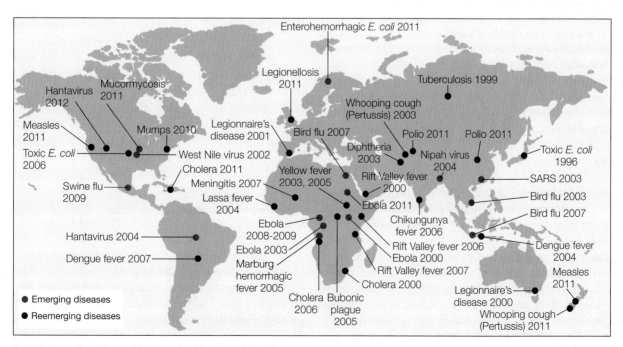

FIGURE 15-13 Emerging and Reemerging Diseases, 1996–2012.

world. Victims suffered from chills, fevers, sore throat, muscle pain, severe headaches, coughing, and general discomfort—as is common in the flu. However, this new strain also caused many deaths worldwide, especially among children. Vaccines against the virus were made available, starting fairly late in 2010. Where did the virus come from?

The H1N1 virus, dubbed the *swine virus*, came from pigs. Although some strains of the H1N1 virus are found only in human beings, others are endemic in pigs and birds. Those found in pigs create swine influenza; those in birds cause avian influenza.

Scientists found that the 2009 global flu pandemic was due to a new strain in humans that originated in pigs, and then spread to humans, hence the name *swine flu*. Interestingly, pigs are an intermediary in viral transmission. That is, many viruses can be transmitted from chickens to pigs to humans, but not directly from chickens to humans.

Although swine influenza is quite common among pigs, the transmission to humans does not occur very often. Even then, swine influenza that is transmitted to humans does not always lead to human influenza. Those who are most likely to contract a swine influenza are individuals who raise and process pigs. If cooked properly, the meat poses very little, if any, risk of infection.

One of the major reasons for the appearance of new diseases is the expanding world population. As populations grow and expand into previously uninhabited areas, humans are exposed to insects and other animals that harbor infectious agents. Human contact with infected animals may result in a direct transmission of pathogens. Such a scenario probably explains how HIV "jumped" to humans from chimpanzees, which suffer a similar disease, and how the Ebola virus results in sporadic outbreaks of hemorrhagic fever in parts of Africa.

Another reason for the appearance of new diseases is the increased worldwide transport of animals—especially animals for the pet trade. The SARS virus probably was transmitted to humans as a result of animal transport in Asia. In fact, the trade in exotic animals into the United States caused the 2003 outbreak of monkeypox.

Increased international travel also can spread diseases to new geographical areas. It is believed that the West Nile virus that emerged in New York City in 1999 came from an individual or animal that was infected in the Middle East, where the virus is endemic. The virus has now spread throughout the entire United States.

Finally, changes in food handling or processing can also be the cause for an emergent disease. As Health Note 15-1 explains, prions, which are responsible for mad cow disease, spread from "infected" beef carcasses that were used to make cattle feed.

Reemerging infectious diseases are ones that have existed in the past but are now showing a resurgence in frequency or geographic range. Some of the more prominent reemerging diseases are cholera, tuberculosis, and dengue fever. One reason diseases are reemerging is antibiotic resistance. Antibiotic-resistant strains of the bacterium that cause tuberculosis, for example, have evolved over time. Because they are resistant to many of the antibiotics used to treat the disease, the disease is spreading. Yet another cause for the reemergence of pathogenic organisms is that large segments of the human population, notably AIDS victims, suffer from lowered immunity to infectious disease, allowing many formerly rare diseases to reappear.

Diseases are also reemerging because of lax public health programs. In the 1980s, the collapse of the Soviet Union and subsequent economic depression that hit many of the newly formed countries resulted in a decline in public health programs. This left many members of the population unvaccinated and thus susceptible to infectious disease. In one case, diphtheria vaccinations disappeared in parts of the former Soviet Union. Within a 3-year period, diphtheria became epidemic in those areas.

The emergence of new diseases and the reemergence of others once thought to be under control may also be due to climate change. Scientific studies have shown that warming in several regions has resulted in the spread of vector-borne infectious diseases such as dengue fever into neighboring areas, which were previously too cool to support the vectors. Although the magnitude of such infections is yet to be determined, global warming may become a prime factor in the emergence and reemergence of infectious disease in the decades ahead. Because additional emerging or reemerging infectious diseases are inevitable in the future, public health systems around the world need to be vigilant and join together to combat emerging infectious diseases and prevent their spread.

> **KEY CONCEPTS**
>
> Some infectious agents, thought to be under control, are reemerging because of antibiotic resistance and lax public health policies; new infectious agents are emerging because of the spread of people into virgin territories, global travel, and transport of animals throughout the world.

Bioterrorism

As you have seen, infectious disease in the human population results from the presence of infectious disease agents and many human factors, such as increasing population density, increased travel, and settlement of previously uninhabited areas. Infectious agents are also being used by terrorists to cause fear or inflict pain and suffering—and even death—in large populations. This threat is called **bioterrorism**. A bioterrorism attack, as defined by the CDC, is the deliberate release of viruses, bacteria, or other germs (agents) used to cause illness or death in people, animals, or plants. The anthrax attacks that occurred in the eastern United States in October 2001 demonstrate the potential for such agents to cause fear and anxiety.

Potential bioterrorists could rely on a large number of infectious agents, including pathogenic bacteria, fungi, viruses, bacterial toxins, and plant toxins like Ricin. The potential impact of the biological agents depends on their virulence and the ease with which they can be disseminated. The pathogens of most concern, called the *Category A Select Agents*, are those that can be spread via aerosols (Figure 15-14).

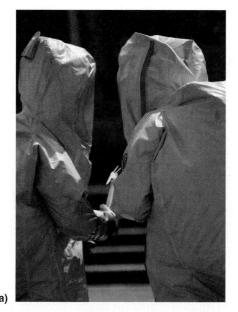
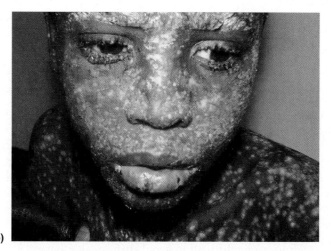

(a)   **(b)**

FIGURE 15-14 **Bioterrorism** (a) Combating the threat of bioterrorism often requires special equipment and protection because many organisms seen as possible bioweapons are spread through the air. (© Photodisc.) (b) This photo, taken in 1967, shows a smallpox patient. (Courtesy of Dr. J. Noble, Jr./CDC.)

This group includes anthrax and smallpox. Category A also includes bacterial toxins, poisonous substances that can be added to food or water supplies, such as the toxin that causes botulism.

Biological weapons offer rogue nations and terrorist groups or individuals several advantages over conventional weapons of mass destruction and terror. Perhaps most important, biological weapons are much cheaper to produce than chemical and nuclear weapons. In addition, they provide a weapon every bit as dangerous and deadly as the nuclear weapons of the more developed nations.

In May 2000, Ken Alibek, a scientist who once worked in the Soviet bioweapons program before defecting to the United States, testified before the U.S. House Armed Services Committee that the best defense against biological weapons is to develop appropriate medical defenses such as vaccines. These could minimize the impact of bioterrorism agents, rendering them useless.

The threat of biological weapons in the United States is being addressed, in part, by careful monitoring of sudden and unusual disease outbreaks. That way, health officials hopefully can react quickly to stop them from spreading. Extensive research is being carried out to determine the effectiveness of various antibiotic treatments. Research is also being carried out on the best ways to develop effective vaccines.

Vaccination of the general public probably offers the best defense against bioterrorism. As of 2003, the United States had stockpiled sufficient smallpox vaccine to vaccinate the entire population if a smallpox bioterrorist event occurred. Vaccines for other agents are being developed. In addition, scientists are trying to develop instruments that detect microbiological weapons before they can cause disease.

KEY CONCEPTS

Bioterrorism is the use of infectious pathogenic agents to frighten and inflict pain and suffering in populations.

15-6   Health and Homeostasis—Staying Healthy

In this chapter, I have described several human diseases that result from microbial pathogens in our environment. Such diseases represent a major disturbance of homeostasis. Just how susceptible are we to contracting an infectious disease? The answer depends on many factors, including our age, our health status, nutrition, and where we live. In the tropical regions of the world, for example, where malaria and cholera are endemic, a person is much more likely to contract these diseases. In the United States, these diseases pose little or no threat.

In general, a person's risk of contracting one of the emerging diseases, such as SARS and West Nile fever, is relatively low—especially if he or she is in good health. A diet rich in plant-based micronutrients, phytochemicals, studies show, can reduce the incidence of many diseases or lesson their impact.

All in all, it is much more likely that illnesses will result from pathogens found in the home or neighborhood. You are much more likely to get sick from contaminated surfaces in your kitchen than from the bite of an infected mosquito.

As an example, one of the most common foodborne illnesses is salmonellosis, caused by *Salmonella* bacteria found in contaminated foods. Salmonellosis is caused by eating meat that wasn't cooked long enough or that was cooked at a lower-than-optimal temperature. It is also caused by consuming food that has been improperly stored after cooking.

Salmonellosis is characterized by diarrhea, fever, and abdominal cramps. Nearly 40,000 cases are reported annually in the United States. However, the Centers for Disease Control and Prevention (CDC) estimate that the actual number of salmonella illnesses each year is greater than 1 million. That's because most mild cases are not reported to physicians. Careful attention to food preparation—such as making sure eggs and meats such as poultry are thoroughly cooked and food preparation surfaces are cleaned with soap and hot water first—can reduce the risk of food-borne pathogens.

Although it can be difficult to avoid surfaces that have been contaminated with flu or cold viruses, some simple steps can help minimize the chances of infection. Washing one's hands frequently with ordinary soap (no need to use antibacterial soaps), for example, is one of the easiest and most effective means of preventing the spread of infectious disease. As a precaution, you should wash your hands thoroughly before preparing or eating food, after coughing or sneezing, after changing a diaper, and after using the toilet. Be sure to wash thoroughly. Some experts recommend that handwashing take as long it takes to sing "Happy Birthday." You can find instructions on proper and washing on the Internet. When soap and water are not readily available, use a hand-sanitizing gel.

Do not handle an animal that looks ill or appears to have an infection, and do not share a drinking glass or eating utensils with someone who is sick. It is important to also stay up to date on infectious diseases that might have the greatest impact in your area, such as West Nile fever.

Certain medicines can keep you from contracting an infectious disease. For example, if you are traveling in an area where malaria is common, you may want to ask your doctor for an antiparasitic medication. In areas where you will encounter mosquitoes, ticks, and other biting insects that carry disease, doctors recommend the use of insect sprays containing the repellent DEET.

Health Tip 15-2

Don't waste your money on antibacterial soaps.

Why?

Antibacterial soaps and other cleaning products that contain antibacterial agents don't protect us like manufacturers would have you believe. One recent study showed that they don't lower the incidence of colds or flu in households that use them.

Antibacterial compounds in these products may be contributing to higher levels of asthma and allergies. Moreover, the antibacterial compounds are ending up in our waterways and in sewage sludge, much of which is applied to farmland. Although researchers are currently trying to determine what effect, if any, they might have, these antibacterial compounds could affect beneficial bacteria in the soil and could end up in the food supply.

Some infectious diseases, such as the common cold, usually do not require a visit to the doctor. However, if you think you have encountered a serious infectious disease, contact your doctor. He or she can perform the necessary tests to determine if you are infected, the seriousness of the infection, and how best to treat it. Even if you only have a virus, physicians may prescribe antibiotics if you develop a secondary bacterial infection. Remember, however, that although antibiotics may be a short-term answer for some bacterial diseases, long-term or indiscriminate use of antibiotics may result in the appearance of more resistant, harder-to-treat strains of bacteria.

Vaccination is the best line of defense for many infectious diseases. Over time, the list of vaccine-preventable diseases has continued to expand. Currently, there are more than a dozen vaccines available. These include vaccines for childhood diseases such as chickenpox, measles, mumps, and rubella. The newest is a vaccine against a virus (human papilloma virus) that can lead to cervical cancer in women. In addition, many vaccines or boosters are available for adults to prevent illnesses such as tetanus and diphtheria. Yearly flu shots usually provide a high degree of protection for you, your family, and the population in general. They are especially recommended for youngsters and adults over 50 who are more likely to die from the flu.

> **KEY CONCEPTS**
>
> Good nutrition and good health are keys to helping prevent infectious diseases or lessen their impacts.

SUMMARY

The Infectious Agents of Human Disease

1. Microorganisms are found in nearly every environment on Earth. Microbes range from submicroscopic viruses to macroscopic parasitic worms. Although they are found everywhere, only a very small percentage of the microbes found in our environment is capable of causing human disease.

2. Pathogens are the microbes capable of causing disease. If they can be transmitted from one organism to another, they are considered contagious; if they invade and cause disease, they are considered infectious.

3. Our present-day understanding of pathogens as the cause of human infectious disease arose largely from the work of Louis Pasteur and Robert Koch, who formulated the germ theory of disease in the late 1800s.

4. Viruses are composed of genetic information (DNA or RNA) and a protein capsid. The capsid may be surrounded by a viral envelope.

5. Viral replication requires the virus to (1) adhere to the appropriate host cells;

(2) be brought into the cell; (3) replicate its genetic information and manufacture capsid proteins; (4) assemble genetic information and capsids into new virus particles; and (5) be released from the host cell.

6. Most bacteria, which can be seen with a light microscope, are composed of rods, spheres, or spirals. They reproduce by replicating the DNA and then splitting to form two separate cells. Such reproduction can occur at a very rapid rate.

7. Eukaryotic microorganisms also can cause infectious disease. Fungi can cause athlete's foot, oral thrush, and serious respiratory infections. Protozoa are responsible for waterborne diseases. Malaria, one of the major infectious disease killers, takes the lives of more people worldwide than any other protozoan disease. The helminths consist of flatworms that cause diseases such as schistosomiasis and roundworms that cause pinworm disease and others.

The Course of a Human Disease

8. The ability of a pathogen to enter the body and cause disease is called pathogenicity. It varies from one pathogen to another. Some pathogens are opportunistic, only causing disease when the host's immune system is suppressed or unable to mount a defense. Virulence describes the degree of pathogenicity.

9. Most pathogens produce a set of signs and symptoms that are characteristic of the disease. Signs are what the physician can detect (mild fever and tissue swelling) and symptoms are what the patient feels (headache and sore throat). A syndrome is a specific set of signs and symptoms characteristic of some diseases, such as AIDS.

10. Most diseases go through a five-stage course, including: (1) the incubation period, (2) the prodromal phase, (3) the acme period, (4) the period of decline, and (5) the period of convalescence. Some convalescent patients may still be carriers for the disease from which they are recovering.

How Pathogens Cause Disease

11. Most pathogens must adhere to specific cells or tissues for an infection to progress. Adhesive proteins may be found on many bacteria and virus cell surfaces.

12. Many bacterial pathogens depend on enzymes to increase their virulence. Enzymes may protect the pathogen from destruction by the immune system, destroy immune cells, or facilitate the penetration of deeper host tissues.

13. Several bacterial pathogens produce toxins that also increase virulence. Exotoxins are products of cell metabolism that can destroy host cell structures or interfere with their metabolism. Their effects vary depending on the specific toxin ingested.

14. Endotoxins are parts of the cell wall in some bacteria. Their effects require a higher dose of toxin than exotoxins. The signs and symptoms are similar for all endotoxins.

How Infectious Diseases Are Transmitted

15. Infectious diseases can be endemic, that is, they remain localized within a small number of individuals. Infectious agents can also create epidemics, explosive increases in the numbers of diseased individuals. An explosive but localized increase in numbers of infected individuals is called an outbreak. A disease that produces explosive numbers of ill individuals worldwide results in a pandemic.

16. Diseases can be transmitted by direct contact of pathogens between an infected individual and another person, for example, from a mother to her unborn child.

17. Animals can carry human diseases, such as rabies and toxoplasmosis, and transmit them directly to humans. Direct contact transmission also occurs through respiratory droplets that can carry an infectious disease a few feet in the air.

18. Indirect contact is also responsible for the transmission of disease. Inanimate objects, for example, can become contaminated with disease agents. Contact with such objects can result in the transmission of the disease to another individual.

19. Aerosols, consisting of moisture droplets and dust contaminated with pathogens, can carry disease farther than respiratory droplets.

20. Arthropods, such as mosquitoes, fleas, and ticks, can carry infectious diseases. Infectious disease also can be spread through food or water.

Emerging Infectious Diseases and Bioterrorism

21. Emerging infectious diseases are those that have appeared in a population for the very first time, such as AIDS in the 1980s and SARS in 2003. Causes for such emergence include a growing world population that has become exposed to insects and animals harboring such infectious diseases; the worldwide transport of animals and exotic animals that harbor infectious disease; and the increased numbers of individuals who travel internationally.

22. Reemerging infectious diseases are those that were under control but now are increasing in incidence. The increase in antibiotic resistance, a lack of a strong public health system, reductions in vaccinations, and global warming are all responsible for the increasing numbers of such diseases.

23. Bioterrorism is a strategy used by terrorist organizations and rogue states to cause fear, inflict harm, or cause death in a population using pathogens or other harmful biological agents. Pathogens that can be dispersed relatively easily, such as the smallpox virus or anthrax bacteria, or bacterial toxins that can be placed in water supplies are some of the most dangerous agents. Minimizing such threats requires strong medical defenses (vaccines) and improved methods of detection.

Health and Homeostasis—Staying Healthy

24. The chance of contracting an emerging infectious disease is not high for individuals who have good, healthy immune systems. It is much more likely that an individual will contract an infection or illness from pathogens in his or her kitchen. Therefore, staying healthy means practicing those methods that eliminate and reduce the chances for disease transmission: keeping one's kitchen and bathroom clean, washing one's hands often, and making sure all vaccinations are up to date.

THINKING CRITICALLY ANALYSIS

This analysis corresponds to the Thinking Critically scenario presented at the beginning of this chapter.

There are several important points concerning you friend's decision to keep and eat the "potentially contaminated" luncheon meat. First, the disease is an example of an outbreak; it appears to be limited (at the time of the newscast) to a few individuals in the town. Second, because your friend is in apparently good health, he is not at great risk. Remember that poor health makes an individual more susceptible to many potentially harmful microbes.

Another point to remember is that if the packaged meat is contaminated with *Listeria*, this disease is not contagious. A person with the disease cannot transmit it directly to anyone else. However, the meat does represent a mechanism for indirect transmission. You should advise your friend not to eat it or offer it to anyone. If he objects, you should note that it is possible that the luncheon meat is contaminated with *Listeria*, but that the number of bacteria at the time he ate his first sandwich might have been low. Leaving the meat in the refrigerator longer provides time for the bacteria to grow and produce significant numbers (a dose) that could be dangerous if ingested at a later date.

Tell your friend no thanks for the lunch invitation! And advise him not to eat it or offer it to anyone else. It's best if he disposed of it immediately.

KEY TERMS AND CONCEPTS

Acme period, p. 341
Aerosol, p. 344
Bacterium, p. 332
Biofilm, p. 342
Bioterrorism, p. 346
Capsid, p. 331
Clinical disease, p. 341
Contagious, p. 329
Direct transmission, p. 343
Disease, p. 338
Emerging infectious disease, p. 345
Endemic, p. 343
Endotoxin, p. 342
Epidemic, p. 343
Exotoxin, p. 341
Flatworm, p. 336
Fungus, p. 334

Germ theory of disease, p. 329
Helminths, p. 336
Host, p. 331
Incubation period, p. 340
Indirect transmission, p. 344
Infection, p. 338
Infectious, p. 329
Microorganism, p. 328
Opportunistic, p. 338
Outbreak, p. 343
Pandemic, p. 343
Parasite, p. 335
Pathogen, p. 329
Pathogenicity, p. 338
Period of convalescence, p. 341
Period of decline, p. 341
Prion, p. 336

Prodromal phase, p. 340
Protozoan, p. 335
Reemerging infectious diseases, p. 346
Roundworm, p. 336
Sign, p. 340
Subclinical disease, p. 340
Symptom, p. 340
Syndrome, p. 340
Tapeworm, p. 336
Toxin, p. 341
Vector, p. 344
Viral envelope, p. 331
Virulence, p. 338
Virus, p. 331

CONCEPT REVIEW

1. List and briefly describe the five steps that must take place for a virus to infect a cell, replicate its kind, and spread to other cells. p. 332.
2. Identify several diseases caused by fungi, protozoa, and helminths. Have you had any of them, or do you know anyone who has? pp. 333–335.
3. Define the terms pathogenicity and virulence. pp. 338–339.
4. List and describe the five stages in the course of an infectious disease. When is an individual likely to spread the disease? pp. 340–341.

5. How are infectious diseases transmitted? Give examples of each major type. pp. 343–345.
6. What is a vector? Describe the importance of vectors in disease transmission. pp. 344–345.
7. Describe the difference between an epidemic and an outbreak. p. 343.
8. How do enzymes help pathogens overcome host defenses? pp. 341–342.
9. What are exotoxins and endotoxins? pp. 341–342.
10. Identify and explain the reasons for the emergence of "new" infectious diseases

and the reemergence of infectious diseases once thought to be under control. pp. 345–346.
11. Make a list of the five most likely areas in your home where illness and disease could arise. Why did you select these areas? pp. 347–348.
12. Define bioterrorism and identify a viral disease and two bacterial diseases that are considered potential bioterror agents. Why are these agents of particular concern? pp. 346–347.

SELF-QUIZ: TESTING YOUR KNOWLEDGE

1. Microbes that cause disease are called _____. They are _____ diseases if they can be transmitted easily between humans. p. 329.
2. Two scientists, _____ and _____, were responsible for proposing and developing the germ theory of disease. p. 330.
3. The viral genetic information is surrounded by a protein coat, called a/an

_____, which in turn may be surrounded by a/an _____. p. 331.
4. In order to replicate, a virus must infect a/an _____ cell. p. 331.
5. The three variations in bacterial shape are _____, _____, and _____. p. 332.
6. Among the eukaryotic pathogens, yeast-like infections are caused by _____, the _____ often

are parasites of the intestinal tract, and the _____ are parasitic worms. pp. 334–336.
7. A _____ is a change from the healthy state of the body, which depends on the _____ of the infecting organism. p. 338.
8. Those pathogens that only cause disease when the host's immune system is suppressed are called _____. p. 338.

9. The term _____ is used to describe those changes in body function that a patient experiences while _____ is used to identify those changes that can be detected and measured by a physician. p. 340.

10. The period between a person's exposure to a pathogen and the appearance of the first symptoms is called the _____ period, while the _____ period is the stage characterized by the development of very specific symptoms characteristic of the disease. pp. 340–341.

11. The two different types of bacterial toxins can be recognized by the fact that the _____ are produced usually only after the death or disintegration of bacterial cells while the _____ are produced by live bacteria in the host tissues. pp. 341–342.

12. The occurrence of rabies in several dozen residents in a city would be considered a/an _____ because the disease had not spread throughout the state or neighboring states. p. 343.

13. Being exposed to the flu virus from touching a contaminated door knob would be an example of _____ transmission while being exposed to the virus from respiratory droplets in the air would be an example of _____ transmission. pp. 343–344.

14. Examples of recently emerging infectious diseases would include _____ and _____; reemerging infectious diseases are represented by _____ and _____. pp. 345–346.

15. The best defense against the potential use of a bioterrorism agent is through _____ of the population. p. 347.

biology.jbpub.com/chiras/8e/

The site features eLearning, an online review area that provides quizzes, chapter outlines, and other tools to help you study for your class. You can also follow useful links for in-depth information, research the differing views in the Point/Counterpoints, or keep up on the latest health news.

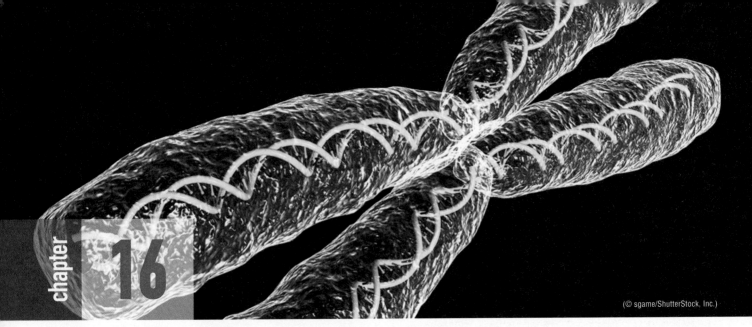

(© sgame/ShutterStock, Inc.)

Chromosomes, Cell Division, and the Cell Cycle

Electricity flowing through electrical wires in your house and in your neighborhood produces a magnetic field that spreads out from the wire. In 1979, two researchers from the University of Colorado discovered that magnetic fields such as these could produce serious health effects. Their studies suggested that magnetic fields generated by high-voltage power lines like those shown in Figure 16-1, may increase the incidence of leukemia (a cancer of the white blood cells) in children who live nearby. Cancer death rates in these children were twice what would have been expected in the general public.

THINKING CRITICALLY

In the 1970s, scientists began warning people about the potential dangers from the release of a class of chemicals called *chlorofluorocarbons* (CFCs) into the atmosphere. Once used in spray cans, refrigerators, freezers, and as blowing agents to make rigid foam like Styrofoam®, CFCs escape from these products and drift into the upper atmosphere. There, sunlight causes them to break apart, producing a chemical that destroys ozone molecules (O_3) in the ozone layer. The ozone layer filters out 99% of the incoming ultraviolet radiation. Exposure to ultraviolet radiation causes skin cancer, cataracts, and other problems. Without it, life on planet Earth could not exist, except in the oceans, lakes, and other waterways. As the ozone layer thinned, environmentalists and others warned that skin cancer rates and mortality from skin cancer could increase dramatically.

Against this gloom and doom, a loud dissenting voice was raised on the airways of America, led in large part by conservative radio talk show host Rush Limbaugh. He called the ozone threat a "scam." He said scientists were behind it, in part, because they viewed it as a way to enhance their funding. (These are the same claims he is currently making to try to discredit assertions about global warming and global climate change.) He and others also alleged that claims by environmentalists and scientists about ozone depletion and its effects were wrong.

Environmentalists countered that Limbaugh and others like him were spreading misinformation. How would you go about analyzing this debate?

FIGURE 16-1 **Too Close for Comfort** Some early studies suggested that extremely low frequency radiation from power lines may increase the incidence of cancer, especially childhood leukemia, in nearby residents. A recent analysis of the research, however, indicates that this link between cancer and electrical lines is false. (© Frank Anusewicz/ShutterStock, Inc.)

In 1986, researchers from the University of North Carolina published reports showing a fivefold increase in childhood cancer (particularly leukemia) in families living 25 to 50 feet from such wires. Experimental work in the laboratory also suggests a link between magnetic fields, known as extremely low frequency (ELF) fields, and cancer. For example, one researcher discovered that ELF fields increased the growth rate of cancer cells in tissue culture. They also rendered the cells 60% to 70% more resistant to the immune system's naturally occurring killer cells, which attack cancer cells.

Despite these and other studies, a panel of scientists convened by the prestigious National Research Council concluded in 1986 that electromagnetic fields associated with power lines, appliances, and other sources do not pose a health risk to humans. Their conclusion was based on an exhaustive analysis of over 500 studies on the subject performed over a 17-year period. They concluded that although some studies do show an effect, the bulk of the evidence shows no credible link. In those studies that demonstrated a link, scientists suggested that other unmeasured factors may have been responsible.

Not everyone agrees with this assessment. Doctors in Europe, for instance, recognize hypersensitivity to electromagnetic fields as a distinct disease. They treat patients with the disease in a way similar to the way they treat people who have allergies. This chapter provides basic information about chromosomes and cell division that will help you understand this debate as well as the information on cancer.

16-1 The Cell Cycle

All cells arise from preexisting cells via cell division. This process explains why you grew from a fertilized ovum to an embryo and then to an adult. It also explains why many tissues and organs of your body can repair themselves when damaged. Simply put, damaged cells are replaced by new cells by cell division.

As shown in Figure 16-2, cell division is one part of a cell's life cycle, called the **cell cycle**. As shown in the diagram, the cell cycle of actively dividing cells consists of two parts: cell division and interphase. During **cell division**, the nucleus and the cytoplasm divide, splitting the cell more or less equally and thus forming two new cells. Interphase is the period between cell divisions. Let's take a look at each part of the cell cycle, starting with interphase.

> **KEY CONCEPTS**
>
> The cell cycle is divided into two parts: cell division and interphase.

Interphase

Once thought to be a cellular resting period, **interphase** is now known to be a time of intense metabolic activity. During interphase, for example, cells replicate their DNA. They also replicate many cytoplasmic components (including organelles).

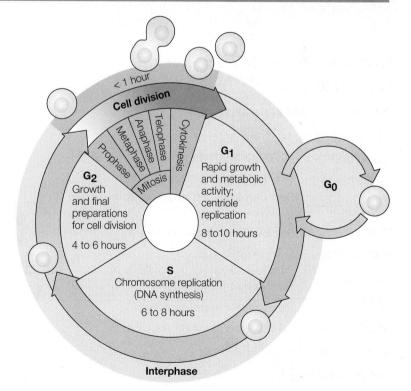

FIGURE 16-2 **The Cell Cycle** The cell cycle consists of two parts: interphase and cell division. Interphase is divided into three parts. The times given here represent the approximate time mammalian cells spend in each phase when grown in the lab. Cells that do not divide enter the G₀ phase.

The replication of their DNA and their cellular organelles prepares cells for division. Thus, when a cell divides, the products of this process—called **daughter cells**—receive the nuclear material (genes) and cellular organelles needed to survive and function.

As shown in Figure 16-1, interphase is divided into three phases: G_1, S, and G_2 (Figure 16-2 and Table 16-1). G_1 (**gap 1**) begins immediately after a cell divides. You'll also notice G_o. G_o is unique to nondividing cells. It is a kind of resting phase in cells that have "left" the cell cycle—that is, stopped dividing (more on this shortly.)

In actively dividing cells, G_1 is a period during which the cell manufactures RNA, proteins, and other key molecules. These molecules are vital for cell function and are stockpiled during G_1 to ensure that each daughter cell receives an adequate supply of nutrients.

The length of G_1 varies considerably from one cell to another. Some cells spend only a few minutes or hours in G_1; others spend weeks or even months. Cells such as nerves that never divide enter G_o (Figure 16-2), a metabolically active period that does not involve DNA replication. In general, mammalian cells in culture spend 8 to 10 hours in G_1.

In the next phase of the cell cycle—**the S phase** or **synthesis phase**—the genetic material, DNA, replicates. **DNA** is shorthand for **deoxyribonucleic acid**. This complex molecule is a polymer of smaller molecules called **nucleotides**. Each nucleotide consists of a sugar, a phosphate, and a nitrogen-containing base. During DNA synthesis, nucleotides covalently bond to another forming a long polynucleotide chain. As shown in Figure 16-3, two nucleotide chains intertwine like snakes, forming a double helix. That is, the intertwined chains form a single spiral-like molecule of DNA called the **double helix** (Figure 16-3). It's a little like a spiral staircase.

As a general rule, mammalian cells grown in the laboratory in containers with proper nutrients spend 6 to 8 hours in the S phase. Immediately following this phase is a period known as G_2 (**gap 2**), which averages 4 to 6 hours in cultured mammalian cells. During this phase, mitochondria divide. The cell also produces spindle fibers, tiny microtubular

TABLE 16-1	Phases of the Cell Cycle
Phase	**Characteristics**
Interphase	
G_1 (gap 1)	Stage begins immediately after mitosis.
	RNA, protein, and other molecules are synthesized.
S (synthesis)	DNA is replicated.
	Chromosomes become double stranded.
G_2 (gap 2)	Mitochondria divide; precursors of spindle fibers are synthesized.
Mitosis	
Prophase	Chromosomes condense.
	Nuclear envelope disappears.
	Centrioles divide and migrate to opposite poles of the dividing cell.
	Spindle fibers form and attach to chromosomes.
Metaphase	Chromosomes line up on equatorial plate of the dividing cell.
Anaphase	Chromosomes begin to separate.
Telophase	Chromosomes migrate or are pulled to opposite poles.
	New nuclear envelope forms.
	Chromosomes uncoil.
Cytokinesis	Cleavage furrow forms and deepens.
	Cytoplasm divides.

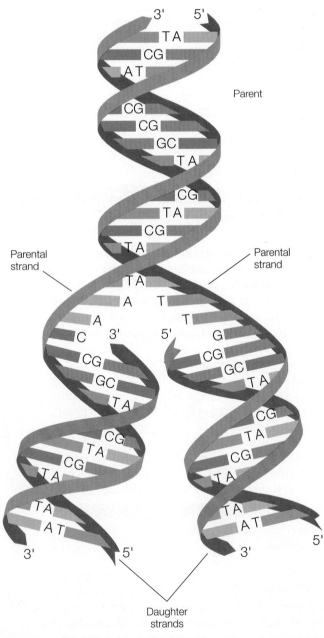

FIGURE 16-3 **DNA Replication** During the S phase of the cell cycle, the DNA molecule unwinds and duplicates, forming two strands, each containing a DNA double helix.

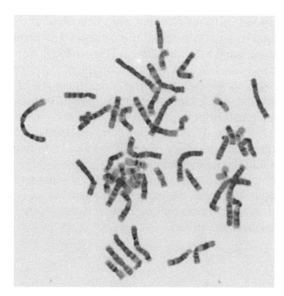

FIGURE 16-4 Metaphase Chromosomes Taken from a human cell during metaphase, these chromosomes are clearly visible under a light microscope. (© Donna Beer Stolz, Ph.D., Center for Biologic Imaging, University of Pittsburgh Medical School.)

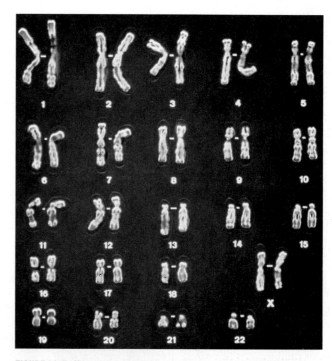

FIGURE 16-5 Karyotype Chromosomes from a normal human female arranged in order of decreasing size. Note that there are 46 chromosomes—or 23 pairs—in somatic cells. Each pair is similar in size, banding pattern, and location of the centromere. (© SPL/Custom Medical Stock Photo.)

structures that are essential to nuclear division. The DNA also begin to condense during G_2, forming distinct blocklike structures called **chromosomes**.

> **KEY CONCEPTS**
>
> Interphase is a period of intense activity between cell divisions and is divided into three parts.

Cell Division

Interphase is followed by cell division, a rapid event that lasts only about 1 hour. During cell division, the nucleus and cytoplasm both divide. However, nuclear division, commonly known as mitosis (my-TOE-siss), precedes cytoplasmic division.

Mitosis involves a series of dramatic structural changes in the genetic material, the chromosomes. In order for the replicated chromosomes to divide, the DNA must first condense. During this process, the chromosomes become compact structures clearly visible via light microscopic preparations (Figure 16-4). Condensation facilitates chromosomal segregation. Without it, cell division would be impossible. Dividing the 46 unraveled chromosomes in humans would result in a tangled mass of broken chromosomes. It would be like trying to separate a plate of spaghetti into two equal piles without breaking a single strand.

During mitosis the chromosomes line up in the center of the cell. As shown in Figure 16-5, each chromosome consists of two parts, called **chromatids**. After lining up in the center of the cell, the two chromatids making each chromosome are drawn apart. One half of each chromosome goes to each daughter cell. This alignment in the center of the cell allows chromosomes to be equally divided.

Cytoplasmic division, or **cytokinesis** (literally, "cell movement," pronounced SITE-toe-ki-NEE-siss), occurs independently of nuclear division (mitosis). It typically happens toward the end of mitosis.

Before we study this process in more detail, let's take a moment to become more familiar with chromosomes.

> **KEY CONCEPTS**
>
> Nuclear and cytoplasmic division occur separately.

16-2   The Chromosome

Chromosomes contain the genetic information of cells, DNA. As you most likely know by now, DNA controls many cellular functions. (For a discussion of the scientific work that led to the discovery of DNA, see **Scientific Discoveries 16-1**; Table 16-2 reviews terminology.) As you also probably know by now, all organisms have a set number of chromosomes. Each cell in your body and every other human, for example, contains 46 chromosomes, except for the gametes (or germ cells—the sperm and egg). Gametes contain half as many cells. In our case, they contain 23 chromosomes.

Figure 16-4 shows the chromosomes in a human being. Take a moment to study the figure. What do you notice? Probably not much, other than the fact that the chromosomes are banded. Now take a moment to examine Figure 16-6. This

TABLE 16-2	Review of Terminology
Chromatin	General term referring to a strand of DNA and associated histone protein
Chromatin fiber	Strand of chromatin
Chromosome	Structure consisting of one or two chromatin fibers
Chromatid	Generally used to refer to one of the chromatin fibers of a replicated chromosome
Centromere	Region of each chromatid to which a sister chromatid attaches

photo shows the same chromosomes but paired or matched according to size, location of the centromere (the point of attachment), and banding patterns (Figure 16-5). How many pairs do you see? Go ahead and count them.

What you should have found is that there are 23 sets or pairs of structurally similar chromosomes in this photo. As you may recall from high school biology, one chromosome of each pair comes from the mother; the other comes from the father. Each member of a pair is called a **homologous** (ha-MOLL-ah-GUS) **chromosome**.

Homologous chromosomes contain genes that control the same inherited traits. For example, if a gene for hair color is located on one chromosome of a pair, it is also located on the other member of the pair.

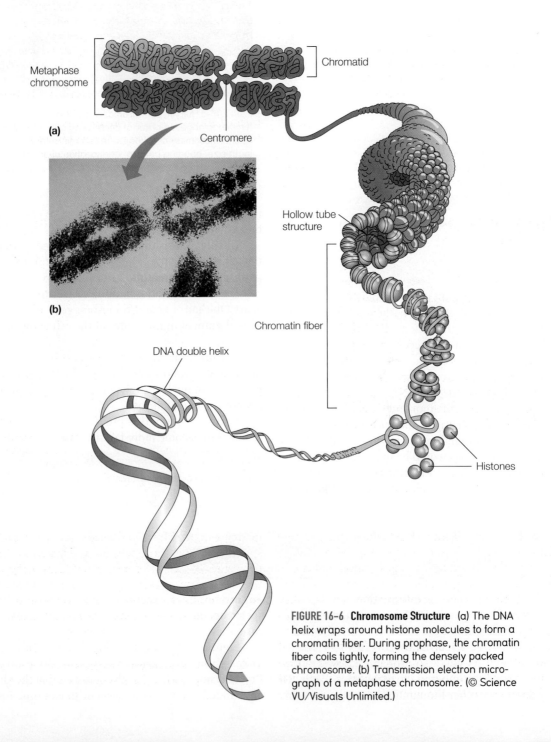

FIGURE 16-6 Chromosome Structure (a) The DNA helix wraps around histone molecules to form a chromatin fiber. During prophase, the chromatin fiber coils tightly, forming the densely packed chromosome. (b) Transmission electron micrograph of a metaphase chromosome. (© Science VU/Visuals Unlimited.)

Scientific Discoveries that Changed the World

16-1 Unraveling the Structure and Function of DNA
Featuring the Work of Watson, Crick, Wilkins, and Franklin

In 1953, two scientists, James Watson and Francis Crick published a brief paper that proposed a model (hypothesis) for the structure of DNA (Figure 1). They suggested that DNA was a double helix and explained how the nucleotides fit together.

Like many other scientific advances that changed our understanding of the world and, in some cases, our way of life, the discovery of the structure of DNA resulted from the efforts of many scientists over many years. This work began in the mid-1940s and culminated in the publication of Watson and Crick's paper in 1953. Watson and Crick used available information from other researchers. One vital piece came from Rosalind Franklin, who worked in the laboratory of British researcher Maurice Wilkins. Franklin produced x-ray photographs of crystals of highly purified DNA (Figure 2). Her remarkable photographs suggested that the DNA molecule was helical.

Another vital piece of evidence came from the laboratory of Erwin Chargaff and his colleagues, who had spent years studying the chemical nature of DNA. Chargaff's lab showed that the amount of purine in DNA always equals the amount of pyrimidine and that the ratios between adenine and thymine and between cytosine and guanine are always 1:1.

Armed with these and other vital details, Watson and Crick devised an ingenious model that fit the chemical and physical data. They published their results in 1953 in the journal *Nature*, and 2 months later they published a second paper proposing that the complementary strands of their model could explain replication. In addition, Watson and Crick hypothesized that genetic information could be stored in the sequence of bases and that mutations might result from a change in that sequence.

In the ensuing years, numerous scientists have provided hard evidence that supports the model proposed by Watson and Crick. And in 1962, Watson, Crick, and Wilkins were awarded the Nobel Prize for their discovery. Interestingly, although much of the x-ray data on which the Watson-Crick model was based had come from Franklin, she was not included in the celebrations. She couldn't be. Unfortunately, Rosalind Franklin died of cancer in 1958, and Nobel Prizes are awarded only to living scientists. Had she been alive, it is likely that she too would have shared in this coveted prize and been a household name as well.

FIGURE 1 The Double Helix Model for DNA Watson and Crick and their double helix model. (© A. Barrington Brown/Science Source, Inc.)

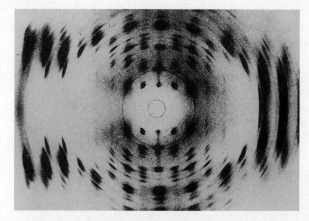

FIGURE 2 Discovering DNA's Helical Shape The famous X-ray diffraction photograph of DNA by Rosalind Franklin. (© Science VU/NIHLBL/Visuals Unlimited.)

The presence of homologous chromosome pairs results from sexual reproduction—the uniting of a sperm and an ovum. During sexual reproduction, a new individual is created when the sperm and egg combine. The sperm and the egg each contain 23 single chromosomes. When they combine, they produce an offspring whose cells contain 46 chromosomes—23 from each parent.

In humans, cells that contain the full number of chromosomes—that is, 46 chromosomes—are said to be **diploid** (diploid = twofold). All body cells, like those of skin and muscle, are diploid. Such cells are often called **somatic cells** to distinguish them from **germ cells**, the sperm and ovum. The sperm and ovum carry paternal and maternal chromosomes, respectively. Germ cells are **haploid** because they contain half the chromosomes of somatic cells. Germ cells are produced by a special kind of cell division known as *meiosis* (my-OH-siss), which occurs in the gonads (ovaries and testes).

> **KEY CONCEPTS**
> Human body cells contain 46 chromosomes that contain all of the genetic information required to control cellular activity.

Chromosomal Condensation

Soon after they duplicate during interphase, the chromosomes of the cell begin to coil, compacting in much the same way that a stretched phone cord shortens and compacts when the tension is removed. Unlike a phone cord, which functions just as well when it is stretched or compacted, chromosomes in the condensed state are metabolically inactive—that is, unable to produce either RNA or DNA.

The structure of a condensed chromosome is shown in the photo in Figure 16-6 and the drawing just above it. To understand this structure, let's begin at the bottom and work up. As illustrated, DNA molecule consists of two strands wrapped around each other to form a double helix. The DNA double helix is associated with protein molecules inside the nucleus called histones (HISS-tones). As shown, **histones** are globular proteins. Histones are extremely important; they play a key role in controlling DNA's activities.

As illustrated in Figure 16-6, loops of the double helix of DNA encircle the histones, forming small clusters. The small clusters along the DNA molecule align in such a way as to form hollow tubules. These tubules, in turn, form larger coils like those of a phone cord. When the coils are compacted, they form the body of the chromosomes.

Chromosomal condensation is vital to cell replication. It is also vital to gene expression—how the genes operate. The fact that chromosomes condense prior to nuclear division also has many practical applications. For example, it enables geneticists to spot potential chromosomal defects. This is especially useful to expectant parents who are concerned about possible genetic defects in their unborn babies.

To test for chromosomal abnormalities, physicians extract fluid from the liquid-filled cavity surrounding the growing fetus through a long needle inserted through the mother's abdomen (Figure 16-7). This process is called **amniocentesis** (AM-knee-oh-cen-TEE-siss).

Fetal skin cells present in the fluid are separated from it, and then grown in culture dishes. There, they proliferate, dividing by mitosis. After the number of cells has increased, the tissue culture is then treated with a chemical substance that halts cell division. The cells are then removed from the culture, placed on a glass slide, and stained. A small piece of glass or plastic is placed over them and compressed, causing the cells to flatten out. The cells containing condensed chromosomes are then photographed through a camera mounted on the microscope. Using a computer program, chromosomes are arranged in homologous pairs with the largest first, as shown in Figure 16-5. The resulting display is called a **karyotype** (CARE-ee-oh-TYPE).

Lining the chromosomes up like this helps geneticists locate obvious structural defects—for example, extra chromosomes or missing segments. More subtle changes in the DNA itself (like mutations) cannot be detected this way. Gross defects in chromosomes can have serious consequences. It can result in children with Down syndrome, a genetic disorder characterized by stunted mental and physical growth. In many cases, chromosomal analysis assures parents that their child has no obvious genetic defects, alleviating worry during this period of heightened anxiety. Chromosomal analysis has the added benefit of helping parents learn the sex of their baby.

> **KEY CONCEPTS**
> Chromosomes condense after replication, which facilitates nuclear division (mitosis).

Chromatids and Chromosomes

The study of genetics can be quite confusing at first. Part of this confusion arises from the fact that chromosomes may contain one or two chromatids, depending on the stage of the cell cycle. For example, as noted earlier, during G_1 of interphase, each chromosome consists of a single DNA molecule

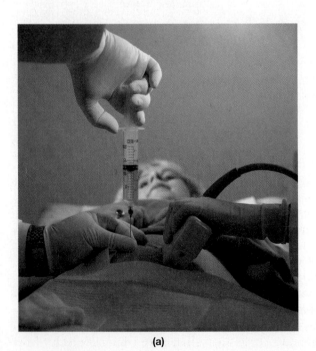

(a)

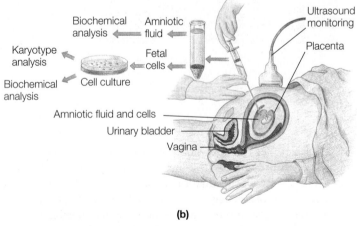

(b)

FIGURE 16-7 **Amniocentesis** (a) A patient undergoing amniocentesis. (© Yoav Levy/Phototake/Alamy Images.) (b) A needle is inserted into the fluid-filled space surrounding the fetus. Fluid containing fetal cells is then withdrawn. Cells are subjected to chromosomal and biochemical analyses. Ultrasound monitors are used to direct the needle and prevent injury to the fetus.

(a double helix) and associated protein. The 46 unreplicated chromosomes in the human cell are packed loosely in the nucleus. During the S phase, however, the DNA strands replicate (Figure 16-3). Each chromosome now consists of two identical chromatids held together at the centromere. Each chromatid is a molecule of DNA, arranged in a double helix, with its associated histone proteins. When a cell divides,

these double-stranded chromosomes are split in half, with one chromatid going to each daughter cell.

> **KEY CONCEPTS**
>
> Chromatids are strands of genetic material containing DNA and histone proteins. The number of chromatids in a chromosome depends on the stage of the cell cycle.

16-3 Cell Division: Mitosis and Cytokinesis

With this information in mind, let's explore cell division, beginning with mitosis, nuclear division.

Mitosis

The human body contains many actively dividing cells, including those of the skin, hair follicles, and the lining of the intestine (Figure 16-2). Cell division, as you learned earlier consists of two parts: nuclear division, called mitosis, and cytoplasmic division, called cytokinesis. Let's begin with mitosis.

As you will learn in this section, **mitosis** involves four stages: prophase, metaphase, anaphase, and telophase shown in Figure 16-8.

Prophase. **Prophase** begins immediately after interphase. Several important changes occur during prophase. As shown in Figure 16-8a and b, the centrioles separate and begin to migrate to opposite sides of the nucleus. They are organelles that are essential to cell division. As shown in Figure 16-9, each **centriole** consists of two small, cylindrical structures identical to the basal bodies. Like other organelles, centrioles replicate during interphase.

As the centrioles migrate to opposite sides of the nucleus, another interesting development takes place—an elaborate array of microtubules forms between the centrioles. This is known as the **mitotic spindle** (Figure 16-10). Its fibers attach to the chromosomes, providing a way to pull the chromatids from chromosome to opposite sides of the cell, which, in turn, enables the nucleus to divide into two equal parts. The formation of the mitotic spindle is shown in Figure 16-9. Take a moment to see how the centrioles separate and the spindle forms.

Figure 16-10 shows the completed spindle and the position of the centrioles. Each spindle fiber consists of a bundle of microtubules composed of tubulin molecules. These molecules are derived from the disassembly of the cell's cytoskeleton during mitosis. As you can see from all three figures, associated with the centrioles in animal cells is a starburst of microtubules called the **aster**. (Aster is a type of flow with radiating petals.) Its role in cell division remains unclear, although it may anchor the spindle to the plasma membrane.

During prophase, the replicated chromosomes condense—that is, shorten and thicken—as you can see by studying Figure 16-8a and b. In addition, the nucleoli disappear. The **nucleoli** are regions of active ribosomal RNA (rRNA) synthesis.

As shown in Figure 16-8c, late in prophase, the nuclear envelope begins to break down. As it disintegrates, mitotic microtubules of the mitotic spindle begin to attach to the chromosomes, setting the stage for the next step, metaphase.

Metaphase. Before chromosomes can be divided equally, they are lined up in the center of the cell with their centromeres located along the equatorial plane (Figure 16-8d). This alignment occurs during **metaphase** and greatly facilitates the separation of chromatids.

Anaphase. When the chromatids of each chromosome begin to separate, the cell enters **anaphase**. The shortest part of mitosis, anaphase lasts only a few minutes (Figure 16-8e). During anaphase, the chromatids of the homologous chromosomes are drawn toward opposite poles of the cell by the mitotic spindle. As shown in Figure 16-8e, the chromosomes are dragged centromere-first, with their arms trailing behind. All of the chromosomes begin to separate simultaneously.

Telophase. The final stage of nuclear division is **telophase** (Figure 16-8f). Telophase begins when the chromosomes complete their migration to the poles. This stage involves a series of changes in the nuclei that is essentially the reverse of prophase. Take a moment to make a mental list of those changes.

Here's what you should have come up with: during telophase, the nuclear membrane reforms. Interestingly, membranes first form around each of the chromosomes of the forming daughter cells. These membranes soon fuse forming one all-inclusive nuclear envelope. Also during telophase, the spindle fibers disappear. They're no longer needed. The chromosomes uncoil, regaining a threadlike appearance. This allows the genes to begin functioning again. Nucleoli also reappear. Telophase ends when the nuclei of the daughter cells appear to be in interphase.

> **KEY CONCEPTS**
>
> Mitosis is divided into four stages.

Cytokinesis

Cytokinesis is the division of the cytoplasm of a cell. It usually begins late in anaphase or early in telophase. Cytokinesis is brought about by a dense network of contractile fibers, called microfilaments, which lie beneath the plasma membrane. These microfilaments, which are part of the

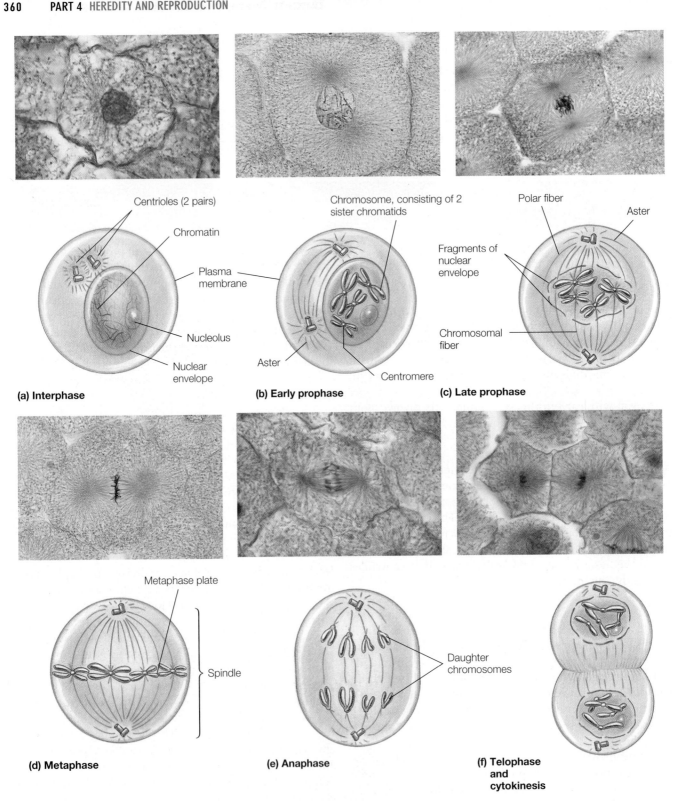

(a) Interphase

Centrioles (2 pairs)
Chromatin
Plasma membrane
Nucleolus
Nuclear envelope

(b) Early prophase

Chromosome, consisting of 2 sister chromatids
Aster
Centromere

(c) Late prophase

Polar fiber
Aster
Fragments of nuclear envelope
Chromosomal fiber

(d) Metaphase

Metaphase plate
Spindle

(e) Anaphase

Daughter chromosomes

(f) Telophase and cytokinesis

FIGURE 16-8 **Mitosis** (Photos a and b © Michael Abbey/Visuals Unlimited, c © Ed Reschke/Getty Images, d © John D. Cunningham/Visuals Unlimited, e and f © M. Abbey/Science Source, Inc.)

cytoskeleton of the cell, are composed of the same proteins found in muscle cells.

During cytokinesis, the microfilaments along the midline of the cell contract and pull the membrane inward (Figure 16-11). The membrane furrows as if it were being constricted by a tightening thread tied around the cell. As cytoplasmic division proceeds, the furrow deepens, eventually pinching the cell in two. This results in two daughter cells with more or less equal amounts of cytoplasm and cellular organelles. Each has all the chromosomes and genes it needs to carry out its functions.

KEY CONCEPTS

Cytokinesis is the division of the cytoplasm and is made possible by contractile protein filaments (microfilaments) found just beneath the plasma membrane.

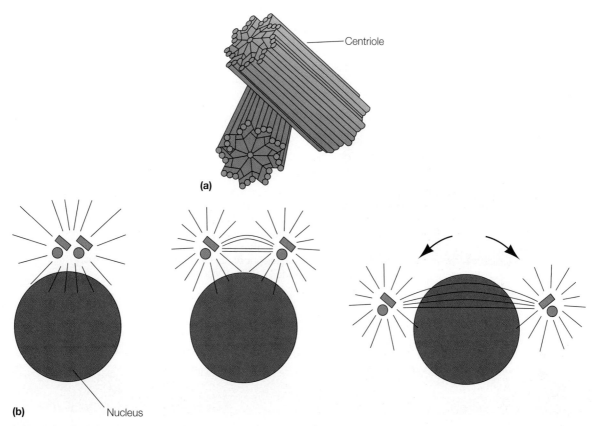

(a)

(b)

Centriole

Nucleus

FIGURE 16-9 Centriole (a) The centriole, like the basal body, consists of nine sets of microtubules with three in each set. Centrioles are found in the cytoplasm and replicate during interphase. (b) They migrate to opposite poles of the nucleus during mitosis.

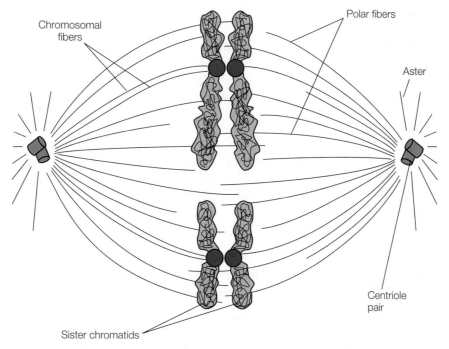

Chromosomal fibers

Polar fibers

Aster

Centriole pair

Sister chromatids

FIGURE 16-10 Mitotic Spindle The mitotic spindle, made of microtubules, forms during prophase in the cytoplasm of the cell. Chromosomal fibers connect to the chromosomes, and polar fibers extend from a pole to the equatorial region of the spindle. In ways not currently understood, the mitotic spindle separates each double-stranded chromosome during mitosis.

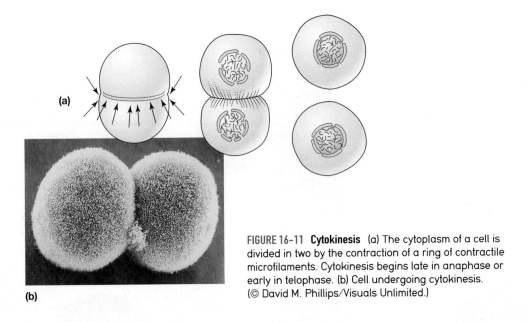

FIGURE 16–11 **Cytokinesis** (a) The cytoplasm of a cell is divided in two by the contraction of a ring of contractile microfilaments. Cytokinesis begins late in anaphase or early in telophase. (b) Cell undergoing cytokinesis. (© David M. Phillips/Visuals Unlimited.)

16-4 Control of the Cell Cycle

The cell cycle controlled by chemical messages arising from within the cell, specifically from the cytoplasm. Research also shows that signals from outside the cell control the cell cycle. For example, studies of cell cultures—that is, cells grown in the laboratory under precise conditions—have shown that normal cells migrate and divide until they spread evenly across the bottom surface of culture dishes. At this point, the cells stop dividing, even though nutrients are plentiful. The contact of cells with one another apparently inhibits further growth; this process is called **contact inhibition**. Contact inhibition also occurs in the tissues of the body.

Interestingly, one of the reasons cancer cells grow uncontrollably is that they lose contact inhibition. Thus, in a culture dish, cancer cells proliferate wildly, growing on top of one another and quickly utilizing nutrients in the nutrient bath they're immersed in. In the body, cancer cells grow in a similar way, forming large growths called **tumors**. Cancer cells in tumors consume large quantities of vital nutrients, weakening and eventually killing host cells. As the cancerous cells explode in number they can destroy vital organs, which leads to major disruptions in homeostasis. Elsewhere in the text you learned about leukemia, a white blood cell (WBC) cancer. As you may recall, cancerous WBCs proliferate in the bone marrow, squeezing out the cells that produce blood-clotting platelets. They also squeeze out the cells that form red blood cells (RBCs), resulting in anemia. In addition, the cancerous WBCs that enter the blood stream do not function normally, so they provide no protection against infectious agents. Loss of clotting, effective WBCs, and RBCs all lead to death.

Hormones affect the cell cycle as well. During each menstrual cycle, the hormone estrogen released from a woman's ovaries stimulates the growth of breast tissue and the uterine lining in preparation for pregnancy. Several other hormones also stimulate the division of body cells.

Growth-promoting and growth-inhibiting factors likewise play a role in controlling the cell cycle. Skin cells, for example, produce a growth-inhibiting factor that prevents cell division. However, damage to the skin—say, a cut or burn—reduces the number of skin cells and presumably reduces the amount of inhibiting factor at the site of injury. This, in turn, releases cells from inhibition. Cells proliferate and repair the damaged area. When the tissue is repaired, the rate of cell division returns to normal.

Researchers who have studied chemical controls of the cell cycle have found that external factors, like hormones, stimulate internal changes that result in changes in the DNA that leads to growth. This sequence of events is extremely complicated and beyond the scope of this book. Researchers have identified two key groups of regulatory molecules: **cyclins** and **cyclin-dependent kinases (CDKs)**. Cyclins are a group of chemicals that bind to CDKs, activating these enzymes. When activated, these enzymes activate or deactivate other enzymes in the cell. They do so by adding phosphates to the target enzymes, a process called **phosphorylation**. Different cyclin-CDK combinations determine which proteins are activated or deactivated. Which cyclin is produced depends on the nature of the external signal. Specific combinations of cyclin-CDK types signal either G_1, S, G_2, or mitosis.

KEY CONCEPTS

Many chemical messages from the cytoplasm control the cell cycle.

SUMMARY

The Cell Cycle

1. The life cycle of a cell is called the cell cycle and consists of two parts: interphase and cell division.
2. Interphase is a period of active cellular synthesis and is divided into three phases in actively dividing cells: G_1, S, and G_2.
3. During G_1, the cell carries out its day-to-day activities. It also produces RNA, proteins, and other molecules and is a period of preparation for division.
4. During the S phase, the DNA replicates. After replication, each chromosome in the nucleus of the cell contains two chromatin fibers, or chromatids.
5. G_2 is a much shorter period and is relatively inactive.
6. During interphase, the cell replicates its organelles and molecules needed by the two daughter cells.
7. Cell division follows interphase. It requires two separate but related processes: mitosis, or nuclear division, and cytokinesis, or cytoplasmic division.
8. Nondividing cells such as muscle and nerve cells do not proceed through G_1, S, and G_2, but rather enter into G_o, a relatively quiescent stage during which the cells carry out the functions required by the cell to continue functioning.

The Chromosome

9. Each organism has a set number of chromosomes. Human cells contain 46 chromosomes.

10. All cells of the body, except the germ cells, are called somatic cells; they contain a full set of chromosomes and are described by geneticists as diploid. In humans, the diploid number of chromosomes is 46.
11. Germ cells or gametes contain half the number of chromosomes of somatic cells and are referred to as haploid cells. In humans, gametes contain 23 chromosomes.
12. Gametes are produced by a special type of cell division known as meiosis, which occurs in the gonads (ovaries and testes).
13. The chromosomes are loosely arranged in the nucleus during interphase, but they condense during prophase.
14. Condensation facilitates chromosome separation and also allows for the study of chromosomes.
15. Chromosomes arranged according to size and other features form a karyotype. The karyotype is used by geneticists to count the chromosomes and to locate potential abnormalities such as missing chromosomes or missing parts of chromosomes.

Cell Division: Mitosis and Cytokinesis

16. During cellular division, the cell divides its chromosomes equally and distributes its cytoplasm and organelles more or less evenly between the two daughter cells.

17. Cell division is essential to reproduction and growth. It is the basis of growth, tissue repair, and provides replacements for cells lost through normal wear and tear.
18. Mitosis, nuclear division, is divided into four stages, prophase, metaphase, anaphase, and telophase, and is summarized in Table 16-1.
19. Successful mitosis depends on the presence of the mitotic spindle, an array of microtubules that forms during prophase. Some of the spindle fibers attach to the chromosomes and help pull them apart during anaphase.
20. Cytokinesis, the division of the cytoplasm, begins in late anaphase or early telophase. Cytokinesis results from the contraction of microfilaments lying beneath the plasma membrane.

Control of the Cell Cycle

21. The cell cycle is controlled by substances known as cyclins produced in the cytoplasm. These chemicals bind to proteins known as cyclin-dependent kinases (CDKs). They initiate changes in the enzymes in a cell.
22. External controls, such as hormones, growth regulators, cell contact inhibition, and others, are imposed on the cell but act through cyclins and CDK.

THINKING CRITICALLY ANALYSIS

This analysis corresponds to the Thinking Critically scenario that was presented at the beginning of this chapter.

When analyzing statements, one should first consider the source and look for bias. Of course, Rush Limbaugh is not a scientist, and critics say he has little knowledge of the atmospheric science necessary to understand ozone depletion. Furthermore, Limbaugh is a powerful advocate of free enterprise and takes a dim view of government regulation. In other words, he's not an unbiased observer—far from it. He has an agenda and advocates it vocally.

Furthermore, if you read his work, you'll find that he gets most of his facts from the late Dixie Lee Ray, also an avowed anti-environmentalist. According to author Robert Boyle, she, in turn, obtained much of her information from Rogelio Maduro, an associate editor of *21st Century*. Maduro coauthored a book entitled, *The Holes in the Ozone Scare*. It asserts that efforts to ban CFCs stem in large part from a corporate plot led by Du Pont, the world's leading manufacturer of CFCs. According to Maduro, they are eager to profit from substitutes for CFCs, which they developed.

This characterization is in error. Unfortunately, Du Pont and other chemical companies actually opposed bans on ozone-depleting CFCs for many years. Only when the evidence that CFCs were damaging the ozone layer became irrefutable did Du Pont back down and support a ban.

Articles in two reputable scientific journals, *Science* and *Chemical and Engineering News*, have systematically refuted the claims made by Limbaugh, Ray, Maduro, and others. Sadly, few people will ever read these journals to explore the scientific treatment of this subject.

KEY TERMS AND CONCEPTS

Amniocentesis, p. 358
Anaphase, p. 359
Aster, p. 359
Cell cycle, p. 353
Cell division, p. 353
Centriole, p. 359
Chromosome, p. 355
Chromotid, p. 355
Contact inhibition, p. 362
Cyclins, p. 362
Cyclin-dependent kinases, p. 362
Cytokinesis, p. 355

Daughter cells, p. 354
Deoxyribonucleic acid (DNA), p. 354
Diploid, p. 357
Double helix, p. 354
G_1 (gap 1), p. 354
G_2 (gap 2), p. 354
Germ cell, p. 357
Haploid, p. 357
Histone, p. 358
Homologous chromosome, p. 356
Interphase (G_1, S, and G_2), pp. 353–354
Karyotype, p. 358

Metaphase, p. 359
Mitosis, p. 355
Mitotic spindle, p. 359
Nucleoli, p. 359
Nucleotides, p. 354
Phosphorylation, p. 362
Prophase, p. 359
S phase (synthesis phase), p. 354
Somatic cells, p. 357
Synthesis phase, p. 354
Telophase, p. 359
Tumor, p. 362

CONCEPT REVIEW

1. Describe the cell cycle. What is it? pp. 353–355.
2. The cell cycle is divided into two major parts. What are they? What takes place during each part of the cycle? pp. 353–355.
3. Draw a diagram of the cell cycle, showing all parts. p. 353.
4. What's the difference between mitosis and cytokinesis? p. 355.

5. During what part of the cell cycle does the DNA duplicate? p. 355.
6. How many chromosomes are found at each stage of the cell cycle? Note whether they are single- or double-stranded. p. 355.
7. List the stages of mitosis, and describe the major nuclear changes occurring in each stage. p. 359.

8. When does cytokinesis occur? How does the cell divide itself? pp. 359–360.
9. Discuss the factors that control the cell cycle. List some factors that control the cell cycle externally? p. 362.
10. What are cyclins and cyclin-dependent kinases? How do they control the cell cycle? p. 362.

SELF-QUIZ: TESTING YOUR KNOWLEDGE

1. The cell cycle is divided into two major parts: cell division and _____. p. 353.
2. Cell division is divided into two parts: _____ and _____. p. 355.
3. In actively dividing cells, ____ is a period during which the cell manufactures RNA, proteins, and other key molecules that ensures that each daughter cell receives an adequate supply of nutrients. pp. 353–354.
4. During the _____ phase of interphase, the cell replicates its DNA. p. 354.
5. During ____ mitochondria divide and the cell produces spindle fibers, tiny

microtubular structures that are essential to nuclear division. p. 354.
6. Nondividing cells like many neurons enter into _____, a phase that allows them to carry out their functions, but does not prepare them for cell division. p. 354.
7. Cytoplasmic division is also known as _____. p. 355.
8. Cytoplasmic division is brought about by _____ proteins lying beneath the plasma membrane of cells. They are part of the _____ network. p. 359.
9. Two chromosomes that contain the same kinds of genes are known as _____ chromosomes. p. 356.

10. A germ cell is called _____ because it contains half the number of chromosomes of somatic cells. p. 357.
11. The proteins associated with the DNA are known as _____. p. 358.
12. During _____ in mitosis, the condensed chromosomes line up in the center of the cell. p. 359.
13. Microtubules that help to separate chromosomes during mitosis make up the _____. p. 359.
14. Chromosomes split and migrate to opposite poles of the cell during _____. p. 359.
15. During _____, a new nuclear envelope forms around the chromosomes. p. 359.

biology.jbpub.com/chiras/8e/

The site features eLearning, an online review area that provides quizzes, chapter outlines, and other tools to help you study for your class. You can also follow useful links for in-depth information, research the differing views in the Point/Counterpoints, or keep up on the latest health news.

(© Monkey Business Images/ShutterStock, Inc.)

Principles of Human Heredity

For most of us, emotions tend to fluctuate with changing circumstances. Some events bring happiness, even elation, while others evoke irritation, anger, fear, or depression. Such peaks and troughs are normal. But for some people, the emotional roller coaster ride is intense—their feelings swing erratically for little or no reason. One minute, they are in a state of elated overactivity (mania); the next, they fall into a state of depressed inactivity. This condition, once known as *manic depression*, but now called *bipolar disease*, is characterized by cyclic mood swings that are unrelated to external events.

THINKING CRITICALLY

Nearly 150 years after his death, President Abraham Lincoln has become the subject of a scientific controversy. The debate among scientists is not over who killed him or why, but rather, over the possibility that he suffered from a genetic disorder called Marfan's syndrome.

Marfan's syndrome is an extremely rare genetic disease that affects the skeletal system, the eyes, and the cardiovascular system. It is difficult to diagnose and is usually identified only after its victims die.

Marfan's victims typically have exceedingly long arms and legs. Nearsightedness and lens defects are also common among Marfan's patients. The most serious problem, however, is enlargement and weakening of the aorta, the large artery that carries blood away from the heart. Weakening occurs in the aortic arch, the very first segment, and is caused by a degeneration of the connective tissue in the wall of this vessel. If untreated, the aorta can burst, causing instant death.

Some evidence suggests that President Abraham Lincoln had Marfan's syndrome. He was tall and lanky, for example, and he wore glasses. And, his great-great-grandfather had the disease, making it possible that Abe had received the Marfan gene. What additional evidence might you look for to confirm this hypothesis? What critical thinking rule(s) do(es) this exercise illustrate?

Bipolar disease varies widely from one person to the next. In some, it persists for only a short time. In others, it continues for years. During the animated and highly energetic state, bipolar patients often act irrationally or obnoxiously. Such behavior can derail college education, ruin relationships, and can lead to financial disaster, especially if it interferes with decision making. When in a state of deep depression, victims sometimes threaten suicide, although their lack of energy often prevents them from following through. In some cases, however, the self-destructive yearning persists after the period of depression and results in suicide.

Bipolar disease occurs in about 3% of the U.S. population and tends to run in families, suggesting a genetic link. Researchers, in fact, have located a gene near the tip of chromosome 11 that may be responsible.

Genes on chromosomes determine the structure and function of cells and organs. They also influence behavior, sometimes profoundly. This chapter examines how genes are passed from parent to offspring. We begin with an overview of meiosis, a special kind of cell division responsible for gamete formation in humans. We then focus our attention on some basic principles of heredity.

17-1 Meiosis and Gamete Formation

> **KEY CONCEPTS**
> During gamete formation, germ cells undergo a special kind of cellular division known as meiosis, which reduces the number of chromosomes by half.

Human body cells or somatic cells contain 46 chromosomes. The male and female gametes have half that number. Thus, when the male and female gametes unite, the offspring will have 46 chromosomes. The mechanism responsible for reducing the number of chromosomes in sex cells is called meiosis (my-OH-siss).

Meiosis is a type of nuclear division that occurs in germ cells and yields haploid gametes, that is, gametes with half the number of chromosomes of a normal (diploid) body cell. Germ cells are formed in the reproductive organs, the ovaries (OH-varees) of women and the testes (TESS-tees) of men. When haploid germ cells unite, they form a diploid zygote with 46 chromosomes, 23 from the mother and 23 from the father. The zygote continues to divide and eventually forms an embryo, then a fetus, and, after birth, a baby that grows to become an adult. Each of the body or somatic cells contains 46 chromosomes.

As you may recall, chromosomes occur in pairs, called *homologous pairs*, one member of each pair from the father and the other from the mother. Homologous pairs are the same length, have the same appearance, and contain the same genes. Meiosis resembles mitosis. If you haven't read that section or have forgotten some of the details, you may want to read or review it now. Unlike mitosis, meiosis involves two nuclear divisions. The two nuclear divisions are referred to as meiosis I and II. Let's consider the main details of each one.

> **KEY CONCEPTS**
> During gamete formation, germ cells undergo a special kind of cellular division known as meiosis, which reduces the number of chromosomes by half.

Meiosis I

Meiosis I is illustrated in Figure 17-1. Take a moment to look at the figure. How does it compare to mitosis?

One of the first things you probably noticed is that meiosis consists of two cell divisions. Each one contains the same four steps as mitosis: prophase, metaphase, anaphase, and telophase. Take a moment to check this out.

Another important similarity is that, just as in mitosis, chromosomes duplicate during interphase. As a result, each chromosome contains two chromatids. During prophase I, the chromosomes condense, as they do in mitosis. However, something unusual occurs in prophase I of meiosis: homologous chromosomes "pair up." In other words, chromosome 1 from the mother and chromosome 1 from the father pair up—that is, become attached. Chromosome 2 from the mother pairs with chromosome 2 of the father, and so on.

During metaphase I of meiosis I, the homologous pairs line up along the equatorial plate. (In mitosis, you may recall all 46 chromosomes line up in the center of the cell during metaphase.)

During anaphase I, the pairs separate, with one member of the pair (containing two chromatids) going to each daughter cell. As a result, each daughter cell ends up with half the number of chromosomes (23) of the parent cells (46). Because the number of chromosomes decreases by half during this process, meiosis I is called a **reduction division**.

> **KEY CONCEPTS**
> Meiosis I is the first of two cell divisions that occur in gamete formation; it results in a reduction in the number of chromosomes from diploid to haploid.

Meiosis II

The second division, **meiosis II**, also resembles mitosis, except for one difference: The cells start out as haploid. As illustrated in Figure 17-1, the chromosomes of the haploid cells condense during prophase II, then line up single file along the equatorial plate during metaphase II. During anaphase II, the chromatids of each chromosome dissociate, one going to each pole. The result of meiosis is four cells, each containing a haploid number of single-stranded chromosomes. It is these cells that give rise to the haploid gametes.

If this is confusing, you may want to take a few minutes to review mitosis. Then go over the steps in meiosis, paying attention primarily to two things: the number of chromosomes and the number of chromatids in each one. Here's a recap of what you'll find: In meiosis I, the cells begin with 46 chromosomes—each with two chromatids. During meiosis I, each daughter cell ends up with 23 chromosomes, and each chromosome has two chromatids. During meiosis II, the

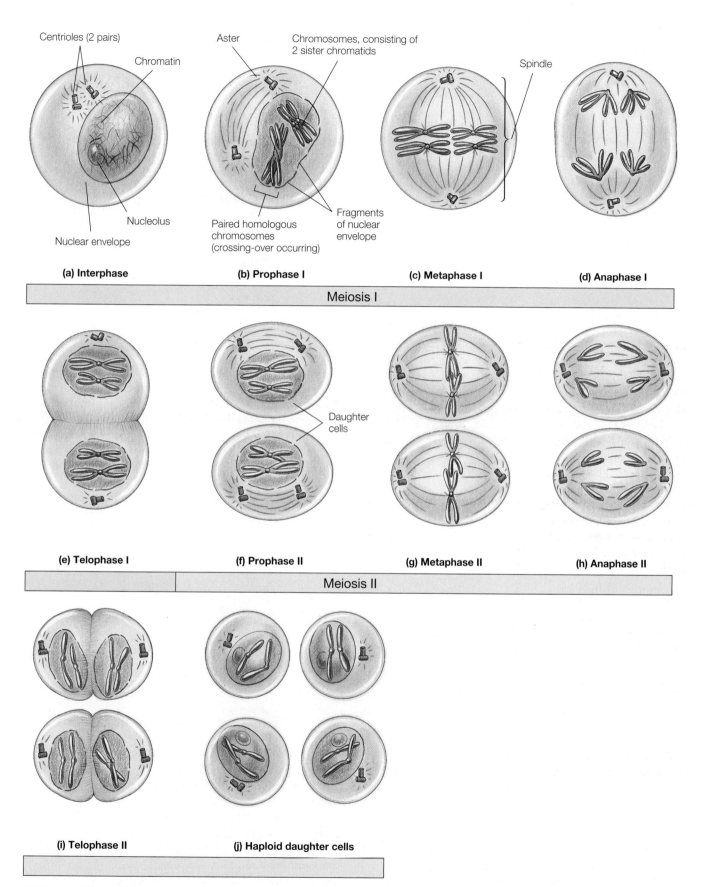

(a) Interphase (b) Prophase I (c) Metaphase I (d) Anaphase I

Meiosis I

(e) Telophase I (f) Prophase II (g) Metaphase II (h) Anaphase II

Meiosis II

(i) Telophase II (j) Haploid daughter cells

FIGURE 17-1 Meiosis Note that meiosis is divided into two phases: meiosis I and meiosis II.

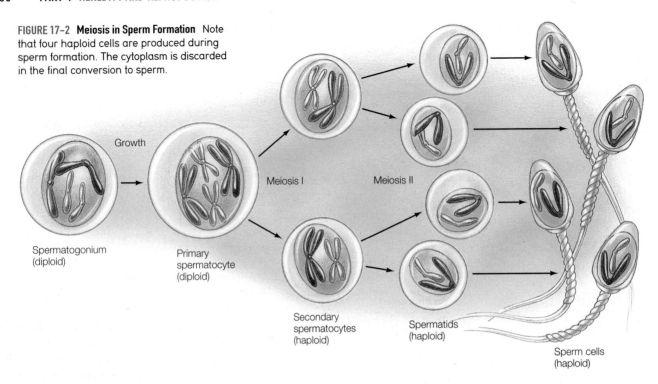

FIGURE 17–2 **Meiosis in Sperm Formation** Note
that four haploid cells are produced during
sperm formation. The cytoplasm is discarded
in the final conversion to sperm.

23 chromosomes split apart, and each daughter cell ends up
with 23 chromosomes—each with one chromatid.

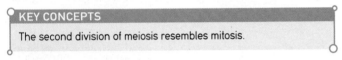

KEY CONCEPTS

The second division of meiosis resembles mitosis.

Gamete Production

Meiosis differs between males and females. In males, the
process is much as I have described above. Thus, a single
diploid cell in the gamete-producing organ, the testis (singular; pronounced TESS-tiss), gives rise to four sperm cells.
During this process, the cytoplasm of the original cells that
give rise to the sperm is divided equally among the four sperm
cells. However, during the final stages of sperm development
(spermiogenesis), much of the cytoplasm is discarded to produce a highly streamlined sperm capable of swimming up the
female reproductive tract (Figure 17-2).

In females, the single diploid cell in humans gives rise
to only one gamete, the egg or ovum (Figure 17-3). Thus, all of

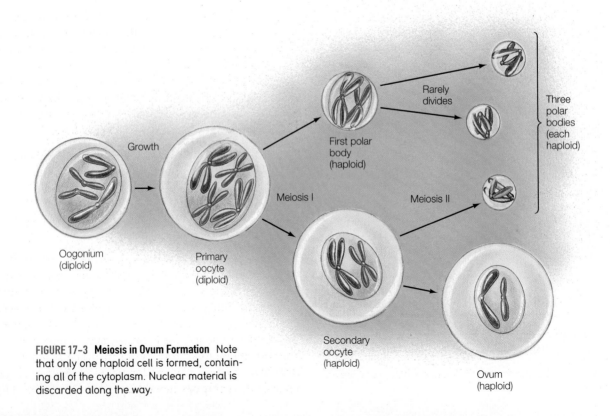

FIGURE 17–3 **Meiosis in Ovum Formation** Note
that only one haploid cell is formed, containing all of the cytoplasm. Nuclear material is
discarded along the way.

the cytoplasm is retained in one cell. The extra chromosomal material is systematically "discarded" during the two meiotic divisions. This difference reflects the fact that the egg is fairly sedentary. It's the sperm that must do most of the work during fertilization! The retention of the ovum's cytoplasm also ensures that the fertilized ovum will contain an adequate amount of nutrients and cellular organelles.

> **KEY CONCEPTS**
>
> In males, meiosis produces four gametes; in females, it produces one gamete.

17-2 Principles of Heredity: Mendelian Genetics

With these basics in mind, we turn our attention to **heredity**—the transmission of genes from parents to offspring. Heredity is one aspect of the study of genetics. The formal study of genetics began with an examination of heredity at the organismic level that was carried out by a nineteenth-century monk, Gregor Mendel, working in the garden at his monastery (Figure 17-4).

Born in Eastern Europe in 1821, Mendel entered an Augustinian monastery in Brno, in what is now the Czech Republic, at the age of 21. He later enrolled at the University of Vienna, where he studied botany, mathematics, and other sciences. Returning to the monastery, he began a series of experiments on garden peas—studies that would reveal several fundamental principles of genetics.

During Mendel's time, many scientists believed that the traits of a child's parents were blended in the offspring, producing a child with intermediate characteristics. In addition, because the ova of sexually reproducing organisms are much larger than the sperm, some scientists believed that the female had a greater influence on the characteristics of the offspring than the male. Over a 10-year period, Mendel designed experiments to examine these assumptions, using the garden pea.

> **KEY CONCEPTS**
>
> Much of our basic understanding of heredity comes from the work of one man, Gregor Mendel, who studied genetics of garden peas.

(a)

(b)

FIGURE 17-4 Mendel and His Garden (a) Gregor Mendel worked out some of the basic rules of inheritance in the mid-1800s in his research on garden peas. (© National Library of Medicine.) (b) Mendel's garden at the monastery in Czechoslovakia as it appears today. (Courtesy of Derek Chen.)

Blending of Traits

When Mendel bred a pea with a white flower to a pea with a purple flower, he did not produce pea plants with intermediate pink flowers as many would have predicted, based on existing assumptions. Instead, these crosses produced plants with purple flowers, for reasons discussed shortly. Experiments that Mendel performed on the inheritance of several other traits produced similar results—refuting the notion that parental traits blend in offspring.

> **KEY CONCEPTS**
>
> Mendel discovered that traits he was studying did not result from a blending of traits.

Principle of Segregation

Mendel's research led him to conclude that adult plants contain pairs of hereditary factors for each trait (for example, seed color) and that each pair governed the inheritance of a single trait. Today, we refer to Mendel's hereditary factors as **genes**, which are located in the DNA of chromosomes.

Mendel reasoned that because a male gamete and a female gamete combine to form a new organism and an adult contains only two hereditary factors for any given trait, each gamete must contain only one hereditary factor (gene) for each trait.

Mendel's studies therefore suggested that the paired hereditary factors of the mother and father must separate during gamete formation so that each gamete contributes one and only one hereditary factor to the zygote. The separation of hereditary factors during gamete formation is known as the **principle of segregation**.

Because the gametes of the parents combine to produce an offspring and because each gamete contains one hereditary factor for each trait, Mendel concluded that the contributions of the parents must be equal. That is, each parent contributes one hereditary factor, or gene, for each trait.

These conclusions may seem unremarkable to us today, especially with our knowledge of meiosis and gamete formation. However, bear in mind that in the 1850s and 1860s when Mendel ran his experiments, chromosomes had not yet been discovered!

> **KEY CONCEPTS**
>
> Mendel's studies showed that parents contribute equally to the characteristics of their offspring; this results from the segregation of hereditary factors (genes) during gamete formation.

Dominant and Recessive Genes

Mendel postulated that hereditary factors (genes) are either dominant or recessive (Table 17-1). When a dominant factor and a recessive factor are present in a pea, Mendel concluded that the dominant factor is always expressed. A recessive factor is expressed only when the dominant factor is missing.

An alternative form of the same gene is called an **allele** (ah-LEEL). Thus, dominant and recessive factors are known today as dominant and recessive alleles. Dominant alleles are designated by capital letters; recessive alleles are signified by lowercase letters. In Mendel's peas, for instance, the gene for flower color is the P gene. The dominant form is P (purple flowers), and the recessive form is p (white flowers). The P and p genes are both alleles of the flower-color gene.

For genes with two alleles, three combinations are possible during fertilization. The first consists of two dominant genes—for example, PP. This individual is said to be **homozygous dominant** (hoe-moe-ZYE-gus) for that particular trait. The second consists of two recessive alleles—in this example, pp—and the individual is said to be **homozygous recessive** for that trait. The third occurs when both dominant

and recessive alleles are present (Pp), and the individual is **heterozygous**.

> **KEY CONCEPTS**
>
> Mendel discovered the principle of dominance, that some hereditary factors (genes) are dominant over others, which are known as recessive traits.

Genotype and Phenotype

As noted earlier, Mendel determined that blending did not occur in the inheritance of traits, such as flower color. Thus, a pea plant with purple flowers (PP) bred with a pea plant with white flowers (pp) produced offspring with purple flowers and not pink flowers, which would occur if blending took place. Mendel proposed that the purple-flowered offspring of this cross were heterozygous—Pp. In such cases, he argued, the recessive hereditary factors (genes) are masked by the dominant hereditary factors (genes). Because of dominance, Mendel asserted, the outward appearance of a plant—its **phenotype** (FEEN-oh-type)—does not always reflect its genetic makeup, or **genotype** (JEAN-oh-type). As a general rule, a homozygous dominant individual and a heterozygous individual are indistinguishable on the basis of outward appearance or phenotype.

> **KEY CONCEPTS**
>
> The genotype of an organism is its genetic makeup; the phenotype is its appearance.

Determining Genotypes and Phenotypes

To track the genotypes and phenotypes of breeding experiments, such as those that Mendel performed, geneticists use a relatively simple tool known as the **Punnett square** (Figure 17-5). To illustrate the process, consider one of the traits that Mendel studied—seed-coat texture. The gene for seed-coat texture has two alleles: one that codes for smooth seeds (S) and one that codes for wrinkled seeds (s). As indicated by the capital S, the smooth-seed allele is dominant.

Suppose that you bred a homozygous-dominant plant (SS) with a homozygous-recessive plant (ss). Figure 17-5 lists the genotypes of each parent and the possible genotypes of the gametes. Determining the genotype is a rather simple affair. If a diploid parent is SS, its haploid gametes are all S. If a parent is ss, the gametes are all s.

To determine the outcome of crossing these two plants, all of the possible gametes produced from one plant are listed along the top of the Punnett square, and all of the possible gametes from the other are listed along the side. The gametes are then combined as shown, producing all of the possible genotypes present in the offspring.

As this example illustrates, a cross between a homozygous-dominant (SS) individual and a homozygous-recessive (ss) individual produces only type of offspring: heterozygous (Ss). As a result, all the offspring are phenotypically uniform. That is, they all resemble the homozygous-dominant parent—they have a smooth seed coat.

In the language of genetics, when one organism is bred to another to study the transmission of a single trait, as in

TABLE 17-1	Genetic Traits Studied by Mendel	
Studied Structure	**Dominant Trait**	**Recessive Trait**
Seeds	Smooth	Wrinkled
	Yellow	Green
Pods	Full	Constricted
	Green	Yellow
Flowers	Axial (along stems)	Terminal (top of stems)
	Purple	White
Stems	Long	Short

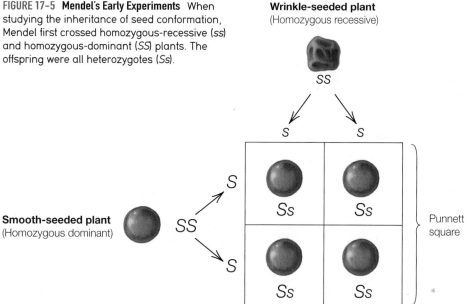

FIGURE 17–5 **Mendel's Early Experiments** When studying the inheritance of seed conformation, Mendel first crossed homozygous-recessive (*ss*) and homozygous-dominant (*SS*) plants. The offspring were all heterozygotes (*Ss*).

Wrinkle-seeded plant
(Homozygous recessive)

Punnett square

All offspring have smooth seeds

the previous example, the procedure is called a **monohybrid cross** (MON-oh-HYE-brid).

In a monohybrid cross, the parents are referred to as the P_1 generation (P for parents). The first set of offspring constitutes the F_1 generation (first filial generation; *filialis* is Latin for "of a son or daughter"). In this example, the F_1 offspring could be bred to produce additional offspring. These offspring are known as the F_2 generation (second filial generation).

To determine the genotypes and phenotypes of such a cross, the Punnett square can be used once again. As **Figure 17-6**

shows, the gametes of the F_1 generation are *S* and *s* because all of the parents are heterozygous (*Ss*). All of the possible gametes from the father are placed along the top of the Punnett square, and all of the possible gametes of the mother are placed along the side. The gametes are then combined as before to determine the genotype of the F_2 generation. Take a moment to make your own Punnett square to predict the outcome.

When finished, take a look at Figure 17-6 to compare your results. As illustrated in Figure 17-6, this monohybrid cross produces three different genotypes: *SS*, *Ss*, and *ss*. The ratio of genotypes of the F_2 generation is one *SS* to two *Ss* to one *ss*, or 1:2:1. Thus, if 100 offspring are produced, you would expect 25 of them to be *SS* (homozygous dominant), 50 to be *Ss* (heterozygous), and 25 to be *ss* (homozygous recessive). Because of dominance, though, this cross yields only two phenotypes: plants with smooth seeds (*SS* and *Ss*) and plants with wrinkled seeds (*ss*). The ratio of phenotypes, or the phenotypic ratio, is three smooth to one wrinkled, or 3:1.

> **KEY CONCEPTS**
>
> The genotypes of offspring can be determined by using the Punnett square; it allows one to estimate the probability of various genetic combinations that can then be used to determine the probability of various phenotypes.

The Principle of Independent Assortment

In his later experiments, Mendel tracked two traits at the same time, a procedure geneticists refer to as a **dihybrid cross** (DIE-HIGH-brid; **Figure 17-7**). This work led to the idea that genes located on different chromosomes are inherited independently during meiosis. To understand what this principle means and how Mendel arrived at it, consider another example that examines seed-coat texture and seed color.

As noted above, peas contain a gene for seed-coat texture, the *S* gene. The dominant form is *S* (smooth), and the recessive form is *s* (wrinkled). Peas also contain a gene for seed color;

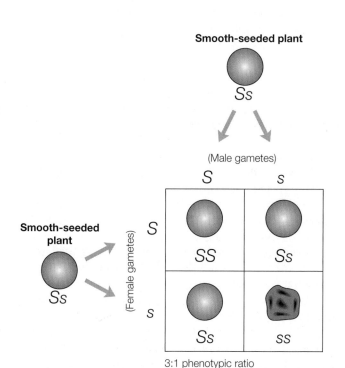

Smooth-seeded plant

Ss

(Male gametes)

Smooth-seeded plant

Ss

(Female gametes)

3:1 phenotypic ratio

FIGURE 17–6 **Monohybrid Cross** In Mendel's early work, he crossed offspring from his F_1 generation (*Ss*) and found a variety of genotypes and phenotypes.

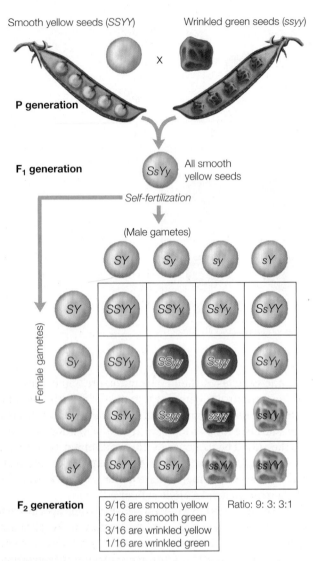

Smooth yellow seeds (*SSYY*)   Wrinkled green seeds (*ssyy*)

X

P generation

F₁ generation   *SsYy*   All smooth yellow seeds

Self-fertilization

(Male gametes)

SY   Sy   sy   sY

(Female gametes)

SY   SSYY   SSYy   SsYy   SsYY

Sy   SSYy   SSyy   Ssyy   SsYy

sy   SsYy   Ssyy   ssyy   ssYy

sY   SsYY   SsYy   ssYy   ssYY

F₂ generation   9/16 are smooth yellow   Ratio: 9: 3: 3:1
3/16 are smooth green
3/16 are wrinkled yellow
1/16 are wrinkled green

FIGURE 17-7 Dihybrid Cross The dihybrid cross examines the inheritance of two traits. Independent assortment of genes produces a large number of genotypes.

the dominant form (*Y*) yields yellow seeds, and the recessive form (*y*) yields green seeds.

To study the inheritance of these two genes, Mendel first crossed homozygous-dominant plants with smooth, yellow seeds (*SSYY*) with homozygous-recessive plants (*ssyy*). This cross is shown in the top of Figure 17-7. The cross resulted in the F₁ generation, consisting of peas with the genotype *SsYy*. What is the phenotype of these plants? If you say they were all phenotypically identical and all displayed the dominant characteristics (smooth, yellow seeds), you are correct.

Next, Mendel crossed two of the F₁ offspring (genotype *SsYy*), as shown in the Punnett square in 17-7. As you can see, he ended up with a mixture of genotypes and phenotypes (Figure 17-7). To produce this combination of offspring, Mendel concluded, the hereditary factors *S* and *s* must have segregated independently of *Y* and *y* during gamete formation. Translated into modern terms, this means that the *Y* and *S* genes (and their alleles *y* and *s*) were on different chromosomes that separated independent of one another during gamete formation (meiosis). Because of independent assortment, the gametes of the F₁ generation contain all combinations of the alleles in equal proportions: *SY, Sy, sY,* and *sy.* Fertilization also occurs at random, giving rise to 16 possible genetic combinations (see F₂ genotypes in Figure 17-7). If the *S* and *Y* genes were on the same chromosome, far fewer combinations would occur.

This research led Mendel to propose the **principle of independent assortment**, which states that the segregation of the alleles of one gene on one chromosome during gamete formation is independent of the segregation of the alleles of another gene on a second chromosome.

KEY CONCEPTS

Mendel discovered that genes on different chromosomes segregate independently during gamete formation; this is known as the principle of independent assortment.

17-3   Mendelian Genetics in Humans

What do studies on garden peas have to do with humans? Interestingly, much of what Mendel discovered in garden peas pertains to human genes and human inheritance. Table 17-2 lists a number of human traits and diseases whose inheritance follows basic Mendelian principles. This section describes the inheritance of those traits, using several common diseases as examples.

Autosomal-Recessive Traits

Human cells contain 23 pairs of chromosomes. They can be divided functionally into one pair of sex chromosomes and 22 pairs of autosomes (AU-toe-zomes).

The **sex chromosomes** are involved in determining whether we turn out male or female—as well as a few other traits. Two types of sex chromosomes exist, X and Y. Females have two X chromosomes; their genotype is XX. Males have one X chromosome and one Y chromosome, so their genotype is XY.

The remaining 22 pairs of chromosomes are called the **autosomes**. The autosomes contain numerous genes that control a variety of traits. This section examines several **autosomal-recessive traits**. As you can surmise from Mendel's experiments, these are traits expressed only when both recessive alleles are present.

Over 600 traits in humans have been identified as autosomal recessive, and another 800 are strongly suspected of being autosomal recessive. Some autosomal-recessive traits are the cause of abnormalities, among them albinism and cystic fibrosis.

Albinism. Albinism is a fairly common genetic defect. It occurs in 1 of every 38,000 Caucasian births and in 1 of every 22,000 African-American births.

Albinism in humans occurs when two recessive genes are inherited from one's parents. The recessive genes result

TABLE 17-2	Traits and Diseases Carried on Human Chromosomes
Autosomal recessive	
Albinism	Lack of pigment in eyes, skin, and hair
Cystic fibrosis	Pancreatic failure, mucus buildup in lungs
Sickle-cell anemia	Abnormal hemoglobin leading to sickle-shaped red blood cells that obstruct vital capillaries
Tay-Sachs disease	Improper metabolism of a class of chemicals called gangliosides in nerve cells, resulting in early death
Phenylketonuria	Accumulation of phenylalanine in blood; results in mental retardation
Attached earlobe	Earlobe attached to skin
Hyperextendable thumb	Thumb bends past 45° angle
Autosomal dominant	
Achondroplasia	Dwarfism resulting from a defect in epiphyseal plates of forming long bones
Marfan's syndrome	Defect manifest in connective tissue, resulting in excessive growth, aortic rupture
Widow's peak	Hairline coming to a point on forehead
Huntington's disease	Progressive deterioration of the nervous system beginning in late twenties or early thirties; results in mental deterioration and early death
Brachydactyly	Disfiguration of hands, shortened fingers
Freckles	Permanent aggregations of melanin in the skin

FIGURE 17-8 **Albinism** Albinism is an autosomal-recessive trait. An albino man with his wife in India. (© Joe McDonald/Visuals Unlimited.)

the respiratory system, and ducts in the pancreas. Defective sweat glands, for example, release excess amounts of salt, a marker that helps physicians diagnose the disease. The most significant symptoms occur in the pancreas and lungs.

In individuals with cystic fibrosis, the ducts that drain digestive enzymes from the pancreas into the small intestine become clogged. This not only impairs digestion but also causes a buildup of enzymes in the organ that results in the formation of cysts. Over time, the pancreas begins to degenerate, and fibrous tissue replaces glandular tissue—hence, the name cystic fibrosis (Figure 17-9).

Despite adequate nutritional intake, individuals with this disease often show signs of malnutrition. To enhance digestion, patients must eat powdered or granular extracts of animal pancreases containing digestive enzymes. Massive doses of vitamins and nutrients must also be taken.

The respiratory systems of most victims of cystic fibrosis produce copious amounts of mucus. Mucus blocks the passages, making breathing difficult. Patients must be treated several times a day to remove the mucus and prevent infections from bacteria trapped in the mucus (Figure 17-10). Patients are also commonly given oxygen to help maintain normal oxygen levels in the blood. Despite treatment, most cystic fibrosis patients live only into their late teens or early twenties.

Cystic fibrosis is one of the most common genetic diseases. One of every 29 Caucasians carries a gene for this disease, and approximately 1 of every 2,000 Caucasians born in the United States suffers from it. In the African-American population, only about 1 in 100,000 to 150,000 individuals is a carrier.

KEY CONCEPTS

Autosomal recessive genes are expressed only when both alleles in the offspring's genome are recessive.

in a deficiency in the metabolic pathways leading to the production of melanin. **Melanin** is the brown pigment responsible for coloration of the eyes, skin, and hair. Individuals who are homozygous-recessive for albinism have either no melanin or reduced levels of melanin (Figure 17-8). Consequently, the skin of an albino is pale, and the hair is white. The eyes of albinos are pink, because there is no pigment in the iris or retina.

Melanin in the skin protects against the effects of ultraviolet radiation. Its absence in albinos makes them highly susceptible to sunburn and skin cancer. Moreover, the lack of pigment in the eyes may result in damage to the light-sensitive region, the retina, which is essential to vision. This, in turn, may result in blindness.

Cystic Fibrosis. Cystic fibrosis (SISS-tick fie-BRO-siss) is an autosomal-recessive disease that leads to early death. The presence of two recessive alleles for the disease alters the function of sweat glands of the skin, mucous glands in

Autosomal–Dominant Traits

Many human traits are **autosomal dominant**. These traits are carried on the autosomes and are expressed in heterozygous (*Aa*) and homozygous-dominant (*AA*) individuals. To date, nearly 1,200 human traits have been identified as autosomal

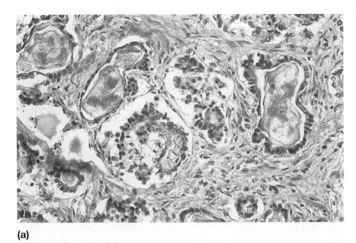
(a)

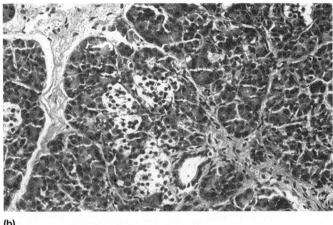
(b)

FIGURE 17-9 **Comparison of a Pancreas from Patient with Cystic Fibrosis with a Normal Pancreas** Cystic fibrosis is a disease caused by an autosomal-recessive gene. (a) It results in a blockage of the ducts draining the pancreas, leading to cysts in the pancreatic tissue. The tissue degenerates and is replaced by fibrous connective tissue. (b) Normal pancreas. (a and b © Science VU/Daniel V. Schidlow/Visuals Unlimited.)

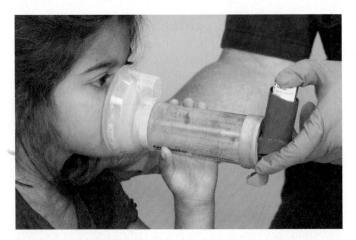

FIGURE 17-10 **Cystic Fibrosis** Inhalants, antibiotics, and physical therapy are used to treat patients with cystic fibrosis. To remove mucus from the lungs, parents or physical therapists must treat the patient two or three times a day. Pounding on the rib cage with a cupped hand (clopping) loosens the mucus. (© Jones & Bartlett Learning. Courtesy of MIEMSS.)

dominant, and 1,000 others are suspected. Freckles, cleft chin, and drooping eyelids are all autosomal-dominant traits, as is Marfan's syndrome, discussed in the Thinking Critically section in this chapter. Let's consider an autosomal-dominant trait called *widow's peak*.

Take a moment to examine the hairlines of your friends and classmates. In some individuals, the hairline runs straight across the forehead. In others, it juts forward in the center, forming a "widow's peak" (Figure 17-11). Widow's peak is determined by an autosomal-dominant gene, indicated by W. Because the W allele is dominant, this phenotype is expressed in both homozygous dominant individuals (*WW*) and heterozygotes (*Ww*). Individuals with the genotype *ww* have a straight hairline.

> **KEY CONCEPTS**
>
> Many traits are autosomal dominant; they are expressed in heterozygous and homozygous phenotypes.

(a)

(b)

FIGURE 17-11 **Widow's Peak** (a) The widow's peak is a dominant trait carried on one of the autosomes. (© Mikhail Nekrasov/ShutterStock, Inc.) (b) A straight hairline is a recessive trait. Simple Mendelian genetics can be used to determine the genotype of the offspring. (© Photodisc.)

17-4   Variations in Mendelian Genetics

Since Mendel's time, researchers have uncovered additional modes of inheritance and gene expression, discussed below.

Incomplete Dominance

One example of another form of inheritance is a phenomenon called incomplete, or partial, dominance. **Incomplete dominance** occurs when heterozygous offspring exhibit intermediate phenotypes. In other words, incomplete dominance results in offspring with phenotypes intermediate to the parental phenotypes. This is the kind of blending that Mendel said didn't exist!

Incomplete dominance occurs in the snapdragon flower. Figure 17-12 shows a cross between a plant with red flowers (genotype: *II*) and one with ivory flowers (genotype: *ii*). Even though the red color is dominant, this cross produces offspring with pink flowers, an intermediate phenotype (Figure 17-12). That's because the *I* gene does not exert complete dominance. Incomplete dominance also occurs in a number of human disorders, including sickle-cell disease.

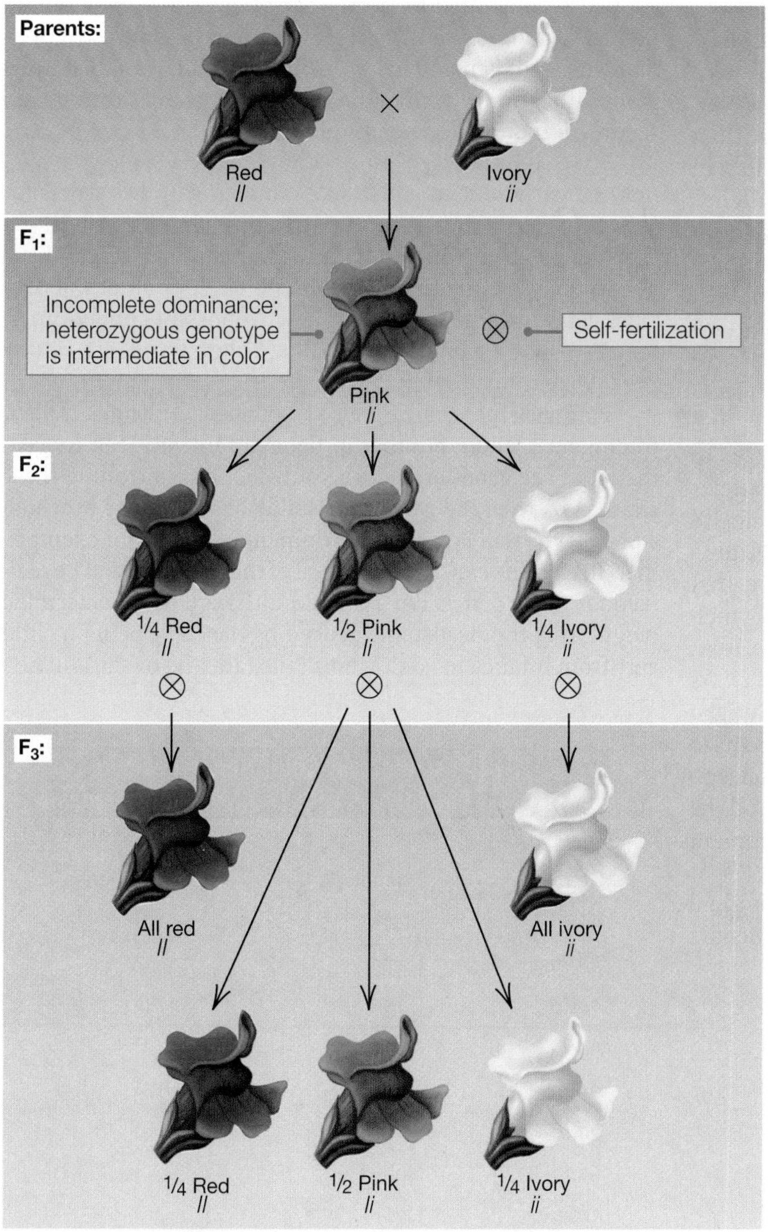

Sickle-Cell Disease. Sickle-cell disease affects the blood, and chiefly afflicts African Americans and Caucasians of Mediterranean descent. Sickle-cell disease is caused by a genetic defect that leads to abnormal hemoglobin. As you may, hemoglobin is a protein found in red blood cells. It carries oxygen inside the cell from the lungs to body tissues.

Sickle-cell disease occurs in individuals who are homozygous recessive for this trait. The disease is therefore often classified as an autosomal recessive disorder. In individuals with sickle cell anemia, the abnormally formed hemoglobin causes red blood cells to become sickle-shaped cells when they encounter low oxygen levels in the blood—for example, when blood cells flow through capillaries in metabolically active tissues.

Sickle cells are larger and less flexible than red blood cells. As a result, they clog capillaries, reducing oxygen flow to brain cells, heart cells, and cells of other organs. In homozygous-recessive individuals, sickle-cell disease is usually lethal. In fact, most individuals with the disease die by their late twenties.

Individuals who are homozygous recessive for sickle cell anemia suffer the ill effects of the disease. Individuals who are heterozygous for the trait live fairly normal lives and are referred to as **carriers**, because they can pass the gene on to their children. Although carriers do not generally suffer from symptoms of sickle cell anemia, moderate sickling occurs when their RBCs are exposed to low oxygen levels in the blood. That's because their RBCs contain 50% normal and 50% abnormal hemoglobin. Moderate sickling is why sickle-cell disease is considered an example of incomplete dominance.

Approximately 1 in every 500 African Americans born in the United States is homozygous recessive, and about 1 of every 12 is heterozygous (a carrier) for the sickle-cell trait. Why is this allele so prevalent?

In tropical climates like much of Africa, the sickle-cell allele protects carriers and homozygous-recessive individuals from malaria, a deadly disease prevalent in humid, tropical regions. Malaria is caused by a microscopic parasite called *Plasmodium* (plaz-MOW-dee-um), which is transmitted from one person to the next by certain mosquitoes. It picks up the parasite when it draws blood from an infected person and then transfers the parasite to other uninfected people when it bites them (drawing blood).

FIGURE 17-12 **Incomplete Dominance** Incomplete dominance involves two alleles, neither of which is dominant over the other. The result is an intermediate phenotype as shown here in the snapdragon flower.

Inside the body, the parasites invade and colonize the liver, where they multiply rapidly. New parasites leave the liver and enter the bloodstream, where they attack and destroy red blood cells. The presence of altered hemoglobin of people with sickle cell anemia changes the plasma membrane of RBCs and prevents the parasite from entering. Interestingly, although homozygous-recessive individuals die earlier, the selective advantage that carriers enjoy has caused the allele to increase in the population.

> **KEY CONCEPTS**
>
> Although Mendel found that traits in peas are not blended, in humans incomplete dominance does result in a blending of certain traits.

Multiple Alleles and Codominance

Blood types are an example of multiple genes that control a trait, the type of glycoprotein found in the plasma membranes of RBC, but they are also an example of codominance.

Mendel studied seven characteristics of peas. Each characteristic is determined by a gene with two alleles—two alternative forms. In humans and other species, however, researchers have shown that some genes can have more than two alleles. What that means is that some genes have different DNA sequences. Such genes are said to have **multiple alleles**.

Let's consider blood types. Blood type is controlled, in part, by a gene known as the *I* gene (for isoagglutinin; pronounced EYE-sa-AH-glue-TEH-nin). The **I gene** is located at a particular site (or locus) on one pair of chromosomes. But unlike the genes Mendel studied, which had only two possible alleles (forms: dominant and recessive) for each trait, the *I* gene has three possible alleles. The three alleles are I^A, I^B, and I^O. However, even though there are three possible alleles in human beings, an individual can have only two of the alleles in his or her genes, one on each homologous chromosome. It's like having three mittens. You can only wear two at a time.

The main result of multiple alleles for blood type is that humans can have multiple blood types, depending on which genes they inherit. As you learned elsewhere in the text, humans have four possible blood types: A, B, AB, and O. As you may recall, blood type is determined by the type of antigen present on the plasma membranes of RBCs. You may want to take a minute or two to refamiliarize yourself with this concept.

Take a look at Table 17-3. It lists the four blood types and the six different genotypes that give rise to them. Take a moment to study them.

As shown, the A blood type occurs when an individual has two *A* alleles of the *I* gene or an A and an O. Type B occurs when an individual has two *B* alleles of the *I* gene or a B and an O. Type AB has both *A* and *B* alleles. Type O occurs when two *O* alleles of the *I* gene are present. The *I* gene is responsible for the production of glycoproteins (proteins with carbohydrate attached) that project from the surface of red blood cells.

Interestingly, both *A* and *B* are dominant genes, and the *O* allele is recessive. Because they are both dominant, the I^A and I^B genes are referred to as codominant. **Codominant genes** are expressed fully and equally. That's why an individual with AB blood has both A and B glycoproteins.

> **KEY CONCEPTS**
>
> Some traits are determined by multiple alleles.

Polygenic Inheritance

For many years, geneticists believed that each gene controls a single trait. In humans and other animals, however, many traits are controlled by a number of genes—from a few to perhaps hundreds. Skin color, for example, is controlled by as many as eight genes. This type of inheritance is called **polygenic inheritance** (polly-JEAN-ick). Polygenic inheritance results in incredible phenotypic variation.

To see how polygenic inheritance leads to such wide phenotypic variation, consider an example using two genes for skin color, designated *A* and *B*. In this example, we will say that the genotype of an African American is *AABB* and the genotype of a Caucasian is *aabb*. Table 17-4 lists all of the possible genotypes in this example with possible phenotypes; Figure 17-13 shows what they look like. Take a moment to study the table and the figure.

Polygenic inheritance is also responsible for height, weight, intelligence, and a number of behavioral traits. But in each instance, the genes are not the only factors controlling these traits. As a rule, the genotype establishes the range in which a phenotype will fall, but environmental factors (for example, diet) determine exactly how much of the potential will be realized for many genes. For example, although genes determine height, this trait is also influenced by diet. The better a child eats from infancy to adolescence, the taller he or she will be.

TABLE 17-3	Phenotype and Genotype in ABO System
Phenotype (blood type)	**Genotypes**
Type A	$I^A I^A$, $I^A I^O$
Type B	$I^B I^B$, $I^B I^O$
Type AB	$I^A I^B$
Type O	$I^O I^O$

TABLE 17-4	Possible Skin-Color Genotypes and Phenotypes with Two Skin-Color Genes	
Genotype	**Phenotype**	**Number of Recessive Genes***
AABB	Black	0
AABb	Dark	1
AaBB	Dark	1
AaBb	Mulatto	2
AAbb	Mulatto	2
aaBB	Mulatto	2
Aabb	Light	3
aaBb	Light	3
aabb	White	4

*Skin color probably involves many more genes.

FIGURE 17-13 Polygenic Inheritance Skin color is probably determined by at least two, perhaps as many as eight, genes, resulting in a wide range of phenotypes: (a) black, (b) dark, (c) mulatto, (d) light, and (e) white. (a © Photos.com. b © LiquidLibrary. c © Photodisc/Getty Images. d © Photos.com. e © LiquidLibrary.)

Health Tip 17-1

Learn to read food labels and read them carefully!

Why?

Food labels contain a great deal of useful information about the food you are eating, but you need to understand food labels to know exactly what you are getting. For example, food labels always include the number of calories per serving. Don't stop here. Check out how many servings are in a package. You may be surprised to find that a package often contains two or more servings.

Be sure to check out the saturated fat content. The higher the saturated fat content, the worse it is for you.

Health Tip 17-2

Heed those food labels!

Why?

A recent study showed that nearly 80% of all Americans read nutrition labels on the foods they buy, but overall, 44% admit to buying food items even though they list unhealthy ingredients. Younger people (ages 18 to 29) are even more likely to ignore the labels. Sixty percent admit to buying foods with unhealthy ingredients.

KEY CONCEPTS

Some traits like skin color are controlled by more than one gene pair; this is known as polygenic inheritance.

Crossing-Over

As pointed out earlier, Mendel found that during gamete formation, the genes under study in his pea plants were segregated independently. Also noted earlier, independent assortment generally occurs only when the genes under study are on different chromosomes. As a rule, if two genes are on the same chromosome, they do not segregate independently.

According to the most recent estimates, humans have an estimated 20,000 to 25,000 genes on their 46 chromosomes. (For years, scientists believed that humans had about 100,000 genes.) Those genes located on the same chromosome tend to be inherited together and are said to be **linked**. However, linkage can be disrupted by a phenomenon called **crossing-over**.

Crossing-over occurs during meiosis, when the homologous chromosomes come together, or "pair up," in prophase I. During this process, the chromatids of homologous chromosomes fuse at one or more points, as shown in Figure 17-14. Later, when the chromosomes separate, arms of one chromosome may end up with the other. In this process, then, chromosomes exchange segments of their chromatin with chromatin on homologous chromosomes. When this occurs, genes that were once said to be linked are not inherited together.

Crossing-over "unlinks" genes and increases genetic variation in gametes; this, in turn, leads to genetic variation in offspring. The more genetic variation, the more possible phenotypes in a population. Genetic variation may lead to characteristics that give one organism an advantage over another. This advantage can be passed to an organism's offspring, resulting in evolutionary changes.

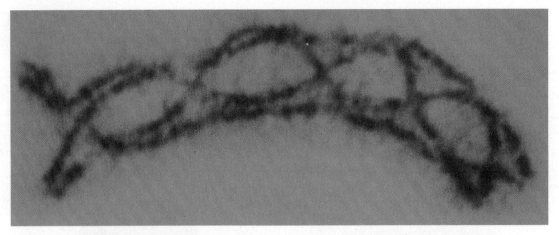

FIGURE 17-14 **Crossing-Over** The crossing-over shown here increases genetic variation in gametes and offspring and occurs during meiosis. (© B. John/Cabisco/Visuals Unlimited.)

Crossing-over can occur anywhere along the length of a chromosome. However, the greater the distance between two genes on the same chromosome, the more likely it is that a crossover will occur between them. This fact helped scientists map the human genome.

> **KEY CONCEPTS**
>
> Genes can be transferred from one chromatid to another during meiosis.

The Human Genome Project

In the 1990s, scientists the world over embarked on what is conceivably the largest unified biological research project of its kind in history—known as the **Human Genome Project**. Coordinated by the U.S. Department of Energy and the National Institutes of Health, this massive research project involving scientists from many different countries was designed to determine the location of all of the genes on all 46 human chromosomes, as well as the sequence of bases of the human genes and intervening DNA in the chromosomes. The Project was also designed to transfer genetic technologies to industry and research facilities throughout the world, thus facilitating the advancement of genetic knowledge. The program was also designed to address key social, psychological, ethical, and economic issues that could arise from the Project.

With an estimated 20,000 to 25,000 genes and three billion base pairs in the human DNA, costs of this monumental task were estimated to be around $3 billion. Time to complete the task was estimated to be 15 years or longer.

To their surprise, scientists found that the first part of the project—mapping the location of the 20,000 to 25,000 genes—occurred at a much faster pace than anticipated. By 2000, they had completed this task—6 years ahead of schedule! The project was terminated in 2003, after 13 years. Researchers stored the information on computers and made it available to researchers in colleges and universities and in private businesses via the Internet. Even you can log on to the website, if you are so inclined. Why was this Project so important?

The Human Genome Project catalyzed the formation of the multibillion-dollar U.S. biotechnology industry and fostered the development of many new medical applications. Knowing the location of genes on human chromosomes adds

to our knowledge of human genetics. By locating genes and determining the sequence of bases in them, scientists can begin to synthesize specific genes—for example, the gene that controls the production of blood-clotting factors. Synthesized in the lab in massive quantities, these and a host of other genes could be used to treat various illnesses. They could be spliced into the DNA of individuals suffering from a number of genetic diseases. A parent, faced with the possibility of having a severely physically impaired child with a genetic disorder, for example, might opt to have gene replacement therapy on the unborn fetus in hopes of having a healthy child.

Researchers are currently experimenting with ways to replace mutated tumor suppressor genes. Tumor suppressor genes are normal genes that hold cell division in check. When mutated, cells can undergo rapid division. By replacing the defective gene in tumor cells with intact tumor suppressor genes, researchers hope someday to tame malignancies, bringing the cells under control.

Although the Human Genome Project could result in many improvements in health care, some critics argue that knowledge gained from it could lead to a situation in which a physician could diagnose an illness through genetic means long before a cure is available. Some individuals are concerned that insurance companies could access a person's records, and, after examining them, deny coverage. Other critics worry about the possibility that the knowledge gained from the Project could result in manipulation of human genes. You can find excellent summaries of the social, economic, and ethical issues on the Internet at the Human Genome Project Information page.

One outcome of this project has been the mapping of chromosomes in nonhuman species, a step that could help bring important genetic improvements in the case of livestock; in the case of disease organisms, chromosome mapping could help scientists find better tools to combat them. Mapping genes of nonhuman species also greatly furthers our understanding of evolution, a process of genetic change that results in new species and improvements in existing species.

> **KEY CONCEPTS**
>
> The Human Genome Project was an intensive effort to determine the sequence of nucleotides in the DNA of the chromosomes of humans and the location of all our genes.

17-5   Sex-Linked Genes

Some genes are carried on the sex chromosomes. These are known as sex-linked genes. In this section we will examine the sex chromosomes and sex-linked traits.

Determination of Sex

Many of us grew up believing that the sex chromosomes determine whether an individual is a male or female. Those with XX chromosomes are females. Those with XY are males.

In recent years, though, genetic scientists refined this view. They asserted that, in truth, sex is determined by the Y chromosome. If the Y chromosome is present, an individual becomes a male. If absent, the individual is a female.

Geneticists found that the Y chromosome exerted its effect during embryonic development. Interestingly, early in the embryo's development, the gonads of males and females are identical. If the Y chromosome is present, however, the embryonic gonad becomes a testis—thanks, at least in part, to the presence of the *SRY* gene (testis-determining gene). The testes, in turn, produced testosterone and other androgens that were responsible for the male secondary sex characteristics such as male hair growth, body shape and size, and so on. Geneticists also found that the *Sry* gene activated other genes that resulted in a phenotypic male. These are referred to as pro-male genes. The absence of the Y chromosome results in the development of an ovary with its hormones and the resultant female characteristics.

It turns out that the sex determination pathway is more complicated. Here's what geneticists think happens. The *SRY* does indeed activate pro-male genes, however, it also inhibits some anti-male genes. So instead of a simple mechanism in which pro-male genes "go all the way" to make a male, there is a balance between pro-male genes and anti-male genes. What is more, if there are too many of anti-male genes, a female is born. If there are more pro-male genes then the offspring is a male.

Researchers now believe that sex is determined by a balance between pro-male, pro-female, anti-males, and anti-female genes. Each year, a small number of babies is born in the United States and other countries who do not conform anatomically to male or female "standards." They are referred to as **intersexuals**. Intersexuals result from the interaction of the genes to produce intermediary sex forms—for example, women born with male genitalia. This phenomenon occurs when an XX baby is exposed to high levels of androgen before birth.

By current estimates, one in every 1,500 to 2,000 live births results in an intersexual baby. Moreover, there are far more individuals who display subtler variations of their assigned sex—characteristics that are not always present at birth. Surgery is usually performed within the first 24 hours to assign a sex to the infant.

Two forms, male and female, are referred to as sexual dimorphism. Because of the fairly frequent occurrence of intersexuality, some researchers propose that gender be viewed as a spectrum rather and a binary—that is, either male or female. The X and Y chromosomes also carry genes that determine many other traits. Genes situated on the X and Y chromosomes are therefore known as **sex-linked genes**. So far, the majority of the sex-linked genes are located on the X chromosome.

These are known as **X-linked genes**. The following sections describe the inheritance of some common sex-linked genes.

> **KEY CONCEPTS**
>
> Sex is determined by the balance between a number of different genes that either promote or inhibit maleness or promote or inhibit femaleness.

Recessive X-Linked Genes

With hundreds of genes, the X chromosome plays an important role in determining our appearance. It also is responsible for certain diseases. Some diseases, like color blindness and certain forms of hemophilia are recessive traits carried on the X chromosome. These are known as **recessive X-linked genes**.

In order for a female (XX) to display a recessive sex-linked trait, each of her X chromosomes must carry the recessive gene. For males (XY), however, only one recessive allele is required. Why? Because the Y chromosome is not genetically equivalent to the X chromosome. Thus, in males, only one recessive gene is needed to exhibit a recessive X-linked trait.

Figure 17-15 shows four possible genetic combinations leading to color blindness. This illustration also introduces you to a genetic-tracking system used to follow traits in families, known as a pedigree. In a **pedigree**, squares represent men and circles represent women. The horizontal line linking a square to a circle indicates a mating. Offspring are shown below their parents. When a square is lightly shaded, the individual is a carrier of the gene. He or she does not have the disease, but carries the recessive gene and can pass it on to his or her offspring. A darkly shaded square or circle indicates the person has the trait. He or she exhibits the trait and is a carrier.

In Figure 17-15a, for example, a man and a woman have four children, two boys and two girls. The woman (light blue circle) is a carrier of color blindness. Her cells contain one X chromosome with a recessive allele for color blindness (designated X^c) and another X chromosome with the normal allele. Consequently, half of her ova will contain the recessive allele. As shown, the woman's husband (white square) is not color-blind. He produces sperm containing either an X or a Y chromosome. When one of his X-bearing sperm unites with one of her ova bearing an X chromosome with the allele for color blindness, the result is a daughter who is a carrier, indicated by a light blue circle. When one of his X-bearing sperm unites with an ovum carrying a normal X chromosome, the result is a daughter who is neither a carrier nor color-blind (white circle).

Males are produced when a Y-bearing sperm unites with an ovum carrying an X chromosome. If the X chromosome carries the recessive allele for color blindness, the boy is color-blind (green square). If the X chromosome is normal, the boy's color vision is unimpaired (white square). Take a moment to study the other possibilities in Figure 17-15.

> **KEY CONCEPTS**
>
> In order for an X-linked recessive gene to be expressed, it must be present on both X chromosomes of females but only on one in males.

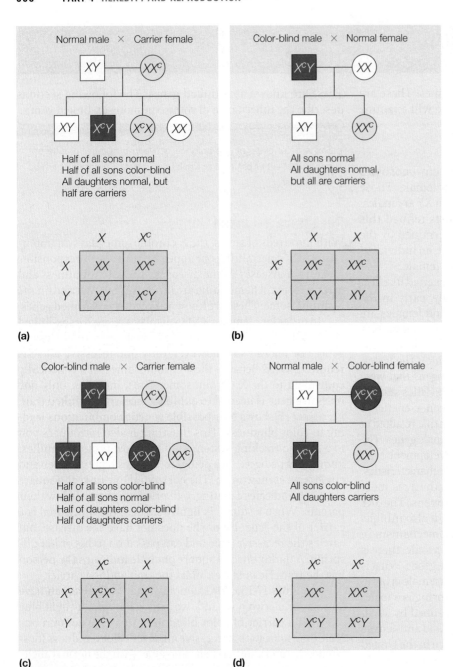

FIGURE 17–15 Inheritance of Color Blindness Four possible genetic ways a sex-linked recessive gene such as color blindness can be passed to offspring. Males are indicated by boxes and females by circles. In this scheme, green boxes and circles represent men and women with color blindness. Light blue boxes represent men and women who are carriers. White boxes and circles represent men and women without the color blindness gene. The notation X^c indicates an X chromosome carrying the gene for color blindness.

Dominant X-Linked Genes

Recessive X-linked genes are the most common sex-linked trait in humans. However, there are a few noteworthy examples of dominant X-linked genes. One of the best understood is a disorder with a tongue-twisting name of **hypophosphatemia** (high-poe-FOS-fuh-TEEM-ee-uh). This disease is characterized by low phosphate levels in the blood and tissues of the body. This, in turn, results in a form of rickets (RICK-its), or bowleggedness (Figure 17-16). Rickets usually results from a dietary deficiency of vitamin D or insufficient exposure to sunlight. Alleviating the dietary deficiency usually solves the problem. In this genetic disease, however, vitamin D cannot reverse the symptoms.

Hypophosphatemia occurs when either the male (XY) or the female (XX) has one X chromosome containing the dominant gene, indicated as X′ (X prime). Figure 17-17 illustrates the pattern of inheritance when a woman who is heterozygous (X′X) for the trait mates with a man who does not carry the trait (XY). Take a moment to study it. As you can see and as you would expect, whenever the dominant gene is present, it is expressed—regardless of the sex of the individual.

> **KEY CONCEPTS**
> Dominant X-linked genes are always expressed in males and females.

Y-Linked Genes

The Y chromosome contains the genes that control the sex of an individual, as noted earlier. The Y chromosome also controls sperm production and male secondary sex characteristics. Because only males have Y chromosomes, Y-linked traits only appear in males. In addition, Y-linked genes can only be transmitted from fathers to sons. Because the X and Y chromosomes are not homologous, each gene on the Y chromosome has only one allele. Thus, both dominant and recessive Y-linked genes are always expressed.

> **KEY CONCEPTS**
> Because the Y chromosomal genes are not homologous, dominant and recessive genes are always expressed when present.

Sex-Influenced Genes

Although genes determine traits in men and women, some autosomal genes behave differently in the two sexes. In one sex, for example, an allele will be dominant; in the other sex, it will be recessive. Because they are affected by the sex of the carrier, these genes are known as **sex-influenced genes**.

The best-known example of a sex-influenced gene is the gene for pattern baldness, now more often referred to as androgenic alopecia. **Androgenic alopecia** occurs in both men and women. In men, the loss of hair often begins in one's twenties, though some cases begin at the end of puberty. The word androgenic refers to the fact that baldness, alopecia, is caused by a male sex hormone known as dihydrotestosterone (DHT). DHT causes hair follicles to shrink, producing very fine hairs. The term **pattern baldness** results from the fact that affected individuals (mostly men) do not go completely bald; they retain a rim of hair on the temples and back of the head.

In men, androgenic alopecia behaves as if it were autosomal dominant. That means that the gene is expressed in both heterozygous and homozygous-dominant individuals. Because the gene is dominant and widespread in many

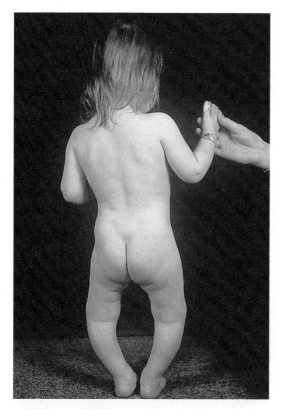

FIGURE 17–16 **Hypophosphatemia** People with hypo-phosphatemia, a dominant X-linked genetic disorder, resemble this child with rickets, which usually results from inadequate vitamin D intake. (© Biophoto Associates/Science Source, Inc.)

populations, many men experience androgenic alopecia. In the United States, by age 35 two-thirds of all men experience noticeable hair loss. By age 50, approximately 85% of men have significantly thinning hair. Unfortunately, about 25% of men who suffer from male pattern baldness begin the painful process before they reach 21. Some begin shortly after puberty.

In women, the allele for androgenic alopecia behaves as if it is autosomal-recessive. Because of this, women are much less commonly afflicted by this genetic trait. In fact, only women who are homozygous recessive for the trait will experience baldness. Hair thinning in women due to androgenic alopecia differs significantly from men. Rather than a distinct pattern baldness, women experience diffuse thinning in all areas of their scalps (Figure 17-18). Some women, however, experience a combination of male and female androgenic alopecia.

As in men, androgenic alopecia in women results from the action of androgens, male hormones. In women, however, male hormones are usually present in very low concentrations. Androgenic alopecia typically results from ovarian cysts that produce androgen, birth control pills that contain large amounts of androgen, pregnancy, and menopause. For a discussion of the causes and cures of baldness, see Health Note 17-1.

KEY CONCEPTS

The activity of some autosomal genes like that for baldness are influenced by the sex of an individual.

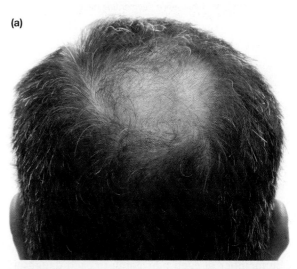

(a)

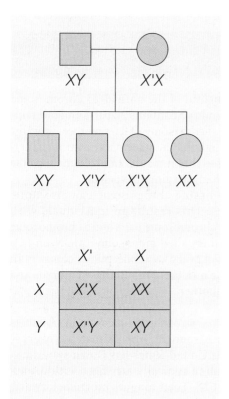

FIGURE 17–17 **Inheritance of a Sex–Linked Dominant Gene** (a) Pedigree. (b) Corresponding Punnett square.

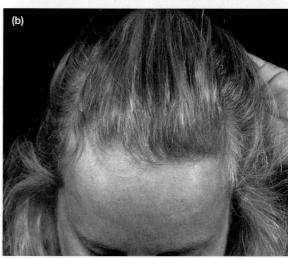

(b)

FIGURE 17–18 **Genetic Expression of Androgenic Alopecia** Hair thinning and loss in (a) a man (© Anastasios Kandris/ShutterStock, Inc.) and (b) a woman (© Mediscan/Visuals Unlimited, Inc.).

health**note**

17–1 The Causes and Cures of Baldness in Men and Women

As we age, we all lose hair. However, some people lose a lot more than others. Doctors recognize three types of hair loss. Partial hair loss involves the loss of small, isolated patches of hair. The second type, pattern baldness, involves the loss of most or all of the hair on the head. Total body hair loss is the complete loss of hair from every part of the body: head, eyelids, eyebrows, arms, legs, and so on.

Hair loss strikes both sexes, all ethnic groups, and all age groups. In fact, two of every three American men will develop some form of balding. An even larger percentage of men and women will lose some hair.

Hair loss, while not life-threatening, can have severe emotional impacts on individuals. Men whose hair begins to thin in their twenties often feel self-conscious and anxious in public. They may feel isolated from their cohorts, for hair loss often makes them look 5 to 10 years older than their peers. Friends may poke fun at them, further adding to their dismay. Even in older men, whose friends are balding, hair loss can result in depression, feelings of low self-esteem, and a general sense of inadequacy. In women, hair loss can have even more devastating effects.

A surprising number of factors can cause hair to fall out. One of the most widely recognized is chemotherapy—chemical treatments used to fight cancer. These drugs attack rapidly dividing cells in the body such as cancer cells and the cells in hair follicles. High fever, severe infections, and severe cases of the flu also cause hair loss, starting 1 to 12 months after the illness. Many prescription drugs also cause hair loss. Birth control pills, iron deficiencies, and protein-deficient diets can have a similar effect in some patients. Many women report excessive hair loss after pregnancy. But the most common cause of all is genetic.

Hereditary balding (pattern baldness) results from genes that arise from either the mother's or the father's side of the family. In men, this condition results in near total or complete loss of hair. Women, on the other hand, tend to experience excessive thinning, but rarely go completely bald. What can be done to combat balding?

Two lines of attack are possible: medicinal and surgical. Consider the medicinal approach first. Several medicines, some of which are sold over the counter, are available. One of the most widely publicized is Rogaine. In a study of 2,300 patients with male pattern baldness, Rogaine treatment resulted in moderate to marked hair growth in 39% of the patients, compared to 11% of those in the control group. For Rogaine to be effective, however, this drug must be applied twice a day every day of one's life. Stopping treatment causes bald spots to reappear. Costing $600 to $1,000 per year, Rogaine is recommended for younger men from ages 20 to 30 who have begun to lose hair within the last 5 years. Men who are completely bald are poor candidates for treatment.

17-6 Chromosomal Abnormalities and Genetic Counseling

Besides defective genes, some defects occur as a result of an abnormal number of chromosomes. This occurs during meiosis, which is a special type of cell division that occurs in the gonads during the formation of gametes. As you may remember, during meiosis I, homologous chromosomes pair up and gather in the center of the cells. The pairs then separate—with one member of each pair going to each daughter cell (Figure 17-19a). If a homologous pair fails to separate during meiosis, however, one of the new cells will end up with an extra chromosome (Figure 17-19b). The other cell will be short one chromosome. The failure of homologous chromosomes to separate is called **nondisjunction** (non-diss-JUNK-shun).

Nondisjunction can also occur in the second meiotic division (Figure 17-19c). In this division, you may recall, the 23 replicated chromosomes split apart, with one chromatid going to each daughter cell. If a chromosome fails to separate into its two chromatids, the result is the same as nondisjunction in meiosis I—a daughter cell with an extra chromosome and another daughter cell missing one chromosome.

When a gamete with one extra chromosome unites with a normal gamete, the zygote produced will contain 47 chromosomes. The zygote may be able to divide successfully by mitosis, producing an embryo whose cells have an additional chromosome. Thus, instead of the normal 23 chromosome pairs, each cell in the embryo contains 22 pairs and one triplet. This condition is called **trisomy** (TRY-sew-mee; literally, "three bodies"). Although that may seem like a minor difference, you shall soon see that it results in profound adverse changes in the individual carrying this defect.

Gametes with a missing chromosome can also unite with normal gametes. This results in individuals with 45 chromosomes—22 chromosome pairs and a chromosome singlet. This condition is called **monosomy** (MON-oh-SO-me). Surprisingly, one of every two conceptions contains an abnormal chromosome number. Most of these embryos and fetuses die inside the mother.

Down Syndrome (Trisomy 21). One of the most common trisomies is **Down syndrome**, or trisomy 21. Approximately 1 of every 700 babies born in the United States has Down syndrome. Down syndrome children typically are short with round, moonlike faces (Figure 17-20). Their tongues protrude forward, forcing their mouths open, and their eyes slant upward at the corners. Their IQs are typically around 50 and rarely over 70. A significant number of Down syndrome babies die from heart defects and respiratory infections in the first year

Antiandrogens can be given to women. These drugs inhibit the binding of androgens (male hormones found in women's blood) to the hair follicles. For reasons not well understood, this treatment can result in a complete reversal of hair loss, but only if it is begun within 2 years of the onset of hair loss. Men can also be treated with anti-androgens, but not without significant problem, for these drugs cause a loss of sex drive and an undesirable elevation of the voice.

For those who do not respond to drugs, surgery is an option. But surgery can be quite expensive, costing as much as $15,000. Despite its potentially high price tag, an estimated 250,000 American males elect to have one of several different types of surgical procedures each year.

One of the most common is hair transplantation (Figure 1). In this operation, small plugs of hair are taken from the sides and back of the scalp where hair grows thickly. These plugs are placed in the bald spot.

Another common procedure is scalp reduction. In this procedure, surgeons remove well-defined bald spots (up to 2 by 7 inches) in the top of the scalp. The edges of the incision are then drawn together, thus reducing the area of baldness. Hair transplants may also be performed in conjunction with this procedure to fill in the remaining area.

Another option is a wig or toupee. Although wigs or toupees have improved dramatically and well-crafted hair pieces made from real or synthetic hair are very difficult to distinguish from the real thing, there are enough bad ones around that this option is often looked upon with disfavor by many men. Hair pieces can even be sutured in place.

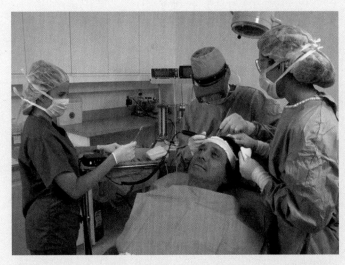

FIGURE 1 Patient undergoing hair transplant surgery. (© Mauro Fermariello/Science Source, Inc.)

People who are experiencing baldness should seek medical attention quickly to determine the causes. With their doctor's help, they can plot a strategy, if they so desire, to ward off balding. Or, they can simply accept their fate and learn to live with—or even appreciate—it.

of infancy. Modern medical care, especially antibiotics, has reduced early death, and many individuals with Down syndrome live to age 20 or beyond.

The incidence of Down syndrome (and many other monosomies and trisomies) increases with maternal age. As Figure 17-21 shows, a woman's chances of having a Down syndrome baby increase dramatically after age 35. For this reason, many couples choose to have their children at an earlier age.

Nondisjunction of the Sex Chromosomes. Nondisjunction can occur in both autosomes and sex chromosomes. Nondisjunction of the sex chromosomes can lead to a variety of nonlethal genetic disorders. One of the most common is Klinefelter syndrome.

Klinefelter syndrome occurs when an ovum with an extra X chromosome is fertilized by a Y-bearing sperm, resulting in an XXY genotype (Figure 17-22). This condition occurs in about 1 of every 700 to 1,000 newborn males. Although Klinefelter syndrome patients are males, masculinization is incomplete. The males' external genitalia and testes are unusually small, and about 50% of them develop breasts. Spermatogenesis is abnormal, and Klinefelter patients are generally sterile.

Another common disorder resulting from nondisjunction of the sex chromosomes is Turner syndrome, a monosomy. **Turner syndrome** may result when an ovum lacking the X chromosome is fertilized by an X-bearing sperm. It may also result when a genetically normal ovum is fertilized by a sperm lacking an X or a Y chromosome (Figure 17-23). The result in both

cases is an offspring with 22 pairs of autosomes and a single, unmatched X chromosome, denoted by XO (Figure 17-22).

Turner syndrome patients look like females and are characteristically short with wide chests and a prominent fold of skin on their necks. Because their ovaries fail to develop at puberty, Turner syndrome patients are sterile, have low levels of estrogen, and have small breasts. For the most part, they lead fairly normal lives. Mental retardation is not associated with the disorder. Turner syndrome occurs in 1 of every 10,000 female births. The rarity of this condition, compared with Klinefelter's syndrome, is due to the fact that the XO embryo is more likely to be spontaneously aborted.

KEY CONCEPTS
Some important chromosomal abnormalities occur as a result of the failure of chromosomes to separate during meiosis during gamete production.

Defects in Chromosome Structure

Another defect in chromosomes involves alterations in chromosome structure, the most common of which are (1) deletions, the loss of a piece of chromosome, and (2) translocations, breakage followed by reattachment elsewhere.

Deletions. A **deletion** is a genetic defect that occurs when a part of a chromosome or a sequence of DNA turns up missing.

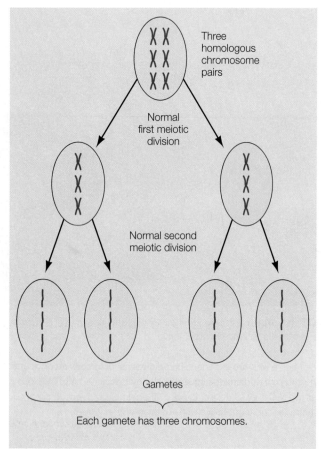

(a)

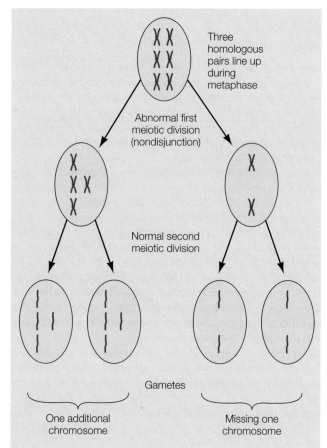

(b)

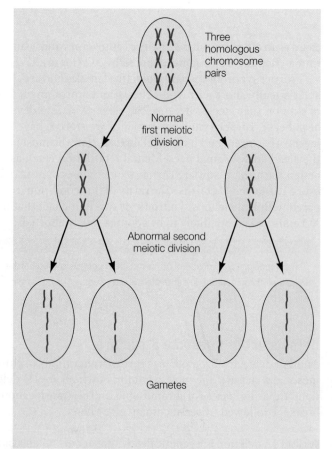

(c)

FIGURE 17–19 Meiosis and Abnormal Chromosome Numbers (a) A simplified version of meiosis. During the first meiotic division, the homologous pairs line up, then separate, producing daughter cells with one-half the number of chromosomes. In the second meiotic division, the chromosomes line up single file and separate, with one chromatid going to each daughter cell. (b) Nondisjunction in the first meiotic division. A chromosome pair may fail to separate during meiosis I, resulting in abnormal gametes. Half are missing a chromosome, and the other half have an extra chromosome. (c) Nondisjunction in the second meiotic division, resulting in two normal gametes, one gamete with two chromosomes, and one gamete with four.

Deletions may result from the loss of one nucleotide or the loss of many—for example, the loss of an entire piece of a chromosome. Deletions most frequently occur during meiosis as a result of crossing over, described earlier in this chapter.

Most deletions of genetic material are deleterious to the health of humans. So harmful are they that embryos whose cells contain chromosomes with deletions are usually eliminated early in pregnancy—naturally aborted. Nevertheless, some embryos whose cells contain deletions do survive. Table 17-5 lists a few disorders caused by deletions. One of the more striking is called Praeder-Willi syndrome (PRAY-der Will-ee).

Praeder-Willi syndrome occurs in 1 in 10,000 to 25,000 births. It is caused by the loss of one of the arms of chromosome 15 during gamete formation and is characterized by slow infant growth, compulsive eating, and obesity. Babies born with the syndrome have a poor suckling reflex and do not feed well. By age 5 or 6, however, these children become compulsive eaters.

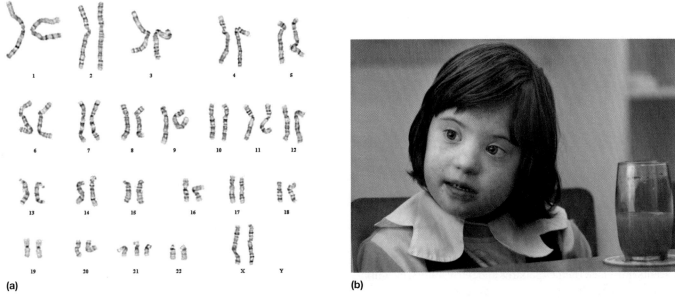

(a)

(b)

FIGURE 17-20 **Down Syndrome** (a) Karyotype of Down syndrome girl. Note trisomy of chromosome 21. (Courtesy of Viola Freeman, Associate Professor, Faculty of Health Sciences, Dept. of Pathology and Molecular Medicine, McMaster University.) (b) Notice the distinguishing characteristics described in the text. (© PhotoCreate/ShutterStock, Inc.)

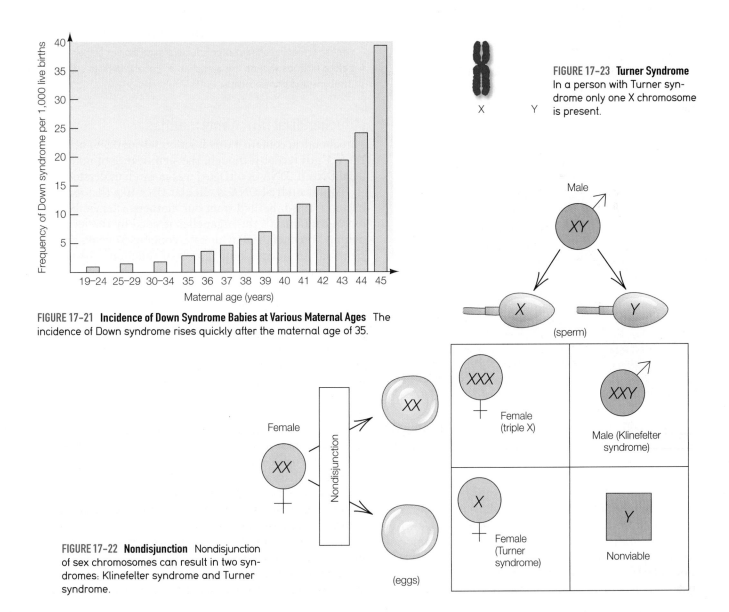

FIGURE 17-21 **Incidence of Down Syndrome Babies at Various Maternal Ages** The incidence of Down syndrome rises quickly after the maternal age of 35.

FIGURE 17-23 **Turner Syndrome** In a person with Turner syndrome only one X chromosome is present.

FIGURE 17-22 **Nondisjunction** Nondisjunction of sex chromosomes can result in two syndromes: Klinefelter syndrome and Turner syndrome.

TABLE 17-5	Chromosome Deletions
Syndrome	Phenotype
Wolf-Hirschhorn syndrome	Growth retardation, heart malformation, cleft palate; 30% die within 24 months
Cri-du-chat syndrome	Infants have catlike cry, some facial anomalies, severe mental retardation
Wilm's tumor	Kidney tumors, genital and urinary tract abnormalities
Retinoblastoma	Cancer of eye, increased risk of other cancers
Praeder-Willi syndrome	Infants are weak and grow slowly; children and adults are obese and are compulsive eaters

Parents must lock their cupboards and refrigerators. Neighbors must be warned to discourage begging and must keep their garbage cans under lock and key. The urge to eat results in obesity, which often leads to late-onset diabetes. If food intake is not restricted, victims can literally eat themselves to death. Researchers believe that the eating disorder may result from an endocrine imbalance caused by the deletion.

Translocations. **Translocations** occur when a segment of a chromosome breaks off and reattaches either to another site on the same chromosome or to another chromosome. Although this might at first seem innocuous, movement of a segment of a chromosome to another site can upset the delicate balance of gene expression and alter homeostasis. Translocations, for example, may be the cause of certain forms of leukemia.

Health Tip 17-3

Not worried about what you smoke or drink or which drugs you take? Better think twice.

Why?

Researchers are finding—much to the surprise of conventional geneticists—that our chemical environment can have profound effects on our DNA. Certain chemicals in our diet, for example, can alter our DNA. DNA in offspring can also be modified while in utero, for example, as a result of exposure to chemicals in a mother's diet. More important, the chemical changes produced in the DNA of the offspring affect their characteristics and—again to the surprise of geneticists—can be transmitted to their offspring. Some changes in the DNA of a person's offspring could have beneficial effects on them; others may not.

> **KEY CONCEPTS**
> Genetic disorders may also result from structural defects in chromosomes like missing pieces.

Genetic Screening

Elsewhere in the text you found that parents can now find out the sex of their child well before birth via amniocentesis. This procedure also permits geneticists to search for abnormal chromosome numbers, as well as deletions and translocations. Furthermore, biochemical tests can also be

run to pinpoint metabolic diseases, although only a dozen or so are routinely screened.

Amniocentesis is usually recommended only if one or more of the following conditions are met: (1) a woman is over 35; (2) she has already delivered a baby with a genetic defect; (3) she is a known carrier of an X-linked genetic disorder; or (4) she or the father has a known chromosomal or genetic abnormality.

As a rule, amniocentesis is usually not performed until the sixteenth week of pregnancy. Before this time, there is not enough fluid surrounding the fetus, and the needle could damage the fetus. Analysis of the fetal cells withdrawn from the amnion requires an additional 10 to 15 days.

To permit earlier detection of genetic defects, a new procedure, known as *chorionic villus biopsy* (CORE-ee-on-ick VILL-us BYE-op-see), has been developed. Chorionic villi are composed of embryonic tissue that forms the fetal portion of the placenta (plah-SEN-tah), a structure that nourishes the growing fetus. To perform this procedure, physicians remove a small sample of a villus from the uterus through the vagina. The cells of the villus are then examined, in much the same way as those removed during amniocentesis. Although chorionic villus biopsy allows for earlier detection, it poses a slightly higher risk to the mother and her fetus.

> **KEY CONCEPTS**
> Genetic screening allows parents to determine the sex of their child before birth as well as the presence of genetic defects such as deletions and translocations.

Mitochondrial DNA Abnormalities

Mitochondria contain a tiny fraction (about 0.3%) of a cell's DNA. Until recently, though, the significance of mitochondrial DNA (mDNA or mtDNA) was not well understood.

Mitochondrial DNA is circular DNA like that found in bacteria. It is inherited from our mothers. (Remember, the ovum contains all the organelles needed by the fertilized ovum.) Mitochondrial in humans contains 37 genes that are essential for normal mitochondrial function. Thirteen of these genes, for instance, control the production of enzymes required in oxidative phosphorylation. This process results in the production of cellular energy—adenosine triphosphate (ATP). The rest of a mitochondrion's genes control the production of RNA molecules needed for protein synthesis.

Although mDNA constitutes only a small portion of a cell's DNA, defects in mitochondrial DNA can result in serious problems. For example, medical researchers have found that a rare form of blindness results from a defect in the mitochondrial DNA. The defective gene in the mitochondria that leads to blindness codes for a protein required in the first step of ATP production. The absence of this protein in the neurons of the optic nerve results in their death. This, in turn, leads to blindness, usually by age 20.

Because mitochondria are passed on by the mother, all of the children of a woman with the defective gene will inherit it. Only a small fraction of the children who inherit the defective gene actually go blind, so other factors must also contribute to blindness.

Several rare genetic diseases may also result from mitochondrial DNA defects. Researchers suggest that even some of the more common diseases may be caused by genetic defects in mitochondrial DNA. Some cases of heart, kidney, and central nervous system failure, whose causes are now unknown, may one day be linked to defective mitochondrial DNA.

Mitochondrial DNA is prone to somatic mutations (mutations in somatic cells) that can lead to certain forms of cancer. Somatic mutations occur in the DNA of certain cells during a person's lifetime and are typically are not passed to one's offspring. Some studies suggest that somatic mutations in mitochondrial DNA may be involved with certain types of cancer, including breast, colon, stomach, liver, and kidney tumors. They may also be associated with leukemia and cancer of immune system cells (lymphoma).

Researchers currently speculate that somatic mutations in mitochondrial DNA increase the production of potentially harmful molecules known as reactive oxygen species (ROS). These are free radicals. Mitochondrial DNA is particularly vulnerable to ROS. Making matters worse, mRNA cannot repair itself as easily as other DNA in the body. This results in further damage—an increase in somatic mutations. Researchers are currently trying to determine whether these mutations can result in uncontrolled cell division that occurs in the growth of cancers.

Interestingly, many phytochemicals in the food we eat—especially the cruciferous vegetables like cabbage, broccoli, kale, cauliflower, kohlrabi, and radishes as well as onions, mushrooms, berries, seeds, and nuts—contain chemicals that help to eliminate free radicals or fight cancer in other ways. Numerous well-conducted scientific studies show that eating a diet rich in these foods can substantially lower one's risk of many forms of cancer. Here's a short list to help you remember them taken from Dr. Joel Fuhrman's work on nutrition: greens, onions, mushrooms, berries, beans, and seeds.

> **KEY CONCEPTS**
>
> Mitochondria contain functional DNA that can be linked to some serious diseases.

17-7   Health and Homeostasis

Throughout this book, the idea that homeostasis and human health are profoundly influenced by our environment has emerged repeatedly. In this section, we will look at genes and behavior and the role of our environment in determining our personality and behavior and, perhaps, our health.

At one time, psychologists viewed a baby as a blank slate. A child's personality, they said, develops through interaction with its environment—its parents, friends, teachers, and so on. However, extensive research now suggests that our personalities are also influenced by our genes. Thus, our genes and our environment operate together in determining personality.

Michael Lewis, a researcher at the Robert Wood Johnson Medical School, studies infant response to stress. His research shows that newborn babies differ markedly in how they respond to the stress of a blood test performed well before their environment could have affected their personality. Lewis found that some children wail when poked with a needle during routine blood tests in the first few days of life; others hardly seem to notice. Of the newborns who cry, some quickly dampen their response. Others seem to go on forever.

Lewis believes that the difference in response to stress, also seen later in life, is genetically based and that the inherent differences will persist. Lewis has also found that babies differ in how they react to frustration. He performed a series of experiments to test infant response to a frustrating situation. Most children responded with anger. Some, however, showed no response at all, and others displayed sadness. These innate differences in behavior, occurring too early to stem from differences in upbringing, also probably result from genetic differences. They may help account for the profound differences in responses to stress seen in adults.

Psychologist Nathan Fox has performed some intriguing studies on shy and extroverted children. His studies show that shy children cling to their mothers when a clown suddenly appears; extroverted children eagerly engage the clown in play. Fox has found that shy children show greater electrical activity in the right part of the brain; extroverted children show a higher level of activity on the left side. Genetics, he suspects, is the reason for this result.

Allison Rosenberg, a researcher at the National Institutes of Health, has also studied shy and outgoing children. Her work shows marked differences between these two groups in both heart rate and the release of the hormone cortisol, further supporting the notion that there are inherent physiological differences in children from the outset. These are related to early differences in personality and, very possibly, to differences in genetic makeup.

Interestingly, geneticists have recently discovered a gene dubbed "the Prozac gene." It is not responsible for the production of Prozac, an antidepressant, but rather is believed to be responsible for mild-mannered behavior in those who have it. Similar findings have been reported in infant monkeys. Important as these genes are, environmental effects, especially events very early in life, can modify genetically programmed behavior.

Another example of the genetic basis of behavior and the role of one's environment in modifying genetically predisposed behavioral patterns comes from research on thrill seeking. Psychologists believe that some individuals are naturally born thrill seekers, or type T people (Figure 17-24). Some researchers believe that thrill seeking may be genetically based. One theory is that risk takers, the type T or "big T" individuals, may be hard to excite and, therefore, may attempt extraordinary feats for excitement. At the opposite end of the spectrum are "small t" people, risk avoiders, who are easily aroused. They seek to avoid stimulation. Presumably, there's a whole spectrum of folks in the middle.

Some researchers believe that type T individuals have an imbalance of a neurotransmitter known as monoamine oxidase (MAO) in the brain. Thrill seeking supposedly increases the MAO levels in the brain, creating a feeling of exhilaration.

FIGURE 17-24 **Type T Behavior** Thrill seeking, or type T behavior, is thought to have a genetic basis. (© Photodisc.)

Type T behavior can be modified by an individual's upbringing and turned in a negative or positive direction, say some psychologists. A positive direction might lead an individual to play for the Green Bay Packers football team. A negative direction might lead that same individual, under different environmental conditions, into gang fighting and crime in the streets.

Peers, teachers, relatives, ministers, parents, and others make up our environment. Their influence may turn the type T child to healthy constructive opportunities or unhealthy, destructive ends. Environmental influences, then, may affect health in a roundabout way. Children who respond abnormally to stress, for instance, may be steered toward mechanisms that reduce stress, resulting in a healthier lifestyle. Homeostasis is thus broadly affected by the psychological environment.

KEY CONCEPTS

Genes profoundly influence the way we behave, starting as newborns.

SUMMARY

Meiosis and Gamete Formation

1. Meiosis is cellular division that occurs in the gonads and results in the production of the male and female gametes, the sperm and ovum.
2. Meiosis involves two nuclear divisions. During the first division, known as meiosis I, the number of chromosomes is reduced by half.
3. In males, meiosis produces four gametes; in females it produces only one.

Principles of Heredity: Mendelian Genetics

4. Gregor Mendel, a nineteenth-century monk, derived several important principles of inheritance from his work on garden peas.
5. Mendel determined that, at least in garden peas, traits do not blend.
6. Mendel's work led him to the conclusion that each adult cell has two hereditary factors for a given trait—one from the mother and the other from the father. These factors are called genes today.
7. Mendel hypothesized that hereditary factors (genes) separate during gamete formation. This is known as the principle of segregation.
8. Mendel also postulated that hereditary factors for a particular trait are either dominant or recessive.
9. The dominant and recessive genes are alternative forms of the gene, or alleles.
10. A dominant factor masks a recessive factor. A recessive factor is expressed only when the dominant factor is missing.
11. Three genetic combinations are possible for a given trait: heterozygous, homozygous dominant, and homozygous recessive.
12. The genetic makeup of an organism is called its genotype. The physical appearance, which is determined by the genotype and the environment, is the phenotype.
13. From his studies, Mendel concluded that the hereditary factors were separated independently of one another during gamete formation. This is the principle of independent assortment and holds true only for nonlinked genes.

Mendelian Genetics in Humans

14. Human cells contain 23 pairs of chromosomes: 22 pairs of autosomes and 1 pair of sex chromosomes.
15. Chromosomes carry dominant and recessive traits, and inheritance of these traits is consistent with Mendel's principles of inheritance, although additional mechanisms are at work in humans and other organisms.
16. Sickle-cell disease, cystic fibrosis, and albinism are autosomal-recessive traits and are expressed only in homozygous-recessive individuals.
17. Widow's peak and Marfan's syndrome are autosomal-dominant traits and are expressed in heterozygous and homozygous-dominant genotypes.

Variations in Mendelian Genetics

18. Genetic research since Mendel's time has turned up several additional modes of inheritance. One mode is incomplete dominance. It occurs when an allele exerts only partial dominance, producing intermediate phenotypes.

19. Some genes have more than two possible alleles. This is referred to as multiple alleles.
20. Multiple alleles result in more possible genotypes and phenotypes in a population.
21. Codominance occurs in multiple-allele genes–genes that exist in more than two forms. Codominant genes are expressed fully and equally.
22. Some traits like skin color in humans are controlled by many genes. This phenomenon is referred to as polygenic inheritance.
23. Genes that are found on the same chromosome are said to be linked. If crossing-over does not occur, these genes are inherited together.

Sex-Linked Genes

24. The X and Y chromosomes are commonly referred to as the sex chromosomes. Studies suggest that the determinant of sex is the Y chromosome. As a rule, individuals with two X chromosomes are females; individuals with an X and a Y chromosome are males.
25. Genetic research shows that sex is determined by many genes and that it is the balance of promale, antimale, profemale, and antifemale genes that determines one's sex.
26. Some individuals are born with intermediate sexual characteristics, a phenomenon known as intersexuality.
27. Intersexuals result from the interaction of the genes to produce intermediary sex forms.

28. The sex chromosomes also carry genes that determine physical traits. A trait determined by a gene on a sex chromosome is a sex-linked trait. Most sex-linked traits occur on the X chromosome.

Chromosomal Abnormalities and Genetic Counseling

29. Abnormalities in the human genome arise from mutations (changes in DNA structure), abnormalities in chromosome number, and alterations in chromosome structure.

30. Alterations in the number of chromosomes result chiefly from errors in gamete formation when chromosomes fail to separate during meiosis, a process called nondisjunction.

31. Variations in chromosome structure result from two occurrences: deletions, or the loss of a piece of chromosome, and translocations, or breakage followed by reattachment elsewhere.

32. Embryos with abnormal chromosome numbers or abnormal chromosome structure are likely to die and be aborted spontaneously.

Health and Homeostasis

33. At one time, psychologists thought that a child's personality developed principally through interaction with the environment—parents, friends, and teachers, etc. New research, however, suggests that our personalities are also influenced by our genetic makeup, starting at birth.

34. Environmental factors can alter patterns of behavior determined by one's genes.

THINKING CRITICALLY ANALYSIS

This analysis corresponds to the Thinking Critically scenario that was presented at the beginning of this chapter.

To begin, reread the previous material in this exercise and make a list of the symptoms characteristic of Marfan's syndrome. They are aortic weakening, abnormally long legs and arms, and nearsightedness. Although you won't be able to find any information about aortic weakening, you could possibly find information on Lincoln's health—perhaps old medical files. In fact, Lincoln's medical records are available, and they show that the former president displayed no signs of cardiovascular disease. That doesn't really prove anything by itself. However, Lincoln's lanky limbs were within the normal dimensions of tall people. As for the president's eyesight, it turns out that Lincoln was farsighted, not nearsighted. This evidence strongly suggests that Abe Lincoln did not have this genetic disorder. This exercise illustrates the importance of a close examination of the facts.

KEY TERMS AND CONCEPTS

Allele, p. 370
Androgenic alopecia, p. 380
Autosomal-dominant traits, p. 373
Autosomal-recessive traits, p. 372
Autosome, p. 372
Carrier, p. 375
Codominant gene, p. 376
Crossing-over, p. 377
Deletion, p. 383
Dihybrid cross, p. 371
Down syndrome, p. 382
Gene, p. 369
Genotype, p. 370
Heredity, p. 369
Heterozygous, p. 370
Homozygous dominant, p. 370
Homozygous recessive, p. 370

Human Genome Project, p. 378
Hypophosphatemia, p. 380
Intersexual, p. 379
I gene, p. 376
Incomplete dominance, p. 375
Klinefelter syndrome, p. 383
Linked, p. 377
Meiosis, p. 366
Meiosis I, p. 366
Meiosis II, p. 366
Melanin, p. 373
Monohybrid cross, p. 371
Monosomy, p. 382
Multiple alleles, p. 376
Nondisjunction, p. 382
Pattern baldness, p. 380
Pedigree, p. 379

Phenotype, p. 370
Polygenic inheritance, p. 376
Principle of independent assortment, p. 372
Principle of segregation, p. 370
Punnett square, p. 370
Recessive X-linked gene, p. 379
Reduction division, p. 366
Sex chromosome, p. 372
Sex-influenced gene, p. 380
Sex-linked gene, p. 379
Sickle cell disease, p. 375
Translocation, p. 386
Trisomy, p. 382
Turner syndrome, p. 383
X-linked gene, p. 379

CONCEPT REVIEW

1. Explain the process of meiosis in general terms. Where does it occur? What does it accomplish? p. 366.

2. Draw a diagram showing the various stages of meiosis I and meiosis II. p. 367.

3. Make a note of the number of chromosomes at each stage and their condition—that is, whether they have one chromatid or two. Which division is the reduction division? p. 366.

4. How is mitosis different from meiosis? How is it similar? p. 366.

5. Mendel's research was designed to answer two basic questions. What were the questions, and what were his findings? p. 369.

6. Define the following terms: principle of segregation and principle of independent assortment. pp. 369–370.

7. What is an allele? How can some traits be influenced by more than one allele? p. 370.

8. Define the terms phenotype, genotype, heterozygous, and homozygous. p. 370.

9. What is a monohybrid cross? What is a dihybrid cross? Give an example of each. pp. 370–372.

10. Freckles are an autosomal-dominant trait. A woman with freckles (*Ff*) marries and has a baby by a man without freckles (*ff*). What are the chances that their children will have freckles? pp. 373–374.

11. Attached earlobes (*A*) are an autosomal-dominant trait. The *A* allele is dominant over the *a* allele, which produces unattached earlobes in homozygous-recessive individuals. A woman with freckles and attached earlobes (*FfAa*) marries a man who has freckles and attached earlobes (*FfAa*). Draw a Punnett square showing the various gametes as well as the genotypes of the offspring. List all possible phenotypes and the genotypes that correspond to them. pp. 370–371.

12. What is sickle-cell disease? What causes it? Why can a person be a carrier of the disease but not display outward symptoms? pp. 375–376.

13. How do incomplete dominance and codominance differ? Give examples of each. pp. 375–376.

14. Assuming that two genes (*A* and *B*) control height, list all of the possible genotypes, and indicate the phenotype associated with each. p. 376.

15. What is the Human Genome Project? Why was it so important to our understanding of human genetics and the treatment of disease? p. 378.

16. Describe how crossing-over works. What effect does it have on the genotype of a person's gametes? pp. 377–378.

17. How does crossing over affect evolution? p. 377.

18. Color blindness results from a recessive, X-linked gene. A color-blind man and his wife have four children, two boys and two girls. One boy and one girl are color-blind, and the other two are normal. What is the genotype of the woman? p. 379.

19. What is a sex-influenced gene? Give an example. pp. 380–381.

20. What is androgenic alopecia? How does it differ in men and women? What causes it? p. 380–381.

21. Why is genetic screening of newborns performed? p. 386.

SELF-QUIZ: TESTING YOUR KNOWLEDGE

1. Meiosis I is also known as a _____ division. During this process, human cells go from _____ to _____ chromosomes (give a number). p. 366.

2. During meiosis II, the _____ chromosomes each contain _____ chromatids. p. 366.

3. In males, meiosis results in the formation of _____ gametes, each with _____ (number) single-stranded chromosomes. p. 366.

4. Mendel postulated that hereditary factors are either _____ or recessive. p. 370.

5. The physical appearance of an organism is known as its _____ while its genetic makeup is known as its _____. p. 370.

6. The principle of independent assortment only applies to genes on _____ chromosomes. pp. 371–372.

7. The human somatic cell contains _____ autosomes and two sex chromosomes. p. 372.

8. The sex chromosomes of women are _____. p. 372.

9. The sex chromosomes in men are _____. p. 372.

10. When is the sex of an individual determined? How is it determined? p. 379.

11. What is intersexuality? p. 379.

12. After researching the topic of sex determination on the Internet, do you think that it is more appropriate to think about sex as a range of manifestations rather than male or female? Why? What evidence do you have to back up this claim? p. 379.

13. Pattern baldness is also known as _____ alopecia. p. 380.

14. Pattern baldness is a sex-_____ trait. p. 380.

15. In an autosomal recessive disease such as cystic fibrosis, both alleles must be _____ for the disease to appear. p. 373.

16. Blending of traits does occur as a result of _____ dominance. p. 375.

17. The ABO blood types are the result of three different alleles. The *A* and *B* alleles are expressed fully and equally. This phenomenon is called _____. p. 376.

18. The mapping of the human genome was the result of the _____ Project. p. 378.

19. A recessive gene on the Y chromosome of a man is _____ expressed. p. 380.

20. The failure of two chromosomes to separate during meiosis is known as _____. p. 382.

21. A _____ occurs when a segment of a chromosome breaks off and attaches to another part of the chromosome. p. 382.

22. Down syndrome is an example of a _____, caused by nondisjunction of chromosome 21 during meiosis. p. 382.

23. Klinefelter syndrome occurs when an ovum with an extra X chromosome is fertilized by a Y-bearing sperm; the resultant genotype is _____. p. 383.

24. Praeder-Willi syndrome occurs by the loss of one arm of chromosome 15; this defect is known as a _____. p. 384.

25. A _____ occurs when a piece of chromosome breaks off and attaches to another part of the same chromosome or to another chromosome altogether. p. 386.

biology.jbpub.com/chiras/8e/

The site features eLearning, an online review area that provides quizzes, chapter outlines, and other tools to help you study for your class. You can also follow useful links for in-depth information, research the differing views in the Point/Counterpoints, or keep up on the latest health news.

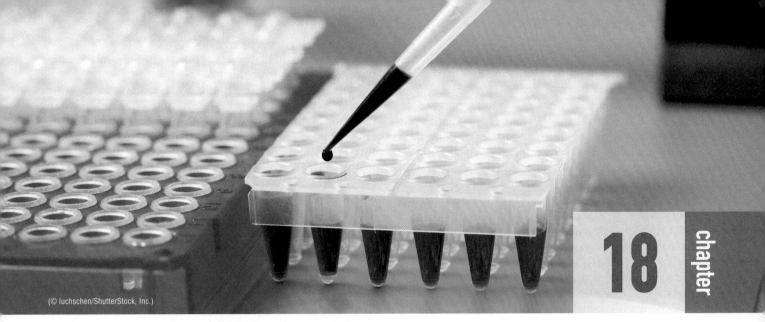
(© luchschen/ShutterStock, Inc.)

How Genes Work and How Genes Are Controlled

In 1988, a U.S. military court sentenced a serviceman in Korea to 45 years in prison for rape and attempted murder. Ten days later, a Florida court convicted a man on two counts of first-degree murder. In 2009, another Florida court ordered the release of a man who had spent 23 years of a life sentence in prison. What makes these three cases important is that courts relied heavily on the results of a technique called *DNA fingerprinting*.

THINKING CRITICALLY

One of America's leading drug companies spent $300 million to construct a facility to produce synthetic growth hormone to be sold to dairy farmers. The hormone is produced by genetically engineered bacteria that contain growth-hormone genes from cattle. The genes were transplanted into the bacteria, which then began to produce growth hormone. Administered to cows, the hormone dramatically increases milk production. Business economists believe that an increase in milk production will reduce the cost of milk to the consumer. Does introducing synthetic growth hormone seem like a good idea? You may want to go on the Internet to study the issue in more detail. Be sure to seek out reliable information. Where would that information most likely come from?

In this procedure, criminologists analyze the composition of the genetic material DNA in samples of hair, semen, or blood recovered from crime scenes. Then they compare the DNA in these samples with the DNA of the accused. If it matches, prosecutors can make a strong case for guilt. Without this procedure, prosecutors believe, the two convictions cited above could not have been won. Likewise, the wrongly convicted murderer would not have been set free. In this and other similar cases, DNA analyses of blood or semen years after the conviction have shown that convicted criminals were not guilty party.

DNA fingerprinting has certainly added a valuable tool to criminal investigations. Many police departments are now routinely collecting DNA samples from all convicted sex offenders and performing DNA analysis, the results of which are kept on file. Like the massive database of fingerprints, they will be readily available for future cases. As a case in point, in 2009 a prisoner released from a California jail after completing his sentence was immediately reincarcerated for another crime he had committed. His DNA fingerprint from a sample collected while in jail for the first crime, was used to convict him of the second offense. To date, there have been over 300 post-conviction exonerations in the United States thanks to DNA fingerprinting.

DNA fingerprinting holds great promise in fighting crime, creating a much higher degree of certainty than previous methods. It is widely used by state and local law enforcement as well as the FBI. One of the advantages of DNA fingerprinting is that there is only 1 chance of a mistaken identity in 4 or 5 trillion. In contrast, methods such as blood typing, which have been used for many years, run the risk of error in about 1 in every 1,000 cases.

Although DNA fingerprinting sounds promising, some geneticists believe that the probabilities (noted above) are based on improper assumptions (violating one of the critical thinking rules) and comparatively little scientific data on the genetic composition of populations.

DNA fingerprinting is one of many offshoots of research in genetics. This chapter covers some of the basic research that led to the discovery of this technique. You will learn about the structure of DNA and RNA, how genes work, and how genes are controlled in body cells, including some of the most recent work on gene control that has revolutionized the way we think about heredity.

18-1   DNA and RNA: Macromolecules with a Mission

In 1953, James Watson, an American biologist, and Francis Crick, a British biologist, proposed a model for the structure of the DNA molecule. This model was based on research by Rosalind Franklin, Maurice Wilkins, and many other scientists. Their work opened the doors to a new field of research known as **molecular genetics**, the study of the structure and function of RNA and DNA at the molecular level. We begin this chapter by looking first at the structure of DNA.

DNA is a molecule that consists of two separate strands that intertwine to form a double helix, a structure that resembles a spiral staircase (Figure 18-1). Each strand of the double helix contains millions of small molecules known as **nucleotides**. Like several other important biological molecules, then, DNA is a polymer, a molecule made up of many smaller ones.

Each nucleotide in turn consists of three smaller molecules: a nitrogen-containing base, a phosphate group, and a simple sugar, a monosaccharide known as deoxyribose (Figure 18-2). As shown in the figure, these molecules are linked together by covalent bonds. The nucleotides, in turn, are joined by additional covalent bonds to form long polynucleotide chains we call DNA.

In the DNA molecule, the two polynucleotide chains are held in place by hydrogen bonds. As shown in Figure 18-1, the hydrogen bonds that form between the bases of the nucleotides of each chain project inward and therefore

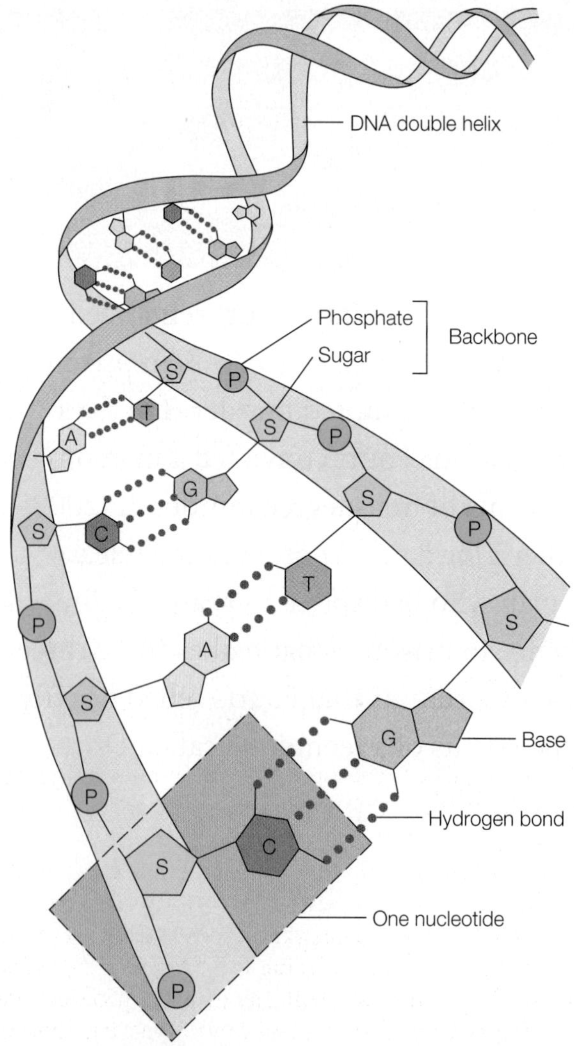

FIGURE 18-1 **DNA** DNA consists of two intertwined polynucleotide chains that form a double helix. Sugars and phosphates form the backbone of each chain, with the bases projecting inward. The bases on opposite strands are connected by hydrogen bonds.

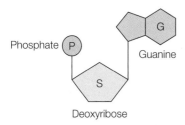

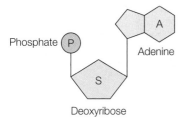

(a) DNA nucleotides containing purine bases

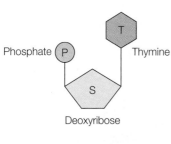

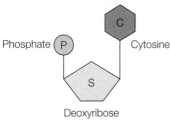

(b) DNA nucleotides containing pyrimidine bases

FIGURE 18–2 Nucleotides Containing Purine and Pyrimidine Bases All nucleotides consist of three subunits: a phosphate group; an organic, nitrogen-containing base; and a five-carbon sugar. DNA nucleotides contain the sugar deoxyribose. Two types of bases are found: purines and pyrimidines. (a) The purines are adenine and guanine. (b) The pyrimidines are cytosine and thymine.

lie inside the helix. Hydrogen bonds between the bases are indicated by the dotted lines in Figure 18-1. These bonds are much weaker than covalent bonds and can be easily broken. This, in turn, allows the molecule to unzip for replication.

DNA contains two types of nitrogen-containing bases. They are the **purines** (PURE-eens) and **pyrimidines** (pa-RIM-a-DEENS) (Figure 18-2). The purines consist of two rings. The pyrimidines contain only one ring. In DNA, you will find two purine bases. They are **adenine** (A) (AD-ah-neen) and **guanine** (G) (GUAN-neen). The pyrimidines in DNA are **cytosine** (C) (sigh-toe-seen) and **thymine** (T) (thigh-mean). When I was an undergraduate student, my genetics professor came from the agricultural college of my university. Evidently

proud of his roots, he provided us with a handy mnemonic "pure AG" to remember that the purines consist of adenine and guanine.

Purines on one strand bind (via hydrogen bonds) to pyrimidines on the opposite strand. But the relationship between the two is even more specific. If you look at **Figure 18-3b**, you will see that the adenine binds only to the pyrimidine thymine (A-T). Guanine binds only to the pyrimidine cytosine (G-C). Adenine and thymine are therefore said to be **complementary bases**, as are guanine and cytosine. As you will soon see, this specific coupling, called **complementary base pairing**, ensures the accurate replication of DNA and the reliable transmission of the genetic information from a parent cell to its daughter cells during cell division.

KEY CONCEPTS

DNA consists of two polynucleotide chains linked by hydrogen bonds. The chains are entwined and form a double helix. Each strand of the DNA consists of purines and pyrimidines that align with complimentary bases on the other string.

DNA Replication

Before a cell can divide, it must first make an exact copy of all of its DNA. DNA replication ensures that a cell about to divide has two identical sets of genetic information, one for each daughter cell.

During the S phase of interphase, cells replicate their DNA by unzipping each DNA double helix along the hydrogen bonds that unite the complementary bases. Each polynucleotide strand then serves as a template (described below) on which a new strand is produced (Figure 18-3a). The new strand is called a **complementary strand**. The template is called the **original strand**. Take a moment to study Figure 18-3b to identify the old and new strands.

DNA replication is referred to as a *semiconservative process* because each polynucleotide chain of the DNA double helix serves as a template for the production of a new strand of DNA.

DNA replication during the S phase begins when special enzymes start to pull apart, or unzip, the DNA double helix. As the two strands are separated, the bases of the polynucleotide chains are exposed. They are then free to form hydrogen bonds with complementary nucleotides found in the nucleus (Figure 18-3). Because adenine binds only to thymine (A-T) and guanine binds only to cytosine (G-C), the original strands (the templates) are said to "direct" the synthesis of new strands. Synthesis on the DNA templates occurs one base at a time, and the accuracy of replication is ensured by complementary base pairing.

During replication, incoming nucleotides must first be aligned properly so that the hydrogen bonds can form between complementary bases and so that the phosphate group of the incoming nucleotide can bond to the sugar of the nucleotide already in place, as illustrated in **Figure 18-4**. Alignment and attachment are aided by an enzyme known as **DNA polymerase** (PUL-yi-merr-ace; Figure 18-4).

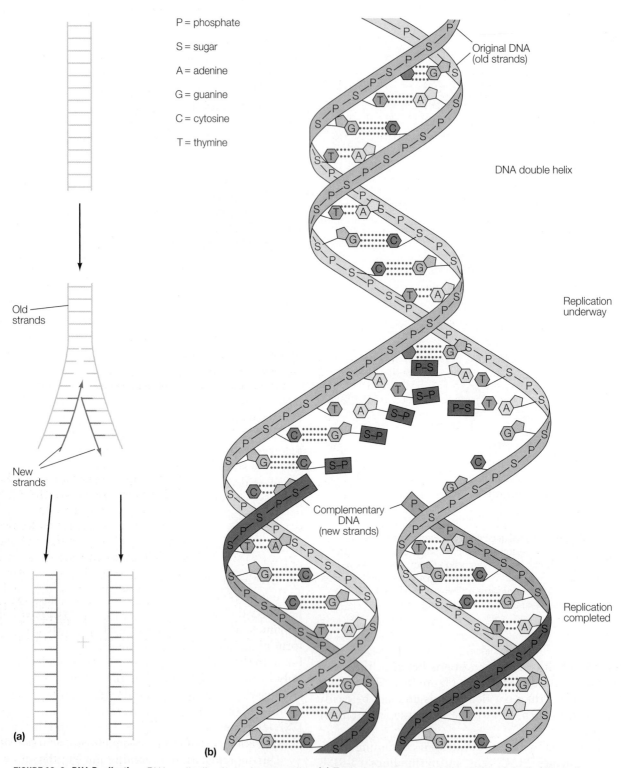

P = phosphate

S = sugar

A = adenine

G = guanine

C = cytosine

T = thymine

FIGURE 18–3 DNA Replication DNA replication is semiconservative. (a) Each double helix unwinds, and each half of the helix serves as a template for the production of a new strand of DNA. When replication is complete, each new helix contains one old and one new strand. (b) Nucleotides attach to the template one at a time and are joined together with the aid of enzymes.

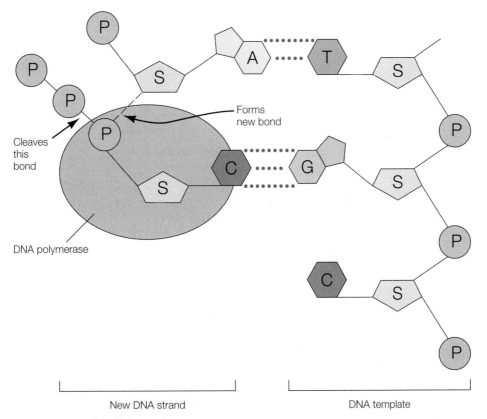

FIGURE 18-4 Role of DNA Polymerase DNA polymerase binds loosely to the DNA template and to the nucleotide, aligning it for insertion. The enzyme catalyzes the formation of the bond between the phosphate of the new nucleotide and the sugar of the one previously inserted. The enzyme also cleaves off two phosphates of each nucleotide.

DNA polymerase slides along the template, aligning nucleotides one at a time, and then linking each new nucleotide to the one already in place. As noted above, the enzyme joins the phosphate group of the new nucleotide to the deoxyribose molecule (sugar) of the nucleotide that's already in place on the new strand.

When DNA synthesis is finished, two new DNA molecules exist, each of which is a double helix. Each double helix, in turn, contains one strand from the original molecule and one new polynucleotide chain. The chromosome, once a single molecule of DNA and protein, now consists of two DNA molecules with associated proteins. As noted in Chapter 16, each strand of DNA and associated protein is called a **chromatid**. The two chromatids of each chromosome are joined at their centromeres after synthesis.

KEY CONCEPTS

DNA unwinds and then serves as a template for the production of new DNA strands.

RNA

DNA contains the genetic information of the cell. DNA therefore determines the structure of the cell and controls most of its functions. Like a commander in the army, DNA does not exert its influence directly. Its work is carried out by other molecules, notably **RNA**. Three types of RNA are involved in this process: ribosomal RNA (rRNA), messenger RNA (mRNA), and transfer RNA (tRNA). Each has a unique function in protein synthesis (Table 18-1). Despite major functional differences in these molecules, all three RNA molecules are biochemically similar. In humans, for example, all RNA molecules are single-stranded, and all are polynucleotides (Figure 18-5). RNA nucleotides consist of three molecules: a sugar, a nitrogenous base, and a phosphate group. RNA nucleotides contain the sugar ribose instead of deoxyribose, which is found in DNA. They also contain the pyrimidine uracil instead of thymine.

TABLE 18-1	Role of RNA Molecules
Molecule	**Role**
Messenger RNA (mRNA)	Carries the genetic information that is needed to make proteins in the cytoplasm from the nucleus
Transfer RNA (tRNA)	Binds to specific amino acids, transports them to the mRNA, and inserts them in the correct location on the mRNA
Ribosomal RNA (rRNA)	Component of the ribosome

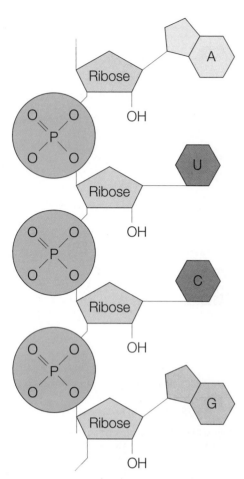

FIGURE 18–5 **RNA** RNA is a single-stranded molecule consisting of many RNA nucleotides. A small section of an RNA molecule is shown here.

TABLE 18-2	Differences Between RNA and DNA in Prokaryotic and Eukaryotic Cells	
DNA		**RNA**
Double-stranded		Single-stranded
Contains the sugar deoxyribose		Contains the sugar ribose
Contains adenine, guanine, cytosine, and thymine		Contains adenine, guanine, cytosine, and uracil
Functions primarily in the nucleus		Functions primarily in the cytoplasm

Table 18-2 summarizes the key differences between RNA and DNA.

KEY CONCEPTS

When it replicates, the DNA double helix unwinds. This allows complementary bases to insert themselves on the existing strand of DNA, ensuring accuracy of replication.

KEY CONCEPTS

Three types of RNA exist, each of which carries out a specific function in protein synthesis.

RNA Synthesis

Unlike DNA, which is self-replicating (meaning it forms on a DNA template), all three types of RNA are synthesized on DNA templates. The synthesis of RNA on a DNA template permits the genetic information coded in the DNA to be transferred to RNA. As you will soon see, the genetic code in DNA needed to control cell structure and function is transferred to one of the three types of RNA, notably messenger RNA. **Messenger RNA** is free to leave the nucleus. It therefore serves as a kind of shuttle (or messenger) that

transfers the genetic information necessary for protein synthesis into the cytoplasm, where proteins are made.

RNA synthesis on a DNA template is called **transcription**. Geneticists chose this name because this process is much like the transcribing process students perform during a lecture where they transfer information from one medium (speech) to another (written form). In this case, genetic information housed in the DNA is simply written in another form, RNA.

Transcription occurs during interphase of the cell cycle. During transcription, small sections of the DNA helix unzip temporarily with the aid of special enzymes. This creates a DNA template on which RNA can be made (Figure 18-6). But RNA is produced on only one of the DNA strands, and only a small portion (less than 1%) of a cell's DNA is used to make RNA. (This helps explain why most mutations in the DNA have no effect on cell function.)

During RNA synthesis, an enzyme called **RNA polymerase** helps align the RNA nucleotides on the DNA template in much the same way that DNA polymerase aligns DNA nucleotides during DNA synthesis. RNA polymerase also catalyzes the formation of covalent bonds between ribose and phosphate groups, thus helping to form the polynucleotide chain. When the synthesis is complete, the RNA molecule is released from the DNA template and the two strands of the DNA molecule reunite, reforming the double helix. RNA is then free to leave the nucleus.

During the synthesis of RNA, base pairing between the DNA template and the RNA nucleotides ensures the proper sequence of nucleotides on the RNA strand. Figure 18-7 shows which bases pair up during RNA synthesis. The only difference in base pairing between RNA synthesis and DNA synthesis is that adenine pairs with uracil on the DNA template.

KEY CONCEPTS

RNA molecules consist of a single strand of nucleotides and are formed on a complementary strand of DNA. Three types of RNA are found in the cell, transfer RNA, messenger RNA, and ribosomal RNA. All three are vital to protein synthesis.

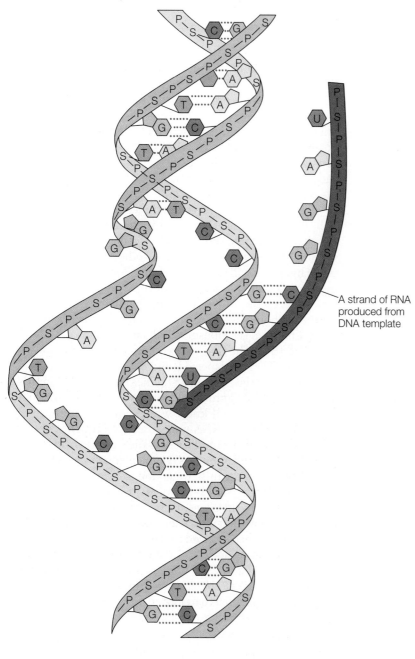

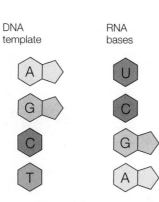

A strand of RNA produced from DNA template

FIGURE 18-6 **RNA Synthesis on a DNA Template** RNA is synthesized on a DNA template. As shown, the DNA double helix unwinds, but only one strand serves as a template for the production of RNA. Complementary base pairing determines the exact sequence of bases in the RNA molecule.

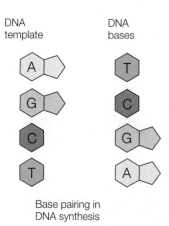

FIGURE 18-7 **DNA and RNA Base Pairing** The sequence of bases on the DNA template determines the sequence of bases on the complementary strands of DNA and RNA.

18-2 How Genes Work: Protein Synthesis

The genetic information required to synthesize protein is transported out of the nucleus as messenger RNA. In the cytoplasm, mRNA serves as a template for protein synthesis. Protein synthesis occurs either on the endoplasmic reticulum or free within the cytoplasm. The synthesis of protein on an mRNA template is called **translation**. That's because the RNA message (the genetic code) is translated into a new "molecular language"—that of the protein. To remember all this, think of transcribing your professor's notes and then handing them to a student from France, who translates them into a new language. Figure 18-8 nicely summarizes these processes.

Protein synthesis requires two additional "players," transfer RNA (tRNA) and ribosomes. **Transfer RNA** molecules are relatively small strands of RNA that bind to specific amino acids in the cytoplasm and transport them to the mRNA. Here the amino acids are incorporated into the protein that's

being constructed on the mRNA molecule (Figure 18-9). A tRNA molecule bound to an amino acid is generally written like this: tRNA–AA.

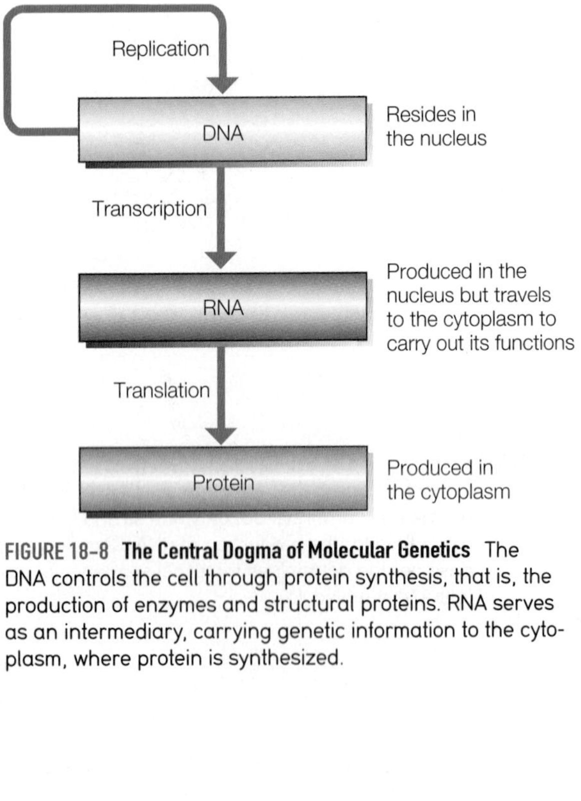

Ribosomes are organelles that appear as small granular structures in the cytoplasm of cells. These organelles play an important role in the production of protein. They are composed of **ribosomal RNA (rRNA)** and protein. Ribosomes consist of two subunits, a small one and a large one. The subunits are often detached in the cytoplasm. They unite on the mRNA during protein synthesis.

Protein synthesis consists of three stages: (1) chain initiation, (2) chain elongation, and (3) chain termination. During chain initiation, shown in the top of Figure 18-10, the small subunit of the ribosome attaches to the mRNA at a specific site. This site is called the initiator codon (COE-dawn). The **initiator codon** is a sequence of three purine and pyrimidine bases on the mRNA molecule that marks where protein synthesis should begin. After the small subunit attaches, the large subunit of the ribosome links up. As shown in the top of Figure 18-10, the ribosome contains two binding sites for tRNA–AA. They are labeled P and A.

Soon after the ribosome attaches to the mRNA, a tRNA bearing a specific amino acid enters the first binding site. But how does the cell know which amino acid to insert? At the bottom of all tRNA molecules is a sequence of three purine and pyrimidine bases. Called an **anticodon**, this three base sequence contains complementary bases that pair with the bases of the initiator codon of the mRNA template.

After the first tRNA–amino acid is in place, a second tRNA–AA enters the scene, inserting itself into the second binding site on the ribosome. The second binding site is located above the next codon, the next three bases on the mRNA template (Table 18-3). In the example shown in Figure 18-10, the codon CGG binds to the tRNA with the anticodon GCC. This tRNA, in turn, binds to only one amino acid, arginine. The cell therefore ensures the proper sequence of amino acids through two mechanisms: (1) complementary base pairing between codons (mRNA) and anticodons (tRNA), and (2) the specificity (matching) of tRNA molecules for (with) amino acids.

FIGURE 18-8 The Central Dogma of Molecular Genetics The DNA controls the cell through protein synthesis, that is, the production of enzymes and structural proteins. RNA serves as an intermediary, carrying genetic information to the cytoplasm, where protein is synthesized.

FIGURE 18-9 Messenger RNA The mRNA, either free within the cytoplasm or bound to the endoplasmic reticulum, serves as a template for protein synthesis. The codons, each consisting of three bases on the mRNA, determine the sequence of amino acids by binding to complementary anticodons on the tRNA molecules. Each tRNA, in turn, delivers a specific amino acid to the mRNA template. The ribosome provides binding sites for tRNA–AA molecules, catalyzes the formation of the peptide bonds, and slides down the mRNA template to permit chain elongation.

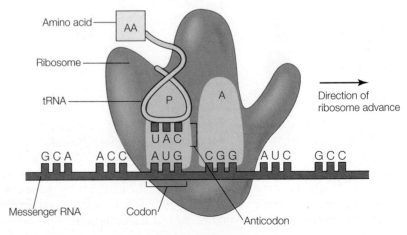

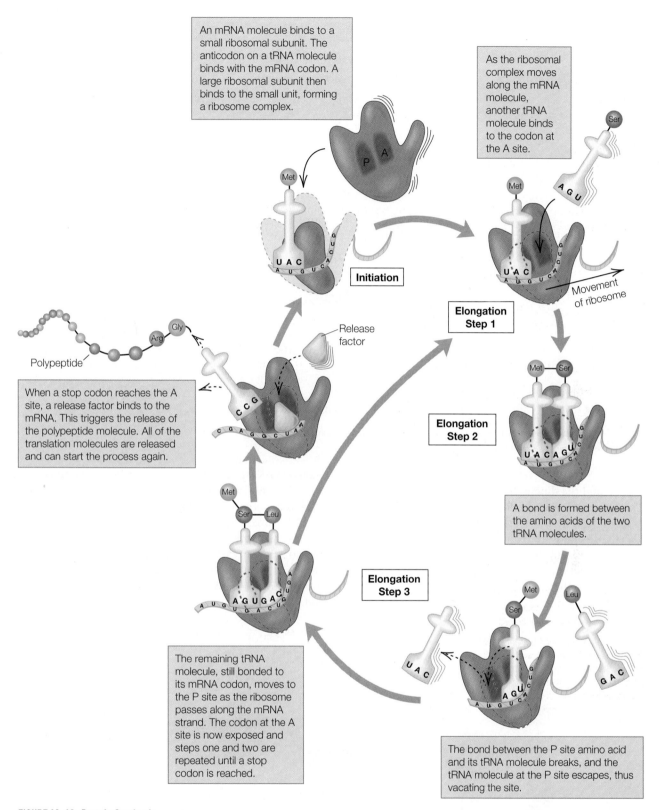

An mRNA molecule binds to a small ribosomal subunit. The anticodon on a tRNA molecule binds with the mRNA codon. A large ribosomal subunit then binds to the small unit, forming a ribosome complex.

As the ribosomal complex moves along the mRNA molecule, another tRNA molecule binds to the codon at the A site.

Initiation

Movement of ribosome

Elongation Step 1

Release factor

Elongation Step 2

When a stop codon reaches the A site, a release factor binds to the mRNA. This triggers the release of the polypeptide molecule. All of the translation molecules are released and can start the process again.

A bond is formed between the amino acids of the two tRNA molecules.

Polypeptide

Elongation Step 3

The remaining tRNA molecule, still bonded to its mRNA codon, moves to the P site as the ribosome passes along the mRNA strand. The codon at the A site is now exposed and steps one and two are repeated until a stop codon is reached.

The bond between the P site amino acid and its tRNA molecule breaks, and the tRNA molecule at the P site escapes, thus vacating the site.

FIGURE 18–10 Protein Synthesis

Once the two amino acids are in place, the next step is to link them via a covalent bond, known as a **peptide bond**. This occurs with the assistance of an enzyme in the ribosome. You can see this in Figure 18-10 in the section labeled "Elongation Step 2." After the peptide bond is formed, the first tRNA (minus its amino acid) leaves the binding site. It is then free to retrieve another amino acid for use later on at another site.

As illustrated in Figure 18-10, "Elongate Step 3," the dipeptide formed during the first reaction is now attached to the second tRNA that is attached at the second binding site.

In order for the chain to grow, however, the ribosome must move down the mRNA strand. This is accomplished with the aid of a contractile protein in the ribosome. This protein permits the ribosome to slide along mRNA one codon at a time.

TABLE 18-3	Codons on mRNA and Their Corresponding Amino Acids						
Codon	Amino Acid	Codon	Amino Acid	Codon	Amino Acid	Codon	Amino Acid
AAU	Asparagine	CAU	Histidine	GAU	Aspartic acid	UAU	Tyrosine
AAC		CAC		GAC		UAC	
AAA	Lysine	CAA	Glutamine	GAA	Glutamic acid	UAA	Terminator codon*
AAG	Threonine	CAG		GAG		UAG	
ACU		CCU	Proline	GCU	Alanine	UCU	Serine
ACC		CCC		GCC		UCC	
ACA		CCA		GCA		UCA	
ACG		CCG		GCG		UCG	
AGU	Serine	CGU		GGU		UGU	Cysteine
AGC	Argenine	CGC	Arginine	GGC	Glycine	UGC	
AGA		CGA		GGA		UGA	Terminator codon*
AGG		CGG		GGG		UGG	Tryptophan
AUU	Isoleucine	CUU		GUU	Valine	UUU	Phenylalanine
AUC	Methionine	CUC	Leucine	GUC		UUC	
AUA		CUA		GUA		UUA	Leucine
AUG		CUG		GUG		UUG	

*Terminator codons signal the end of the formation of a polypeptide chain.

After the first tRNA is released, the ribosome moves down the mRNA, and the dipeptide now occupies the first binding site. This frees up the second site, permitting it to accept another tRNA–AA. A new tRNA–AA then enters the vacant site and is joined to the dipeptide, thus forming a tripeptide. This is shown in Figure 18-10 just to the left of the label "Elongation Step 3."

Chain elongation takes place by the addition of one amino acid at a time, but this does not mean that protein synthesis is a slow process. Quite the contrary, protein synthesis occurs with remarkable speed. In bacteria, many proteins containing numerous amino acids are synthesized in 15 to 30 seconds. In humans, the large subunits of the very large hemoglobin molecule are synthesized in 3 minutes.

As the peptide chain is formed, hydrogen bonds between amino acids on different parts of the chain cause it to bend and twist, forming the secondary structure of the protein or peptide (Figure 18-11). When the ribosome reaches the **terminator codon**, the sequence of bases that signals the end of the pro-

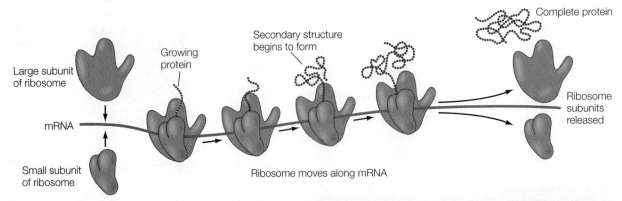

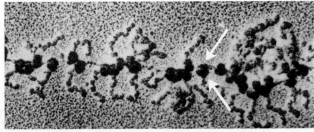

FIGURE 18-11 **Protein Synthesis** As the protein is synthesized on the mRNA, it begins to coil and bend, forming its secondary structure. Several ribosomes may "work" a single strand of mRNA simultaneously. (Photo © E. Kifelva and D. Fawcett/Visuals Unlimited.)

tein chain, the protein is released. If the protein is produced on mRNA on the surface of the rough endoplasmic reticulum, it is transferred into its interior. The protein may then be chemically modified and packaged into transfer vesicles, which pinch off and transport their cargo to the Golgi complex. If protein synthesis occurs on mRNA molecules in the cytoplasm, the protein is released into the cytoplasm.

KEY CONCEPTS

Protein synthesis occurs on mRNA templates in the cytoplasm of cells with the aid of enzyme-containing ribosomes and t-RNA, which retrieve specific amino acids from the cytoplasm and insert them in the proper location on the mRNA strand thanks to base pairing.

18-3 Controlling Gene Expression

In the previous sections, you've seen the fascinating cellular machinery that produces DNA, RNA, and protein. In this section, we look at another intriguing cellular mechanism of extreme importance: the control of genes. This information is useful to geneticists because it helps them discover new ways to treat, or even cure, diseases such as cancer that originate in the genes of an organism.

DNA carries the genetic information needed by the cell to grow, reproduce, and carry out a myriad of functions. Specific segments of DNA, known as genes, are responsible for controlling these functions. Genetic control occurs in at least five levels (Figure 18-12).

Control at the Chromosome Level. As you may recall from your study of chromosomes, the chromatin fibers in the nucleus which contain DNA and histone protein condense and become inactive during prophase of mitosis in preparation for cell division. In the condensed state, they are inactive and cannot produce RNA or new DNA.

Cells also inactivate some of their chromatin during interphase. That's because most cells have far more DNA than they need. Muscle cells, for instance, have all the genes needed by liver cells and brain cells. Because the muscle cells do not need this DNA, they inactivate it by condensing sections of chromatin that carry those genes. Another example of chromosome inactivation occurs in cells of human females. Females contain two sex chromosomes, XX; however, one of the X chromosomes is typically condensed and therefore mostly turned off.

Chromosomal condensation or coiling provides a crude way of regulating genetic expression. It is one form of epigenetic control, discussed shortly. More precise controls occur at other levels.

Control of Transcription. Genetic control also occurs at the transcription phase—that is, at mRNA production—in humans. Control at this level occurs as a result of the interaction of chemical substances with genes. This often occurs when the end products of metabolic pathways, specific chemicals produced by chemical reactions in the cell, either turn on (activate) some genes or turn them off (repress) genes. When the cell turns off a gene, it terminates the production of mRNA needed to make proteins like enzymes required in a metabolic pathway. For more on this process, see Scientific Discoveries that Changed the World 18-1.

Human cells also control gene activity via a third mechanism, known as enhancement. **Enhancement** is a process in which already active genes increase their production of mRNA. This occurs as a result of the action of proteins or enhancers that bind to nearby segments of the DNA. Put simply, **enhancers** bind to enhancer regions of the chromosome near genes and then increase mRNA production on the DNA template. This increases the output of metabolic pathways like putting more workers on an already operating production line at a factor. Thus, enhancers do not turn nearby genes on and off; they amplify their activity.

Control before Translation. Genetic studies show that DNA in humans contains two types of genetic material: sections that produce mRNA and sections that do not. As Figure 18-13 shows, the noncoding segments of DNA are called **introns** and are interspersed within the functional segments of DNA (functional genes), the **exons**. The name *intron* signifies that these are intervening segments of noncoding DNA. Exons are expressed segments—sections of the DNA that can be used to produce the mRNA that is used to make protein.

Both introns and exons are transcribed during the cell cycle, producing a messenger RNA transcript that is

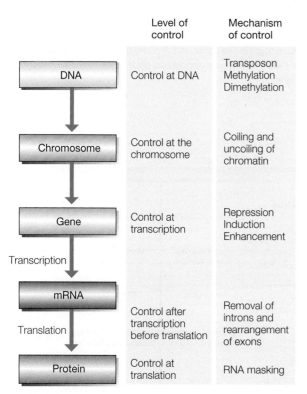

	Level of control	Mechanism of control
DNA	Control at DNA	Transposon Methylation Dimethylation
Chromosome	Control at the chromosome	Coiling and uncoiling of chromatin
Gene	Control at transcription	Repression Induction Enhancement
Transcription		
mRNA	Control after transcription before translation	Removal of introns and rearrangement of exons
Translation		
Protein	Control at translation	RNA masking

FIGURE 18-12 Gene Expression In humans, gene expression is regulated at four levels.

Scientific Discoveries that Changed the World

18-1 Unraveling the Mechanism of Gene Control
Featuring the Work of Jacob and Monod

One of the questions that intrigued geneticists for many years—and still does—is how the genetic information contained in a cell's DNA is controlled. Many experiments have been performed over the years in an attempt to answer this question, but none so important as those of two French molecular geneticists, François Jacob and Jacques Monod, on the common intestinal bacterium *E. coli* (Figure 1). Today, much of what is known about gene regulation in bacteria and humans comes from their studies.

Working at the Pasteur Institute in the late 1950s, Jacob and Monod focused much of their attention on the bacterial uptake and breakdown of lactose, a disaccharide also known as milk sugar. *E. coli* absorb lactose and break it down with the aid of an enzyme known as *galactosidase* (ga-lack-TOSE-uh-DACE). The breakdown of lactose results in the release of energy.

E. coli living in a lactose-free medium, however, contain very little of the enzyme. It is only when lactose is added that the enzyme is synthesized.

The introduction of lactose also increases the production of a carrier molecule (galactoside permease) found in the plasma membrane. It transports lactose into the bacterium. In addition, the presence of lactose increases the intracellular concentration of another enzyme, which may play a role in lactose breakdown but is not essential to the process. The carrier protein and two enzymes are part of what scientists call an *inducible enzyme system*. The production of these chemicals is induced by the presence of lactose.

Mapping studies have shown that three adjacent genes on the bacterial DNA are responsible for the production of these proteins. These studies led Jacob and Monod to hypothesize that the three genes belong to a single unit, which they called an **operon**. They defined an operon as a cluster of genes with related functions that is regulated in such a way that all the genes in the group are activated and inactivated simultaneously.

In a paper published in 1961, the researchers proposed a mechanism by which these genes might be controlled. This intriguing model is "one of the truly important conceptual advances in biology," according to University of Wisconsin cell biologist Wayne Becker.

Extensive studies of bacteria helped Jacob and Monod piece together a picture of gene regulation in *E. coli*. As a testimony to the thoroughness of their work, the original model has undergone very little change in over 60 years.

FIGURE 1 Jacob (left) and Monod (center) (with André Lwoff, recipient of the 1965 Nobel Prize in Medicine), two French scientists who proposed the operon hypothesis. (© Institut Pasteur.)

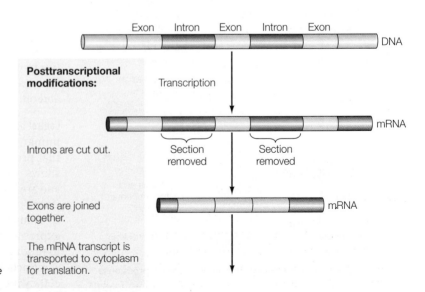

FIGURE 18–13 Posttranscriptional Control of Human Genes
One key mechanism of control occurs after transcription. DNA contains noncoded segments (introns) and useful segments (exons). An mRNA transcript is produced from the DNA, and the segments produced by introns are removed. The final product is a strand of mRNA containing only RNA copies of the exons. The exon copies can be linked differently, producing slightly different products.

a complete "readout" of the DNA. Cells then process the RNA transcript, cutting out the sections produced by introns, and then joining the sections produced by exons together to produce a functional mRNA molecule. This is clearly demonstrated in Figure 18-13. In order for a functional mRNA molecule to be synthesized, then, it must be "cut and pasted" together. Interestingly, the portions of mRNA produced by exons can be linked differently. This produces an mRNA molecule that produces slightly different products needed to meet a cell's needs. In this way, cells obtain another means of controlling genetic expression.

Control at Translation. Eukaryotic cells can also control gene expression at the level of translation (protein synthesis). Messenger RNA, for example, which is produced in the nucleus, may be transferred to the cytoplasm in an inactive state. The mRNA is said to be *masked*. Masking allows the cell to build up large supplies of mRNA in preparation for a sudden burst of protein synthesis. When the time is right, the mRNA is activated and begins producing large quantities of protein. This phenomenon occurs in the human ovum, the female sex cell produced in the ovaries. The ovum produces an enormous amount of masked mRNA as it awaits fertilization. When the sperm arrives and fertilizes the ovum, the mRNA that has accumulated in the cytoplasm is unmasked, and there is an explosion of protein synthesis on the unmasked mRNA, creating a flood of protein necessary for subsequent cell divisions.

Jumping Genes: Transposons. In recent years, geneticists have come to recognize an additional mechanism of genetic control, **transposons** or jumping genes, one of the newest and most revolutionary discoveries in genetics since the discovery of the structure of DNA.

Jumping genes are the discovery of plant geneticist Barbara McClintock, who received a Nobel Prize for her work in 1983 at the age of 81—some 32 years after she first presented her findings. It took that long, in part, because what McClintock discovered flew in the face of genetic science. She found that sequences of DNA in corn moved from one place to another when plants were under stress, for example, when they were faced with extreme drought and water shortages. These segments of DNA were sliced out of the DNA and often inserted near or into active genes, resulting in a kind of genetic cut and paste. In their new locations, these genes could profoundly influence nearby genes or the genes into which they were inserted. What is more, McClintock found that transposons often moved to certain parts of the genome where mutations would most likely result in a beneficial effect. In other words,

their movement wasn't random. It seems to be directed. In the words of geneticist and author of *Survival of the Sickest*, Dr. Sharon Moalem, "The corn plant seemed to be engaged in some sort of intentional mutation—neither random, nor rare."

Since her initial discovery, other scientists have made similar observations of an internal restructuring of the genome in other species such as fruit flies. The phenomenon has also been observed in hundreds of species of bacteria. All of these genetic changes appear to be a response to changes in internal and external conditions. They result in beneficial changes that enhance an organism's ability to survive and reproduce.

Scientists also believe that the immune system's ability to produce hundreds of different antibodies results from jumping genes. To date, researchers have found that about one-fourth of the active genes in the human genome have incorporated DNA from jumping genes, according to Dr. Moalem.

In addition to the genetic cut and paste occurring at the level of the DNA, researchers have found that some genes make copies of complementary RNA. Interestingly, it serves as a template for the production of DNA. That is, it makes copies of the original DNA in the genes. This DNA is then pasted into the genome with similar results, creating a kind of genetic copy and paste.

Transposons reveal yet another way to control the function of cells—a directed genetic mutation in response to stress. What is more, evidence suggests that transposons originate in the noncoding DNA, long referred to as "junk DNA." (These are the introns discussed earlier.) Perhaps even more interesting, it appears as if this "junk DNA" very likely originated from viruses that incorporated themselves into the DNA of humans and other species over the course of evolution.

Transposons not only also radically reshape our understanding of genetics, they also change our understanding of mutation and evolution. It appears that jumping genes may be responsible for the rapid evolutionary change that has occurred in the history of life on Earth. That's because they mutate more rapidly than the rest of our genes.

KEY CONCEPTS

Genetic control occurs at five levels: (1) control at the level of the DNA, (2) control at chromosome level; (3) control at transcription, that is, mRNA production; (4) control after transcription before translation; and (5) control at translation or protein production. They are summarized in Figure 18-12. New research shows that genes can also be snipped from the DNA and placed in new locations, resulting in profound changes in the cells in which this occurs. DNA can also make copies of RNA that are used to make DNA segments that can be inserted into genes, altering a cell's structure and function.

18-4 Epigenetic Control of Gene Expression

In 2000, two researchers from Duke University, Professor Randy Jirtle and his graduate student Robert Waterland, performed an experiment that rocked the scientific community, forcing us to rethink the basic mechanisms of heredity. In their experiment, Jirtle and Waterland studied reproduction in a strain of mice known as *agouti mice*. Adult agouti mice eat ravenously and therefore grow up to be quite overweight (Figure 18-14). These fat-as-a-pincushion mice have a yellow coat and, because they are overweight, are prone to develop diabetes and cancer. When they breed, their offspring exhibit similar characteristics.

FIGURE 18-14 Agouti Mice It is hard to believe, but a simple change in the diet of the mother (left) produced offspring that were amazingly different not only in appearance but also in disease susceptibility. (Courtesy of Randy Jirtle.)

In their experiment, the researchers changed the diet of female mice just before they were bred, presenting the females with a diet rich in chemicals that supply methyl (CH_3) groups to organic chemicals in our bodies. These molecules are commonly found in foods such as onions, beets, and garlic, and also are present in food supplements such as those containing folic acid. In this experiment, the pregnant mice were fed their ordinary diet during pregnancy.

Much to the surprise of the researchers, the babies who were born to these females were markedly different from their parents. Instead of being fat and yellow, they were brown and normal in size and no longer prone to diabetes or cancer, as shown in Figure 18-14.

Even though the offspring of the agouti mice contained the gene that made their parents fat and made them susceptible to cancer and diabetes, their offspring were normal. What is more, when the offspring bred, they gave birth to normal young.

These results suggested that diet could alter inheritance. This conflicted with years of scientific research that showed that the sequence of bases in the DNA in the parents determines the genotype and phenotype of their offspring. The researchers explained this unusual phenomenon by hypothesizing that the methyl-donating chemicals in the diet enter the females' bloodstream, and pass through the placenta into the blood of the babies. The methyl groups, the researchers suggest, then attach to (methylate) the agouti gene and inactivate it. Surprisingly, this relatively modest chemical addition to the DNA, which does not change the sequence of bases, is passed to future generations.

This work was dismissed at first because it violates the precepts of heredity that say that only changes in the sequence of bases in DNA can permanently alter heredity. Further research, however, has shown similar results not just in animals, but also in plants, leading geneticists to alter their beliefs about heredity.

This work has led to a new field of genetics, **epigenetics** ("epi" meaning "on" the genes). Researchers are finding that although DNA contains the information needed to control cellular structure and function, the DNA itself is controlled by an array of chemical switches. Some of these controls are nothing more than methyl groups that switch off genes. Other controls exist in the histone proteins associated with the DNA (discussed earlier in this chapter). These controls lie along the length of the DNA double helix. They form something akin to a complex software code that controls the DNA hardware.

Researchers are learning that vitamins, toxic chemicals, even the behavior of mothers (that cause chemical changes in their offspring) can alter this control code, referred to as the **epigenome**. Scientists are finding that the epigenome is just as critical as DNA to the healthy development of an organism's offspring. Changes can alter the structure, function, and—even more remarkably—behavior of an organism for life can be passed on to offspring, sometimes for several generations. In short, the epigenome is part of the biological information a cell passes on to its daughter cells during mitosis and meiosis. It is, therefore, also part of the biological information a parent passes on to its offspring. The implication to humans, as Ethan Watters observed in *Discover* magazine, is that "as bizarre as it may sound, what you eat or smoke today could affect the health and behavior of your great grandchildren."

Epigenetic factors also have been implicated in cancer. Methylation of genes often occurs during the transformation of a normal cell to a cancer cell. That is to say, it is not just DNA mutations that result in cancer, as scientists have long thought. Simple chemical changes (like methylation) in certain genes—changes that do not alter the sequence of bases—may play an important role in the development of cancer. Research, for example, shows that cells contain genes that suppress cancer cell growth. Methylation of these genes can switch the genes off. This, in turn, leads to runaway growth in the afflicted cell.

Cells also contain genes that promote cancer. They can be activated by removal of their methyl groups, resulting in runaway cell division. (Methyl groups act as off switches in such genes.) Interestingly, abnormal methylation has been found in cancers of the cervix, prostate gland, stomach, colon, thyroid, and breast.

Researchers also speculate that other diseases such as rheumatoid arthritis and diabetes may involve abnormal methylation. In addition, researchers are also studying the effects of methylation on aging.

The possible good news in all of this is that drugs and foods that reduce methylation could someday be used to slow the aging process and prevent cancer and other diseases. At this writing, one such drug has already been approved for use in the treatment of one form of cancer. Numerous other drugs are currently under development. If you go online, you will encounter a huge amount of scientific information—most of it very technical—on methylation and disease.

Certain foods like onions may also be recommended. Careful scientific studies that compare diets of human populations and their intake of onions show that those populations in which individuals eats lots of onions suffer from a much lower incidence of cancer, for example, a 50% lower rate of stomach cancer, 56% lower rate of colon cancer, 71% lower rate of prostate cancer, and an 88% lower rate of esophageal cancer. Researchers have found that a chemical in green tea can prevent methylation, leading some researchers to speculate that scientists may someday be able to recommend an epigenetic diet. (You will probably see an epigenetic diet book in the not too distant future to join the legion of fad diet books!)

Research also shows that methylation of RNA and proteins may also play a role in epigenetics. Studies, for instance, show that small molecules of RNA, called *small RNAs*, are very likely involved in changing the chromatin structure leading to condensation of chromosomes. It may also somehow direct DNA methylation.

Clearly, the new research on epigenetic control changes the way scientists think about the transfer of genetic information from one generation to the other. No longer is DNA the sole determinant of our physical characteristics. We inherit more than DNA from our parents. We inherit chromosomes, half of which are DNA and half of which are proteins, the histones. These proteins carry the epigenetic marks and information. DNA can also be methylated, and these simple chemical changes can apparently affect us and our offspring.

> **KEY CONCEPTS**
>
> Genetic expression may also be controlled through methylation, the addition or subtraction of methyl groups from DNA as well as RNA and protein. Such changes do not alter the structure of the DNA and can be passed from one generation to the next.

18-5   Health and Homeostasis

One of the leading causes of death in humans is a disease called *cancer*. Not one disease, cancer actually consists of over 100 different diseases such as liver cancer, lung cancer, breast cancer, and prostate cancer. Cancer is caused by numerous biological, chemical, and physical agents in a two-step process. One cause of cancer is genetic mutations, that is, alterations of parts of the genetic material.

Researchers have made a startling discovery about such mutations that may explain the cause of most cancers. They have found that humans and other species contain a group of genes called proto-oncogenes (pro-toe-ON-co-JEANS). **Proto-oncogenes** control functions related to cellular replication, among them cell adhesion and the production of the plasma membrane receptors that bind to growth factors or hormones. When mutations in these genes are not repaired, the result can be uncontrolled cellular proliferation, cancer. (For a discussion of methods used to assess the potency of potential cancer-causing agents, see the **Point/Counterpoint**.)

Certain viruses also cause cancer. Cancer-causing viruses fall into two groups. Some viruses stimulate cancer by activating proto-oncogenes in body cells. Others contain cancer-causing genes themselves, which they insert into the DNA of the nuclei of infected cells. These genes are known as **viral oncogenes**. After being inserted into a cell's DNA, **oncogenes** turn on the cell's genes that control cell division.

The discovery of proto-oncogenes and viral oncogenes has resulted in a quantum leap in our understanding of cancer. Some scientists believe that the key to treating cancer may ultimately lie in finding ways to turn off activated proto-oncogenes. Such changes could halt replication of cancer cells. Who knows, maybe someday you or someone you know will benefit directly from this idea?

Thousands of mutations occur in each cell of your body every day, many of which could lead to serious problems, including cancer. As noted earlier, the cells of your body repair much of the damage. This repair is brought about by special enzymes in the nucleus that "snip off" damaged sections of the DNA, then rebuild the molecule, thus helping to ensure normal cell function.

Remaining healthy, though, requires that we not stretch that resiliency to the breaking point. Some researchers fear that modern industrial society may be doing just that. How?

Today, 84,000 chemical substances are in commercial use. The National Academy of Sciences notes that few of these substances have been adequately tested for their ability to cause mutations, cancer, and birth defects. Exposure to any of these chemicals or combinations of chemicals, some fear, may increase our chances of developing cancer.

Others believe that widespread publicity over the ill effects of some chemicals such as pesticides has created a chemical paranoia—sometimes referred to as *chemophobia*—among some members of our society. New research suggests that while the threat of chemical carcinogens is real, many people are overreacting to it.

In fact, numerous studies suggest that naturally occurring chemicals in our food, like one type of mold that grows on peanuts, probably cause more cancer than pesticide residues.

Interestingly, numerous extensive studies in large populations of humans show that diet can have a huge impact on cancer. The standard American diet, for instance, consists of lots of fatty meats, lots of carbohydrates and fats, and tons of snack foods containing fats. This diet promotes cancer. Populations that subsist on this imbalanced diet have much higher rates of cancer than those that consume a more vegetable-based diet.

While eating a fatty, meat-based, low-micronutrient diet often leads to cancer, eating a healthier whole-grain, vegetable, and fruit-based diet full of micronutrients and antioxidants actually lowers ones risk of cancer and a host of other diseases.

All this shows that diet can have a profound effect on health and that chemicals of modern society are probably much less important than an improper diet, except in select areas, something we all can control.

Point/Counterpoint 18-1   Are Current Procedures for Determining Carcinogenesis Valid?

Animal Testing for Cancer Is Flawed *by Philip H. Abelson*

The principal method of determining potential carcinogenicity of substances is based on studies where huge doses of chemicals are administered daily to inbred rodents for their lifetime. Then by questionable models, which include large safety factors, the results are used to extrapolate the effects of minuscule doses in humans. Resultant stringent regulations and attendant frightening publicity have led to public anxiety and chemophobia. If current ill-based regulatory levels continue to be imposed, the cost of cleaning up phantom hazards will be in the hundreds of billions of dollars with minimal benefit to human health. In the meantime, real hazards are not receiving adequate attention.

The current procedures for gauging carcinogenicity are coming under increasing scrutiny and criticism. A leader in the examination is Bruce Ames, who with others has amassed an impressive body of evidence and arguments. Ames and Gold summarized some of their data and conclusions in *Science* (31 August 1990, p. 970). Three articles in the *Proceedings of the National Academy of Sciences* provide an elaboration of the information with extensive bibliographies. The articles also provide data about other pathologic effects of natural chemicals.

Philip Abelson was the Deputy Editor of *Science*. This essay is excerpted from "Testing for Carcinogens with Rodents," *Science* 249 (1990): p. 1357. Reprinted with permission from AAAS. (Photo © Philip Abelson.)

A limited number of chemicals tested, both natural and synthetic, react with DNA to cause mutations. Most chemicals are not mutagens, but when the maximum tolerated dose (MTD) is administered daily to rodents over their lifetime, about half of the chemicals give rise to excess cancer, usually late in the normal life span of the animals. Experiments in which synthetic industrial chemicals were administered in the MTD to both rats and mice resulted in 212 of 350 chemicals being labeled as carcinogens. Similar experiments with chemicals naturally present in food resulted in 27 of 52 tested being designated as carcinogens. These 27 rodent carcinogens have been found in 57 different foods, including apples, bananas, carrots, celery, coffee, lettuce, orange juice, peas, potatoes, and tomatoes. They are commonly present in quantities thousands of times as great as the synthetic pesticides.

The plant chemicals that have been tested represent only a tiny fraction of the natural pesticides. As a defense against predators and parasites, plants have evolved a large number of chemicals that have pathologic effects on their attackers and consumers. Ames and Gold estimate that plant foods contain 5,000 to 10,000 natural pesticides and breakdown products. In cabbage alone, some 49 natural pesticides have been found. The typical plant contains 1% or more of such substances. Compared with the amount of synthetic pesticides we consume, we eat about 10,000 times more of the plant pesticides.

It has long been known that virtually all chemicals are toxic if ingested in sufficiently high doses. Common table salt can cause stomach cancer. Ames and others have pointed out that high levels of chemicals cause large-scale cell death and replacement by division. Dividing cells are much more subject to mutations than quiescent cells. Much of the activity of cells involves oxidation, including formation of highly reactive free radicals that can react with and damage DNA. Repair mechanisms exist, but they are not perfect. Ames has stated that oxidative DNA damage is a major contributor to aging and to cancer. He points out that any agent causing chronic cell division can be indirectly mutagenic because it increases the probability of DNA damage being converted to mutations. If chemicals are administered at doses substantially lower than MTD, they are not likely to cause elevated rates of cell death and cell division and hence would not increase mutations. Thus, a chemical that produces cell death and cancer at the MTD could be harmless at lower dose levels.

Diets rich in fruits and vegetables tend to reduce human cancer. The rodent MTD test that labels plant chemicals as cancer-causing in humans is misleading. The test is likewise of limited value for synthetic chemicals. The standard carcinogen tests that use rodents are an obsolete relic of the ignorance of past decades. At that time, extreme caution made sense. But now tremendous improvements of analytical and other procedures make possible a new toxicology and far more realistic evaluation of the dose levels at which pathological effects occur.

Current Methods of Testing Cancer Are Valid *by Devra Davis*

The vast majority of the scientific community endorses the conduct of experimental studies in order to try to identify those materials that could cause disease and prevent harmful exposures. In the typical toxicologic study of 50 rodents, each animal is a stand-in for 50,000 people. Because rodents live about 2 years on average, they are usually exposed to amounts of the suspect agent that approximate what a human would encounter in an average lifetime of 70 years.

Some have argued that the use of the maximum tolerated dose (MTD) in these studies produces tissue damage and cell proliferation, which, in turn, lead to cancer. This is biological nonsense that ignores a fundamental characteristic of cancer biology that has been known for more than 50 years: cancer is a multi-stage disease, with multiple causes, which arises by a stepwise evolution that involves progressive genetic changes, cell proliferation, and clonal expansion. Thus, swamping of tissues with high doses alone may well kill an animal or damage its tissues, but high doses alone are not sufficient to cause cancer. A 2-year study at the National Institute of Environmental Health Sciences (NIEHS) by David Hoel and colleagues provides good evidence on this point. They looked for signs of damage in tissues taken from rats and mice used in cancer studies. Cancers occurred in organs that did not show apparent damage, and some damaged organs were completely free of tumors.

In addition, most compounds tested do not cause cancer only in the highest dose group tested but typically produce a dose-response relationship, where the amount of cancer developed is proportional to the dose administered. Some chemicals cause toxicity only, others cause only cancer. Not all of those that cause cancer do so through organ toxicity. In fact, almost 90% of the substances shown to cause cancer in the National Toxicology Program do so without producing any increased cellular toxicity, and they also cause cancer at both lower and higher doses.

Animal studies are evolving and being further refined, as is our understanding of differences between species that need to be taken into account in conducting these studies. Every compound known to cause cancer in humans also produces cancer in animals, when adequately tested. For 8 of the 54 known human cancer-causing agents, evidence of carcinogenicity was first obtained in laboratory animals; in many cases, the same target organs and doses have been involved in producing cancer in both animals and humans.

The NIEHS has carried out nearly 400 long-term rodent studies, which have been published following peer review by specialists in the field. Chemicals nominated for testing usually represent a sample of potentially "problematic" materials. About 40% of the "suspect" pesticides evaluated to date have been found to cause cancer. As to the role of so-called natural pesticides, rodent diets are also loaded with many of these materials. Nevertheless, a number of test compounds added to these diets markedly increase the amount of tumors produced. Thus, animals are more sensitive to certain synthetic compounds than to the background level of natural materials. Humans are also likely to have acquired some resistance to these natural materials through evolution.

To date, only about 20% of all synthetic organic chemicals have been adequately tested for their potential human toxicity. Those who must set public policy on the use of chemicals need a rational basis on which to stake their actions. Continued advances in animal testing provide an important contribution to environmental health sciences and to public health efforts to predict, and then prevent, the development of environmentally caused disease.

In summary, the current system should be used until it can be replaced by a demonstrably better one. There is no scientific basis for rejecting the MTD as capable of inducing cancer.

Devra Davis is a senior adviser in the office of the Assistant Secretary for Health, Department of Health and Human Services a Senior Distinguished Visiting Research Scholar in the WHO Collaborating Center for Capacity Building in Public Health at Hebrew University—Hadassah School of Public Health and Community Medicine. (© Devra Davis.)

Sharpening Your Critical Thinking Skills

1. Summarize each author's main points and supporting data. Do you see any inconsistencies or examples of faulty logic in either essay? If so, where?

2. Why is it that two scientists can disagree on an issue such as this?

3. Given the disagreement, what course of action would you recommend?

Visit Human Biology's site for links to websites offering more information on this topic.

SUMMARY

DNA and RNA: Macromolecules with a Mission

1. DNA is the genetic information of the cell.

2. DNA molecules consist of two intertwined polynucleotide strands consisting of numerous nucleotides. DNA polynucleotide chains are joined by hydrogen bonds found between purine and pyrimidine bases on the opposite strands.

3. Each nucleotide in the DNA molecule consists of a purine or pyrimidine base, the sugar deoxyribose, and a phosphate group. The nucleotides are joined by covalent bonds.

4. Complementary base pairing is an unalterable coupling in which adenine on one strand of the DNA molecule always binds to thymine on the other and guanine always binds to cytosine.

5. Complementary base pairing ensures accurate replication of the DNA and accurate transmission of genetic information from one cell to another and from one generation to another.

6. To replicate, DNA must first unwind with the aid of a special enzyme. After unwinding, the original strands provide templates for the production of complementary DNA strands, a process aided by the enzyme DNA polymerase.

7. The cell contains three types of RNA: transfer RNA, ribosomal RNA, and messenger RNA. All play important roles in the synthesis of protein in the cytoplasm.

8. All three RNA molecules are single-stranded polynucleotide chains. RNA nucleotides contain the sugar ribose instead of deoxyribose, which is found in DNA. RNA nucleotides contain four bases: adenine, guanine, cytosine, and uracil (instead of thymine).

9. RNA is synthesized in the nucleus on a template of DNA. The synthesis of RNA is called transcription.

How Genes Work: Protein Synthesis

10. The genetic information contained in the DNA molecule is transferred to messenger RNA. This process is called transcriptions.

11. Messenger RNA molecules carry the genetic information from the nucleus to the cytoplasm, where proteins are synthesized. Messenger RNA serves as a template for protein synthesis.

12. Ribosomes are required to produce proteins on the mRNA template. They bind to the mRNA and contain binding sites for transfer RNA molecules.

13. Transfer RNA molecules deliver amino acid molecules to the mRNA and insert them in the growing chain.

14. Each tRNA binds to a specific amino acid and delivers it to a specific codon, a sequence of three bases on the mRNA. Thus, the sequence of codons determines the sequence of amino acids in the protein. Proteins are synthesized by adding one amino acid at a time. During protein synthesis, the ribosome first attaches to the mRNA at the initiator codon. The large subunit then attaches. A specific tRNA bound to an amino acid binds to the initiator codon and the first binding site of the ribosome. A second tRNA–amino acid then enters the second site.

15. An enzyme in the ribosome catalyzes the formation of a peptide bond between the two amino acids. After the bond is formed, the first tRNA (minus its amino acid) leaves the first binding site.

16. The ribosome moves down the mRNA one codon, shifting the tRNA bound to its two amino acids to the first binding site and opening the second site for another tRNA–amino acid. This process repeats itself many times in rapid succession.

17. As the peptide chain is formed, hydrogen bonds begin to form between the amino acids, and the chain begins to bend and twist, forming the secondary structure of the protein or peptide. When the ribosome reaches the terminator codon, the peptide chain is released.

Controlling Gene Expression

18. Gene control in humans occurs at least five levels: at the level of the DNA, at the chromosome, at transcription, after transcription but before translation, and at translation.

19. At the DNA level: Sections of DNA called transposons can move to new location where they reattach, affecting gene expression. The addition or removal of methyl groups, known as methylation or demethylation, can affect the DNA in profound ways. Such changes are hereditary.

20. At the chromosome level: Control occurs by condensation, or coiling, of the chromatin fibers to inactivate genes. At transcription: Certain chemicals either turn on or turn off genes. These chemicals are either the end products of metabolic reactions or substrates in these pathways. Some segments of the DNA, called enhancers, can greatly increase the activity of nearby genes. Geneticists think that protein molecules bind to enhancer regions of the chromosome and increase gene activity.

21. After the production of RNA before translation: Human DNA contains far more genetic material than it needs. Those segments of the DNA used to produce the RNA that will be used to make protein are called exons. The intervening segments of noncoding DNA are called introns.

22. Both introns and exons are transcribed during the cell cycle. The cell, however, removes the RNA from the introns and joins the RNA fragments produced by exons to create a functional mRNA molecule.

23. The exon-produced RNA can be spliced together in different ways; the resulting mRNAs produce different proteins.

24. At translation: Messenger RNA may be transferred to the cytoplasm in an inactive, or masked, state. Masking permits the cell to build up large supplies of mRNA in preparation for a sudden burst of protein synthesis.

Epigenetic Control of Gene Expression

25. New research shows that our genetic makeup can be modified in ways that do not involve mutations of the DNA. These modifications may be passed to future generations.

26. These studies suggest that diet, chemical toxins, and even the behavior of our parents can alter our genes through the epigenome.

27. The epigenome is an array of chemical markers and switches. Some of these controls are nothing more than methyl groups that switch off genes. Other controls exist in the histone proteins associated with the DNA.

28. These controls lie along the length of the DNA double helix. They form something of a complex software code that controls the DNA hardware.

29. Changes in the epigenome may also be responsible for certain diseases, such as cancer, rheumatoid arthritis, and diabetes.

Health and Homeostasis

30. Humans and other organisms contain specific genes, called proto-oncogenes, which lead to cancer when mutated. Proto-oncogenes are normal genes that code for cellular structures and functions, such as cell adhesion, mitotic proteins, and the production of plasma-membrane receptors for growth factors or hormones.

31. Chemical, physical, and biological agents can mutate these genes, resulting in uncontrolled cellular proliferation (cancer).

32. Viruses can also cause cancer. Some viruses, for example, possess oncogenes, which enter human body cells and are incorporated in the cell's DNA, stimulating uncontrolled cellular division. Other viruses carry genes that stimulate cancer by activating human proto-oncogenes.

THINKING CRITICALLY ANALYSIS

This analysis corresponds to the Thinking Critically scenario that was presented at the beginning of this chapter.

After you have thought about this issue for a while, you may want to consider some additional facts. The U.S. dairy industry already produces an excess of milk. The federal government buys the surplus, dehydrates some of it, and makes cheese out of the rest. The government stockpiles dehydrated milk and cheese, which are given to needy families. Will increasing the surplus produce more food for the poor?

Before you draw any conclusions, consider another issue—the effects on small dairy operations of increasing the surplus of milk. Many owners of small dairy herds believe that the large producers will be the primary beneficiaries of the hormone. It will allow them to increase their milk production and probably drive down the cost of milk and milk products. In the process, it could put many small dairy farmers out of business. This loss could adversely affect the economies of many rural regions.

Some consumer groups are concerned about the potential health effects of using synthetic growth hormone. A trace of growth hormone is present in milk produced normally. Will the use of synthetic growth hormone increase this concentration in milk? What effect will that have on your health or the health of children?

Make a list of the pros and cons of using synthetic growth hormone. Then make a list of questions—those that you can answer and those that you cannot answer. Do you need more information to decide?

If you are able to make up your mind, what factors swayed your opinion? Did the concerns of one group outweigh the concerns of another? What critical thinking rules did you use in this exercise?

KEY TERMS AND CONCEPTS

Adenine (A), p. 393
Anticodon, p. 398
Complementary base, p. 393
Complementary base pairing, p. 393
Complementary strand, p. 393
Cytosine, p. 393
DNA, p. 392
DNA polymerase, p. 393
Enhancement, p. 401
Enhancer, p. 401
Epigenetics, p. 404
Epigenome, p. 404

Exons, p. 401
Guanine (G), p. 393
Initiator codon, p. 398
Introns, p. 401
Messenger RNA, p. 396
Molecular genetics, p. 392
Nucleotide, p. 392
Oncogene, p. 405
Operon, p. 402
Original strand, p. 393
Peptide bond, p. 399
Purine, p. 393

Proto-oncogene, p. 405
Pyrimidine, p. 393
Ribosomal RNA, p. 398
Ribosome, p. 398
RNA polymerase, p. 396
Terminator codon, p. 401
Thymine (T), p. 393
Transcription, p. 396
Transfer RNA, p. 398
Translation, p. 398
Transposon, p. 403
Viral oncogene, p. 405

CONCEPT REVIEW

1. What is DNA? p. 392.
2. What is the genetic code? p. 395.
3. What is meant when we say that the genetic code is housed in the DNA and translated into instructions the cell can understand? pp. 395–397.
4. Describe how DNA is synthesized. How does the cell ensure the accurate replication of DNA? pp. 395–396.
5. List the three types of RNA. What do they have in common? p. 395.
6. Briefly describe the function of each type of RNA in protein synthesis. pp. 398–399.
7. In what ways are RNA and DNA similar? In what ways are they different? p. 396.
8. What are the three major parts of protein synthesis? p. 398.
9. What is a codon? What is an anticodon? Why are they important for the accurate production of protein? p. 398.
10. What is the initiator codon? What is the terminator codon? pp. 398, 400–401.

11. Describe the production of protein on mRNA in detail. Be sure to note the enzymes involved and the role of the ribosome. Be sure to include the terms mRNA, tRNA, ribosome, large subunit of the ribosome, small subunit of the ribosome, binding sites, p. 398.
12. Discuss the various levels at which human genes are controlled. Give an example of each. pp. 401–403.
13. What is the epigenome? How can it be changed through diet or exposure to toxic chemicals? Is this another level at which human genes can be controlled? pp. 403–405.
14. How has knowledge of changes in the epigenome affected our understanding of heredity? pp. 404–405.
15. What is a transposon? Why can it cause genetic change in an organism? What factors can cause a gene or a group of genes to move from one location on a chromosome to another? p. 403.

16. What is a proto-oncogene? How is it affected by a mutagen, a chemical or physical agent that causes mutation? p. 405.
17. Proto-oncogenes control functions related to cell growth. What are they? How are these functions related to cell growth? p. 405.
18. Go on the Internet and see if you can find scientifically accurate articles that describe foods that could fight cancer through methylation. What are they? p. 404.
19. Do some research on the Internet or in books to find out more about operons. What is an operon? How are operons controlled? p. 402.
20. Do some research on the Internet on the importance of antioxidants and micronutrients in our diet to combat disease. What are your conclusions? How could your diet change? p. 405.

SELF-QUIZ: TESTING YOUR KNOWLEDGE

1. The DNA molecule is a double helix; its strands are held together by _____ bonds between bases. p. 392.

2. DNA consists of two _____ strands made of nucleotides. p. 392.

3. DNA's two strands intertwine forming a _____ helix. p. 392.

4. In DNA, the purine bases are adenine and _____. p. 393.

5. IN DNA, the base adenine is joined to the base _____ on the complementary strand. p. 393.

6. Because new DNA is produced on an existing strand of DNA, the process is referred to as _____. p. 393.

7. In DNA, the sugar molecules in the nucleotides are known as _____. p. 392.

8. The exact replication of DNA is made possible by _____ base pairing. p. 393.

9. RNA contains a nitrogen base known as _____ that is not found in DNA. p. 395.

10. RNA molecules are all _____-stranded. p. 396.

11. The formation of RNA on a DNA template is called _____. p. 396.

12. The formation of protein on an mRNA template is known as _____. p. 398.

13. RNA is produced on a DNA template inside the cell's _____. p. 396.

14. _____ attaches to amino acids in the cytoplasm and inserts them in newly forming protein chains. p. 398.

15. In cells during interphase, chromosomal _____ results in a tight coiling of the DNA, inactivating genes in that section. p. 401.

16. _____ is a process in which already active genes increase their production of mRNA. p. 401.

17. _____ are segments of DNA that are not responsible for the production of mRNA. That is, they are not expressed. p. 401.

18. A _____ is a segment of DNA that moves to other parts of the genome, causing profound genetic change in an organism. p. 403.

19. Genes in cells that control functions related to cellular replication, including cell adhesion and the production of plasma membrane receptors that bind to growth factors like certain hormones are known as _____. p. 405.

20. Genes carried by viruses that can cause cancer are known as _____. p. 405.

 biology.jbpub.com/chiras/8e/

The site features eLearning, an online review area that provides quizzes, chapter outlines, and other tools to help you study for your class. You can also follow useful links for in-depth information, research the differing views in the Point/Counterpoints, or keep up on the latest health news.

(© Pressmaster/ShutterStock, Inc.)

Genetic Engineering and Biotechnology

In the movie *Jurassic Park*, scientists removed DNA from dinosaur blood found in the gut of mosquitoes preserved in amber (hardened sap of pine trees) during the age of the dinosaurs. They took this DNA and inserted it into the nuclei of frog cells and produced living, breathing, lawyer-munching dinosaurs. We all loved that part of the movie!

THINKING CRITICALLY

A chef at a famous restaurant in New York City organized fellow chefs in the city to protest the sale of genetically engineered produce, such as tomatoes. His main concern is that the government has refused to require farmers who are growing genetically engineered tomatoes to label their products. Consumers, therefore, won't know if they're getting genetically engineered vegetables or ones produced naturally. How would you analyze this issue? What critical thinking rules might apply? To help you analyze this issue, take some time to review the **Point/Counterpoint** in this chapter.

Although this scenario is pure science fiction, it illustrates the kinds of genetic manipulations that scientists are engaged in. One procedure, in which a segment of the DNA of one species is cut out and transferred to another species, is popularly referred to as **genetic engineering**. For reasons explained shortly, scientists typically refer to it as *recombinant DNA technology*.

19-1 Genetic Manipulation: An Overview

Although we like to think of genetic engineering as a modern invention, it is not new at all. In fact, plants, animals, and microorganisms have been "performing" their own genetic engineering experiments for at least 3.5 billion years. They've been undergoing mutations, exchanging segments of genes (via crossing-over), and combining genes in new ways through sexual reproduction. Even bacteria swap genes via tiny segments of DNA called plasmids, which leads to new genetic combinations, including bacteria that are antibiotic resistant. In addition, a lot of the DNA in we humans appears to have been picked up from viruses that have, over the long course of human evolution, infected human cells. These alterations in the genome are natural and create new genetic combinations essential to the evolution of life on Earth. Over the course of evolution, nature's own genetic engineering has resulted in a diverse array of life forms.

Throughout much of human history, we humans have also actively and intentionally engaged in genetic engineering through a process known as **selective breeding**—intentionally crossing certain plants or animals to produce offspring with desirable traits. Domestic animals, such as dogs, came about by selective breeding, as have cattle, cats, fancy pigeons, and other animals (Figure 19-1). These are all remarkable breeds. Vital food crops—such as corn, wheat, and tomatoes—are here today thanks to the efforts of geneticists who selected plants with desirable traits and then fertilized them with pollen from other

(a)

(b)

(c)

FIGURE 19-1 **Fancy Pigeons** (a) This beautiful bird, a Norwich Cropper, barely resembles its relative the rock dove, or common pigeon. Selective breeding produced this bird as it has many other new genetic combinations. (Courtesy of the Australian National Pigeon Assn.; breeder, Eddie Kloprogge.) (b) A Pygmy Pouter (© Richard Bailey/Corbis.) (c) A Saxon Fairy Swallow. (© Eric Isselee/ShutterStock, Inc.)

plants with similar traits. The goal was to produce high-yield plants that produced good-tasting fruits and seeds. Geneticists have also developed drought-resistant strains and much more.

Some consider modern forms of genetic engineering, the newest method of genetic modification, to be nothing more than an advanced form of the genetic tinkering that occurs in nature and in human societies through selective breeding (Figure 19-2). Others see it as an intrusion into the natural process of evolution. Still others view it as a dangerous gamble that has potential health effects. Some opponents warn that genetically engineered organisms could spread throughout the world, causing disease and environmental problems. (For a debate on genetic engineering, see the Point/Counterpoint in this chapter.)

> **KEY CONCEPTS**
>
> Genetic engineering or recombinant DNA technology is the newest form of genetic tinkering humans have engaged in. It's designed, in large part, to improve the genetics of food crops and livestock to help enhance reproduction and growth as well as other important traits.

FIGURE 19-2 **Short Stalk Wheat** Produced by selective breeding, this new variety helps reduce costly inputs and stands up better to wind. (© ZoranKrstic/ShutterStock, Inc.)

19-2 Cloning

Yet another growing controversy has to do with a process called **cloning**. In this technique, cells from an organism are used to develop another, genetically identical organism—a clone. Nuclei from the cells of prize-winning cattle, for example, can be transferred to enucleated eggs (egg cells whose nuclei have been removed) of the same breed. The artificially fertilized eggs are then transferred to surrogate mothers, cows in whose reproductive tracts the eggs develop. The result is a genetically superior offspring. The goal of cloning is to extract many eggs that are artificially fertilized with nuclei from superior males. In so doing, a breeder can mass produce genetically superior cattle.

In 1997, much furor arose when a Scottish researcher reported the birth of a cloned sheep. The clone, named Dolly, was derived from a cell taken from the mammary gland of another sheep. The nucleus from this cell was placed in the enucleated cell, and inserted in the uterus of a female (ewe) where it developed into Dolly. The announcement stirred considerable controversy, as people began to wonder whether this technique would be used for humans. In 1998, scientists announced yet another clone—this one in mice.

In 2000, the first cloned cow was sold at auction. Produced by a company in Wisconsin, called Infigen, this cow was an exact copy of a prize-winning Holstein. It was produced when an ear cell from the mother was fused with an enucleated egg. The egg with its ear cell nucleus was then activated using a chemical that stimulates the release of calcium in the egg, causing it to divide. The newly "fertilized" egg was transplanted into another cow, where it developed fully.

Researchers at Texas A & M University have used cloning to produce a disease-resistant cow. They used genetic material from a bull that was naturally resistant to several common diseases. These diseases not only affect cattle but also can be passed to humans in uncooked beef or unpasteurized milk. One remarkable aspect of this feat is that the donor cells from the disease-resistant bull had been frozen 15 years earlier!

Cloning has also been used to produce an Asian gaur, an endangered oxlike animal native to the bamboo forests of India and Burma (Figure 19-3). A skin cell from the gaur was fused with an enucleated egg cell. Researchers hope that this technique may help increase populations of endangered wildlife and may even bring back recently extinct species. They are beginning cloning work to help the Giant Panda, another species in danger of extinction. They're hoping that eggs of black bears will prove to be up to the task.

In the past decade, numerous animals have been successfully cloned. In December 2001, for instance, scientists at Texas A&M University created the first cloned cat, appropriately called CopyCat. Who says scientists don't have a sense of humor?

To date over a dozen animals have been successfully cloned, including carp, cats, cattle, dogs, deer, ferrets, frogs, fruit flies,

FIGURE 19-3 **Asian Gaur** Scientists are using cloning to help increase the numbers of this and other endangered species. (© Chen Wei Seng/ShutterStock, Inc.)

horses, mice, monkeys, mules, pigs, rabbits, rats, water buffalo, and wolves. You may want to take a few moments, when you can break away from your studies to go online to see some photos of cloned animals.

In 2003, a group of researchers reported the cloning of a human, although evidence of this feat has not been forthcoming. Even Hollywood got into the act in 2004 with a film called *Godsend*, a movie about human cloning.

This chapter examines genetic manipulation and its real and potential applications. It also outlines the basic controversy over the use of modern genetic engineering techniques.

> **KEY CONCEPTS**
>
> Cloning is a technique in which nuclei from cells of one animal are used to fertilize eggs of another of the same species usually to produce superior offspring, but also to resurrect extinct species.

19-3   Recombinant DNA Technology

In **recombinant DNA technology**, genes are removed from one organism, replicated, and then spliced into the cells of another organism of the same or a different species. Genes can also be artificially produced in the lab and then inserted into the genome of host cells. In recent years, scientists have developed, removed or even modified existing genes in an organism. These techniques also fall within the realm of genetic engineering. Organisms produced by these techniques are referred to as **genetically modified organisms (GMOs)**.

Interestingly, the term genetic engineering was coined in 1951 by science fiction writer, Jack Williamson—quite amazingly a year before the role of DNA in heredity was confirmed! The science of genetic engineering or genetic modification got its start in the early 1970s, over 40 years ago as the result of research by Stanley Cohen at Stanford University and Herbert Boyer at the University of California at San Francisco. In 1973, these scientists cut segments of DNA from one bacterium using a special enzyme and spliced them into the genetic material of another bacterium. The recipient bacterium then multiplied and, in the process, produced many copies of the altered genetic material. The process Cohen and Boyer pioneered still provides the basis for much of the genetic engineering research and development being done today. To understand how it works, let's look at the individual steps, beginning with the slicing of a piece of DNA from a chromosome.

> **KEY CONCEPTS**
>
> Genetic engineering is a process in which scientists remove genes, duplicate them, and then insert them into a new organism, known as a genetically modified organism. It also involves techniques to delete or modify genes.

Removing Genes

Genetic engineering was made possible by the discovery of a group of naturally occurring enzymes that snip off segments of the DNA molecule. Known as **restriction endonucleases**, these enzymes cut through both strands of the DNA, as shown in Figure 19-4.

Restriction endonucleases are found in many species of bacteria. In bacteria, these enzymes serve a protective function. They destroy the DNA of viruses that attack viruses, commandeering the bacterial resources to produce more of their own kind (these viruses are referred to as *bacteriophages*—bacteria eaters). Endonucleases prevent these cellular pirates from taking over and killing infected bacteria. To date, scientists have

discovered hundreds of different restriction endonucleases, each one specific for a particular sequence of bases on the DNA.

Scientists slice segments of DNA from bacteria and other organisms, including humans, using restriction endonucleases. They then use the same enzyme to slice open the recipient cell's DNA. The gene segments removed by endonucleases are then inserted into the gap in the recipient's DNA.

After a slice of DNA is added to the bacterial DNA, another enzyme is used to seal it in place. This enzyme is called **DNA ligase**. At this point, the DNA splicing is complete. Geneticists have formed a molecule that contains DNA from two different organisms (a combination of DNAs)—technically referred to as a recombinant DNA molecule.

> **KEY CONCEPTS**
>
> Scientists use enzymes called nucleases to slice specific genes from one cell and to create gaps in another cell into which the spliced gene can be placed.

Mass Producing Genes

As noted earlier, recombinant DNA technology began with experiments in bacteria. As you may recall, bacteria contain a single, circular "strand" of DNA.

They also contain smaller circular strands, called **plasmids** (PLAZ-mids) (Figure 19-5). Plasmids carry a few genes—such as those that confer antibiotic resistance to bacteria. They also replicate independent of the chromosome. Plasmids are much smaller than bacterial DNA, are relatively easy to extract, and can be used for gene transplants. How is this done?

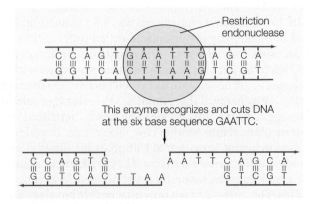

FIGURE 19-4 **DNA Scissors** Restriction endonucleases extracted from bacteria selectively slice through different segments of the DNA double helix, producing a staggered cut as shown here.

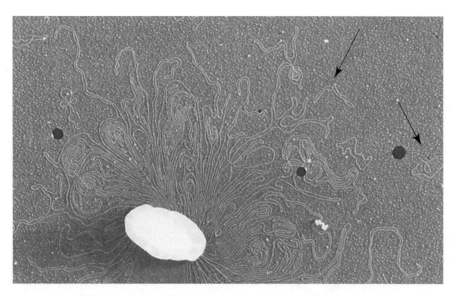

FIGURE 19-5 **Plasmids** Rupturing a bacterium releases circular chromosome and numerous plasmids, indicated by arrows. (© H. Potter and D. Pressler/ Visuals Unlimited.)

To begin, take a look at Figure 19-6. The top left drawing shows a bacterium and its plasmid. As illustrated, plasmids can be removed from bacterial cells, sliced open with restriction endonucleases. Notice the small gap in the plasmid in the drawing. On the right side of the diagram, genes are removed from another cell, for example, a human cell. The gene is then inserted into the opening in the plasmid and sealed in place with DNA ligase.

The recombinant plasmid DNA is then inserted into the bacterium, transferring the new DNA to the bacterium. This is achieved by adding the genetically modified plasmids to culture dishes containing bacteria (or yeast cells). In culture, the plasmids readily incorporated into their new hosts. As the host bacterial cells divide, the plasmids replicate. This produces multiple copies of the gene. This process is known as gene cloning. **Gene cloning** produces numerous identical copies of the foreign gene. Steps 6 and 7 in Figure 19-6 show that the mass produced genes can be used to genetically modify organisms, such as corn, to produce hardier, higher-producing crops or useful byproducts, such as hormones that are needed to treat patients with hormonal deficiencies.

> **KEY CONCEPTS**
> Desirable genes from one organism can be transplanted into plasmids from bacteria where they are biologically reproduced. This allows us to mass produce desirable genes and gene byproducts.

Producing Genes on mRNA

Geneticists can produce multiple copies of DNA in other ways, too. One common way they do this is by extracting messenger RNA (mRNA) from cells whose genes they want to copy. They then make copies of DNA from the mRNA molecules. This process is known as **reverse transcription**. (Remember: normally, mRNA is produced from DNA by the process known as *transcription*.) This genetic slight of hand is made possible by a unique enzyme found in certain viruses known as **reverse transcriptase**.

To make DNA, geneticists extract mRNA from cells and then add reverse transcriptase (Figure 19-7). The enzyme creates single strands of DNA on the mRNA templates. As illustrated, the hybrid DNA/RNA molecules are then treated with a chemical that destroys the RNA, leaving only single strands of DNA. Additional enzymes convert the single-stranded DNA to double-stranded DNA molecules. DNA molecules containing the genes researchers want to transplant into another organism are mass produced from the mRNA and are referred to as *cDNA*.

> **KEY CONCEPTS**
> Desirable genes can also be mass produced from strands of mRNA extracted from cells and treated with reverse transcriptase, an enzyme that creates DNA from templates of RNA.

The Polymerase Chain Reaction

The production of genes from mRNA and plasmids is called **gene amplification**. Another, more widely used method of gene amplification is the **polymerase chain reaction**. This technique allows very rapid production of DNA in mass quantities and was invented by biochemist Kary Mullis.

In this technique, geneticists extract a segment of the DNA double helix containing genes they want to replicate from a cell's nucleus. They then split the DNA molecules into two strands by heating them. Next, they cool the solution and add DNA polymerase enzymes. These enzymes catalyze the formation of complementary DNA molecules.

The DNA molecules are then split again with heat, and more DNA polymerase is added. The new polymerase triggers another round of DNA replication. This procedure, which is carried out by machine, permits scientists to produce a million or more copies of the original DNA.

Mass production permits scientists to produce large quantities of DNA, which are needed to chemically analyze the DNA of various species. In so doing, geneticists can determine the exact sequence of bases that comprise the various genes. This procedure has helped scientists map the genes of the human **genome**. Interestingly, this study has shown that the human genome contains many of the same genes found in other organisms, such as bacteria, mice, and snakes.

> **KEY CONCEPTS**
> DNA can also be mass produced using plasmids, reverse transcriptase, and the DNA polymerase chain reaction.

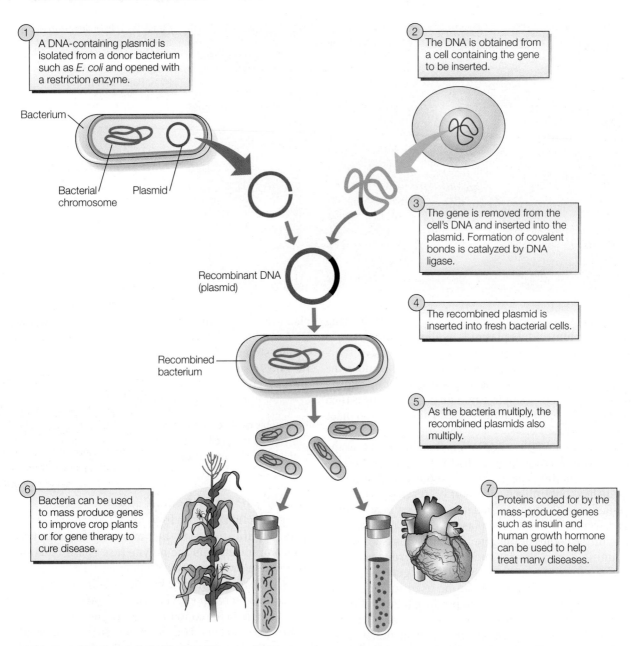

1. A DNA-containing plasmid is isolated from a donor bacterium such as *E. coli* and opened with a restriction enzyme.

2. The DNA is obtained from a cell containing the gene to be inserted.

Bacterium

Bacterial chromosome   Plasmid

Recombinant DNA (plasmid)

3. The gene is removed from the cell's DNA and inserted into the plasmid. Formation of covalent bonds is catalyzed by DNA ligase.

4. The recombined plasmid is inserted into fresh bacterial cells.

Recombined bacterium

5. As the bacteria multiply, the recombined plasmids also multiply.

6. Bacteria can be used to mass produce genes to improve crop plants or for gene therapy to cure disease.

7. Proteins coded for by the mass-produced genes such as insulin and human growth hormone can be used to help treat many diseases.

FIGURE 19–6 Genetic Engineering Simplified Genes can be snipped from chromosomes and inserted in plasmids. The recombinant plasmids can then be reinserted into bacteria, where they are replicated in large quantity. This process is known as cloning.

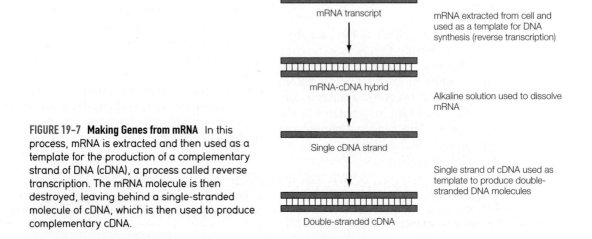

mRNA transcript

mRNA extracted from cell and used as a template for DNA synthesis (reverse transcription)

mRNA-cDNA hybrid

Alkaline solution used to dissolve mRNA

Single cDNA strand

Single strand of cDNA used as template to produce double-stranded DNA molecules

Double-stranded cDNA

FIGURE 19–7 Making Genes from mRNA In this process, mRNA is extracted and then used as a template for the production of a complementary strand of DNA (cDNA), a process called reverse transcription. The mRNA molecule is then destroyed, leaving behind a single-stranded molecule of cDNA, which is then used to produce complementary cDNA.

19-4   Applications of Recombinant DNA Technology

Recombinant DNA technology has resulted in a number of practical, although sometimes controversial, applications.

Treating Diseases

By removing the human genes that code for hormones and then transferring those genes into bacterial cells that are grown in large tanks, drug companies can generate huge quantities of hormones (Figure 19-8). They can also produce other important proteins such as blood clotting factors missing in some individuals. This technique has resulted in a dramatic increase in the supply of protein hormones such as insulin, growth hormone, and oxytocin and other important proteins such as blood clotting factors. One really important product is tPA, a substance that is routinely used to treat heart attacks. It quickly dissolves blood clots and, if given early enough, can dramatically improve one's chances of surviving a heart attack and minimizing damage to the heart. Table 19-1 lists some of the products and their uses.

Consider the benefits of producing human growth hormone. Human growth hormone is used to treat patients who produce abnormally low amounts. In years past, human growth hormone was extracted from the pituitary glands of human cadavers. To extract enough growth hormone for one dose, however, technicians had to remove glands from 50 human cadavers. Treatment was therefore very costly.

Pituitaries from pigs have also been used to obtain growth hormone. Although pig growth hormone acts like human growth hormone, it is different enough chemically to cause an immune reaction in some patients.

TABLE 19-1	Some Useful Products of Genetic Engineering
Products	**Used to**
Growth hormone	Treat patients with dwarfism
Insulin	Treat patients with diabetes
Tumor necrosis factor	Treat patients with cancer
tPA (tissue plasminogen activator)	Dissolves clots formed during heart attacks
Erythropoietin	Stimulates blood cell formation for patients with some forms of anemia
Clotting factors	Treat people with hemophilia
Surfactant	Treat babies with blue baby syndrome, caused by collapse of air sacs in lungs, which is caused by a lack of surfactant
Interferons	Treat some forms of cancer and some viral infections
Vaccines	Prevent hepatitis A, B, and C, as well as AIDS*, malaria*, and herpes*
DNA probes	Map human chromosomes, perform DNA fingerprinting, detect infectious diseases, and detect genetic disorders

*Under development, but not currently available.

FIGURE 19-8   **Mass Producing Genetically Engineered Bacterial Products**   Large vats like this one contain hundreds, sometimes thousands of gallons of culture medium and genetically modified bacteria that produce valuable proteins such as human growth hormone for treating human diseases. (© Visuals Unlimited.)

Recombinant DNA technology offers an opportunity to produce a growth hormone that is much safer than pig growth hormone because it avoids potential immune reactions.

Protein hormones, like growth hormone, can also be produced for other uses, notably for livestock production. Cows, for instance, can be injected with bovine growth hormone produced by genetically engineered bacteria. This treatment greatly increases milk production. (Discussed elsewhere in the text.)

> **KEY CONCEPTS**
>
> Genetic engineering techniques can be used to mass produce many medically important proteins like hormones, blood clotting factors, antiviral agents, cancer-fighting medications, and safer vaccines that can be used to treat diseases and save lives.

Vaccines

Drug companies can also use genetic engineering to mass produce vaccines more economically and more quickly than is possible by conventional means. Vaccines contain bacteria or viruses that have been inactivated, severely weakened, or killed. Vaccines to bacterial toxins contain inactivated toxins and are referred to as *toxoid vaccines.*

When injected in an individual, vaccines stimulate the immune system—just as if it were infected by the real bacteria. The immune system attacks the foreign agent in the vaccine over a period of 7 to 10 days. It also creates memory cells, a cellular reserve like the Army Reserve. Memory cells "remember" the foreign agent and mount a very rapid response to eliminate the foreign infectious agent if you are ever exposed again.

Two types of genetically engineered vaccines are being produced. The first consists of proteins or polypeptides identical to those in the capsids of viruses. These vaccines can be produced from viral genes created by genetic engineering. The advantage of this type of vaccine is that it confers immunity but lacks all the infectious properties of attenuated vaccines produced by traditional methods.

The other type of genetically engineered vaccine consists of genetically modified but inactive viruses. Although the vaccine stimulates an immune reaction, some researchers are concerned that the genetic changes created by genetic engineering could have unpredictable biological effects when the vaccines are administered to humans. Some scientists are concerned that the altered viruses may infect species not currently affected by the viruses. They are also concerned that the genetically altered viruses could exchange genetic material with live viruses—with potentially adverse effects in humans.

At present, numerous vaccines are under development. One vaccine, for an infectious disease that strikes the liver (hepatitis B), has been produced via this method. But geneticists hope to develop others, perhaps even one for the AIDS virus.

> **KEY CONCEPTS**
>
> Genetic engineering can be used to produce safer vaccines than those produced by conventional means, and to produce them more quickly and more economically.

Transgenic Organisms

In 1997, scientists from the Medical College of Georgia announced the successful transplantation of genes that code for a green fluorescing protein into zebrafish, a common type of tropical fish. It turned the red blood cells green under blue light, allowing scientists to study blood cell formation in the zebra fish embryos. The scientists have performed the same process with nerve cells so they can study how nerve cells form connections in the developing brain.

Transferring genes from one species into another results in a **transgenic organism**. This process can be achieved by one of several techniques, two of which are discussed here.

In some cases, foreign genes are injected directly into the eggs of other species. Figure 19-9 shows a microneedle (on right) through which genes are being injected into an egg cell. Genetically modified eggs are then placed in the reproductive tract of females, where they can develop into embryos and then fetuses.

Another much more complicated technique is shown in Figure 19-10. In this technique, fibroblast cells (connective tissue cells) are first removed from a fetus, in this instance, a cow fetus. Foreign genes are then added to a culture dish containing the fibroblasts and are incorporated into some of the fibroblast nuclei. The fibroblasts are then fused whole with eggs whose nuclei have been removed. (These are enucleated cells.) The fused fibroblast-enucleated eggs are then treated with an electrical current. This treatment causes the fibroblast DNA to be incorporated into the eggs. It also stimulates cell division. In this process, a geneticist can create many identical genetically modified embryos. Those embryos are then transplanted into surrogate mothers, producing multiple genetically modified offspring. Take a moment to read the numbered boxes in the diagram and study the drawings to be sure you understand this process.

As a rule, transgenic organisms don't differ much from their original form; in fact, they may display only one new trait, such as a larger body size. In others, the offspring are quite different from their mothers. One example occurs when the human gene that codes for growth hormone is transplanted into pigs and

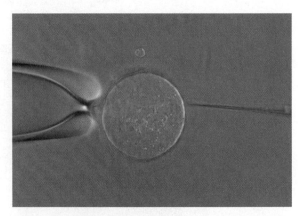

FIGURE 19-9 Injecting Genes Genes can be injected directly into eggs via microneedles (on right). The large structure on the left is a glass rod that holds the egg so the genes can be injected. The egg is then placed in the oviduct of a surrogate mother, where it can develop. (© M. Baret/ Science Source/Science Source, Inc.)

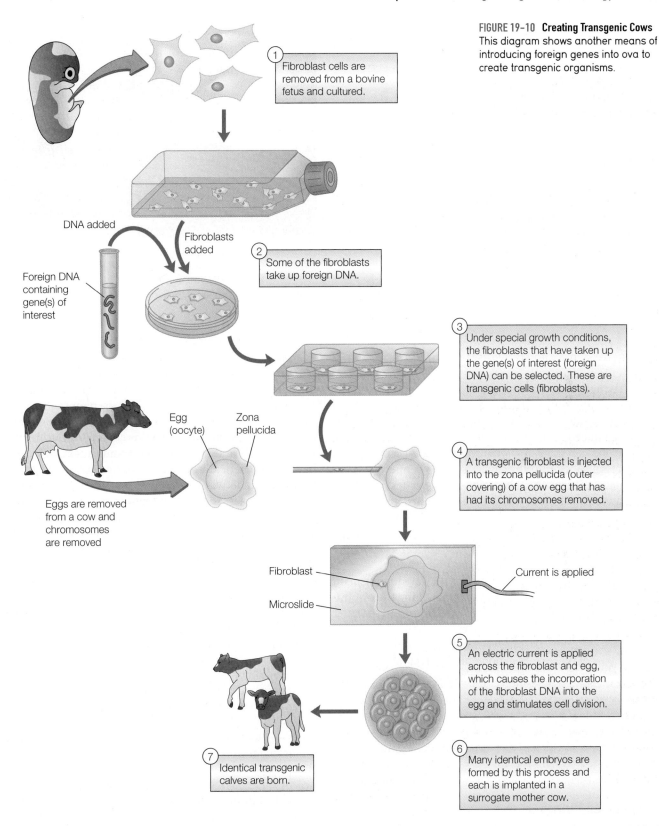

FIGURE 19–10 **Creating Transgenic Cows**
This diagram shows another means of introducing foreign genes into ova to create transgenic organisms.

1. Fibroblast cells are removed from a bovine fetus and cultured.

DNA added

Fibroblasts added

Foreign DNA containing gene(s) of interest

2. Some of the fibroblasts take up foreign DNA.

3. Under special growth conditions, the fibroblasts that have taken up the gene(s) of interest (foreign DNA) can be selected. These are transgenic cells (fibroblasts).

Egg (oocyte)

Zona pellucida

Eggs are removed from a cow and chromosomes are removed

4. A transgenic fibroblast is injected into the zona pellucida (outer covering) of a cow egg that has had its chromosomes removed.

Fibroblast

Microslide

Current is applied

5. An electric current is applied across the fibroblast and egg, which causes the incorporation of the fibroblast DNA into the egg and stimulates cell division.

7. Identical transgenic calves are born.

6. Many identical embryos are formed by this process and each is implanted in a surrogate mother cow.

mice (Figure 19-11). When successful, this results in offspring that are much larger than mice that didn't receive the gene.

Researchers are looking for ways to re-engineer cows so that the milk they produce will be chemically more similar to human milk. Scientists, for example, have transplanted a human gene that codes for the production of a chemical found in mother's milk, called lactoferin, into cows. This substance helps babies fight off infections.

Transgenic organisms can also be used to create important medical advances. Researchers have long relied on animals like rats and mice to study human disease. Genetic engineering allows scientists to create lab animals that suffer from additional human diseases. Genetically modified mice, for instance, are currently being used to study cancer, obesity, heart disease, diabetes, arthritis, substance abuse, anxiety, aging, and Parkinson disease. Researchers

FIGURE 19–11 **Supermice** The mouse on top, pictured with its litter-mate, has received a gene for human growth hormone. (Courtesy of Jacob Panici, Southern Illinois University School of Medicine.)

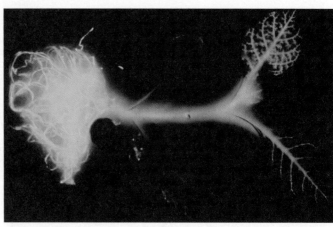

FIGURE 19–12 **Glow Little Tobacco Plant, Glimmer Glimmer** This tobacco plant contains firefly genes for the enzyme luciferase. When the plant is sprayed with luciferin, the enzyme breaks it down, releasing light. (© Keith V. Wood/Science VU/Visuals Unlimited.)

can use these mice to study disease and to test potential treatments.

Although gene transplant techniques sound simple, they're actually quite difficult. Scientists can't just inject a foreign gene into a baby pig or cow. The gene generally has to be introduced at the embryonic stage when the organism is in the one-cell stage. As noted above, this is accomplished by injecting genes into the nuclei of ova via microneedles. Some nuclei incorporate the foreign gene or genes, which are then transmitted from one cell to the next during embryonic development. Even if the gene is incorporated into the genome of the egg cell, however, there's no guarantee that it will end up in the proper location or that the location it ends up in will result in the production of the correct end product.

Scientists are also experimenting with ways to transplant normal genes into cancer cells or the cells lining the respiratory tracts of people suffering from cystic fibrosis. In the latter case, they're inserting genes into viruses that infect the cells lining the respiratory passageways, hoping that the genes will become part of these cells' genome and correct the problems.

KEY CONCEPTS

Transgenic organisms are organisms that contain genes from individuals of other species; they may help improve economically desirable species such as livestock, and can also be used to produce organisms that are vital to better understanding and treating human disease.

Transgenic Plants

Genetic modification of plants, especially food crops, is an area of great interest among agriculturalists and concern among some citizens. One of the first success stories involved the transplantation of a gene from insects that codes for an enzyme called **luciferase** (lew-SIFF-er-ace) into a tobacco plant. This enzyme is found in fireflies and is responsible for catalyzing a light-generating reaction. In the tobacco plant, the enzyme didn't do much unless the plant was sprayed with a chemical called luciferin. At this point, the plant glowed (Figure 19-12).

This experiment was not performed to create tobacco plants that glow in the dark so they could be harvested at night, but only to show that genes could be transplanted from one species to another. It opened up the door to more practical experiments—for instance, transplanting genes that help increase crop production as well as enhancing the nutritional value of seeds and fruit. Soybeans and canola, for instance, have been genetically altered to produce more healthful oils. Efforts are underway to produce commercially valuable crops that thrive in less than optimum soils, for example, salty soils that normally stunt plant growth. Transplanting salt-tolerant genes into commercial crop species could allow farmers to use vast acreages in the United States and other countries idled by the buildup of salts. Salt buildup is caused by the irrigation of poorly drained soils with water high in salts.

Scientists are trying to develop transgenic plants whose roots produce nitrogen from gaseous nitrogen in the air we breathe (much like legumes such as beans and peas). This could help reduce fertilizer use and promote healthier soils. Researchers are also developing ways to make crop plants resistant to herbicides, chemicals applied to crops to control weeds. Although herbicides generally act only on weed species, they sometimes impair the growth of crop species. Proponents hope that herbicide-resistant crops will help farmers increase yields. Critics argue that if herbicide-resistant crops are used, farmers may apply larger amounts of herbicide to their fields. These chemicals, in turn, could be washed into nearby waterways, poisoning aquatic life. They could also poison soils, killing microorganisms essential for the long-term health of the soil. (For a discussion, see the Point/Counterpoint 19-1)

Geneticists are also working on ways to alter plants themselves to increase their resistance to pests. By transferring the genes that protect wild species from insects into crop species, scientists may be able to produce many varieties that require little, if any, chemical pesticides. The benefit to the soils, waterways, and other species could be enormous.

Efforts to produce genetically engineered vegetables such as tomatoes have also proven successful. Genetically modification of crops is big business. In 2012, genetically modified crop seeds were valued at US$15 billion. Just to put that into perspective, that's more than the size of the entire state budget for Kansas.

Critics, however, fear that such experiments may result in unforeseen reactions such as allergies in people who consume the genetically modified organisms (GMOs). There is evidence to suggest that some people do indeed develop allergies to genetically engineered crops.

> **KEY CONCEPTS**
>
> Transgenic plants are being developed through genetic engineering to increase crop production, increase the nutritional value of food crops, protect crops from insects and harmful effects of herbicides, enhance soil quality, help plants grow in marginal soils, and much more.

Transgenic Microorganisms

Genetic engineering began with microorganisms when researchers inserted the human insulin gene in bacteria and found that the bacteria produced human insulin. Four years after their finding, the U.S. Food and Drug Administration approved the use of these bacteria in insulin production, which is used to treat diabetes. Since that time, transgenic microorganisms have been used to produce numerous medically important proteins including blood-clotting factors, anti-cancer agents, and human growth hormone which is used to treat dwarfism.

In recent years, researchers have developed a genetically modified bacterium that could help reduce tooth decay. Cavities are produced by a strain of bacteria, *Streptococcus mutans*. These bacteria feed on sugars in our mouths during and after meals and snacks. These bacteria also secrete a sticky substance known as plaque. Without proper brushing, bacteria become entrapped in the plaque where they secrete lactic acid. It digests tooth enamel, producing cavities. To prevent cavities, scientists have modified the bacterium so that it no longer secretes lactic acid. Researchers hope that the bacteria could be used to replace lactic acid-secreting bacteria in a person's mouth, thus reducing the incidence of cavities.

Transgenic microorganism can also be used to protect plants from harmful insects. Consider efforts to combat root-eating insects in corn. To combat these pesky insects, researchers have successfully transplanted a bacterial gene that produces an insect-killing toxin from one species into another species of bacteria, one that colonizes the roots of certain crop plants. Insects that dine on the roots of these newly inoculated plants subsequently ingest the lethal bacteria and die.

This novel technique could benefit farmers, allowing them to use fewer toxic chemicals to control root-eating insects, thus lowering production costs. It could also benefit the environment, because it reduces the amount of potentially dangerous chemicals that released into soils. These chemicals can poison soil microbes or leach into groundwater or be washed into nearby lakes and streams all of which serve as a source of drinking water for people.

Genetic researchers have also developed a bacterium that retards the formation of frost on plants. The genetic variant is found naturally in the environment but not in sufficient quantities to protect crops. Thus, researchers have cloned the

bacterium and have begun testing its efficiency in the field. Should it prove successful and safe, the bacterium could be sprayed on crops early and late in the growing season when frosts can be devastating, potentially saving farmers millions of dollars a year. Some critics worry that this bacterium could escape into the environment and displace normally occurring bacteria. Who knows what detrimental effects this might have on the function of natural systems?

> **KEY CONCEPTS**
>
> Transgenic bacteria are used to produce proteins such as hormones and blood-clotting factors that can be used to prevent disease and could be used to protect crops from insects and frost.

Gene Therapy

As noted earlier, researchers are also experimenting with ways in which genetic engineering can be used to combat disease directly—by altering the genome. In this process, known as **gene therapy**, genes are used to replace, supplement, or alter genes of an individual's cells to cure disease. The most common use of gene therapy involves genes that replace mutated genes. Other forms of gene therapy involve attempts to repair genetic mutation

The promise of gene therapy is enormous. It could actually cure genetic disorders.

Surprisingly, genetic defects are quite common in newborns. In fact, approximately 1 of every 100 children born in the United States suffers from a serious genetic defect, such as sickle-cell disease, cystic fibrosis, or hemophilia. If successful, gene therapy could be used to replace defective genes with normal genes, allowing thousands of people to live normal lives and saving millions of dollars a year.

One of the biggest challenges in gene therapy is how to place genes into the DNA of body cells in which they are missing. For instance, to cure a genetic defect in brain cells, the genes must be delivered to the cells of the brain. Several mechanisms have been developed to achieve this goal. One of the most successful ways of introducing genes into cells is the use of a **vector**, a gene-carrying agent. The adenovirus is one such vector (Figure 19-13). This virus causes upper respiratory system infections because it attacks and invades the cells lining the respiratory system. Scientists have successfully used the virus as a vector to deliver genes to correct cystic fibrosis.

Another potentially promising technique involves the use of microspheres—tiny lipid spheres. Called *liposomes*, these vectors can be packed with genes and coated with antibodies for specific target cells. If successful, this will allow the liposomes to deliver their contents to the correct cells.

Geneticists have discovered other nonvector means of delivering genes to host cells. One promising means of delivering genes to sites where they're required is the bone marrow transplant. Consider an example. Krabbe's disease is a rare genetic disorder resulting from a deficiency in one enzyme in human brain cells. The absence of this enzyme allows fat to accumulate in the nervous system, causing nerve cells to degenerate. Seizures and visual problems occur early in life, and most victims die within the first 2 years of life.

To test a means of correcting this disease, scientists used a strain of mice that suffers from a similar condition.

Point/Counterpoint Controversy over Herbicide Resistance Through Genetic Engineering

The Benefits of Genetically Engineered Herbicide Resistance *by Charles J. Arntzen*

One of the principal outcomes of plant genetic engineering has been a rapid expansion in our understanding of plant cell biology, genetics, and molecular controls over the structure and function of plants. This understanding is now helping to solve practical problems such as developing insect- and disease-resistant crops, thereby decreasing needs for chemical pesticides. Other approaches include creating herbicide-resistant plants, which will allow farmers more choices in weed control, and to switch to newer, rapidly biodegradable herbicides.

Dr. Charles J. Arntzen is an endowed chair at Arizona State University and CEO of Boyer Thompson Institute. He holds Director positions with several major universities and corporations. Dr. Arntzen was elected to the U.S. National Academy of Sciences in 1983 as a result of pioneering research in photosynthesis and plant molecular biology. He is a fellow of The American Association for the Advancement of Science, received the Award for Superior Service from the U.S. Department of Agriculture, served as chairman of the National Biotechnology Policy Board of the National Institutes of Health, and served for eight years on the Editorial Board of *Science*. This article is adapted by the author from a 1991 publication entitled, *The Genetic Revolution, Scientific Prospects of Public Perceptions* with permission of the publisher, Johns Hopkins University Press. (Photo © Charles J. Arntzen.)

Herbicides are chemicals sprayed on crops to control weeds. Their effect is selective—that is, they kill weeds without significantly harming crops. All herbicides currently available to farmers began with an evaluation of the sensitivity of weeds and crops to experimental chemicals. This evaluation is accomplished by applying potentially useful chemicals to test samples of weed and crop seedlings. Compounds that kill weeds but not crops are then subjected to animal toxicology studies. Those deemed acceptable are developed for farmers' use.

The success of herbicides is based on the fact that crop species contain enzymes that convert chemical compounds, including herbicides, to inactive forms. These enzymes tend to be crop-specific. For instance, the enzymes that render soybeans immune to 30 or more commercial herbicides are different from the enzymes that make corn resistant to other herbicides.

This can be a benefit when corn or soybeans are grown in rotation. In such instances, both the unwanted "volunteer" corn and weeds that germinate in a soybean field can be controlled by the soybean herbicide.

Interestingly, certain weeds have the same enzymes as the crop species they invade. This is often true when the weeds and crops are somewhat related, such as with wild oats in a field of wheat or barley. Thus, the oats are resistant to the same herbicides as the wheat or barley. Consequently, the farmer currently has few or no choices in chemical weed control because herbicides that would kill the weeds would also kill the crop.

Through genetic engineering, researchers have devised ways to make crops resistant to new classes of herbicides. That is, they have found ways to give crops such as soybeans additional enzymes to augment their natural herbicide resistance. By transferring genes that code for enzymes conferring herbicide resistance, scientists can give farmers a broader set of options when selecting herbicides. If successful, these options will solve problems where there is currently no weed control, or where the herbicides available are prohibitively expensive or environmentally damaging.

This line of research may also give the farmer a simpler means to control weeds through the use of a single, more effective herbicide selected on the basis of several traits, including herbicide resistance, reduced cost, and, ideally, greater safety in the environment.

This explanation of the benefits of genetic research aimed at making crops more resistant to herbicides may not satisfy the reader who is asking, "Why don't they simply stop using all herbicides?" The main answer is economics. Weed control is essential for all crops. Left unchecked, weed populations can reduce crop yields to near zero. Prior to the 1950s, weeds were controlled strictly by mechanical means—that is, by cultivation or hand pulling. This was a labor-intensive and costly process. By the 1980s, more than 95% of all major row crops in the United States were produced using herbicides. Farmers based this decision (to apply herbicides rather than using a cultivator or a hoe) on costs. An application of herbicides at the time of planting, costing as little as $10 per acre, is many times cheaper than mechanical cultivation. Weed control through herbicides is a factor in the low food prices we enjoy in the United States. It also saves on energy use and, when properly used, does not harm the environment.

With the continuing development of herbicide-resistant crops as part of an integrated weed-control strategy, crop production costs can be held at low levels, while new methods for weed control are being developed.

The Perils of Genetically Engineered Herbicide Resistance *by Margaret Mellon*

Environmentalists have been active in the debate about genetic engineering since the beginning. The current focus of environmental interest in genetic engineering is based on the prospect of commercial production of a broad range of genetically engineered organisms—for example, bacteria and viruses. These organisms could have direct and adverse impacts on our environment and our health.

Public concern over genetic engineering is also focused on indirect environmental consequences, the best example of which is herbicide resistant plants. The development of major crops that are tolerant of chemical herbicides seems likely to lead to increased or prolonged use of these dangerous agricultural chemicals.

Herbicide-resistant crops, like most genetically engineered agricultural products, are being developed by large, transnational chemical companies. Worldwide, at least 28 enterprises have launched more than 65 research programs to develop herbicide-resistant crops. All of the major crop plants are involved, including cotton, corn, soybeans, wheat, and potatoes. Engineered crops being made resistant to a company's own herbicides will surely increase that company's market share in chemicals. Thus, industry analysts expect herbicide-resistant plants to be big business, mainly for pesticide and herbicide producers.

Herbicide-resistant products run directly counter to the promise of a reduced dependence on agricultural chemicals put forward by biotechnology advocates. Rather than weaning agriculture from chemicals, these crops will be shackled to herbicide use for the foreseeable future. Herbicides represent an estimated 65% of the chemical pesticides used in agriculture. By continuing to promote the use of herbicides, the biotechnology industry greatly restricts its potential for reducing overall chemical use.

While herbicide-resistant crops offer no hope of environmental benefits in agriculture, there are alternatives to these products that do promise substantial reduction in overall pesticide use. These alternatives fall under the rubric of sustainable agriculture.

Sustainable agriculture employs a variety of agricultural practices such as crop rotation, intercropping, and ridge tillage (that produce long ridges of soil on which plants are grown). These practices control pests without the application of synthetic pesticides and fertilizers. Used by knowledgeable farm managers, these techniques work to make farms more profitable and environmentally sound. Growing different crops in successive growing seasons, for example, dramatically reduces pests by sequentially removing the hosts on which the pests depend. With fewer weeds or insects to contend with, the farmers reduce the need for costly chemical pesticides. Once thought impractical, sustainable agriculture is rapidly gaining support in national policy forums, including the National Academy of Sciences.

Dr. Margaret Mellon is a senior scientist with the Food & Environment Program at the Union of Concerned Scientists (UCS). UCS scientists join with the public to use rigorous, independent science to create innovative, practical solutions for a healthy, safe, and sustainable future. Dr. Mellon is one of the nation's most respected experts on biotechnology and food safety. She lectures widely on biotechnology issues and frequently appears on television and radio talk shows. (Photo © Margaret Mellon.)

The environmental advantage of such an approach is clear. Crop rotations that reduce pests reduce the need for pesticides now and into the future. Thus, developing sustainable practices should be the highest priority for agriculture.

Although a minor part of the sustainable agriculture picture up to now, biotechnology could help by developing new crop varieties for use in sustainable systems—for example, faster germinating, cold-tolerant crop varieties could enhance low-input sustainable systems by enabling more effective weed control. Engineered products such as these would fulfill the promise of biotechnology, and would benefit farmers, the environment, and the rural economy alike.

Sharpening Your Critical Thinking Skills

1. State the main thesis of each author in your own words.

2. List the data or arguments used to support each hypothesis and the data used by the authors to refute the other's point of view, if any.

3. Can you determine whether there are any flaws in the reasoning in either essay?

4. Which viewpoint do you agree with? Why?

Visit Human Biology's Internet site for links to websites offering more information about this topic.

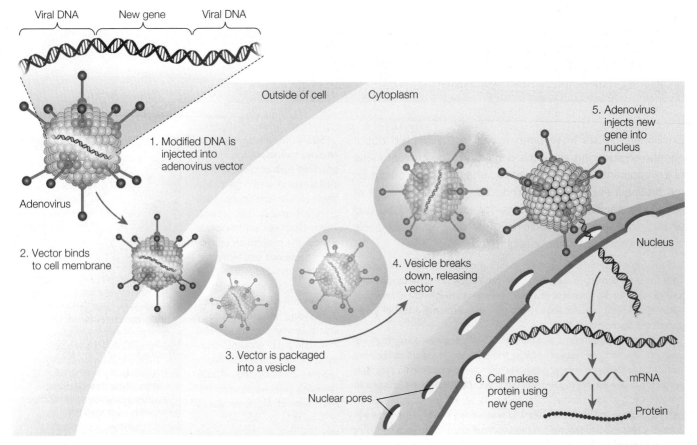

FIGURE 19-13 **Adenovirus as a Vector** In this drawing, genetically engineered adenoviruses are used to deliver genes to genetically defective cells. (Reproduced from the Genetics Home Reference/NLM.)

They injected them with bone marrow cells containing the gene that codes for the missing enzyme. Scientists found that the transplanted cells became established in the lungs and liver, where they restored enzymatic activity. Some even found their way into the brain!

Researchers have also developed a novel transplantation technique that could enable surgeons to introduce genetically engineered cells into specific organs in the human body. In laboratory studies, scientists have injected genetically altered liver cells into a foamlike material. The foam was impregnated with a hormone that stimulates the growth of blood vessels from nearby larger vessels. The foam was then transplanted into rats. Within a week after implantation, the artificial tissue was riddled with a network of blood vessels, allowing the cells to live on. This technique could someday be used to introduce genetically normal cells into numerous tissues and organs of the body, providing a cure for disorders such as diabetes mellitus and Parkinson's disease.

> **KEY CONCEPTS**
> Gene therapy provides hope of curing genetic diseases. Genes are injected into genetically defective cells of the body by one of several means, including vectors.

Mapping the Human Genome

Geneticists have long been interested in mapping human chromosomes—that is, in determining which genes are on which chromosomes. For years, the mapping process relied on studies of patterns of linked traits in families, a rather crude technique.

One useful tool in gene mapping is the **gene probe**, small fragments of single-stranded DNA. Produced by machines called *DNA synthesizers*, gene probes allow scientists to determine the DNA from blood cells left at a crime scene. Probes bind to complementary base pairs of the DNA sample under study. Because scientists know the composition of the **DNA probes**, they can determine the sequence of bases in the DNA under study. This allows technicians to identify DNA in samples of blood, hair, or semen to identify perpetrators of crime. DNA probes can also be used to identify infectious organisms, even anthrax and other potentially lethal biological agents.

Another technique is also used. In this procedure, scientists fuse a human cell with a mouse cell (Figure 19-14). The new hybrid cell containing 96 chromosomes divides, but when the cell divides, it begins to lose human chromosomes. The daughter cells divide again and lose more. Eventually, cells are produced with all mouse chromosomes and one or two human chromosomes. By studying the resulting protein products produced by those cells, geneticists can tell which protein-coding genes belong to each chromosome.

> **KEY CONCEPTS**
> DNA probes and fusion can be used to determine which genes belong on chromosomes.

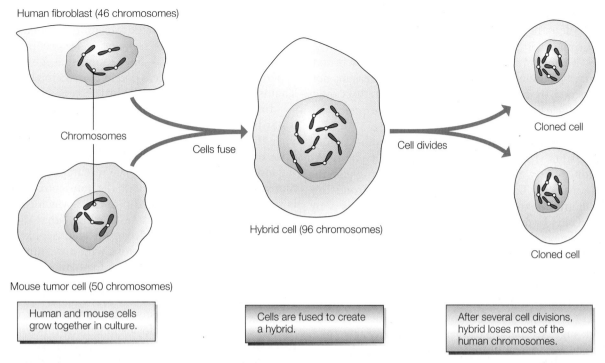

Human fibroblast (46 chromosomes)

Chromosomes

Cells fuse

Mouse tumor cell (50 chromosomes)

Hybrid cell (96 chromosomes)

Cell divides

Cloned cell

Cloned cell

| Human and mouse cells grow together in culture. | Cells are fused to create a hybrid. | After several cell divisions, hybrid loses most of the human chromosomes. |

FIGURE 19-14 Mapping by Fusion This technique permits scientists to determine the chromosomal address of certain genes. Human and mouse cells are fused, creating a new cell with excess chromosomes. This cell divides and, in so doing, loses most of the excess chromosomes. Cells with one or two human chromosomes can be grown in culture. By determining the unique human proteins being synthesized in these cells and the chromosomes present, geneticists can determine the chromosomal "address" of certain genes. More precise techniques are needed to determine the exact location on the chromosome.

19-5   Controversies over Genetic Engineering—Ethics and Safety Concerns

In October 1997, several scientists issued a worldwide alert asking all governments to ban imports of a genetically engineered soybean from the United States. The soybeans were genetically altered to be resistant to the herbicide Roundup. The scientists were concerned that the use of this herbicide would increase the level of plant estrogens in soybeans, thus increasing the levels of these potentially harmful chemicals in humans.

In the year 2000, taco shells, tortillas, and numerous snack foods produced from an unapproved genetically engineered corn were recalled from U.S. markets as a result of health concerns. Approximately 300 types of taco shells, tortillas, and snack chips made by the Texas-based Mission Foods were affected by the recall, with an estimated value of about $10 million. This genetically engineered corn (and the resulting products made from it) had not been approved because of questions about its ability to cause allergic reactions. Interestingly, "normal" corn crops were also affected because pollen from the genetically engineered crops was dispersed by the wind into their fields, thus altering the neighboring "normal" corn crops.

At this time, many safety questions remain unanswered. As noted earlier, perhaps the greatest safety concern is the possibility of unleashing genetically altered bacteria or viruses. At least two studies now indicate that genetically engineered bacteria that are applied to seeds take up residence on the roots but

migrate very little from the site of application in the short term. Critics, however, are concerned with long-term consequences. Some individuals fear that a genetically altered strain could spread through the environment, wreaking havoc on ecosystems and, possibly, human populations. Once unleashed, it would be impossible to retrieve. The genetically altered bacterium that retards frost formation, for example, could enter the atmosphere on dust particles, reducing cloud formation and altering global climate. No one knows for sure how serious this threat really is. The safety of genetically engineered food is discussed in the **Point/Counterpoint** in this chapter.

Other critics object to genetic tinkering, especially the transfer of genes from one species to another, on ethical grounds. Do humans have the right, they ask, to interfere with the course of evolution?

Proponents of genetic tinkering argue that livestock and plant breeders have been selectively breeding hardy animals to produce genetically superior livestock and crops for hundreds, if not thousands, of years. In so doing, people have been performing a kind of genetic engineering—albeit a slow one—for millennia. Genetic engineering, say proponents, merely offers a quicker way of achieving the same goals.

Genetic engineering and selective breeding do indeed achieve the same end points in some instances. Selective breeding simply alters gene frequencies—that is, the percentage of organisms in a population carrying a specific allele. Genetic engineering does the same, as in the case of the frost-retarding bacteria. In other words, geneticists are simply producing large quantities of a mutant already found in nature. But that in and of itself may be reason for concern. In nature, for reasons not well understood, the mutant is found in a small quantity. Could shifting the allele frequency artificially cause some ecological catastrophe?

In other instances, the two techniques produce quite different results. That is, in some cases, genetic engineering introduces new genes into species, producing genetic combinations never before encountered in life. Introducing the human growth hormone gene into cattle embryos is an example of this form of tinkering. This manipulation violates a fundamental law of nature: Different species do not interbreed. Is it ethical to blur nature's naturally maintained boundaries? Who can say? And can we put a lid on our ambition? Will we know when to stop?

Unfortunately, experience with genetic engineering is too limited to answer the safety concerns of critics. Preliminary work suggests that the dangers have been exaggerated and that genetically engineered bacteria are not a threat to ecosystem stability. Still, further research is needed to be certain. Ethical questions, like political questions that have been plaguing some countries for centuries, may never be answered to the satisfaction of everyone.

Another point that is rarely mentioned in the debate is the potential for wrong-doing—for creating superinfectious organisms that could wipe out huge numbers of people. Imagine for instance, the catastrophe that might arise if some mad scientist introduced the deadly AIDS virus genes into the genes of the flu virus, which is rapidly transmitted around the world.

Scientists who have been exploring designer pathogens—highly infectious agents—have found that most potential superbugs turn out to be less potent than their natural forms. That is, virulence was typically lost.

In 1998 and 1999, however, Australian researchers made a startling discovery while working to find a way to render mice infertile, in an effort to protect world grain supplies. Mice and rats ravage grain supplies in many countries, reducing food available for people. The researchers had spliced a mouse immune system gene into the mousepox virus, which is similar to the human smallpox virus. Much to their surprise, the mice did not become infertile—they died—even mice that had been immunized against mousepox virus. The researchers published their work with a warning that this technique could potentially be used to produce deadlier forms of human pathogens for biological warfare.

Whatever the outcome of the debate, genetic engineering is here to stay. Future disasters, should they occur, may compel us to prohibit certain forms of genetic tinkering, but in cases where genetic engineering reduces suffering, advances agriculture, and facilitates environmental cleanup, public support may be hard to contain. For more on the subject see the two point/counterpoints in this chapter.

> **KEY CONCEPTS**
>
> Genetic engineering offers many benefits but poses risks as well. Additional research is needed to determine whether the risk are significant and whether they outweigh the benefits.

SUMMARY

Genetic Manipulation: An Overview

1. Genetic engineering, or recombinant DNA technology, is a procedure by which geneticists remove segments of DNA from one organism and insert them into the DNA of another.

Cloning

2. Cloning is the production of exact replicas of an organism inserting the DNA from one (existing or extinct) organism into an enucleated cell of another. The term cloning can also be used to describe the replication of genes via one of several methods.

Recombinant DNA Technology

3. Geneticists remove and transplant segments of DNA from an organism using a naturally occurring enzyme called restriction endonuclease.
4. After a segment of DNA is transplanted into another cell, a second enzyme, DNA ligase, is used to seal it in place, forming a recombinant DNA molecule—a molecule that contains DNA from two different organisms.
5. Bacteria contain small, circular strands of DNA, called plasmids, which are separate

from the chromosome. Foreign genes can be spliced into plasmids, and the plasmid carrying a foreign gene can be reinserted into bacteria or yeast cells in culture. As these cells divide, the plasmids replicate and produce multiple copies of the gene, which is then said to be cloned.

6. Cloning produces numerous identical copies of the foreign gene, which can be used for genetic studies or to produce useful products, such as hormones.
7. Geneticists can produce multiple copies of DNA by extracting messenger RNA from cells and using it to produce DNA using the enzyme reverse transcriptase.
8. The most widely used method of gene amplification is the polymerase chain reaction. In this technique, DNA is alternately heated to split the double helix, then cooled. DNA polymerase enzymes are added to catalyze the formation of complementary DNA molecules.

Applications of Recombinant DNA Technology

9. Recombinant DNA technology and other related techniques have resulted in a number of practical and sometimes controversial applications.

10. Hormones and other proteins are mass produced in genetically engineered microorganisms and used to treat a variety of disorders.
11. Recombinant DNA techniques are also to produce vaccines.
12. Gene splicing can be used to transfer genes from one species to another, creating transgenic organisms. These organisms rarely differ much from their original form; they may, in fact, only display one new trait, such as a larger body size.
13. Transgenic plants are easier to produce than transgenic animals, and today, efforts are underway to create plants that produce more food as well as more nutritious fruit and seeds, generate soil nitrogen, live in marginal soils, combat insect pests, are more resistant to herbicides and pests.
14. Numerous transgenic microorganisms are used today to produce medically important hormones.
15. Recombinant DNA technology may also be used to cure genetic disease, a treatment called gene therapy. In this procedure, scientists hope to be able to insert normal human genes into genetically defective body cells.

16. The location of genes on chromosomes can be determined by several means. One of them is the DNA probe.
17. DNA probes are tiny segments of genes that bind to complementary base pairs of sample DNA—for example, hair, blood, or semen taken from crime scenes.
18. DNA probes can also be used to detect infectious organisms and genetic disorders.
19. Fusing human cells with mouse cells also allows scientists to determine the location of genes on specific chromosomes.

Controversies Over Genetic Engineering—Ethics and Safety Concerns

20. Although genetic engineering has spawned a great deal of enthusiasm, it does have its critics who are concerned with safety and ethical issues.
21. Perhaps the greatest safety concern is the possibility of intentionally or unintentionally releasing potentially dangerous genetically altered bacteria or viruses into the environment.
22. Some critics object to genetic tinkering, especially the transfer of genes from one

species to another, on ethical grounds, arguing that we do not have the right to interfere with the course of evolution.
23. Unfortunately, experience with genetic engineering is too limited to answer the safety concerns of critics. Preliminary work suggests that the dangers have been exaggerated and that genetically engineered bacteria are not a threat to ecosystem stability. Further research is needed to be certain.

THINKING CRITICALLY ANALYSIS

This analysis corresponds to the Thinking Critically scenario that was presented at the beginning of this chapter.

Before you accept or reject an argument, such as the one made by the chef, it's important to gather all the facts. Reading this chapter will give you some perspective on genetic engineering. I'd also suggest reading some specific articles on genetically engineered tomatoes.

But you'll also learn that geneticists are transplanting bacterial genes into tomato plants. One gene in particular retards the production of ethylene gas, which normally stimulates ripening. When the bacterial gene is present, it slows ripening, allowing farmers to leave the tomato on the vine several extra days. This increases the tomato's flavor, with obvious benefits to anyone who has had the opportunity to compare a store-bought tomato with one grown in a garden.

The questions here might be: Will this harm consumers? Can you think of any other concerns and ways to address them? Are the chefs merely reacting to the phrase "genetic engineering"? Could this be one of those thought-stopping terms?

KEY TERMS AND CONCEPTS

Cloning, p. 413
DNA ligase, p. 414
DNA probes, p. 424
Gene amplification, p. 415
Gene cloning, p. 415
Gene probe, p. 424
Gene therapy, p. 421

Genetically modified organism (GMO), p. 414
Genetic engineering, p. 412
Genome, p. 415
Luciferase, p. 420
Plasmid, p. 414
Polymerase chain reaction, p. 415
Recombinant DNA technology, p. 414

Restriction endonuclease, p. 414
Reverse transcriptase, p. 415
Reverse transcription, p. 415
Selective breeding, p. 412
Transgenic organism, p. 418
Vector, p. 421

CONCEPT REVIEW

1. What is genetic engineering? p. 412.
2. In what ways have humans been involved in genetic engineering over the years? pp. 412–413.
3. Describe the methods of recombinant DNA technology. Your answer should explain how genes are removed from DNA of the donor, spliced into the recipient, and then cloned. pp. 414–416.
4. What is gene amplification? Describe three methods of gene amplification. p. 415.
5. What is a transgenic organism? Give an example of some transgenic animals? What advantages do they have? pp. 418–419.

6. Give a few examples of ways transgenic plants could help increase agricultural production? p. 420.
7. Describe several ways transgenic microorganisms are used to improve medicine. p. 421.
8. What is the main advantage of producing human growth hormone or insulin from genetically engineered bacteria over conventional means of production? p. 421.
9. Describe ways in which transgenic microorganisms are used to improve agriculture. pp. 420–421.
10. List the major concerns for genetic engineering and describe each one. pp. 425–426.

11. Debate the following statement: Genetic engineering is morally wrong and should be discontinued. It violates the laws of nature and gives humans a power beyond their control. pp. 425–426.
12. What is gene therapy? p. 421.
13. Describe ways in which genes are inserted into host cells during gene therapy. p. 421.
14. What is a DNA probe? What are they used for? p. 424.
15. Fusing human cells with mouse cells is one technique that helped geneticists determine which chromosomes occurred on each chromosome. How does this technique work? p. 424.

SELF-QUIZ: TESTING YOUR KNOWLEDGE

1. Genetic engineering is also known as _____ DNA technology. p. 412.

2. Humans have been engineering genomes of crops and livestock for centuries using technique known as _____ breeding. p. 412.

3. _____ is the process of producing an exact replica of an organism. p. 413.

4. The enzyme used to slice off a piece of DNA belongs to a group of enzymes known as restriction _____. p. 414.

5. The enzyme DNA _____ is used to insert segments of DNA into an opening in the structure. p. 414.

6. _____ are small circular strands of DNA in bacteria that can be used to insert genes into bacteria for mass production. p. 414.

7. _____ is the name of the process in which RNA is synthesized on a strand of DNA. p. 415.

8. Genes can be cloned by synthesizing DNA from mRNA. This process is possible thanks to an enzyme known as reverse _____. p. 415.

9. The _____ chain reaction is yet another way to mass produce DNA for genetic studies and other purposes. p. 415.

10. Mass producing genes using plasmids and other techniques is known as gene _____. p. 415.

11. The human genome contains _____ to _____ genes. p. 424.

12. An organism containing genes transplanted from another organism is known as a _____ organism. p. 418.

13. Inserting genes into the cells of diseased individuals to cure diseases is known as _____. p. 421.

14. Approximately _____ of every 100 children born in the United States suffers from some kind of genetic defect. p. 421.

15. A DNA _____ is used to identify DNA from evidence taken from the scene of various crimes to identify potential perpetrators. p. 424.

biology.jbpub.com/chiras/8e/

The site features eLearning, an online review area that provides quizzes, chapter outlines, and other tools to help you study for your class. You can also follow useful links for in-depth information, research the differing views in the Point/Counterpoints, or keep up on the latest health news.

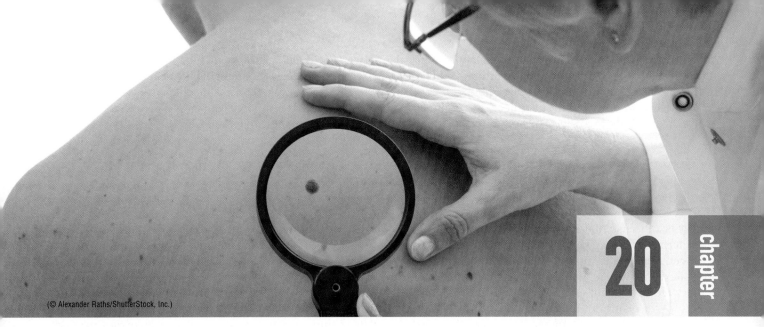

(© Alexander Raths/ShutterStock, Inc.)

Cancer

In December 2009, the National Institutes of Health announced a decline in the incidence of some of the major cancers. From 1999 to 2006, they reported, the incidence rates of all cancers dropped 0.7%. Death rates from cancer fell 1.6% from 2001 to 2006. Nonetheless, cancer kills approximately 560,000 Americans each year and is a disease that most of us will either experience ourselves or through our relatives and friends. In fact, according to national statistics, one of every three Americans will have cancer during his or her lifetime, and one of every four will die from it. Current estimates suggest that about one-third of these deaths result from tobacco use (smoking and chewing tobacco).

THINKING CRITICALLY

Researchers at a local medical school have just completed a two-year study on mice and rats that indicates that a common cleaning agent to which many of us are exposed in our daily lives causes cancer in these laboratory animals. They suggest that the chemical should be carefully scrutinized and perhaps banned from public use to protect human health. What research would you suggest needs to be done before measures are taken to regulate this chemical's sale? Why?

Another third are thought to be related to poor nutrition, lack of physical activity, being overweight, or obesity. The rest arise from environmental and workplace pollutants.

Although we often think of cancer as a human disease, cancer occurs in a wide range of animals and plants (Figure 20-1). In fact, cancer is turning up in many fish, which scientists believe is the result of chemical pollutants in fresh and salt water ecosystems. This chapter looks at cancer in humans. It will familiarize you with terms and concepts that you will encounter throughout your lifetime.

20-1 Benign and Malignant Tumors

Cancer is a disease that afflicts many different body cells characterized by unregulated cell division. Put another way, in cancer, cells divide and grow uncontrollably, often forming cancerous tumors also known as malignant tumors. Tumors require lots of nutrients to support rampant cell growth. As a result, an extensive network of blood vessels forms on malignant tumors. The network of vessels reminded the ancient Greek Physician, Hippocrates, of a crab, so he coined the term cancer around 400 BC (he called these cancerous cells *karkinos*—Greek for crab).

Malignant tumors are dangerous because they do not stop growing. They sap the body of strength and eventually weaken and kill its victims. Malignant tumors can also invade neighboring areas of the body, impairing their function. They can also spread to distant tissues and organs via the lymphatic and circulatory systems. Cancer cells settle in new areas where they multiply, forming new malignant tumors, described shortly.

Bear in mind that not all abnormal cellular proliferation is cancerous. Some cells, in fact, form small masses that reach a certain size, and then stop growing. These are called **benign tumors**. Good examples are moles and growths in the uterus called fibroids. As a rule, benign tumors are just that, mild. They generally pose no significant medical problems. However, that's not always the case. Benign tumors can put pressure on nerves, causing pain. Some may block blood vessels, reducing blood flow to vital organs. Benign tumors may also compress organs like the intestines, blocking the flow of digested food materials through the gastrointestinal (GI) tract. Some benign tumors of endocrine glands result in an overproduction of hormones.

Of great concern, however, are the malignant tumors, which as just noted never stop growing and frequently spread to other parts of the body. The term cancer is reserved for malignant tumors, of which there are over 200 different types found in humans, such as brain cancer, lung cancer, and liver cancer. (Table 20-1 lists other types of cancer.) Within these general categories are subtypes; for example, there are several different types of skin cancer.

Almost every type of cell in the body can become malignant. Notable exceptions include neurons and muscle cells,

FIGURE 20-1 Cancer in Plants? Cancer occurs in many different plant and animal species. This tree suffers from multiple tumors. (Courtesy of Lian Bruno.)

TABLE 20-1	Estimated Numbers of New Cases and Deaths from Common Types of Cancers	
Cancer Type	Estimated New Cases	Estimated Deaths
Bladder	72,570	15,210
Breast (female—male)	232,340—2,240	39,620—410
Colon and rectal (combined)	142,820	50,830
Endometrial	49,560	8,190
Kidney (renal cell) cancer	59,938	12,586
Leukemia (all types)	48,610	23,720
Lung (including bronchus)	228,190	159,480
Melanoma	76,690	9,480
Non-Hodgkin lymphoma	69,740	19,020
Pancreatic	45,220	38,460
Prostate	238,590	29,720
Thyroid	60,220	1,850

Data from American Cancer Society: *Cancer Facts and Figures 2013.* Atlanta: American Cancer Society, 2013; and Chow W.-H., Dong L. M., and Devesa S. S.: Epidemiology and risk factors for kidney cancer. *Nature Reviews Urology* 7(5):245–257, 2010.

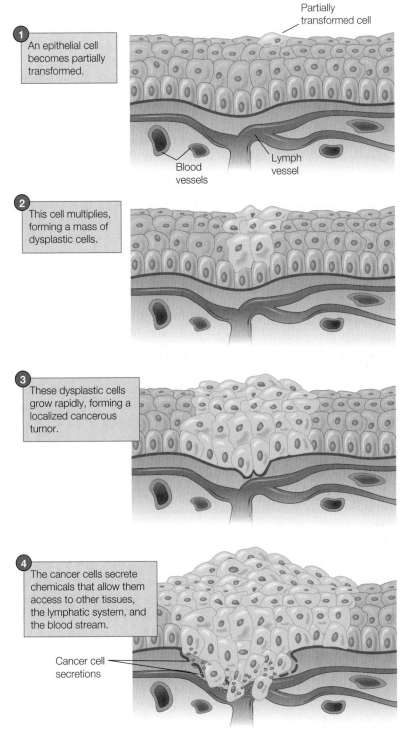

1 An epithelial cell becomes partially transformed.

Partially transformed cell

Blood vessels

Lymph vessel

2 This cell multiplies, forming a mass of dysplastic cells.

3 These dysplastic cells grow rapidly, forming a localized cancerous tumor.

4 The cancer cells secrete chemicals that allow them access to other tissues, the lymphatic system, and the blood stream.

Cancer cell secretions

FIGURE 20–2 Cancer Growth and Metastasis Cancers grow by cell division. Cells can break free from the tumor and spread in the blood and lymphatic systems to other parts of the body where they establish secondary tumors. Secondary tumors often develop in the liver, lungs, and lymph nodes.

two types of cells that cannot divide. What is more, each type of cancer has its own unique properties.

> **KEY CONCEPTS**
>
> Two types of tumors can form in humans, benign and malignant. As a rule, malignant tumors are the most serious because they can spill over into neighboring tissues and spread to other parts of the body.

Metastasis

Malignant tumors spread when individual cells or clusters of cells break loose from the original tumor, known as the **primary tumor** (Figure 20-2). They then can travel through the body in the circulatory system, or blood vessels, and in the lymphatic system, a network of vessels that drains excess fluid from body tissues. These cells may settle in other sites. In their new residence, the cancer cells can divide uncontrollably to form **secondary tumors**. The spread of cells from one region of the body to another is called **metastasis** (ma-TASS-tah-SISS).

The ability of cancer cells to metastasize is one of the main reasons cancers are often so difficult to treat and so lethal. Acquiring the ability to metastasize is, therefore, a key event in the development of most cancers. What allows cancer cells to break away from the primary tumor?

Most cells in the body stay in place. They're joined in tissues and organs by molecules on their surfaces, known as **cell-cell adhesion molecules**. In cancer cells, these molecules are either lacking or altered. Cancer cells therefore lose their anchorage. (Interestingly, researchers found that experimentally restoring these molecules in cancer cells grown in culture dishes can render cancer cells ineffective [unable to spread] when they are transplanted into mice.)

Not all cells that break away from the primary tumor survive to form a secondary tumor, however. In fact, researchers estimate that only 1 in 10,000 cancer cells that enter the circulatory system is successful in establishing a secondary tumor.

The most common site of metastasis is the lungs. Why? After leaving the organ in which the primary tumor is located, the first vascular bed most cancer cells encounter is in the lungs. That's because cancer cells that travel in the bloodstream enter the venous system, and then travel to the heart. From the heart, they travel to the lungs. The lungs have extensive networks of capillaries that are involved in the exchange of oxygen and carbon dioxide. Cancer cells quickly attach

to the lining of the small blood vessels, where they set up residence. The liver is the second most common site for secondary tumors. However, some types of cancer cells show a preference for certain organs other than the lung and liver. The prostate is an organ belonging to the male reproductive system. It produces secretions that compose the bulk of the ejaculate. The prostate is also a common site of cancer in men. Prostate cancer cells tend to settle in the bones. It is believed that molecules on the prostate cancer cells bind specifically to molecules on the cells lining the blood vessels in bone. Cancer cells traveling in the lymphatic system tend to spread to the lymph nodes, small organs located along the lymphatic system. They filter bacteria and viruses from the lymph, helping to fight infections.

Metastasis can be identified by x-rays and computer assisted tomography (CAT) scans. Doctors can also detect metastasis by feel. Metastases to the lymph nodes, for instance, can sometimes be palpated (felt) or visualized, if they are large enough. Metastases to the liver and spleen result in abdominal pain and palpable enlargement of these organs. Metastases to the bone can result in bone fractures and pain.

> **KEY CONCEPTS**
>
> Cancer cells originating from a primary tumor can spread through the bloodstream and lymphatic vessels. Cancer cells can establish a new tumor in a new site, called a secondary tumor. This process is known as metastasis.

20-2   How Cancers Form

A growing body of research suggests that the production of a malignant tumor (cancer) involves two steps that may take place over a long period: (1) conversion or initiation, and (2) development and progression (Figure 20-3).

The conversion of a normal cell to a cancerous cell, the initiation of a cancer, typically begins with a mutation, a change in the DNA of cells. Mutations are caused by chemical, physical, or biological agents (viruses, for example). Good examples are the many highly reactive free radicals found in the bloodstream. Free radicals are atoms or groups of atoms with an unpaired electron in their outer shells. They can be formed when oxygen interacts with certain molecules. Many

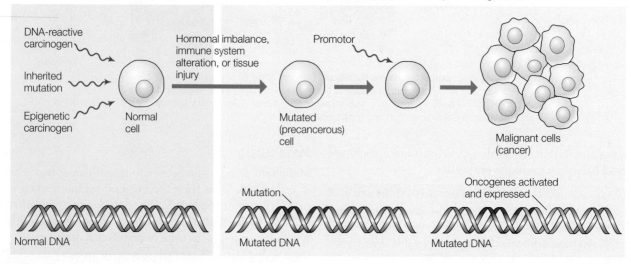

FIGURE 20-3 Stages Leading to Cancer

ous vegetables, including cabbage, broccoli, cauliflower, chard, kale, and others contain isothiocyanates (ITCs). To date, over 150 ITCs have been discovered. Some of them inhibit anti-angiogenesis. Mushrooms and garlic contain similar agents. Onions and garlic contain sulfur-containing compounds that have the same effect.

> **KEY CONCEPTS**
> Shrinking the blood supply to cancers through drugs and phyto-chemicals present in certain fruits and vegetables, can cause cancers to shrink and die.

Attacking Cancer at the Level of the Gene

Research is also underway to treat tumors by attacking genes, including oncogenes, tumor suppressor genes, and others. Oncogenes are cancer-causing genes found in viruses. The term is also used to describe a mutated version of human genes that regulate cell growth, that is, proto-oncogenes (both are discussed earlier in this chapter).

Tumor suppressor genes are naturally occurring genes (proto-oncogenes) that normally keep cells from becoming malignant. Drugs that block mutated forms of these genes, alter the methylation of these genes, or even altering their products, researchers hope to find ways to target cancer cells specifically. Their hope is that such treatments at the genetic and epigenetic level will help them stop cancer dead in its tracks, while reducing toxic side effects of other forms of treatment, notably chemotherapy.

Some researchers are even studying ways to correct the genetic defects in cancer cells that lead to uncontrolled replication. Take the tumor suppressor gene, for example. Tumor suppressor genes code for the production of a suppressor protein that, in normal cells, serves as a brake to DNA replication. Mutations of the gene can render the protein inactive. Cells that are normally held in check are unleashed to proliferate madly. By inserting normal tumor suppressor genes in tumor cells, researchers hope to return them to normalcy. Encouraging results are being produced in laboratory studies.

Research in these areas is growing quickly and new drugs are regularly becoming available. Drugs are now available, for instance, to treat one type of intestinal cancer and one form of leukemia. Numerous clinical trials are currently underway.

> **KEY CONCEPTS**
> Researchers are working on ways to kill cancer at the level of gene, specifically tumor suppressor genes and oncogenes (proto-oncogenes that have become mutated).

Treating Cancer with Laser Light

Scientists have found that certain drugs, when activated by laser light, can kill cancer cells. This is called *photodynamic therapy (PDT)*. In one therapy, patients with tumors in the lining of the esophagus and the bronchi of the lung are given the drug Photofrin. This is a light-sensitive drug. After the drug circulates in the blood and bathes the cancer cells, the researchers insert a small fiberoptic fiber in the affected area. A laser beam is trained on the cancer and, when it is turned on, it activates the chemical. The drug produces a form of oxygen that destroys nearby cancer cells.

PDT treatment has been approved in the United States for certain forms of advanced esophageal cancer and one form of lung cancer. One of the chief advantages of this therapy is that it poses little risk to adjacent tissues and can be repeated several times and combined with other therapies. Another advantage is that PDT is typically performed in outpatient clinics. Patients don't need to be admitted to the hospital. It can also be used in conjunction with other conventional therapies, including surgery, radiation, or chemotherapy.

> **KEY CONCEPTS**
> Photodynamic therapy uses a light-sensitive drug that is introduced into the tumor. When exposed to a specific wavelength of light, the drug produces a form of oxygen that kills tumor cells.

Early Detection

One of the most effective "treatments" of cancer is not a drug or radiation—it is early detection. The earlier a cancer is detected, the greater are one's chances of survival. When a cancer is detected early, there's less time for it to spread to other parts of the body.

Early detection can be achieved by self-examination. Women, for instance, are advised to check for lumps in their breasts every month. Men can also regularly check themselves for testicular cancer. Doctors also provide valuable cancer screening. Individuals can also be on the lookout for potential skin cancers. Any abnormalities in the breasts, testes, or skin should be reported to a physician immediately.

Physician cancer screening is carried out for many types of cancer, including breast, prostate, lung, colorectal, and testicular cancer. During periodic physical exams, for instance, doctors perform a digital examination of the prostate gland in men. Cancer screening for internal cancers such as cancer of the lung or colon require photographic or videographic techniques. Lung cancer, for instance, is detected by X-rays of the lung. Cancers of the large intestine, the colon, are identified by a **colonoscopy** (aka **coloscopy**). Colonoscopy is the endoscopic examination of the large intestine and the connecting portion of the small intestine using a video camera on a flexible tube that is passed into the colon via the anus. It allows the physician to detect precancerous conditions (small growths called *polyps*) and cancer. Polyps can be removed and biopsied, that is, studied under the microscope to determine if they are precancerous growths (Figure 20-7).

Another similar screening is sigmoidoscopy—examination of the lower half of the colon where most tumors occur. Men and women are advised to have a sigmoidoscopy performed every 5 years after age 50, and a colonoscopy every 10 years.

Women's breasts are also routinely examined by physicians via mammograms, low-energy X-rays. X-ray photos taken during these procedures are used to identify possible tumors (Figure 20-8). The American Cancer Society's recommends every three years for women in their 20s and 30s, every year for women 40, and every other year, after age 50.

While mammograms save lives, there's some disagreement about the age at which a woman should start getting mammograms and how often women receive them.

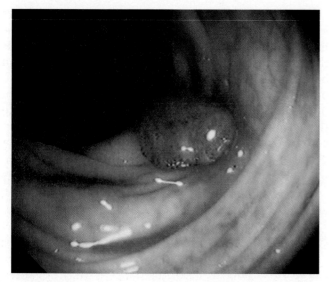

FIGURE 20–7 **Polyp** Image of a cancerous polyp observed during a colonoscopy. (© Dr. Larpent/CNRI/Science Source.)

According to a 2011 study published in *Annals of Internal Medicine*, mammograms should not be done on a one-size fits all basis. Rather, a schedule of mammograms should be personalized based on the age of a woman, family history, the density of breast tissue, and several other factors.

In addition to X-rays to detect cancers, physicians can also perform blood tests. New technologies, for instance, make it possible to detect cancer cells in the blood of patients after the cells have broken away from the primary tumor. Prostate tumor cells release a substance known as **prostate-specific antigen** (**PSA**). PSA is an enzyme in prostate cells. It is produced by cells lining the internal chambers of the prostate gland. PSA liquefies semen, which allows sperm to swim freely. It may also dissolve mucus in a woman's cervix, which allows the sperm ready access to the uterus.

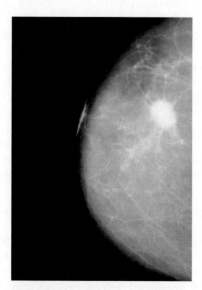

FIGURE 20-8 **Breast Tumor** X-ray of woman's breast shows a small tumor. (© Kings College School of Medicine/Science Source.)

PSA is found in low concentration in the blood of men with healthy prostates, but is often found in elevated concentrations in the blood of men with prostate cancer. Doctors routinely send out men's blood for PSA screening during physical exams, although the United States Preventive Services Task Force, a panel of healthcare experts who routinely evaluate the most recent clinical studies, recommends against it. They argue that too many false positives occur—that is, too many men are diagnosed with prostate cancer when they have none. This leads to overtreatment because "most prostate cancer is asymptomatic for life." Treating a cancer that will not cause problems involves more risks than benefits.

> **KEY CONCEPTS**
>
> The earlier cancer can be detected, the better one's chances of survival. Strategies include self-examination and tests such as sigmoidoscopy and colonoscopy for colon cancer, mammography and clinical breast exam for breast cancer in women, and a PSA for prostate cancer in men.

Cancer Prevention

Despite the promising new developments in cancer treatment, prevention is always the best approach. Here are the most important things you can do to reduce your chances of getting cancer: (1) quit smoking or do not start it in the first place; (2) eat a healthy diet rich in plant nutrients; (3) maintain a healthy weight; (4) drink plenty of water and no or moderate quantities of alcohol; (5) avoid excess exposure to sunlight and X-rays; and (6) choose a safe occupation and dwelling.

Dermatologists recommend we limit our exposure to ultraviolet radiation by wearing sunscreen, hats, and protective clothing. They reduce our chances of contracting skin cancer. Eating lots of fruits and vegetables especially cruciferous vegetables, mushrooms, berries, onions, and garlic is important. As noted earlier in this chapter, substances in fruits and vegetables known as phytochemicals greatly reduce the risk of cancer and can cause cancer remission (see Figure 20-6). Many phytochemicals are antioxidants, which remove free radicals that can trigger early changes in the DNA that may lead to cancer. Other chemicals fight cancer directly, for example, by blocking angiogenesis. Still others support the immune system, allowing the body to mount a more effective line of defense. As a side note, eating organic fruits and vegetables may also help reduce the risk of cancer as doing so dramatically lowers blood levels of potentially cancer-causing pesticides. Dietary fiber in grains and vegetables helps reduce your chances of contracting colon cancer, a leading killer of men. Even certain vitamins present in a healthy diet or through supplements have been shown to reduce one's risk of certain forms of cancer. A recent review of 63 research studies, for instance, showed that vitamin D (found in fortified milk and cereals, but sunlight still is the best way to get vitamin D) lowers risk of developing several forms of cancer, among them cancer of the colon, ovary, breast, and possibly prostate. In this study, however, vitamin D's protective effect was only seen in people consuming 1,000 IU of vitamin D a day, well above the 200 to 600 IU daily requirement (the amount varies with age).

Recent research suggests that a chemical substance called ellagic acid, which is found in raspberries, blackberries, strawberries, and other fruit and nuts, may destroy carcinogenic molecules in the body, reducing the likelihood of developing cancer. Scientists studying carcinogenic agents found in tobacco smoke, auto exhaust, and foods noted that in laboratory animals, ellagic acid reduced DNA damage caused by one prevalent carcinogen by 45% to 70%.

Health Tip 20-3

For optimum health, both now and in the long term, you should eat 13 half-cup servings of fruit and vegetables every day. The simple way to achieve this goal is to be sure one-half of every meal is fruits and vegetables.

Why?

Since the early 1990s, nutritionists have recommended 2 to 4 servings of fruit and 3 to 5 servings of vegetables—that is, 5 to 9 servings of fruit and vegetables every day.

Knowing that most Americans resist eating right, the Centers for Disease Control and an industry group, the Produce for Better Health Foundation, have been advising people to eat "Five a Day." In 2005, however, the U.S. government issued new guidelines recommending 13 half-cup servings a day. Their new slogan is "Fruits and Veggies . . . More Matters!"

Studies show that eating certain fish—cold-water species like salmon, herring, mackerel, and sardines—dramatically decreases one's chances of developing kidney cancer. These species contain much higher levels of omega-3 fatty acids.

Yet another preventive measure is a brand new vaccine, designed to prevent cervical cancer caused by the human papilloma virus.

Researchers believe that the recent drop in cancer is also due to better lifestyle choices—better diets, exercise programs, and adoption of other healthy habits.

Despite the small drop in the number of cancer deaths, approximately over 1.6 million Americans will be diagnosed with cancer in 2012, according to the American Cancer Society. About 577,000 of them will die—or about 1,500 people per day. And, despite gains from years of intensive research and billions of dollars spent studying cancer, modern science may actually be losing the war against many types of cancer. For example, long-term survival rates in cancer patients over the past decade or so have increased only 4.2%, compared with an overall increase of 5.1% in survival rates for all other diseases. Because of this statistic, many agree that the cheapest and most effective cure is prevention.

There are many ways to prevent cancer, among them eating a very healthy diet (especially eating foods that fight cancer), weight control, choice of occupation and residence, and protections against X-rays and UV exposure.

SUMMARY

1. Cancer is a disease characterized by uncontrolled cell division. It occurs in around 200 different cell types in humans.
2. Cancer is a disease that most of us will experience either directly or indirectly. One out of three Americans will contract the disease, and one out of four will die from it.

Benign and Malignant Tumors

3. Cells that divide uncontrollably form tumors. Tumors that fail to grow and spread are benign and rarely cause serious medical problems.
4. Malignant tumors are those that continue to grow, often invading nearby tissues and spreading to other parts of the body where they invade tissues and vital organs.
5. Malignant tumor cells spread (metastasize) through the body from the primary tumor site via the circulatory and lymphatic systems, often becoming established in distant sites, where they form secondary tumors. The lungs are one of the main sites of metastasis as are the kidneys.

6. Metastases can be identified by palpating organs and through X-rays and CAT scans.

How Cancers Form

7. The transformation of a normal cell into a cancerous one is the first stage in the development of a cancer, known as initiation.
8. Most cancers are a result of mutations, change in the DNA of cells resulting from a chemical, biological, or physical agent. These agents are called cocarcinogens.
9. Many reactive free radicals are found in the body and are capable of causing mutations. Free-radicals can be eliminated by eating a diet rich in antioxidants such as green vegetables.
10. The initial mutation caused by a cocarcinogen may or may not result in a cancerous growth. A second factor, a promoter, however, may stimulate the cell to proliferate uncontrollably.
11. In many cancers, a long dormant period occurs between the first stage, initiation, and the second stage, development and progression.

12. Cancers can also result epigenetically—at levels that do not initially involve mutations of the DNA.
13. Epigenetic causes of cancer include hormonal imbalances, immune system suppression, methylation of DNA, and a host of other factors.

Cancer Treatments

14. Tumors can be removed surgically or destroyed with radiation, but successful treatment depends on the location and destruction of secondary tumors.
15. Patients can also be treated with chemotherapeutic agents, chemical substances that attack rapidly dividing cells.
16. Many new treatments are also underway. Medical researchers are, for example, studying ways to target cancer cells specifically, for example, by activating the immune system via anti-cancer vaccines so that it produces antibodies and T cells that attack cancer cells. Monoclonal antibodies designed to attack cancer cells also help combat several types of cancer.

17. Scientists are also researching ways to deliver lethal cancer-fighting agents by microspheres and have developed several successful treatments that are now in use.

18. Some researchers have found that preventing the growth of blood vessels stops cancer growth. They have developed anti-angiogenesis drugs. Many healthful fruits and vegetables also offer these benefits.

19. Other scientists are working on ways to remedy the underlying genetic defects in cancer cells by inserting healthy genes to replace the mutated ones.

20. Despite all of the advances, early detection and prevention remain two of the most important measures in the battle against cancer. Self-exams and physician exams are keys to lowering the incidence of cancer and one's changes of death.

THINKING CRITICALLY ANALYSIS

This analysis corresponds to the Thinking Critically scenario that was presented at the beginning of this chapter.

Just because a chemical causes cancer in rats and mice doesn't mean that it will have the same effect on humans. To study this chemical, it would be wise to examine the scientific literature to see if there are any studies of workers exposed to this particular chemical through their workplace. If so, it would be prudent to determine the level of exposure—that is, how high the concentrations were at their workplaces. If they match those of the experiment and are close to those levels of exposure in our daily lives, perhaps it might be wise to consider regulating the chemical or finding ways of dramatically reducing our exposure. It might also be wise to study the effects of the chemical in our daily lives, by looking for potential health effects in people already exposed to it. Studies such as these can be costly and may take several years to complete, but, combined with other research, they could give us an indication of what steps, if any, need to be taken to protect the public from unnecessary exposure.

KEY TERMS AND CONCEPTS

Angiogenesis, p. 436
Anti-cancer vaccines, p. 435
Benign tumor, p. 430
Cancer, p. 430
Cell-cell adhesion molecules, p. 432
Chemotherapy, p. 435
Cocarcinogens, p. 433
Colonoscopy (coloscopy), p. 437

Development and progression, p. 433
DNA-reactive carcinogens, p. 433
Epigenetic carcinogens, p. 433
Immunotherapy, p. 435
Malignant tumor, p. 430
Metastasis, p. 432
Monoclonal antibodies, p. 435
Oncogene, p. 433

Primary tumor, p. 431
Promoter, p. 433
Prostate-specific antigen, p. 438
Proto-oncogene, p. 433
Radiation treatment, p. 434
Secondary tumor, p. 432
Tissue factor, p. 433
Tumor suppressor gene, p. 433

CONCEPT REVIEW

1. In what ways are cancer cells different from normal cells? pp. 430–431.

2. What is the difference between a benign and a malignant tumor? p. 430.

3. Why are malignant cancers so dangerous? p. 430.

4. Why are nonmalignant cancers sometimes harmful to human health? Describe some effects of nonmalignant or benign tumors. p. 430.

5. Describe the two stages of cancer development. What occurs in each stage? pp. 432–433.

6. Define the terms initiator and promoter in relation to cancer. p. 433.

7. What are DNA-reactive carcinogens? p. 433.

8. What is an epigenetic carcinogen? Can any epigenetic carcinogens lead to mutations? pp. 433–434.

9. Why don't more mutated cells become cancerous? p. 433.

10. Cancer is often considered a disease of aging. Given the fact that cancer is caused by DNA mutation, why is this assertion correct? p. 434.

11. List and describe the four major treatments for cancer today? Why are they often used in conjunction? pp. 434–435.

12. What is immunotherapy? p. 435.

13. Describe the term anti-cancer vaccine. Does it prevent cancer? p. 435.

14. What is a monoclonal antibody? How does it fight cancer? p. 435.

15. Do some online research to create a list of cancers that are currently being treated with monoclonal antibodies. Why are diets rich in phytochemicals also often considered anti-cancer diets? What cancer-causing agents do phytochemical-rich diets avoid or reduce? What cancer-fighting factors do these diets add to your food? p. 435.

SELF-QUIZ: TESTING YOUR KNOWLEDGE

1. A _____ tumor is one that grows uncontrollably. p. 430.
2. The conversion of a normal cell to a pre-cancerous cell is known as _____. p. 432.
3. A chemical substance that stimulates a precancerous cell to divide is known as a _____. p. 433.
4. The period between the initial exposure to a cancer-causing agent and the development of a cancer is known as the _____ period. It can often be 10 to 20 years. p. 434.
5. Mutations of proto-oncogenes and tumor _____ genes can lead to cancer. p. 433.
6. Chemicals that alter the DNA, causing mutations that lead to cancer, are known as DNA-_____ carcinogens. p. 433.
7. _____ is the treatment of cancer using anti-cancer drugs. p. 434.
8. _____ is the treatment of cancer using one's own immune system. p. 435.
9. Anti-cancer _____ destroy cancer cells and are attracted to the surface molecules of cancer cells. p. 435.

biology.jbpub.com/chiras/8e/

The site features eLearning, an online review area that provides quizzes, chapter outlines, and other tools to help you study for your class. You can also follow useful links for in-depth information, research the differing views in the Point/Counterpoints, or keep up on the latest health news.

(© Monkey Business Images/ShutterStock, Inc.)

Human Reproduction

THINKING CRITICALLY

Imagine that the town you live in has two hospitals—one large and the other much smaller. Suppose that 45 babies are born each day, on average, in the larger hospital and 15 babies are born each day, on average, in the smaller hospital. As you probably know, about half of all babies born in this country are girls and half are boys. Over a year's time, the hospitals both recorded days on which more than 60% of the babies born were girls. Which of the two hospitals do you think posted more days?

Sandra Collins woke one day with a pain in her abdomen that persisted throughout the morning. Instead of calling her doctor, she shrugged off the pain and went Christmas shopping with her husband. A few hours later, while she was browsing through a bookstore, the pain grew worse and she blacked out. Her husband rushed her to the emergency room, where doctors discovered that Sandra was suffering from internal bleeding caused by an ectopic (eck-TOP-ick) pregnancy—that is, a fertilized ovum that had developed outside of the uterus.

In Sandra's case, it had embedded in the upper part of her reproductive tract, in one of the Fallopian tubes. Surgeons whisked her into the operating room, where they surgically removed the fetus, placenta, and surrounding tissue and repaired torn blood vessels.

Reproduction is one of the most basic body functions. However, as this account shows, it doesn't always operate smoothly. This chapter examines human reproduction and related topics. Sexually transmitted diseases are covered in A Closer Look at the end of this chapter.

21-1 The Male Reproductive System

The male reproductive system is shown in **Figure 21-1a**. **Table 21-1** summarizes the key components and their functions. As illustrated, the male reproductive system consists of two gonads, the **testes** (TESS-teas). They produce sperm and male sex hormones, mostly testosterone. The testes are suspended in the **scrotum**, a sack of skin attached to the body below the attachment of the penis. The scrotum provides a slightly cooler environment that is necessary for sperm to survive.

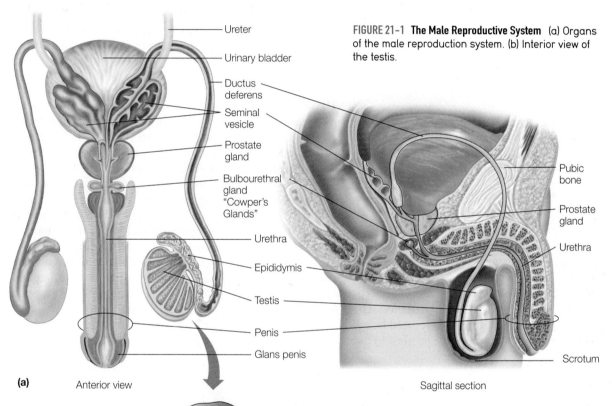

FIGURE 21-1 **The Male Reproductive System** (a) Organs of the male reproduction system. (b) Interior view of the testis.

(a) Anterior view

Sagittal section

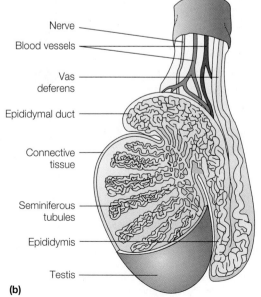

(b)

The Testes

As Figure 21-1b shows, each testis contains numerous highly convoluted **seminiferous tubules** (SEM-in-IF-er-uss), in which sperm are formed.

Sperm produced in the seminiferous tubules empty into a network of connecting tubules in the "back" of the testes. These tubules, in turn, empty into the **epididymal duct** (ep-eh-DID-eh-mal). The epididymal duct forms the **epididymis**. It is a storage site for sperm prior to their release during ejaculation, the ejection of sperm. While being stored, sperm fully mature, a process that makes them capable of swimming.

During ejaculation, the epididymal duct of each testis empties into a duct known as the **vas deferens** (plural, vasa deferentia; pronounced VAH-sah DEAF-er-en-she-ah). These ducts pass from the scrotum into the body cavity through the **inguinal canals**, small openings in the wall of the lower abdomen (**Figure 21-2a**). The inguinal canals are potential weak spots in the abdominal wall. In some men, loops of intestine

TABLE 21-1	The Male Reproductive System
Component	Function
Testes	Produce sperm and male sex steroids
Epididymes	Store sperm
Vasa deferentia	Conduct sperm to urethra
Sex accessory glands	Produce seminal fluid that nourishes sperm
Urethra	Conducts sperm to outside
Penis	Organ of copulation
Scrotum	Provides proper temperature for testes

can bulge through weakened inguinal canals (Figure 21-2b). This condition, known as an **inguinal hernia**, can be corrected surgically.

The vasa deferentia then empty into the urethra. From here, sperm course through the penis to the outside of the body during ejaculation (discussed shortly).

> **KEY CONCEPTS**
>
> Sperm are produced in the seminiferous tubules inside the testes and stored in the epididymis where they mature prior to ejaculation. The epididymal ducts of the testes empties into the vasa deferentia which empty into the urethra.

The Sex Accessory Glands

During ejaculation, sperm are combined with fluids produced by small glands located near the neck of the urinary bladder, the **sex accessory glands**. These include the seminal vesicles, the prostate gland, and the Cowper's glands (Figure 21-1).

The paired seminal vesicles empty into the vasa deferentia and produce the largest portion of the ejaculated fluid. The Cowper's glands are a pair of pea-size glands located below the prostate on either side of the urethra. The prostate gland

surrounds the neck of the bladder and empties its contents directly into the urethra.

The sex accessory glands produce fluid that, when combined with sperm produced by the testis, form **semen**. Sperm, the male sex cells, constitute only 1% of the volume of the semen.

Fluid from the sex accessory glands has many chemical components. For example, it contains the simple sugar fructose, which is used by sperm to generate energy needed to help propel themselves through the female reproductive tract. Semen also contains a chemical buffer, a substance that neutralizes the lethal (to sperm) acidic secretions of the female reproductive tract. Yet another component of the semen is prostaglandin. This is a chemical that causes the muscle of the womb (uterus) to contract. Muscle contractions are believed to be primarily responsible for the movement of sperm up the female tract.

> **KEY CONCEPTS**
>
> Semen released from the penis during ejaculation contains sperm and fluids produced by the sex accessory glands.

Prostate Diseases

The prostate gland is about the size of a walnut, but increases as men grow older. Enlargement of the prostate is known as benign prostatic hypertrophy or hyperplasia (BPH). BPH is a normal part of aging in men. By various estimates, approximately half of all men show signs of BPH by 60. By age 85, 90% of all men show signs of it. BPH results from the formation of small nodules inside the gland and growth of the cells between the glands.

In BPH, the enlarged prostate presses on the urethra, around which the gland is wrapped. This, in turn, makes it difficult for men to initiate urination and to completely stop it (resulting in dribbling). BPH makes it difficult to completely empty the urinary bladder, which results in more frequent urination, a symptom noticed especially at night. In addition, BPH usually results in a weak urine stream.

BPH is more a nuisance than a serious medical problem. Men can take steps to reduce their discomfort, for example,

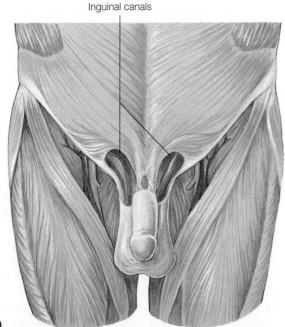

Inguinal canals

(a)

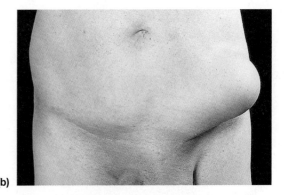

(b)

FIGURE 21–2 The Inguinal Canal and Hernia (a) During development, the testis descends through the inguinal canal, an opening in the musculature in the lower abdominal wall. In adults, the inguinal canal provides a route for the vas deferens, blood vessels, and nerves that supply each testis. (b) Loops of intestine may push through the weakened musculature surrounding the inguinal canal. (© Biophoto Associates/Science Source, Inc.)

spreading out their consumption of liquids during a day, spending extra time voiding their urinary bladder, and avoiding beverages known to increase urination such as coffee, caffeinated soft drinks, caffeinated energy drinks, and alcoholic beverages. BPH can also be treated with drugs. In some cases the nodules grow so large that they block the flow of urine and make urination painful. This condition typically requires surgery.

The prostate is also a common site for cancer in men. Enlarged prostates and prostate cancers are detected by rectal exams that allow doctors to feel the prostate. A blood test is also now commonly used to determine levels of prostate-specific antigen (PSA), circulating levels of antigen produced by cancer cells. This test is not recommended by some health officials because it results in many false negatives.

Patients with early prostate cancer often exhibit no symptoms, although some men complain of discomfort and other symptoms. Typical symptoms are similar to those experienced by patients with BPH. In addition, some men experi-

ence annoying urine discharges that occur when they laugh or cough, an inability to urinate when standing up, painful or burning sensations during urination or ejaculation, and blood in urine or semen.

Over the years, prostate cancer has been treated by surgery, chemotherapy, cryotherapy (freezing the cancerous tissue), hormonal therapy (reducing testosterone levels), and/or radiation. Many prostate cancers grow very slowly so are simply monitored throughout a man's life. If the cancer grows to a size that causes problems, the gland is surgically removed.

> **KEY CONCEPTS**
>
> The prostate gland enlarges with age in men. This disorder is known as benign prostatic hypertrophy or hyperplasia. Enlargement of the prostate can block urinary flow. The prostate is also a common site of cancer in older men.

Sperm Formation

The formation of sperm in the seminiferous tubules is illustrated in Figure 21-3. Sperm are formed from special cells in the periphery of the tubules, known as **spermatogonia** (sper-MAT-oh-GO-nee-ah). The spermatogonia divide by mitosis, producing a constant supply of new cells needed to produce sperm.

Some spermatogonia, however, undergo meiosis, a process that involves two cellular divisions that result in the formation of four cells, called spermatids. Each spermatid contains one-half the number of chromosomes of the spermatogonia. The

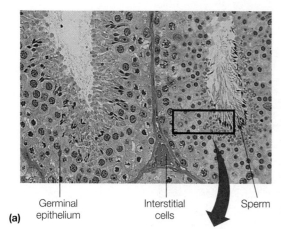
(a) Germinal epithelium · Interstitial cells · Sperm

FIGURE 21-3 Formation of the Sperm (a) Cross section through two seminiferous tubules showing the germinal epithelium where sperm are formed and the interstitial cells where testosterone is produced. (© Fred Hossler/Visuals Unlimited.) (b) Details of sperm formation.

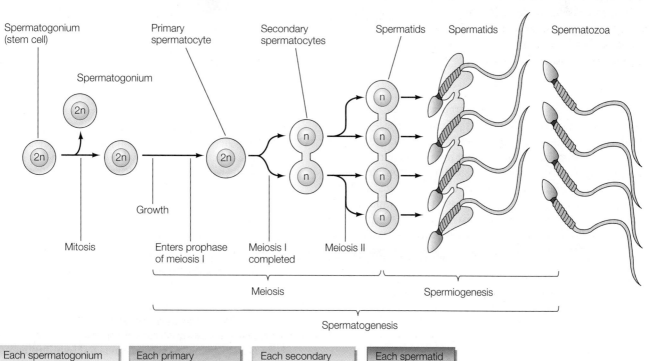
(b)

spermatids then differentiate, becoming **sperm**. During this process, the nucleus condenses and the cytoplasm is eliminated. A long, whiplike tail (a flagellum) forms from the centriole. These changes streamline the cell so it is able to swim and to fertilize an ovum. The mitochondria of the spermatid congregate around the first part of the tail, where they can provide energy for propulsion. The Golgi apparatus enlarges and forms an enzyme-filled cap that fits over the condensed nucleus like a stocking cap. This structure, the **acrosome** (ACK-row-sohm), will help the sperm digest its way through the coatings surrounding the ovum during fertilization.

On average, men produce 200 to 300 million sperm every day. The average 3-milliliter ejaculate contains 240 million or more. Such large numbers evolved because many sperm are required to ensure fertilization. In fact, nearly all sperm are eliminated as they travel through the female reproductive tract.

In humans, each sperm formed during meiosis contains 23 single-stranded (unreplicated) chromosomes—half the number in a normal somatic cell. Thus, when the sperm unites with an ovum (also containing 23 unreplicated chromosomes), they produce a **zygote** containing 46 single-stranded chromosomes. One-half of its chromosomes come from each parent.

> **KEY CONCEPTS**
> Sperm are streamlined germ cells that are formed in the testes from spermatogonia via meiosis. They contain 23 single-stranded chromosomes.

Interstitial Cells

The testes also produce male sex hormones. These hormones are produced in cells found in spaces between the seminiferous tubules. These large, highly visible cells are known as Leydig cells, after the scientist who first described them, or **interstitial cells** (in-ter-STISH-al). They produce a group of sex steroid hormones known as **androgens**, so named because they exert a masculinizing effect. The most important androgen is testosterone (discussed below).

Testosterone

Testosterone is a steroid home that circulates in the blood and stimulates many cells. In the testis, it stimulates the formation of sperm. It also stimulates cellular growth in bone and muscle and accounts, in part, for the fact that men are generally taller and more massive than women. Testosterone promotes facial hair growth and thickening of the vocal cords, typically giving men deeper voices than women. It also tends to make men more aggressive than women. Testosterone is also responsible for pattern baldness in men and causes acne, discussed next.

> **KEY CONCEPTS**
> Interstitial cells are hormone-producing cells that lie between the seminiferous tubules of the testes. They produce androgens, male sex steroids including testosterone.

Acne and Testosterone

Testosterone stimulates the oil glands of the skin in both sexes. (Women secrete lesser amounts of testosterone.) Oil-producing glands, known as **sebaceous glands** (seh-BAY-schuss), secrete oil onto the skin, moisturizing it (Figure 21-4a).

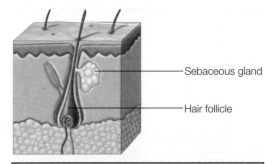

(a) Sebaceous glands associated with hair follicles secrete sebum, an oily substance that lubricates the skin.

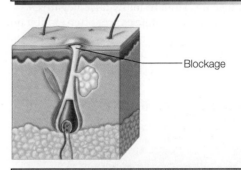

(b) A follicle may become blocked by excess sebum and dead skin cells. Unable to escape, the sebum builds up in the hair follicle.

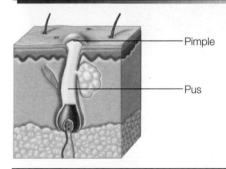

(c) Bacteria present on the skin may infect the sebum, causing inflammation; pus and swelling form an acne pimple.

FIGURE 21-4 **Formation of an Acne Pimple** (a) Testosterone stimulates oil production in the sebaceous glands. (b) If the outlet is blocked, sebum builds up in the gland and (c) the gland may become infected.

During puberty (sexual maturation) in boys, rising levels of testosterone dramatically increase sebaceous gland activity. This itself is not a problem. However, if dead skin cells block the pores that carry the oil to the skin's surface, oil known as *sebum* may collect inside the gland (Figure 21-4b). Bacteria on the skin may invade the gland, where they proliferate. This results in inflammation, pus formation, and swelling. The skin protrudes, forming an acne pimple.

Acne is a skin condition characterized by pimples, commonly referred to as "zits." Individuals with acne develop whiteheads, blackheads, and painful, red, inflamed patches of skin (known as cysts). If the plug is white, it is referred to as a whitehead. If the plug is dark, it is known as a blackhead.

Deep acne results in hard, painful cysts. This condition is referred to as *cystic acne*.

Acne commonly occurs in teenagers. In fact, three-fourths of all teenagers experience some form of acne at some time in their teenage years. Acne can also be found in men and women in their twenties, but is sometimes experienced by older adults and occasionally even babies. Acne tends to run in families and may be triggered by hormonal changes at puberty or during menstrual periods or pregnancy. Hormones in birth control pills and stress can also cause acne. Additional causes include oily cosmetic and hair products, certain drugs such as steroids, testosterone, and estrogen, and high levels of atmospheric humidity and sweating.

Mild acne can be treated by gently washing the skin twice a day with warm water and a mild, unscented soap. The skin should also be treated with an acne cream containing benzoyl peroxide. Moderate and severe acne can be successfully treated with special ointments and antibiotics prescribed by a doctor. In the 1950s and 1960s, some people with severe acne were treated with X-rays, a dangerous practice that could increase an individual's chances of contracting skin cancer.

> **KEY CONCEPTS**
>
> Acne is a skin condition common in teenagers and individuals in their twenties. It is characterized by whiteheads, blackheads, and sometimes, deep, painful, red, inflamed patches of skin and is often caused by hormonal changes.

The Penis

Sperm are deposited in the female reproductive tract with the aid of the male sexual organ, the penis. The **penis** consists of a shaft of varying length and an enlarged tip, the glans penis (Figure 21-5). The glans is covered by a sheath of skin at birth, the foreskin. The foreskin gradually becomes separated from the glans in the first 2 years of life. At puberty, the inner lining of the foreskin begins to produce an oily secretion. Bacteria can grow in the protected, nutrient-rich environment created by the foreskin, so special precautions must be taken to keep the area clean.

Because of potential health problems or religious reasons, many parents opt to have the foreskin removed in the first few days of their son's life. The operation, called *circumcision*, may help reduce penile cancer in men and may also reduce cervical cancer in the wives or sexual partners of circumcised men. It has recently been shown to reduce the incidence of AIDS, a deadly sexually transmitted disease caused by a virus known as HIV.

During sexual arousal, nerve impulses traveling to the penis from the spinal cord cause arterioles in the organ to dilate. Blood flows into a spongy **erectile tissue** (eh-REK-tile) in the shaft of the penis, making it harden. The growing turgidity (swelling) compresses a large vein on the dorsal surface of the penis, blocking the outflow of blood and further stiffening the organ.

Some men lose their ability to achieve or to sustain an erection. This condition is known as **erectile dysfunction** (formerly called *impotence*). Most men experience impotence at some time in their life, but it is usually temporary. Persistent impotence, however, is a more serious condition, and is more common in middle-age and elderly men.

Persistent impotence may be caused by marital conflict, stress, fatigue, and anxiety. Naturally low testosterone levels and nerve damage may also cause impotence. Alcohol ingestion and some medications may contribute to the condition as well. Blockages (due to atherosclerosis) in the arteries leading to the penis may be the cause in some cases. Impotence is especially prevalent in men who smoke, as smoking leads to atherosclerosis of the arteries, including those that deliver blood to the penis, restricting blood flow.

Numerous treatments are possible, depending on the cause of the problem. Two of the most popular treatments are the drugs Viagra and Cialis. These medications relax the muscle in the walls of the arteries supplying the penis. This permits blood to flow into the organ more readily when stimulated.

> **KEY CONCEPTS**
>
> The penis is the male copulatory organ via which sperm are deposited in the female reproductive tract. It consists of erectile tissue that becomes engorged with blood during sexual arousal.

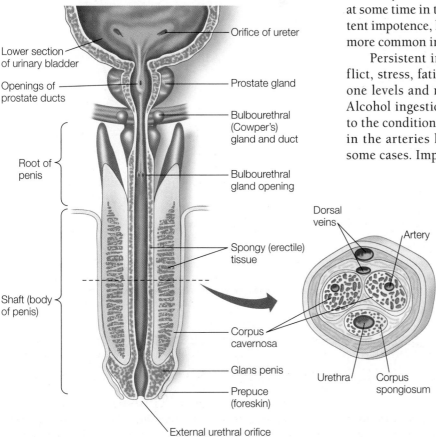

FIGURE 21-5 Anatomy of the Penis The penis consists principally of spongy tissue that fills with blood during sexual arousal. The urethra passes through the penis, carrying urine or semen.

Ejaculation

Ejaculation occurs when the penis is erect. Ejaculation is a reflex that is triggered by sexual stimulation. Sexual stimulation causes motor neurons in the spinal cord to send impulses along nerves to the smooth muscle in the walls of the epididymis, vasa deferentia, and sex accessory glands. These nerve impulses cause the smooth muscle to contract, which propels sperm and fluid into the urethra.

> **KEY CONCEPTS**
>
> Ejaculation is a reflex mechanism that ejects semen from the testis.

Hormonal Control of Male Reproduction

As noted earlier, the testes produce sex steroid hormones—mainly, testosterone. Testosterone secretion, however, is controlled by a hormone from the pituitary gland. This hormone is known as **luteinizing hormone** (LH). In males, LH is also known as **interstitial cell stimulating hormone (ICSH)** because it stimulates the interstitial cells to produce androgen hormones.

ICSH secretion is controlled by a releasing hormone produced by the hypothalamus, known as **gonadotropin releasing hormone (GnRH)**. As Figure 21-6 shows, the secretion of GnRH and ICSH is controlled by testosterone levels in the blood in a negative feedback loop. When testosterone levels in the blood decline, receptors in the hypothalamus detect the change and signal an increase in GnRH secretion. This hormone flows in the blood to the anterior pituitary and stimulates the production and release of ICSH. It then travels in the bloodstream to the testes, where it stimulates testosterone production. When testosterone levels return to normal, GnRH release subsides, as does ICSH secretion.

The pituitary also produces a gonadotropin known as **follicle-stimulating hormone (FSH)**. Like testosterone, FSH stimulates sperm formation. FSH secretion is controlled by GnRH and a peptide hormone called *inhibin*, produced by the testes. Inhibin gets its name from the fact that it inhibits the production of FSH in the anterior pituitary.

> **KEY CONCEPTS**
>
> Sperm formation is controlled by the hormones testosterone from the testis and FSH from the anterior pituitary. These hormones, in turn, are controlled by negative feedback mechanisms.

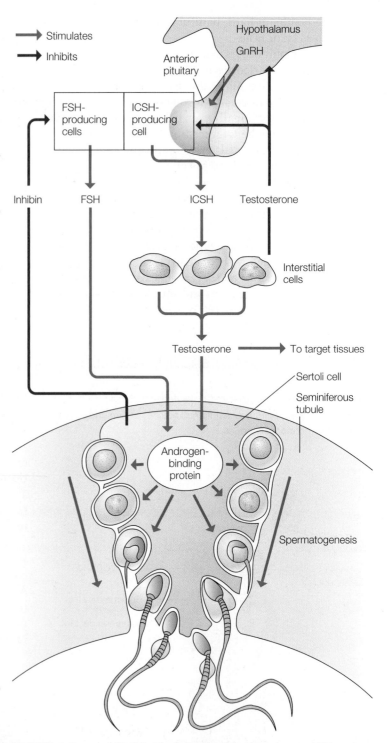

FIGURE 21-6 **Hormonal Control of Testicular Function** Testosterone, FSH, and ICSH participate in a negative feedback loop. The testes also produce a substance called inhibin, which controls GnRH secretion.

21-2 The Female Reproductive System

Figure 21-7 illustrates the female reproductive system, which consists of two parts: the reproductive tract (internal organs) and the external genitalia. Let's start with the structures of the female reproductive tract. Table 21-2 summarizes the role of each.

Anatomy of the Female Reproductive System

The **ovaries** are paired, almond-shaped organs located in the pelvic cavities of women. They produce the female gametes, the ova or eggs. The ovaries also produce several important reproductive hormones, discussed shortly.

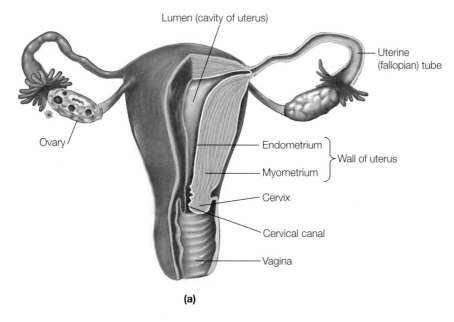

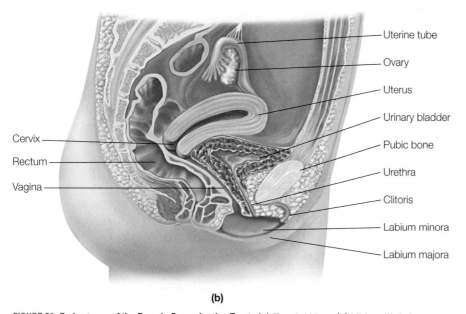

FIGURE 21–7 **Anatomy of the Female Reproductive Tract** (a) Frontal View. (b) Midsagittal view.

of the uterine tubes draw the egg inside and down the tubes to the uterus.

Fertilization occurs in the upper third of the uterine tubes. The fertilized ovum is then transported down the uterine tubes to the uterus. Inside the uterus, the fertilized ovum attaches to the lining, or the **endometrium** (EN-doh-MEE-tree-um), and embeds itself there, remaining for the duration of pregnancy.

The womb or **uterus** is a pear-shaped organ about 7 centimeters (3 inches) long and about 2 centimeters (less than 1 inch) wide at its broadest point in nonpregnant women. The wall of the uterus contains a thick layer of smooth muscle cells. The uterus houses and nourishes the developing fetus.

At birth, the baby is expelled from the uterus through the **cervix** (SIR-vix), the lowermost portion of the uterus. As Figure 21-7b shows, the cervix protrudes into the vagina (vah-GINE-ah). Although the canal running through the cervix is quite narrow, the cervix stretches considerably at birth to allow the passage of the baby into the vagina.

The **vagina** is a distensible, 3-inch, tubular organ that leads to the outside of the body. Its walls contain a considerable amount of muscle that allow it to expand during birth. The vagina also serves as the receptacle for sperm during sexual intercourse. To reach the ovum, sperm must travel through a tiny opening and narrow canal of the cervix that leads into the uterus. From here, sperm move up both uterine tubes.

An egg released from an ovary is taken up by one of two hollow, muscular tubes, known as the **oviducts** (OH-va-ducts) or **uterine tubes** (YOU-ter-in) or **Fallopian tubes**. As Figure 21-7a shows, the ends of the uterine tubes are widened and fit loosely over the ovaries. Currents created by cilia in the lining

KEY CONCEPTS

The female reproductive system consists of the ovaries, which produce female gametes, the ova; the oviducts, the site of fertilization takes place; the uterus, the site in which fertilized ova develop into a fetus; and the vagina, which serves as a receptacle for semen during copulation, and as a path from the uterus to the outside of a woman's body during childbirth.

The External Genitalia

The external genitalia are known as the **vulva**. It consists of two flaps of skin on either side of the vaginal and urethral openings (Figure 21-8). The outer folds are known as the **labia majora** (LAY-bee-ah ma-JOR-ah). These large folds of skin extend from the **mons pubis**, a mound of fatty tissue lying over the pubic bone. The labia are covered with hair on the outer surface and contain numerous sebaceous glands on the inside. The inner

TABLE 21-2	The Female Reproductive System
Component	**Function**
Ovaries	Produce ova and female sex steroids
Uterine tubes	Transport sperm to ova; transport fertilized ova to uterus
Uterus	Nourishes and protects embryo and fetus
Vagina	Site of sperm deposition, birth canal

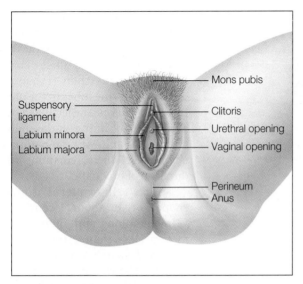

FIGURE 21–8 **Female External Genitalia**

flaps are the **labia minora** (meh-NOR-ah). Anteriorly, they meet to form a hood over a small knot of tissue called the **clitoris** (CLIT-er-iss). The clitoris is a highly sensitive organ involved in female sexual arousal. It consists of erectile tissue and becomes filled with blood during sexual arousal. It is formed from the same embryonic tissue as the penis. In fact, some women are born with greatly elongated clitorides (plural).

KEY CONCEPTS

A woman's external genitalia consists of the labia, flaps of skin that flank the vaginal opening, and the clitoris, an organ that consists of erectile tissue that is involved in sexual arousal and derived from the same tissue as the penis of males.

Sexual Arousal and Orgasm

The sexual response of women involves four stages. During the first phase, the excitement phase, cervical glands produce a secretion that lubricates the vagina. Erectile tissue in the labia fills with blood, causing them to expand. The nipples become erect. Heart rate, muscle tension, and blood pressure increase.

During the next phase, breathing, heart rate, blood pressure, and muscle tension increase more. Blood vessels in the wall of the vagina and clitoris fill with blood, becoming engorged. Glands near the opening of the vagina are activated, producing a lubricating fluid.

Continued sexual arousal can lead to orgasm, rhythmic contractions of the uterus and vagina. These contractions are often accompanied by intense physical pleasure. Women often report feelings of warmth throughout their bodies after orgasm.

In the fourth stage, blood drains from the clitoris and labia. Blood pressure, heart rate, and respiration return to normal.

KEY CONCEPTS

Sexual arousal in women occurs in four stages, starting with excitement, culminating in orgasm, and ending with a return to previous physiological conditions.

The Ovaries

During each menstrual cycle, one ovary releases an **ovum**, the female gamete, commonly referred to as an egg. This process is called **ovulation** (OV-you-LAY-shun). The release of an ovum occurs approximately once a month in women during their reproductive years—from puberty (ages 11 to 15) to menopause (ages 45 to 55). Ovulation is temporarily halted when a woman is pregnant and may be suppressed by emotional and physical stress.

The structure of an ovary is shown in Figure 21-9. As illustrated, several structures are visible. One is the female germ cells—immature ova. They are surrounded by varying numbers of cells. Together, the immature ova (also known as oocytes) and the cells surrounding them are called **follicles**. As illustrated, follicles vary in size. The earliest ones contain only a few follicle cells. The more developed follicles contain many layers. Each month, a dozen or so follicles begin to develop. Figure 21-10 (right side) illustrates the growth and development of follicles.

In the largest follicles, a clear liquid begins to accumulate between the follicle cells. Eventually, so much liquid accumulates that one central cavity is formed. At this point, the follicle is called an **antral follicle** (AN-tril). Although a dozen or so follicles begin developing during each cycle, as a rule, only one makes it to ovulation. The rest stop growing and degenerate.

The follicle (or follicles) that survives continues to enlarge by accu-

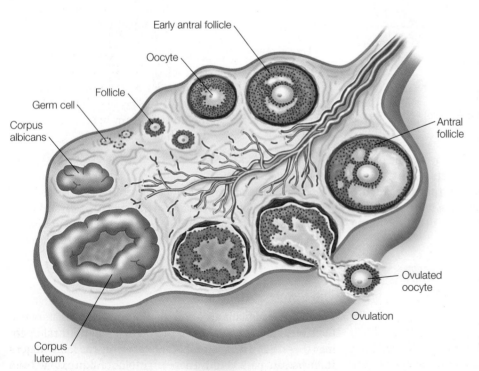

FIGURE 21–9 **Structure of the Ovary** This drawing illustrates the phases of follicular development and also shows the formation and destruction of the corpus luteum (CL). Antral follicles give rise to the CL. A fully formed CL and antral follicle would not be found in the ovary at the same time.

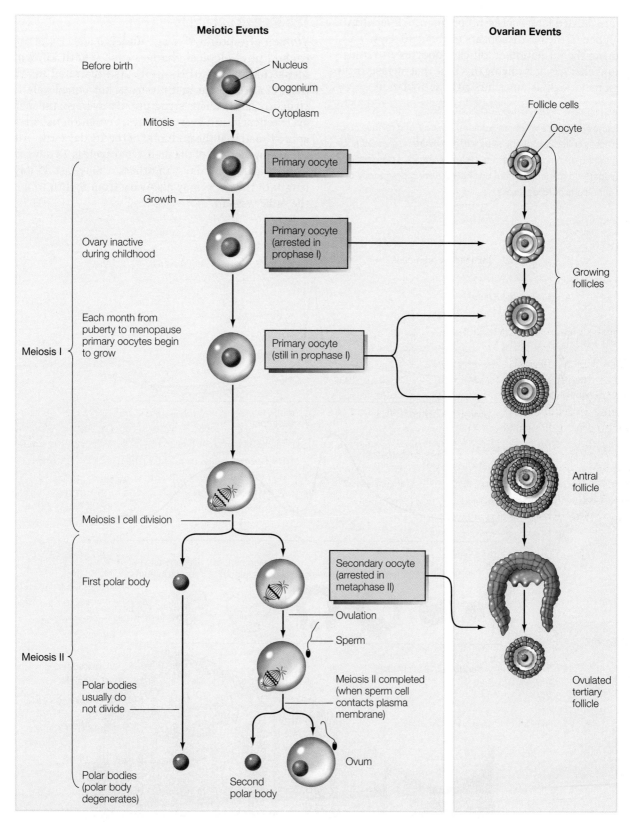

Meiotic Events

Ovarian Events

FIGURE 21–10 **Oogenesis and Follicle Development**

mulating more fluid. As the fluid builds up, the antral follicle begins to bulge from the surface of the ovary. The pressure exerted on the outside of the ovary causes the ovary's surface to stretch and eventually burst. The oocyte is then released and the cells of the collapsed follicle begin to enlarge and secrete hormones. These cells form a structure called **corpus**

luteum (CORE-puss LEU-tee-um; "yellow body") or CL for short—so named because of the yellow pigment it contains in cows and pigs (Figure 21-9). The CL produces two sex hormones, estrogen and progesterone.

If the oocyte is fertilized, the CL remains active for several months, producing estrogen and progesterone, both of which are needed for a successful pregnancy. If fertilization does not occur, the CL disappears in about 10 days.

During the formation of follicles, oocytes also undergo important changes. It is during this time that meiosis begins. The first meiotic division occurs right at ovulation.

KEY CONCEPTS

The ovaries produce the female gametes, the ova, and release them in a process called ovulation; as a rule, one ovum is released each month during a woman's menstrual cycle. Menstrual cycles continue from puberty to menopause but are halted during pregnancy and other conditions such as stress.

The Menstrual Cycle

Women of reproductive age undergo a series of anatomical and physiological changes each month known as the **menstrual cycle** (MEN-strell). As illustrated in Figure 21-11, these changes occur in three areas: hormone levels, ovarian structure, and uterine structure. On average, the menstrual cycle repeats itself every 28 days. Ovulation usually occurs approximately at the midpoint of the 28-day cycle. Although the average length of the menstrual cycle is 28 days, in some women it lasts 25 days; in others it may last 35 days. The length of the cycle may also vary from month to month in the same woman.

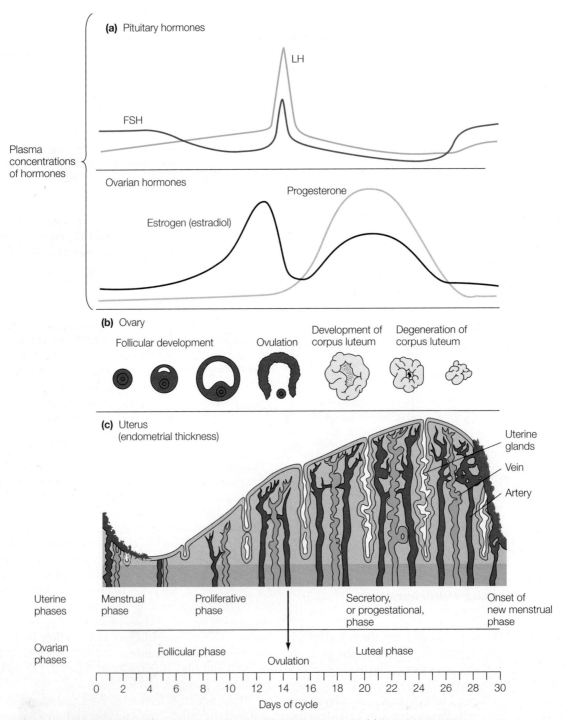

FIGURE 21–11 The Menstrual Cycle (a) Hormonal cycles. (b) The ovarian cycle. (c) The uterine cycle.

Figure 21-11, starting with the top panel, shows the changes in pituitary hormones, ovarian hormones, the ovary, and the uterus.

> **KEY CONCEPTS**
>
> The menstrual cycle consists of changes in pituitary and ovarian hormones, follicles inside the ovary, and the uterine lining.

Hormonal Control of Female Reproduction

As illustrated in the top panel of Figure 21-11, two pituitary gonadotropic hormones, FSH and LH, are released during the first half of the menstrual cycle. These hormones peak in the middle of the menstrual cycle, just before ovulation.

As its name implies, FSH stimulates follicular development. It does so by stimulating the mitotic division of the follicle cells. LH stimulates estrogen production by follicles and is responsible for the increase in estrogen during the first half of the menstrual cycle. My Ph.D. research at the University of Kansas School of Medicine showed that estrogen, like FSH, stimulates mitotic division of follicle cells, promoting follicle growth. Because the first half of the menstrual cycle is a time of follicular growth, it is referred to as the *follicular phase.*

FSH and LH secretion are controlled in a negative feedback mechanism involving estrogen and GnRH produced by the hypothalamus. Just before ovulation, however, something unusual happens: FSH and LH secretion surge. This peak in production is the result of one of the body's rarest events, a positive feedback loop, caused when estrogen levels reach a certain critical level. At this point, the hypothalamus responds with a sudden outpouring of GnRH. The pituitary responds with a sudden massive release of LH and FSH, which stimulates ovulation.

After ovulation, FSH and LH release decline sharply. The corpus luteum, shown in the third panel, forms from the collapsed follicle. It releases both estrogen and progesterone. The second half of the menstrual cycle is called the **luteal phase** (LU-tee-al), so named because the corpus luteum forms during this time.

> **KEY CONCEPTS**
>
> Two pituitary hormones stimulate growth of the follicles. FSH stimulates growth directly. LH stimulates growth by stimulating the production and release of estrogen that in turn, stimulates follicle growth.

Uterine Changes

Changes in ovarian hormones that occur during the menstrual cycle also have profound effects on the uterine lining, as shown in the bottom panel in Figure 21-11. As illustrated, the lining or endometrium thickens throughout much of the cycle in preparation for pregnancy. In the absence of fertilization, however, most of the thickened endometrium is shed, a process called **menstruation** (MEN-strew-A-shun).

To understand how the endometrium responds to hormonal changes, we begin on day 1 of the average 28-day menstrual cycle. Day 1 of the cycle marks the first day of menstruation. During the next 4 or 5 days, the uterine lining is shed—that is, the tissue that formed in the previous menstrual cycle sloughs (pronounced sluffs) off from the lining. It then passes out of the uterus into the vagina along with a considerable amount of blood—on average, about 50 to 150 milliliters. (This loss of blood explains why women are more prone to

develop anemia than men, as well as why women should take iron supplements or eat iron-rich foods such as spinach, clams, corn flakes cereal, and bran cereals.)

As soon as the endometrium has been shed, the lining of the uterus begins to rebuild—to prepare for the possibility of a pregnancy in the new cycle. Initial regrowth is stimulated by estrogen released by the ovaries. Estrogen stimulates the growth of glands in the endometrium and causes them to fill with a nutritive secretion that will nourish an embryo should fertilization occur. It also promotes cell division in the deepest layer of the endometrium, which causes it to grow thicker.

After ovulation, the endometrium continues to thicken under the influence of estrogen and progesterone in preparation for pregnancy. The uterine glands become swollen with a nourishing glycogen-rich secretion.

If fertilization does not occur, however, the uterine lining starts to shrink approximately 4 days before the end of the cycle. It then begins to slough off, starting menstruation. The shedding of the uterine lining is triggered by a decline in estrogen and progesterone concentrations in the blood.

When progesterone levels fall, the uterus begins to undergo periodic contractions. That's because progesterone inhibits smooth muscle contraction in the myometrium. These contractions propel the detached tissue out of the uterus and are responsible for the cramps many women experience during menstruation.

> **KEY CONCEPTS**
>
> The endometrium, the lining of the uterus thickens throughout much of the menstrual cycle in preparation for pregnancy. If fertilization does not occur, the thickened endometrium is shed in a process known as menstruation.

Effects of Fertilization

If fertilization occurs, the fertilized egg passes into the uterus and embeds in its thickened lining, from which the fertilized egg derives its nutrients early in pregnancy. If pregnancy is to continue, ovarian hormones must continue to be secreted. Ovarian hormone production, however, is now stimulated by a hormone produced not from the pituitary but from the newly formed embryo. This hormone is known as **human chorionic gonadotropin** (KO-ree-ON-ick), or HCG.

HCG has the same effect as LH—that is, it stimulates estrogen production by the ovary (in particular by the CL). Continued estrogen production keeps the uterine lining from deteriorating. HCG shows up in detectable levels in a woman's blood and urine about 10 days after fertilization. If the ovum is fertilized and the zygote successfully embeds in the uterine lining, HCG will maintain the CL for approximately 6 months.

Pregnancy tests available through a doctor's office or drugstore detect HCG in a woman's urine. The home pregnancy tests are relatively inexpensive, fairly reliable, and provide results almost immediately.

> **KEY CONCEPTS**
>
> If an ovum is fertilized, the newly formed embryo releases HCG. It stimulates estrogen production by the corpus luteum formed from the collapsed follicle. Estrogen prevents sloughing of the uterine lining and thus ensures the newly formed embryo has a suitable environment in which it can develop.

Estrogen

Like testosterone in boys, estrogen secretion in girls increases dramatically at puberty. As the level of estrogen in the blood increases, the hormone begins to stimulate follicle development in the ovaries. Menstrual cycles begin. Estrogen also stimulates the growth of the external genitalia and the breasts. It also stimulates growth of internal structures including the uterus, uterine tubes, and vagina.

Estrogen's influence extends far beyond the reproductive system. For example, estrogen promotes rapid bone growth in the early teens. Because estrogen secretion in girls usually occurs earlier than testosterone secretion in boys, girls typically experience a growth spurt before similarly aged boys. However, estrogen also stimulates the closure of the growth zones of the bones, putting an end to growth in females much earlier than in boys. Thus, most girls reach their full adult height by the ages of 15–17. Boys continue growing until the ages of 20–21. Besides promoting bone growth, estrogen stimulates the deposition of fat in women's hips, buttocks, and breasts, giving the female body its characteristic shape.

> **KEY CONCEPTS**
>
> The hormone estrogen plays a key role in follicle development, preparation of the uterus for pregnancy, growth of the breasts and external genitalia, and body growth. Estrogen secretion is responsible for the early growth spurt in teenage girls.

Health Tip 21-1

Eating organic fruits and vegetables, which have been grown without pesticides, may reduce a woman's risk of developing breast cancer.

Why?

Breast cancer has several known causes. Studies show that women with breast cancer are five to nine times more likely to have pesticide residues in their blood than those who do not.

Progesterone

Progesterone is sometimes referred to as the "hormone of pregnancy." As just noted, it promotes growth of the uterine lining that prepares it for implantation. It also calms muscle of the uterus during pregnancy, to prevent expulsion of embryos and fetuses. As you shall soon see, women experience a slight increase in body temperature during ovulation. This is caused by progesterone.

Like estrogen, progesterone affects many other body functions outside the female reproductive tract. Progesterone appears to act as an anti-inflammatory agent and also helps to regulate the immune system. It helps maintain homeostasis by regulating blood clotting and vascular tone as well as levels of zinc copper, and cellular oxygen levels. Some studies suggest that it protects women against endometrial cancer by opposing the actions of estrogen. The list goes on. And, it signals insulin release. It even affects fetal development. Fetuses, for instance, convert maternal progesterone into adrenal steroids.

> **KEY CONCEPTS**
>
> Progesterone plays a key role in reproduction in women, but also controls or assists in controlling many other important processes.

Premenstrual Syndrome

For reasons not yet fully understood, many women suffer from irritability, depression, fatigue, and headaches just before menstruation. Many also complain of bloating, tension, joint pain, and swelling and tenderness of the breasts. These complaints are symptomatic of a condition known as **premenstrual syndrome** (**PMS**).

PMS is a clinically recognizable condition characterized by one or more of the symptoms noted above. Four of every 10 women of reproductive age experience PMS in varying degrees. While most women of child-bearing age experienced physical symptoms, such as bloating or breast tenderness, a diagnosis of PMS is limited to those who consistently experience emotional and physical symptoms during the second half of the menstrual cycle (luteal phase) that are sufficiently severe to interfere with some aspects of their lives.

Symptoms of PMS vary from one woman to the next; however, each woman's symptoms are predictable. They occur consistently as well, during a 10-day period prior to menstruation. Symptoms disappear shortly before or shortly after the beginning of menstruation.

The cause of PMS is still unknown, but scientists have identified numerous factors that increase a woman's likelihood of experiencing it. These include high caffeine intake, stress, increasing age, a family history of depression, and dietary factors. Dietary factors include low levels of Vitamin E and D, as well as low levels of magnesium, manganese, and zinc.

Although scientists are still researching the causes of PMS, dozens of cures or partial cures have been suggested. They range from aerobic exercise; reductions in caffeine, sugar, and sodium intake; an increase in fiber consumption; adequate rest and sleep; hormonal therapy; antidepressants; and vitamins B_6; and E. Hormonal therapy consists of oral contraceptives combined with wearing a contraceptive patch. Buyers beware, however, for very little good scientific evidence is available to indicate which, if any, of these "cures" really work.

Fortunately, work is now underway to test various treatments to see if any of them consistently bring relief. In the meantime, physicians recommend that women suffering from PMS see their family doctor to be certain that the symptoms are not caused by some other medical problem. Doctors recommend relaxation and avoidance of stress to help avoid or relieve PMS. Light exercise may help, as may warm baths. More frequent light meals with plenty of carbohydrates and fiber may be beneficial, too. Reducing salt intake and avoiding excess chocolate consumption are also recommended. Cutting out caffeine drinks and taking vitamin B_6 supplements are also generally advised.

> **KEY CONCEPTS**
>
> PMS is characterized by numerous adverse mental and physical symptoms including irritability, bloating, stress, and depression that occur in women during the last half of the menstrual cycle.

Menopause

The menstrual cycle continues throughout the reproductive years. As a woman ages, however, estrogen levels begin to decline. This decline is brought about by a decrease in the number of ovarian follicles. (Remember that growing follicles produce estrogen.) As estrogen levels decline, ovulation and menstruation become increasingly erratic. Between the ages of 45 and 55, most of the follicles that were in the ovary at puberty have been stimulated to grow and have either degenerated or ovulated. This results in complete cessation of the menstrual cycle and is called **menopause**.

The decline in estrogen production results in several important body changes. For example, it causes the breasts and internal reproductive organs to begin to shrink. Vaginal secretions often decline, and, in some women, sexual intercourse becomes painful without artificial lubrication.

The decline in estrogen levels may also result in behavioral disturbances. Many women, for instance, become more irritable and suffer bouts of depression. Headaches, insomnia, and sleepiness may occur in some women. Very noticeable physical changes may also occur. Three-quarters of all women suffer "hot flashes"—that is, there are times in the day where they suddenly feel very hot. Three-quarters of all women suffer "night sweats"—that is, they sweat profusely at night while sleeping. Hot flashes and night sweats are induced by massive vasodilation (expansion of arterioles feeding capillary beds) in the skin. For some women, however, the symptoms are quite mild. Fortunately for all people concerned, these symptoms usually pass.

Declining estrogen levels also accelerate osteoporosis, a softening of the bone due to a reduction in calcium. To counter osteoporosis and relieve other unwelcome changes taking place during menopause, physicians can prescribe small amounts of estrogen or a combination of estrogen and progesterone. This is known as **hormone replacement therapy**. Historically, women have been treated for 5 to 10 years after the onset of menopause. New data suggest that hormone replacement therapy should not last longer than 5 years in order to maximize its benefits and minimize a woman's risk of developing breast cancer. Osteoporosis can also be prevented, even reversed, with new drugs and exercise. As noted in **Health Tip 21-2**, vitamin D can also help protect women from developing osteoporosis.

> **KEY CONCEPTS**
>
> Menopause is the cessation of ovulation and menstrual cycles that typically occurs in women between the ages of 45 and 55. Caused by declining estrogen levels, menopause results in temporary behavioral changes such as irritability and depression as well as temporary physical changes such as night sweats. It can also lead to osteoporosis.

Health Tip 21-2

As you get older, increase your intake of vitamin D.

Why?

Studies show that consuming vitamin D reduces the risk of bone fractures among elderly men and women. However, a recent study showed that only those who consumed 300 to 400 IU (international units) of vitamin D above the recommended daily intake of 400 to 600 IU enjoyed the benefit. They suffered 25% fewer fractures. That is, the elderly should consume 700 to 1,000 IU of vitamin D every day.

21-3   Birth Control

Birth control is any method or device that prevents births. Birth control measures fit into two broad categories: (1) contraception, ways of preventing pregnancy, and (2) induced abortion, the deliberate expulsion of a fetus.

Contraception

Figure 21-12 summarizes the effectiveness of the most common means of contraception. Effectiveness is expressed as a percentage. A 95% effectiveness rating means that 95 women out of 100 using a certain method in a year will not become pregnant.

> **KEY CONCEPTS**
>
> Contraception is any technique that reduces one's chances of pregnancy. It involves measures taken by both men and women.

Abstinence and Surgical Sterilization

The most effective means of contraception is **abstinence**, refraining from sexual intercourse. This form of birth control not only reduces unwanted pregnancy, it prevents the transmission of sexually transmitted diseases (see **A Closer Look: Sexually Transmitted Diseases**).

The next most effective means of preventing unwanted pregnancies is **surgical sterilization**. It is also the leading method of contraception practiced by married couples in the United States.

In women, sterilization is performed by cutting or cauterizing the uterine tubes. This technique is called **tubal ligation** and is shown in **Figure 21-13a**. Tubal ligations, carried out in a hospital, are performed by making tiny incisions in the abdomen to access the uterine tubes.

Male sterilization is known as a **vasectomy** (vah-SECK-toe-me), and can be carried out in a physician's office under local anesthesia (Figure 21-13b). To perform a vasectomy, a physician makes a small incision in the scrotum. Each vas deferens is exposed, cut, and the free ends are tied off or cauterized.

Vasectomies only prevent the sperm from passing into the urethra during ejaculation. They do not impair sex drive, and because they do not block the flow of the sex accessory glands, which produce 99% of the volume of the ejaculate, they have virtually no effect on ejaculation. Vasectomy and tubal ligation can be reversed through microsurgery, although the operation is not always successful.

> **KEY CONCEPTS**
>
> Two of the most effective means of contraception are abstinence and surgical sterilization. In men, sterilization is achieved by cutting and tying off the vas deferens. In women, it is achieved by tubal ligation, cutting and sealing the oviducts.

Hormonal Implants, IUDs, and Birth Control Pills

The next most effective means of birth control are implants and shots, both of which contain a synthetic form of progesterone. **Hormonal implants** are matchstick-size capsules that

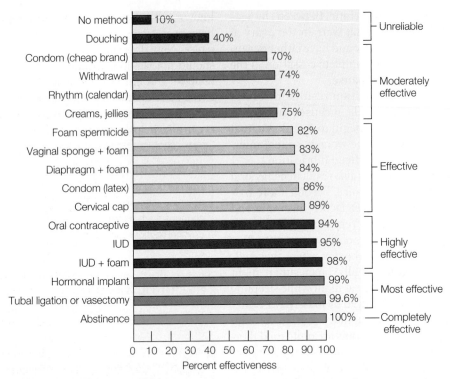

FIGURE 21–12 **Effectiveness of Contraceptive Measures** Percent effectiveness is a measure of the number of women in a group of 100 who will not become pregnant in a year.

contain a potent form of progesterone (Norplant). They are inserted under the skin of a woman's arm (Figure 21-14). They are effective in preventing pregnancy for up to 3 years. **Shots** also include a potent synthetic progesterone and are effective for 12 to 13 weeks.

Progesterone inhibits ovulation but also causes the mucus in the cervix to thicken, preventing sperm from entering the uterus.

Another highly effective contraceptive device is the **intrauterine device** (**IUD**; Figure 21-15). The IUD consists of a small plastic or metal object inserted into the uterus by a physician. Two types are used. One type contains copper that is slowly released into the uterine cavity. Copper impairs sperm movement through the vagina and uterus, thus preventing fertilization. The other type releases a synthetic form of progesterone called progestin. It causes the cervical mucus to thicken,

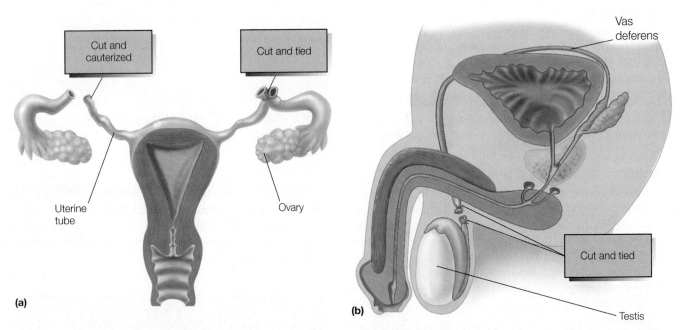

FIGURE 21–13 **Sterilization Methods** (a) In a tubal ligation, the uterine tubes are cut, then tied off or cauterized. (b) In a vasectomy, the vasa deferentia are cut, and then tied off.

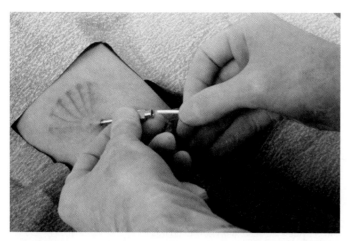

FIGURE 21-14 Subcutaneous Progesterone Implant Inserted under the skin, this tiny device releases a steady stream of progesterone, blocking ovulation for months. (© Hattie Young/Science Source, Inc.)

impeding sperm movement into the uterus. Progestin also alters the endometrium in ways that prevent implantation of fertilized eggs. They are left in place either 5 or 10 years, depending on the type.

Like all forms of contraception, the IUD has adverse impacts. In some cases, the uterus expels the device, leaving a woman unprotected. IUDs may also cause abdominal cramps in some women—ranging from mild to severe—and increase menstrual bleeding. In rare instances, they can cause uterine infections and perforation (a penetration of the uterine wall by an IUD). The IUD is slightly more effective when combined with spermicidal foams inserted in the vagina right before intercourse.

The **birth control pill** is yet another highly effective method—and is one of the most popular. Birth control pills come in several varieties, but the most common contains a mixture of synthetic estrogen and progesterone. These hormones inhibit the release of LH and FSH by acting on the pituitary and hypothalamus. The lack of LH and FSH, in turn, inhibits follicle development and ovulation. Progesterone

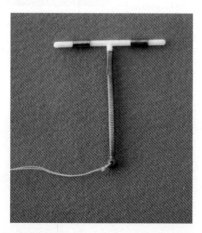

FIGURE 21-15 The IUD IUDs come in a variety of shapes and sizes and are inserted into the uterus, where they prevent implantation. Only one type is currently legal in the United States. (© Jones & Bartlett Learning. Photographed by Kimberly Potvin.)

in the birth control pill also thickens the mucus in a woman's cervix, blocking sperm from entering the uterus. Birth control pills can also prevent pregnancy by rendering the lining of the uterus inhospitable to a fertilized egg. Birth control pills are nearly 100% effective in preventing pregnancy. Unfortunately, skipping even a few days may release the pituitary and hypothalamus from the inhibitory influences of estrogen and progesterone, resulting in ovulation and possible pregnancy.

A minipill containing progesterone alone is also available. Even though it is less effective than the combined pill (95% vs. 99.9%), the minipill is more suitable for some women because it has fewer side effects.

Birth control pills come packaged in a thin case containing either 21 or 28 pills. Twenty-one-day packages contain 21 active pills—that is, pills containing hormones. Women take the pills for 21 days, and then wait 7 days to start the next pack. During that period, they menstruate. Packages with 28 pills contain 21 active pills and 7 pills with no hormones to help women keep track of them. During the time they are taking placebos, women menstruate.

Pharmaceutical companies have developed a new option for hormonal contraceptives: the **extended-cycle pill**. The first approved extended-cycle pill, Seasonale, contains low doses of progesterone and estrogen taken continuously for 12 weeks, followed by one week of inactive pills. With this regime, yearly menstrual periods are reduced from 13 a year to four.

Although effective, birth control pills may have adverse health effects in some women. Even though the incidence of these adverse side effects is small, a woman considering different birth control options should study them carefully before making a decision.

One very rare side effect is death. Deaths from the use of birth control pills may result from heart attacks, strokes, or blood clots. To reduce this risk, pharmaceutical companies have dramatically lowered the estrogen content of the combined pill, because estrogen is responsible for most of the adverse effects.

Early studies showed a positive correlation between the use of birth control pills and cancers of the breast and cervix (see Health Note 21-1). More recent studies suggest that the new generation of low-estrogen pills is less likely to cause cancer of the breast or cervix.

Even with reduced estrogen levels, however, women who take birth control pills are more likely to develop cervical cancer than women who do not. Consequently, physicians recommend annual Pap smears for women on the pill.

During a Pap smear, the cervical lining is swabbed. The swab picks up cells sloughed off by the lining of the cervix. The cells are then examined under a microscope for signs of cervical cancer. Early diagnosis increases a woman's chances of survival.

KEY CONCEPTS

Hormonal implants, IUDs, and birth control pills are fairly effective means of contraception. Hormonal implants and birth control pills contain hormones that block ovulation, thus preventing pregnancy; IUDs impairs the implantation of a fertilized egg in the uterine lining.

health**note**

21–1 Breast Cancer: Early Detection Is the Key to Survival

Breast cancer is one of the most common cancers affecting women in the United States. Each year, over 200,000 cases of breast cancer are detected in this country. Breast cancer is a fairly lethal disease. One of every five women who contracts breast cancer will die from it.

Doctors and medical researchers have uncovered many risk factors associated with this disease. Numerous studies, for instance, show that the risk of breast cancer increases as a woman ages. In fact, the risk doubles every 10 years of a woman's life. Most women who are diagnosed with breast cancer are in their 50s; the disease is very rare in women under 30 years of age.

Studies also indicate that estrogen is a risk factor for breast cancer. Generally speaking, the longer a woman is exposed to estrogen, the higher her chances of contracting the disease. Because of this, the early onset of menstruation is another risk factor, as is the late onset of menopause. Both of these risk factors are thought to be related to overall estrogen exposure.

Estrogen in birth control pills may also slightly increase a woman's chances of developing a cancerous breast tumor. Estrogen given to post-menopausal women for 10 years or more can elevate a woman's risk, too.

Obesity also increases a woman's risk. Studies show that excess body fat causes an increase in estrogen. For reasons not completely understood, the age of a woman at the time of her first child's birth also affects her risk of developing breast cancer. If a woman's first child is born after her thirtieth birthday, her chances of developing breast cancer are two times higher than for a woman whose first child was born before her twentieth birthday. If she never has children, her chances of developing breast cancer are also twice as high as a woman who gives birth to a child before age 20.

Some women inherit a higher risk for breast cancer. There are at least two abnormal genes that can be linked to this disease. Women who inherit these genes often elect to have their breasts removed as a kind of pre-emptive strike.

Researchers have also discovered factors that help reduce the risk of breast cancer. A low animal-fat diet is one of them. Exercise helps as well, as noted in Health Tip 21-3. Breast feeding for 2 years or more also seems to lower risk.

Early detection is a key to beating this disease. The earlier it is detected, the more likely a woman is to survive. Because of this, for years many doc-

tors have recommended that women over 40 have a routine mammogram, an X-ray of the breast, every 1 to 2 years. Women over 50 should have a mammogram every year (Figure 1). In 2009, however, a study of the risks and benefits of mammography suggested that routine mammography should not begin until age 50. Women between 50 and 70, the researchers stated, should receive routine mammograms every other year. Women in the 40-to-50-year-old age group should begin screening based on individual risk factors. In other words, those in the highest risk groups should receive mammograms.

Women should also perform regular breast self-exams. The best time to perform a self-exam is immediately after a menstrual period. Doctors will provide instructions on how to do this.

Most lumps women detect are harmless, but they should be checked out immediately by a physician.

Breast cancer is typically treated by surgery. If the tumor is small, the surgeon will remove it along with some of the surrounding tissue. This procedure is known as a lumpectomy. If the tumor is large, or if there are multiple tumors, all of the tissue in the breast will be removed. This procedure is known as a mastectomy (mass-teck-toe-me). The surgeon usually removes a few lymph nodes from the armpits as well, because breast cancer frequently spreads to the lymph nodes.

Surgery is typically followed up by radiation therapy, especially if the tumor was large or if the cancer had spread to the lymph nodes. Women with estrogen-sensitive tumors are typically treated with the drug tamoxifen. This drug blocks estrogen's effect and causes the tumor to shrink. In other instances, doctors may recommend chemotherapy.

Studies show that a combination of drug treatment and surgery enhance a woman's chances of survival. Nine of every 10 women whose tumors are detected early and treated immediately survive 10 years or more. If a tumor has spread to other organs, however, chances of survival are much lower.

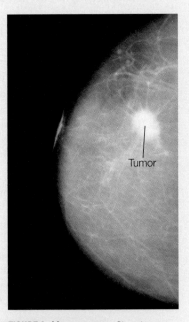

FIGURE 1 **Mammogram** Showing a small tumor. (© Kings College School of Medicine/Science Source.)

Barrier Methods: Cervical Caps, Condoms, Diaphragms, and Vaginal Sponges

The next most effective means of birth control are the barrier methods—the diaphragm, cervical cap, condom, and vaginal sponge—all of which prevent the sperm from entering the uterus.

The **diaphragm** (DIE-ah-FRAM) is a rubber cup that fits over the end of the cervix (Figure 21-16). To increase its effectiveness, a spermicidal (sperm-killing) jelly, foam, or cream should be applied to the rim and inside surface of the cup.

Health Tip 21-3

Regular exercise is essential to good cardiac health and also reduces a woman's chances of developing breast cancer. Try to get at least 30 minutes of aerobic exercise like running or riding a bike at least 3 times a week.

Why?

Exercise helps a woman maintain proper weight, which reduces the risk of many diseases from heart attack and stroke to late-onset diabetes to breast cancer.

FIGURE 21-16 **The Diaphragm** Worn over the cervix, the diaphragm is coated with spermicidal jelly or cream and is an effective barrier to sperm. (© Jones & Bartlett Learning. Photographed by Kimberly Potvin.)

Smaller versions of the diaphragm, called **cervical caps**, are also available. Fitting over the end of the cervix, the cervical cap most often used is held in place by suction.

Condoms are thin, latex rubber sheaths that fit onto the erect penis (Figure 21-17). Sperm released during ejaculation are trapped inside (in a small reservoir at the tip of the condom) and are therefore prevented from entering the vagina. Besides preventing fertilization, condoms also protect against sexually transmitted diseases, a benefit not offered by any other birth control measure except abstinence.

Yet another barrier method is the **vaginal sponge** (Figure 21-18). This small absorbent piece of foam is impregnated with spermicidal foam. Inserted into the vagina, the sponge is positioned over the end of the cervix. The sponge is effective immediately after placement and remains effective for 24 hours. Cervical sponges can be purchased without a doctor's prescription. Like the condom, one size fits all.

As mentioned earlier, spermicidal jellies, creams, foams, and films contain chemical agents that kill sperm but are

FIGURE 21-18 **The Vaginal Sponge** Infused with a spermicidal chemical, the vaginal sponge is inserted into the vagina and is effective for up to 24 hours. (Courtesy of Tekoa King.)

apparently harmless to the woman. Spermicidal preparations are most often used in conjunction with diaphragms, condoms, and cervical caps. Foams are about 82% effective, whereas spermicidal creams and jellies are about 75% effective.

> **KEY CONCEPTS**
>
> Chemical and physical barriers that prevent sperm from entering the uterus thus preventing fertilization include cervical caps, condoms, diaphragms, vaginal sponges, and spermicidal preparations.

Withdrawal and the Rhythm Method

One of the oldest—but least successful—means of birth control is **withdrawal**, removing the penis prior to ejaculation. This method requires tremendous willpower and frequently fails, for three reasons: because couples often throw caution to the wind in the heat of passion, because the penis is withdrawn too late, or because of pre-ejaculatory leakage—the release of a few drops of sperm-filled semen before ejaculation.

Abstaining from sexual intercourse around the time of ovulation—the **rhythm**, or **calendar**, **method**—also can help couples reduce the likelihood of pregnancy. If a couple knows the exact time of ovulation, they can time sexual intercourse to prevent pregnancy.

To determine when ovulation occurs, women can keep records on the length of their menstrual cycles, sample cervical secretions, or take body temperature just after awakening each morning. In most women, body temperature rises one-half to one degree right after ovulation as a result

FIGURE 21-17 **The Condom** Worn over the penis during sexual intercourse, it prevents sperm from entering the vagina. (© Andriy Rovenko/ShutterStock, Inc.)

of rising levels of progesterone. By keeping a temperature record over several menstrual cycles, a woman can determine the length of her cycle and the time of ovulation. Once the length of the cycle and the time of ovulation have been determined, a couple can determine the days they should practice abstinence.

Another method used to time ovulation involves taking samples of the cervical mucus. Cervical mucus varies in consistency during the menstrual cycle. By testing its thickness on a daily basis, a woman can tell fairly accurately when she has ovulated. Because ova remain viable 12 to 24 hours after ovulation and sperm may remain alive in the female reproductive tract for up to 3 days, abstinence 4 days before and 4 days after the probable ovulation date should provide a margin of safety. This minimizes the chances of a viable sperm reaching a viable ovum. Unfortunately, some women experience the greatest sexual interest around the time of ovulation. Moreover, sexual intercourse after a period of abstinence may also advance the time of ovulation.

> **KEY CONCEPTS**
>
> The least effective means of contraception are withdrawal and the rhythm method.

Male Contraceptives

For years, most contraceptive measures have been developed for women, primarily because of the ease with which female reproductive processes can be blocked or halted. Considerable effort is underway to develop male contraceptives beyond the three traditional approaches: vasectomy, condoms, and withdrawal. Some methods are in the early or middle stages of research and development. That is, they have been tested on lab animals and are now being studied in human subjects. Others are close to being ready to market, yet at this writing (January 2013) none are currently available in the United States.

One method that is being studied in China is **vas occlusion**. Unlike vasectomies that seal or remove part of the vas deferens, vas occlusion methods block the vas deferens, preventing sperm from reaching the urethra. Scientists have developed injectable silicon plugs that leave the duct intact. Other methods of blocking the vas deferens under study are via clips, plugs, or even tiny valves. These devices are referred to as **intra vas devices** (IVDs; Figure 21-19). In India, scientists have created an injectable compound that is now in clinical trials. Like the IUD, the IVD is a reversible form of contraception. It is, however, not yet approved for use in any country.

Two pharmaceutical companies, working with a dozen universities and non-governmental organizations, are developing male contraceptive pills that are analogous to female hormonal contraceptive pills—that is, they contain a combination of hormones designed to suppress sperm production. Studies show that a combination of progestin and androgen is safe and effective. Male contraception may come in the form of a male hormone pill or a patch or gel applied to the skin.

Currently, researchers are working with numerous non-hormonal pharmaceuticals that have contraceptive effects. Two researchers in the United Kingdom, for instance, have

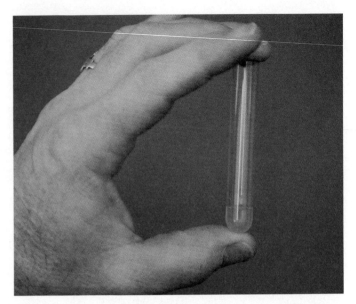

FIGURE 21-19 Intra Vas Device Material such as reversible inhibition of sperm under guidance (RISUG) (in the tube) inactivates the sperm before ejaculation. (© Male Contraception Information Project.)

found that two common medications used to treat high blood pressure and schizophrenia (phenoxybenzamine or thioridazine) disrupt the transport of sperm out of the testes. They do so by changing the way the smooth muscles of the vasa deferentia behave during orgasm. These drugs clamp the vasa shut, so that sperm are not released and mixed with the secretions of the sex accessory glands. These drugs cause infertility in men being treated for high blood pressure and schizophrenia (although the drug thioridazine has been discontinued because of extreme side effects). But these drugs may also have other effects that contribute to infertility. The researchers have identified a number of other drugs that produce similar results but with fewer side effects. They hope to develop a drug that will take effect within 2 to 3 hours of ingestion and will wear off within 24 hours. A man could take this pill only as needed before having sex.

Researchers are also studying drugs that disrupt sperm maturation in the testes. The drug Adjudin, for instance, changes the way Sertoli cells interact with developing sperm. In normal sperm production, Sertoli cells remain connected with immature sperm through a series of microscopic bridges and channels. These bridges provide materials and information needed to direct the development of spermatids.

As you learned earlier, spermatids undergo a series of changes in order to become functional sperm. These changes include condensation of the sperm's DNA, streamlining of the cell for fast swimming, and preparation of the cell membrane to recognize and fuse with an egg. When rats are treated with Adjudin, the spermatids are released from the Sertoli cells before the maturation processes are complete. The prematurely released sperm are therefore unable to fertilize an egg.

> **KEY CONCEPTS**
>
> Numerous efforts are underway to create more male contraceptive options including intra vas devices (IVDs) and chemical and hormonal contraceptives.

Abortion

Some couples may elect to terminate pregnancy through abortion.

Abortion is not suitable or morally acceptable to all people. Pro-life advocates argue that abortion should be outlawed or severely restricted—that is, allowed only in cases of rape, incest, and threat to the life of the mother. These individuals advise unmarried women to abstain from sexual intercourse, or, if they become pregnant, to give birth and either keep the baby or put it up for adoption.

Pro-choice advocates, on the other hand, support abortion. They argue that women should have the freedom to choose whether to terminate a pregnancy or have a child. Abortion, they say, reduces unwanted pregnancies and untold abuse and suffering among unwanted infants, especially in poor families, and gives women more options than motherhood. Nonetheless, pro-choice advocates often point out that abortion should not be practiced as a primary means of birth control. Abstinence and contraception are less costly, less traumatic, and more morally acceptable.

In the first 12 weeks of pregnancy, abortions can be performed surgically in a doctor's office via vacuum aspiration. In this procedure, the cervix is first dilated by a special instrument. Next, the contents of the uterus are drawn out through an aspirator tube. Vacuum aspiration is a fairly simple and relatively painless procedure. Usually no anesthesia is given. Although women bleed for a week or so after the procedure, they generally experience few complications.

From weeks 13 to 16, aspiration is supplemented by physical scraping of the lining of the uterus to ensure complete removal of the fetal tissue. After 16 weeks, abortions are more difficult and more risky. Solutions of salt, urea, or prostaglandins, which stimulate uterine contractions, are injected into the sac of fluid surrounding the fetus to induce premature labor. The hormone oxytocin may be administered to the woman with the same effect.

Implantation of an embryo can also be prevented by use of a pill called RU486, incorrectly dubbed by opponents as "the abortion pill." In the United States, doctors also recommend another birth control pill. Although it is called the "morning after pill," it actually consists of two to four pills that can be taken one to several days after unprotected sex. One product (Preven) consists of four pills, two taken within 3 days of sex, and the other two taken 12 hours later. Containing synthetic progesterone, this pill prevents implantation and is about 85% effective.

> **KEY CONCEPTS**
>
> Abortion is a controversial form of birth control that involves the intentional removal or expulsion of an embryo from the uterus.

21-4   Infertility

A surprisingly large percentage (about one in six) of U.S. couples cannot conceive a child. The inability to conceive is called **infertility**. According to some statistics, in about 50% of the couples, infertility results from problems occurring in the woman. Approximately 30% of the cases are due to problems in the man alone, and about 20% are the result of problems in both partners.

If after a year of actively trying to conceive, a couple is unsuccessful, they can consult a fertility specialist, who will first check obvious problems such as infrequent or poorly timed sex. If timing is not the problem, and it rarely is, the physician tests the man's sperm count. A low sperm count is one of the most common causes of male infertility and is easy to test.

A low sperm count may result from overwork, emotional stress, and fatigue. Excess tobacco and alcohol consumption can also contribute to the problem. Tight-fitting clothes and excess exercise, both of which raise the scrotal temperature, also tend to reduce the sperm count. The testes are also sensitive to a wide range of chemicals and drugs that reduce sperm production.

If infertility appears to be caused by a low sperm count, a couple may choose to undergo artificial insemination, using sperm from a sperm bank. These sperm are generally acquired from anonymous donors and are stored frozen. When thawed, the sperm are reactivated, then deposited in the woman's vagina or cervix around the time of ovulation.

If sperm production and ejaculation appear normal, a physician checks the woman's reproductive tract. Physicians first assess ovulation. If ovulation is not occurring, fertility drugs may be administered. Several kinds of drugs are available. One of the more common is HCG, which, as noted earlier, is an LH-like hormone that induces ovulation. Unfortunately, fertility drugs often result in superovulation (the ovulation of many fertilizable ova), leaving couples with a "litter" of 4 to 6 babies, instead of the one child they had hoped for. Most of the multiple births you hear about on the news are the result of fertility drugs.

If tests show that ovulation is occurring normally, the physician examines the uterine tubes to determine whether they are obstructed. In some instances, a previous sexually transmitted disease (for example, a gonorrheal or chlamydial infection) that spread into the tubes may have caused scarring that obstructed the passageway. In such instances, couples may be advised to adopt a child or to try *in vitro* (in VEE-trow) fertilization.

During in vitro fertilization, ova are surgically removed from the woman and fertilized by the partner's sperm outside her body. The fertilized ovum is then implanted in the uterus of the woman, where it can grow and develop successfully. Besides being expensive and time-consuming, this procedure has a low success rate.

> **KEY CONCEPTS**
>
> Approximately one in six U.S. couples suffers from infertility. Infertility results from numerous causes, most important of which are low sperm count, failure of ovulation, and blockages of the uterine tubes.

21-5   Health and Homeostasis

The human reproductive system is not essential to either homeostasis or the survival of an individual. It is, however, essential to the survival of our species.

Like other systems, the reproductive system is susceptible to numerous factors, including medicines like antibiotics, pollution, diet, and stress.

Numerous studies show that since the 1950s sperm count has been declining in young men in the United States. Prior to 1950, the average sperm count was about 110 million per ml. By 1980 and 1981, sperm counts had dropped to about 60 million per ml. Further studies show a continuing decline to fewer than 50 million in the late 1980s.

Statistical studies suggest that the decline may be related to growing pesticide use, air pollution, and other factors, including diet. A recent study in Denmark, for example, showed a strong correlation between low sperm counts and the consumption of saturated fats from meat and other fat-rich foods. Brazilian researchers found that men whose diets contained more grains had higher sperm counts.

Especially harmful may be chemicals in the environment that mimic the female sex hormone estrogen, such as dioxins, generated in paper production and waste incineration. Phthalates (pronounced THAL-ates), which are added to plastics and are present in many foods and beverages, may also be responsible.

Research suggests that exposing males to such chemicals early *in utero*, when they are extremely chemically sensitive, could have dramatic effects on their testes. At least 60 such chemicals have been identified to date.

Ten commonly prescribed antibiotics can also reduce sperm count. Tagamet, a drug that is used to relieve stress and treat excess stomach acidity, reduces sperm count by over 40%. By one estimate, at least 40 commonly used drugs depress sperm production. Thousands of other drugs and environmental pollutants have not been tested.

These facts do not necessarily mean that the United States is in a sperm crisis, but they do suggest the need for further research to determine the potential impacts, if any, of the many thousands of chemicals now commonly used or released into the environment. Research may prove that we need to clean up our act, or it may show that these fears are unwarranted.

> **KEY CONCEPTS**
>
> Reproduction is not essential to homeostasis, but it is essential to the continuation of our species. Numerous medications such as antibiotics and environmental pollutants such as pesticides and plasticizers contribute to low sperm count in men, a problem of epic proportions.

SUMMARY

The Male Reproductive System

1. The male reproductive tract consists of seven components: (1) testes, (2) epididymis, (3) vasa deferentia, (4) sex accessory glands, (5) urethra, (6) penis, and (7) scrotum.
2. The testes reside in the scrotum, which provides a suitable temperature for sperm development.
3. Sperm produced in the seminiferous tubules are stored in the epididymis. During ejaculation, sperm pass from the epididymis to the vas deferens, then to the urethra. Secretions from the sex accessory glands are added to the sperm during ejaculation, forming semen.
4. The interstitial cells of the testes lie between the seminiferous tubules and produce the hormone testosterone.
5. Testosterone secretion is controlled by a luteinizing hormone (LH) from the anterior pituitary.
6. Testosterone stimulates the formation of sperm as well as facial hair growth, thickening of the vocal cords, sebaceous gland secretion, and bone and muscle development.
7. The penis is the organ of copulation. It contains erectile tissue, which fills with blood during sexual arousal, making it rigid.

8. Ejaculation is under reflex control. Motor neurons in the spinal cord send impulses to the smooth muscle in the walls of the epididymis, the vasa deferentia, the sex accessory glands, and the urethra, causing ejaculation.
9. The prostate gland enlarges as a man ages, a condition known as benign prostatic hypertrophy or hyperplasia. BPH blocks the flow of urine but is usually not a serious condition.
10. Prostate cancer is common in older men but often grows very slowly.

The Female Reproductive System

11. The female reproductive system consists of two parts: the reproductive tract and the external genitalia.
12. The reproductive tract consists of (1) the uterus, (2) the two uterine tubes, (3) the two ovaries, and (4) the vagina.
13. The external genitalia consist of two flaps of skin on both sides of the vaginal opening, the labia majora and the labia minora.
14. Female germ cells are housed in follicles in the ovary.
15. A dozen or so follicles enlarge during each menstrual cycle, but most follicles degenerate. In humans, usually only one follicle makes it to ovulation during each cycle.

16. The oocyte is released during ovulation and then drawn into the uterine tubes.
17. The menstrual cycle consists of a series of changes occurring in the ovaries, uterus, and endocrine system of women.
18. The first half of the menstrual cycle is called the follicular phase. It is during this period that FSH from the pituitary stimulates follicle growth and development. LH stimulates estrogen production.
19. Estrogen levels rise slowly during the follicular phase, and then trigger a positive feedback mechanism that results in a surge of FSH and LH, which initiates ovulation.
20. The oocyte is expelled from the antral follicle at ovulation. The follicle then collapses and is converted into a corpus luteum (CL), which releases estrogen and progesterone.
21. In the absence of fertilization, the CL degenerates. If fertilization occurs, however, HCG from the embryo maintains the CL for approximately 6 months.
22. The CL produces estrogen and progesterone, essential to maintaining pregnancy.
23. During the menstrual cycle, ovarian hormones stimulate growth of the uterine lining, which is necessary for successful

implantation. If fertilization does not occur, the uterine lining is sloughed off during menstruation, which is triggered by a decline in ovarian estrogen and progesterone.

24. Like testosterone levels in boys, estrogen levels in girls increase at puberty. Estrogen promotes growth of the external genitalia, the reproductive tract, and bone. It also stimulates the deposition of fat in women's hips, buttocks, and breasts.

25. Progesterone works with estrogen to stimulate breast development. It also promotes growth of the uterine lining and inhibits uterine contractions.

26. Many women suffer from premenstrual syndrome (PMS), which is characterized by irritability, depression, tension, fatigue, headaches, bloating, swelling and tenderness of the breasts, and joint pain.

27. The menstrual cycle continues throughout the reproductive years, but as a woman ages, estrogen levels decline. Ovulation and menstruation become erratic as a woman approaches 45. Between the ages of 45 and 55, ovulation and menstruation cease. The end of reproductive function in women is known as menopause.

28. The decline in estrogen levels results in the atrophy of the reproductive organs and other symptoms such as irritability, depression, and hot flashes and night sweats induced by intense vasodilation of vessels in the skin.

Birth Control

29. Birth control refers broadly to any method or device that prevents births and includes two general strategies: contraception and induced abortion.

30. The most effective form of birth control is abstinence. Another highly effective measure is sterilization—vasectomy in men and tubal ligation in women.

31. Progesterone implants (under the skin) also prevent pregnancy and offer long-lasting protection up to 3 years.

32. Progesterone shots are also effective contraceptives, providing protection for 12 to 13 years.

33. The intrauterine device or IUD is a plastic or metal coil that is placed inside the uterus, remaining in place for 5 to 10 years, depending on the product.

34. The pill is a highly effective means of birth control. The most common pill in use today contains a mixture of estrogen and progesterone that inhibits ovulation. In some women, however, estrogen causes adverse health effects.

35. Birth control pills come in packs containing either 21 or 28 pills that are taken each month. Pharmaceutical companies have developed a new option for hormonal contraceptives, the extended-cycle pill that is taken continuously for 12 weeks, followed by one week of inactive pills.

36. The diaphragm, condom, and vaginal sponge are less effective measures of birth control.

37. The condom is a thin, latex rubber sheath worn over the penis that prevents sperm from entering the vagina during sexual intercourse.

38. The vaginal sponge is a tiny, round sponge worn by the woman. It is infused with a spermicidal chemical.

39. One of the oldest, but least effective, methods of birth control is withdrawal—removing the penis before ejaculation. Spermicidal chemicals used alone are about as effective as withdrawal.

40. Abstaining from sexual intercourse around the time of ovulation, known as the rhythm method, is another way to prevent pregnancy. Statistics on effectiveness show that the natural method is one of the least successful of all birth control measures.

41. Scientists are currently testing numerous male contraceptives, including oral contraceptives and IVDs, devices that block sperm entry into the vasa deferentia.

Infertility

42. Infertility, the inability to conceive, can result from a variety of problems in men and women, such as poorly timed or infrequent sex, low sperm count in men, and obstruction of the uterine tubes in women.

Health and Homeostasis

43. A variety of drugs and chemical pollutants affect sperm development and may be causing a decline in the sperm count of U.S. men.

THINKING CRITICALLY ANALYSIS

This analysis corresponds to the Thinking Critically scenario that was presented at the beginning of this chapter.

When asked, most people say that both hospitals would report the same number of days on which 60% of the children born were female. After all, half of all births each year are boys and half are girls. But in reality, we should expect more 60% days in the smaller hospital. Why would this be so?

As you learned, when performing experiments, scientists like a large sample size. The larger the sample size, the more reliable the outcome. That's because there's more variability of outcomes in smaller samples. That is, there's more chance that seemingly unrepresentative events will occur in a smaller sample, according to Thomas Kida, author of *Don't Believe Everything You Think.* Most of us don't recognize this fact. Small samples are not as representative of reality as large samples.

KEY TERMS AND CONCEPTS

CONCEPT REVIEW

1. You are a fertility specialist. A young woman arrives in your office complaining that she has been trying to get pregnant for 2 years, but to no avail. Describe how you would go about determining whether the problem was with her, her husband, or both of them. p. 461.
2. Describe the anatomy of the male reproductive system. List each organ and its role. pp. 443–445.
3. Where are sperm produced? Where are they stored? p. 443
4. What is semen? What structures produce the semen? pp. 444–446.
5. List the hormones that control testicular function. Where are they produced, and what effects do they have on the testes? p. 446.

6. How are male hormone levels controlled? p. 448.
7. Trace the pathway for a sperm from the seminiferous tubule to the site of fertilization. pp. 443–446.
8. What is an ovarian follicle? p. 450.
9. Describe the process of ovulation and the hormones that stimulate it? pp. 449–452.
10. What is FSH and where is it produced? Where is estrogen produced? p. 453.
11. What is the corpus luteum? How does it form? What does it produce? Why does it degenerate at the end of the menstrual cycle if fertilization does not occur? pp. 451–452.
12. What is menstruation? What triggers its onset? pp. 452–453.

13. Describe the effects of estrogen and progesterone on the reproductive tract and the body. p. 454.
14. A 50-year-old female friend of yours complains of irritability and depression. She says that she wakes up in the middle of the night in a sweat. Would you give her the name of a psychiatrist? Why or why not? If not, then what would you do? pp. 454–455.
15. Describe each of the following birth control measures, explaining what they are and how they work: the pill, IUD, diaphragm, cervical cap, condom, spermicidal jelly, hormonal implants, and natural method. pp. 455–461.

SELF-QUIZ: TESTING YOUR KNOWLEDGE

1. Sperm are produced in the testes inside the _____ tubules. p. 443.
2. The bulk of the ejaculate consists of fluids produced by the sex _____ glands. p. 444.
3. Sperm are stored in the _____, where they mature before ejaculation. p. 443.
4. The hormone _____ produced by the pituitary stimulates testosterone production in the testes. p. 446.
5. Testosterone is produced by the _____ cells in the testes. p. 446.
6. The hormone _____ produced by Sertoli cells inhibits FSH secretion in men. p. 448.

7. Interstitial cell stimulating hormone is produced by the _____ _____ gland. p. 448.
8. _____ tissue in the penis becomes engorged with blood during sexual arousal. p. 447.
9. The inability to achieve an erection is known as _____ dysfunction. p. 447.
10. In women, ova are produced by the _____, located in the pelvic cavity. p. 448.
11. Fertilization takes place in the upper one-third of the _____. p. 449.
12. The fertilized egg implants in the lining of the _____, where it develops. p. 447.

13. Female germ cells are found in structures inside the ovaries called _____. p. 450.
14. Ovulation is stimulated by the release of two hormones from the pituitary, LH and _____. p. 453.
15. A collapsed follicle is converted into a structure called a _____, which produces estrogen and progesterone. p. 453.
16. Human _____ gonadotropin is an LH-like hormone released by the newly developed embryo. p. 453.
17. Hormone replacement therapy in women helps treat symptoms of _____, including osteoporosis. p. 455

18. The most effective contraceptive method is _____. p. 455.

19. Surgical sterilization in women is called _____ ligation; in men it is known as _____. p. 455.

20. Hormonal implants contain a synthetic form of the hormone _____. p. 455.

21. The _____ is a device that is inserted into the uterus. One type contains _____ that is slowly released into the uterus where it impairs sperm movement through the vagina and uterus, preventing fertilization. The other type releases a synthetic progesterone called hormone progestin. p. 456.

22. Birth control pills come in several varieties, but the most common contains a mixture of synthetic _____ and _____. p. 457.

23. These hormones in birth control pills inhibit the release of _____ and _____, which inhibits follicle development and ovulation. p. 457.

24. During a _____ smear, the cervical lining is swabbed and the cells are examined under the microscope for signs of cervical cancer. p. 457.

25. The least effective means of contraception are _____ and the _____ method. p. 459.

26. Scientists have developed silicon plugs that are injected into the vasa deferentia. These are known as _____. p. 460.

biology.jbpub.com/chiras/8e/

The site features eLearning, an online review area that provides quizzes, chapter outlines, and other tools to help you study for your class. You can also follow useful links for in-depth information, research the differing views in the Point/Counterpoints, or keep up on the latest health news.

Sexually Transmitted Diseases

Certain bacteria and viruses are transmitted from one person to another by sexual contact. These organisms penetrate the lining of the reproductive tracts of men and women and thrive in the moist, warm environment of the body. Here these organisms often create local infections with symptoms such as painful urination and pus-like discharges. Some organisms primarily affect other parts of the body with profound consequences, even death. Syphilis and the HIV virus, for instance, are primarily systemic diseases—that is, they affect entire body systems. HIV impairs the immune system while syphilis affects the nervous and cardiovascular systems.

Diseases caused by bacteria and viruses that are transmitted by sexual contact are called **sexually transmitted diseases** (**STDs**). Most of the infectious agents that cause STDs are spread by vaginal intercourse, but other forms of sexual contact are also responsible for the transmission of these microbes. Syphilis, for example, is an STD caused by a bacterium that is spread by oral, anal, and vaginal sex.

One complicating factor in controlling STDs is that occasionally some diseases such as gonorrhea produce no obvious symptoms in many men and women. As a result, the disease can be transmitted without a person knowing he or she is infected. In other STDs, such as AIDS, symptoms may not appear for weeks or even years after the initial infection. Thus, sexually active individuals who are not monogamous can transmit the AIDS virus to many people before they are aware that they are infected. New blood tests are being used to detect HIV earlier and may help to reduce the spread of this deadly disease. In this section we will examine the most common STDs, except AIDS.

Gonorrhea

Gonorrhea (GON-or-REE-ah; referred to colloquially as the "clap") is an extremely common STD. The bacterium thrives in warm, moist areas of the male and female reproductive tracts, including the cervix, uterus, and Fallopian tubes of women, and the urethras of women and men. It can also proliferate in the mouth, throat, eyes, and anus of both men and women.

Gonorrhea often causes no symptoms. In men, when they do appear symptoms may include painful urination and a pus-like discharge from the urethra. Women may experience a cloudy vaginal discharge and lower abdominal pain. If a woman's urethra is infected, urination may be painful. Symptoms of gonorrhea usually appear about 1 to 14 days after sexual contact.

Gonorrhea is spread via anal, vaginal, and oral sex. It can be transmitted even if a man does not ejaculate. Gonorrhea can be transferred from a mother to her offspring during birth—that is, as the baby passes through the vagina during birth.

Gonorrhea is treated with antibiotics and clears up quickly, usually within 3 to 4 days, if treatment begins early. If left untreated, however, gonorrhea in men can spread to the prostate gland and the epididymis, which can be more difficult to treat. Infections in the urethra lead to the formation of scar tissue that may narrow the urethra, making urination difficult. In some women, bacterial infection spreads to the uterus and uterine tubes, causing the buildup of scar tissue. Scar tissue in the uterine tubes may block the passage of sperm and ova, resulting in infertility.

Gonorrheal infections can spread into the abdominal cavity through the opening of the uterine tubes. If the infection enters the bloodstream in men or women, it can travel throughout the body. Interestingly, if left untreated gonorrhea may increase a person's risk of either contracting or transmitting HIV.

The Centers for Disease Control and Prevention (CDC) notes that there were over 300,000 reported cases of gonorrhea in 2011, but estimates that 700,000 people contracted gonorrhea. (About half the cases were not reported to the CDC.) Although any sexually active person can contract this STD, the populations at highest risk are sexually active teenagers, young adults, and African Americans. Gonorrhea can be detected by microscopic examination of urine samples or swabs of infected areas, such as the oral cavity.

> **KEY CONCEPTS**
>
> Gonorrhea is an extremely common sexually transmitted disease that can spread to many organs.

Syphilis

Syphilis is an STD caused by a bacterium that penetrates the linings of the oral cavity, vagina, and penile urethra. It may also enter through breaks in the skin. It can also be transmitted from a mother through the placenta.

If untreated, syphilis proceeds through three stages. In stage 1, between 1 to 8 weeks after exposure, a small, painless red sore develops, usually in the genital area. Easily visible

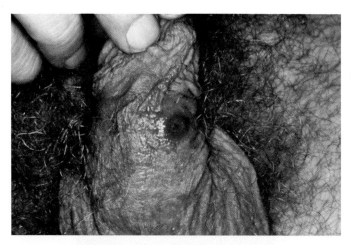

FIGURE 1 Stage 3 Syphilis Sores like this are common during stage 1 of syphilis. They disappear, sometimes leaving little evidence. (Courtesy of Dr. Gavin Hart and Dr. N. J. Fiumara/CDC.)

when on the penis, these sores often go unnoticed when they occur in the vagina or cervix (Figure 1). The sore heals in 1 to 5 weeks, leaving a tiny scar.

Approximately 6 weeks after the sore heals, individuals complain of fever, headache, and loss of appetite. Patients often display prominent nonitchy rashes that appear throughout the body, even on the palms of the hands and soles of the feet. Lymph nodes in the neck, groin, and armpit swell as the bacteria spread throughout the body. This is stage 2, and it lasts for about 4 to 12 weeks.

As a rule, the symptoms of stage 2 syphilis disappear for several years. In some individuals, the disease progresses no further. That is, it spontaneously resolves. In others, the disease flares up again, creating additional symptoms that persist until the individual dies. This is stage 3.

During stage 3, an autoimmune reaction occurs. Patients experience a loss of their sense of balance and a loss of sensation in their legs. As the disease progresses, individuals experience paralysis, senility, and even insanity. Some patients go blind. Huge ulcers may develop on the skin or on internal organs (Figure 2). In some cases, the bacterium weakens the walls of the aorta, causing aneurysms.

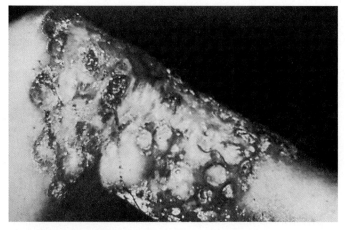

FIGURE 2 Stage 3 Syphilis This large open sore appeared on the skin of the patient during Stage 3 syphilis. These ulcerated sores may also appear on the internal organs. (Courtesy of Susan Lindsley/CDC.)

Children infected *in utero* may be stillborn or may be born blind, with numerous structural birth defects.

Syphilis is diagnosed by the patient's symptoms and by examining the pus under a microscope. Blood tests are also useful in stages 1 and 2. Syphilis can be successfully treated with antibiotics, but only if the treatment begins early. Suspicious sores in the mouth and genitals should be brought to the attention of a physician. In stage 3, antibiotics are useless. Tissue or organ damage is permanent.

Syphilis is one of the most serious STDs, though infection rates in the United States are much lower than gonorrhea. According to the CDC, there are over 46,000 new cases of syphilis each year. The majority of the cases occur in men who have sex with men.

As in gonorrhea, syphilis increases one's chances of transmitting and contracting HIV through oral, anal, and vaginal sex. That's because syphilis sores make it easier for an individual to transmit as well as acquire the HIV. In fact, the CDC estimates that a person is two to five times more likely to contract HIV if exposed sores are present. Like other sexually transmitted diseases, syphilis can be prevented by using condoms. Early diagnosis and treatment are also vital to preventing the spread. The CDC recommends,

> Because untreated syphilis in a pregnant woman can infect and kill her developing baby, every pregnant woman should receive prenatal care and be tested for syphilis during pregnancy and at delivery.

KEY CONCEPTS

Syphilis, caused by a bacterium, can be extremely debilitating if left untreated. There are three stages of the STD. Pregnant women should be tested for syphilis during pregnancy and at delivery to protect her neonate from infection or death.

Chlamydia

The most common sexually transmitted disease is **chlamydia** (clam-ID-ee-ah).

Caused by a bacterium, this STD is often referred to as "silent" infection. That's because most people infected by the bacterium show no symptoms. If symptoms do occur, they often do not appear until several weeks after infection. In men chlamydia is characterized by a burning sensation during urination and a discharge from the penis. Women also experience a burning sensation during urination and a vaginal discharge. If the bacterium spreads, it can cause more severe infection and infertility.

Like most other STDs, untreated chlamydial infections can spread to the uterus and uterine tubes, resulting in *pelvic inflammatory disease (PID)*. PID can be silent, that is, without symptoms, or can cause symptoms such as abdominal and pelvic pain. It is important to point out that even when chlamydia causes no symptoms, it can damage the reproductive organs of a woman, resulting in infertility. Damage to the uterine tubes can also result in a potentially lethal ectopic pregnancy—that is, a pregnancy outside the uterus.

Chlamydia can be transmitted from one individual to another during vaginal, anal, or oral sex. Like other STDs, many people experience no symptoms at all and therefore risk spreading the disease to others. Women risk being reinfected if their partner is not treated. Chlamydia can also be spread from a mother to her baby during childbirth. Children born to mothers with chlamydia can develop eye infections and pneumonia.

Chlamydia bacteria often migrate to the lymph nodes, where they cause considerable enlargement and tenderness in the affected area. Blockage of the lymph nodes may result in tissue swelling in the surrounding tissue.

Chlamydia is four times more prevalent than gonorrhea. The CDC estimates that there are approximately 2.8 million new cases of chlamydia each year! Although any sexually active man or woman can be infected with the bacterium, chlamydia is most common among young people—high school and college students. As in other STDs, men who have sex with other men are also at risk for chlamydial infection. Chlamydia is easily treated by doxycycline and other antibiotics.

> **KEY CONCEPTS**
>
> Chlamydia is the most common of the STDs, and can spread throughout the body because most people infected by the bacterium show no symptoms. Left untreated, chlamydial infections can spread to a woman's uterus and uterine tubes, resulting in pelvic inflammatory disease and infertility.

Genital Herpes

Genital herpes (HER-peas) is another common sexually transmitted disease caused by two closely related viruses, herpes simplex viruses type 1 (HSV-1) and type 2 (HSV-2).

HSV-1 can cause sores in the genital area as well as infections of the mouth and lips, which are typically referred to as *fever blisters*. HSV-1 is spread by oral sex and vaginal sex. As a rule, HSV-2 is transmitted via sexual contact with someone who has a genital HSV-2 infection. Unfortunately, the virus can be transmitted from one person to another even if the former exhibits no visible sores. In some cases, an individual who spreads the disease may not even know that he or she is infected. As in other STDs, herpes increases the transmission of the HIV virus because open sores allow HIV to spread more readily.

Herpes can also be spread from one part of the body to another. For instance, if a person with genital herpes touches his or her sores or the fluids released from them, he or she can transfer the virus to other parts of the body. Infections in sensitive location such as the eyes can be troublesome. As a result, individuals with genital herpes are advised not to touch the sores or fluids and to wash their hands thoroughly and immediately after contact to reduce the likelihood of transfer.

The viruses that cause this STD enter the body and remain there for life. The first sign of viral infection is pain, tenderness, or an itchy sensation on the penis or female external genitalia. These symptoms usually occur 6 days or so after contact with someone infected by the virus. Soon afterward, painful blisters appear on the external genitalia, thighs, buttocks, and cervix, or in the vagina (Figure 3).

The blisters break open and become painful ulcers that last for 1 to 3 weeks, and then disappear. Unfortunately, the herpes virus is a lifelong resident of the body. Continued outbreaks of genital herpes are fairly common, especially during the first year of infection. They often occur when an individual is under stress. Repeat outbreaks are typically less severe and shorter than the first one. Even though the herpes

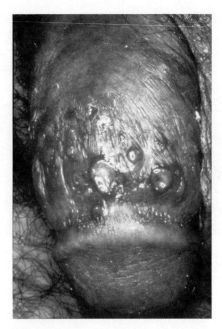

FIGURE 3 Genital Herpes Blisters on the external genitalia. (Courtesy of Dr. N. J. Flumara and Dr. Gavin Hart/CDC.)

viruses remain in the body for life, the number of outbreaks usually decreases over time and may cease altogether.

Transmission from an infected male to his female partner is more likely than from an infected female to her male partner. Herpes can be transmitted to other individuals during sexual contact only when the blisters are present or (as recent research suggests) just beginning to emerge. Even if a person does not exhibit symptoms, he or she can infect others. When the virus is inactive, sexual intercourse can occur without a partner becoming infected.

Although herpes cannot be cured, physicians can suppress outbreaks with antiviral drugs such as *acyclovir* (A-sigh-CLOE-ver). These drugs not only reduce the incidence of outbreaks but also accelerate healing of the blisters.

Herpes is not a particularly dangerous STD, except in pregnant women, who run the risk of transferring the virus to their infants at birth. Because the virus can be fatal to newborns, these women are often advised to deliver by cesarean section (an incision made just above the pubic bone) if the virus is active at the time of birth.

According to the CDC, "The surest way to avoid transmission of sexually transmitted diseases, including genital herpes, is to abstain from sexual contact, or to be in a long-term mutually monogamous relationship with a partner who has been tested and is known to be uninfected." Herpes can be prevented by the use of condoms; however, the virus can be transmitted to areas not covered by a condom.

Individuals infected with the herpes virus should abstain from all forms of sex when sores or other symptoms of herpes are present, advises the CDC. Antiviral medications can prevent or shorten outbreaks and lessen their severity. In addition, notes the CDC, "daily suppressive therapy (i.e., daily use of antiviral medication) for herpes can reduce the likelihood of transmission to partners."

According to the U.S. Department of Health and Human Services, an estimated 45 million Americans are infected with the virus, and up to 1 million new infections occur each year. Surprisingly, one of every five adolescent and adults is currently infected with genital herpes. Infection is a bit more common in women (approximately one in four) than in men (almost one out of five).

KEY CONCEPTS

Genital herpes is caused by the herpes simplex viruses type 1 (HSV-1) and type 2 (HSV-2) that remain in the body for life. Infection with the virus primarily results in painful blisters in the genital area, and surrounding areas.

Genital Warts

Genital warts are caused by any of 40 strains of **human papillomavirus** (HPV), which is common in adults.

According to the CDC, approximately 79 million Americans are infected with the HPV. They estimate that an additional 20 million people are infected each year and note that "HPV is so common that at least 50% of sexually active men and women get it at some point in their lives."

HPV is transmitted most often during vaginal and anal sex. It may also be transmitted during oral sex and genital-to-genital contact.

HPV-infected skin cells can develop into warts or cancers. Fortunately, in most cases, the immune system fights off HPV so infected areas can return to normal. In instances where the immune system cannot combat the infection, HPV can lead to genital warts or cancer. Genital warts may first appear within a few weeks or several months after being infected with HPV. Cancer typically takes many years to develop.

Genital warts are benign growths that appear on the external genitalia and around the anuses of men and women (Figure 4). Warts also grow inside the vagina of women. In rare instances, warts may form inside the mouth. Warts generally occur in individuals whose immune systems are suppressed, for example, after long periods of stress.

These warts can remain small or can grow to cover large areas, creating cosmetically unsightly growths. They may cause mild irritation, and certain strains of HPV are associated with cervical cancer in women. According to the CDC, the types of HPV that cause genital warts are not the same as strains of virus that cause cancers.

The CDC estimates that each year in the United States, approximately 12,000 women develop cervical cancer and that almost all of these cancers are caused by HPV. This virus can also cause cancer of other areas including the penis, vagina, anus, and oropharynx (back of the mouth). Fortunately, scientists have developed two vaccines against HPV that can prevent HPV infections: bivalent vaccine (Cervarix) and quadrivalent vaccine (Gardasil). Gardasil can also prevent the HPV types that cause most genital warts.

Genital warts can be treated with chemicals or removed surgically—although rates of recurrence are quite high. In 20% to 30% of the cases, genital warts disappear spontaneously. Getting rid of the virus, however, is impossible, for it resides in the body forever.

KEY CONCEPTS

Genital warts are benign growths caused by the human papilloma virus, which can appear in infected people with suppressed immune systems.

Trichomoniasis

Trichomoniasis (or "trich") is yet another very common STD. This one, however, is caused by a parasite, a protozoan known as *Trichomonas vaginalis*. Unfortunately, most individuals infected with the parasite are unaware of it.

According to the CDC, an estimated 3.7 million people in the United States are infected with this parasite; however, only about 30% develop any symptoms. Also according to the CDC, infection is more common in women than men, and older women are more likely than younger women to have been infected.

As in other STDs, the parasite is transmitted during sex. In women, the parasite commonly infects the lower genital tract, the external genitalia, vagina, and urethra. In men, the parasite most commonly invades the urethra in the penis. Unlike some other STDs, the parasite rarely infects other body parts.

Interestingly, about 70% of people infected with the parasite exhibit no signs or symptoms. When they do occur, symptoms range from mild irritation to severe inflammation. Symptoms may develop soon after infection or may appear much later. Symptoms may also reappear from time to time.

As in many other STDs, men feel an itching sensation or irritation in the urethra, a burning sensation after urination or ejaculation, and may report some discharge from the penis. Women may experience itching, burning, redness or soreness in the genitals and some discomfort during urination. Some experience a thin, unusually smelling discharge. Pregnant women may give birth earlier and may give birth to an underweight child.

According to the CDC, "trichomoniasis can increase the risk of getting or spreading other sexually transmitted infections," such as AIDS.

Trichomoniasis is considered the most readily cured STDs. It can be eliminated with a single dose of antibiotic medication (either metronidazole or tinidazole).

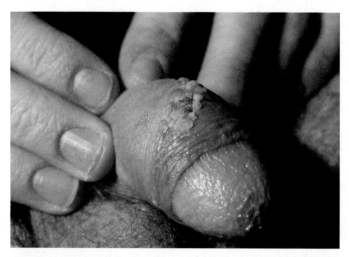

FIGURE 4 Genital Warts Genital warts on the penis are caused by infection of the skin by papilloma viruses. Genital warts can be removed by a variety of treatments but sometimes recur. (© Dr. P. Marazzi/Science Source, Inc.)

KEY CONCEPTS

Trichomoniasis, caused by a protozoan, is a curable STD.

Genetically Resistant STDs

Like other infectious microorganisms, those that cause STDs can become resistant to antibiotics. Gonorrhea, for instance, was once successfully treated with penicillin and more recently with another group of antibiotics (cephalosporin). In 2008, however, Japanese health officials discovered an antibiotic-resistant strain. Cephalosporin-resistant microbes have now reached in sizable numbers, according to a study published online in *The Journal of the American Medical Association*. This and other genetically resistant strains may leave millions of people worldwide with fewer treatment options, according to the World Health Organization.

At this writing, gonorrhea is the only STD that has developed genetic resistance to antibiotics in human populations. Should others occur, the health consequences could be quite serious.

Fortunately, the bacterium that causes the very common STD chlamydia are relatively stable bacteria—that is, they are slow to mutate. Even so, according to the journal *Future Microbiology*, scientists have discovered that drug-resistant chlamydia can develop in the lab.

The bacterium that causes syphilis is also slow to mutate, although some strains have developed resistance to erythromycin, according to Lola Stamm, an associate professor of epidemiology at the University of North Carolina's Gillings School of Global Public Health. Currently, penicillin is an effective treatment. "If it ever gets resistant to penicillin, that would be a scary day," Stamm says. "We don't have a lot of other options."

> **KEY CONCEPTS**
> Infectious microorganisms that cause STDs can become resistant to antibiotics.

Preventing STDs

Sexually transmitted diseases caused by bacteria can, for the most part, be treated with antibiotics, eliminating the disease organism. Even syphilis can be eliminated if it is treated early. STDs caused by viruses, however, are another story. Although drugs are available to suppress the outbreak of viruses like the one that causes genital herpes, the virus remains in the body forever.

One of the best medicines for sexually transmitted diseases is prevention through abstinence. Sexual relations within the confines of a monogamous relationship, in which both partners are STD-free, are also effective means of preventing the spread of STDs. Those who chose to partake in sex outside of a monogamous relationship are advised to use condoms. Condoms are thin latex sheaths worn over the penis during sexual intercourse. They prevent the exchange of body fluids and direct contact of the penis with the vaginal or anal canal lining, in the case of anal intercourse, preventing disease organisms from passing from one individual to another. Bear in mind, as noted earlier, skin not protected by the condom can pick up some forms of virus, such as the herpes virus.

Careful selection of partners is also a means of reducing the likelihood of contracting an STD. Those who engage in sex freely with other promiscuous individuals, for instance, are much more likely to contract a sexually transmitted disease than those who engage in sex with individuals who are not sexually promiscuous.

> **KEY CONCEPTS**
> Abstinence and safe sex practices like using condoms can help prevent the spread of sexually transmitted disease.

SUMMARY

1. Certain viruses and bacteria can be transmitted from one individual to another during sexual contact. Infections spread in this way are called sexually transmitted diseases (STDs).
2. STDs affect the reproductive organs of men and women, but can also spread to other organs, causing serious health effects, even death.
3. Many STDs are spread because individuals carrying the infectious agent are unaware they are infected. That's because symptoms may not appear immediately after infection or, in some cases, symptoms do not appear at all.
4. Gonorrhea is caused by a bacterium that commonly infects the urethra in men and the cervical canal in women. Overt symptoms of the infection are frequently not present, so people can spread the disease without knowing it. If left untreated, gonorrhea can spread to other organs, causing considerable damage.
5. Syphilis is a serious STD caused by a bacterium that penetrates the linings of the oral cavity, vagina, and penile urethra. If untreated, syphilis proceeds through three stages. It can be treated with antibiotics during the first two stages, but in stage 3, when damage to the brain and blood vessels is evident, treatment is ineffective.
6. Chlamydial infections, an extremely common bacterial STD, resemble gonorrhea.
7. Genital herpes is also a very common STD. It is caused by a virus. Once the virus enters the body, it remains for life. Blisters form on the genitals and sometimes on the thighs and buttocks. The blisters break open and become painful ulcers. At this stage, an individual is highly infectious. New outbreaks of the virus may occur from time to time, especially when an individual is under stress.
8. Genital warts are caused by HPV and occur in men and women whose immune systems are suppressed.
9. Trichomoniasis is a very common and curable STD caused by a protozoan parasite known as *Trichomonas vaginalis*. Unfortunately, most individuals infected with the parasite are unaware of it.
10. STDs can be prevented by abstinence, refraining from sex outside of monogamous relationships in which both partners are STD-free, careful selection of partners, and the use of condoms.
11. Early detection and treatment is also vital to preventing the spread of STDs.

KEY TERMS AND CONCEPTS

Chlamydia, p. 467
Genital herpes, p. 468
Genital warts, p. 469

Gonorrhea, p. 466
Human papillomavirus (HPV), p. 469
Sexually transmitted disease (STD), p. 466

Syphilis, p. 466
Trichomoniasis, p. 469

(© Monkey Business Images/ShutterStock, Inc.)

Human Development and Aging

THINKING CRITICALLY

Thanks to improvements in medicine, people are living longer. You may have heard this statement dozens of times in one form or another. It is repeated so often that most of us believe it implicitly. But is it true? Are people really living longer than they used to?

In a hospital room in a major city, two parents bring a newborn child into the world with the assistance of a nurse midwife. The wife holds their new daughter in the hospital bed, while three floors up in the geriatric ward, an elderly man quietly passes away from pancreatic cancer.

Birth and death are two of life's most dramatic acts. They're the subject of this chapter. To begin our journey, we start with the remarkable process that marks the beginning of human life, fertilization. We then turn our attention to the development of the fertilized ovum into an embryo and then into a fetus. We will conclude with a look at aging and death.

22-1   Fertilization

Fertilization, the uniting of sperm and egg, is the first step in human development. This process occurs in the upper third of the oviducts. To get there, sperm must travel from the vagina, the site of ejaculation, through the cervical canal. They must then travel through the uterus into the oviducts. Thirty minutes after ejaculation, those sperm that have successfully found their way into the cervical canal arrive at the junction of the uterus and oviducts or uterine tubes. Sperm then travel up the oviducts, where they may encounter a secondary oocyte (commonly referred to as an egg) released from the ovary during ovulation.

> **KEY CONCEPTS**
>
> Sperm are released into the vagina and travel through the reproductive tract to the upper third of the oviducts where fertilization takes place.

Passage through the Female Reproductive Tract

Although many millions of sperm are deposited in the vagina, only a tiny fraction make it into the oviducts (Figure 22-1). The rest are killed by the acidic secretions of the vagina or get lost along the way. Some sperm, for instance, get lost in the vagina and fail to find their way into the cervix. Many sperm that enter the cervix get lost in the folds of the inner lining. The same fate faces sperm that enter the uterus.

Studies based on laboratory animals suggest that in humans only 1 in 3 sperm makes it through the cervical canal and 1 in 1,000 makes it to the uterine tubes (Figure 22-1). Those that are killed by acidic secretions of the vagina or get lost in the vagina, cervix, uterus, or oviducts are eventually engulfed by cells in the lining of these organs.

> **KEY CONCEPTS**
>
> Tens of millions of sperm are deposited in the vagina but only a small number makes it to the site of fertilization.

Sperm Mobility

Sperm travel by swimming and by muscular contraction of the female reproductive tract. Sperm travel through the reproductive tract of women partly on their own, by swimming. This action is made possible by the whip-like tail of the sperm, known as the *flagellum*. Most transport, however, is the result of muscular contractions occurring in the walls of the uterus and uterine tubes. These contractions are stimulated by hormone-like substances in the semen, prostaglandins.

> **KEY CONCEPTS**
>
> Sperm movement through the female reproductive tract occurs as a result of action of the flagella and muscular contractions of the wall of the tract.

Sperm Penetration

After swimming around the secondary oocyte for a while, a process that dissolves away the protective coating on the sperm cells, sperm begin to wiggle their way through the layer of cells surrounding the female gamete. The cells surrounding the oocyte form the *corona radiata* (radiating crown). Getting through the closely attached cells is made possible by enzymes released from the acrosome, a stocking cap-like structure on the head of the sperm (Figure 22-2).

After passing through the *corona radiata*, sperm must digest their way through another barrier, the *zona pellucida* (the clear zone), a clear gel-like layer immediately surrounding the oocyte. Sperm penetrate it with the aid of additional acrosomal enzymes.

Sperm cells that penetrate the *zona pellucida* enter the space between the zona pellucida and the plasma membrane of the oocyte. As a rule, the first sperm cell to come in contact with the plasma membrane of the oocyte will fertilize it; all other sperm are excluded.

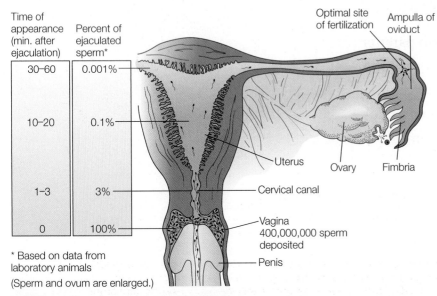

Time of appearance (min. after ejaculation)	Percent of ejaculated sperm*
30–60	0.001%
10–20	0.1%
1–3	3%
0	100%

* Based on data from laboratory animals

(Sperm and ovum are enlarged.)

Optimal site of fertilization — Ampulla of oviduct

Uterus

Ovary   Fimbria

Cervical canal

Vagina 400,000,000 sperm deposited

Penis

FIGURE 22-1   Sperm Transport in the Female Reproductive System Sperm move rapidly up the female reproductive tract of humans and other vertebrates principally as a result of contractions in the muscular walls of the uterus and uterine tubes. Notice the rapid decline in sperm number along the way.

22-2 Pre-Embryonic Development

Human development consists of three stages: pre-embryonic, embryonic, and fetal. **Pre-embryonic development** includes all the changes that occur from fertilization to the time just after an embryo implants in the uterine wall. During this phase, the zygote undergoes rapid cellular division and is soon converted into a solid ball of cells not much bigger than the fertilized ovum (Figure 22-4). This structure is called a **morula** from the Latin *morus* meaning mulberry. The morula is nourished by secretions produced by the lining of the uterine tubes. Approximately 3 to 4 days after ovulation, the morula enters the uterus.

Fluid soon begins to accumulate in the morula, converting it into a **blastocyst**, a hollow sphere of cells as shown in Figure 22-4. The blastocyst consists of a clump of cells, the inner cell mass, which will become the **embryo**, and a ring of flattened cells, the trophoblast ("to nourish the blastocyst"). The trophoblast gives rise to the embryonic portion of the **placenta**, an organ that supplies nutrients to and removes wastes from the embryo and fetus. The blastocyst remains unattached inside the uterus for 2 to 3 days. During this period, it is nourished by secretions of uterine glands.

> **KEY CONCEPTS**
>
> During pre-embryonic development the zygote is first converted to a morula, and then into a blastocyst consisting of the inner cell mass that will become the embryo and the trophoblast that will become the embryonic portion of the placenta.

Implantation

The blastocyst eventually attaches to the uterine lining and then digests its way into this thickened layer using enzymes released by its cells. This process is called **implantation**. It occurs 6 to 7 days after fertilization (Figure 22-5).

Most blastocysts implant high on the back wall of the uterus. The cells of the trophoblast first contact the endometrium, then adhere to it but only if the uterine lining is healthy and properly primed by estrogen and progesterone. If the endometrium is not ready or is "unhealthy"—for example, because of the presence of an IUD, use of a "morning after pill," or an endometrial infection—the blastocyst cannot implant. Blastocysts may also fail to implant if their cells

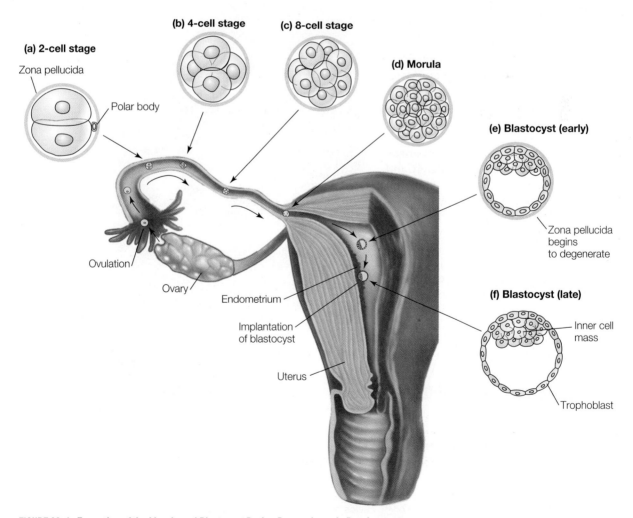

FIGURE 22-4 **Formation of the Morula and Blastocyst During Pre-embryonic Development**

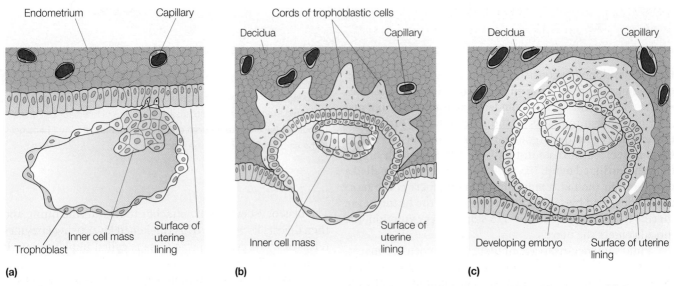

(a)   **(b)**   **(c)**

FIGURE 22-5 Implantation (a) The blastocyst fuses with the endometrial lining. Endometrial cells proliferate, forming the decidua. (b) The blastocyst digests its way into the endometrium. Cords of trophoblastic cells invade, digesting maternal tissue and providing nutrients for the developing blastocyst. (c) The blastocyst soon becomes completely embedded in the endometrium.

contain certain genetic mutations. Unimplanted blastocysts are absorbed (phagocytized) by the cells of the uterine lining, the endometrium, or are expelled during menstruation.

When implantation occurs, the cells of the endometrium at and around the point of contact enlarge, causing the lining of the uterus to thicken. Enzymes released by the cells of the trophoblast digest a small hole in the thickened endometrial lining, and the blastocyst "bores" its way into the deeper tissues of the uterine lining (Figure 22-5b). During this process, the blastocyst feeds on nutrients released from the cells it digests.

By day 14, the uterine endometrium grows over the blastocyst, enclosing it completely and walling it off from the uterine cavity. Endometrial cells respond to the invasion of the blastocyst by producing certain **prostaglandins**. These substances stimulate an increase in the development of uterine blood vessels, which ensures an ample supply of blood and nutrients for the blastocyst.

Soon thereafter, endometrial and embryonic tissue combine to form a structure known as the **placenta** (Figure 22-6). The placenta performs many important functions. It nourishes the embryo and fetus. It also provides oxygen and removes the wastes produced by embryos and fetuses throughout their residence inside the womb.

KEY CONCEPTS

The blastocyst embeds in the hormone-primed uterine lining. The blastocyst first digests its way into the thickened lining and then is completely surrounded by it.

Ectopic Pregnancy

Fertilized ova do not always make their way to the uterus for implantation. They may end up

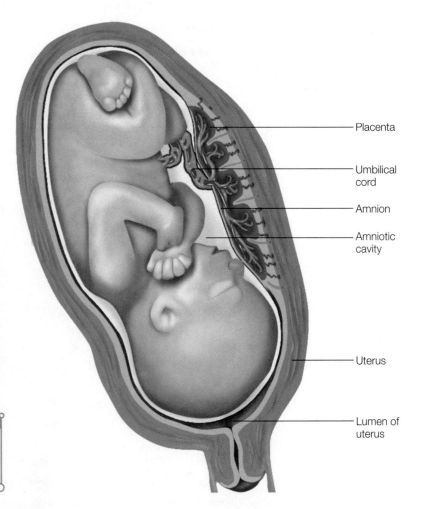

FIGURE 22-6 The Placenta This organ, made from maternal and fetal tissue, helps nourish the developing fetus and remove wastes. It also produces important hormones.

implanting in the uterine tubes, for example, because scar tissue caused by infections from one of several sexually transmitted disease organisms like gonorrhea blocks their passage into the uterus. They may also implant outside the reproductive tract altogether in the body cavity. That is, they may escape through the open end of the uterine tube. Implantation outside of the body of the uterus is known as an **ectopic pregnancy**. About 1 in every 200 pregnancies is ectopic.

In a tubal pregnancy (where the embryo implants in the uterine tubes), placental development occurs, but typically damages the uterine tubes. This causes internal bleeding and severe abdominal pain. Because the uterine tube cannot sustain the embryo, a tubal pregnancy cannot generally proceed to term. Surgery is required to remove the embryo.

KEY CONCEPTS

Blockages of the upper portion of the female reproductive tract often caused by infections from STDs can result in ectopic pregnancy, implantation of a blastocyst in the uterine tubes or body cavity.

Placental Hormones

The placenta also produces several hormones vital to a successful pregnancy. This section discusses three of the placenta's hormones: human chorionic gonadotropin (HCG), estrogen, and progesterone (Table 22-1). Figure 22-7 shows blood levels of these three hormones during pregnancy.

HCG is a hormone produced by the embryo early in pregnancy. Like LH from the pituitary of women, HCG stimulates estrogen and progesterone production by the corpus luteum or CL, a hormone-producing "gland" that forms in the ovary from the collapsed follicle after ovulation. These hormones are essential to maintaining pregnancy. However, as illustrated, HCG secretion from the embryonic tissue lasts

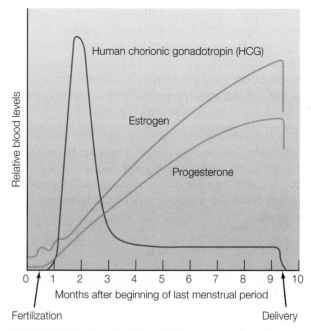

FIGURE 22-7 Blood Levels of Placental Hormones Human chorionic gonadotropin levels peak in the second month of pregnancy, then drop off by the end of the third month. Levels of estrogen and progesterone, produced chiefly by the placenta, continue to rise.

only about 10 weeks. Because of the natural decline in HCG, the CL degenerates. At this time, estrogen and progesterone production shifts to the placenta.

Estrogen and progesterone secreted by the placenta during the remainder of the pregnancy serve a number of essential functions. Estrogen, for example, stimulates growth of the smooth muscle cells of the uterus. This allows the organ to expand to many times its original size during pregnancy.

TABLE 22-1	Hormones Produced by the Placenta
Hormone	**Function**
Human chorionic gonadotropin (HCG)	Maintains corpus luteum of pregnancy
	Stimulates secretion of testosterone by developing testes in XY embryos
Estrogen (also secreted by corpus luteum of pregnancy)	Stimulates growth of myometrium, increasing uterine strength for parturition (childbirth)
	Helps prepare mammary glands for lactation
Progesterone (also secreted by corpus luteum of pregnancy)	Suppresses uterine contractions to provide quiet environment for fetus
	Promotes formation of cervical mucous plug to prevent uterine contamination
	Helps prepare mammary glands for lactation
Human chorionic somatomammotropin	Helps prepare mammary glands for lactation
	Believed to reduce maternal utilization of glucose so that greater quantities of glucose can be shunted to the fetus
Relaxin (also secreted by corpus luteum of pregnancy)	Softens cervix in preparation of cervical dilation at parturition
	Loosens connective tissue between pelvic bones in preparation for parturition

The additional muscle mass will also provide the additional propulsive force that's needed to expel the baby at birth.

Progesterone calms the uterine musculature during pregnancy, preventing the embryo and fetus from being expelled. It also stimulates the production of cervical mucus, which plugs the cervical canal, preventing bacteria from entering the uterus and infecting the growing embryo.

> **KEY CONCEPTS**
>
> The placenta produces three hormones essential for maintaining pregnancy: HCG, estrogen, and progesterone from the ovary. HCG stimulates the production of estrogen and progesterone. Estrogen allows the uterus to grow and progesterone calms the uterus.

The Amnion

As the placenta begins to form, the *inner cell mass (ICM)* of the blastocyst undergoes some remarkable changes. Early in development, a layer of cells separates from the ICM to form the **amnion** (AM-knee-on). A small cavity, the amniotic cavity, forms between the ICM and the amnion. The amniotic cavity fills with a liquid called **amniotic fluid**. It will eventually form a cushion around the baby during development, helping to protect it from injury (Figure 22-6).

> **KEY CONCEPTS**
>
> The amnion forms from the inner cell mass. It creates a cavity that fills with fluid that helps protect the embryo and fetus during development.

22-3   Embryonic Development

After the amnion forms: the cells of the inner cell mass differentiate, forming three distinct germ cell layers, the ectoderm, mesoderm, and endoderm. These are known as the *primary germ layers*. The formation of the primary germ layers marks the beginning of **embryonic development**.

The primary germ layers of the embryo give rise to the organs in a process called **organogenesis**. Organogenesis therefore is the main event of embryonic development. Table 22-2 shows the organs that form from each layer.

One of the first events of organogenesis is the formation of the central nervous system (the spinal cord and brain). It arises from the **ectoderm** of the ICM. As shown in Figure 22-8a, early in embryonic development, the ectoderm along the back of the embryo folds inward. This creates a long trench, the *neural groove*, that runs the length of the back surface of the embryo (Figure 22-8b). Over the next few weeks, the neural groove deepens and eventually closes off, creating the neural tube (Figure 22-8d). The walls of the neural tube thicken and form the *spinal cord*. In the head region, the neural tube expands to form the *brain*.

As you may recall from your study of the nervous system, numerous nerves attach to the spinal cord and brain. They're known as the *spinal nerves* and *cranial nerves*, respectively. These nerves develop from small aggregations of ectodermal cells, the **neural crest**, lying on either side of the *neural tube*, as shown in Figure 22-8f. These cells give rise to axons that grow into the body and attach to organs, muscle, bone, and skin. The ectoderm also gives rise to the outer layer of skin (the epidermis).

The middle germ layer, the **mesoderm**, gives rise to deeper structures—the muscle, cartilage, bone, and others. Much of the mesoderm first aggregates in blocks, called the *somites* (SO-mights; Figure 22-8d). The somites form the backbone and the muscles of the neck and trunk. Mesoderm lateral to the somites becomes the dermis of the skin, connective tissue, and the bones and muscles of the limbs.

The **endoderm**, the "lowermost" germ layer of the ICM, forms a large pouch under the embryo called the **yolk sac** (Figures 22-8a and 22-9). The uppermost part of the yolk sac becomes the lining of the intestinal tract. The yolk sac also gives rise to blood cells and primitive germ cells. During organogenesis, the germ cells migrate from the wall of the yolk sac to the developing testes and ovaries, which are located near the kidneys. These cells become spermatogonia in males or oogonia in females.

> **KEY CONCEPTS**
>
> The formation of organs occurs during embryonic development, which begins with the formation of the primary germ layers, the ectoderm which gives rise to the skin and nervous system, the mesoderm, which gives rise to the muscles and bone, and the endoderm, which gives rise to the lining of the intestinal tract.

TABLE 22-2	**End Products of Embryonic Germ Layers**	
Ectoderm	**Mesoderm**	**Endoderm**
Epidermis	Dermis	Lining of the digestive system
Hair, nails, sweat glands	All muscles of the body	Lining of the respiratory system
Brain and spinal cord	Cartilage	Urethra and urinary bladder
Cranial and spinal nerves	Bone	Gallbladder
Retina, lens, and cornea of eye	Blood	Liver and pancreas
Inner ear	All other connective tissue	Thyroid gland
Epithelium of nose, mouth, and anus	Blood vessels	Parathyroid gland
Enamel of teeth	Reproductive organs	Thymus
	Kidneys	

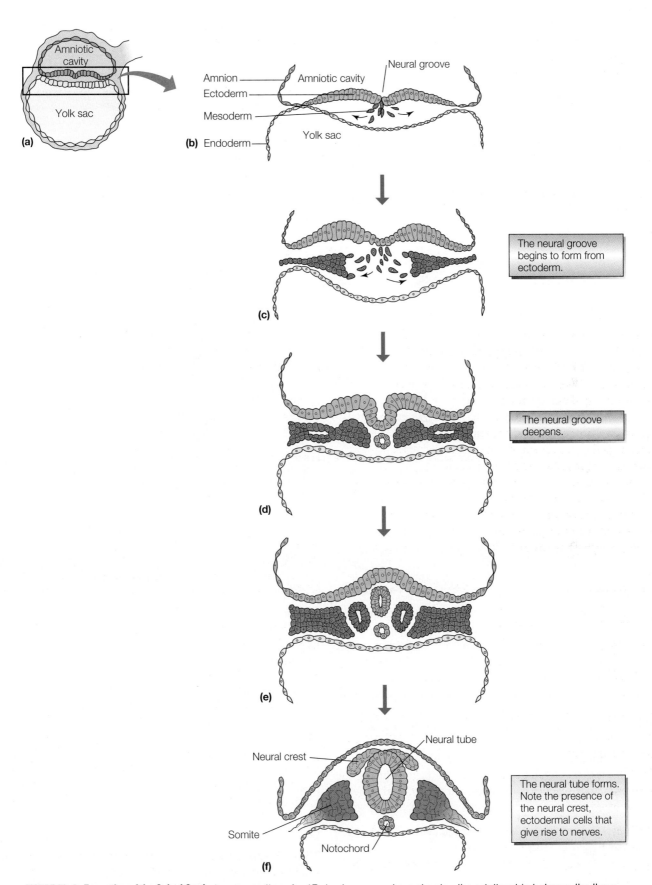

FIGURE 22–8 Formation of the Spinal Cord A cross section of a 17-day human embryo showing the relationship between the three embryonic tissues as well as the amniotic cavity and the yolk sac.

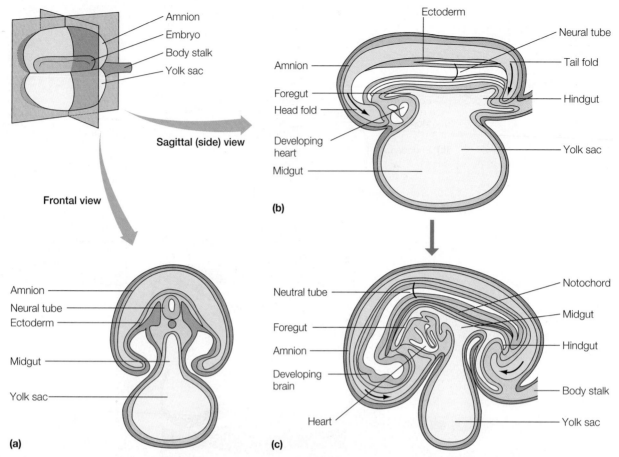

FIGURE 22–9 The Yolk Sac   (a) A cross section of the embryo. The yolk sac forms from embryonic endoderm. (b) A longitudinal section showing how the upper end of the yolk sac forms the embryonic gut, (c) which will become the lining of the intestinal tract.

22-4   Fetal Development

Fetal development begins in the eighth week of pregnancy and ends at birth. It involves (1) continued organ development and growth, and (2) changes in body proportions—for example, elongation of the limbs.

The fetus grows rapidly during the fetal period. It increases in length from about 2.5 centimeters (1 inch) to 35 to 50 centimeters (14 to 21 inches) and increases in weight from 1 gram to 3,000 to 4,000 grams (6.6 to 8.8 pounds) during this 28-week period. The fetus also undergoes considerable change in physical appearance, becoming more human-like as each month passes (Figure 22-10). Organ development that begins during the embryonic stage reaches completion during the fetal stage.

> **KEY CONCEPTS**
>
> Fetal development involves continued organ development and growth and changes in body proportions.

The Fetal Circulatory System

One organ system that does not quite reach completion during fetal development is the circulatory system. As shown in Figure 22-11a, the fetal circulatory system is much like the circulatory system of newborns (and adults) except for three bypasses—pathways that divert blood around two organs that are still growing and not yet functional: the lungs and liver.

Fetal blood circulates to and from the placenta in the umbilical cord. The cord contains two umbilical arteries and a single, large umbilical vein. The umbilical vein, however, carries oxygen- and nutrient-rich blood from the placenta to the fetus. Some of the blood flows into and through the fetal liver, as shown in Figure 22-11a. From the liver, the blood passes into the inferior vena cava and on to the heart. Because the liver is not yet functional, most of the blood bypasses the organ. It does so via a small shunt, a blood vessel known as the *ductus venosus*. It connects the umbilical vein directly to the inferior vena cava. Take a moment to locate it in Figure 22-11a.

Blood from the inferior vena cava flows into the right atrium of the heart. In an adult, all of the blood flows from the right atrium to the right ventricle, then is pumped to the lungs where it is oxygenated. In a fetus, however, only a small

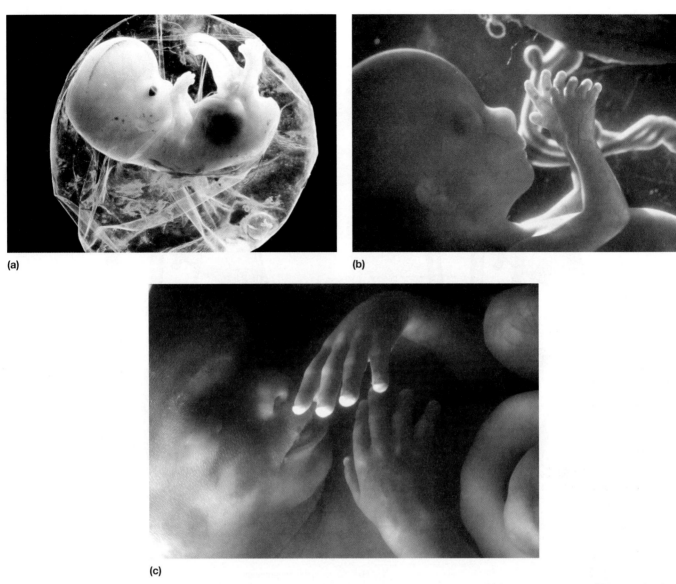

(a)

(b)

(c)

FIGURE 22-10 **Fetal Development** (a) Fetus at 5 to 6 weeks. (© Biophoto Associates/Science Source, Inc.) (b) Fetus at 4 months. (© Nestle/Petit Format/Science Source, Inc.) (c) Fetus at 5 months. (© Neil Bromhall/Science Source, Inc.)

portion of the blood goes into the right atrium, and then to the lungs where it supplies nutrients required for growth and development. The majority of the blood bypasses the lungs. It flows directly from the right ventricle to the left ventricle through a small hole in the wall between these two chambers. The hole is called the *foramen ovale*—literally, oval hole. It is shown in Figure 22-11a. Blood entering the left atrium is then pumped to the left ventricle. From there, it is pumped to the head and the rest of the body through the aorta and its many branches.

A third shunt also exists. Known as the *ductus arteriosus*, it lies between the pulmonary artery and the aorta (Figure 22-11a). Like the foramen ovale, it helps to divert blood away from the lungs.

Blood pumped throughout the body of the fetus via the aorta returns to the heart by the inferior and superior vena cavae. However, a large percentage of the blood that has been stripped of oxygen and filled with wastes of cellular activities returns to the placenta via the umbilical arteries.

In the placenta, the blood is rid of its waste and recharged with oxygen and vital nutrients picked up from the mother's bloodstream.

At birth, the fetal circulatory system quickly changes to the adult pattern. The foramen ovale normally closes at birth, establishing the adult circulation pattern. The two other shunts shrivel and close up, then eventually disappear. In some children, however, the foramen ovale remains open. Babies born with this defect usually appear blue because blood is unable to get to the lungs to be oxygenated. Blood deprived of oxygen appears slightly blue. Surgery is required to repair the defect.

KEY CONCEPTS

The fetal circulatory system receives oxygen and nutrients from and rids itself of wastes via the maternal blood system through the placenta. The fetal circulatory system contains three temporary shunts that allow blood to bypass organs that will not become functional until a baby is born.

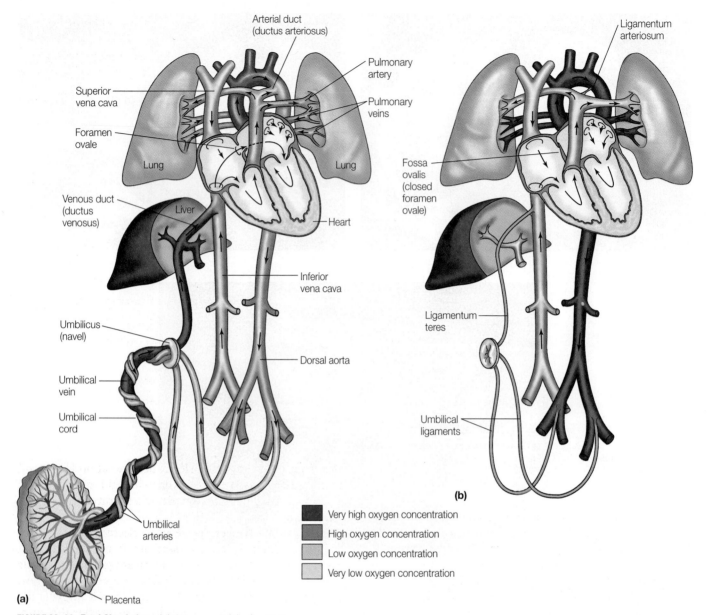

Arterial duct (ductus arteriosus)
Superior vena cava
Foramen ovale
Lung
Venous duct (ductus venosus)
Liver
Umbilicus (navel)
Umbilical vein
Umbilical cord
Umbilical arteries
Placenta
Pulmonary artery
Pulmonary veins
Lung
Heart
Inferior vena cava
Dorsal aorta

Ligamentum arteriosum
Fossa ovalis (closed foramen ovale)
Ligamentum teres
Umbilical ligaments

(a)   **(b)**

- Very high oxygen concentration
- High oxygen concentration
- Low oxygen concentration
- Very low oxygen concentration

FIGURE 22-11 Fetal Circulation   (a) Before and (b) after birth.

22-5 Birth Defects

Health Tip 22-1

When the time comes to bear children, women should get plenty of folic acid through their diet and supplements.

Why?

Folic acid deficiencies, which are common among women, can lead to birth defects. Deficiencies can be corrected by supplements and by eating foods rich in folate such as all-bran cereal, wheat-bran cereal, oranges, grapefruit, strawberries, raspberries, and dark leafy vegetables such as Romaine lettuce, spinach, and kale. The list also includes asparagus, broccoli, enriched spaghetti, and enriched white rice. Cooking destroys folic acid, so uncooked folate-rich fruits and vegetables should be high on the list of preferred food items.

By several estimates, 31% of all fertilizations end in a miscarriage—that is, a spontaneous abortion. Two of every three of these miscarriages occur before a woman is even aware that she is pregnant. Why such a high rate? Biologists hypothesize that early miscarriage is nature's way of "discarding" defective embryos.

Despite this culling that helps our species remain genetically fit, many children are born each year with **birth defects**, physical or physiological abnormalities. By various estimates, 10% to 12% of all newborns have some kind of birth defect, ranging from minor biochemical or physiological problems, which are not even noticed at birth, to gross physical defects (**Figure 22-12**). Most birth defects arise from chemical, biologi-

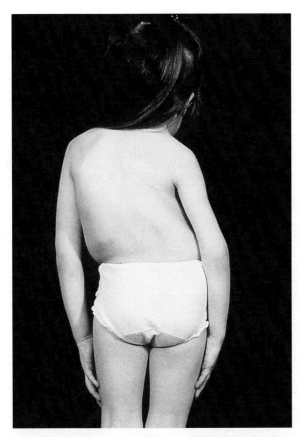

FIGURE 22–12 A Common Birth Defect Scoliosis is a lateral curvature of the spine. (© SIU/Visuals Unlimited.)

TABLE 22-3	Known and Suspected Human Teratogens
Known Agents	**Possible or Suspected Agents**
Progesterone	Aspirin
Thalidomide	Certain antibiotics
Rubella (German measles)	Insulin
Alcohol	Antitubercular drugs
Irradiation	Antihistamines
	Barbiturates
	Iron
	Tobacco
	Antacids
	Excess vitamins A and D
	Certain antitumor drugs
	Certain insecticides
	Certain fungicides
	Certain herbicides
	Dioxin
	Cortisone
	Lead

cal, and physical agents, although a dietary deficiency of the nutrient folic acid is also linked to birth defects. Table 22-3 lists known and suspected agents.

The effect of a birth-defect causing agent on the developing embryo is related to (1) the time of exposure, (2) the nature of the agent, and (3) the dose. Consider time first.

Because the organ systems develop at different times, the timing of exposure determines which systems are affected by a given agent. Organ systems are usually most sensitive to potentially harmful agents early in their development, as indicated by the pink bars in Figure 22-13. The central nervous system (CNS), for example, begins to develop during the third week of pregnancy. Because most women do not know they are pregnant for 3 to 4 weeks after fertilization, the central nervous system is often at risk—that is, it is exposed to potentially harmful agents, especially alcohol. In contrast, the teeth, palate, and genitalia do not begin to form until about the sixth or seventh week of pregnancy. Exposure during the seventh week might therefore affect them, but have little effect on the CNS, which has entered a less sensitive phase.

The nature of a birth-defect causing agent also influences the outcome of exposure. Some chemicals, for instance, affect a variety of developing systems. Others affect only one system. Ethanol in alcoholic beverages, for example, is a broad-spectrum chemical—one that affects many systems. Children who are born to alcoholic mothers—or even women who have consumed one or two drinks early in pregnancy—may

have a variety of physical defects. Behavioral problems and learning disabilities are also common in children of alcoholic mothers. These symptoms are part of a condition called *fetal alcohol syndrome.*

In contrast, other agents are more selective, targeting only one system. Methyl mercury found in fish and shell-fish, for instance, damages the CNS, creating abnormalities in the brain and spinal cord, but it has little effect on other systems.

Finally, like most toxic substances, chemicals generally follow a dose-response relationship: The greater the dose, that is, the amount an embryo or fetus is exposed to, the greater the effect.

One of the best-known birth-defect causing agents is not a chemical substance, but the virus that causes German measles or rubella (rue-BELL-ah). If a pregnant woman contracts the disease during the first 3 months of pregnancy, she has a one in three chance of giving birth to a baby with a serious birth defect. Deafness, cataracts, heart defects, and mental retardation are common. Fifteen percent of all of these babies die before the age of one.

Many birth defects can be avoided. German measles vaccine given to young girls usually protects them throughout their childbearing years. Vaccinating boys also reduces the risk to society. Other precautions also help prevent birth defects. Pregnant women or women who are trying to get pregnant should carefully control what they eat or drink.

Once the period of high sensitivity has passed, good nutrition and a healthy environment remain essential. This is because fetal development is also influenced by a number of physical and chemical agents in our homes and places of work. Toxic chemicals can stunt fetal growth and, in higher

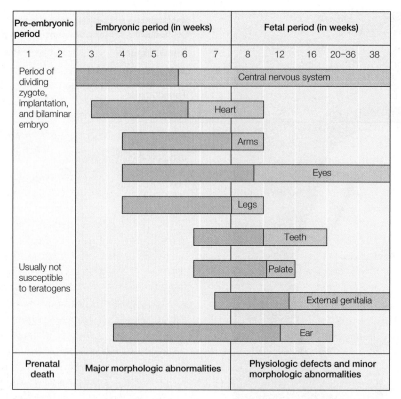

| Pre-embryonic period | Embryonic period (in weeks) | | | | | Fetal period (in weeks) | | | | |

FIGURE 22-13 Organogenesis Human development is divided into three periods or stages: pre-embryonic, embryonic, and fetal. Organogenesis occurs during the embryonic stage. Each bar indicates when an organ system develops. The pink-shaded area indicates the periods most sensitive to teratogenic agents (agents that can cause birth defects).

quantities, can even kill fetuses, resulting in stillbirths. Proper nutrition is essential because nutrient, vitamin, and mineral deficiencies can retard growth.

> **KEY CONCEPTS**
>
> Birth defects include structural and functional defects that appear in the newborn. Birth defects may be caused by chemical, biological, and physical agents. The effect they have depends on (1) the time of exposure, which determines which organ system is affected; (2) the dose, that is, how much a baby is exposed to; and (3) the nature of the agent, that is, whether that agent is a narrow or broad-spectrum teratogen.

22-6 Childbirth and Lactation

Childbirth begins with mild uterine contractions. During labor, contractions increase in strength and frequency until the baby is born.

Several factors stimulate uterine contractions. A rise in estrogen levels late in pregnancy is thought to stimulate the production of oxytocin receptors in the smooth muscle cells of the uterus. As you may recall from your study of the endocrine system, oxytocin is a hormone produced by the posterior pituitary of women. This hormone causes contraction of the smooth muscle of the uterus. The increase in the number of receptors in the smooth muscle cells renders them increasingly responsive to the small amounts of oxytocin.

Oxytocin first comes from the fetal pituitary gland, not the mother. The fetal pituitary releases small quantities right before birth. Fetal oxytocin travels across the placenta and circulates in the mother's bloodstream. When it arrives at the sensitized uterine musculature, it stimulates muscle contraction. Fetal oxytocin also stimulates the release of placental prostaglandins that act on smooth muscle. Together, these hormones stimulate more frequent and powerful uterine contractions.

Maternal oxytocin also plays an important role in childbirth. Scientists believe that uterine contractions and discomfort create stress that triggers the release of maternal oxytocin. It, in turn, augments muscle contractions caused by fetal oxytocin and prostaglandins (**Figure 22-14**). As uterine muscle contraction increases, maternal oxytocin release increases. This stimulates even stronger contractions, resulting in additional oxytocin release, a positive feedback loop that continues until the baby is delivered.

Research also suggests that the high levels of estrogen at the end of pregnancy block the "calming" influence of placental progesterone. Consequently, the uterus begins to contract at irregular intervals.

Physicians (and expectant mothers) recognize two types of labor contractions. The first are false labor contractions. These irregular uterine contractions usually begin a month or two before childbirth and are also known as **Braxton-Hicks contractions**. As the due date approaches, false labor contractions occur with greater frequency, causing many anxious couples to race to the hospital, only to be told to go home and wait for a few weeks for the real thing. In contrast, true labor contractions are more intense than those of false labor and occur at regular intervals.

Contractions are not all that's needed for childbirth. Some changes must also take place to enable the large baby to emerge from the uterus and pass through the uterus and vagina without causing damage. Much of the success of childbirth depends on the hormone **relaxin**. Produced by the ovaries and the placenta, relaxin is released near the end of pregnancy. As its name implies, relaxin softens the fibrocartilage uniting the pubic bones. This, in turn, allows the pelvic cavity to widen and thus greatly facilitates childbirth. Relaxin also softens the cervix, allowing it to expand so the baby can pass through it without causing damage.

KEY CONCEPTS

Childbirth is caused by uterine contractions that are stimulated by levels of the hormone oxytocin from the fetus and the mother. Estrogen also plays a key role in childbirth by increasing the number of oxytocin receptors and by blocking the calming influence of the hormone progesterone. The hormone relaxin allows the cervix to expand, permitting childbirth.

Stages of Childbirth

Childbirth consists of the three stages of: dilation, expulsion, and placental, which are shown in **Figure 22-15**.

Stage 1, the *dilation stage*, gets its name from the dilation of the cervix. This phase begins when uterine contractions start and generally lasts 6 to 12 hours, but it can last much longer.

At the beginning of stage 1, uterine contractions typically last only 30 seconds and may come every half hour or so. As time passes, however, contractions become more frequent and powerful. Uterine contractions generally rupture the amnion early in the dilation phase, causing the release of the amniotic fluid, an event commonly referred to as "breaking the water."

Uterine contractions push the infant's head against the relaxin-softened cervix, causing it to stretch and become thin (Figure 22-15b). By the end of stage 1, the cervix has dilated to about 10 centimeters (4 inches), approximately the diameter of a baby's head. During this phase, the baby is pushed downward by the uterine contractions and descends into the pelvic cavity. When its head is "locked" in the pelvis, the baby is said to be engaged.

Stage 2, the *expulsion stage*, begins after the cervix is dilated to 10 centimeters and the baby is engaged (Figure 22-15c). At this time, uterine contractions usually occur every 2 or 3 minutes and last 1 to 1.5 minutes each. For most first-time mothers, stage 2 lasts 50 to 60 minutes. If a woman

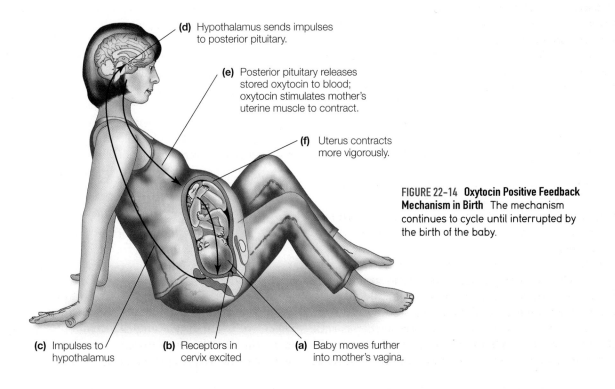

(d) Hypothalamus sends impulses to posterior pituitary.

(e) Posterior pituitary releases stored oxytocin to blood; oxytocin stimulates mother's uterine muscle to contract.

(f) Uterus contracts more vigorously.

FIGURE 22-14 Oxytocin Positive Feedback Mechanism in Birth The mechanism continues to cycle until interrupted by the birth of the baby.

(c) Impulses to hypothalamus

(b) Receptors in cervix excited

(a) Baby moves further into mother's vagina.

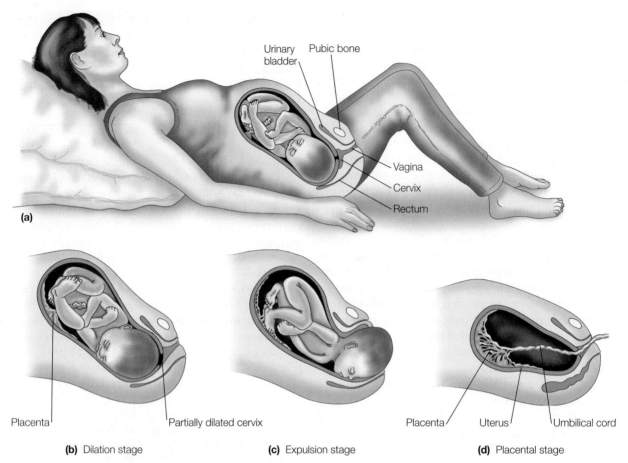

Urinary bladder
Pubic bone
Vagina
Cervix
Rectum

(a)

Placenta
Partially dilated cervix

Placenta
Uterus
Umbilical cord

(b) Dilation stage

(c) Expulsion stage

(d) Placental stage

FIGURE 22–15 **Stages of Labor** (a) Position of the fetus near birth. (b) Dilation stage. Uterine contractions push the fetal head lower in the uterus and cause the relaxin-softened cervix to dilate. (c) Expulsion stage. Fetus is expelled through the cervix and vagina. (d) Placental stage. Placenta is delivered.

is delivering her second child, it lasts about 20 to 30 minutes on average. The expulsion stage ends when the child is pushed through the vagina into the waiting hands of a doctor, midwife, proud father, or bewildered cab driver.

To facilitate the delivery, an incision is often made in the skin to widen the vaginal opening. The incision prevents unnecessary tearing and allows the infant to pass quickly. The incision is sutured immediately after the baby is born.

Once the baby's head emerges from the vagina, the rest of the body slips out quickly. However, the baby is still attached to the placenta inside the mother via the **umbilical cord**. The cord pulsates for a minute or more, continuing to deliver blood to the newborn. When the pulsating stops, the cord is tied off and cut.

Most babies (95%) are delivered head first with their noses pointed toward the mother's tailbone (Figure 22-15c). Occasionally, however, babies may be oriented in other positions—making delivery more difficult, time-consuming, and hazardous to both mother and baby. The most common alternative delivery is the breech birth, in which the baby is expelled rear-end first. Because breech births require more time, they can cause extreme fatigue in the mother and brain damage in the baby. In breech babies, the umbilical cord sometimes wraps around the infant's neck during birth. This can cut off the fetus's blood supply, causing brain damage or death.

To avoid such complications, physicians often "turn" breech babies before birth by applying pressure to the wom-an's abdomen. If a fetus cannot be turned, the baby is usually delivered by a **cesarean section**, a horizontal incision through the abdomen just at the pubic hair line and then through the uterus. Cesarean sections are also performed if labor is prolonged or if the mother has an active infection caused by herpes or some other sexually transmitted disease that might be transferred to her child as it passes through the vagina.

The final stage of delivery is the *placental stage* (Figure 22-15d). The placenta is expelled by uterine contractions, usually within 15 minutes of childbirth. The uterine blood vessels then clamp shut, preventing hemorrhage, although the mother continues to lose some blood for 3 to 6 weeks after delivery. After delivery, the uterus gradually returns to its normal size.

KEY CONCEPTS

Childbirth occurs in three stages: (1) dilation, during which the fetus is pushed downward by uterine contractions against the cervix; (2) expulsion, during which the fetus is pushed through the cervix and vagina, and (3) placental during which the placenta is expelled.

Lactation

A baby emerges into a novel environment at birth. No longer connected to the placenta, which has provided nutrients from the maternal bloodstream, the newborn must acquire food from another source, its mother's breasts in the form of

Scientific studies show that proper diet and exercise can increase life span. Laboratory studies, for instance, show that by manipulating the chemical environment of cells through diet, one can produce more cell divisions than normal. For example, the presence of vitamin E in large quantities increases the number of cell divisions in tissue cultures and a cell's life span. Vitamin E is an antioxidant that destroys harmful free radicals that attack cells. Whether vitamin E extends a person's life is not yet known, however.

Research shows that the better one lives, the longer one lives. Diet, exercise, and stress management are keys to a healthy life.

Health Tip 22-4

Closely monitor what you eat and the amount of exercise you get.

Why?

Studies show that most people overestimate the amount of exercise they get and underestimate the number of calories they consume. Try this: Estimate how much you eat each day and how much exercise you get. Then, record what you eat for a week or two and how much you exercise to see if your estimates are accurate.

Aging also results from a decrease in or deterioration of cellular function. Studies show that the decline in cell function results from problems arising in the DNA, RNA, and proteins. Cells are exposed to many potentially harmful factors in the course of a lifetime. Natural and human-produced radiation and other potentially harmful agents (such as pesticides and cleaning agents in the home and at work) may damage cells and impair their function. The cells of the body also produce oxygen-free radicals, which are oxygen atoms that carry unpaired electrons. These chemical species are natural byproducts of cellular metabolism and can react with (oxidize) proteins, DNA, lipids, and other important molecules, causing damage. As we age, our cells accumulate oxidized lipids and proteins, including important enzymes, which cause cell function to deteriorate.

Additional evidence shows that cells may lose the ability to repair DNA as they age, making matters worse. As a result, cells would be unable to fix mutations in genes that produce important structural and functional proteins such as enzymes. Cancer, the incidence of which increases with age, could result. As the damage accumulates, body function gradually deteriorates.

Some researchers think that aging is largely the result of a gradual decline in immune system function caused by a reduction in the number of cells and a loss of cell function. Although the immune system does falter, this decline surely cannot explain the other signs of aging.

KEY CONCEPTS

Aging is caused by a decrease in the number of cells and a deterioration of cell function, caused by exposure to harmful chemicals, radiation, and free radicals that, in turn, leads to changes in the structure and function of the body, including the immune system. A healthy lifestyle can slow the aging process.

Healthy Aging

Poor health and loss of independence are not inevitable consequences of aging. According to the U.S. Centers for Disease Control, the following strategies have proven effective in improving the health of older adults and thus allowing one to thrive in old age:

- *Healthy lifestyles.* Research has shown that healthy lifestyle behaviors, such as being physically active, eating a healthy diet, and not smoking are more influential than genetic factors in helping older people avoid the deterioration traditionally associated with aging. Starting early—right now!—can pay huge dividends in the long run. If you smoke, quit or cut down. If you drink alcohol, consider the same strategy. If you eat poorly and never exercise, start eating much better and add regular exercise to your daily routine.
- *Early detection of diseases.* Screening to detect chronic diseases early in their course, when they are most treatable, can save lives; however, many older adults have not had all of the recommended screenings. Consider obtaining a type of health insurance that covers or defrays the cost of annual physicals, health screening, and preventive care.
- *Immunizations.* About 36,000 people aged 65 or older die each year of influenza and invasive pneumococcal disease. Immunizations can reduce a person's risk for hospitalization and death from these diseases.
- *Injury prevention.* Falls are the most common cause of injuries to older adults. More than one-third of adults aged 65 or older fall each year, and of those who fall, 20% to 30% suffer moderate to severe injuries that decrease mobility and independence. Staying healthy and getting exercise improves one's balance and strength and reduces the likelihood of falls.
- *Self-management techniques.* Programs to teach older adults self-management techniques can reduce both the pain and the costs of chronic disease. For example, people with arthritis can learn practical skills such as how to manage their pain, how to deal with fatigue and stress, and how to develop a personal exercise program.

KEY CONCEPTS

Declining health and loss of independence can be alleviated in old age by maintaining a healthy lifestyle, including a healthy diet, regular physical exams to detect diseases early on, immunization, steps to prevent injury and self-management skills.

Preventing Diseases of Old Age

Although body function declines with age, many people succumb to disease in the later years. Many diseases are associated with old age. Osteoporosis, arthritis, atherosclerosis, Alzheimer's disease, and cancer are examples. It is important to note that *although these diseases are more prevalent in older people, they are not the inevitable consequence of aging.* That is, they are not unavoidable signs of aging. As previous chapters have shown, the likelihood of developing most of these diseases can be greatly reduced by exercise, diet, and other lifestyle adjustments. Although our bodies will

healthnote

22-2 Can We Reverse the Process of Aging?

Suppose a drug were invented that would help you live to be 120. Would you take it?

If you are like most people, you probably would, as long as your additional years would be healthy and fairly trouble free. Because most of us would like to beat aging, we secretly hope for a medical breakthrough that would prolong our lives.

The search for ways to reverse aging is extensive and typically involves nonhuman organisms, such as mice, rats, and fruit flies. Unfortunately, the search for a "cure" for old age in humans has been fraught with frustration. Over the years, numerous treatments have been tried and have failed. There is some encouraging news, however. Scientists recently discovered a protein called *stomatin* (stow-MA-tin), which they isolated from fibroblasts that have stopped dividing in tissue culture. Scientists hope that stomatin will stop other cells from dividing as well. If this turns out to be true, researchers may have discovered an important clue in the aging puzzle: the signal that ends cell division. Moreover, they may have found a way to prolong human life.

If the stomatin gene can be located, researchers may be able to find a way to inactivate it. A drug, for example, could be injected in people to block the gene's effect. The tissues of these people might continue to regenerate beyond the genetically programmed life span.

Efforts to reverse the aging process have met with some success in another arena, too, that of slowing down the aging of skin. Because age-related changes in the skin mirror those occurring in other aging tissues, scientists have long studied the skin to expand their understanding of the aging process. Their studies have shown that skin aging results from two processes: intrinsic or chronological aging—which may be genetically programmed—and extrinsic aging, which results from accumulated environmental damage from sunlight and other factors.

Many skin scientists doubt whether anything can be done about intrinsic aging. Their studies suggest that after a skin cell has lived out its lifetime, its plasma membrane receptors become insensitive to growth factors that stimulate DNA replication and cell division.

Research suggests, however, that extrinsic aging, especially that induced by sunlight, is preventable and even reversible. Many individuals do not realize that sunlight damages epithelial cells of the epidermis and the fibroblasts in the underlying dermis. The plasma membranes of skin cells, in fact, can be damaged within a few minutes of exposure to sunlight (Figure 1). Sunlight energy can also be absorbed

---

eventually fail, we do not have to suffer from the chronic illnesses that plague so many people. Health Note 22-2 offers some additional information on efforts now underway to slow the process of aging.

> **KEY CONCEPTS**
>
> Although many people experience diseases such as osteoporosis as they age, arthritis, and cancer, they are not the inevitable consequence of aging. A healthy lifestyle can help you live a long, healthy life.

Health Tip 22-5

For long-term mental health, eat green leafy vegetables and citrus fruit.

Why?

Studies show that folate, a B vitamin found in leafy green vegetables and citrus fruits, may help keep the brain sharp as we age. Older men, ages 50 to 85, who have higher levels of folate in their blood score better on certain tests of mental ability. It appears that folate may protect against the toxic effect of a chemical known as homocysteine, also found in our blood. Folate is added to flour (fortified) and grain products such as breakfast cereal to help reduce the incidence of birth defects. This vitamin may also protect against Alzheimer's disease.

Death

Death may be defined by the loss of brain or cardiovascular function.

Aging results in a deterioration of function that eventually leads to death. But death can also result from traumatic injury to the body—for example, severe damage to the brain or a sudden loss of blood—and many other causes. What is death?

Although we can all agree that death is the cessation of life, the trouble begins when people try to define it in instances requiring life-support systems. In 25 states, a person is considered dead when his or her heart stops beating and breathing ceases. In the remaining 25 states, however, death is defined as an irreversible loss of brain function. A woman in a coma, for example, who shows little or no brain activity other than that needed to keep the heart beating and the lungs working is alive in some states and legally dead in others.

Maintaining a person on life support is extremely costly to a family and to society if, for example, the patient is receiving federal benefits through Medicare. The money spent on maintaining a life could arguably be used to save dozens of other lives through preventive measures. These issues inevitably lead to another biomedical dilemma, the controversy over euthanasia, a word derived from the Greek meaning "easy death."

Euthanasia (you-thin-A-zah) refers to any act or method of causing death painlessly and may be either passive or active. *Passive euthanasia* consists of decisions and actions to withhold treatment that might prolong life. *Active euthanasia* consists of decisions and actions that actively shorten a person's life—for example, injecting a lethal substance to terminate a patient's life. The late Michigan physician,

by oxygen in the tissue, resulting in the formation of oxygen-free radicals, highly destructive chemicals that are primarily responsible for extrinsic aging.

In 1988, medical researchers found that Retin-A, a derivative of vitamin A used to treat severe cases of acne, also reduces wrinkles, age spots, and roughness as long as it is applied regularly. Studies show that Retin-A stimulates the growth of new blood vessels in the dermis. This, in turn, may nurture the regeneration of damaged skin cells. Retin-A also detoxifies oxygen-free radicals in tissues and can inhibit the destruction of collagen fibers in the dermis.

Despite the discovery of stomatin and Retin-A, there is as yet no evidence that medical scientists can prevent or even retard the overall rate of aging. The best route to a long and healthy life is to live it well: learn ways to reduce stress, eat well, exercise regularly, get enough sleep, take alcohol in moderation or not at all, and avoid harmful practices such as smoking.

FIGURE 2 Beware: Skin Damage A gorgeous tan may lead to premature aging. (© Photos.com.)

 Visit Human Biology's Internet site for links to websites offering more information on this topic.

Dr. Jack Kevorkian, was engaged in active euthanasia but was convicted of second degree homicide and sentenced to 10 to 25 years in prison in 1999. Kevorkian was paroled in 2007, in part because he was suffering from a terminal illness himself.

Opinions on euthanasia vary widely, and the controversy will, no doubt, be debated for many years. (For a debate on physician-administered euthanasia, see the Point/Counterpoint in this chapter.) As a final note, individuals who want to save their relatives emotional and legal turmoil can sign a living will. This is a legal document stipulating conditions under which doctors should allow a person to die.

> **KEY CONCEPTS**
>
> Death is the final stage of our lives. It is legally defined by some as the loss of brain function and by others as the loss of cardiovascular function.

22-8 Health and Homeostasis

Computer monitors, or video display terminals (VDTs), produce numerous types of radiation with frequencies ranging from x-rays to radio waves. Fortunately, manufacturers have installed protective shields that prevent most of this radiation from escaping and bombarding the user. It is the lower-frequency radiation (radio frequencies) that has people concerned. Why?

Laboratory experiments show that electromagnetic fields generated by extremely low-frequency radiation can alter fetal development in chickens, rabbits, and swine. In addition, researchers in Oakland, California, recently published results of a medical study on the incidence of miscarriage in nearly 1,600 women. The study showed that women who sit in front of a computer monitor for more than 20 hours a week during the first 3 months of pregnancy are nearly twice as likely to miscarry as women in similar jobs who do not use computers.

Researchers suspect that miscarriages result from radiation emitted from the VDTs, but believe that other factors may also be involved—for example, the stress of using a computer. Preferring to err on the conservative side, some scientists advise pregnant women to minimize their exposure to radiation from computer monitors. Should the link between radiation from VDTs and miscarriages be substantiated by further research, it would provide yet another link in the growing case for the importance of a healthy environment to human health.

Point/Counterpoint Physician-Assisted Euthanasia

The Right of a Physician Who Will Not Kill *by Rita L. Marker*

Should physician-administered euthanasia—the direct and intentional killing of a patient by a physician—be legalized? Common sense has always said NO. And laws have wisely banned euthanasia.

Now, however, an effort is underway to change these laws. Those who seek to legalize euthanasia have framed their arguments in terms of personal rights and freedom. They've clearly recognized the importance of words in molding public opinion; carefully crafted verbal engineering is being employed to transform the appalling crime of mercy killing into an appealing matter of patient choice.

Rita L. Marker has been the executive director of the Patients Rights Council since it began in 1987 and is the author of *Deadly Compassion.* (© Rita L. Marker.)

Proposals appearing in state after state have such benign titles as, "Death With Dignity Act," and the deceptively soothing term "aid-in-dying" is often substituted for "euthanasia."

As efforts of euthanasia activists increase, careful examination of what is at stake has become increasingly important. This examination should include answers to key questions.

Who will be the recipients of new "rights" if euthanasia is legalized? The law would benefit doctors, not patients. Competent adults already have the legal right to refuse medical treatment as well as the right to make their wishes known about such interventions for the future. Additionally, neither attempting nor carrying out the very personal and tragic decision to commit suicide is a criminal offense. Bluntly put, people can refuse medical care or can kill themselves without a doctor's help.

Doctors, however, are prevented by law from killing their patients. Legalization of euthanasia would give doctors the right to directly and intentionally kill patients. An action that is now considered homicide would become a "medical service."

What about safeguards? Efforts to legalize euthanasia have failed in Washington and California. Opponents pointed out the real dangers inherent in the proposals. Advocates, however, insisted that their measures contained sufficient safeguards. Yet, after each version failed, proponents acknowledged the lack of necessary safeguards. It's as though each and every attempt to legalize euthanasia is treated as a dress rehearsal. Yet we're not dealing with a play or musical event. A "learn as you go" attitude in law and public policy could lead to the deaths of millions of vulnerable people.

What changes in law are now being promoted? Recognizing that the dangers of their initial proposals were too apparent, euthanasia proponents have repackaged their agenda, calling for changes that would allow physician "assisted" death. Often referred to as "assisted suicide" measures, these new proposals would permit a physician to prescribe medication with the express purpose that it be used to end a patient's life. Such a change would lead to classification of such an action as a medical intervention. The reality would be simple: The physician would have prescribed medication to kill.

Logic and common sense make it readily apparent that "prescribing" encompasses far more than writing a prescription that is taken to the corner drug store. Medication prescribed for a patient in a hospital or care facility is often provided by health professionals. The current "change" allowing "only" for the death prescription is yet another deception intended to mask reality in meaningless safeguards.

What has happened where euthanasia has been practiced? One need only look to the experience in Holland for a glimpse of where the euthanasia road leads. Although it is widely practiced in Holland, euthanasia remains technically illegal. Yet, even with "safeguards," which are far tighter than those which have been proposed in the United States, a 1991 Dutch government study released horrifying information. The study found that in tiny Holland—a country with only one-half the population of the state of California—Dutch physicians deliberately and intentionally end the lives of more than 11,000 people each year by lethal overdoses or injections. And more than half of those killed had not requested euthanasia.

What will happen if euthanasia is legalized? With increasing emphasis being placed on cost containment, individuals who are disabled are particularly vulnerable. Leaders in the disability rights community are becoming increasingly alarmed by suggestions such as that of Jack Kevorkian that their "choice" of death would "enhance public health and welfare."

Euthanasia proponents have been clear in describing their goals. For example, Derek Humphry has called for expansion of euthanasia to include those who are physically and mentally disabled. He has also written that there would be a means for handling the dilemma of "terminal old age" once euthanasia is legalized. Jack has outlined plans for designated death zones with special clinics where "planned death" could be carried out.

The Right to Choose to Die *by Derek Humphry*

If we are truly free people and if our bodies belong to ourselves and not to others, then we have the right to choose when and how to die. Death comes to us all in the end, although modern medicine can often help us improve and extend our life spans. Thus, given today's high technology and ruinous cost of medicine, it is wise if we all give advance thought to the manner of our dying. Such decisions should be transferred to paper (a living will and durable power of attorney for health care), because 46 states now legally recognize one's wishes in respect to the withdrawal of life-support equipment.

Advance-declaration documents, of course, deal only with the legal and ethical problems of medical equipment and treatments. Less than half of dying people are connected to such equipment. For many more patients, technology does nothing for their terminal illness, so—in effect—there is no "plug to pull." Therefore, the cutting edge of the right-to-die debate in the twenty-first century centers on assisted death and voluntary euthanasia.

Hemlock Society supporters feel not only that hopelessly sick people should have an unfettered right to accelerate their end to save them pain, distress, and indignity, but also that willing doctors should be able to help them die. [To clarify some terms: self-deliverance (suicide) is ending your own life to be free of suffering; assisted suicide is helping another die; euthanasia is the direct ending of another's life by request. Currently, suicide in any form, for any reason, is legal, but assisted suicide and euthanasia are technically crimes.]

Most people at the close of life do not wish to suffer, desiring a quick and painless demise. The sophisticated pain-management drugs now available, when properly administered, control some 90% of terminal pain. But they do nothing to alleviate the indignities, the psychic pain, and the loss of quality of life associated with some debilitating terminal illnesses.

A complaint that any hastening of the end interferes with God's authority can be answered for many through a belief that their God is tolerant and charitable and would not wish to see them suffer. Other people have no faith in God. Pious people differ with this view, of course, and I respect that.

Numerous opinion polls in the United States and other Western countries indicate that two-thirds of people want the right to have a doctor lawfully assist them to die. People in the states of Washington, Oregon, Maine, Michigan, and California have engaged in political action to achieve this law reform. The proposed law in these states was called the Death With Dignity Act, and the broad outline of its purpose is as follows:

1. The patient wanting physician-assisted dying would have to be a mature adult suffering from a terminal illness likely to cause death within about 6 months.

2. People with emotional or mental illness (especially depression) could not get help to die under this law.

3. The request would have to be in writing, and the signature would have to be witnessed by two independent persons.

4. The physician could decline to help the patient die on grounds of conscience but would then cease to be the treating physician. The patient could then seek a physician who was willing.

Derek Humphry was the principal founder of the Hemlock Society and its executive director from 1980 to 1992. He is the author of several books on euthanasia, including the best-seller *Final Exit* (1991). Website: www.finalexit.org. (© Derek Humphry.)

5. The family would have to be informed and its views, if any, taken into account. But the family could neither promote nor veto the patient's request to die.

6. The physician would have to be satisfied that the patient was fully aware of his or her condition, had been informed of all possible alternatives, and was competent to make this request.

7. If the physician was unsure of the patient's mental state, a mental-health professional could be called to make an evaluation.

8. The time and manner of the assisted dying would have to be negotiated between patient and physician, with the patient's wishes paramount.

9. At any time the patient could orally or in writing revoke the request for assisted dying.

10. Any person who pressured another person to get assistance in dying, who forged such a request, or who ignored a revocation would be subject to prosecution.

11. After helping the patient die, the physician would have to report the action in confidence to a state health agency.

The right-to-die movement feels that these conditions, plus others in the Death With Dignity Act too numerous to describe here, are an intelligent and humane approach to euthanasia with the necessary safeguards against abuse. Only Oregon has passed such a law.

Dying on one's own terms is not only an idea whose time has come, but also the ultimate civil and personal liberty.

Sharpening Your Critical Thinking Skills

1. Summarize the key points of each author and their supporting arguments.

2. Which viewpoint corresponds to yours? Why?

Visit Human Biology's Internet site for links to websites offering more information about this topic.

SUMMARY

Fertilization

1. The sperm and egg of humans unite during fertilization, which usually takes place in the upper third of the uterine tube.
2. Sperm deposited in the vagina reach the site of fertilization partly on their own (by swimming), but mostly with the aid of muscular contractions in the walls of the uterus and uterine tube.
3. Sperm dissolve away the cells surrounding the oocyte and then bore through the zona pellucida and contact the plasma membrane. The first one to contact the membrane fertilizes the oocyte. Further sperm penetration is blocked. The chromosomes of the sperm and oocyte duplicate and merge in the center of the cell, where mitosis begins.
4. Human development is divided into three stages: pre-embryonic, embryonic, and fetal.

Pre-Embryonic Development

5. During pre-embryonic development (from fertilization to implantation), the zygote undergoes rapid cellular division, forming a morula.
6. The morula is then converted into a blastula, a structure with a hollow cavity, called a blastocyst in humans. The blastocyst consists of a clump of cells, the inner cell mass (ICM)—which becomes the embryo, and the trophoblast, which gives rise to the embryonic portion of the placenta.
7. While the placenta is forming, a layer of cells from the ICM of the blastocyst separates from it and forms the amnion. The amnion fills with fluid and enlarges during embryonic and fetal development, eventually surrounding the entire embryo and fetus, protecting them during development.

Embryonic Development

8. The ICM differentiates into the three germ cell layers: ectoderm, mesoderm, and endoderm. The formation of the three primary germ layers marks the beginning of embryonic development.

The organs develop from these three primary tissues during organogenesis. Table 22-2 lists the organs and tissues formed from each of the layers.

Fetal Development

9. Fetal development begins 8 weeks after fertilization and is primarily a period of fetal growth because most of the organ systems have developed or are under development.
10. The placenta produces several hormones that play an important part in reproduction. HCG maintains the corpus luteum during pregnancy. Progesterone and estrogen stimulate uterine growth and the development of the glands and ducts of the breast.
11. The placenta delivers oxygen and nutrients to fetal blood and removes waste products from it.
12. The fetal circulatory system is very much like the adult circulatory system, except that it contains three pathways that allow blood to bypass the lungs and liver, which are developing but not functioning. At birth these bypasses close up and the adult pattern of circulation develops.

Birth Defects

13. Birth defects arise from chemical, biological, and physical agents. The effect of these agents is related to the time of exposure, the nature of the agent, and the dose. A defect is most likely to arise if a woman is exposed during the embryonic period when the organs are forming.

Childbirth and Lactation

14. Labor consists of intense and frequent uterine contractions believed to be caused by the release of small amounts of fetal oxytocin prior to birth. Fetal oxytocin stimulates the release of prostaglandins by the placenta. Oxytocin and prostaglandins stimulate contractions in the sensitized uterine musculature.
15. Maternal oxytocin is also released, augmenting muscle contractions. As uterine contractions increase, they cause more maternal oxytocin to be released, which stimulates stronger contractions and more oxytocin release, a positive feedback loop that continues until the baby is born.
16. Labor consists of the dilation, expulsion, and placental stages.
17. The breasts consist primarily of fat and connective tissue interspersed with milk-producing glandular tissue and ducts.
18. During pregnancy, the glands and ducts proliferate under the influence of placental and ovarian estrogen and progesterone as well as human placental lactogen.
19. Milk production is induced by maternal prolactin. Before milk production begins, the breasts produce small quantities of a protein-rich fluid, colostrum. A newborn can subsist on colostrum for the first few days; the baby derives antibodies from colostrum that help protect it from bacteria.
20. Suckling causes a surge in prolactin secretion. Each surge stimulates milk production needed for the next feeding.

Aging and Death

21. Aging is the progressive deterioration of the body's homeostatic abilities and the gradual deterioration of the function of body organs. These changes result from at least two factors: a decrease in the number of cells in the organs and a decline in the function of existing cells.
22. Death results from aging, traumatic injury, or infectious disease.
23. Although declines in body function occur with age, we can age in a healthy fashion, living out our latter years in good health, free of disease.
24. To avoid disease and maintain health, we must adopt healthy lifestyles, get medical help early when diseases do occur, receive regular immunizations, protect against falls, and learn self-management techniques to reduce both the pain and costs of chronic disease.

THINKING CRITICALLY ANALYSIS

This analysis corresponds to the Thinking Critically scenario that was presented at the beginning of this chapter.

Medical scientists measure longevity by a statistic called life expectancy at birth. Technically, it is the number of years, on average, a person lives after he or she is born. Life expectancy at birth has increased dramatically in the last century. In 1900, for example, on average, white American females lived only 50 years. Today, life expectancy is 81 years. For males, a similar trend is observed. In 1900, for example, the life expectancy of a white American male was 47 years; today, it is 74 years.

Most people take these statistics to mean that men and women are actually living to a much older age. In reality, something very different is happening: the increase in the life expectancy at birth is largely the result of declining infant mortality. That is, thanks to improvements in medicine, more children are living through the first year of life and this has dramatically increased life expectancy. In the early 1900s, 100 of every 1,000 babies died during the first year of life. Today, that number has fallen to 12. The impact on average life expectancy is incredible.

This is not to say that all of the gain in life expectancy results from a decline in infant mortality. Medical advances have certainly increased life expectancy after infancy, but these changes are small in comparison to those brought about by lowering infant mortality. In fact, about 85% of the increase in life span in the last century is the result of decreased infant mortality.

KEY TERMS AND CONCEPTS

Aging, p. 488
Amnion, p. 478
Amniotic fluid, p. 478
Birth defect, p. 433
Blastocyst, p. 475
Braxton-Hicks contractions, p. 485
Breast, p. 487
Cesarean section, p. 486
Colostrum, p. 487
Ectoderm, p. 478
Ectopic pregnancy, p. 477

Embryo, p. 475
Embryonic development, p. 478
Endoderm, p. 478
Euthanasia, p. 492
Fertilization, p. 472
Fetal development, p. 480
Human placental lactogen, p. 487
Implantation, p. 475
Lactation, p. 487
Mesoderm, p. 478
Morula, p. 475

Neural crest, p. 478
Organogenesis, p. 478
Ovum, p. 473
Placenta, p. 476
Pre-embryonic development, p. 475
Prostaglandin, p. 476
Relaxin, p. 485
Umbilical cord, p. 486
Yolk sac, p. 478
Zygote, p. 473

CONCEPT REVIEW

1. Give an overview of the process of development in humans. Describe what happens during each stage. pp. 472–481.
2. Why is the formation of endoderm, mesoderm, and ectoderm so important during embryonic development? pp. 478–479.
3. Describe the process of fertilization in humans. Use drawings to elaborate your points, and label all drawings. pp. 472–474.
4. Define the terms inner cell mass and trophoblast. pp. 475–478.
5. Where does the human embryo acquire nutrients before the placenta forms? p. 496.
6. What are the major functions of the human placenta? pp. 476–478.

7. Describe the flow of blood from the placenta to the fetus and back. What is the role of the various bypasses? pp. 480–481.
8. What are birth defects? What causes them? How can they be prevented? pp. 482–483.
9. Describe the factors that trigger labor. pp. 484–485.
10. What factors are responsible for cervical dilation during labor? p. 485.
11. Define the term aging. Describe five of the major changes that occur as we age. p. 488.
12. What is the difference between intrinsic and extrinsic aging? p. 492.
13. All people age, but some do so without developing common diseases of aging. What are these diseases, and how can they be prevented? pp. 488, 491–492.

14. List and discuss some of the hypotheses that attempt to explain why we age. pp. 490–491.
15. Make a list of the five major strategies to age in a healthy fashion. p. 491.
16. Thanks to modern medicine, men and women are living longer, says a friend. Using your understanding of average life expectancy data and the reason for an increase in life expectancy in the last century, explain why this statement is not quite true. pp. 471 and 497.
17. Take a moment to peruse the Internet for successful ways aging has been reversed in humans and other species. Based on your studies, do any of these measures hold great promise for slowing aging in humans?

SELF-QUIZ: TESTING YOUR KNOWLEDGE

1. Fertilization occurs in the upper third of the _____. p. 472.
2. Enzymes in the head of the sperm in a structure called the _____ help it dissolve its way through the layers of cells and the zona pellucida surrounding the egg. p. 472.
3. The fertilized egg or _____ divides and forms a solid ball of cells known as a _____. p. 473.
4. The blastocyst is a hollow sphere of cells that gives rise to the embryo and the trophoblast, which forms the embryonic portion of the _____. p. 475.
5. The blastocyst attaches to the lining of the uterus and digests its way into it; this process is called _____. p. 475.
6. The _____ provides oxygen and nutrients to the developing fetus. p. 475.
7. The placenta produces three hormones, HCG, estrogen, and _____. p. 477.

8. The _____ is filled with fluid that protects the embryo and fetus from damage. p. 478.
9. Organogenesis, the formation of organs, occurs primarily during _____ development. p. 478.
10. An _____ pregnancy occurs when a fertilized ovum implants in the oviduct or inside the body cavity, not the uterus. p. 477.
11. Milk production in the breast is stimulated by the hormone _____ from the pituitary gland. p. 487.
12. A mother's milk provides nutrients and early _____ protection. p. 487.
13. _____ is a protein-rich milk that contains antibodies. p. 487.
14. Milk production stimulates the release of _____ from the anterior pituitary gland of the mother. p. 487.

15. Milk expulsion from the breasts is stimulated by suckling which causes the release of _____ from the posterior pituitary. p. 487.
16. Aging may occur, in part, from an accumulation of toxic wastes inside the body, known as _____. p. 488.
17. Vitamin ____ increases the number of cell divisions cells are capable of in vitro. p. 491.
18. Active _____ is the act of causing death by injecting a lethal substance into a patient. p. 492.
19. Death is legally defined as a loss of _____ function or heart function. p. 493.
20. _____ aging may be genetically programmed. p. 490.

biology.jbpub.com/chiras/8e/

The site features eLearning, an online review area that provides quizzes, chapter outlines, and other tools to help you study for your class. You can also follow useful links for in-depth information, research the differing views in the Point/Counterpoints, or keep up on the latest health news.

(© Hintau Aliaksei/ShutterStock, Inc.)

Evolution

Scientists believe that the atoms that make up both the living world and the nonliving world (rocks, for example) once existed in outer space as a giant cloud of cosmic dust and gas—a solar nebula that gave rise to the sun and our solar system. About 5 billion years ago, the materials in this huge cloud began to condense. No one knows what triggered this process, but many believe that a nearby exploding star (supernova) may have been the catalyst. Scientists speculate that the explosion forced gases (mostly hydrogen) and matter in the vicinity to begin to coalesce. Rotation of the gas cloud and gravity also very likely played a role in this process.

THINKING CRITICALLY

The timing and origin of the emergence of modern humans is a matter of considerable debate among scientists. Two basic hypotheses exist. The first, the Multiregion Evolution model, says that modern humans derive from small populations living in many regions of Europe, Africa, China, and Indonesia. These populations gave rise to present-day populations. Thus, modern Chinese came from an ancient Chinese population, and so on.

The second hypothesis holds that although archaic regional populations existed, they were displaced by a single population of humans that arose in Africa approximately 200,000 years ago. This new species supposedly spread into new territories, replacing regional human populations.

The second hypothesis, the African Origins model, is based on genetic studies of DNA in mitochondria. In these analyses, scientists have found that only the African population exists in an unbroken line. Genetic lineages of archaic peoples (regional populations) seem to have faded into oblivion.

Using your critical thinking skills, make a list of questions that might be useful in analyzing the African Origins model.

As the huge cloud condensed, scientists speculate that gas and dust particles in the center compacted more rapidly than outlying particles. This region would eventually give rise to the sun. Scientists hypothesize that when the mass of the forming sun reached a critical density, heat and pressure began to cause hydrogen and helium in it to chemically combine to form larger atoms. This process, known as *fusion*, not only forms larger atoms but also releases large amounts of energy. This energy is found in many forms, including heat and light. The process continues today in our sun.

For many years, scientists believed that cosmic dust lying outside the forming sun condensed to form many of the rocky planets (Greek for "wanderers"), including the Earth, Mercury, and Mars. They speculated that a rapid inward collapse of cosmic dust, caused by gravity, led to the formation of these planets.

In the 1960s, the Apollo space program changed this view—and quite dramatically. Evidence from this program, along with the work of geologists, suggested that the Earth and other rocky planets were formed by a different process, one referred to as *accretion*, the gradual accumulation of material. Studies suggest that cosmic dust in outer space lumped together to form larger particles. These particles collided with others, forming ever larger and larger masses. They, in turn, collided to form some of the planets. Over a period of at least 100 million years, and after numerous collisions, the rocky planets like Earth formed.

When the Earth first formed, about 4.6 billion years ago, the planet was a solid mass of rock and ice. However, scientists believe that the continued bombardment of the Earth by large bodies, some the size of the moon or even Mars, produced immense amounts of heat in the interior, melting the rock. Bombardment resulted in the formation of Earth's molten core, the source of our planet's magnetic field, and may have lasted for 1 billion years.

Geologists believe that the Earth's crust and first early atmosphere formed about 1 billion years ago. Cooling of the Earth caused lighter silicate rocks to rise to the surface, forming the crust. Intense heat in the Earth's interior, however, caused frequent volcanic eruptions. These eruptions released gases such as carbon dioxide that formed the Earth's first atmosphere. Water in the atmosphere also arose from the volcanic eruptions. Later, as the Earth continued to cool, water in the atmosphere began to rain down from the skies, creating lakes and oceans. Today, the oceans cover nearly 70% of the planet. It is in these oceans, specifically on their shores or perhaps even deep on the ocean's floor, that many scientists believe life began.

In this chapter we trace the emergence of life on Earth and study evolution. We conclude by looking at human evolution.

23-1   The Evolution of Life

In the most general terms, **evolution** is a process in which existing life forms change in response to changes in their environment. In some cases, they change so dramatically that they produce entirely new species.

The evolution of life on Earth can be divided into three phases. During the first phase, organic molecules formed. During the second phase, cells evolved. And during the third phase, multicellular organisms arose. This section outlines the key events in each phase.

> **KEY CONCEPTS**
> Evolution is responsible for the diversity of life on planet Earth and has occurred on three different levels: molecular, cellular, and organismic.

Chemical Evolution

When they first formed, the seas were lifeless bodies of water. The land masses, so richly carpeted today with plants, were barren rock. How could life have formed from such an unpromising start?

Research suggests that chemical evolution probably began about 4 billion years ago—or 500 to 600 million years after the Earth formed. At that time, the Earth's atmosphere contained a mixture of simple chemicals such as water vapor, methane, ammonia, and hydrogen. Water vapor condensed to form rain and washed many of these molecules from the sky.

In the shallow waters of the newly formed seas, sunlight, heat from volcanoes, or lightning energized these molecules. This energy caused them to react with one another, producing a variety of simple organic molecules, including simple sugars (monosaccharides) and amino acids.

According to the **theory of chemical evolution**, these organic molecules, in turn, began to react with one another. In the process, they formed small polymers—large molecular weight molecules consisting of many smaller molecules. As a result, primitive proteins and simple RNA or DNA molecules formed. Slowly but surely, over tens of thousands of years, all of the organic molecules necessary for life began to emerge.

But that's not the end of the process. Scientists believe that polymers, in turn, combined to form the very first, albeit primitive, cells.

> **KEY CONCEPTS**
> The formation of organic molecules that are essential for life from inorganic molecules is known as chemical evolution.

Evidence for Chemical Evolution Theory

Although the theory of chemical evolution may seem like the stuff of science fiction, many studies have been performed over the years demonstrating that it is possible. The first direct evidence in support of the theory of chemical evolution came from an American graduate student, Stanley Miller. While studying for his Ph.D. in chemistry at the University of Chicago in the early 1950s, Miller devised an apparatus to test part of the theory (Figure 23-1). To his closed, sterilized glass apparatus, Miller added three gases thought to have existed in the Earth's primitive atmosphere: methane (CH_4), ammonia (NH_3), and hydrogen (H_2). A sparking device simulated

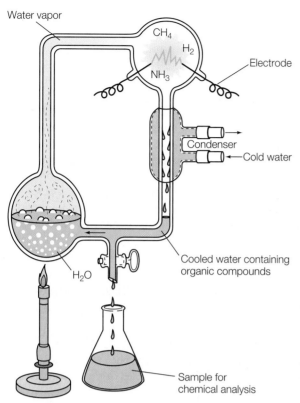

FIGURE 23-1 Miller's Apparatus This device showed that organic molecules could be produced from the chemical components of the Earth's early atmosphere. (Photo courtesy UC San Diego.)

lightning and provided energy. Boiling water created steam that circulated through the device, carrying with it the reactant gases and the products of the reactions occurring inside.

Several days later, Miller found that the water in the apparatus had turned brown. A chemical analysis of the liquid revealed the presence of several biologically important organic compounds, including amino acids, urea, and lactic acid (Table 23-1). Other researchers soon found that a variety of organic molecules, including the building blocks of DNA and RNA, could be created in conditions similar to those believed to have been present early in the Earth's history. Further experiments showed that these smaller molecules spontaneously assembled into long-chain molecules, giving birth to proteins, polysaccharides, DNA, and RNA.

> **KEY CONCEPTS**
>
> Several scientific experiments have demonstrated the plausibility of chemical evolution.

TABLE 23-1	Chemical Products Produced Abiotically in Miller's Experiment	
Glycine	Glutamic acid	Urea
Alanine	Acetic acid	Succinic acid
Aspartic acid	Formic acid	Aldehyde
Butyric acid	Lactic acid	Hydrogen cyanide

The Panspermia Hypothesis

While most scientists believe that molecules essential to life arose by chemical evolution, some hypothesize that they arrived on Earth by another route, notably in meteorites and comets. This idea is known as the **Panspermia theory** (although use of the word **theory** here is wrong in this context. It is a *hypothesis*). Studies show the presence of amino acids in the core of a meteorite that had drifted in space for 4.5 billion years. Studies suggest that the amino acids existed inside the meteorite before it struck the Earth. Although other evidence has also been found to support this idea, many scientists still believe that chemical evolution on Earth was the main source of the building blocks of life. However, they believe that additional organic matter came from outer space.

> **KEY CONCEPTS**
>
> Some scientists believe that organic molecules essential for the evolution of life on Earth arrived on meteorites and comets.

The First Cells

Scientists hypothesize that the earliest cells were simple aggregations of the polymers and lipids formed in water. (Lipids and proteins may have formed a primitive cell membrane.) Scientists also hypothesize that some of the proteins and perhaps even some of the RNA molecules in these primitive cells, called **protocells**, may have served as primitive enzymes. These enzymes may have allowed these primitive cells to synthesize some of their own molecules and break down others to generate energy. In other words, they may have made metabolism possible. Researchers also hypothesize that small molecules of DNA that formed during chemical evolution may have been incorporated into protocells, providing a primitive mechanism of heredity. It's very likely that numerous types of protocells formed in the seas, each with special characteristics. Thus, there was a diverse and rather large community of primitive cells, as illustrated in Figure 23-2. Over time, they became more and more complex. Like living organisms, they were probably subject to the forces of evolution that "pick" the most successful (more on this shortly).

A number of biologists believe that many of the different types of protocells received nourishment from organic molecules such as glucose that they absorbed from their environment. As such, these organisms were **heterotrophs** (literally, nourished by others). Some may have evolved simple mechanisms to acquire energy through fermentation (the breakdown of carbohydrates in the absence of oxygen) because oxygen was not yet present.

During the early evolution of life, the oceans first became populated by vast numbers of protocells. They evolved in many different ways. How life proceeded from this point on is subject to some debate. New research suggests that these primitive cells, now containing genetic information, engaged in extensive gene swapping. That is, these single-celled organisms derived from the first protocells transferred genes freely, creating newer forms of life, many of which were better able to survive and reproduce.

Over time, bacteria-like heterotrophs gave rise to numerous species of bacteria, as shown in Figure 23-2. They also gave rise to another group of single-celled organisms, the

Archaea. Together, the bacteria and Archaea form a group known as **prokaryotes**.

Archaea are organisms similar to bacteria in most aspects of cell structure and metabolism. However, they also possess some very distinct differences. Most notably, these organisms differ in genetic transcription and translation—that is, how they transcribe the DNA into RNA and make protein from their RNA. In this respect, they are more similar to nucleated cells, the eukaryotes.

Many Archaea live in extreme environments, for example, a number thrive in relatively high temperatures, often above 100° C (212° F), for example, in geysers. Some species live in extremely cold habitats. Others live in highly saline, acidic, or alkaline environments. Not all Archaeans are extremophiles. Some live in milder conditions such as marshlands and soil. Those that produce methane are found in the digestive tracts of animals such as humans.

> **KEY CONCEPTS**
>
> The earliest cells were probably aggregations of organic polymers and lipids known as protocells that acquired nutrients directly from the environment. They eventually acquired genetic material and gave rise to prokaryotes, primitive bacteria and Archaea, cells that on some levels resemble truly nucleated cells, the eukaryotes.

The Emergence of Eukaryotes

As shown in Figure 23-2, the protocells and their prokaryotic descendants also gave rise to the eukaryotes. As you may

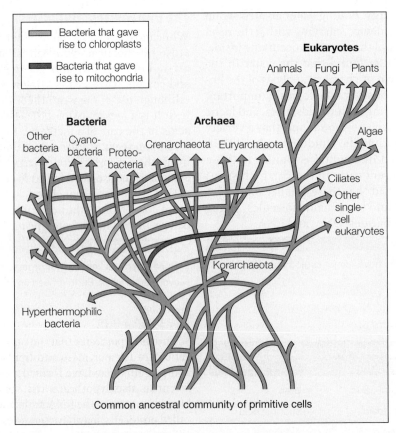

FIGURE 23-2 Summary of the Evolution of the Earth and Life (Reproduced from *Uprooting the Tree of Life* by W. Ford Doolittle. Copyright © 2000 by Scientific American, Inc. All rights reserved.)

recall from elsewhere in the text, **eukaryotes** are nucleated cells with distinct cellular organelles. Some eukaryotes, like amoebas, are single-celled organisms. Others, like animals and plants, are multicellular.

The earliest eukaryotes were single-celled organisms, which biologists believed evolved from some of the Earth's first protocells. Biologists think that they may have also received genes from primitive bacteria and Archaea. But that's not all they received.

As shown in Figure 23-2, biologists believe—and with good scientific evidence to support their beliefs—that eukaryotes, which evolved about 1.2 billion years ago, developed organelles, in part, by incorporating entire bacteria within their interiors. That is, some eukaryote precursors actually engulfed bacteria that were capable of breaking down glucose. These bacteria remained within the eukaryotes forever and became the eukaryotes' mitochondria. They lived in a permanent symbiotic relationship. As cells divided, so did their internal partners, which were then passed on to future generations. Other primitive eukaryotes incorporated photosynthetic bacteria. They became the cells' chloroplasts. Nuclei and membranous organelles like the endoplasmic reticulum may have formed by the infolding of the cell membrane of this line of aspiring eukaryotic cells.

This process is known as **endosymbiotic evolution**. It gave rise to eukaryotic cells of animals, fungi, and plants (For a timetable of these events, see Figure 23-3.) What drove the evolution of eukaryotic cells?

One factor was the emergence of oxygen in the previously oxygen-free environment. As oxygen began to be produced by the earliest photosynthetic bacteria, cells that had incorporated bacteria that could handle oxygen were more likely to survive and pass on their genes. Cells that incorporated photosynthetic bacteria and later gave rise to plants also enjoyed an enormous advantage over other cells. They were able to make their own food from carbon dioxide, water, and sunlight via photosynthesis.

> **KEY CONCEPTS**
>
> Protocells and prokaryotes gave rise to nucleated cells with cellular organelles, the eukaryotes, via a process known as endosymbiotic evolution.

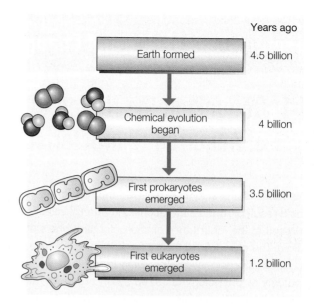

FIGURE 23-3 **Timeline for the Evolution of Life on Earth**

Evolution of Plants and Animals

Over many millions of years, eukaryotic organisms gave rise to a variety of multicellular plants and animals. They evolved first in the oceans, protected from ultraviolet radiation from the sun. Over time, oxygen produced by photosynthesis created the ozone layer. Ozone contains three atoms of oxygen (O_3). The ozone layer blocked harmful ultraviolet radiation from striking the Earth. This, in turn, allowed life to evolve on land.

The first species to invade the land from the oceans were very likely plants. They thrived in moist environments. Soon after the land plants invaded, animals came ashore. (The animals now had a food source, so terrestrial life was possible.) The very first of these animals were probably scorpion-like creatures. In time, millions of new plant and animal species evolved.

Eukaryotes gave rise to animals, fungi, and plants. This process was driven in large part by the emergence of oxygen in the environment.

23-2 How Evolution Works

Many new life forms have emerged over several billion years of the Earth's history. And many have changed, or evolved. As you shall soon see, the theory of evolution explains the emergence of these new life forms and changes in existing life forms. This theory is quite different from a once-competing theory, discussed in Scientific Discoveries that Changed the World 23-1.

Genetic Variation

Evolution occurs as a result of several forces or factors. One factor that plays a key role is **genetic variation** in populations. It leads to three key differences in organisms in a population

in structure (anatomical), function (physiological), and behavior.

Basically, here's how evolution works. In all populations, organisms share many similar characteristics. However, some members of the population differ from the others with respect to structure or function or behavior—sometimes combinations of them. These differences are caused by genetic variation—slight differences in the genes of these organisms.

Genetic variations in members of populations may result from mutations that occur naturally and randomly. Mutations occur in the body cells (somatic cells) and germ cells. In the evolution of multicellular organisms, somatic mutations are

Scientific Discoveries that Changed the World

23-1 Debunking the Notion of the Inheritance of Acquired Characteristics

In the 1800s, some scientists believed that life forms evolved not as a result of random mutations and natural selection, but as a result of the inheritance of characteristics one's parents acquired. This now-defunct theory was known as the inheritance of acquired characteristics. According to the theory, changes in anatomy or physiology acquired by an organism, such as muscle enlarged through heavy exercise, are transmitted to one's offspring.

For many years, this theory of inheritance of acquired characteristics was incorrectly attributed to an eighteenth-century French naturalist Jean-Baptiste Lamarck. Through textbooks and lectures, writers and teachers have repeatedly asserted that the theory was the brainchild of a French naturalist by the name of Jean-Baptiste Lamarck. More careful study of the historical roots of theory show that Lamarck did not come up with the idea. It was widely discussed at the time but had been proposed in ancient times by Hippocrates and Aristotle. In Lamarck's time, the idea was commonly accepted. Although Lamarck promoted it in his writings, he also promoted the theory of evolution. This led none other than Charles Darwin to praise the French naturalist for his role in popularizing the concept.

History aside, let's look at the theory of acquired characteristics in a bit more detail. To understand the idea, let's examine the supposed evolution of the giraffe. According to supporters of the inheritance of acquired characteristics, the earliest giraffes were short-necked animals. Over time, however, the giraffes' necks lengthened as a result of the simple act of stretching—a feat required to feed on leaves that were out of the reach of other animals. The act of stretching, said supporters, resulted in changes in the germ cells. As a result, theorists proposed, the slightly stretched neck that an adult giraffe had acquired would be transmitted to its offspring. The offspring, in turn, stretched their necks to reach food, causing further elongation.

The result?

According to supporters of the theory of inheritances of acquired characteristics, generation after generation of giraffes, each stretching to find food, led to the modern giraffe.

Scientists proposed an evolutionary theory based on the "use and disuse" of organs. It stated that an individual acquired traits during its lifetime and that such traits were in some way incorporated into the hereditary material and passed to the next generation. This explained how a species could change over time. Conversely, not using an organ could lead to its disappearance.

The inheritance of acquired characteristics was doubted by many scientists, in large part because of a lack of evidence from direct observations or experiments. This concept is based on the mistaken belief that all of the organs of the parents produce hereditary factors that control the formation of corresponding parts in the offspring. This hereditary factor would somehow then be transmitted to the sperm of the male or egg of a female. Thus, when a new offspring was formed, it would have the acquired characteristics.

August Weismann, a German scientist in the late 1800s, attempted to bring some sense to the debate in a series of extraordinary essays. Although Weismann did not perform experiments to disprove the inheritance of acquired characteristics, his essays were based on the research of others. In his writings, he argued that the germ cells contained the hereditary information responsible for the transmission of traits from one generation to the next. He also argued that the body cells of an organism did not dispatch small hereditary particles to germ cells.

Weismann concluded that the cells in mammals that determine heredity (the *germline*) became isolated before birth from the body cells (soma). Since then, no mechanism had been determined by which changes in the *soma* could affect the *germline*. This conclusion is known as *Weismann's barrier* and has helped debunk previous erroneous views of genetics and evolution.

Twenty years later, two scientists, W. E. Castle and John C. Phillips, published the results of studies that helped to support Weismann's conclusions. They transplanted ovaries from black guinea pigs (homozygous dominant) into albino guinea pigs (homozygous recessive) whose ovaries had been removed. Later, the albino females were bred to albino males.

Previous studies had shown that matings between albino males and females resulted in 100% albino offspring. Therefore, if acquired characteristics were inherited, Castle and Phillips argued, the offspring of these matings should be white, having acquired their coloration from the host. When bred to an albino male, however, the females produced only black offspring. The young, they said, "are such as might have been produced by the black guinea pig herself, had she been allowed to grow to maturity and been mated with the albino male used in the experiment."

Although the inheritance of acquired characteristics is defunct, new evidence suggests that the transmission of genetic traits is much more complex than geneticists have long believed. For instance, you learned that genes of an embryo can be influenced by the mother's diet, especially exposure to methylating agents.

As you may recall, methylating agents are chemicals in the diet that are rich in methyl groups. They can attach to DNA and alter our genes. Exposure of body cells to methylating agents may result in dramatic changes in the genes of a female's offspring. These changes, in turn, may dramatically alter structure, function, and behavior of the offspring.

What is more, as you've learned, these changes are heritable. That is, they can be passed on to subsequent generations. In order to do so, the methylation must also occur in the germ cell DNA.

Interestingly, exposure of a pregnant woman to methylating agents can change the DNA of eggs that are present in a female fetus prior to birth. These changes would be felt not only in the daughter but also the granddaughters and grandsons of the pregnant woman.

Although these results don't negate the Weismann barrier, they do show that heredity is much more complex than once thought. There is evidence to suggest that certain types of viruses can carry DNA from somatic cells to germ cells. As geneticist and author Sharon Moalem points out, the theory of the inheritance of acquired traits "isn't exactly right, but it may not be exactly wrong."

meaningless. They cannot be passed to future generations. It is germ cell mutations that are of importance.

Beneficial germ cell mutations produce characteristics in an organism's offspring that may give it an advantage over other members of the population. Genetically based characteristics that increase an organism's ability to survive and reproduce and thus pass on its genes are called **adaptations**. For example, a random mutation in the germ cell of a fish may make its offspring more efficient in catching prey. This, in turn, increases the likelihood that their offspring will survive and reproduce. Members of the same population that do not share this trait are less likely to survive and reproduce, and will very likely produce fewer offspring. Over time, the better adapted fish will leave a larger number of offspring. These offspring are also more likely than others to survive and reproduce. Their success results in a shift in the gene pool so that future populations contain proportionately more offspring of the fish with the beneficial mutation. On a genetic level, then, the gene that gave some fish an advantage over others increases in frequency in the population. Beneficial mutations, which produce genetic variation in populations, are often called the raw material of evolution.

Genetic variation comes from other processes as well. For example, when a sperm and an egg unite, the genes from the parents combine to produce a new offspring that may have characteristics that result in an advantage over other members of the population. If the new combination of genes produces advantageous adaptations, the genes may increase in frequency in the population.

> **KEY CONCEPTS**
>
> Genetic variation in populations can lead to differences in structure, function, and behavior that give some organisms a selective advantage over others; those that enjoy an advantage will more likely survive and reproducing, which can cause a shift in the genetic makeup of a population.

Natural Selection

Sociologist Andrew Schmookler once wrote that "evolution employs no author, only an extremely patient editor." What he meant is that evolution is not directed. There is no author. Instead, genetic variation arises through mutations and genetic variants resulting from sexual reproduction, as just noted. Mutations and sexual reproduction may produce adaptations that may persist over time, creating organisms better suited to a particular environment. In some cases, new species may evolve.

The shifts in gene pools are brought about by evolution's patient "editor," in a process called **natural selection. Charles Darwin**, a nineteenth-century British naturalist, and **Alfred Wallace** independently proposed the idea of natural selection. Darwin described it as a process in which slight variations, if useful, are preserved (Figure 23-4). Thus, natural selection is a process by which organisms often become better adapted to their environment.

Two principal factors contribute to natural selection: **biotic factors**, or other living organisms, and **abiotic factors**, the physical and chemical environment (temperature, rainfall, and so on). Abiotic and biotic factors influence survival and reproduction

by "selecting" the fittest—those best able to reproduce and pass on their genes. If these conditions change, those organisms best adapted to the new conditions tend to remain and pass their genes on to subsequent generations. This process causes a shift in the frequency of certain genes in a population.

In Darwin's time, evolution was widely discussed among naturalists and other scientists, but the mechanism by which it occurred remained a mystery. Darwin dedicated many years to the search for an answer, even traveling around the world by ship and cataloging species (Figure 23-4b). It was on this lengthy voyage that Darwin reportedly came up with the idea of natural selection.

A careful scientist, though, Darwin spent the next 20 years looking for flaws in his own reasoning. In 1858, much to his surprise, Darwin received a paper from Wallace, a respected naturalist who had proposed the same concept. Darwin sent Wallace's paper to some of his colleagues and suggested that they publish it. Fortunately, Darwin's colleagues, who were aware of his own work, encouraged him to write a paper of his own. In 1858, Wallace's and Darwin's papers were both presented to the Linnaean Society. In 1859, Darwin published his now-famous book, On the *Origin of Species by Means of Natural Selection*.

As in the case of Mendel, it took many years for Darwin's and Wallace's ideas to be understood and accepted. Not until the 1940s, about 80 years after the men presented their ideas to the world, did natural selection become widely accepted. Today, it is one of the central tenets of biology.

> **KEY CONCEPTS**
>
> Natural selection is a process in which slight variations in populations, if useful, are preserved. Natural selection is therefore a process by which organisms often become better adapted to their environment. Numerous abiotic and biotic factors influence are agents of natural selection.

Survival of the Fittest

Darwin used the phrase "survival of the fittest" to describe how natural selection worked. Survival of the fittest is commonly interpreted as "survival of the strongest." To a biologist, however, **fitness** is a measure of reproductive success, which can be attributed to many different features, not just strength. Fitness does not necessarily result from speed either. The ability to hide, digest food more effectively, or use water more efficiently can be just as important, if not more important, than standard measures of strength.

> **KEY CONCEPTS**
>
> Survival of the fittest describes how natural selection works; fitness is a measure of reproductive success and results from many factors, strength, ability to hide, ability to catch prey, and so on.

Jumping Genes and Evolution

Although scientists agree on the mechanisms by which evolution takes place—notably, through mutation and natural selection—mutations may have occurred much more rapidly and perhaps not as randomly as previously thought. Even more startling, evolution of species like our own may have been aided by viruses.

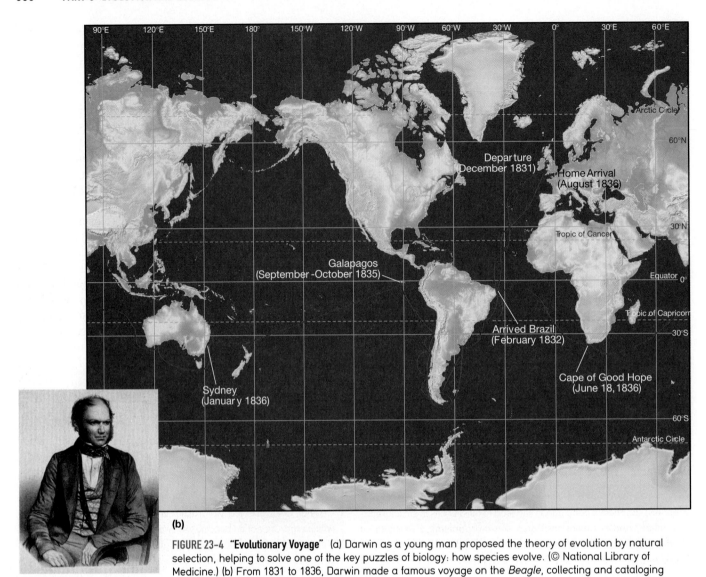

(b)

(a)

FIGURE 23-4 "**Evolutionary Voyage**" (a) Darwin as a young man proposed the theory of evolution by natural selection, helping to solve one of the key puzzles of biology: how species evolve. (© National Library of Medicine.) (b) From 1831 to 1836, Darwin made a famous voyage on the *Beagle*, collecting and cataloging thousands of diverse species.

Here's an overview of the new thinking: Many geneticists believe that during evolution of life on Earth many viruses have been incorporated into the DNA of living organisms, humans included. In the process, their DNA has become part of the genome of all living things. Even more important, geneticists believe that viral DNA may have formed new genes that gave organisms an advantage over others. In addition, because viruses mutate more quickly than DNA in the cells of organisms, they may have also given organisms an opportunity to change rapidly in response to changes in the environment. This led one scientist to refer to viruses as "the ultimate genetic creators" capable of "inventing new genes in large numbers." As geneticist Sharon Moalem writes in his book, *The Survival of the Sickest*, "With all that mutating power" viruses . . . "are bound to happen on useful genes far faster than we could without their help." In essence, "this partnership with viruses may have helped us evolve into complex organisms much faster than we could have on our own." He adds, "Over the last few millions of years, perhaps we have given them the ride of their life and, in return, they've given

us the chance to borrow some code from their huge genetic library." But where are these viruses now?

They're in the so-called junk DNA, the DNA that gives rise to jumping genes—segments of DNA that can move about in the nucleus, creating new genetic combinations in response to internal and external change. Evidence suggests that about half of the junk DNA is composed of jumping genes. Studies suggest that jumping genes play a role in development of the brain and personality. Their ability to rearrange the genetic deck is responsible, in part, for the diversity of personalities that emerge. They may also be responsible for our ability to make a diverse array of antibodies. Clearly, they're riding along with us, continuing to provide benefits.

KEY CONCEPTS

Evolution of humans and other organisms has very likely been aided by genes acquired from viruses, a biological agent that is quick to mutate. Beneficial genes acquired from viruses and jumping genes can result in new genetic combinations that give organisms a selective advantage.

23-3   The Evidence Supporting Evolution

Darwin's theory of evolution by natural selection can be summarized as follows. Natural variations exist in all species. Those inherited (genetic) variations that provide an advantage to certain members of a species survive and reproduce. The factors that determine which of these organisms survive and reproduce is referred to as natural selection—in essence, it's nature's way of selecting. While evolution is commonly referred to as a theory, like cell theory and many other key tenets of modern biology, there is a great deal of evidence that supports it.

The Fossil Record

One of the most important sources of information on evolution is the fossil record. **Fossils** consist of the remains of organisms that lived on the Earth, such as the bones of dinosaurs, shells, or teeth. Impressions are also part of the fossil record. For example, some creatures such as dinosaurs left footprints in the mud that hardened into stone (Figure 23-5a). Imprints of leaves make up part of the fossil record (Figure 23-5b). Some ancient organisms (frogs, insects, and flowers) were preserved in amber, resins released by ancient trees (Figure 23-5c).

The fossil remains of species no longer in existence provide evidence that supports the theory of evolution. Because fossils and the rocks they come from can be dated using radioactive techniques, scientists can determine when different species lived.

Studying when a species lived and tracking the types of life forms present at different times during the Earth's history have permitted scientists to piece together an evolutionary history of the planet. These studies show that the oldest rock contains relatively simple single-celled organisms, and that successively younger rock houses fossils representing increasingly more complex life forms.

To date, scientists have discovered fossils belonging to about 250,000 species, most of them from rock formed within the last 600 million years. In this still-growing record, some lineages are nearly complete.

The fossil record not only shows the changes that occur to a particular animal—for example, how horses evolved—they also demonstrate the origin of new species. Studies of the fossil record show that, during evolution, a primitive fish probably gave rise to the amphibians. Amphibians gave rise to the reptiles. Birds and mammals evolved from reptiles.

> **KEY CONCEPTS**
>
> The evolutionary history of life on Earth can be recreated by studying fossils.

Homologous Structures

Today's organisms are adapted to a wide range of conditions and exhibit a dazzling diversity of appearance. Despite the diversity of life, many organisms exhibit similar anatomical features. Structures thought to have arisen from common

(a)

(b)

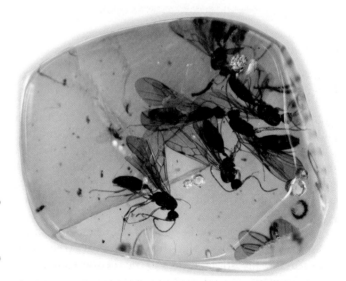

(c)

FIGURE 23–5 Fossils (a) Dinosaur tracks made in a Texas streambed about 120 million years ago. (Courtesy of Jerry Sintz/Bureau of Land Management.) (b) Imprint of a leaf. (© Falk Kienas/ShutterStock, Inc.) (c) Insects embedded in amber about 40 million years ago. (© Ismael Montero Verdu/ShutterStock, Inc.)

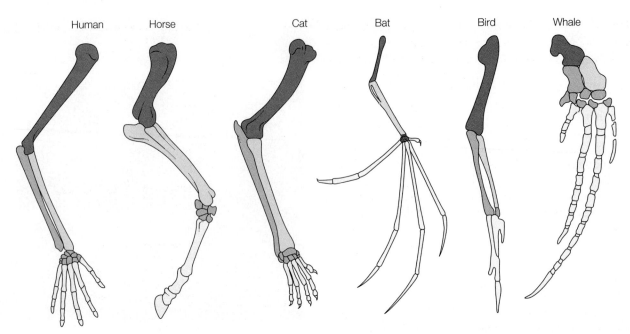

FIGURE 23-6 Homologous Structures among Vertebrates The presence of homologous structures in vertebrates and other groups supports the theory of evolution.

ancestors are known as **homologous structures**. The presence of homologous structures among life's diverse forms represents yet another piece of evidence in support of evolution. **Figure 23-6** illustrates some homologous structures: wings, flippers, arms, and legs. As shown, the bones in the wing of a bird and bat, the flipper of a whale, the arm of a human being, and the leg of a horse and cat are quite similar, even though these appendages perform quite different functions. Similarities such as these correspond to what you would expect if birds and mammals evolved from a common ancestor.

> **KEY CONCEPTS**
> Common anatomical features in different species suggest a common ancestry of all organisms and clearly support the theory of evolution.

Biochemical Similarities

Organisms as distantly related as roses and rhinos are made of the same basic biochemicals—ATP, DNA, RNA, and protein. All living organisms store genetic information in DNA, and all use ATP to capture, store, and transport energy. In addition, many biochemical pathways in organisms (such as those involved in producing energy) bear a remarkable similarity.

To many biologists, the common biochemistry of the Earth's diverse organisms lends further support to the notion of a common ancestry and thus to the theory of evolution. Although many species have similar enzymes and other proteins, differences do exist. Hemoglobin, for example, is found in vertebrate red blood cells, that is, the red blood cells of animals with backbones. However, differences in the composition of hemoglobin are common among vertebrates. These differences have proven quite useful to researchers.

By comparing the amino acid sequence of certain proteins (such as hemoglobin) from different species, evolutionary biologists can determine how closely related they are. Chimpanzees and humans, for instance, share about 98% of the same genes, suggesting that they are indeed very closely related. The more differences there are, the more distant the evolutionary relationship.

Analysis of the amino acid sequence of proteins and the composition of the DNA has enabled evolutionary biologists to create their own evolutionary trees. Not surprisingly, the evolutionary trees they have developed correspond well to those created by comparative anatomists.

Biochemical analyses of DNA and protein have recently been expanded to include extinct organisms (museum specimens) and even fossils. DNA extracted from the cells of extinct species can be cloned via genetic engineering. This process provides sufficient amounts of DNA to determine its sequence and make comparisons with other organisms, both living and dead, thus allowing evolutionary biologists to study the relationships among many more species.

> **KEY CONCEPTS**
> The fact that organisms have many biochemical similarities lends support to the notion of a common ancestry and thus to the theory of evolution.

Embryologic Development

Studies show that the embryos of many different groups of organisms develop similarly. Thus, the embryos of chickens, turtles, mice, fishes, and humans—all vertebrates—bear a remarkable resemblance during their early stages of development. For example, each of them has a tail and gill slits, even though only adult fishes and amphibian larvae have gills.

One plausible explanation for the similarity of vertebrate embryos is that all of these species contain the genes that control the development of tails and gills and that these genes were passed on from a common ancestor. In humans, however, these genes are active only during embryonic development. The structures they produce either become inconspicuous, as in the case of the tail (all that is visible is the tailbone), or become other structures, as in the case of the tissue lying between the gill slits. It forms various structures in the neck.

> **KEY CONCEPTS**
>
> The embryos of many different groups of vertebrates bear a remarkable resemblance during their early stages of development, suggesting a genetic similarity and an evolutionary relationship.

Experimental Evidence

Humans have been "evolving" organisms for centuries by artificially selecting for desired traits. Dogs and fancy pigeons are two of our most remarkable success in creating new organisms through selective breeding—breeding for certain characteristics. This work shows that desired characteristics can be selected for, at least artificially.

A long list of experiments supports the theory of evolution by natural selection. An exhaustive series of field and laboratory studies, for example, shows how genetic differences in wild populations of fruit flies can be attributed to natural selection.

> **KEY CONCEPTS**
>
> Artificial selection and numerous experiments strongly support Darwin's theory of natural selection, the key operative principle behind evolutionary change.

Biogeography

Biogeography is the study of the distribution of plants and animals throughout the world. It seeks to understand why certain species occur in specific places and not in others. Why, for example, are the egg-laying mammals found in Australia, Tasmania, and New Guinea but nowhere else in the world? It also seeks to understand why very similar species are found in distant lands—often separated by oceans.

The answer lies in understanding that 195 to 240 million years ago, all the continents were part of a very large land mass called Pangaea (**Figure 23-7**, top). This land mass split apart, however, and the pieces of this once singular land mass began to migrate to new positions, creating distinct continents (Figure 23-7, middle and bottom). Movement of continents is made possible because continents are part of large plates in the Earth's crust, called *tectonic plates*. Tectonic plates are capable of moving around. This occurs as a result of the motion of molten rock circulating in the Earth's core.

Tectonic plate movement continues today and is responsible for earthquakes and volcanic eruptions. It continues to move the continents around, promising a very different world 100 million years from today.

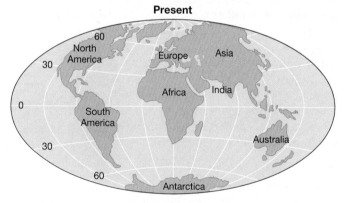

FIGURE 23-7 The Restless Planet The continents move about the planet like pieces of a giant jigsaw puzzle as a result of current of molten rock beneath them. Pangaea, shown in the top panel, was a supercontinent that broke apart to form the world as we know it today. But that's not the end of the story. The continents continue to move, reshaping our world.

Tracing the origin of continents helps scientists understand why two very similar species exist in different parts of the world—for example, South America and Africa. You can see from Figure 23-7 that these two continents fit together like two pieces of a puzzle, having been joined in Pangaea. This explains why two similar species exist today in these very distant continents and has helped scientists study natural selection and other processes of evolution.

> **KEY CONCEPTS**
>
> Biogeography is the study of the distribution of species and ecosystem on various continents in relation to the Earth's geologic history.

23-4 Early Primate Evolution

Humorist Will Cuppy once quipped that, "all modern men are descended from a worm-like creature, but it shows more in some people." In reality, humans belong to a group, or more correctly, an order known as the *primates*. The **primates** are believed to have evolved not from worms, but from an insect-eating mammal that probably resembled the modern-day tree shrew of Southeast Asia (Figure 23-8).

To understand human evolution, we need to briefly examine the evolution of primates, beginning with the mammalian insectivores, shown in Figure 23-9. This animal, in turn, gave rise to a group called the **prosimians**, the very first primates to inhabit

FIGURE 23-8 Look Familiar? An organism resembling this tree shrew is believed to have been the early ancestor of the primates. (© Dmitrijs Mihejevs/ShutterStock, Inc.)

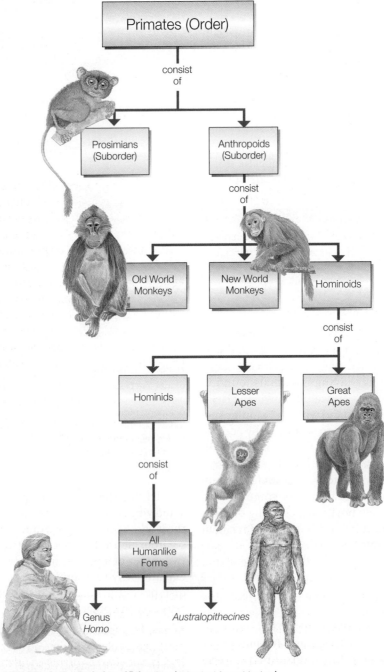

FIGURE 23-9 The Evolution of Primates (Adapted from Mader.)

(a)   (b)

FIGURE 23-10 **Prosimians** The prosimians (premonkeys) were probably the first primates. The earliest ones probably resembled modern-day (a) tarsiers and (b) lemurs. (a, © Gordon Galbraith/ShutterStock, Inc.; b, © Christian Riedel/ShutterStock, Inc.)

the Earth. Today, the prosimians consist of tree shrews, lemurs, and tarsiers. Figure 23-10 shows a tarsier and a lemur. During the course of evolution, the prosimians gave rise to the **anthropoids**. This group includes monkeys, apes, and humans. As shown in Figure 23-9, humans evolved from an ape-like animal.

Primates are characterized by several features, including grasping hands, which permit them to pick up objects, and forward-directed eyes, which enhance three-dimensional vision. Primates also have the largest brains of all mammals in proportion to their body size.

> **KEY CONCEPTS**
>
> Humans are members of the order primates. Primates include humans, monkeys, and apes. All primates evolved from insect-eating vertebrates known as the prosimians.

Fossil Evidence

Unlike the dinosaurs, early primates did not live in habitats conducive to fossil preservation. Whereas the mud in swamps

or along rivers preserved the bones of many a dinosaur, the forests and grasslands in which primates lived were not environments where bones of a dead animal would be readily covered with sediment and preserved. Their bones were scattered widely and even eaten by predators or scavengers. Because of this, the primate fossil record, including the record of early hominids, is less than complete.

Because the incompleteness of the primate fossil record, the origin of modern humans remains in question. In the past 30 years, however, new light has been shed on our origins by the discovery of a number of new fossils. This has completely changed the view of human evolution.

> **KEY CONCEPTS**
>
> Scientists have pieced together evidence suggesting how modern humans evolved, a process that has been difficult because the forests and grasslands in which primates lived were not environments where bones were readily covered with sediment and preserved as fossils.

23-5   Evolution of the Australopithecines

For years, archaeological evidence suggested that the first **hominids**, the first human-like primates, belonged to the genus *Australopithecus* (awe-STRAY-loh-PITH-eh-cuss; meaning "southern apeman"; Figure 23-11). They were believed to have evolved, as noted above, from a then-unidentified ape-like ancestor.

At the time, the oldest known australopithecine skeleton—about 3.2 million years old—was unearthed in Africa. This specimen, named Lucy, was one of the most complete fossils of early hominids yet found (Figure 23-12). Known as *Australopithecus afarensis* (*afarensis* means from the Afar region of Ethiopia), these hominids stood only about 3 feet

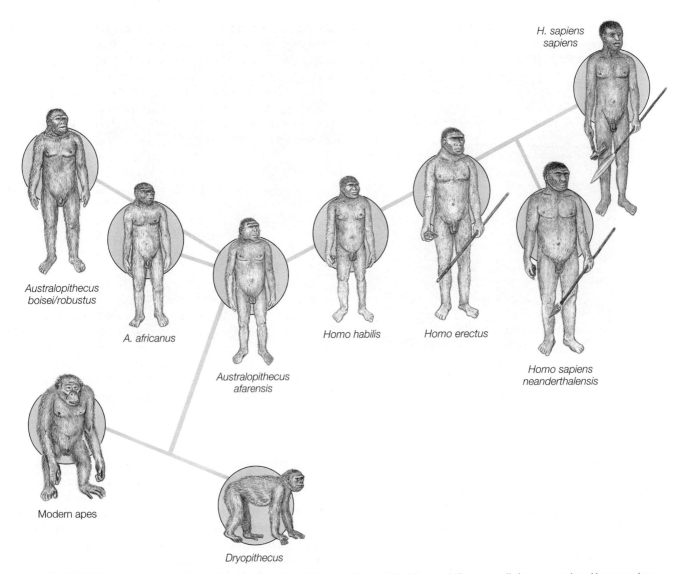

*H. sapiens
sapiens*

*Australopithecus
boisei/robustus*

A. africanus

*Australopithecus
afarensis*

Homo habilis

Homo erectus

*Homo sapiens
neanderthalensis*

Modern apes

Dryopithecus

FIGURE 23-11 Old Thinking on the Evolution of *Homo sapiens* Scientists once thought that the evolutionary path to our species, *Homo sapiens*, was pretty straightforward (linear) as depicted here. New studies show that the evolution of our kind is much more complicated than this. While the details are still being hammered out, it is clear that human evolution is more like a meandering process involving many different species.

(a)

(b)

FIGURE 23-12 *Australopithecus afarensis* (a) This skeleton, estimated to be about 3.5 million years old, is popularly known as Lucy. (© Science VU/Visuals Unlimited.) (b) Reconstruction of what Lucy may have looked like. (Courtesy of the Saint Louis Zoo, www.stlzoo.org.)

high and had brains only slightly larger than those of apes. The skulls of *A. afarensis* have many ape-like features, including massive brow ridges, low foreheads, and forward-jutting jaws. The shape of its pelvis suggests that *A. afarensis* walked upright, however. Researchers believe they originated many years earlier, as shown in the timeline in **Figure 23-13** (this graph shows researchers' latest thinking on when various hominid species existed).

More recent work, however, shows that Lucy's kind were not the first *Australopithecus*. As illustrated in **Figure 23-14**, *A. afarensis* was preceded by another Australopithecine, known as *A. anamensis*. However, they were preceded by yet another hominid, known as *Ardipithecus ramidus*, which appeared about 200,000 years before *A. anamensis* (Figure 23-14).

Even more recent work has uncovered a much earlier ancestor, known as *Sahelanthropus tchandensis*. This ape-like creature lived 6 million to 7 million years ago in the Sahel region (southern range of the Sahara) in northern Chad (in Africa), just after humans split from the great apes. It may represent the earliest human ancestor.

As the early fossil record is starting to become more detailed, human origins are becoming clearer and clearer, although one can expect much more clarification (and debate) in the years to come. With this in mind, let's return to *A. afarensis*, the small-brained biped. Although it was once assumed that the line from this species to modern humans was pretty straightforward, as shown in Figure 23-11, new findings show that early human evolution was much more complex. In fact, studies show that *A. afarensis*, which vanished about 3 million years ago, was not the only Australopithecine to live in Africa. It appears that several other species of *Australopithecus* lived in Africa at the same time as *A. afarensis*. More details on these species will be forthcoming in years to come as scientists sift through the evidence and make new discoveries.

Eventually, *A. afarensis* disappeared and was replaced by another species, *A. africanus*, which lived for about 1 million years (Figure 23-14). *A. africanus* was slightly taller than its predecessor and had a slightly larger brain. Studies show that it was not alone, either. At least one other species of *Australopithecus* lived alongside it.

Members of the genus *Australopithecus* had many common features. For example, they were all *bipedal* (buy-PED-al)—that is, they walked on two legs—and they all ranged in height from 1 meter to 1.7 meters (3.3 to 5.6 feet). Their brains were larger than those of chimpanzees but considerably smaller than those of modern humans. The differences among the species are relatively minor, mostly a matter of degree. As the genus evolved, their brains got larger and their height

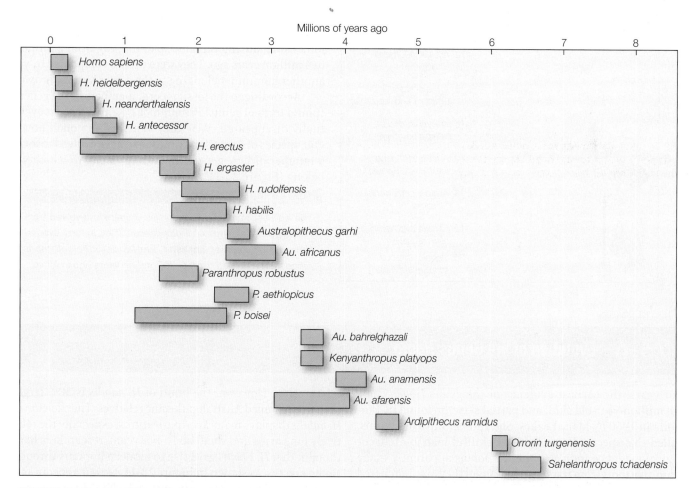

FIGURE 23-13 Timeline of Human Evolution This graph shows when various hominid species existed over the past 7 million years. Note that several species existed at the same time. The earliest hominid was *Sahelanthropus tchadensis*, which lived approximately 6 to 7 million years ago. (Adapted from Woody, B., *Nature* 418 [2002]: 133–135.)

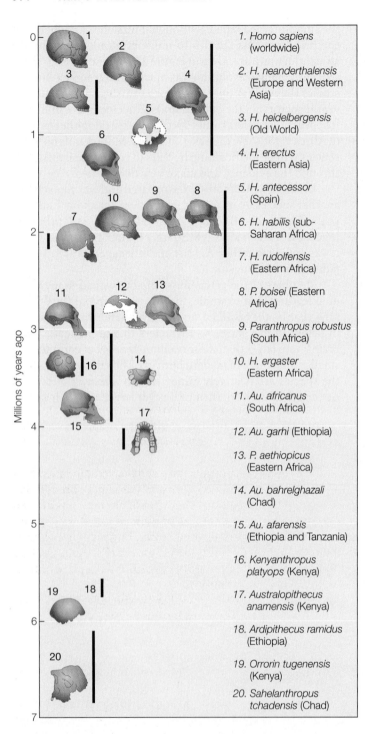

Millions of years ago

1. *Homo sapiens* (worldwide)

2. *H. neanderthalensis* (Europe and Western Asia)

3. *H. heidelbergensis* (Old World)

4. *H. erectus* (Eastern Asia)

5. *H. antecessor* (Spain)

6. *H. habilis* (sub-Saharan Africa)

7. *H. rudolfensis* (Eastern Africa)

8. *P. boisei* (Eastern Africa)

9. *Paranthropus robustus* (South Africa)

10. *H. ergaster* (Eastern Africa)

11. *Au. africanus* (South Africa)

12. *Au. garhi* (Ethiopia)

13. *P. aethiopicus* (Eastern Africa)

14. *Au. bahrelghazali* (Chad)

15. *Au. afarensis* (Ethiopia and Tanzania)

16. *Kenyanthropus platyops* (Kenya)

17. *Australopithecus anamensis* (Kenya)

18. *Ardipithecus ramidus* (Ethiopia)

19. *Orrorin tugenensis* (Kenya)

20. *Sahelanthropus tchadensis* (Chad)

FIGURE 23–14 Tracing Our Roots This diagram is one author's (Ian Tattersall) speculation on how modern humans evolved. There are gaps in the evolutionary lineage, which is to say that scientists are not sure how our species evolved. Note that several hominid species lived at the same time. Some evolutionary paths were dead ends. (Adapted from Tattersall, I., "Once We Were Not Alone," *Sci. Am.* [May 2003]: 285–292.)

increased. The enlarged brain is believed to be the result of natural selection.

The fossil record indicates that about 2.3 million years ago, another species of *Australopithecus* emerged in eastern Africa, *A. robustus*—so named because it was heavier and taller and had a larger brain than *A. africanus*. About 2.2 million years ago, the fourth species, *A. boisei*, appeared. It was even more robust than *A. robustus*. Because they have unique features that distinguish them from the other Australopithecines, these species have been reclassified since their discovery. They are now considered members of a new genus, *Paranthropus*. As shown in Figure 23-14, they may have arisen from *A. afarensis* through an intermediary, *Paranthropus aethiopicus*, which lived in eastern Africa.

Figure 23-14 indicates that several species of *Australopithecus* and *Paranthropus* coexisted in Africa, at least for a while, during the time span ranging from 2 million to 4 million years ago. They were even accompanied by yet another hominid, belonging to the genus *Kenyanthropus*.

According to the fossil record, members of the genera (plural form of genus) *Paranthropus*, *Kenyanthropus* eventually disappeared. Why they vanished, no one knows. One species of Australopithecus may have evolved to form a number of new genera, including *Homo*, to which we belong (Figure 23-14).

> **KEY CONCEPTS**
>
> The earliest humanlike organisms, hominids, evolved from members of the genus *Australopithecus*. They, in turn, evolved from a much earlier ancestor, known as *Sahelanthropus tchadensis* that lived 6 million to 7 million years ago in Africa.

23-6 Evolution of the Genus *Homo*

For years, the earliest evidence of the genus *Homo* was a 1.8 million-year-old skull and partial skeleton found in Tanzania in 1960 by Mary Leakey. She and her husband, Louis, called this species **Homo habilis** ("skillful man"). Archaeological evidence indicates that *H. habilis* apparently made primitive tools from fractured rocks and they also butchered large animals.

What was *H. habilis* like physically? The skull of *H. habilis* is like that of *Australopithecus*, suggesting an evolutionary

relationship. However, the brain of *H. habilis* is 50% larger than its presumed Australopithecine relatives. The skeleton of *H. habilis* displays many ape-like features, especially the relatively long arms and small body. For years, researchers have thought that *H. habilis* gave rise to modern humans through **Homo erectus**, as shown in Figure 23-11. Now it appears as if *H. habilis* was a dead end in evolution. So how did we evolve?

More recently, researchers have unearthed an even-earlier hominid, known as *Homo ergaster*. As shown in Figure 23-14, it probably evolved from a species of *Australopithecus*. Which one, scientists don't know.

H. ergaster may have given rise to *H. erectus*, which may have also vanished. Contrary to what scientists have long thought, *H. erectus* did not evolve into modern humans (Figure 23-15). In fact, many scientists believe that *H. ergaster* may be the species that gave rise to *Homo sapiens* via several other species also shown in Figure 23-14, which are discussed next.

> **KEY CONCEPTS**
>
> Scientists are still piecing together the evolutionary history of modern humans, *Homo sapiens*. It appears as if we evolved from *H. ergaster*, not *H. erectus* as long thought.

Homo neanderthalensis and *Homo sapiens*

Hominids moved out of Africa, presumably in the form of *H. ergaster* or a close relative, reaching China and Java about 1.8 million years ago.

H. erectus moved out, too, and was well established in these regions one million years ago. Members of the genus *Homo* also began to appear in Europe at a later time. One of them was *Homo antecessor*, who lived in Spain about 800,000 years ago. It was later followed by *Homo heidelbergensis*, a species found in Europe and Africa. It is believed that in Europe, *H. heidelbergensis* gave rise to a group of hominids whose best-known member was *Homo neanderthalensis*. This species thrived in Europe and western Asia between 230,000 and 300,000 years ago.

H. heidelbergensis may have also given rise to *Homo sapiens*. Like our ancestors, we may have arisen in Africa.

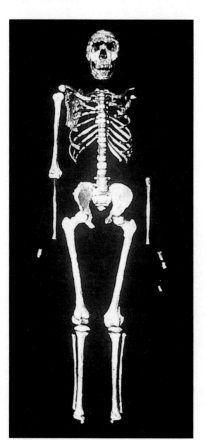

FIGURE 23-15 **Skeleton of *Homo erectus*** This, the most complete fossilized skeleton of *H. erectus* ever found, belonged to a boy who lived about 1.6 million years ago. (© National Museum of Kenya/Visuals Unlimited.)

Before we look at modern humans, let's take a look at one of our most famous relatives, *Homo neanderthalensis*, the Neanderthals (Figure 23-16). Widely distributed in Europe and Asia, the Neanderthals lived in caves and camps. Their name comes from the Neander Valley in Germany, where the first specimens of this group were discovered. Archaeological evidence shows that the Neanderthals gathered fruits, berries, grains, and roots, and also hunted animals with weapons. They cooked some of their food on fires. Neanderthals stood erect and walked upright. Evidence also indicates that they lived in small clans and buried their dead in elaborate rituals.

Although the skeletons of the Neanderthals resemble those of humans, they differ in several ways. For example, the skulls of Neanderthals were, on average, larger than those of modern humans. Their skeletons were also more massive and heavily muscled than those of modern humans, and they had rather short lower limbs, much like Inuits (who inhabit the Arctic regions) and other cold-adapted people.

Many people think of the Neanderthals as dimwitted and brutish, shuffling along bent over like an ape. This view is based on an interpretation of a Neanderthal skeleton found in 1908 in France. At the time of the discovery, however, the researchers failed to recognize that the skeleton under study was bent over because the individual suffered from arthritis of the hip and had diseased vertebrae. Unfortunately, this hasty conclusion from one observation spawned such a long-standing myth.

For years, anthropologists considered Neanderthals as a subspecies of *Homo sapiens, Homo sapiens neanderthalensis*. Others argued that they should be in a class of their own because the differences in their physical appearance are quite striking. Their broad faces, large projecting brow ridges, and heavily built bodies, gave credence to the idea (Figure 23-16). Today, they've been afforded a place of their own, *Homo neanderthalensis*.

Neanderthals disappeared approximately 30,000 years ago for reasons still not understood. Some archaeologists

FIGURE 23-16 **Neanderthal** Notice the projecting brow ridges of this reconstruction of a Neanderthal man. (© Cabisco/Visuals Unlimited.)

believe that they were replaced by the earliest known members of *Homo sapiens*. Where did these **H. sapiens** come from?

Modern forms of *H. sapiens* first appear in the fossil record about 195,000 years ago in Africa. About 40,000 years ago, with the appearance of the **Cro-Magnon** culture in Africa, tools became much more sophisticated. These people used a wider variety of raw materials than their *H. sapiens* ancestors, such as bone and antler. They also produced implements to make clothing. Decorated tools, beads, ivory carvings, clay figurines, musical instruments, and spectacular cave paintings appeared over the next 20,000 years. Archaeological evidence shows that Cro-Magnons also used sophisticated weapons, including the bow and arrow, and were highly skilled nomadic hunters, following great herds of animals during their seasonal migrations. They may have had a well-developed language. Cro-Magnons lived in caves and rock shelters in groups of 50 to 75 people; they are best known for the elaborate artwork that adorned the walls of their caves (Figure 23-17).

Approximately 10,000 years ago, truly modern humans emerged, but the changes were not great—mostly less robustness in the face, jaws, and teeth. Over the past 40,000 years, however, evolution has produced little noticeable change in the physical appearance of humans. A Cro-Magnon on the streets of Los Angeles, in fact, would probably go unnoticed. Those 40,000 years have not been without change, however, for during this period, *H. sapiens* had developed a rich and varied culture, complex language, and extraordinarily sophisticated tools.

> **KEY CONCEPTS**
>
> Humans are *Homo sapiens* that may have evolved from *H. heidelbergensis*, not *Homo neanderthalensis* in Africa.

FIGURE 23-17 **Cro-Magnon Art** (© Pixtal/age fotostock.)

Human Races

Studies in evolution show that species that become subdivided into isolated populations undergo changes in response to their environment. If the changes are profound, the subpopulations often lose the ability to interbreed and produce new species.

Over the course of time, the human population has wandered far and wide on the planet and has fragmented into distinct subpopulations, living in very different environments. This fragmentation has resulted in the formation of a number of phenotypically distinct subpopulations, or **races** (Figure 23-18). It is important to remember, however, that races

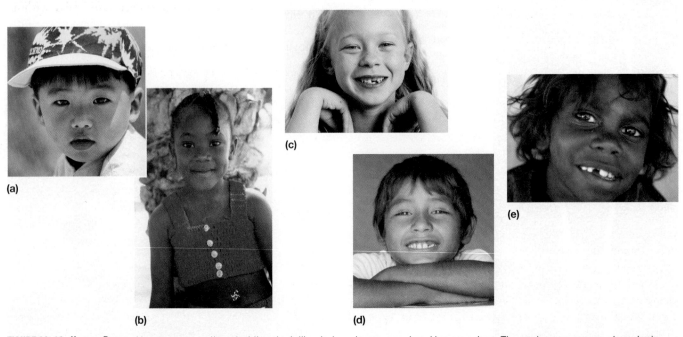

FIGURE 23-18 **Human Races** Humans, no matter what they look like, belong to one species, *Homo sapiens*. The various races are phenotypic variants of that species. Many scientists recognize five distinct races: (a) Mongoloid, (b) Negroid, (c) Caucasian, (d) Native American, and (e) Australian Aborigine. (a–d © Photos.com, e © Adaps Picture Library t/a apl/Alamy Images.)

are not distinct species, just regional variations of the one species, *Homo sapiens*.

Most of us are familiar with regional characteristics. We know, for instance, that Mexicans tend to have brown eyes, dark skin, and black hair and that Scandinavians tend to have blue eyes and blond hair. Many differences found in the races are adaptations to differing environmental conditions. The darker skin that evolved in tropical and subtropical populations, for instance, may be an adaptation that protects people from harmful ultraviolet radiation.

How many races are there? Some scientists recognize three main ones: Mongoloids, Negroids, and Caucasians.

Others recognize two others: Native Americans and Australian Aborigines (Figures 23-18d and 23-18e). Still others recognize 30 different races. What this tells us is that the races are arbitrary categories. As people move about the planet and interbreed, the races are mixing, blurring the lines among the world's peoples. As time goes on, then, the distinction among various races will inevitably decrease.

> **KEY CONCEPTS**
>
> Human races are a reflection of phenotypic differences (adaptations to differing environmental conditions) created by geographic isolation and selected for by abiotic factors of the environment.

23-7 Health and Homeostasis

Evolution has resulted in the presence of many homeostatic mechanisms—in the bodies of organisms like us and in the environment on which we depend. As pointed out previously, humans have adopted practices that unknowingly upset those mechanisms. One of these ways—the use of pesticides—has potentially far-reaching impacts.

Chemical pesticides kill insects and other organisms that damage crops. Unfortunately, the use of pesticides has been shown to create genetic resistance in pest populations. **Genetic resistance** results from naturally occurring genetic variation—the presence of a mutation that makes a small percentage (about 5%) of any insect population resistant to chemical pesticides.

When farmers spray their fields with chemical pesticides, they kill 95% of the pests. The survivors are primarily those that are naturally resistant to the pesticide. They, in turn, breed and, over time, produce new populations that are even more resistant to pesticides. To kill these pests, farmers must apply higher doses of pesticides or switch to another type. However, a small segment of the population is genetically resistant to the higher dose or the new chemical preparation. That group survives the spraying, and is selected for and produces an even more resistant population, thus continuing an ever-escalating cycle often referred to as the "pesticide treadmill."

The pesticide treadmill is an excellent example of artificial selection at work. Some scientists warn farmers that they can never win in the battle against pests. Thanks to genetic variation in a population, no matter what pesticide they use, or how much they use, there will always be a resistant strain. Ever more powerful and more frequent applications at higher doses will be necessary just to stay even. Today, over 550 species of insects are resistant to one or more chemical pesticides. Twenty of the worst pests are now resistant to all types of insecticide.

Don't despair. There are ways to control pests without the use of pesticides and without creating genetically resistant strains. Organic farmers are practicing these methods quite successfully. One technique is crop rotation, which requires farmers to alternate the crops they plant in a given field. One year, corn might be grown in the field, the next year, beans, and the third year, alfalfa. How does it work?

Many insect pests are specialists, preferring one crop to another. A field of corn, for example, provides an abundant supply of food for corn borers. At the end of the season, corn borers lay their eggs in the soil or in organic material on the surface. If the same crop is planted the next year, the newly hatched insects will have an abundant food supply. The insect population will increase quickly, causing considerable damage. If a different crop is planted the second year, however, the corn borer population will decline. Continued rotation helps to hold down pests year after year without pesticides.

> **KEY CONCEPTS**
>
> The use of pesticides is resulting in the evolution of pesticide-resistant insects by a process of natural selection. Fortunately, there are more environmentally sound ways to control pests that do not result in pesticide resistance.

Health Tip 23-1

When buying poultry, eggs, and dairy products, choose the organic variety.

Why?

The Consumers Union recommends organic varieties to avoid exposure to hormones that may cause health effects in humans and antibiotics that contribute to the creation of antibiotic-resistant strains of bacteria. Organic beef also lowers your risk of being exposed to the agent (prions) that causes mad cow disease and may also contribute to the growing incidence of Alzheimer's and related diseases.

healthnote

23-1 Well Done, Please: The Controversy over Antibiotics in Meat

More than 60 years ago, livestock growers began adding antibiotics to cattle and pig feed to protect animals confined to pens from disease as they were being fattened for market (Figure 1). Antibiotics were first given to control disease, which can run rampant under crowded conditions. However, farmers soon found that the drugs had an unanticipated beneficial effect: For reasons still not fully understood, antibiotics accelerated the rate of body growth.

Not surprisingly, today 70% of all cattle and 90% of all veal calves and pigs are reared on feed treated with penicillin or tetracycline. Nearly half of the antibiotics sold in the United States, in fact, are used for livestock feed. This practice is also popular in Europe.

The addition of antibiotics to feed has been criticized by microbiologists and health officials. They fear that it could promote the evolution of superstrains of bacteria immune to antibiotics. In an editorial in the *New England Journal of Medicine*, a Tufts University microbiologist, Stuart Levy, noted that "every animal . . . taking an antibiotic . . . becomes a factory producing resistant strains" of bacteria. Resistant bacteria, in turn, could transfer their resistance to other bacteria, creating highly lethal strains that could infect humans.

Scientists are also concerned that some resistant bacteria in cattle (such as *Salmonella*) could be transmitted directly to people via meat or milk. The effects could be grave.

Medical researchers are also worried that antibiotics in livestock grains may also end up in the meat of the animal. It may then be passed on to the meat-eating public. In humans, the antibiotics could result in the

FIGURE 1   Feed Lot Cattle Antibiotics reduce infections in cattle in tight quarters and accelerate growth. Microbiologists, however, worry that they may be stimulating the evolution of super strains of bacteria. (Courtesy of USDA.)

production of antibiotic-resistant bacteria, adding to the already growing problem caused by antibiotics used to treat infections,

In recent years, studies have shown that early concerns are valid. Antibiotic-resistant and potentially harmful bacteria are indeed present in supermarket meat from antibiotic-treated chickens and cattle. Andrew Gunther, program director for Animal Welfare, writes: "The problem for humans is that by allowing intensive livestock farms to routinely expose bacteria to regular sub-therapeutic levels of antibiotics . . . we are actually providing the ideal conditions for bacteria to mutate and become resistant to their effects." What is more, antibiotic resistance can "jump" from animal pathogens to human pathogens. As a result, many scientists and food advocacy groups are pushing for tighter regulations on the industry and more details about how it uses antibiotics

SUMMARY

The Evolution of Life

1. Scientists believe that the Earth and the Sun—and the rest of the system—came from an enormous cloud of cosmic dust and gas.
2. The Earth formed about 4.5 billion years ago.
3. The evolution of life probably began in the sea and is divided into three phases: chemical evolution, cellular evolution, and the evolution of multicellular organisms.
4. Chemical evolution, scientists hypothesize, began about 4 billion years ago. Simple inorganic chemicals that were dissolved in the seas combined to form small organic molecules. Some organic molecules may also have come from

outer space in meteors that struck the Earth. Over time, the organic molecules combined to form polymers—small proteins and nucleic acids. These polymers combined to form aggregates that may have been the precursors of cells.

5. The first true cells arose from the primitive aggregates. They contained primitive enzymes and simple genes, and may have derived nourishment from organic molecules they absorbed from the environment. Many different types of primitive cells could have existed simultaneously.
6. During early cellular evolution, considerable gene swapping could have occurred among the early cells, producing more complex assortments of ever more complex cells.

7. Over time, the primitive cells gave rise to prokaryotes, primitive bacteria and Archaea. Prokaryotes emerged about 3.5 billion years ago, and eukaryotes (cells with nuclei) evolved about 1.2 billion years ago. The evolution of eukaryotes opened the door for the evolution of multicellular organisms.
8. A variety of multicellular plants and animals evolved from single-celled eukaryotes in the oceans. As the ozone layer developed, life on land became possible.

How Evolution Works

9. Evolution has produced a great diversity of organisms.
10. Evolution takes place because of genetic variation and natural selection.

Despite these concerns, efforts to reduce the use of antibiotics in feed have been defeated since the late 1970s. Livestock producers, feed producers, and the multimillion-dollar drug industry, however, argued back that microbiologists' concerns had not been proven. (In 2007, Tyson announced that it will no longer sell chicken treated with antibiotics.)

Since then, scientific evidence has begun to accumulate, confirming the suspicions of the medical profession. A study by researchers at the Centers for Disease Control and Prevention, published in *Science* in 1984, showed that antibiotic-resistant bacteria artificially (created by adding antibiotics to cattle feed) could be transferred directly from meat to humans. The study also showed that 20% to 30% of the *Salmonella* outbreaks in the United States involved antibiotic-resistant strains. About 4% of the people contracting the antibiotic-resistant bacteria died, compared with only 0.2% of the victims of the nonresistant *Salmonella*.

In 1999, a team of Minnesota researchers published results of a study that provided additional evidence. The researchers found that only 1.3% of bacteria isolated from humans in 1992 were resistant to a certain type of antibiotic. In 1998, 3 years after the drug was approved for use in animal feed, 10.2% of the bacteria were resistant to the antibiotic. The researchers concluded that the practice of giving the drug to poultry "created a reservoir" of resistant bacteria. In 2000, Danish researchers found that antibiotic-resistant bacteria could be transmitted from meat to humans.

Proponents of antibiotic use in feed believe that the link between antibiotics and human disease is still weak and that further research is needed. Even if these findings are substantiated by further research, proponents believe that the benefits of using antibiotics outweigh the potential health effects. Banning antibiotics or cutting back on their use, they say, could have enormous economic impacts that must be weighed against sickness and loss of life. But health experts hope that the United States, like Europe, which strictly limited the use of antibiotics in animal feed in the early 1970s, will find the political will to end this activity.

Individuals can take direct action, too. You can buy "organically grown" beef—that is, beef fed a diet that contains no hormones or antibiotics. Grass-fed organic beef may be even healthier, as these meats contain less fat than corn-fed beef. Some species of grass-fed beef, for instance, Belted Galloways, are even leaner and healthier (Figure 2). You can also change your eating habits. Instead of asking for rare meat or a half-cooked burger, you may want to consider asking for a well-done piece of meat. If ground beef has been heated in the center to over 160° F (71° C), you are pretty safe; chicken and poultry need to be heated to more than 170° F (77° C) or until the fluids run clear.

FIGURE 2 **Belted Galloways** These grass-fed Belted Galloways, sometimes fondly called "Oreo cows," raised by the author on his farm in Missouri (evergreenbelties.com), have thick furry coats that keep them warm in the winter and reduce fat deposits that help insulate other breeds. (Courtesy Dan Chiras.)

 Visit Human Biology's Internet site for links to websites offering more information about this topic.

11. Genetic variation in a species arises from many factors such as mutations and from new genetic combinations resulting from sexual reproduction.
12. Genetic variation results in variations in traits that may offer some organisms an advantage over others, giving them a better chance of surviving and reproducing and passing their genes on to future generations.
13. Beneficial traits are preserved in a population by natural selection.

The Evidence Supporting Evolution

14. The scientific knowledge in support of evolution is rich and varied. The fossil record, anatomical similarities in groups of organisms, the common biochemical makeup of organisms, similar embryological development among many groups of organisms, and experimental evidence all support the theory of evolution by natural selection.

Early Primate Evolution

15. Humans belong to the order called primates. Today, most primate species live in tropic and subtropic forests. The main exception is humans, who inhabit a wide range of habitats.
16. Primates are characterized by grasping hands, forward-looking eyes, and large brains (in proportion to body size).
17. Based on fossil evidence, it appears that the primates evolved from a mammalian insectivore that resembled the modern-day tree shrew and lived about 80 million years ago.
18. Humans evolved from an early ape-like creature.

Evolution of the Australopithecines

19. The first hominid, *Sahelanthropus tchadensis*, may have lived 6 to 7 million years ago in Africa. It may have evolved into the first australopithecines.
20. One of the oldest known australopithecine skeletons was unearthed in Africa and is believed to be about 3.5 million years old. It belongs to a group called *Australopithecus afarensis*. It stood about 3 feet high and had a brain only slightly larger than an ape's, but it probably walked erect.

21. *A. afarensis* may have evolved to form *A. africanus*, which was slightly taller than its predecessor and had a slightly larger brain.

22. About 2.3 million years ago *Paranthropus robustus* emerged. It was taller and heavier and had a larger brain than its predecessors. About 2.2 million years ago, the fourth species, *Paranthropus boisei*, appeared. These species appear to be dead ends in human evolution.

Evolution of the Genus *Homo*

23. Many paleontologists once thought that *A. afarensis* gave rise to the genus *Homo*, the ancestors of modern humans, *Homo sapiens*. They thought that one member of that group, *Homo habilis*, gave rise to *Homo erectus*, which, in turn, evolved to form modern humans.

24. More recent findings suggest that modern humans arose from *Homo ergaster*. It probably evolved from a species of *Australopithecus*. Which one, scientists don't know.

25. Hominids moved out of Africa, presumably in the form of *Homo ergaster* or a close relative, reaching China and Java about 1.8 million years ago. *Homo erectus* moved out, too, and was well established in these regions 1 million years ago.

26. Members of the genus *Homo* began to appear in Europe at a later time. One of them was *Homo antecessor*, which lived in Spain about 800,000 years ago. It was later followed by *Homo heidelbergensis*, a species found in Europe and Africa.

27. In Europe, *H. heidelbergensis* gave rise to a group of hominids whose best-known member was *Homo neanderthalensis*. This species thrived in Europe and western Asia between 230,000 and 300,000 years ago.

28. *H. heidelbergensis* may have also given rise to our kind, *Homo sapiens*. Like our ancestor, we may have arisen in Africa.

29. The Neanderthals lived in caves and camps in Europe and Asia until approximately 30,000 years ago, when they disappeared. Some archaeologists believe that they were replaced by modern humans, the Cro-Magnons, members of *H. sapiens*.

30. *Homo sapiens* appeared in Africa about 195,000 years ago, and then migrated to other continents. The Cro-Magnons, a culturally more advanced form of *Homo sapiens*, first appeared in Africa and then spread across Europe and northern Asia, perhaps wiping out the Neanderthals or possibly interbreeding with them.

31. Over the course of time, the human population has wandered far and wide on the planet and has come to inhabit a wide range of climatic zones. This has resulted in the formation of a number of distinct subpopulations, or races. Many differences found in the races are thought to be adaptations to differing environmental conditions.

Health and Homeostasis

32. Homeostasis at the organismic and environmental levels can be upset by pesticides. Pesticides can create a form of artificial selection because they select resistant species.

33. Farmers who spray their fields to kill insects leave behind a genetically resistant subpopulation that breeds and repopulates farm fields. A second application at a higher dose or an application of another pesticide kills off more susceptible insects, but this leaves behind another subset that often becomes a further pest, forcing farmers to use higher doses or switch to another pesticide. This escalation in the war against pests is called the pesticide treadmill.

34. Chemical resistance is a rather common occurrence in the modern world. Weeds can become resistant to herbicides, and even microorganisms develop resistance to antibiotics.

THINKING CRITICALLY ANALYSIS

This analysis corresponds to the Thinking Critically scenario that was presented at the beginning of this chapter.

One of the first questions you might ask is this: Does the archaeological data support the genetic data? In other words, does the fossil record support the African Origins model? Or does it support the Multiregion Evolution model?

Some archaeological evidence suggests that the Multiregion Evolution model might be the most valid hypothesis. For example, studies of facial features of the Chinese skulls illustrate commonalities in modern and archaic Chinese. More importantly, these features are not present in the skulls of early modern Africans. One would expect the skulls of early modern Africans to resemble the Chinese skulls if the modern Chinese populations did indeed come from an ancestral African population. A similar study of skulls of modern and archaic European populations shows no similarities with either the earliest modern Africans or archaic African populations.

Another question that might be useful is this: Is there any archaeological evidence of the spread of African populations into other areas at the proper time? Again, the answer is no. Nor is there evidence of the spread of African culture (as witnessed by stone tools) as one might expect.

Given these facts, can you think of any reason to question the Multiregion Evolution model?

One common criticism of the Multiregion Evolution model is that archaeologists make many subjective judgments and often base their conclusions on small numbers of skulls. Geneticists wonder if such small samples are truly representative of an entire population. In contrast, the genetic data offers quantitative indicators of evolutionary change that are relatively free of biases.

Given the controversy, it is clear that neither hypothesis can be ruled out. More research is needed, and many years will no doubt pass before scientists can agree on which is correct.

KEY TERMS AND CONCEPTS

Abiotic factor, p. 505
Adaptation, p. 505
Alfred Wallace, p. 505
Anthropoid, p. 511
Archaea, p. 502
Australopithecus, p. 511
Biogeography, p. 509
Biotic factor, p. 505
Charles Darwin, p. 505
Cro-Magnons, p. 516
Endosymbiotic evolution, p. 503

Eukaryote, p. 503
Evolution, p. 500
Fitness, p. 505
Fossil, p. 507
Genetic resistance, p. 517
Genetic variation, p. 503
Heterotroph, p. 502
Hominids, p. 511
Homo erectus, p. 514
Homo habilis, p. 514
Homo sapiens, p. 516

Homologous structure, p. 508
Natural selection, p. 505
Panspermia theory, p. 501
Primate, p. 510
Prokaryote, p. 502
Prosimian, p. 510
Protocells, p. 501
Race, p. 516
Theory, p. 501
Theory of chemical evolution, p. 500

CONCEPT REVIEW

1. Define the term, evolution. What are the end products of evolutionary change? p. 500.
2. Describe how the first cells may have arisen during evolution. What critical requirements must have been met for life to begin? pp. 501–503.
3. The development of photosynthesis and the emergence of eukaryotes were pivotal events in evolution. Why? p. 503.
4. How do scientists believe the first organelles appeared in single-celled organisms? What evidence supports this hypothesis? p. 502.
5. Numerous plants and animals evolved in the sea before life emerged on land. Why? p. 503.
6. A critic of evolution says, "Life forms are too diverse to have come from a common ancestor." How would you respond? pp. 500–503.
7. How do random mutations in germ cells contribute to the evolutionary process? pp. 503–505.
8. Define the following terms: adaptation, variation, natural selection, biotic factors,

abiotic factors, selective advantage, and fitness. p. 505.
9. What are the sources of genetic variation in a population? pp. 503, 505.
10. What is meant by survival of the fittest? What is fitness? Why does fitness not always mean strength? p. 505.
11. Explain what is meant by the theory of evolution by natural selection. p. 505
12. Discuss the following statement: Natural selection is nature's editor. pp. 505–506.
13. What factors in an organism's environment contribute to natural selection? p. 505.
14. Check this out on the Internet: What was Charles Darwin's contribution to thinking about evolution in 1859? What was Alfred Wallace's contribution? In what ways were these scientists' views similar and in what ways were they different? p. 505.
15. How may viral genes have facilitated the evolution of humans? p. 506.
16. Discuss in detail the various lines of evidence supporting the theory of

evolution. Give examples of each one. pp. 507–509.
17. Why is the primate fossil record incomplete? What problems does this gap create in tracing the evolutionary history of primates? pp. 510–511.
18. Briefly describe the evolution of primates. In other words, where did primates come from? Then, briefly describe the evolution of modern humans, *Homo sapiens*, touching on what scientists think are our main ancestors pp. 510–515.
19. Describe the term pesticide treadmill. What is it and how can farmers avoid this problem? p. 517.
20. Based on your understanding of the development of pesticide resistance in insects, what predictions might you make about the widespread use of antibacterial soaps and cleaning agents now widely used in homes and commercial establishments in North America. p. 517.

SELF-QUIZ: TESTING YOUR KNOWLEDGE

1. Inorganic molecules in the oceans may have reacted chemically to produce the very first _____ molecules. p. 500.
2. Some scientists speculate that the first organic molecules essential for life on Earth came from _____ and comets. This is known as the _____ hypothesis. p. 501.
3. The very first cells were known as _____ and probably consisted of aggregations of polymers and other molecules formed during chemical evolution. p. 501.
4. Cells with nuclei belong to a group known as _____. p. 503.
5. _____ evolution is the name of the process by which single-

celled organisms acquired organelles like mitochondria, forming modern cells. p. 503.
6. Life first evolved in the _____ because intense UV radiation impaired evolution of land plants and animals. p. 503.
7. Genetic variation leads to behavioral, functional, and _____ changes in organisms that may better their chances of survival and reproduction. p. 503.
8. Sexual reproduction results in new combinations of genes that can confer a selective _____ in organisms. p. 505.
9. A genetically based characteristic that increases an organism's chance of

surviving and reproducing is called a(n) _____. p. 505.
10. Abiotic and biotic factors influence survival and reproduction by "selecting" the fittest—those best able to reproduce and pass on their genes. This process is called natural _____. p. 505.
11. Two scientists are noted for their explanation of the mechanism behind evolution, Charles Darwin and _____ _____. p. 505.
12. _____ is a measure of reproductive success that can be attributed to a variety of factors such as greater strength or cunning. p. 505.
13. _____ genes have been incorporated into the DNA of humans and other organisms over the years and

may have helped accelerate evolution. p. 506.

14. A _____ gene is a segment of DNA that comes from so-called junk DNA. p. 506.

15. Fossils have helped scientists trace the path of evolution because the rocks they come from can be _____ using radioactive techniques. p. 507.

16. Structures thought to have arisen from a common ancestor like the arm of a human and the fin of a whale are known as _____ structures. p. 508.

17. At one time, all the continents were joined to form a supercontinent known as _____. p. 509.

18. Humans belong to the order called ___ _____. p. 510.

19. Primates are believed to have evolved from a(n) _____ mammal that belongs to a group known as the prosimians (premonkeys). p. 510.

20. The prosimians gave rise to _____, a group consisting of monkeys, apes, and hominids. p. 511.

21. The first hominid, *Sahelanthropus tchadensis*, may have lived 6 to 7 million years ago in Africa. It may have evolved into the first _____. p. 513.

22. Human races are the result of _____ separation. p. 505.

23. In Europe, *H. heidelbergensis* gave rise to a group of hominids whose best-known member was *Homo _____*. This species thrived in Europe and western Asia between 230,000 and 300,000 years ago. p. 515.

24. *H. heidelbergensis* may have also given rise to our kind, *Homo sapiens*. Like our ancestor, we may have arisen in _____. p. 515.

25. _____ resistance refers to naturally occurring resistance among pest species, bacteria, and other organisms to pesticides, herbicides, and antibiotics. p. 517.

biology.jbpub.com/chiras/8e/

The site features eLearning, an online review area that provides quizzes, chapter outlines, and other tools to help you study for your class. You can also follow useful links for in-depth information, research the differing views in the Point/Counterpoints, or keep up on the latest health news.

(Courtesy of Department of Energy)

Ecology and the Environment

Most of us live our lives seemingly apart from nature. We make our homes in cities and towns surrounded by concrete and steel and drown out the sound of birds with our noise. We spend countless hours indoors, watching television, playing computer games, and talking with or texting friends. The closest many of us get to nature is a romp with the family dog on the grass in our backyard or an occasional nature show.

THINKING CRITICALLY

A newspaper article notes that "on the issue of global warming the scientific community is divided." In support of this assertion, the author quotes two scientists. One says that he's "convinced the Earth's temperature is increasing and this increase is being caused by pollution, notably carbon dioxide from human activities such as energy production and transportation." The other argues that "there's not enough evidence to support this conclusion." The article goes on to say that because of the uncertainty among the scientific community, it makes no sense to launch a global effort to reduce carbon dioxide emissions. Can you detect any problem in this reporting? What critical thinking rules were helpful to you in this examination?

Raymond Dasmann, a world-renowned ecologist, wrote that despite what many of us may think, a human apart from nature is an abstraction. No such being exists. Human life depends on the environment. The clothes we wear, the food we eat, the energy we consume, and even the oxygen we breathe are all products of nature.

Nature provides other free services as well. For example, plants protect the land, preventing flooding and erosion. Swamps help purify the water we drink. Birds control insect populations. Clearly, nature "serves" us well. Therefore, although we may have isolated ourselves from nature, we are extremely dependent on it. This chapter will help you understand why.

24-1 An Introduction to Ecosystems

This chapter discusses ecology and environmental issues. **Ecology** is a branch of science that focuses on the many ways organisms interact with one another. Ecology also examines how organisms are affected by the physical and chemical environment. In addition, ecology studies the ways organisms affect their environment.

Ecology, like all disciplines in science, is a body of knowledge and a process of inquiry. However, ecology probably ranks as one of the most misused words in the English language. Banners proclaim, "Save Our Ecology." Speakers argue that "our ecology is in danger," and others talk about the "ecological movement." These common uses of the word ecology are incorrect. Why?

Ecology is a discipline of science. It is not synonymous with the word environment. It does not mean the web of interactions in the environment. We can save our ecology department and ecology textbooks, but we cannot save our ecology. Our ecology is not in danger; our environment is. You cannot join the ecology movement, but you can join the environmental movement.

> **KEY CONCEPTS**
> Ecology is the study of interactions that occur in ecosystems between organisms and between organisms and their chemical and physical environment.

The Biosphere

The science of ecology, unlike many other branches of scientific endeavor, often focuses on systems. The largest biological system on Earth is the **biosphere** (BUY-oh-sfear). The biosphere is the thin skin of life on the planet. As shown in Figure 24-1 the biosphere forms at the intersection of air, water, and land. Organisms consist of components derived from all three. The carbon atoms in the proteins in your body, for instance, come from carbon dioxide in the atmosphere captured by plants. The minerals in your bones come from the soil in which plants grow. Water comes from streams and lakes.

The biosphere extends from the bottom of the ocean to the tops of the highest mountains. Although that may seem like a long way, it's not—at least, in comparison with the size of the Earth. In fact, if the Earth were the size of an apple, the biosphere would be about the thickness of its skin. Although life exists throughout the biosphere, it is rare at the extremes, where conditions for survival are marginal. Ecologists refer to the biosphere as a closed system. In other words, the biosphere is much like a sealed terrarium. That is, it receives no materials from the outside. (The one exception is cosmic dust that settles on the Earth on a daily basis.) Because the

Earth is a closed system, all materials necessary for life must be recycled over and over, as you shall soon see.

Just like a terrarium, the only outside contribution to the biosphere is sunlight. As noted previously in this book, sunlight is the source of energy for virtually all life. Even the energy released when coal, oil, and natural gas are burned owes its origin to the sunlight that fell on the Earth several hundred million years ago. Coal, for instance, is made from ancient plants that lived in and along swamps. Leaves and other plant matter fell into the water and were covered by sediment. Over time, pressure and heat converted the plant matter to coal. Oil is made from ancient marine algae that settled to the bottom and were covered by sediment. Over time, heat and pressure converted the organic matter into oil, a complex mixture of organic compounds. Natural gas is an organic compound associated with coal and oil deposits. It is produced from once-living organisms.

> **KEY CONCEPTS**
> The biosphere is the thin skin of life on the planet, encompassing ecosystems on land and in the oceans, lakes, and rivers.

Biomes and Aquatic Life Zones

Viewed from outer space, the Earth resembles a giant jigsaw puzzle, consisting of large landmasses and vast expanses of ocean. The landmasses, or continents, can be divided into large biological subregions or biomes. A **biome** is a region characterized by a distinct climate, along with specific plants and animals adapted to it. Some of the most familiar biomes include the desert, grassland, and tundra (Figure 24-2).

The oceans also can be divided into subregions, known as **aquatic life zones**. Aquatic life zones are the aquatic equivalent of biomes. Like their land-based counterparts, each of these regions has a distinct environment and characteristic plant and animal life adapted to conditions of the zone. Four major aquatic life zones exist: coral reefs, estuaries (the mouths of rivers where fresh- and saltwater mix), the deep ocean, and the continental shelf.

Humans inhabit all biomes on Earth and are active in many aquatic life zones, especially estuaries. Humans benefit in numerous ways from the microbes, plants, animals, soil, water, and air that comprise the biomes and aquatic life zones. We often tend to think of biomes and aquatic life zones in terms of their natural resources (coal, timber, oil, and fish). Although vital to human society's economic health, these resources are but a fraction of the services we receive from the living Earth. Trees, for instance, not only provide lumber to build homes and make paper, but they also provide oxygen

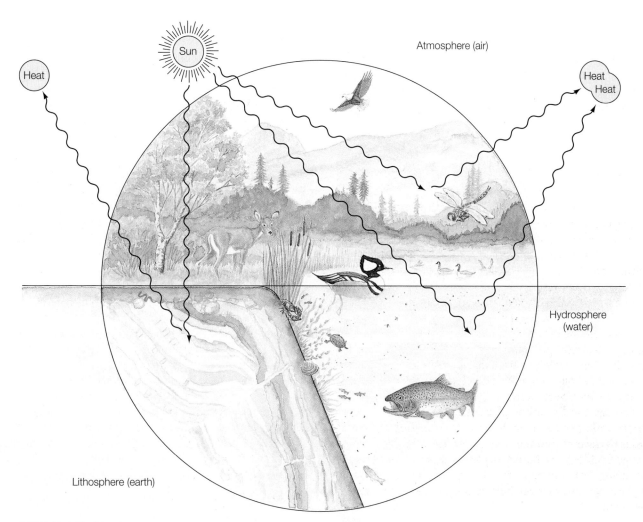

FIGURE 24-1 **The Biosphere** Life exists at the intersection of land, air, and water.

and protect watersheds. The oxygen we breathe, which makes glucose metabolism and energy production in cells possible, comes from plants like trees. In addition, trees may remove certain pollutants from the air. Millions of species besides humans also depend on these services.

> **KEY CONCEPTS**
>
> The Earth consists of aquatic and terrestrial regions known as aquatic life zones and biomes, respectively. They provide natural resources and many ecological services vital to all life on the planet.

Ecosystems

The biosphere is a global **ecological system**, or ecosystem for short. The term **ecosystem** is used to describe a community of organisms and their physical and chemical environment. Numerous interactions are possible within an ecosystem.

Ecosystems consist of two basic components: abiotic and biotic. The **abiotic components** of an ecosystem are the physical and chemical factors necessary for life. They include sunlight, precipitation, temperature, and nutrients.

Humans often alter the physical and chemical components of the environment. Time and again, scientists are finding that these changes can result in serious biological impacts. If conditions change drastically, for instance, some species perish. Some changes can even affect our own survival.

The **biotic components** of an ecosystem are the organisms that live there—the plants, animals, and microorganisms. Within biomes and aquatic life zones, species often occupy very specific regions to which they are well adapted. A group of organisms of the same species occupying a specific region constitutes a **population**. In any given ecosystem, several populations exist together and form a **biological community**, an interdependent network of plants, animals, and microorganisms.

> **KEY CONCEPTS**
>
> Ecosystems are communities of organisms and their environment. The environment consist of abiotic and biotic components.

Habitat

If asked to give a brief description of yourself, you would probably begin by describing the place where you live. You would then likely discuss the work you do, the friends you have, and other important relationships that describe your place in society. A biologist would do much the same when describing an organism. He or she would start with a description of the place an organism lives—that is, its **habitat**.

> **KEY CONCEPTS**
>
> An organism's habitat is where it lives.

(a)

(b)

(c)

(d)

(e)

FIGURE 24–2 **North American Biomes** (a) Tundra, (b) northern coniferous forests, (c) temperate deciduous forest, (d) prairie, and (e) desert. (a © Zastavkin/ShutterStock, Inc.; b © Caleb Foster/ShutterStock, Inc.; c © Corbis; d courtesy of Tim McCabe/USDA NRCS; e © LouLouPhotos/ShutterStock, Inc.)

Niche

Next, the biologist would describe how the organism "fits" into the ecosystem, its **ecological niche**, or simply **niche** (pronounced *nitch*). An organism's niche includes all of its relationships with its environment. For example, the niche includes what an organism eats, what eats it, and other important facts.

The concept of the niche is very important. For example, successful control of an insect pest is best achieved through an understanding of its niche. Such an analysis might show that a particular species of bird or insect feeds on the pest. As another example, to protect a plant or animal you have to protect the ecosystem it lives in. Such efforts help protect the species' niche, which, in turn, helps protect the species itself.

24-2 Ecosystem Function

Producers

Life on land and in water is possible principally because of the existence of a group of organisms known as **producers** (Figure 24-3). Producers primarily include the algae and plants. These organisms absorb sunlight. They then use its energy to make organic molecules from atmospheric carbon dioxide and water. This process is called **photosynthesis**. Organic molecules generated by photosynthesis nourish the producers and also feed all the other organisms in ecosystems.

Another large group of organisms is the **consumers**. They reap the benefit of photosynthesis in producers. Ecologists place consumers into four general categories, depending on the type of food they eat. Some consumers, such as deer and cattle, feed directly on plants. They are called **herbivores**. Others, such as wolves, feed on herbivores and other animals and are known as **carnivores**. Humans and many other animal species consume both plants and animals and are known as **omnivores**. The final group feeds on animal waste or the remains of plants and animals. They are **detrivores** (DEE-treh-vores), or **decomposers**. This important group includes many bacteria, fungi, and insects.

> **KEY CONCEPTS**
>
> The biotic components of an ecosystem consist of producers, which make much of the food consumed by consumers, and decomposers that are responsible for recycling waste and the dead remains of organisms so they can be used again by producers.

Food Chains and Food Webs

In ecosystems, each organism is a food source for some other organism. As a result, all organisms belong to one or more food chains. Technically, a **food chain** is a series of organisms, each one feeding on the organism preceding it (Figure 24-3). Biological communities consist of numerous food chains.

Biologists recognize two general types of food chains: grazer and decomposer. **Grazer food chains** begin with plants and algae, the producers. These organisms are consumed by herbivores, or grazers; hence, the name grazer food chain. Herbivores, in turn, may be eaten by carnivores. **Decomposer food chains** begin with dead material—either animal wastes (feces) or the remains of plants and animals (Figure 24-4).

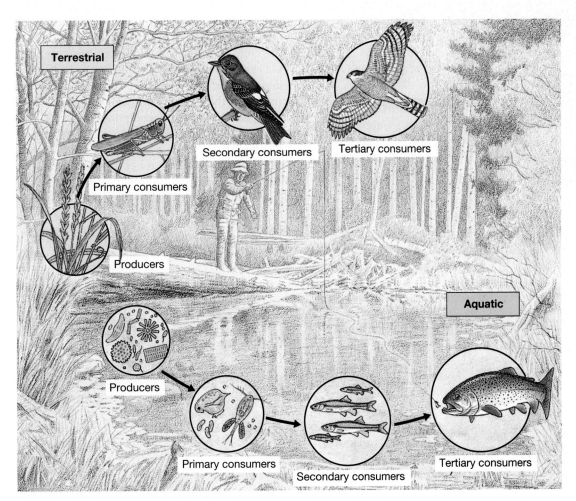

FIGURE 24-3 **Simplified Food Chains** Two grazer food chains are shown, a terrestrial one and an aquatic one.

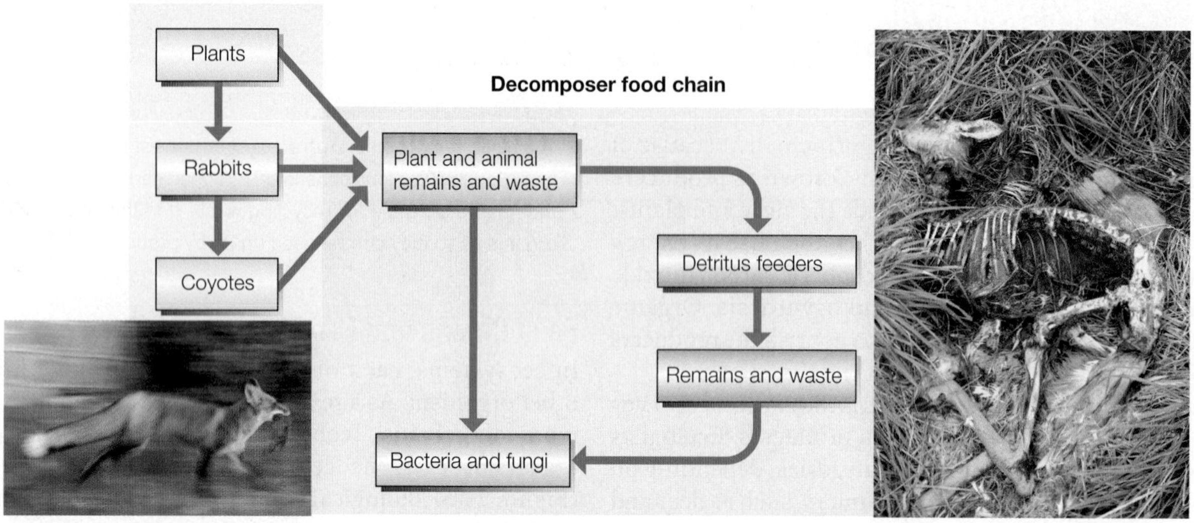

Grazer food chain

FIGURE 24-4 **Food Chains** A grazer food chain and a decomposer food chain, showing the connection between the two. (Coyote photo (© Izzat Bakhadyrov/ShutterStock, Inc.; bones photo © Perihelion/Alamy Images.)

As in most things in nature, decomposer and grazer food chains typically function together (Figure 24-4). For example, waste products from the grazer food chain enter the decomposer food chain. Nutrients liberated by the decomposer food chain enter the soil and water and are reincorporated into plants at the base of the grazer food chain.

In a biological community, food chains are woven together into more complex networks, called **food webs** (Figure 24-5). Food webs present a complete picture of the feeding relationships in any given ecosystem.

> **KEY CONCEPTS**
>
> A food chain is a series of organisms, each one feeding on the organism preceding it. Two types of food chains are found: grazer and decomposer. Biological communities consist of numerous interwoven food chains, creating a food web.

Energy and Nutrient Flows

Although we tend to think of food chains simply as a map of feeding relationships, they are much more. Food chains are avenues for the flow of energy and the cycling of nutrients through the environment. Let's consider energy first.

In the biosphere, almost all of the energy needed by organisms comes from the Sun. Solar energy is captured by plants and algae and is used to produce organic food molecules. This energy is stored in the molecules of organisms. In the food chain, organic molecules pass from plants to animals, where they are broken down in their mitochondria. They release stored solar energy, which is then used to power numerous cellular activities.

During cellular respiration, much of the energy stored in organic food molecules is lost as heat. Heat escaping from plants and animals is radiated into the atmosphere and then into outer space. It cannot be recaptured and reused by organisms. Because all solar energy is eventually converted to heat, energy is said to flow one way through food webs. Energy cannot be recycled.

In sharp contrast, nutrients are recycled over and over. Nutrients in the soil, air, and water are first incorporated into plants and algae, and then passed from plants to animals. Nutrients in the food chain eventually reenter the environment by one of three paths. They may enter via the excretion of wastes from animals. Urine from animals, for instance, contains nutrients that are incorporated into the soil, and then into plants. Nutrients may also reenter the environment via the decomposition of solid waste such as feces or leaves deposited on the soil by trees in the fall. And, lastly, they may reenter by the decomposition of dead organisms—plants and animals that die and decompose.

One way or another, all nutrients eventually make their way back to the environment for reuse. Each new generation of organisms therefore relies on the recycling of matter. The atoms in your body, for instance, have been recycled many times since the beginning of life on Earth. Who knows? Some of those atoms may have been part of the very first cells.

> **KEY CONCEPTS**
>
> Food chains are avenues for the one-way flow of energy and the cycling of nutrients through ecosystems.

Trophic Levels

Ecologists classify the organisms in a food chain according to their position, or **trophic level** (TROE-fic; literally, "feeding" level). The producers comprise the base of the grazer food chain and are therefore members of the first trophic level. The grazers are members of the second trophic level. Carnivores that feed on grazers are members of the third trophic level, and so on.

Most terrestrial food chains are limited to three or four trophic levels. In fact, longer terrestrial food chains are quite rare. Food chains generally do not have a large enough producer base to support many levels of consumers.

When plotted, the biomass (the amount of organic matter) at the various trophic levels forms a pyramid, the **biomass**

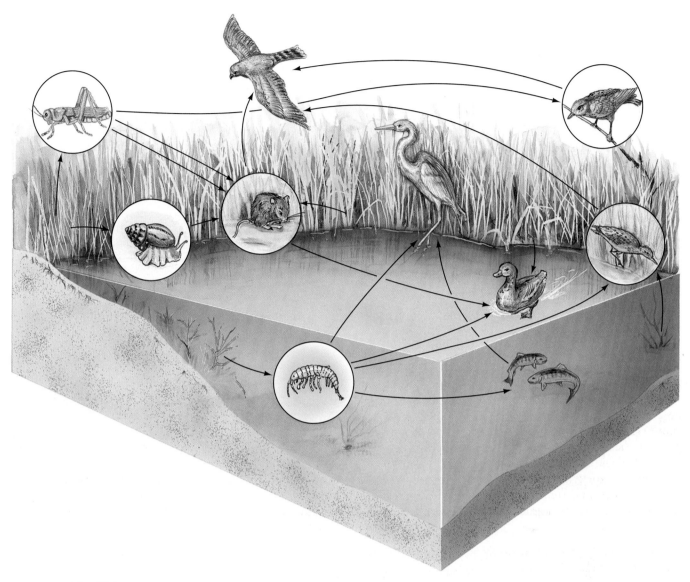

FIGURE 24–5 A Food Web

pyramid (Figure 24-6). Because biomass contains energy (stored in the chemical bonds), the biomass pyramid can be converted into a graph of the chemical energy in the various trophic levels. This graph is called an **energy pyramid**. In most food chains, the number of organisms also decreases with each trophic level, forming a **pyramid of numbers**.

> **KEY CONCEPTS**
>
> Organisms are classified by their trophic level in food chains. Producers comprise the first trophic level. Grazers are members of the second trophic level. Carnivores are on the third trophic level.

Nutrient Cycles

As noted earlier, nutrients flow from the environment through food webs, and then are released back into the environment. This circular flow constitutes a **nutrient cycle**, or a **biogeochemical cycle**.

Nutrient cycles can be divided into two phases: environmental and organismic. In the environmental phase, a nutrient exists in the air, water, or soil, or in two or more of them simultaneously. In the organismic phase, nutrients are found in living organisms.

Dozens of global nutrient cycles operate continuously to ensure the availability of chemicals vital to all organisms—humans included. Unfortunately, many human activities seriously disrupt nutrient cycles and can profoundly influence the survival of species that share this planet with us. Some threaten our own survival. This section looks at three important nutrient cycles.

> **KEY CONCEPTS**
>
> The circular flow of a nutrient through ecosystems is known as a biogeochemical cycle or nutrient cycles. Nutrients are cycled through the environment and organisms in food webs.

The Water Cycle. Water is part of a global recycling network known as the **hydrological cycle**, or **water cycle**. The hydrological cycle runs day and night, collecting, purifying, and distributing water throughout the planet. Along its way, it serves humans and other living creatures in a multitude of ways.

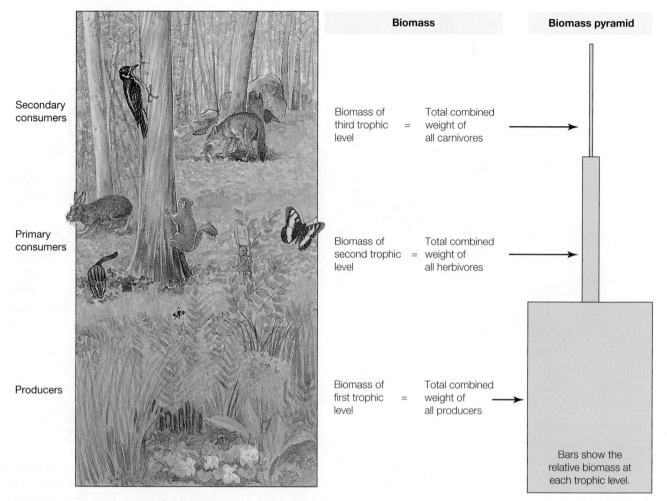

	Biomass		Biomass pyramid
Secondary consumers	Biomass of third trophic level	= Total combined weight of all carnivores	
Primary consumers	Biomass of second trophic level	= Total combined weight of all herbivores	
Producers	Biomass of first trophic level	= Total combined weight of all producers	

Bars show the relative biomass at each trophic level.

FIGURE 24-6 Biomass Pyramid In most food chains, biomass decreases from one trophic level to the next higher one.

The water cycle is driven by two basic processes, evaporation and precipitation. Evaporation occurs when water molecules escape from surface waters, soils, and plants and become suspended in air (Figure 24-7). When water molecules depart, they leave behind impurities. Thus, evaporated water is free of contamination until it mixes with atmospheric pollutants from human and natural sources.

In the atmosphere, water is suspended as fine droplets, creating water vapor. Although you may not usually be aware of it, you can detect its presence on a cold day when moisture beads up on cold windows, or on a warm day, when moisture collects on the outside of a glass of ice-cold lemonade.

The amount of moisture (water vapor) air can hold depends on the temperature of the air. The warmer the air, the more moisture it can hold.

Water that has evaporated eventually returns to Earth through precipitation. Here's how it happens. On a warm summer day, sunlight warms the Earth. This causes water to evaporate from surface waters, land, and plants. Warm, moisture-laden air then rises. As it rises, the air expands because atmospheric pressure decreases. As it expands, it cools. As it cools, its ability to hold water decreases, and clouds begin to form. (This same phenomenon occurs when moisture-laden air is pushed upward by mountain ranges.)

Clouds form as water molecules in air begin to attach onto various small particles suspended in the air such as

salts from the sea, dusts, or particulates from factories, power plants, and vehicles.

In tropical and semitropical regions, fine cloud droplets collide and combine, forming raindrops. Over a million fine water droplets must come together to produce a single drop of rain. In temperate climates and at the poles, the temperature of the air in clouds is often well below the freezing point, even in the warmer months of the year. As a result, minute ice crystals tend to form in the clouds. When the crystals reach a certain critical mass, they fall from the sky as snowflakes. In the spring, summer, and fall, however, the snowflakes generally melt as they fall, producing rain.

Clouds move about on the winds, and deposit their moisture throughout the globe as rain, drizzle, snow, hail, or sleet. This process, called precipitation, returns water to lakes, rivers, oceans, and land.

Water that falls on the land may evaporate again, or it may flow into lakes, rivers, streams, or groundwater. Eventually it returns to the ocean, from which it may once again evaporate.

At any single moment, 97% of the Earth's water is in the oceans. The remaining 3% is freshwater. Of this, most (99%) is locked up in polar ice, in glaciers, and in deep, inaccessible aquifers—underground reservoirs usually in some permeable material such as sandstone. This leaves a paltry 0.003% available for use by humans; in addition, most of this water is hard

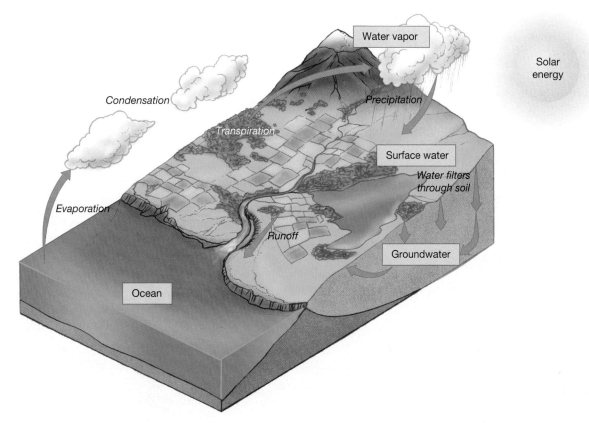

FIGURE 24–7 **The Hydrological Cycle**

to reach and much too costly to be of any practical value. At present, human civilization and all land-based forms of life are maintained by only a tiny fraction of the Earth's water supply, a fact that underscores the importance of treating our freshwater supplies with care.

KEY CONCEPTS

Water moves cyclically through the biosphere in the hydrological cycle that is driven by evaporation and precipitation.

The Carbon Cycle. The **carbon cycle** is illustrated in Figure 24-8. Let's begin with free carbon dioxide. In the environmental phase of the cycle, carbon dioxide resides primarily in the atmosphere and surface waters (oceans, lakes, and rivers). As illustrated, atmospheric carbon dioxide is absorbed by plants and other photosynthetic organisms. These organisms convert carbon dioxide into organic food materials, which travel along the food chain from one trophic level to the next.

Carbon dioxide reenters the environmental phase via cellular energy production (cellular respiration) of the organisms in the grazer and decomposer food chains and from the decomposition of waste and dead organisms.

For tens of thousands of years, the carbon cycle was in balance. With the advent of the Industrial Revolution, however, production of carbon dioxide began to exceed the planet's ability to absorb it. This was due in large part to two factors: (1) the widespread combustion of fossil fuels, which releases carbon dioxide, and (2) deforestation. Because trees absorb enormous amounts of carbon dioxide for photosynthesis, deforestation reduced the amount of atmospheric carbon dioxide they could incorporate.

Today, about 8 billion tons of carbon—in the form of carbon dioxide—is added to the atmosphere each year. Three-quarters of the carbon dioxide comes from the combustion of fossil fuels such as coal to make electricity and gasoline to power our cars. The remaining quarter stems from deforestation. As noted later in the chapter, carbon dioxide traps heat escaping from Earth and reradiates it to the Earth's surface. As carbon dioxide levels increase, global temperature rises.

KEY CONCEPTS

Carbon is vital to life. It cycles through the biosphere, starting in the atmosphere as carbon dioxide, then to photosynthetic organisms where it is used to make carbon-based food molecules then to consumers that feed on plants and other animals that feed on them. It then flows back to the atmosphere as carbon dioxide.

The Nitrogen Cycle. Nitrogen is an element essential to many important biological molecules, including amino acids, DNA, and RNA. The Earth's atmosphere contains enormous amounts of it. However, atmospheric nitrogen exists as nitrogen gas (N_2), which is unusable to all but a few organisms. To be incorporated into living organisms, atmospheric nitrogen must first be converted to a usable form—either nitrate or ammonia—in the **nitrogen cycle**.

The conversion of nitrogen to ammonia is known as **nitrogen fixation**. As Figure 24-9 shows, nitrogen fixation partly occurs in the roots of certain plants called legumes. These include peas, beans, clover, alfalfa, and others. Inside small nodules in their roots live symbiotic bacteria that convert atmospheric nitrogen to ammonia. (Ammonia is also produced by certain bacteria that live in the soil.)

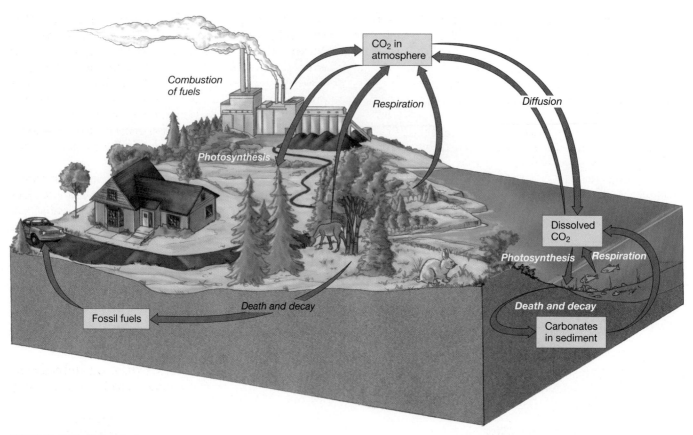

FIGURE 24-8   The Carbon Cycle

Once ammonia is produced, other soil bacteria convert it to nitrite and then to nitrate. Nitrates are incorporated by plants and are used to make amino acids and nucleic acids. All consumers—ourselves included—therefore ultimately receive the nitrogen they require from plants.

Nitrate in soil is also produced indirectly from the decay of animal waste and the remains of plants and animals.

Humans alter the nitrogen cycle in at least four ways: (1) by applying excess nitrogen-containing fertilizer on farmland, much of which ends up in waterways; (2) by disposing of nitrogen-rich municipal sewage in waterways; (3) by raising cattle in feedlots adjacent to waterways; and (4) by burning fossil fuels, which release nitrogen oxides into the atmosphere. The first three activities increase the

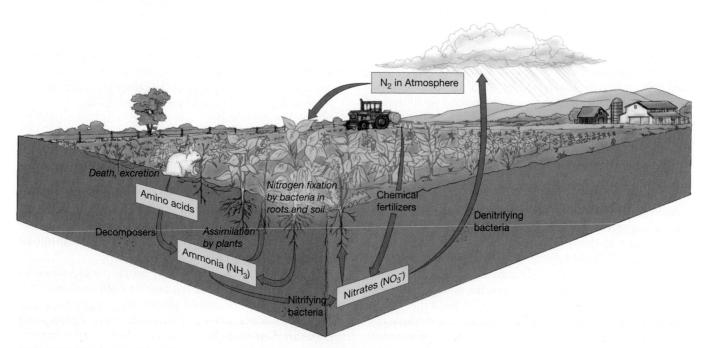

FIGURE 24-9   The Nitrogen Cycle Nitrogen in the atmosphere is converted to NH_3 (ammonia) by bacteria and cyanobacteria in soil. Ammonia is converted to nitrates and taken up by plants.

concentration of nitrogen in the soil or water, upsetting the ecological balance. Nitrogen oxides released into the atmosphere by power plants, automobiles, and other sources are converted to nitric acid, which falls with rain or snow. Known as **acid rain and snow**, it can have devastating effects on terrestrial and aquatic ecosystems, as described shortly.

Nitrogen is a plant nutrient. It stimulates the growth of aquatic plants. As a result, nitrogen-polluted rivers and lakes may become congested with dense mats of vegetation, making them unnavigable. Sunlight penetration to deeper water levels is also impaired by the growth of plants. This causes oxygen levels in deeper waters to decline. In the autumn, when aquatic plants die and decay, oxygen levels can fall further, killing aquatic life.

> **KEY CONCEPTS**
>
> Nitrogen is essential to life. It cycles through the biosphere from the atmosphere to the soil and then plants and animals.

Biological Succession

Now that you understand ecosystems, let's take a look at the development of biological communities. Biological communities, containing complex food webs and nutrient cycles, may develop in areas devoid of life or in areas that have been severely disturbed. Because the development of new communities takes place over time in a series of changes, biologists refer to this process as **succession**. Succession is a process of change in which one community is gradually replaced by another until a mature ecosystem is formed. A mature ecosystem is one that has reached a state of long-term dynamic balance. Two types of succession exist: primary and secondary.

Primary succession occurs where no biotic community previously existed—for example, when deep-sea volcanoes erupt and form islands. In the tropics, a rich paradise can form on the barren volcanic rock, but it will take tens of thousands of years. Seeds for plants may be carried to the islands by waves or may be dropped by birds. Over time, the plants take root, then spread to cover the entire island. New species may arise as a result of genetic changes in these species. Birds may settle on the island as well. In the ensuing years, new species may evolve. The process of primary succession on rock exposed by the retreat of glaciers is shown in Figure 24-10.

Secondary succession occurs when a disturbed area, such as a farm field, is no longer used. As in primary

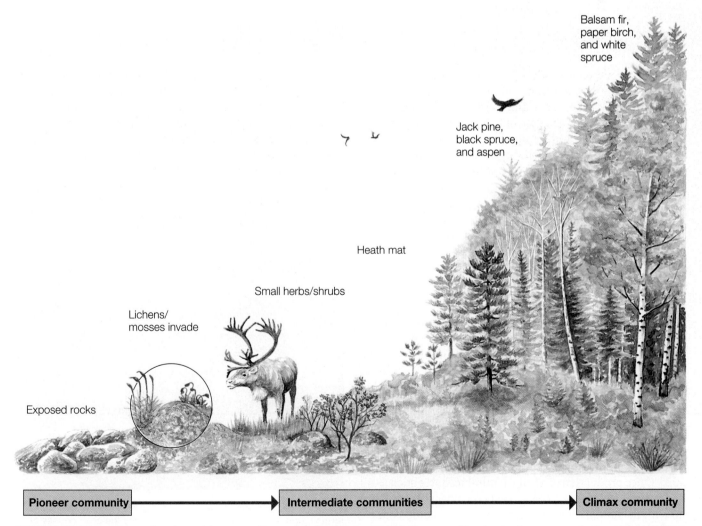

Balsam fir, paper birch, and white spruce

Jack pine, black spruce, and aspen

Heath mat

Small herbs/shrubs

Lichens/ mosses invade

Exposed rocks

| Pioneer community | → | Intermediate communities | → | Climax community |

FIGURE 24-10 **Primary Succession** As rock is exposed by a retreating glacier, biological communities develop in a process that can take hundreds or thousands of years.

succession, it also goes through a series of changes that can take decades to complete. Eventually the original ecosystem is restored.

Secondary succession occurs much more rapidly than primary succession because soil is already present. In primary succession, soil must be formed from rock, gravel, or sand, and this can take 100 to 1,000 years.

> **KEY CONCEPTS**
>
> Biological succession is a process of change in which one community is gradually replaced by another until a mature ecosystem is formed. Primary succession occurs on virgin ground and secondary occurs on land whose biological communities were destroyed by natural or human causes.

24-3 Overshooting the Earth's Carrying Capacity

Environmental problems of significance occur in both the rich, industrialized nations and the poor, less-developed nations. Although the problems vary from one nation to another, they all are signs of a common root cause: human society is exceeding the Earth's carrying capacity.

Carrying capacity is the number of organisms an ecosystem can support indefinitely—that is, the number it can sustain. Carrying capacity for all organisms, including humans, is determined by three factors: (1) food production; (2) resource supply; and (3) the environment's ability to assimilate pollution.

> **KEY CONCEPTS**
>
> Carrying capacity is the number of organisms an ecosystem can support based on food supplies, resource supplies, and waste assimilation.

Exceeding the Carrying Capacity

As the global human population grows, many nations are finding it more and more difficult to meet rising demands for food. Starvation abounds in the less-developed nations. There, an estimated 12 million people, many of them children, perish each year from malnutrition and starvation or from diseases worsened by hunger, a sure sign of populations living beyond the local carrying capacity (Figure 24-11).

> **KEY CONCEPTS**
>
> Populations are living beyond their local carrying capacity.

Resources

Human populations require many resources such as fuel, fiber, and building materials. Those resources that are finite such as oil, natural gas, and minerals are known as **nonrenewable resources**. Resources that replenish themselves via natural biological and geological processes are called **renewable resources**. Wind, hydropower, trees, and fishes are examples.

Today, some important nonrenewable resources such as oil and natural gas are on the decline. In fact, global oil production may have already peaked or may peak soon. Global natural gas production is expected to peak between 2015 and 2025. Once production of a finite natural resource such as oil and natural gas peaks, production is no longer able to keep up with demand. Serious economic problems could ensue.

While some nonrenewable resources are beginning to decline, others are just about used up.

Some renewable resources are also in danger. Tropical rain forests, for example, are being depleted faster than they are being replanted. The rapid destruction of renewable and nonrenewable resources is another sign that we're living beyond the Earth's carrying capacity.

> **KEY CONCEPTS**
>
> Human populations require many renewable and nonrenewable resources, many of which are on the decline, a clear indication that we are living beyond the Earth's carrying capacity.

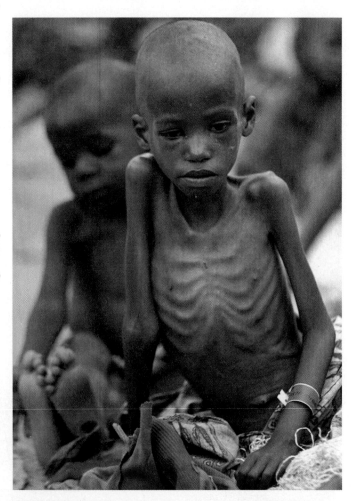

FIGURE 24-11 Food Shortages Overpopulation, drought, political turmoil, and mismanagement of farmland are the chief causes of hunger and starvation. Children die by the thousands every day. (© David Guttenfelder/AP Photos.)

Pollution

Carrying capacity is also determined by the environment's ability to assimilate and degrade pollutants. In natural eco-systems, wastes are usually diluted to harmless levels or broken down and recycled in nutrient cycles. Unfortunately, human populations often produce amounts of wastes that overwhelm these cycles, resulting in pollution of air and water that can be detrimental to life.

> **KEY CONCEPTS**
>
> Local, regional, and global air pollution are also signs that humans are exceeding the Earth's carrying capacity.

24-4   Overpopulation: Problems and Solutions

One reason people are living beyond the carrying capacity, say many scientists, is overpopulation. **Overpopulation** can be explained in six words: too many people, reproducing too rapidly. In March 2012, the world population topped the 7 billion mark and was increasing at a rate of 1.1% per year. Although the growth rate may seem small, it translates into around 77 million new people every year, or about 210,000 people being added to the world population every day. If the current rate continues, world population could easily reach 9.4 billion people by the year 2050.

Where Is Overpopulation Occurring?

Today, the most rapid growth is occurring in three areas: Africa, Asia, and Latin America. In Europe, the population is shrinking. In the United States, population growth is among the fastest of all more-developed nations.

Most people view overpopulation as a problem of the less-developed nations. Actually, overpopulation is a problem in all countries. It is as serious in the United States as it is in Bangladesh. The reason for this apparent irony is the high standard of living in the rich, industrialized countries. The higher the standard of living, the greater the energy and resource consumption and the greater the impact on the environment. In fact, each baby born in the United States will use 20 to 40 times as many resources as a newborn in India. This means the approximately 313 million Americans alive in 2012 caused as much environmental damage as 6 to 12 billion people in the Third World!

The human population has not always been so large, nor has it always grown so rapidly. As Figure 24-12a shows, it was not until the last 200 years that global human population began to skyrocket. This growth was stimulated by better sanitation, improvements in medicine, and advances in technology. Pollution, resource depletion, hunger, starvation, and many other environmental problems have also grown in concert with the skyrocketing human population.

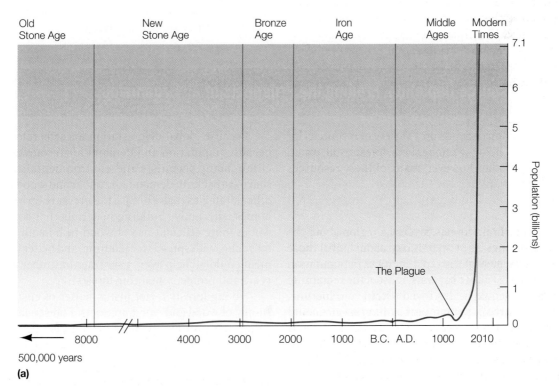

(a)

FIGURE 24-12 **Exponential Growth of the World Population** (a) World population.

(continued)

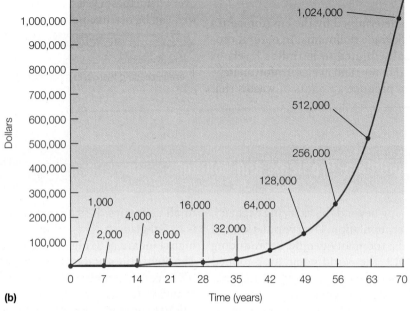

(b)

FIGURE 24-12 *(continued)* **Exponential Growth of the World Population** (b) Exponential growth of a bank account starting with $1,000 at 10% interest.

KEY CONCEPTS

Population growth and overpopulation are problems in less-developed nations but also more-developed nations like the United States, in large part because of the high rate of resource consumption and pollution.

Benefits of Reducing Population Growth

Most world leaders agree that reducing the rate of population increase will help both less-developed and more-developed nations solve many pressing social, economic, and environmental problems such as hunger, environmental deterioration, habitat loss, species extinction, and resource shortages.

Slowing the growth of the human population may not be sufficient to create an enduring human presence, however. Over the next 50 years, some experts believe that it will very likely be necessary to reduce the size of the human population. The principal means of reducing population size is through attrition—reducing the birth rate so that it falls below the death rate. Under such conditions, populations will begin to shrink.

KEY CONCEPTS

Stabilizing and perhaps reducing the size of the human population may be required to solve many of the most pressing social, economic, and environmental problems of today.

24-5   Resource Depletion: Eroding the Prospects of All Organisms

The human population depends on a variety of resources for its survival and well-being—among them, forests, soil, water, minerals, and oil. Today, however, many of these resources are in danger.

Forests

At one time, tropical rain forests covered a region about the size of the United States (Figure 24-13). Today, about half of those forests are gone. Because of the rapid increase in population size and the resulting timber harvests, most of the remaining tropical rain forests could be destroyed within your lifetime. Along with them, perhaps as many as 1 million species could vanish. Making matters worse, in the tropics, only 1 tree is replanted for every 10 trees that are cut down. In tropical Africa, this ratio is 1 to 29.

The decline of the world's forests is not limited to the tropics. In the United States, for example, much of the forested land has been cut down and converted to other uses.

Because global deforestation exceeds reforestation and because population and demand for resources continue to climb, many scientists and environmentalists are urging sharp cutbacks in demand for timber and wood products and other strategies that will ease the pressure on forests. Smaller homes, alternative building materials that don't depend on wood, more efficient use of wood in home building, paper recycling, widespread tree planting, and better forest management could all help avert a shortage of timber in the coming years and protect remaining forests.

Saving forests is not just a matter of ensuring a steady supply of wood and wood products. Forest conservation also helps purify air and reduces carbon dioxide buildup in the atmosphere. Forests protect watersheds, reduce soil erosion, and maintain recreational opportunities as well. They also ensure adequate oxygen levels. Tropical forests are a potential source of new medicines and food plants that could help feed the world's people. Forest conservation also protects the habitat

FIGURE 24-13 **Deforestation** A family in Brazil has settled a deforested plot where they hope to eke out an existence by raising food for themselves. Widespread deforestation by farmers, logging operations, and other activities are taking a toll on the rainforests, wiping out many species and contributing to greenhouse warming. (© Jany Sauvanet/Science Source.)

FIGURE 24-14 **Soil Erosion** Millions of acres of farmland are destroyed each year because of poor land management practices that lead to severe soil erosion. (Courtesy of Lynn Betts/NRCS USDA.)

of wild species, many of which are endangered. (For a discussion of extinction, see the **Point/Counterpoint** in this chapter.)

KEY CONCEPTS

The world's forests have been greatly reduced and continue to be depleted. Forests provide numerous important economic resources, including wood, recreation, water purification, and oxygen.

Soil

Although soil erosion occurs naturally, it is often greatly accelerated by human activities such as farming, construction, and mining (Figure 24-14). In the past 100 years, about one-third of the fertile topsoil on U.S. farmland has been eroded away by water and wind, largely because of poor land management. Although efforts have been made to reduce soil erosion in the United States and other countries, high levels of erosion still occur.

Globally, annual erosion rates are 18 to 100 times greater than the rates at which soil is regenerated. (The average renewal time is 500 years, but it ranges from 200 to 1,000 years.) Making matters worse, millions of acres of farmland are lost to urban and suburban sprawl, highway construction, and other human activities worldwide (Figure 24-15). In the United States, approximately 3,500 acres of rural land are lost every day. That's equivalent to 1.25 million acres a year, or a strip 0.3 mile wide extending from New York City to San Francisco.

These figures paint a rather grim picture for the long-term future of food production both here and abroad, especially when viewed in light of increasing populations. The loss of topsoil can be stopped, and soils can be replenished. However,

FIGURE 24-15 **Farmland Conversion** Cities are often surrounded by excellent farmland soils that drain well and are flat and highly productive. Many of the features that make soils suitable for farmland also make them suitable for building. This scene, unfortunately, is all too common in expanding urban areas. (Courtesy of Lynn Betts/NRCS USDA.)

Point/Counterpoint   Why Worry About Extinction?

Humans Are Accelerating Extinction *by David M. Armstrong*

Evolution is the process of change in gene pools. When one gene pool becomes reproductively independent of another, a new species has formed. Such speciation generates species; extinction takes them away. Simply put, extinction is a failure to adapt to change, the termination of a gene pool, and the end of an evolutionary line.

Extinction is a natural process. Most of the species that have lived on this planet are now extinct. The 3 to 30 million species on Earth today are no more than 1% to 10% of the species that have evolved since life began about 3.5 billion years ago. Given these facts, why are thoughtful people concerned about endangered species? After all, history makes it clear that—given enough time—all species will become extinct.

David M. Armstrong teaches science for non-scientists at the University of Colorado at Boulder and has written several books on the mammals and ecology of the Rocky Mountain region. (© David M. Armstrong.)

The basis for concern is that today the natural process of extinction is proceeding at an unnatural rate. Let us estimate by how much human activity has accelerated rates of extinction. The lifespan of species seems to average from 1 million to 10 million years. Assume (to be conservative) that the average longevity of a species of higher vertebrates is 1 million years. In round numbers (to make calculations easy), there are 10,000 species of birds and mammals. So, on average, one species ought to go extinct each century. However, between 1600 and 1980, at least 36 species of mammals and 94 species of birds became extinct. That is about 0.29 species per year, 29 times the natural rate.

What does it mean to increase a rate by 29 times? The speed limit is 55 miles per hour. Exceed the speed limit by 29-fold, and you are moving 1,595 miles per hour, over twice the speed of sound. The difference between natural rates of extinction and present, human-influenced rates is analogous to the difference between a casual drive and Mach 2! Is that a problem? You decide: Concern is a moral construct, not a scientific one.

Several human activities have contributed—mostly inadvertently—to accelerating rates of extinction. The dodo and the passenger pigeon were extinguished by overhunting. Wolves and grizzly bears were exterminated over much of their ranges as threats to livestock. The black-footed ferret was driven to the verge of extinction because prairie dogs, its staple food, were poisoned as agricultural pests. The smallpox virus was exterminated in the "wild" (but survives in a half-dozen laboratories).

Habitat change is the most important cause of endangerment and extinction. Clearing forests for agriculture has decimated the lemurs of Madagascar. Chemical pesticides led to the decline of the peregrine falcon. Introducing exotic species (like goats on the Galapagos and mongooses in Hawaii) displaces native animals and plants. Developing the Amazon Basin is a habitat alteration, and a cause of extinction, on an unprecedented scale.

Many urge saving species for their aesthetic value. Whooping cranes are beautiful, and part of the beauty is that they are products of a marvelous evolutionary process. Most concern about accelerated extinction, however, stresses economic value. A tiny fraction of Earth's seed plants are used commercially. Perhaps an obscure plant like jojoba will become a source of oil more reliable than that beneath the sands of Saudi Arabia. Wild grasses have furnished genes that improved disease resistance in wheat. Numerous wild animals (like musk ox, kudu, and whales) could contribute protein to the human diet. Wild species may have medical value; penicillin, after all, was once merely an obscure mold on citrus fruit. Some sensitive species are useful monitors of environmental quality, and the presence of healthy populations of many species may promote greater stability or resilience of ecosystems. Naturalist Aldo Leopold noted that we humans have a way of "tinkering" with the ecosphere to see how it works. Given that, we ought to remember the first rule of tinkering: Never throw away any of the parts.

"Extinction is forever," and extinction impoverishes both Earth's ecosystems and the potential richness of human life. Borrowing again from Aldo Leopold, I believe we should be concerned about unnaturally rapid extinction because such concern is part of a "right relationship" between people and the landscapes that nurture and inspire them.

Biologist Sir Julian Huxley noted that "we humans find ourselves, for better or worse, business agents for the cosmic process of evolution." We hold power over the future of the biosphere, the power to destroy or to preserve. German philosopher George Hegel noted that freedom (including the power to destroy species) implies responsibility (to preserve them). I agree. The question of human-accelerated extinction boils down to a simple ethical question, "Does posterity matter?" Some of us have ethics that are human-centered. We ask simply, "Do my children deserve a life as rich, with as much opportunity, as mine?"

Extinction Is the Course of Nature *by Norman D. Levine*

Evolution is the formation of new species from preexisting ones by a process of adaptation to the environment. Evolution began long ago and is still going on. During evolution, those species better adapted to the environment replaced the less well adapted. It is this process, repeated year after year for millennia, that has produced the present mixture of wild species. Perhaps 95% of the species that once existed no longer exist.

Human activities have eliminated many wild species. The dodo is gone, and so is the passenger pigeon. The whooping crane, the California condor, and many other species are on the way out. The bison is still with us because it is protected, and small herds are raised in semicaptivity. The Pacific salmon remains because we provide fish ladders around our dams so it can reach its breeding places. The mountain goat survives because it lives in inaccessible places. But some thousands of other animal species, to say nothing of plants, are extinct, or soon will be. Some nature lovers weep at this passing and collect money to save species. They make lists of animals and plants that are in danger of extinction and sponsor legislation to save them.

I don't. What the species preservers are trying to do is to stop the clock. It cannot and should not be done.

Extinction is an inevitable fact of evolution, and it is needed for progress. New species continually arise, and they are better adapted to their environment than those that have died out.

Extinction comes from failure to adapt to a changing environment. The passenger pigeon did not disappear because of hunting alone, but because its food trees were destroyed by land clearing and farming. The prairie chicken cannot find enough of the proper food and nesting places in the cultivated fields that once were prairie.

And you cannot necessarily introduce a new species, even by breeding it in tremendous numbers and putting it out into the wild. Thousands of pheasants were bred and set out year after year in southern Illinois, but in the spring of each year there were none left. Another bird, the capercaillie, is a fine, large game bird in Scandinavia, but every attempt to introduce it into the United States has failed. An introduced species cannot survive unless it is preadapted to its new environment.

A few introduced species are preadapted and some make spectacular gains. The United States has received the English sparrow, the starling, and the house mouse from Europe, and also the gypsy moth, the European corn borer, the Mediterranean fruit fly, and the Japanese beetle. The United States gave Europe the gray squirrel and the muskrat, among others. The rabbit took over in Australia, at least for a time.

The rabbit and the squirrel were successful on new continents because their requirements are not as narrow as those of species that failed. Today, adjustment to human-made environments may be just as difficult as adjustment to new continents. The rabbit and the squirrel have succeeded in adjusting to the backyard habitat, but most wild animals have disappeared.

Norman D. Levine was a professor emeritus at the College of Veterinary Medicine and Agricultural Experiments Station, University of Illinois at Urbana. His research interests covered parasitology, protozoology, and human ecology. This essay is from "Evolution and Extinction." *Bioscience* 39 (1989): p. 38, and is courtesy of the American Institute of Biological Sciences. (© Norman D. Levine.)

Human-made environments are artificial. People replace mixed grasses, shrubs, and trees with rows of clean-cultivated corn, soybeans, wheat, oats, or alfalfa. Variety has turned into uniform monotony, and the number of species of small vertebrates and invertebrates that can find the proper food to survive has become markedly reduced. But some species have multiplied in these environments and have assumed economic importance; the European corn borer in this country is an example.

Would it improve Earth if even half of the species that have died out were to return? A few starving, shipwrecked sailors might be better off if the dodo were to return, but I would not be. The smallpox virus has been eliminated, except for a few strains in medical laboratories. Should it be brought back? Should we bring panthers back into the eastern states? Think of all the horses that the automobile and tractors have replaced, and of all the streets and roads that have been paved and the wild animals and plants killed as a consequence. Before people arrived in the United States about 10,000 years ago, the animal-plant situation was quite different. What should we do? Should we all commit suicide?

Sharpening Your Critical Thinking Skills

1. Summarize the key points of both authors.

2. Do you see any flaws in the reasoning of either author?

3. Which viewpoint do you adhere to? Why?

Visit Human Biology's Internet site for links to websites offering more information about this topic.

TABLE 24-1 Some Solutions to Alleviate World Hunger
Reduce population growth.
Reduce soil erosion.
Reduce desertification.
Reduce farmland conversion.
Improve yield through better crop strains.
Improve yield through better soil management.
Improve yield through fertilization.
Improve yield through better pest control.
Reduce spoilage and pest damage after harvest.
Use native animals for meat production.
Tap farmland reserves available in some countries.

worldwide conservation efforts are needed, as are measures to reduce population growth, which help to reduce the conversion of farmland to other uses such as housing. More compact development in cities and towns can also reduce the incursion on valuable farmland and protect the world's food supplies. Additional solutions shown in Table 24-1 could alleviate world hunger and could also reduce the loss of farmland and rangeland.

> **KEY CONCEPTS**
>
> Soils are vital to the long-term economic welfare of human society and the survival of humans, however, soils throughout the world have been badly eroded and continue to be destroyed by poor agricultural practices.

Water

Water shortages are becoming acute worldwide. Many of the most populous nations, including China, India, and the United States, are facing severe shortages in urban and agricultural areas. Water shortages generally result because too many people are drawing on limited water supplies. Global warming may also be causing shortages.

Many steps can be taken to conserve water and stretch available supplies. In the agricultural sector, for instance, lining irrigation ditches with concrete or using pipes rather than open ditches to transport water to fields can dramatically reduce losses due to evaporation. More efficient sprinklers, computerized systems that monitor soil moisture so that farmers know exactly how much irrigation water is required, and other measures can also help. Residential water conservation can also help. Water-efficient toilets, showerheads, dishwashers, and clothes washers can have a huge impact on reducing the average water consumption in our homes.

> **KEY CONCEPTS**
>
> Clean, fresh water is vital to humans, agriculture, and industry, but pollution, overuse, and global warming are causing severe shortages in many parts of the world.

Oil

Oil is the lifeblood of modern society. In the United States, for example, oil supplies 42% of our annual energy demand. But the supply of oil is finite. No one knows how much oil is left in Earth's crust, but as noted earlier, most experts agree that oil production worldwide will soon peak or may have already peaked. When oil extraction peaks, demand for oil products including gasoline and jet fuel will outstrip production. This, in turn, could cause major economic problems, including runaway inflation and economic stagnation.

The first step in meeting future demand is energy efficiency. The efficient use of oil and its many important by-products, such as gasoline, diesel fuel, and home heating oil, can help us stretch current oil supplies considerably. More efficient vehicles and greater reliance on mass transit in urban areas can also extend oil supplies.

Renewable fuels can help meet future demand. Ethanol produced from corn, wheat, and other crops can power automobiles and trucks. Ethanol can be made more efficiently from sugar cane and cellulose from wood and corn plants. Production from these sources could make it much more economical than corn ethanol production, the main source in the United States. Another potentially important fuel is biodiesel, a fuel produced from vegetable oils and hydrogen. Hydrogen is widely viewed as a potential new energy resource because it can be made from water. Hydrogen can be burned directly, but can also be fed into fuel cells. In fuel cells, it is converted to electricity, which is used to run electric motors in cars, trucks, and busses. The process is very clean, as the only waste product is water, the original source of hydrogen. Unfortunately, the production of hydrogen requires a significant amount of energy in the form of electricity. In fact, studies show that it is three to four times more efficient to use that electricity to power a car directly than it is to use the electricity to split water to power a hydrogen fuel cell in a car to make electricity to run an electric motor. Electric cars may meet much of our future transportation demand, as 50% of all Americans drive only 20 to 30 miles a day and 90% drive fewer than 60 miles—well within the range of electric vehicles. Several electric cars are now available like the Nissan Leaf that can travel about 70 miles on a single charge.

> **KEY CONCEPTS**
>
> Oil supplies are in serious decline but many options are available to satisfy energy needs supplied by oil, including energy efficiency, renewable energy fuels, and alternative vehicles like electric cars.

24-6 Pollution

As noted earlier, waste and pollution from human society is overwhelming many nutrient cycles, poisoning other species (and ourselves). This section recaps four of the most serious waste problems.

Global Warming and Climate Change

Carbon dioxide is produced during the combustion of all organic materials—most importantly, fossil fuels. In normal concentrations in the atmosphere, carbon dioxide has a

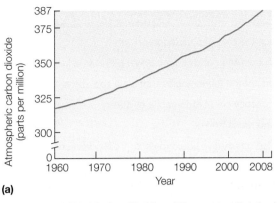

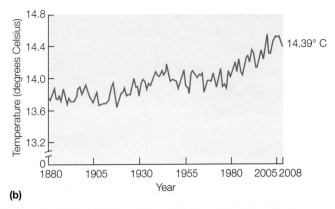

FIGURE 24-16 **Global Carbon Dioxide and Temperature Trends** (a) Carbon dioxide levels in the atmosphere have risen dramatically since 1958. (b) Graph of average global temperature since 1950.

warming effect on the planet. Acting much like the glass in a greenhouse, it traps heat escaping from the Earth and radiates it back to the surface. Carbon dioxide is, therefore, known as a **greenhouse gas**. A little bit of carbon dioxide is essential to life on Earth. In fact, without carbon dioxide in the atmosphere, the planet would be about 55° F (30° C) cooler than it is. Too much, however, may lead to overheating, a phenomenon called **global warming**. Global warming can lead to changes in the climate, which can have many adverse effects on us and the millions of species that share this planet with us.

Since 1880, global carbon dioxide levels have increased from 285 ppm to 400 ppm, or about 71%, principally as a result of industrialization powered by the combustion of fossil fuels and deforestation (Figure 24-16a). Several other pollutants also contribute to global warming, however. These include methane from livestock and ozone-destroying chlorofluorocarbons (CFCs) and their replacements, the hydrochlorofluorocarbons (HCFCs), both of which are used in refrigerators, air conditioners, and freezers.

Many scientists believe that the dramatic increase in greenhouse gas emissions and deforestation is causing the global temperature to increase and that further increases resulting from the expanding use of fossil fuels may cause the temperature to rise even more dramatically in the coming years, causing an extraordinary shift in climate. Figure 24-16b shows average global temperature since 1880.

> **KEY CONCEPTS**
>
> Carbon dioxide is a greenhouse gas, so named because it traps heat escaping from the Earth's surface. Rising levels of carbon dioxide in the atmosphere are warming the Earth's surface, resulting in a phenomenon known as global warming that is causing dramatic and costly changes in global climate.

Impacts of Global Warming

According to latest projections based on the present rate of increase in greenhouse gases, global temperatures could be 5° to 10° F (215° to 212° C) hotter by 2100. The models suggest that global rainfall patterns will shift dramatically as a result of warming. Computer simulations of U.S. climate suggest that the Midwest and much of the western United States will be drier and hotter than they are today. If that happens, many Midwestern farmers will be driven out of business. As rainfall declines in this agriculturally productive region, farming

may intensify in the northern states. Overall agricultural productivity in the United States may fall, however, because northern soils are not as rich as those in the Midwest. Food shortages caused by the decline in U.S. agriculture in the Midwest and rising food prices could affect the economy in profound ways. Computer climate models predict that the southern United States and Pacific Coast may be wetter but hotter.

A rise in global temperature is melting glaciers throughout the world as well as the polar ice caps (Figure 24-17). Studies show that polar bear populations that rely on Arctic ice to hunt are already experiencing problems. Melting ice is resulting in less success in hunting, leading to leaner adults and reduced infant survival rates that could cause this beloved species to become extinct within the not-too-distant future.

Warmer temperatures would also expand the volume of the seas. Together, melting ice and expanding oceans would cause the sea level to rise. In the past 50 years, sea level has risen 10 to 12 centimeters (4 to 6 inches). By 2100, the computer models suggest that the sea level will rise 50 centimeters (2 feet) because of global warming.

In the United States, approximately half of the population lives within 50 miles of the ocean, and many large cities such as Miami are located only a few feet above sea level (Figure 24-18). A rising sea level would flood many low-lying regions. Storms could cause more damage than they do now because high water would be able to move farther inland. Expensive dikes

FIGURE 24-17 **Glacier Bay, Alaska** This enormous glacier and others like it could melt as the Earth gets warmer, raising the sea level. (Courtesy of John Bortniak/NOAA.)

FIGURE 24-18 **Miami Underwater?** A rising sea level due to global warming could flood many coastal cities the world over. (© Elias H. Debbas II/ShutterStock, Inc.)

TABLE 24-2	Measures to Reduce Global Climate Change
Reduce the rate of population growth.
Switch from coal- and oil-fired power plants to solar and wind energy
Implement the technologies that burn coal more efficiently.
Dramatically boost automobile efficiency.
Expand mass transit.
Develop alternative liquid fuels for transportation like biodiesel and vegetable oil.
Dramatically improve the efficiency of industry.
Make new and existing homes much more energy-efficient.
Build new homes much more efficiently and so that they use solar energy for space heating.
Reduce global deforestation.
Begin a massive global reforestation effort.
Reduce consumption of unnecessary items.
Expand recycling efforts.

and levees would be needed to protect cities such as Miami. Other coastal cities would have to be rebuilt on higher ground, at a staggering cost. A rising sea level would be particularly hard on Bangladesh and other Asian countries with extensive lowland rice paddies.

In 2000, a report by the National Research Council predicted that rising temperatures will cause the tropical climate of equatorial regions to shift northward into the lower-tier states. Their climate, in turn, will shift to the mid-tier states and the climate of the top-tier states would move into Canada. The effect of such a rapid shift on vegetation and wildlife could be devastating. Signs of this change are already present.

Rising global temperatures are causing the spread of tropical diseases into neighboring areas. Studies show that insects that carry malaria are moving to higher altitudes and higher latitudes in Africa and Central America.

Global warming could cause considerable human suffering, with many deaths attributed to heat waves. In 2003, an estimated 40,000 Europeans, mostly elderly individuals, died in a record-breaking heat wave. Global warming is triggering more violent weather, especially hurricanes and tornados that cause billions of dollars worth of damage and loss of human life. A recent study of severe storms in the United States showed that violent downpours are on the rise. The incidence of tornados in the United States doubled from the 1980s to the 1990s. The strength of hurricanes is also on the rise.

> **KEY CONCEPTS**
>
> Global climate change will result in many costly impacts ranging from a marked decrease in agricultural production to rising sea levels to more violent weather and increased death rate.

Solutions to Global Warming

While scientists and politicians debate global warming, the Earth appears to be getting hotter. In fact, the 1990s was the hottest decade on record in the past 100 years. That is, until the 2000s came around. The last decade's temperatures are shattering records as well.

Reducing the many impacts of global warming, say experts, will require sharp reductions in fossil-fuel consumption by energy efficiency and conservation, and the use of alternative fuels will be required. Global reforestation is essential. Additional strategies are shown in Table 24-2. We as individuals can also help (Table 24-3).

> **KEY CONCEPTS**
>
> Solving climate change will require many actions from energy efficiency in all sectors of society to a massive increase in the use of renewable energy to power our homes, vehicles, and businesses.

Acid Deposition

Throughout the world, thousands of lakes have turned acidic, killing fishes and other aquatic organisms (Figure 24-19). Acidification is most prevalent in the northeastern United States, southeastern Canada, and in Sweden and Norway, where dying lakes number in the thousands.

The acidification of lakes is a result of acids falling from the skies in rain or snow. They are produced from two atmospheric pollutants: sulfur dioxide and nitrogen dioxide. These pollutants arise chiefly from the combustion of fossil fuels: coal, oil, gasoline, natural gas, and jet fuel. In the atmosphere, these gases combine with water and oxygen to form sulfuric and nitric acids, which are responsible for acid rain and acid snow, collectively referred to as **acid deposition**.

In the United States and Europe, acid deposition has been worsening for over six decades, largely as a result of increased fossil-fuel combustion. Today, acid rain and snow are commonly encountered downwind from virtually all major population centers. The acids come from pollutants produced by power plants, motorized vehicles, factories, and homes.

Acid deposition changes the pH of lakes and streams, killing fishes and other aquatic organisms. Acids falling on land also dissolve toxic minerals such as aluminum from the soil and wash them into surface waters, killing fish.

TABLE 24-3 | **Individual Actions That Can Reduce Global Climate Change**

Automobile energy savings

    Buy energy-efficient vehicles.

    Reduce unnecessary driving.

    Carpool, take mass transit, walk, or bike to work.

    Combine trips.

    Keep your car tuned and your tires inflated to the proper level.

    Drive at or below the speed limit.

Home energy savings

    Increase your attic insulation to R30 or R38.

    Caulk and weather-strip your house.

    Add storm windows and insulated curtains.

    Install an automatic thermostat.

    Turn the thermostat down a few degrees in winter, and wear warmer clothing.

    Replace furnace filters when needed.

    Lower water heater setting to 120° to 130° F (49° to 54° C).

    Insulate water heater and pipes, install a water heater insulation blanket, and repair or replace all leaky faucets.

Home energy savings (*continued*)

    Take shorter showers.

    Use cold water as much as possible.

    Avoid unnecessary appliances.

    Buy energy-efficient appliances.

    Use low-energy light bulbs.

Reducing waste and resource consumption

    Recycle at home and at work.

    Avoid products with excessive packaging.

    Reuse shopping bags.

    Refuse bags for single items.

    Use a diaper service instead of disposable diapers.

    Reduce consumption of throwaways.

    Donate used items to Goodwill, Disabled American Veterans, the Salvation Army, or other charities.

    Buy durable items.

    Give environmentally sensitive gifts.

Figure 24-20 shows the areas of North America that are most susceptible to acid deposition. These regions are generally mountainous and contain soils with little capacity to neutralize acids. Consequently, acids that fall on the land quickly wash into nearby lakes and streams.

Acids also damage crops and trees, directly and indirectly. Some scientists believe that massive forest diebacks occurring throughout the world may be the result of acidic rainfall and acidic fog that often blanket forests (Figure 24-21). Finally, acid deposition also damages buildings, statues, and other structures. The estimated cost in the United States is about $5 billion per year.

FIGURE 24-19 Victims of Modern Society These fish were killed by acids from acid deposition. (© Christian Draghici/Dreamstime.com.)

Reducing the deposition of acids will require a dramatic reduction in the release of sulfur dioxide and nitrogen dioxide from power plants, factories, and automobiles. One way of reducing sulfur dioxide is the smokestack scrubber, a device that traps sulfur dioxide gas escaping from power plants, removing up to 95% of this pollutant. Installing and operating a scrubber is rather expensive, and some utility companies have objected to this strategy. So far, most have chosen the cheaper low-sulfur coal strategy.

Another even cheaper strategy is energy efficiency. By using fossil-fuel energy more efficiently, we reduce the overall rate of fossil-fuel combustion. This reduces sulfur dioxide and nitrogen dioxide release. Using renewable energy is another more important strategy for reducing acid deposition. Solar and wind energy, for example, meet our needs, but release no acid precursors. They can also help us dramatically reduce the carbon dioxide emissions that lead to global climate change.

KEY CONCEPTS

The deposition of acids produced by pollution from the combustion of many types of fossil fuel is causing significant damage to crops, buildings, and aquatic and terrestrial ecosystems that could be reduced by energy efficiency and renewable energy.

The Ozone Layer

Encircling the Earth, 20 to 30 miles above its surface, is a region of the atmosphere, the **ozone layer**, which contains a slightly elevated level of ozone gas. The ozone layer shields the Earth from harmful ultraviolet radiation from the sun. Early in the evolution of life, in fact, the formation of the ozone layer probably allowed the colonization of land by plants and animals.

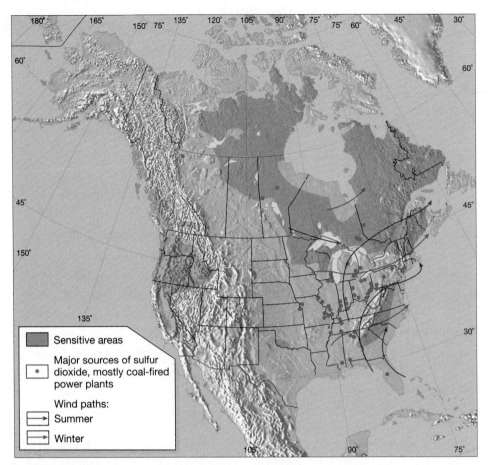

FIGURE 24-20 **Sensitive Areas** This map of North America shows areas experiencing acid deposition and the most geologically vulnerable regions.

FIGURE 24-21 **Forest Die-Off** Ghostly remains of trees killed by acid deposition in the Northeastern United States. (© JoLin/ShutterStock, Inc.)

Today, however, the ozone layer is being destroyed by chemicals released into the atmosphere by modern society. The most potent destroyer of ozone is a class of chemicals known as the *chlorofluorocarbons (CFCs)*. Once used as spray-can propellants, refrigerants, blowing agents for plastic foam, and cleansing agents, CFCs are highly stable molecules.

In the early 1970s, scientists discovered that CFCs released from human activities gradually drift into the upper atmosphere. There, they are broken down by sunlight, where their byproducts react with and destroy ozone molecules.

In 1988, a panel of atmospheric scientists reviewing 20 years of satellite data on ozone levels concluded that the ozone layer is on the decline (Figure 24-22). Ozone depletion was particularly evident at the poles. Scientists found that each year a giant hole in the ozone layer about the size of the United States formed over Antarctica. Ozone levels in the hole declined by as much as 50%.

Declining ozone levels increase ultraviolet radiation striking the Earth. Like many natural components of our environment, ultraviolet radiation is beneficial. In small amounts, ultraviolet radiation tans light skin and stimulates vitamin D production in the skin. However, excess ultraviolet exposure can cause problems. In humans, it can cause serious skin burns, cataracts (clouding of the eye's lens), skin cancer, and premature aging. Over the next five decades, the Environmental Protection Agency (EPA) estimates that ozone depletion will result in approximately 200,000 cases of skin cancer in the United States.

Studies of skin cancer show that light-skinned people are much more sensitive to ultraviolet radiation than more heavily pigmented individuals. Land- and water-dwelling plants could also suffer. Intense ultraviolet radiation is usually lethal to plants. Smaller, nonfatal doses damage leaves

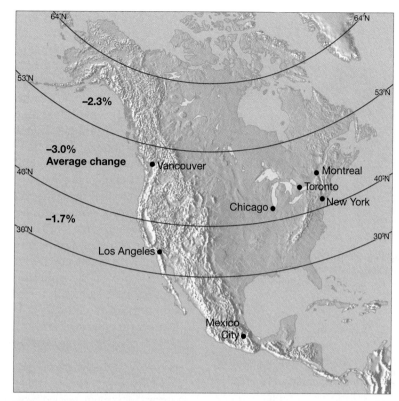

FIGURE 24-22 **Decline in the Ozone Layer** This map of North America shows the ozone depletion at different latitudes between 1969 and 1988.

and inhibit photosynthesis. They can also cause mutations and stunt growth. Declining ozone and increasing ultraviolet radiation could also cause dramatic declines in commercial crops, costing billions of dollars a year. It may also damage certain commercially valuable tree species.

Finally, ultraviolet light is harmful to paints, plastics, and other materials. Losses from further decreases in the ozone layer could cost society enormous amounts of money. Fortunately, ozone depletion has been addressed head on by the nations of the world. They have signed three treaties to ban ozone-depleting chemicals. Despite the bans, it could take the ozone layer 50 to 100 years to recover.

> **KEY CONCEPTS**
>
> The release of chlorofluorocarbons has resulted in a significant decrease in the ozone layer that has resulted in an increase in the amount ultraviolet radiation reaching the Earth's surface. This, in turn, kills plants, causes skin cancer and cataracts.

24-7 Health and Homeostasis

The trends and problems discussed in this chapter are, say many experts, spawning serious problems throughout the world. What many people forget when they deal with such issues is that these problems could combine to produce impacts far greater than anticipated. That is, they produce a synergistic effect. For example, soil erosion, global warming, farmland conversion, acid deposition, ozone depletion, and population growth together could combine to produce massive famine.

Solving these problems will require tougher laws and regulations, reductions in population growth, more efficient technologies, and changes in human behavior.

Many people believe that broader societal changes are required. One suggestion is that we build a sustainable society, one that lives within the carrying capacity of the environment. Living within the limits of nature means achieving a population size and a way of life that do not exceed the planet's ability to supply food and other resources and to handle wastes. It means promoting human societies that manage their affairs in ways that protect the Earth's homeostatic mechanisms—ultimately creating a population that lives within the Earth's carrying capacity.

Building a sustainable society will require a shift to the patterns seen in nature. It will require that we become more efficient in our use of resources; recycle to the maximum extent possible; shift to renewable resources, especially energy; restore ecosystem damage; and stabilize, if not reduce, world population.

Building a sustainable society will require a lifetime of commitment on the part of businesses, governments, and individuals. Some changes required to create an enduring

FIGURE 24-23 Energy–Efficient Lighting Many new lightbulbs use only 24% of the energy of standard lightbulbs and last 10 times longer, saving $20 to $40 over their lifetime. (Courtesy of Osram Sylvania. Used with permission.)

human presence may come as a result of legislative actions—new laws and regulations.

Technological innovations will also help us become a sustainable society. Such household items as compact fluorescent (CFLs) and light emitting diodes (LEDs) have replaced incandescent light bulbs (Figure 24-23). Improved photovoltaics—thin silicon wafers that produce electricity from sunlight—are helping us make the transition to solar energy. That's how I have powered my home in Colorado since 1976. I currently power my new home, farm, and business in east-central Missouri entirely on solar and wind energy.

But new technologies are not the only, or even the most important, answer. Many existing technologies simply need to be installed. Insulation, weather stripping, efficient shower-heads, and efficient appliances already on the market can make tremendous inroads into waste reduction.

Many proponents also believe that we will eventually need to reduce consumption and alter our lifestyles. Taking shorter showers, shaving without the water running, turning off lights, eliminating plastic bags and drinking bottles, and hundreds of other small actions on the part of individuals, when combined with similar actions by millions of other people, can result in significant cuts in resource demand.

Ending the waste and pollution—and soon—is crucial if we are to protect the environment and ourselves. The balance of nature is, after all, the balance that sustains us. We risk upsetting it at our own peril. Protecting the planet is the ultimate form of health care.

> **KEY CONCEPTS**
>
> Many experts believe that solving the problems that face the world will require the creation of a sustainable society, one that lives in harmony with nature, helping to maintain homeostatic conditions on planet Earth essential for the health and well-being of humans and all other species.

SUMMARY

An Introduction to Ecosystems

1. Humans depend on the environment for many free services and resources such as food, oxygen, clothing, materials, and much else.
2. Ecology is the study of ecosystems. It examines the relationship of organisms to their environment and the many interactions between the abiotic and biotic components of ecosystems.
3. The living "skin" of the planet is called the biosphere. It extends from the bottom of the oceans to the tops of the highest mountains.
4. The biosphere is a closed system in which materials are recycled over and over. The only outside contribution is sunlight, which powers virtually all biological processes.
5. The Earth's surface is divided into large biological regions, or biomes, each with a characteristic climate and characteristic plant and animal life. The oceans can also be divided into biological regions, known as aquatic life zones.
6. An ecosystem consists of a community of organisms, its environment, and all of the interactions between them.
7. A group of organisms of the same species living in a specific region constitutes

a population. Two or more populations occupying that region form a community.
8. The physical space a species occupies is called its habitat. A species' niche includes its habitat, its position in the food chain, and so on.

Ecosystem Function

9. Virtually all life on Earth depends on plants and other producers, organisms that synthesize organic materials from sunlight, carbon dioxide, and water via photosynthesis.
10. Organisms dependent on producers and other organisms for food are called consumers. Four types of consumers are present: herbivores, carnivores, omnivores, and detrivores.
11. All organisms are part of one or more food chains. Food chains that begin with plants consumed by grazers (herbivores) are known as grazer food chains. Those that begin with animal waste or the remains of dead organisms are decomposer food chains. Food chains are parts of larger food webs.
12. The position of an organism in a food chain is called its trophic level.
13. Nutrients necessary for life flow through nutrient cycles. Humans can interrupt nutrient cycles, locally and globally.

14. The term succession refers to a series of changes in an ecosystem in which one community replaces another until a mature ecosystem is produced. Primary succession occurs where no community previously existed. Secondary succession occurs where a community was destroyed by natural or human events.

Overshooting the Earth's Carrying Capacity

15. Although environmental problems vary from one nation to another, they are all the result of human populations exceeding the carrying capacity of the environment.
16. Carrying capacity is the number of organisms an ecosystem can support indefinitely and is determined by food and resource supplies and by the capacity of the environment to assimilate or destroy waste products of organisms.

Overpopulation: Problems and Solutions

17. Overpopulation occurs any time a population exceeds the carrying capacity. It is manifest in shortages of food, lack of other resources, or excessive pollution—sometimes all three.
18. The most rapid growth in human population is occurring in the developing

world in parts of Africa, Asia, and Latin America. Resource depletion and food shortages are the most common problems in these regions. Many experts believe that the industrialized countries are also overpopulated. Judging from the quality of our air, water, and soils, we are clearly exceeding the Earth's carrying capacity.

19. The human population problem can be summed up in six words: too many people, reproducing too rapidly, which results in resource shortages, excessive pollution, and poverty.

Resource Depletion: Eroding the Prospects of All Organisms

20. The human population requires a variety of renewable and nonrenewable resources for survival, and many of these resources are being depleted.
21. In many parts of the world, forests are being cut down faster than they can be replaced. To prevent the destruction of the world's forests, tree planting, paper recycling, and other strategies are needed.
22. Worldwide, agricultural soils are being eroded from rangeland and farmland at an unsustainable rate, and millions of acres of farmland are being destroyed by human encroachment. The destruction of productive soils threatens the long-term prospects for food production.
23. Soil conservation and population control measures can help ensure an adequate supply of soil, but serious efforts must begin soon.

24. Many areas of the world suffer water shortages. The rise in population and the rise in demand are likely to cause further shortages in the future. The growth of the human population must be reduced and strict water conservation is needed.
25. Oil supplies, like mineral supplies, are also finite. Global oil production may have already peaked or may peak very soon. When oil production peaks, production may not be able to meet demand.
26. By using oil much more efficiently and seeking clean, renewable alternative fuels, modern society can make a smooth transition to a sustainable energy supply.

Pollution

27. Pollution from human activities is overwhelming nutrient cycles, poisoning other species (and ourselves), and destroying the planet's homeostatic mechanisms.
28. One of the most serious threats from pollution comes from carbon dioxide. Carbon dioxide is a greenhouse gas, trapping heat in the Earth's atmosphere. This, in turn, could increase the planet's surface temperature, altering its climate, shifting rainfall patterns and agricultural zones, flooding low-lying regions, and destroying many species that cannot adapt to the sudden change in temperature.
29. To slow global warming, many scientists recommend massive reforestation projects and dramatic improvements in

the efficiency of fossil-fuel combustion as well as the development of alternative fuels.

30. Sulfur dioxide and nitrogen dioxide are two gaseous pollutants released from power plants, factories, automobiles, and other sources. In the atmosphere, they are converted to sulfuric and nitric acid, respectively.
31. Acids fall from the sky in rain and snow. Acids alter the pH of lakes and streams, killing fish and other organisms. They also leach (dissolve and remove) toxic metals from soils that kill fishes. They destroy trees and crops and deface buildings, costing society billions of dollars a year.
32. Acid rain and snow can be reduced by installing pollution-control devices in factories and power plants and by employing preventive measures such as energy conservation and alternative fuels.
33. The ozone layer encircles the Earth, trapping ultraviolet light. It is being destroyed by chlorofluorocarbons and other pollutants.

Health and Homeostasis

34. Solving our problems will not only require tougher laws and tighter regulations but also a profound change in the way we live and conduct business.
35. A sustainable society is built on five operating principles: conservation, recycling, renewable resources, restoration, and population control.

THINKING CRITICALLY ANALYSIS

This analysis corresponds to the Thinking Critically scenario that was presented at the beginning of this chapter.

Critical thinking rules encourage us to question sources of information and their conclusions. In this example, we find that the author of this article is grossly in error when he claims that the scientific community is divided on the issue. About 99% of the nation's 700 atmospheric scientists believe that global warming is a reality; only a handful embrace the opposite view. Critics say those detractors are on the payroll of oil companies. The evidence the author introduces to support his assertion—a quote from each side of the issue—is terribly misleading.

This type of reporting is quite common in newspapers, television, magazines, and books. Although it is intended to give a balanced view of issues, in reality, it provides an extremely unbalanced view.

What would have been a more accurate conclusion? Perhaps that not all scientists agree that global warming is occurring, but that the vast majority does. It's important to note that even though the majority of the world's atmospheric scientists agree that global warming is happening, this doesn't mean they're right.

Digging a little deeper, finding out more, often throws conclusions into question.

KEY TERMS AND CONCEPTS

CONCEPT REVIEW

1. Define the term ecology, and give examples of its proper and improper use. p. 524.
2. The Earth is a closed system. What does that mean? What are the implications of this statement? p. 524.
3. Define the following terms: biosphere, biome, aquatic life zone, and ecosystem. pp. 524–525.
4. Define the following terms: habitat, niche, producer, consumer, trophic level, food chain, and food web. pp. 525–528.
5. Outline the flow of carbon dioxide through the carbon cycle, and describe ways in which humans adversely influence the carbon cycle. p. 531.
6. With the knowledge you have gained, explain why it is beneficial to set aside habitat areas to protect endangered species. pp. 525–534.
7. Look back over all you have just learned about the science of ecology. In what

ways is this information important to the long-term prosperity of human society. p. 524.
8. Human populations in both the industrialized and nonindustrialized countries are overshooting the Earth's carrying capacity. What does that mean? Do you agree or disagree with this statement? Why or why not? Give examples. pp. 534–540.
9. Why is overpopulation as much a problem in the United States as it is in Bangladesh? p. 535.
10. Describe key trends in the use of forests, soils, water, minerals, and oil that suggest that humanity is on an unsustainable course. List and discuss solutions to each of the problems. pp. 536–540.
11. Given trends in natural resource use, many experts believe that global population growth must stop. Do you agree or disagree? Why? Is your position

supported by scientific fact or based more on a general feeling? pp. 535–540.
12. Describe the cause and impacts of global warming. How do you contribute to this problem? pp. 540–542.
13. Describe the causes and impacts of acid deposition. pp. 542–543.
14. Describe the causes and impacts of stratospheric ozone depletion wastes. pp. 543–545.
15. How could conservation (efficiency), recycling, renewable resources, and population control help address problems from global warming and acid deposition? pp. 542–545.
16. Check out the terms sustainable development and sustainable society on the Internet. After studying them, do you think efforts to achieve both of these goals will help preserve global biospheric homeostasis? pp. 545–546.

SELF-QUIZ: TESTING YOUR KNOWLEDGE

1. _____ is a discipline in the sciences that studies the relationship of organisms to their environment. p. 524.
2. An organism's environment consists of two components: _____ and _____. p. 524.
3. The _____ is a region of the planet on which life can be found. p. 524.
4. A biologically distinct region such as a grassland is called a/an _____. p. 524.
5. A biologically distinct region of the ocean is known as a/an _____ _____ zone. p. 524.
6. The term _____ is used to describe a community of organisms and their physical and chemical environment. p. 525.
7. A group of organisms belonging to the same species and occupying a specific

area is called a/an _____. p. 525.
8. All of the organisms in an ecosystem form a biological _____. p. 525.
9. The area where an organism lives is known as its _____. p. 525.
10. The _____ of an organism includes all of its relationships to the biotic and abiotic environment in which it lives. p. 526.
11. Organisms like plants and algae are known as _____ because they can create organic food molecules from carbon dioxide, water, and energy. p. 527.
12. An animal that feeds entirely on other organisms is called a/an _____. p. 527.
13. A _____ food chain begins with detrivores feeding on dead material. p. 527.

14. The position of an organism in a food chain is known as its _____ level. p. 528.
15. Most food chains contain no more than _____ to _____ levels. p. 528.
16. Numerous food chains in a biological community are interlinked, thus forming a _____ _____. p. 528.
17. Carbon dioxide is returned to the atmosphere as a result of the decay of waste and also _____. p. 531.
18. The conversion of atmospheric nitrogen into ammonia in the roots of plants and elsewhere is known as nitrogen _____. p. 531.
19. The development of a biological community in a previously barren environment is known as _____ succession. p. 533.

20. The _____ capacity is the number of organisms an ecosystem can support indefinitely. p. 534.
21. The accumulation of _____ in the atmosphere, primarily from the combustion of fossil fuels and deforestation, is resulting in a warming trend known as _____ warming. p. 541.
22. Carbon dioxide is a _____ gas that traps heat in the Earth's atmosphere. p. 541.
23. Sulfuric and nitric acids falling with the rain is known as _____ rain. p. 542.
24. Ozone is present in the upper layer of the atmosphere. It is being depleted by chemicals once used in spray cans and cooling systems known as _____. p. 544.
25. Ozone depletion could result in an increase in the amount of _____ radiation striking the Earth, which could increase skin cancer among humans. p. 544.

biology.jbpub.com/chiras/8e/

The site features eLearning, an online review area that provides quizzes, chapter outlines, and other tools to help you study for your class. You can also follow useful links for in-depth information, research the differing views in the Point/Counterpoints, or keep up on the latest health news.

Periodic Table of Elements

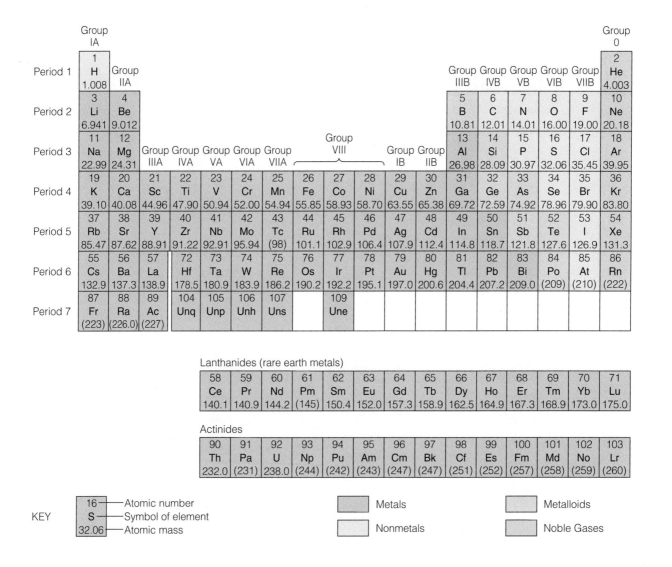

Group IA																	Group 0
1 H 1.008	Group IIA											Group IIIB	Group IVB	Group VB	Group VIB	Group VIIB	2 He 4.003
3 Li 6.941	4 Be 9.012											5 B 10.81	6 C 12.01	7 N 14.01	8 O 16.00	9 F 19.00	10 Ne 20.18
11 Na 22.99	12 Mg 24.31	Group IIIA	Group IVA	Group VA	Group VIA	Group VIIA	Group VIII			Group IB	Group IIB	13 Al 26.98	14 Si 28.09	15 P 30.97	16 S 32.06	17 Cl 35.45	18 Ar 39.95
19 K 39.10	20 Ca 40.08	21 Sc 44.96	22 Ti 47.90	23 V 50.94	24 Cr 52.00	25 Mn 54.94	26 Fe 55.85	27 Co 58.93	28 Ni 58.70	29 Cu 63.55	30 Zn 65.38	31 Ga 69.72	32 Ge 72.59	33 As 74.92	34 Se 78.96	35 Br 79.90	36 Kr 83.80
37 Rb 85.47	38 Sr 87.62	39 Y 88.91	40 Zr 91.22	41 Nb 92.91	42 Mo 95.94	43 Tc (98)	44 Ru 101.1	45 Rh 102.9	46 Pd 106.4	47 Ag 107.9	48 Cd 112.4	49 In 114.8	50 Sn 118.7	51 Sb 121.8	52 Te 127.6	53 I 126.9	54 Xe 131.3
55 Cs 132.9	56 Ba 137.3	57 La 138.9	72 Hf 178.5	73 Ta 180.9	74 W 183.9	75 Re 186.2	76 Os 190.2	77 Ir 192.2	78 Pt 195.1	79 Au 197.0	80 Hg 200.6	81 Tl 204.4	82 Pb 207.2	83 Bi 209.0	84 Po (209)	85 At (210)	86 Rn (222)
87 Fr (223)	88 Ra (226.0)	89 Ac (227)	104 Unq	105 Unp	106 Unh	107 Uns		109 Une									

Period 1, Period 2, Period 3, Period 4, Period 5, Period 6, Period 7

Lanthanides (rare earth metals)

58 Ce 140.1	59 Pr 140.9	60 Nd 144.2	61 Pm (145)	62 Sm 150.4	63 Eu 152.0	64 Gd 157.3	65 Tb 158.9	66 Dy 162.5	67 Ho 164.9	68 Er 167.3	69 Tm 168.9	70 Yb 173.0	71 Lu 175.0

Actinides

90 Th 232.0	91 Pa (231)	92 U 238.0	93 Np (244)	94 Pu (242)	95 Am (243)	96 Cm (247)	97 Bk (247)	98 Cf (251)	99 Es (252)	100 Fm (257)	101 Md (258)	102 No (259)	103 Lr (260)

KEY

16 — Atomic number
S — Symbol of element
32.06 — Atomic mass

Metals Metalloids
Nonmetals Noble Gases

The Metric System

Most Common English Units and the Corresponding Metric Unit

Measure	English Unit	Metric Unit
Weight	tons	metric tons
	pounds	kilograms
	ounces	grams
Length	miles	kilometers
	yards	meters
	inches	centimeters
Square Measure	acres	hectares
	square miles	square kilometers
Volume	quarts and gallons	liters
	fluid ounces	milliliters

Converting English Units to Metric Units

Measure	English Unit	Metric Unit
Weight	1 ton (2000 pounds)	0.9 metric tons
	1 pound	0.454 kilograms
	1 ounce	28.35 grams
Length	*1 mile	1.6 kilometers
	1 yard	0.9 meters
	*1 inch	2.54 centimeters
Square Measure	1 acre	0.4 hectares
	1 square mile	2.59 square kilometers
Volume	1 quart	0.95 liters
	*1 gallon	3.78 liters
	1 fluid ounce	29.58 milliliters

*Most useful conversions to know.

Converting Metric Units to English Units

Measure	Metric Unit	English Unit
Weight	*1 metric ton	2204 pounds
	1 metric ton	1.1 tons
	*1 kilogram	2.2 pounds
	1 gram	28.35 ounces
Length	*1 kilometer	0.6 miles
	1 meter	1.1 yards
	1 centimeter	0.39 inches
Square Measure	*1 hectare	2.47 acres
	1 square kilometer	0.386 square miles
Volume	1 liter	1.057 quarts
	1 liter	0.26 gallons
	1 milliliter	0.0338 fluid ounces

*Most useful conversions to know.

Acetylcholine Neurotransmitter substance in the central and peripheral nervous systems of humans.

Acetylcholinesterase Enzyme that destroys the neurotransmitter acetylcholine in the synaptic cleft.

Acid Any chemical substance that adds hydrogen ions to a solution, such as sulfuric acid.

Acrosome Enzyme-filled cap over the head of a sperm. Helps the sperm dissolve its way through the corona radiata and zona pellucida of the ovum.

Actin microfilaments Contractile filaments made of protein and found in cells as part of the cytoskeleton. Especially abundant in the microfilamentous network beneath the plasma membrane and in muscle cells.

Action potential Recording of electrical change in membrane potential when a neuron is stimulated.

Active immunity Immune resistance gained when an antigen is introduced into the body either naturally or through vaccination.

Active site The region of an enzyme molecule that binds substrates and performs the catalytic function of the enzyme.

Active transport Movement of molecules across membranes using protein molecules and energy supplied by ATP. Moves molecules and ions from regions of low to high concentration.

Adaptation Genetically based characteristic that increases an organism's chances of passing on its genes to its offspring.

Adenine One of the purine bases found in nucleotides such as ATP.

Adenosine diphosphate (ADP) Precursor to ATP; consists of adenine, ribose, and two phosphates. *See also* Adenosine triphosphate.

Adenosine triphosphate (ATP) A molecule composed of ribose sugar, adenine, and three phosphate groups. The last two phosphate groups are attached by "high-energy bonds" that require considerable energy to form but release that energy again when broken. ATP serves as the major energy carrier in cells.

Adipose tissue Type of loose connective tissue containing numerous fat cells. Important storage area for lipids.

Adrenal cortex Outer portion of the adrenal gland. Produces a variety of steroid hormones, including cortisol and aldosterone.

Adrenal gland Endocrine gland located on top of the kidney. Consists of two parts: adrenal cortex and medulla, each with separate functions.

Adrenal medulla Inner portion of the adrenal glands. Produces epinephrine (adrenalin) and norepinephrine (noradrenalin).

Adrenalin (epinephrine) Hormone secreted under stress. Contributes to the fight-or-flight response by increasing heart rate, shunting blood to muscles, increasing blood glucose levels, and other functions.

Adrenocorticotropic hormone (ACTH) Polypeptide hormone produced by the anterior pituitary. Stimulates the cells of the adrenal cortex, causing them to synthesize and release their hormones, especially glucocorticoids.

Aerobic exercise Exercise, such as swimming, that does not deplete muscle oxygen. Excellent for strengthening the heart and for losing weight.

Agglutination Clumping of antigens that occurs when antibodies bind to several antigens.

Aging Inevitable and progressive deterioration of the body's function, especially its homeostatic mechanisms.

AIDS Acquired immunodeficiency syndrome. Fatal disease caused by the HIV virus, which attacks T helper cells, greatly reducing the body's ability to fight infection.

Albinism Genetic disease resulting in a lack of pigment in the eyes or the eyes, skin, and hair. An autosomal recessive trait.

Aldosterone Steroid released by the adrenal cortex in response to a decrease in blood pressure, blood volume, and osmotic concentration. Acts principally on the kidney.

Alga (algae, plural) Heterogeneous group of aquatic plants consisting of three major groups: green, red, and brown. Important producer essential to aquatic food chains.

Alleles Alternative form of a gene.

Allergen Antigen that stimulates an allergic response.

Allergy Extreme overreaction to some antigens, such as pollen or foods. Characterized by sneezing, mucus production, and itchy eyes.

Allosteric inhibition Enzyme regulation in which an inhibitor molecule binds to an enzyme at a site away from the active site, changing the shape or charge of the active site so that it can no longer bind substrate molecules.

Allosteric site Region of an enzyme where products of metabolic pathways bind, changing the shape of the active site. In some enzymes this prevents substrates from binding to the active site; in others, it allows them to bind. Thus, allosteric sites can either turn on or turn off enzymes.

Alveoli Tiny, thin-walled sacs in the lung where oxygen and carbon dioxide are exchanged between the blood and the air.

Ameboid motion Cellular locomotion common in single-celled organisms and some cells in the human body. The cells send out slender cytoplasmic projections that attach to the substrate "ahead" of the cell. The cytoplasm flows into the projections, or pseudopodia, advancing the organism.

Amino acid A small organic molecule that serves as a building block for proteins. Approximately 20 different amino acids are found in the proteins in the human body.

Amniocentesis Procedure whereby physicians extract cells and fluid from the amnion surrounding the fetus. The cells are examined for genetic defects, and the fluid is studied biochemically.

Amnion Layer of cells that separates from the inner cell mass of the embryo and eventually forms a complete sac around the fetus.

Amniotic fluid Liquid in the amniotic cavity surrounding the embryo and fetus during development. Helps protect the fetus.

Amylase Enzyme in saliva that helps break down starch molecules.

Anabolic steroids Synthetic androgen hormones that promote muscle development.

Analogous structures Anatomical structures that function similarly but differ in structure; for example, the wing of a bird and the wing of an insect.

Anaphase Phase of mitosis during which the chromatids of each chromosome begin to uncouple and are pulled in opposite directions with the aid of the mitotic apparatus.

Androgens Male sex steroids such as testosterone produced principally by the testes.

Anemia Condition characterized by an insufficient number of red blood cells in the blood or insufficient hemoglobin. Often caused by insufficient iron intake.

Aneurysm Ballooning of the arterial wall caused by a degeneration of the tunica media.

Anoxia Lack of oxygen.

Antagonistic Refers to hormones or muscles that exert opposite effects.

Anterior pituitary Major portion of the pituitary gland, which is controlled by hypothalamic hormones. Produces seven protein and polypeptide hormones.

Anthropoids Monkeys, the great apes, and humans.

Antibodies Proteins produced by immune system cells that destroy or inactivate antigens, including pollen, bacteria, yeast, and viruses.

Anticodon Sequence of three bases found on the transfer RNA molecule. Aligns with the codon on messenger RNA and helps control the sequence of amino acids inserted into the growing protein.

Anticodon loop Part of the transfer RNA molecule that bears three bases that bind to the three bases of the codon on messenger RNA.

Antidiuretic hormone (**ADH**) Hormone released by the posterior pituitary. Increases the permeability of the distal convoluted tubule and collecting tubules, increasing water reabsorption.

Antigens Any substance that is detected as foreign by an organism and elicits an immune response. Most antigens are proteins and large molecular weight carbohydrates.

Aorta Largest artery in the body; carries the oxygenated blood away from the heart and delivers it to the rest of the body through many branches.

Aquatic life zones Ecologically distinct regions in fresh water and salt water.

Aqueous humor Liquid in the anterior and posterior chambers of the eye.

Arteries Vessels that transport blood away from the heart.

Arteriole Smallest of all arteries; usually drains into capillaries.

Arthroscope Device used to examine internal injuries, such as in joints.

Association cortex Area of the brain where integration occurs.

Association neurons Nerve cells that receive input from many sensory neurons and help process them, ultimately carrying impulses to nearby multipolar neurons.

Aster Array of microtubules found in the cell in association with the spindle fibers during cell division.

Asthma Respiratory disease resulting from an allergic response. Allergens cause histamine to be released in the lungs. Histamine causes the air-carrying ducts (bronchioles) to constrict, cutting down airflow and making breathing difficult.

Astigmatism Unequal curvature of the cornea (sometimes the lens) that distorts vision.

Atom Smallest particles of matter that can be achieved by ordinary chemical means, consisting of protons, neutrons, and electrons.

Atomic mass unit Unit used to measure atomic weight. One atomic mass unit is 1/12 the weight of a carbon atom.

Atomic weight Average mass of the atoms of a given element, measured in atomic mass units.

Atrioventricular bundle Tract of modified cardiac muscle fibers that conduct the pacemaker's impulse into the ventricular muscle tissue.

Atrioventricular node (**AV node**) Knot of tissue located in the right ventricle. Picks up the electrical signal arriving from the atria and transmits it down the atrioventricular bundle.

Atrioventricular valves Valves between the atria and ventricles.

Auditory (**eustachian**) **tube** Collapsible tube that joins the nasopharynx and middle ear cavities and helps equalize pressure in the middle ear.

Auricle (or **pinna**) Skin-covered cartilage portion of the outer ear.

Autoimmune reaction Immune response directed at one's own cells.

Autonomic nervous system That part of the nervous system not under voluntary control.

Autosomal dominant trait Trait that is carried on the autosomes and is expressed in heterozygotes and homozygote dominants.

Autosomal recessive trait Trait that is carried on the autosomes and is expressed only when both recessive genes are present.

Autosomes All human chromosomes except the sex chromosomes.

Autotrophs Organisms such as plants that, unlike animals, are able to synthesize their own food.

Axial skeleton The skull, vertebral column, and rib cage. Contrast with appendicular skeleton.

Axon Long, unbranched process attached to the nerve cell body of a neuron. Transports nerve impulses away from the cell body.

Basal body Organelle located at the base of the cilium and flagellum. Consists of nine sets of microtubules arranged in a circle. Each "set" contains three microtubules.

Basilar membrane Membrane that supports the organ of Corti in the cochlea.

Benign tumor Abnormal cellular proliferation. Unlike a malignant tumor, the cells in a benign tumor stop growing after a while, and the tumor remains localized.

Bile Fluid produced by the liver and stored and concentrated in the gallbladder.

Bile salts Steroids produced by the liver, stored in the gallbladder, and released into the small intestine where they emulsify fats, a step necessary for enzyme digestion.

Biome Terrestrial region characterized by a distinct climate and a characteristic plant and animal community.

Biorhythms (biological cycles) Naturally fluctuating physiological process.

Biosphere Region on Earth that supports life. Exists at the junction of the atmosphere, lithosphere, and hydrosphere.

Bipedal Refers to the ability to walk on two legs.

Birth control Any method or device that prevents conception and birth.

Birth defects Physical or physiological defects in newborns. Caused by a variety of biological, chemical, and physical factors.

Blastocyst Hollow sphere of cells formed from the morula. Consists of the inner cell mass and the trophoblast.

Blood clot Mass of fibrin containing platelets, red blood cells, and other cells. Forms in walls of damaged blood vessels, halting the efflux of blood.

B-lymphocytes Type of lymphocyte that transforms into a plasma cell when exposed to antigens.

Bowman's capsule Cup-shaped end of the nephron that participates in glomerular filtration.

Brain stem Part of the brain that consists of the medulla and pons. Houses structures, such as the breathing control center and reticular activating systems, that control many basic body functions.

Braxton-Hicks contractions Contractions that begin a month or two before childbirth. Also known as false labor.

Breathing center Aggregation of nerve cells in the brain stem that controls breathing.

Breech birth Delivery of a baby feet first.

Bronchi Ducts that convey air from the trachea to the bronchioles and alveoli.

Bronchioles Smallest ducts in the lungs. Their walls are largely made of smooth muscle that contracts and relaxes, regulating the flow of air into the lungs.

Bronchitis Infection or persistent irritation of the bronchi characterized by cough.

Buffer A chemical substance that helps resist changes in acidity.

Bulimia Eating disorder characterized by recurrent binge eating followed by vomiting.

Calcitonin (thyrocalcitonin) Polypeptide hormone produced by the thyroid gland that inhibits osteoclasts and stimulates osteoblasts to produce bone, thus lowering blood calcium levels.

Cancer Any of dozens of diseases, all characterized by the uncontrollable replication of various body cells.

Capillaries Tiny vessels in body tissues whose walls are composed of a flattened layer of cells that allow water and other molecules to flow freely into and out of the tissue fluid.

Capillary bed Branching network of capillaries supplied by arterioles and drained by venules.

Capsid Protein coat of a virus.

Carbohydrate Organic compound consisting of carbon, hydrogen, and oxygen. A structural component of plant cells, it is used principally as a source of energy in animal cells.

Carcinogens Cancer-causing agents.

Cardiac muscle Type of muscle found in the walls of the heart; it is striated and involuntary.

Carnivores Meat eaters; organisms that feed on other animals.

Carrier proteins Class of proteins that transport smaller molecules and ions across the plasma membrane of the cell. Are involved in facilitated diffusion.

Carriers Individuals who carry a gene for a particular trait that can be passed on to their children, but who do not express the trait.

Carrying capacity The number of organisms an ecosystem can support on a sustainable basis.

Cartilage Type of specialized connective tissue. Found in joints on the articular surfaces of bones and other locations such as the ears and nose.

Catabolic reaction A reaction in which molecules are broken down; for example, the breakdown of glucose is a catabolic reaction.

Catalyst Class of compounds that speed up chemical reactions. Although they take an active role in the process, they are left unchanged by the reaction. Thus, they can be used over and over again. Catalysts lower the activation energy of a reaction. *See also* Enzyme.

Cataracts Disease of the eye resulting in cloudy spots on the lens (and sometimes the cornea) that cause cloudy vision.

Cell body Part of the nerve cell that contains the nucleus and other cellular organelles; the center of chemical synthesis.

Cell cycle Repeating series of events in the lives of many cells. Consists of two principal parts: interphase and cellular division.

Cell division Process by which the nucleus and the cytoplasm of a cell are split, creating two daughter cells. Consists of mitosis and cytokinesis.

Cellular respiration The complete breakdown of glucose in the cell, producing carbon dioxide and water. Composed of four separate but interconnected parts: glycolysis, the intermediate reaction, the citric acid cycle, and the electron transport system.

Central nervous system The brain and spinal cord.

Centriole Organelle consisting of a ring of microtubules, arranged in nine sets of three. Structurally identical to basal bodies but associated with the spindle apparatus. Gives rise to the basal body in ciliated cells.

Centromere Region on each chromatid that joins with the centromere of its sister chromatid.

Cerebellum Structure of the brain that lies above the pons and medulla. It has many important functions including synergy.

Cerebral cortex Outer layer of each cerebral hemisphere, consisting of many multipolar neurons and nerve cell fibers.

Cerebral hemisphere Convoluted mass of nervous tissue located above the deeper structures, such as the hypothalamus and limbic system. Home of consciousness, memory, and sensory perception; originates much conscious motor activity.

Cerebrum The largest part of the brain; consists of two cerebral hemispheres.

Cervix Lowermost portion of the uterus; it protrudes into the vagina.

Cesarean section Delivery of a baby via an incision through the abdominal and uterine walls.

Chemical equilibrium The point at which the "forward" reaction from reactants to products proceeds at the same rate as the "backward" reaction from products to reactants, so that there is no net change in chemical composition.

Chemical evolution Formation of organic molecules from inorganic molecules early in the history of the Earth.

Chlorofluorocarbons (CFCs) Chemical substances once used as spray can propellants, refrigerants, and blowing agents for foam products. These substances drift to the upper stratosphere and dissociate. Chlorine released by CFCs reacts with ozone, thus eroding the ozone layer.

Chlorophyll A green pigment that acts as the primary light-trapping molecule for photosynthesis.

Cholecystokinin (CCK) Hormone produced by cells of the duodenum when chyme is present; causes the gallbladder to contract, releasing bile.

Chromatid Strand of the chromosome consisting of DNA and protein.

Chromatin Long, threadlike fibers containing DNA and protein in the nucleus.

Chromosome Structure found in the nucleus of cells that houses the genetic information. Consists of DNA and protein.

Chromosome-to-pole fibers Microtubules of the spindle that extend from the centriole to the chromosome where they attach. They play a crucial role in separating the double-stranded chromosomes during mitosis.

Chronic bronchitis Persistent irritation of the bronchi, which causes mucous buildup, coughing, and difficulty breathing.

Chyme Liquified food in the stomach.

Cilia A cellular organelle that protrudes from the surface of many cells. Cilia beat in unison to cause fluids to move over their surface.

Circadian rhythm Biorhythm that occurs on a daily cycle.

Circulatory system Organ system consisting of the heart, blood vessels, and blood.

Circumcision Operation to remove the foreskin of the penis. Generally performed on newborns.

Citric acid cycle A cyclic series of reactions in which the pyruvates produced by glycolysis are broken down to CO_2, accompanied by the formation of ATP and electron carriers. Occurs in the matrix of mitochondria.

Clitoris Small knot of tissue located where the labia minora meet. Consists of erectile tissue.

Cloning Technique of genetic engineering whereby many copies of a gene are produced.

Cochlea Sensory organ of the inner ear that houses the receptor for hearing.

Codominant Refers to two equally expressed alleles.

Codon Three adjacent bases in the messenger RNA that code for a single amino acid.

Color blindness Condition that occurs in individuals who have a deficiency of certain cones. The most common form involves difficulty in distinguishing between red and green.

Colostrum Protein-rich product of the breast, produced for two to three days immediately after delivery and prior to milk production.

Community All of the plants, animals, and microorganisms in an ecosystem.

Compact bone Dense bony tissue in the outer portion of all bones.

Complement Group of blood proteins that circulate in the blood in an inactive state until the body is invaded by bacteria; then they help destroy the bacteria.

Complementary base pairing Unalterable coupling of the purine adenine to the pyrimidine thymine, and the purine guanine to the pyrimidine cytosine in DNA. Responsible for the accurate transmission of genetic information from parent to offspring. Also functions in RNA production.

Cone Type of photoreceptor that operates in bright light; is responsible for color vision.

Connective tissue One of the primary tissues. It contains cells and varying amounts of extracellular material. Holds cells together, forming tissues and organs.

Connective tissue proper Name referring to loose and dense connective tissue; supports and joins various body structures.

Consumers Organisms that eat plants and algae (producers) and other consumers.

Contact inhibition Cessation of growth that results when two or more cells contact each other. A feature of normal cells but absent in cancer cells.

Contraceptive Any measure that helps prevent fertilization and pregnancy.

Control group A group of subjects (plants, animals, etc.) that is used in an experiment; this group is identical to the experimental group in all regards except that it does not receive the experimental treatment.

Cornea Clear part of the wall of the eye continuous with the sclera; allows light into the interior of the eye.

Coronary bypass surgery Surgical technique used to reestablish blood flow to the heart muscle by grafting a vein to shunt blood around a clogged coronary artery.

Corpus luteum (CL) Structure formed from the ovulated follicle in the ovary; produces estrogen and progesterone.

Cortisol Glucocorticoid hormone that increases blood glucose by stimulating gluconeogenesis. Also stimulates protein breakdown in muscle and bone.

Cranial nerves Nerves arising from the brain and brain stem.

Critical thinking Process in which one seeks to determine the validity of conclusions based on sound reasoning. This often requires one to examine hidden bias, the appropriateness of experimental design, and other factors.

Crossing-over Exchange of chromatin by homologous chromosomes during prophase I of meiosis. Results in considerably more genetic variation in gametes and offspring.

Cushing's syndrome Disease that results from pharmacologic doses of cortisone usually administered for rheumatoid arthritis or allergies.

Cystic fibrosis Autosomal recessive disease that leads to problems in sweat glands, mucous glands, and the pancreas. Pancreas may become blocked, thus reducing the flow of digestive enzymes to the small intestine. Mucus buildup in the lungs makes breathing difficult.

Cytokinesis Cytoplasmic division brought about by the contraction of a microfilamentous network lying beneath the plasma membrane at the midline. Usually begins when the cell is in late anaphase or early telophase.

Cytoplasm Material occupying the cytoplasmic compartment of a cell. Consists of a semifluid substance, the cytosol, containing many dissolved substances, and formed elements, the organelles.

Cytoskeleton A network of protein tubules in the cytoplasmic compartment of a cell. Attaches to many organelles and enzyme molecules and thus helps organize cellular activities, increasing efficiency.

Cytotoxic T cells Type of T cell (T-lymphocyte) that attacks and kills virus-infected cells, parasites, fungi, and tumor cells.

Daughter cells Cells produced during cell division.

Decomposer food chain Series of organisms that feed on organic wastes and the dead remains of other organisms.

Defibrillation Procedure to stop fibrillation (erratic electrical activity) of the heart.

Dendrite Short, highly branched fiber that carries impulses to the nerve cell body.

Dense connective tissue Type of connective tissue that consists primarily of densely packed fibers, such as those found in ligaments and tendons.

Dermis Layer of dense irregular connective tissue that binds the epidermis to underlying structures.

Detrivores Organisms that feed on animal waste or the remains of plants and animals.

Diabetes insipidus Condition caused by lack of the pituitary hormone ADH. Main symptoms are excessive drinking and excessive urination.

Diabetes mellitus Insulin disorder either resulting from insufficient insulin production or decreased sensitivity of target cells to insulin. Results in elevated blood glucose levels unless treated.

Dialysis Procedure used to treat patients whose kidneys have failed. Blood is removed from the body and pumped through an artificial filter that removes impurities.

Diaphragm (birth control) Birth control device consisting of a rubber cup that fits over the end of the cervix. Used in conjunction with spermicidal jelly or cream.

Diaphragm (muscle) Dome-shaped muscle that separates the abdominal and thoracic cavities.

Diaphysis Shaft of the long bones. Consists of an outer layer of compact bone and an inner marrow cavity.

Diastolic pressure The pressure at the moment the heart relaxes. The lower of the two blood pressure readings.

Differentiation Structural and functional divergence from the common cell line. Occurs during embryonic development.

Diffusion Movement of molecules from a region of high concentration to a region of low concentration through a semipermeable membrane.

Dihybrid cross Procedure where one plant is bred with another to study two traits.

DNA polymerase Enzyme that helps align the nucleotides and join the phosphates and sugar molecules in a newly forming DNA strand.

Dominant Adjective used in genetics to refer to an allele that is always expressed in heterozygotes. Designated by a capital letter.

Double helix Describes the helical structure formed by two polynucleotide chains making up the DNA molecule.

Down syndrome Genetic disorder caused by an additional chromosome 21 that results in distinctive facial characteristics and mental retardation. Also known as Trisomy 21.

Dryopithecus Genus of apelike creatures that is thought to have given rise to the gibbons, gorillas, orangutans, and chimpanzees.

Duodenum First portion of the small intestine; site where most food digestion and absorption takes place.

E. coli Common bacterium that lives in the large intestine of humans and other mammals. Digests leftover glucose and other materials from food. Used in much genetic research.

Ecological niche An organism's habitat and all of the relationships that exist between that organism and its environment.

Ecological system (ecosystem) System consisting of organisms and their environment and all of the interactions that exist between these components.

Ecology Study of living organisms and the web of relationships that binds them together in the economy of nature. The study of ecosystems.

Ecosystem *See* ecological system.

Ecosystem balance Dynamic equilibrium in ecosystems. Maintained by the interplay of growth and reduction factors.

Ectoderm One of the three types of cells that emerges in human embryonic development. Gives rise to the skin and associated structures, including the eyes.

Edema Swelling resulting from the buildup of fluid in the tissues.

Effector General term for any organ or gland that is controlled by the nervous system.

Ejaculation Ejection of semen from the male reproductive tract.

Elastic arteries Arteries that contain numerous elastic fibers interspersed among the smooth muscle cells of the tunica media.

Elastic cartilage Type of cartilage containing many elastic fibers found in regions where support and flexibility are required.

Electron Highly energetic particle carrying a negative charge that orbits the nucleus of an atom.

Electron carrier A molecule that can reversibly gain and lose electrons. Electron carriers generally accept high-energy electrons produced during an exergonic reaction and donate the electrons to acceptor molecules that use the energy to drive endergonic reactions.

Electron transport system Series of protein molecules in the inner membrane of the mitochondrion that pass electrons from the citric acid cycle from one to another, eventually donating them to oxygen. The electrons come from the citric acid cycle. During their journey along this chain of proteins, the electrons lose energy, which is used to make ATP.

Elements Purest form of matter; substances that cannot be separated into different substances by chemical means.

Emphysema Progressive, debilitating disease that destroys the tiny air sacs in the lung (alveoli), caused by smoking and air pollution.

Endocrine glands Glands of internal secretion that produce hormones secreted into the bloodstream.

Endocrine system Numerous, small, hormone-producing glands scattered throughout the body.

Endocytosis Process by which cells engulf solid particles, bacteria, viruses, and even other cells.

Endoderm One of the three types of cells that emerges during embryonic development. Gives rise to the intestinal tract and associated glands.

Endolymph Fluid inside the semicircular canals that deflects the cupula, signaling rotational movement of the head and body.

Endometrium Uterine endothelium or lining.

Endoplasmic reticulum Branched network of channels found throughout the cytoplasm of many cells. Formed from flattened sheets of membrane derived from the nuclear membrane. *See* rough and smooth endoplasmic reticulum for functions.

Endothelium Single-celled lining of blood vessels.

End-product inhibition The inhibition of an enzyme by a product of the chemical reaction it catalyzes or by the product of a series of chemical reactions of which the enzyme is a part; may result from binding of the end product to the allosteric site or active site of the enzyme.

Energy carrier A molecule that stores energy in "high-energy" chemical bonds and releases the energy again to drive coupled endergonic reactions. ATP is the most common energy carrier in cells; NAD and FAD are others.

Energy pyramid Diagram of the amount of energy at various trophic levels in a food chain or ecosystem.

Enzyme A protein catalyst that speeds up the rate of specific biochemical reactions.

Epidermis Outermost layer of the skin that protects underlying tissues from drying out and from bacteria and viruses.

Epididymis Storage site of sperm. Located on the testis, it consists of a long, tortuous duct, the epididymal duct.

Epigenome That part of the chromosome outside the DNA that controls the function of DNA, notably methyl groups that can chemically bond to DNA and the histone proteins that surround the DNA.

Epiglottis Flap of tissue that closes off the trachea during swallowing.

Epiphyseal plate Band of cartilage cells between the shaft of the bone and the epiphysis. Allows for bone growth.

Episiotomy Surgical incision that runs from the vaginal opening toward the rectum. Enlarges the vaginal opening, easing childbirth.

Epithelium One of the primary tissues. Forms linings and external coatings of organs.

Erectile tissue Spongy tissue of the penis that fills with blood during sexual excitement, making the penis turgid.

Erythropoietin Hormone produced by the kidney when oxygen levels decline. Stimulates red blood cell production in the bone marrow.

Esophagus Muscular tube that transports food to the stomach.

Essential amino acid One of nine amino acids that must be provided in the human diet.

Euchromatin Metabolically active chromatin.

Eukaryote Any cell containing a distinct nucleus and organelles. They are found in single-celled organisms of the kingdom Protista and all multicellular organisms of the kingdoms Plantae, Animalia, and Fungi.

Evolution Process that leads to structural, functional, and behavioral changes in species, making them better able to survive in their environment; also leads to the formation of new species. Results from natural genetic variation and environmental conditions that select for organisms best suited to their environment.

Exhalation Expulsion of air from the lungs.

Exocrine gland Gland of external secretion; empties its contents into ducts.

Exocytosis Process by which cells release materials stored in secretory vesicles. The reverse of endocytosis.

Exon Expressed segment of DNA.

Experiment Test performed to support or refute a hypothesis.

Experimental group One of two groups in many scientific experiments (the other is the control group). Treated like the control except for one variable, the experimental variable.

Extension Movement of a body part (limbs, fingers, and toes) that opens a joint.

External auditory canal Channel that directs sound waves to the eardrum.

External genitalia External portion of the female reproductive system consisting of the clitoris, labia minora, and labia majora.

Extrinsic eye muscles Six muscles located outside the eye that are responsible for eye movement.

Facilitated diffusion Process in which carrier proteins shuttle molecules across plasma membranes. The molecules move in response to concentration gradients.

Feces Semisolid material containing undigested food, bacteria, ions, and water; produced in the large intestine.

Feedback mechanism A mechanism in which the product of one process regulates the rate of another process, either turning it on or shutting it off.

Fermentation Process occurring in eukaryotic cells in the absence of oxygen, during which pyruvic acid is converted to lactic acid. Also occurs in those prokaryotes that live in oxygen-free environments.

Fertilization Union of sperm and ovum.

Fiber Any of the indigestible polysaccharides in fruits, vegetables, and grains.

Fibrillation Cardiac muscle spasms occurring during heart attacks due to a loss of synchronized electrical signals.

Fibrin Fibrous protein produced from fibrinogen, a soluble plasma protein. Helps form blood clots.

Fibrinogen Protein in plasma that forms fibrin.

Fibroblast Connective tissue cell, found in loose and dense connective tissues that produces collagen, elastic fibers, and a gelatinous extracellular material; responsible for repairing damage created by cuts or tears to connective tissue.

Fibrocartilage Type of cartilage whose extracellular matrix consists of numerous bundles of collagen fibers. Principally found in the intervertebral disks.

Fight-or-flight response An automatic response of an organism to danger that enables it to flee or stand and fight; it is activated by the autonomic nervous system and results in an increase in heart rate and breathing and an increase in blood flow to the muscles.

Fitness Measure of reproductive success of an organism and, therefore, the genetic influence an individual has on future generations.

Flagellum Long, whiplike extension of the plasma membrane of certain protozoans and sperm cells in humans. Used for motility.

Flexion Movement of a limb, finger, or toe that involves closing a joint.

Follicle (ovary) Structure found in the ovary. Each follicle contains an oocyte and one or more layers of follicle cells that are derived from the loose connective tissue of the ovary surrounding the follicle.

Follicle (thyroid) Structure found in the thyroid gland. Consists of an outer layer of cuboidal cells surrounding thyroglobulin, a proteinaceous material from which thyroxine is formed.

Follicle-stimulating hormone (FSH) Gonadotropic hormone from the anterior pituitary that promotes gamete formation in both men and women.

Food chain Series of organisms in an ecosystem in which each organism feeds on the organism preceding it.

Food vacuole Membrane-bound vacuole in a cell containing material engulfed by the cell.

Food web All of the connected food chains in an ecosystem.

Foreskin Sheath of skin that covers the glans penis.

Fossil Remains or imprints of organisms that lived on Earth many years ago, usually embedded in rocks or sediment.

Gallbladder Sac on the underside of the liver that stores and concentrates bile.

Gastrin Stomach hormone that stimulates HCl production and release by the gastric glands.

Gene Segment of the DNA that controls cell structure and function.

Gene pool All the genes of all of the members of a population or species.

Gene therapy The use of artificially produced genes to treat, even cure, diseases.

Genetic engineering The artificial manipulation of genes in which certain genes from one organism are removed and transferred to another organism of the same or a different species. This permits scientists to transfer important genes to improve species—for example, to increase resistance to disease.

Genome All of the genes of an organism.

Genotype Genetic makeup of an organism.

Germ cell Refers to the sperm or ovum (egg) and the cells from which they are derived; germ cells contain half the chromosomes of somatic cells.

Germinal epithelium Germ cells in the wall of the seminiferous tubule that give rise to sperm.

Glans penis Slightly enlarged tip of the penis.

Glaucoma Disease of the eye caused by pressure resulting from a buildup of aqueous humor in the anterior chamber.

Glomerulus Tuft of capillaries that make up part of the nephron; site of glomerular filtration.

Glucagon Hormone released by the pancreas that stimulates the breakdown of glycogen in the liver and the release of glucose molecules, thus increasing blood levels of glucose.

Glucocorticoids Group of steroid hormones produced by the adrenal cortex that stimulate gluconeogenesis.

Glycolysis Metabolic pathway in the cytoplasm of the cell, during which glucose is split in half, forming two molecules of pyruvic acid. The energy released during the reaction is used to generate two molecules of ATP.

Glycoproteins Proteins that have carbohydrate attached to them.

Goiter Condition in which the thyroid gland enlarges due to lack of dietary iodide.

Golgi complex Organelle consisting of a series of flattened membranes that form channels. It sorts and chemically modifies molecules and repackages its proteins into secretory vesicles.

Golgi tendon organs Special receptors found in tendons that respond to stretch. Also known as neurotendinous organs.

Gonadotropin General term for FSH and LH, which are produced by the anterior pituitary and target male and female gonads.

Gonadotropin-releasing hormone (GnRH) Hormone produced by the hypothalamus that controls the release of FSH (ICSH in males) and LH.

Gray matter Gray, outermost region of the cerebral cortex.

Grazer Herbivorous organism.

Grazer food chain Food chain beginning with plants and grazers (herbivores).

Greenhouse gas Gas, such as carbon dioxide and chlorofluorocarbons, that traps heat escaping from the Earth and radiates it back to the surface.

Growth hormone A protein hormone produced by the anterior pituitary that stimulates cellular growth in the body, causing cellular hypertrophy and hyperplasia. Its major targets are bone and muscle.

Habitat Place in which an organism lives.

Helper T cell Type of T-lymphocyte that stimulates the proliferation of T and B cells when antigen is present.

Hemoglobin Protein molecules inside red blood cells; binds to oxygen.

Hemophilia Disease caused by a gene defect occurring on the X chromosome. Results in absence of certain blood-clotting factors.

Herpes One of the most common sexually transmitted diseases; caused by a virus.

Heterochromatin Inactive chromatin that is slightly coiled or compacted in the interphase nucleus.

Heterotrophs Organisms such as animals that, unlike plants, are unable to synthesize their own food. Consume plants and other organisms.

Heterozygous Adjective describing a genetic condition in which an individual contains one dominant and one recessive gene in a gene pair.

High-density lipoproteins (HDLs) Complexes of lipid and protein that transport cholesterol to the liver for destruction.

Histamine Potent vasodilator released by certain cells in the body during allergic reactions.

Histone Globular protein thought to play a role in regulating the genes.

Homeostasis A condition of dynamic equilibrium within any biological or social system. Achieved through a variety of automatic mechanisms that compensate for internal and external changes.

Hominid First humanlike creatures.

Hominoids Subgroup of anthropoids.

Homologous structures Structures thought to have arisen from a common origin.

Homozygous Adjective describing a genetic condition marked by the presence of two identical alleles for a given gene.

Hormone Chemical substance produced in one part of the body that travels to another, typically through the bloodstream, where it elicits a response.

Human chorionic gonadotropin (HCG) Hormone produced by the embryo that stimulates the corpus luteum in the mother's body to produce estrogen.

Humoral immunity Immune reaction that protects the body primarily against viruses and bacteria in the body fluids via antibodies produced by plasma cells.

Hyperglycemia High blood glucose levels.

Hypertension High blood pressure.

Hypothalamus Structure in the brain located beneath the thalamus. It consists of many aggregations of nerve cells and controls a variety of functions aimed at maintaining homeostasis.

Hypothesis Tentative and testable explanation for a phenomenon or observation.

Immune system Diffuse system consisting of trillions of cells that circulate in the blood and lymph and take up residence in the lymphoid organs, such as the spleen, thymus, lymph nodes, and tonsils, as well as other body tissues. Helps protect the body against invasion by foreign cells, such as bacteria and viruses, and protects against cancer cells.

Immunity Term referring to the resistance of the body to infectious disease.

Immunocompetence Process in which lymphocytes mature and become capable of responding to specific antigens.

Immunoglobulins Antibodies.

Implantation Process in which the blastocyst embeds in the uterine lining.

Impotence Inability of a male to achieve an erection.

Incomplete dominance Partial dominance. Occurs when an allele exerts only partial dominance over another allele, resulting in an intermediate trait.

Incontinence Inability to control urination.

Induced abortion Deliberate expulsion of a fetus or embryo.

Infectious mononucleosis White blood cell disorder caused by a virus. Characterized by a rapid increase in the number of monocytes and lymphocytes in the blood.

Inferior vena cava Large vein that empties deoxygenated blood from the body below the heart into the right atrium of the heart.

Infertility Inability to conceive; can be due to problems in either the male or the female or both partners.

Inflammatory response Response to tissue damage including an increase in blood flow; the release of chemical attractants, which draw monocytes to the scene; and an increase in the flow of plasma into a wound.

Inhalation Process of air being drawn into the lungs.

Inhibiting hormone Hormone from the hypothalamus that inhibits the release of hormones from the anterior pituitary.

Initiator codon Codon found on a messenger RNA strand that marks where protein synthesis begins.

Inner cell mass Cells of the blastocyst that become the embryo and amnion.

Insulin Hormone that stimulates the uptake of glucose by body cells, especially muscle and liver cells. Stimulates the synthesis of glycogen in liver and muscle cells.

Insulin-dependent diabetes Type of diabetes that can only be treated with injections of insulin. May be caused by an autoimmune reaction. Also known as early-onset diabetes.

Insulin-independent diabetes Type of diabetes that often occurs in obese people. In most patients, it can be controlled by diet. Also known as late-onset diabetes.

Integral protein Large protein molecules in the lipid bilayer of the plasma membrane.

Integration Process of making sense of various nervous inputs so that a meaningful response can be achieved.

Interferon Protein released from cells infected by viruses that stops the replication of viruses in other cells.

Interphase Period of cellular activity occurring between cell divisions. Synthesis and growth occur in preparation for cell division.

Interstitial cells Testosterone-producing cells located in the loose connective tissue between the seminiferous tubules of the testes.

Interstitial cell stimulating hormone (ICSH) Luteinizing hormone in males. Stimulates testosterone secretion.

Interstitial fluid Fluid surrounding cells in body tissues. Provides a path through which nutrients, gases, and wastes can travel between the capillary and the cells.

Intervertebral disks Shock-absorbing material between the bones of the spine.

Intron Segment of DNA that is not expressed. Lies between exons (expressed segments).

Ion Atom that has gained or lost one or more electrons. May be either positively or negatively charged.

Ionic bond Weak bond that forms between oppositely charged ions.

Iris Colored segment of the middle layer of the eye visible through the cornea.

Isotope Alternative form of an atom; differs from other atoms in the number of neutrons found in the nucleus.

Joint capsule Connective tissue that connects to the opposing bones of a joint and forms the synovial cavity. The inner layer of the joint capsule produces synovial fluid.

Kidney Organ that rids the body of wastes and plays a key role in regulating the chemical constancy of blood.

Kingdom A large grouping of organisms; scientists typically recognize five major kingdoms: Monera, Protista, Animalia, Plantae, and Fungi.

Klinefelter syndrome Genetic disorder that results from an XXY genotype.

Krebs cycle *See* citric acid cycle.

Labor The process or period of childbirth.

Lactation Milk production in the breasts.

Laparoscope Instrument used to examine internal organs through small openings made in the skin and underlying muscle.

Laryngitis Inflammation of the lining of the larynx, resulting in hoarseness. Caused by bacterial and viral infection and also excessive use of the voice.

Larynx Rigid but hollow cartilaginous structure that houses the vocal cords and participates in swallowing.

Lens Transparent structure that lies behind the iris and in front of the vitreous humor. Focuses light on the retina.

Leukemia Cancer of white blood cells.

Leukocytosis An increase in the concentration of white blood cells, which often occurs during a bacterial or viral infection.

Ligament Connective tissue structure that runs from bone to bone, located alongside and sometimes inside the joint. Offers support for joints.

Limbic system Array of structures in the brain that work in concert with centers of the hypothalamus. Site of instincts and emotions.

Lipid Commonly known as fats. Water-insoluble organic molecules that provide energy to body cells, help insulate the body from heat loss, and serve as precursors in the synthesis of certain hormones. A principal component of the plasma membrane.

Liver Organ located in the abdominal cavity that performs many functions essential to homeostasis. It stores glucose and fats, synthesizes some key blood proteins, stores iron and certain vitamins, detoxifies certain chemicals, and plays an important role in fat digestion by producing bile.

Loose connective tissue Type of connective tissue that serves primarily as a packing material. Contains many cells among a loose network of collagen and elastic fibers. Often contains cells that help protect the body from foreign organisms.

Low-density lipoproteins (LDLs) Complexes of protein and lipid that transport cholesterol, depositing it in blood vessels.

Lungs Two large saclike organs in the thoracic cavity where the blood and air exchange carbon dioxide and oxygen.

Luteinizing hormone (LH) Hormone produced by the anterior pituitary that stimulates gonadal hormone production. In men, LH stimulates the production of testosterone, the male sex steroid. In women, LH stimulates estrogen secretion.

Lymph Fluid contained in the lymphatic vessels. Similar to tissue fluid, but also contains white blood cells and may contain large amounts of fat.

Lymph node Small nodular organ interspersed along the course of the lymphatic vessels. Serves as a filter for lymph.

Lymphatic system Network of vessels that drains extracellular fluid from body tissues and returns it to the circulatory system.

Lymphocyte Type of white blood cell. *See also* B-lymphocyte and T-lymphocyte.

Lymphoid organs Organs, such as the spleen and thymus, that belong to the lymphatic system.

Lysosome Membrane-bound organelle that contains enzymes. Responsible for the breakdown of material that enters the cell by endocytosis. Also destroys aged or malfunctioning cellular organelles.

Macronutrients Nutrients required in relatively large amounts by organisms. Includes water, proteins, carbohydrates, and lipids.

Macrophage Phagocytic cell derived from monocytes that resides in loose connective tissues and helps guard tissues against bacterial and viral invasion.

Malignant tumor Structure resulting from uncontrollable cellular growth. Cells often spread to other parts of the body.

Marrow cavity Cavity inside a bone containing either red or yellow marrow.

Mast cell Cell found in many tissues, especially in the connective tissue surrounding blood vessels. Contains large granules containing histamine.

Matrix Extracellular material found in cartilage. Also the material in the inner compartment of the mitochondrion.

Matter Anything that has mass and occupies space.

Medulla Term referring to the central portion of some organs; for example, the adrenal medulla.

Megakaryocyte Large cell found in bone marrow that produces platelets.

Meiosis Type of cell division that occurs in the gonads during the formation of gametes. Requires two cellular divisions (meiosis I and meiosis II). In humans, it reduces the chromosome number from 46 to 23.

Meiosis I First meiotic division.

Meiosis II Second meiotic division.

Memory cells T or B cells produced after antigen exposure. They form a reserve force that responds rapidly to antigen during subsequent exposure.

Menopause End of the reproductive function (ovulation) in women. Usually occurs between the ages of 45 and 55.

Menstrual cycle Recurring series of events in the reproductive functions of women. Characterized by dramatic changes in ovarian and pituitary hormone levels and changes in the uterine lining that prepare the uterus for implantation. Ovulation occurs at the midpoint of the menstrual cycle.

Menstruation Process in which the endometrium is sloughed off, resulting in bleeding. Occurs approximately once every month.

Mesoderm One of the three types of cells that emerge in human embryonic development. Lies in the middle of the forming embryo. Gives rise to muscle, bone, and cartilage.

Messenger RNA (mRNA) Type of RNA that carries genetic information needed to synthesize proteins to the cytoplasm of a cell.

Metabolism The chemical reactions of the body, including all catabolic and anabolic reactions.

Metaphase Stage of cellular division (mitosis) in which chromosomes line up in the center of the cell.

Metastasis Spread of cancerous cells throughout the body, through the lymph vessels and circulatory system or directly through tissue fluid.

Methylation The addition of methyl groups to DNA that influences their function. Methyl groups are "donated" by foods we eat and profoundly affect heredity.

Microfilament Solid fiber consisting of contractile proteins that is found in cells in a dense network under the plasma membrane. Forms part of the cytoskeleton.

Micronutrients Nutrients required in small quantities. They include two broad groups, vitamins and minerals.

Microtubules Hollow protein tubules in the cytoplasm of cells that form part of the cytoskeleton. Also form spindles.

Microvilli Tiny projections of the plasma membranes of certain epithelial cells that increase the surface area for absorption.

Middle ear Portion of the ear located within a bony cavity in the temporal bone of the skull. Houses the ossicles.

Mineralocorticoids Group of steroid hormones produced by the adrenal cortex. Involved in electrolyte or mineral salt balance.

Mitochondrion Membrane-bound organelle where the bulk of cellular energy production occurs in eukaryotic cells. Houses the citric acid cycle and electron transport system.

Mitosis Term referring specifically to the division of a cell's nucleus. Consists of four stages: prophase, metaphase, anaphase, and telophase.

Mitotic spindle Array of microtubules constructed in the cytoplasm during prophase. Microtubules of the mitotic spindle connect to the chromosomes and help draw them apart during mitosis.

Molecule A structure formed by two or more atoms.

Monocyte White blood cell that phagocytizes bacteria and viruses in body tissues.

Monohybrid cross Procedure in which one plant is bred with another to study the inheritance of a single trait.

Morning sickness Nausea that often occurs in the first two to three months of pregnancy.

Morula Solid ball of cells produced from the zygote by numerous cellular divisions.

Motor unit Muscle fibers supplied by a single axon and its branches.

Mucus Thick, slimy material produced by the lining of the respiratory tract and parts of the digestive tract. Moistens and protects them.

Multipolar neuron Motor neuron found in the central nervous system. Contains a prominent, multiangular cell body and several dendrites.

Muscle fiber Long, unbranched, multinucleated cell found in skeletal muscle.

Muscle spindles Stretch receptors found in skeletal muscle. Also known as neuromuscular spindles.

Muscle tissue A contractile tissue found in varying amounts in all organs of the body. Consists of three types: skeletal, cardiac, and smooth.

Mutation Technically, a change in the DNA caused by chemical, physical, and biological agents. Also refers to a wide range of chromosomal defects.

Myelin sheath Layer of fatty material coating the axons of many neurons in the central and peripheral nervous systems.

Myofibril Bundle of contractile myofilaments in skeletal muscle cells.

Myometrium Uterine smooth muscle.

Myosin Protein filament found in many cells in the microfilamentous network. Also found in muscle cells.

Naked nerve ending Unmodified dendritic ending of the sensory neurons. Responsible for at least three sensations: pain, temperature, and light touch.

Natural selection Evolutionary process in which environmental abiotic and biotic factors "weed" out the less fit—those organisms not as well adapted to the environment as their counterparts.

Nephron Filtering unit in the kidney. Consists of a glomerulus and renal tubule.

Nerve Bundle of nerve fibers. May consist of axons, dendrites, or both. Carries information to and from the central nervous system.

Nervous tissue One of the primary tissues. Found in the nervous system and consists of two types of cells: conducting cells (neurons) and supportive cells.

Neuroendocrine reflex A reflex involving the endocrine and nervous systems.

Neuron Highly specialized cell that generates and transmits nerve impulses from one part of the body to another.

Neurosecretory neurons Specialized nerve cells of the hypothalamus and posterior pituitary that produce and secrete hormones.

Neurotransmitter Chemical substance released from the terminal ends (terminal boutons) of axons when a nerve impulse arrives. May stimulate or inhibit the next neuron.

Neutron Uncharged particle in the nucleus of the atom.

Neutrophil Type of white blood cell that phagocytizes bacteria and cellular debris.

Nitrogen fixation Process in which bacteria and a few other organisms convert atmospheric nitrogen to nitrate or ammonia, forms usable by plants.

Node of Ranvier Small gap in the myelin sheath of an axon; located between segments formed by Schwann cells. Responsible for saltatory conduction.

Nondisjunction Failure of a chromosome pair or chromatids of a double-stranded chromosome to separate during mitosis or meiosis.

Noradrenalin (norepinephrine) Hormone produced by adrenal medulla and secreted under stress. Contributes to the fight-or-flight response.

Nuclear envelope Double membrane delimiting the nucleus.

Nuclear pores Minute openings in the nuclear envelope that allow materials to pass to and from the nucleus.

Nucleic acids Refers to DNA and RNA.

Nucleoli Temporary structures in the nuclei of cells during interphase. Regions of the DNA that are active in the production of RNA.

Nucleotides The building blocks of DNA and RNA. Consist of a nitrogenous base, a sugar, and a phosphate group.

Nucleus (atom) Dense, center region of an atom that contains neutrons and protons.

Nucleus (cell) Cellular organelle that contains the genetic information that controls the structure and function of the cell.

Nutrient cycle Circular flow of nutrients from the environment through the various food chains back into the environment.

Olfactory membrane Receptor for smell; found in the roof of the nasal cavity.

Olfactory nerve Nerve that transmits impulses from the olfactory membrane to the brain.

Optic nerve Nerve that carries impulses from the retina to the brain.

Organ Discrete structure that carries out specialized functions.

Organ of Corti Receptor for sound; located in the inner ear within the cochlea.

Organ system Group of organs that participate in a common function.

Organogenesis Organ formation during embryonic development.

Osmosis Diffusion of water across a selectively permeable membrane.

Osmotic pressure Force that drives water across a selectively permeable membrane. Created by differences in solute concentrations.

Ossicles Three small bones inside the middle ear that transmit vibrations created by sound waves to the organ of Corti.

Osteoarthritis Degenerative joint disease caused by wear and tear that impairs movement of joints.

Osteoblast Bone-forming cell; secretes collagen.

Osteoclast Cell that digests the extracellular material of bone. Stimulated by the parathyroid hormone.

Osteocyte Bone cell derived from osteoblasts that has been surrounded by calcified extracellular material.

Osteoporosis Degenerative disease resulting in the deterioration of bone. Due to inactivity in men and women and loss of the ovarian hormone estrogen in postmenopausal women.

Outer ear External portion of the ear.

Oval window Membrane-covered opening in the cochlea where vibrations are transmitted from the stirrup to the fluid within the cochlea.

Ovary Female gonad. Produces ova (eggs) and steroid hormones, estrogen and progesterone.

Overpopulation Condition in which a species has exceeded the carrying capacity of the environment.

Ovulation Release of the oocyte from the ovary. Stimulated by hormones from the anterior pituitary.

Ovum (ova, plural) Germ cell containing 23 single-stranded chromosomes. Produced during the second meiotic division.

Oxaloacetate Four-carbon compound of the citric acid cycle. It is involved in the very first reaction of the cycle and is regenerated during the cycle.

Oxidation The loss of hydrogens or electrons from a substance.

Oxytocin Hormone from the posterior pituitary hormone. Stimulates contraction of the smooth muscle of the uterus and smooth-muscle-like cells surrounding the glandular units of the breast.

Ozone (O_3) Molecule produced from molecular oxygen. Accumulates in the stratosphere (upper layer of the atmosphere). Helps screen out incoming ultraviolet radiation. *See also* Ozone layer.

Ozone layer Region of the atmosphere located approximately 12 to 16 miles above the Earth's surface where ozone molecules are produced. Helps protect the Earth from ultraviolet light.

Pancreas Organ found in the abdominal cavity under the stomach, nestled in a loop formed by the first portion of the small intestine. Produces enzymes needed to digest foodstuffs in the small intestine and hormones that regulate blood glucose levels.

Pap smear Procedure in which cells are retrieved from the cervical canal to be examined for the presence of cancer.

Parasympathetic division (of the autonomic nervous system) Portion of the autonomic nervous system responsible for a variety of involuntary functions.

Parathyroid glands Endocrine glands located on the posterior surface of the thyroid gland in the neck. Produce parathyroid hormone.

Parathyroid hormone (PTH) Hormone that helps regulate blood calcium levels. Stimulates osteoclasts to digest bone, thus raising blood calcium levels. Also known as parathormone.

Passive immunity Temporary protection from antigen (bacteria and others) produced by the injection of immunoglobulins.

Penis Male organ of copulation.

Pepsin Enzyme released by the gastric glands of the stomach. Breaks down proteins into large peptide fragments.

Pepsinogen Inactive form of pepsin.

Perichondrium Connective tissue layer surrounding most types of cartilage. Contains blood vessels that supply nutrients to cartilage cells.

Periodic table of elements Table that lists elements by ascending atomic number. Also lists other vital statistics of each element.

Peripheral nervous system Portion of the nervous system consisting of the cranial and spinal nerves and receptors.

Peristalsis Involuntary contractions of the smooth muscles in the wall of the esophagus, stomach, and intestines, which propel food along the digestive tract.

Peritubular capillaries Capillaries that surround nephrons. They pick up water, nutrients, and ions from the renal tubule, thus helping maintain the osmotic concentration of the blood.

Pharynx Chamber that connects the oral cavity with the esophagus.

Phenotype Outward appearance of an organism.

Photoreceptors Modified nerve cells that respond to light. Located in the retina of humans and other animals.

Photosynthesis The series of chemical reactions in which the energy of light is used to synthesize high-energy organic molecules, usually carbohydrates, from low-energy inorganic molecules, usually carbon dioxide and water.

Pituitary gland Small pea-sized gland located beneath the brain. It produces numerous hormones and consists of two main subdivisions: anterior and posterior pituitary.

Placenta Organ produced from maternal and embryonic tissue. Supplies nutrients to the growing embryo and fetus and removes fetal wastes. Also produces hormones that help maintain pregnancy.

Plasma Extracellular fluid of blood. Comprises about 55% of the blood.

Plasma cell Cell produced from B-lymphocytes (B cells); synthesizes and releases antibodies.

Plasma membrane Outer layer of the cell. Consists of lipid and protein and controls the movement of materials into and out of the cell.

Plasmids Small circular strands of DNA found in bacterial cytoplasm separate from the main DNA.

Platelet Cell fragment produced from megakaryocytes in the red bone marrow. Plays a key role in blood clotting.

Polygenic inheritance Transmission of traits that are controlled by more than one gene.

Polyploidy Term referring to a genetic disorder caused by an abnormal number of chromosomes. Includes tetraploidy and triploidy.

Polyribosome Also known as polysome. Organelle formed by several ribosomes attached to a single messenger RNA. Synthesizes proteins used inside the cell.

Polysaccharide A carbohydrate molecule such as glycogen and starch that is made of many smaller molecules (monosaccharides).

Population Group of like organisms occupying a specific region.

Portal system Arrangement of blood vessels in which a capillary bed drains to a vein, which drains to another capillary bed.

Posterior chamber Posterior portion of the anterior cavity of the eye.

Posterior pituitary Neuroendocrine gland that consists of neural tissue and releases two hormones, oxytocin and antidiuretic hormone.

Predation Process in which an individual kills and feeds on another smaller individual.

Premenstrual syndrome (PMS) Condition that occurs in some women in the days before menstruation normally begins. Characterized by a variety of symptoms such as irritability, depression, fatigue, headaches, bloating, swelling and tenderness of breasts, joint pain, and tension.

Premotor area Region of the brain in front of the primary motor area. Controls muscle contraction and other less voluntary actions (playing a musical instrument).

Primary motor cortex Ridge of tissue in front of the central sulcus. Controls voluntary motor activity.

Primary oocyte Germ cell produced from oogonium in the ovary. Undergoes the first meiotic division.

Primary response Immune response elicited when an antigen first enters the body.

Primary sensory cortex Region of the brain located just behind the central sulcus. The point of destination for many sensory impulses traveling from the body into the spinal cord and up to the brain.

Primary succession Process of sequential change in which one community is replaced by another. Occurs where no biotic community has existed before.

Primary tissue One of the major tissue types, including epithelial, connective, muscle, and nervous tissue.

Primary tumor Cancerous growth that gives rise to cells that spread to other regions of the body.

Primates An order of the kingdom Animalia. Includes prosimians (premonkeys), monkeys, apes, and humans.

Primordial germ cells Cells that originate in the wall of the yolk sac and eventually become either spermatogonia or oogonia.

Principle of independent assortment Mendel's second law. Hereditary factors are segregated independently during gamete formation. Occurs only when genes are on different chromosomes.

Principle of segregation Mendel's first law, which states that hereditary factors separate during gamete formation.

Producers Generally refers to organisms that can synthesize their own foodstuffs. Major producers are the algae and plants that absorb sunlight and use its energy to synthesize organic foodstuffs from water and carbon dioxide.

Prokaryote A cell that has no nucleus and no organelles such as a bacterium. All prokaryotes are members of the Kingdom Monera.

Prolactin Protein hormone secreted by the posterior pituitary. In humans, it is responsible for milk production by the glandular units of the breast.

Prophase First phase of mitosis during which chromosomes condense, the nuclear membrane disappears, and the spindle forms.

Proprioception Sense of body and limb position.

Prosimians Premonkeys; tarsiers and lemurs.

Prostaglandins Group of chemical substances that have a variety of functions. Act on nearby cells.

Prostate gland Sex accessory gland that is located near the neck of the bladder. Produces fluid that is added to the sperm during ejaculation. Empties into the urethra.

Protein A polymer consisting of many amino acids.

Proton Subatomic particle found in the nucleus of the atom. Each proton carries a positive charge.

Proto-oncogenes Genes in cells that, when mutated, lead to cancerous growth.

Puberty Period of sexual maturation in humans.

Pulmonary circuit (or circulation) Short circulatory loop that supplies blood to the lungs and transports it back to the heart.

Pulmonary veins Veins that carry oxygenated blood from the lungs to the left atrium.

Pupil Central opening in the iris that allows light to penetrate deeper into the eye.

Purine Type of nitrogenous base found in DNA nucleotides. Consists of two fused rings.

Purkinje fiber Modified cardiac muscle fiber that conducts nerve impulses to individual heart muscle cells.

Pus Liquid emanating from a wound. Contains plasma, many dead neutrophils, dead cells, and bacteria.

Pyramid of numbers Diagram of the number of organisms at various trophic levels in a food chain or ecosystem.

Pyrimidine One of two types of nitrogen base found in DNA nucleotides. Consists of one ring.

Pyrogen Chemical released primarily from macrophages that have been exposed to bacteria and other foreign substances. Responsible for fever.

Radioactivity Tiny bursts of energy or particles emitted from the nucleus of some unstable atoms. Results from excess neutrons in the nuclei of some atoms.

Receptor Any structure that responds to internal or external changes. Three types of receptors are found in the body: encapsulated, nonencapsulated (naked nerve endings), and specialized (e.g., the retina and semicircular canals).

Recessive Term describing an allele of a gene that is expressed when the dominant factor is missing.

Recombinant DNA technology Procedure in which scientists take segments of DNA from an organism and insert them into DNA from other organisms.

Red blood cells (RBCs) Enucleated cells in blood that transport oxygen in the bloodstream.

Red marrow Tissue found in the marrow cavity of bones. Site of blood cell and platelet production.

Reflex Automatic response to a stimulus. Mediated by the nervous system.

Relaxin Hormone produced by the corpus luteum and the placenta. It is released near the end of pregnancy and softens the cervix and the fibrocartilage uniting the pubic bones, thus facilitating birth.

Releasing hormone Any of a group of hypothalamic hormones that stimulates the release of other hormones by the anterior pituitary.

Resting potential Minute voltage differential across the membrane of neurons. Also known as the membrane potential.

Restriction endonuclease Enzyme used in recombinant DNA technology. Cuts off segments of the DNA molecule for cloning and splicing.

Reticular activating system (RAS) Region of the medulla that receives nerve impulses from neurons transmitting information to and from the brain. Impulses are transmitted to the cortex, alerting it.

Retina Innermost, light-sensitive layer of the eye. Consists of an outer pigmented layer and an inner layer of nerve cells and photoreceptors (rods and cones).

Retrovirus Special type of RNA virus that carries an enzyme enabling it to produce complementary strands of DNA on the RNA template.

Reverse transcriptase Enzyme that allows the production of DNA from strands of viral RNA.

Rheumatoid arthritis Type of arthritis in which the synovial membrane of the joint becomes inflamed and thickens. Results in pain and stiffness in joints. Thought to be an autoimmune disease.

Ribosomal RNA (rRNA) RNA produced at the nucleolus. Combines with protein to form the ribosome.

Ribosome Cellular organelle consisting of two subunits, each made of protein and ribosomal RNA. Plays an important part in protein synthesis.

RNA polymerase Enzyme that helps align and join the nucleotides in a replicating RNA molecule.

Rod Type of photoreceptor in the eye. Provides for vision in dim light.

Root nodule Swelling in the roots of certain plants (legumes) containing nitrogen-fixing bacteria.

Rough endoplasmic reticulum (RER) Ribosome-coated endoplasmic reticulum. Produces lysosomal enzymes and proteins for use outside the cell.

Saccule Membranous sac located inside the vestibule. Contains a receptor for movement and body position.

Salivary gland Any of several exocrine glands situated around the oral cavity. Produces saliva.

Saltatory conduction Conduction of a nerve impulse down a myelinated neuron from node to node.

Sarcomere Functional unit of the muscle cell. Consists of the myofilaments, actin, and myosin.

Sarcoplasmic reticulum Term given to the smooth endoplasmic reticulum of a skeletal muscle fiber. Stores and releases calcium ions essential for muscle contractions.

Schwann cell Type of neuroglial cell or supportive cell in the nervous system. Responsible for the formation of the myelin sheath.

Science Body of knowledge on the workings of the world and a method of accumulating knowledge. *See also* Scientific method.

Scientific method Deliberate, systematic process of discovery. Begins with observation and measurement. From observations, hypotheses are generated and tested. This leads to more observation and measurement that supports or refutes the original hypothesis.

Sclera Outermost layer of the eye.

Scrotum Skin-covered sac containing the testes.

Sebum Oil excreted by sebaceous glands onto the surface of the skin.

Secondary sex characteristics Distinguishing features of men and women resulting from the sex steroids. In men, includes facial hair growth and deeper voices. In women, includes breast development and fatty deposits in the hips and other regions.

Secondary succession Process of sequential change in which one community is replaced by another. It occurs where a biotic community previously existed, but was destroyed by natural forces or human actions.

Secondary tumor Cancerous growth formed by cells arising from a primary tumor.

Secretin Hormone produced by the cells of the duodenum. Stimulates the pancreas to release sodium bicarbonate.

Secretory vesicles Membrane-bound vesicles containing protein (hormones or enzymes) produced by the endoplasmic reticulum and packaged by the Golgi complex of some cells. They fuse with the membrane, releasing their contents by exocytosis.

Selective permeability Control of what moves across the plasma membrane of a cell.

Semen Fluid containing sperm and secretions of the secondary sex glands.

Semicircular canal Sensory organ of the inner ear. Houses the receptors that detect body position and movement.

Seminal vesicles Sex accessory glands that empty into the vas deferens. Produce the largest portion of ejaculate.

Seminiferous tubule Sperm-producing tubule in the testis.

Sex accessory gland One of several glands that produce secretions that are added to sperm during ejaculation.

Sex chromosomes X and Y chromosomes that help determine the sex of an individual. They also carry a few other traits.

Sex-linked gene A gene that is on a sex chromosome.

Sex steroids Steroid hormones produced principally by the ovaries (in women) and testes (in men). Help regulate secretion of gonadotropins and determine secondary sex characteristics.

Sexually transmitted diseases (venereal diseases) Bacterial and viral infections that are transmitted by sexual contact.

Sickle-cell disease Genetic disease common in African Americans that results in abnormal hemoglobin in red blood cells, causing cells to become sickle shaped when exposed to low oxygen levels. Sickling causes cells to block capillaries, restricting blood flow to tissues.

Sinoatrial node The heart's pacemaker. Located in the wall of the right atrium, it sends timed impulses to the heart muscle, thus synchronizing muscle contractions.

Skeletal muscle Muscle that is generally attached to the skeleton and causes body parts to move.

Skeleton Internal support of humans and other animals. Consists of bones joined together at joints.

Sliding filament mechanism Sliding of actin filaments toward the center of a sarcomere, causing muscle contraction.

Smooth endoplasmic reticulum (SER) Endoplasmic reticulum without ribosomes. Produces phosphoglycerides used to make the plasma membrane. Performs a variety of different functions in different cells.

Smooth muscle Involuntary muscle that lacks striations. Found around circulatory system vessels and in the walls of such organs as the stomach, uterus, and intestines.

Somatic cell A body cell such as those of bone, muscle, liver, brain, and blood.

Species Group of organisms that is structurally and functionally similar. When members of the group breed, they produce viable, reproductively competent offspring. Also a subgroup of a genus.

Specificity The property of an enzyme allowing it to catalyze only one or a few chemical reactions.

Spermatozoan Sperm. Male germ cell.

Spinal nerve Nerve that arises from the spinal cord.

Spongy bone Type of bony tissue inside most bones. Consists of an irregular network of bone spicules.

Starch A polysaccharide found in plants and made of many glucose units.

Subatomic particles Electrons, protons, and neutrons. Particles that can be separated from an atom by physical means.

Substrate Molecule that fits into the active site of an enzyme.

Succession Process of sequential change in which one community is replaced by another until a mature or climax ecosystem is formed. *See also* Primary succession and Secondary succession.

Sulcus Indented region or groove in the cerebral cortex between ridges.

Surfactant Detergent-like substance produced by the lungs. Dissolves in the thin watery lining of the alveoli; helps reduce surface tension, keeping the alveoli from collapsing.

Suspensory ligament Zonular fibers that connect the lens to the ciliary body.

Sympathetic division Division of the autonomic nervous system that is responsible for many functions, especially those involved in the fight-or-flight response.

Synapse Juncture of two neurons that allows an impulse to travel from one neuron to the next.

Synaptic cleft Gap between an axon and the dendrite or effector (e.g., gland or muscle) it supplies.

Synergy Coordination of the workings of antagonistic muscle groups.

Systemic circulation System of blood vessels that transports blood to and from the body and heart, excluding the lungs.

Systolic pressure Peak pressure at the moment the ventricles contract. The higher of the two numbers in a blood pressure reading.

Taste bud Receptor for taste principally found in the surface epithelium and certain papillae of the tongue.

T cell *See* T-lymphocyte.

Telophase Final stage of mitosis in which the nuclear envelope reforms from vesicles and the chromosomes uncoil.

Tendons Connective tissue structures that generally attach muscles to bones.

Terminal boutons Small swellings on the terminal fibers of axons. They lie close to the membranes of the dendrites of other axons or the membranes of the effectors, and transfer nerve impulses from one cell to another.

Terminator codon Codon found on each strand of messenger RNA that marks where protein synthesis should end.

Testes Male gonads. They produce sex steroids and sperm.

Testosterone Male sex hormone that stimulates sperm formation and is responsible for secondary sex characteristics, such as facial hair growth and muscle growth.

Theories Principles of science—the broader generalizations about the world and its components. Theories are supported by considerable scientific research.

Theory of Natural Selection A theory proposed by Charles Darwin to explain how evolution works; it states that the fittest organisms of a population survive and reproduce, thus passing their genes on to future generations. Over time, this results in a shift in the genetic makeup of the population.

Thyroid gland U- or H-shaped gland located in the neck on either side of the trachea just below the larynx. Produces three hormones: thyroxin, triiodothyronine, and calcitonin.

Thyroid-stimulating hormone (TSH) Hormone produced by the pituitary gland. Stimulates production and release of thyroxine and triiodothyronine by the thyroid gland.

Thyroxine Hormone produced by the thyroid gland that accelerates the rate of mitochondrial glucose catabolism in most body cells and also stimulates cellular growth and development.

Tissue Component of the body from which organs are made. Consists of cells and extracellular material (fluid, fibers, and so on). *See* primary tissue.

T-lymphocyte Type of lymphocyte responsible for cell-mediated immunity. Attacks foreign cells, virus-infected cells, and cancer cells directly. Also known as T cell.

Trachea Duct that leads from the pharynx to the lungs.

Transcription RNA production on a DNA template.

Transfer RNA (tRNA) Small RNA molecules that bind to amino acids in the cytoplasm and deliver them to specific sites on the messenger RNA.

Transition reaction Part of cellular respiration in which one carbon is cleaved from pyruvic acid, forming a two-carbon compound, which reacts with Coenzyme A. The resulting chemical enters the citric acid cycle.

Translation Synthesis of protein on a messenger RNA template.

Translocation Process in which a segment of a chromosome breaks off but reattaches to another site on the same chromosome or another one.

Transposons Also known as jumping genes. Sequences of DNA that move from one location on the DNA to another in response to stress and other factors. Transposons may be inserted into existing genes, altering their structure and function or near existing genes where they influence their function.

Triiodothyronine Hormone produced by the thyroid gland. Nearly identical in function to thyroxin.

Trophic hormones Hormones that stimulate the production and secretion of other hormones. Also known as tropic hormones.

Trophic level Feeding level in a food chain.

Trophoblast Outer ring of cells of the blastocyst that form the embryonic portion of the placenta.

Tubal ligation Sterilization procedure in women. Uterine tubes are cut, preventing sperm and ova from uniting.

Tumor Mass of cells derived from a single cell that has begun to divide. In malignant tumors, the cells divide uncontrollably and often release clusters of cells or single cells that spread in the blood and lymphatic systems to other parts of the body. Benign tumors grow to a certain size, then stop.

Turner syndrome Genetic disorder in which an offspring contains 22 pairs of autosomes and a single, unmatched X chromosome. Phenotypically female.

Tympanic membrane (eardrum) Membrane between the external auditory canal and middle ear that oscillates when struck by sound waves.

Type I diabetes Form of diabetes that occurs mainly in young people and results from an insufficient amount of insulin production and release. Brought on by damage to insulin-producing cells of the pancreas. Also known as early-onset diabetes.

Type II diabetes Form of diabetes that occurs chiefly in older individuals (around the age of 40) and results from a loss of tissue responsiveness to insulin. Also known as late-onset diabetes caused by obesity.

Umbilical cord A structure that connects the fetus to the placenta and contains two blood vessels that carry blood to and from the embryo.

Umbilical vein Vein in the umbilical cord that carries blood from the placenta to the fetus.

Ureter Hollow, muscular tube that transports urine by peristaltic contractions from the kidney to the urinary bladder.

Urethra Narrow tube that transports urine from the urinary bladder to the outside of the body. In males, it also conducts sperm and semen to the outside.

Urinary bladder Hollow, distensible organ with muscular walls that stores urine. Drained by the urethra.

Urine Fluid containing various wastes that is produced in the kidney and excreted out of the urinary bladder.

Uterus Organ that houses and nourishes the developing embryo and fetus.

Utricle Membranous sac containing a receptor for body position and movement. Located inside the vestibule of the inner ear.

Vaccine Preparation containing dead or weakened bacteria and viruses that, when injected in the body, elicits an immune response.

Vagina Tubular organ that serves as a receptacle for sperm and provides a route for delivery of the baby at birth.

Variation Genetically based differences in physical or functional characteristics within a population.

Vas deferens Duct that carries sperm from the testis to the urethra. Contracts during ejaculation.

Vasectomy Contraceptive procedure in men in which the vas deferens is cut and the free ends sealed to prevent sperm from entering the urethra during ejaculation.

Vein Type of blood vessel that carries blood to the heart.

Vena cava One of two large veins that empty into the right atrium of the heart.

Venule Smallest of all veins. Empties into capillary networks.

Villi Fingerlike projections of the lining of the small intestine that increase the surface area for absorption.

Virus Nonliving entity consisting of a nucleic acid—either DNA or RNA—core surrounded by a protein coat, the capsid. Viruses are cellular parasites, invading cells and taking over their metabolic machinery to reproduce.

Vitamin Any of a diverse group of organic compounds. Essential to many metabolic reactions.

Vocal cords Elastic ligaments inside the larynx that vibrate as air is expelled from the lungs, generating sound.

White blood cells (WBCs) Cells of the blood formed in the bone marrow. Principally involved in fighting infection.

White matter The portion of the brain and spinal cord that appears white to the naked eye. Consists primarily of white, myelinated nerve fibers.

Yellow marrow Inactive marrow of bones in adults containing fat. Formed from red marrow.

Yolk sac Embryonic pouch formed from endoderm. Site of early formation of red blood cells and germ cells.

Zygote Cell produced by a sperm and ovum during fertilization. Contains 46 chromosomes.

Index

Note: Page numbers followed by *f* or *t* indicate material in figures or tables, respectively.

SHERLOCK HOLMES
THE COMPLETE WORKS

A Wilco Book
Outstanding works of universal interest

© *Wilco* 2013
This edition reprinted in 2016

ISBN NO : 978-81-8252-644-0

SHERLOCK HOLMES - THE COMPLETE WORKS

Published by
Wilco
Publishing House
Mumbai - India
Tel: (91-22) 2204 1420 / 2284 2574
Fax: (91-22) 2204 1429
E-mail: wilco@wilcobooks.com
Website: www.wilcobooks.com
www.wilcopicturelibrary.com

Printed and bound in India